D1986035

INTERNATIONAL SERIES
OF
MONOGRAPHS ON PHYSICS

SERIES EDITORS

J. BIRMAN	CITY UNIVERSITY OF NEW YORK
S. F. EDWARDS	UNIVERSITY OF CAMBRIDGE
R. FRIEND	UNIVERSITY OF CAMBRIDGE
M. REES	UNIVERSITY OF CAMBRIDGE
D. SHERRINGTON	UNIVERSITY OF OXFORD
G. VENEZIANO	CERN, GENEVA

INTERNATIONAL SERIES OF MONOGRAPHS ON PHYSICS

138. I. M. Vardavas, F. W. Taylor: *Radiation and climate*
137. A. F. Borghesani: *Ions and electrons in liquid helium*
136. C. Kiefer: *Quantum gravity, Second edition*
135. V. Fortov, I. Iakubov, A. Khrapak: *Physics of strongly coupled plasma*
134. G. Fredrickson: *The equilibrium theory of inhomogeneous polymers*
133. H. Suhl: *Relaxation processes in micromagnetics*
132. J. Terning: *Modern supersymmetry*
131. M. Mariño: *Chern-Simons theory, matrix models, and topological strings*
130. V. Gantmakher: *Electrons and disorder in solids*
129. W. Barford: *Electronic and optical properties of conjugated polymers*
128. R. E. Raab, O. L. de Lange: *Multipole theory in electromagnetism*
127. A. Larkin, A. Varlamov: *Theory of fluctuations in superconductors*
126. P. Goldbart, N. Goldenfeld, D. Sherrington: *Stealing the gold*
125. S. Atzeni, J. Meyer-ter-Vehn: *The physics of inertial fusion*
123. T. Fujimoto: *Plasma spectroscopy*
122. K. Fujikawa, H. Suzuki: *Path integrals and quantum anomalies*
121. T. Giamarchi: *Quantum physics in one dimension*
120. M. Warner, E. Terentjev: *Liquid crystal elastomers*
119. L. Jacak, P. Sitko, K. Wieczorek, A. Wojs: *Quantum Hall systems*
118. J. Wesson: *Tokamaks, Third edition*
117. G. Volovik: *The Universe in a helium droplet*
116. L. Pitaevskii, S. Stringari: *Bose-Einstein condensation*
115. G. Dissertori, I.G. Knowles, M. Schmelling: *Quantum chromodynamics*
114. B. DeWitt: *The global approach to quantum field theory*
113. J. Zinn-Justin: *Quantum field theory and critical phenomena, Fourth edition*
112. R. M. Mazo: *Brownian motion - fluctuations, dynamics, and applications*
111. H. Nishimori: *Statistical physics of spin glasses and information processing — an introduction*
110. N. B. Kopnin: *Theory of nonequilibrium superconductivity*
109. A. Aharoni: *Introduction to the theory of ferromagnetism, Second edition*
108. R. Dobbs: *Helium three*
107. R. Wigmans: *Calorimetry*
106. J. Kübler: *Theory of itinerant electron magnetism*
105. Y. Kuramoto, Y. Kitaoka: *Dynamics of heavy electrons*
104. D. Bardin, G. Passarino: *The Standard Model in the making*
103. G. C. Branco, L. Lavoura, J. P. Silva: *CP Violation*
102. T. C. Choy: *Effective medium theory*
101. H. Araki: *Mathematical theory of quantum fields*
100. L. M. Pismen: *Vortices in nonlinear fields*
99. L. Mestel: *Stellar magnetism*
98. K. H. Bennemann: *Nonlinear optics in metals*
97. D. Salzmann: *Atomic physics in hot plasmas*
96. M. Brambilla: *Kinetic theory of plasma waves*
95. M. Wakatani: *Stellarator and heliotron devices*
94. S. Chikazumi: *Physics of ferromagnetism*
91. R. A. Bertlmann: *Anomalies in quantum field theory*
90. P. K. Gosh: *Ion traps*
88. S. L. Adler: *Quaternionic quantum mechanics and quantum fields*
87. P. S. Joshi: *Global aspects in gravitation and cosmology*
86. E. R. Pike, S. Sarkar: *The quantum theory of radiation*
83. P. G. de Gennes, J. Prost: *The physics of liquid crystals*
82. B. H. Bransden, M. R. C. McDowell: *Charge exchange and the theory of ion-atom collision*
73. M. Doi, S. F. Edwards: *The theory of polymer dynamics*
71. E. L. Wolf: *Principles of electron tunneling spectroscopy*
70. H. K. Henisch: *Semiconductor contacts*
69. S. Chandrasekhar: *The mathematical theory of black holes*
51. C. Møller: *The theory of relativity*
46. H. E. Stanley: *Introduction to phase transitions and critical phenomena*
32. A. Abragam: *Principles of nuclear magnetism*
27. P. A. M. Dirac: *Principles of quantum mechanics*
23. R. E. Peierls: *Quantum theory of solids*

Ions and Electrons in Liquid Helium

A.F. BORGHESANI

Department of Physics, University of Padua, Italy

OXFORD
UNIVERSITY PRESS

OXFORD
UNIVERSITY PRESS

Great Clarendon Street, Oxford OX2 6DP

Oxford University Press is a department of the University of Oxford.
It furthers the University's objective of excellence in research, scholarship,
and education by publishing worldwide in

Oxford New York

Auckland Cape Town Dar es Salaam Hong Kong Karachi
Kuala Lumpur Madrid Melbourne Mexico City Nairobi
New Delhi Shanghai Taipei Toronto

With offices in

Argentina Austria Brazil Chile Czech Republic France Greece
Guatemala Hungary Italy Japan Poland Portugal Singapore
South Korea Switzerland Thailand Turkey Ukraine Vietnam

Oxford is a registered trade mark of Oxford University Press
in the UK and in certain other countries

Published in the United States
by Oxford University Press Inc., New York

© A.F. Borghesani 2007

The moral rights of the authors have been asserted
Database right Oxford University Press (maker)

First published 2007

All rights reserved. No part of this publication may be reproduced,
stored in a retrieval system, or transmitted, in any form or by any means,
without the prior permission in writing of Oxford University Press,
or as expressly permitted by law, or under terms agreed with the appropriate
reprographics rights organization. Enquiries concerning reproduction
outside the scope of the above should be sent to the Rights Department,
Oxford University Press, at the address above

You must not circulate this book in any other binding or cover
and you must impose this same condition on any acquirer

British Library Cataloguing in Publication Data
Data available

Library of Congress Cataloging in Publication Data
Data available

Printed in Great Britain
on acid-free paper by Biddles Ltd. www.biddles.co.uk

ISBN 978–0–19–921360–3 (Hbk)

1 3 5 7 9 10 8 6 4 2

Sic unumquicquid paulatim protrahit aetas
in medium ratioque in luminis erigit oras.
Namque alid ex alio clarescere corde videbant,
artibus ad summum donec venere cacumen.

Titus Lucretius Caro
De rerum natura
V, 1454–1457

PREFACE

Liquid helium with its fascinating properties is a subject that has strongly attracted the interest of scientists. The appearance of superfluidity as the temperature is lowered below the λ-point at $T = 2.17\,\mathrm{K}$ has been considered as a distinctive sign of the occurrence of Bose–Einstein condensation, although the fact that the helium atoms interact with each other substantially modifies the properties of the condensate with respect to an ideal Bose system.

The superfluidity of liquid helium can be tested in several different experiments, but electrons and ions have proved to be among the most used and successful probes. In liquid helium they are actually complex entities, whose structure, which is either a solvation cluster surrounding a positive He ion or an empty bubble encompassing an electron, is determined by the interaction between the different types of charge and the He atoms.

The transport of charges in the liquid in different ranges of electric field or velocity sheds light on different aspects of the superfluidity. At small electric fields and at low temperature the drift velocity of ions depends on the collisions with the elementary excitations of the superfluid: phonons, rotons, and ^{3}He atomic impurities.

At higher electric fields and drift velocities, ions produce and interact with hydrodynamic structures of the superfluid: vortex rings and vortex lines. One of the major scientific achievements is the acknowledgment that these hydrodynamic patterns are quantized.

The importance of ions in the study of liquid helium, however, is not limited to the investigation of transport. Many other scientific issues have been settled by using ions in the liquid. Among these it suffices here to remember the study of the liquid–vapor interface or of the spectroscopic properties of the electron trapped in the bubble.

In spite of the importance of ions in connection with liquid helium, there are no comprehensive monographs dealing with this specific topic. Pieces of information are scattered throughout the specialized literature. Thus, the goal of this book is to critically describe and analyze the vast amount of scientific data on ions in liquid helium collected over the years.

A large part of the book is devoted, *a fortiori*, to the problem of charge transport in the superfluid He II because in this field the use of ions has produced the most important results. Low-field, low-temperature drift velocity data yield information on the quasiparticle–ion interaction within the frame of reference of kinetic theory. High-drift velocity data give information on the quantized hydrodynamic structures of the superfluid, how vortices are nucleated, and how they interact with ions and with the elementary excitations. However, other experimental results, not related to transport, are also described.

Liquid He, however, is not a synonym for superfluid only. As for all other liquids, He can be found in its normal liquid phase. The λ-transition and the normal liquid–vapor transition are examples of continuous transitions, showing long-range critical fluctuations of the appropriate order parameter that can be investigated by using ions because they couple to them via electrostriction. Ions are also best suited to probing classical hydrodynamics on a microscopic scale.

Moreover, as a fortunate case, a fermionic counterpart of the bosonic ^{4}He liquid exists. Liquid ^{3}He, that liquefies at a lower temperature than ^{4}He, consists of fermions and is considered a paradigm for a Fermi liquid. Ions, in this case, are used to investigate how an impurity couples with a Fermi liquid.

At a temperature much lower than its bosonic counterpart, ^{3}He becomes superfluid too. The superfluid transition is the result of a p-wave atom pairing mechanism closely related to the mechanism leading to superconductivity in metals. The particular type of atom pairing is responsible for the very rich topological structure of the superfluid phases of ^{3}He that influences charge transport. For this reason, experiments with ions in liquid ^{3}He have efficiently supplemented NMR measurements in shedding light on the structure of superfluid ^{3}He.

Among the ions used in all of the experiments described in this book, the negative ion deserves a special place because it is a universal probe consisting of a cavity surrounding the electron. The cavity forms as the result of the competition between the short-range repulsive exchange interaction among the excess electron and the electron shell of the surrounding atoms, the increase of the electron kinetic energy due to localization, and the expansion of the liquid to form the bubble.

Whereas electrons injected into the liquid, either ^{3}He or ^{4}He, very rapidly form a fully-fledged electron bubble, electrons injected into He gas may form bubbles only when the thermodynamic parameters of the gas are favorable. Thus, electron transport experiments in dense He gas give important pieces of information on the dynamics and thermodynamics of the formation of the electron bubble. These experiments are interesting because they allow the scientists to continuously monitor the formation of the ultimate negative charge in the liquid as a function of the gas density and temperature.

The study of physics is a mixture of experiment and theory. Owing to my background as an experimentalist, I have also described in some detail the experimental set-ups and techniques. There are sometimes such clever and elegant experiments that they are immediately perceived as “beautiful”, no less elegant than a good theory. Therefore, this book attempts to give some “flavor” of the experiments. Nonetheless, the theories underlying the interpretation of the experimental facts are quite extensively described.

The interplay of the physical processes ruling the behavior of charges in the superfluid also manifests itself in the succession of chapters. For instance, the transport behavior of the “bare” ions at high electric fields in superfluid He II has been necessarily postponed to the description of the ion–vortex interaction because the former cannot be understood if the latter has not been described previously. In spite of this, I believe that the reader will not encounter any difficulty by abandoning for a while the topic of the low-field transport of charges, switch to quantum hydrodynamics and ion–vortex interaction, and come back to the topic of high-field ion transport.

For the same reason, most chapters rely on what has been written in the previous ones. However, some chapters, e.g., that about the techniques of ion production, can be skipped without impairing the understanding of the others.

The book is divided into three parts. The first one deals with liquid ^{4}He, the second one with liquid ^{3}He, and the third part is about dense He gas. In the first part, the transport of ions in superfluid He is treated first and the normal fluid only thereafter because of the overwhelming importance of superfluidity in this material. The order is reversed in the case of liquid ^{3}He for the converse reason that more studies are devoted to investigating the normal fluid than the superfluid.

The book is intended for graduate and postgraduate students, beginning research in this low-temperature field, as well as for the specialists who wish to have at hand a complete review on this topic. The required background includes basic statistical mechanics, kinetic theory, hydrodynamics, and cryogenics. In any case, all the theoretical developments are explained in detail.

Particular care has been devoted to compiling a rich and, as far as possible, complete bibliography. All figures and diagrams have been redrawn from the originals by myself for the sake of uniformity and greater clarity, and all units have been changed so as to conform to the SI system. In any case, the reader is advised to consult the original papers in the quest for the highest accuracy.

I am indebted to several colleagues. First of all, to Prof. M. Santini and Prof. L. Bruschi, with whom I have collaborated here in Padua for several years. As the reader can spot from the bibliography, they were heavily involved in this field in the past and many of the most interesting results are due to them and to the people of the group of Prof. Careri in Rome. I had the privilege to listen to them first hand.

I am also indebted to Prof. Dr W. F. Schmidt, of the Freie Universitaet Berlin, who first asked me to write a review on this subject. Actually, the project of writing this book originally started as a chapter I contributed to the book *Electronic excitations in liquefied rare gases*, edited by Prof. W. F. Schmidt and Prof. E. Illenberger and published by ASP in 2005.

A. F. Borghesani
Padua, April 2007

CONTENTS

Part I

Liquid helium-4

1

SUPERFLUIDITY AND CHARGE CARRIERS

Liquid ^{4}He is unique among the liquids of rare gases because of the relevance of quantum effects related to its small mass that keep it liquid down to absolute zero. Below the critical temperature, $T_c \approx 5.2\,\mathrm{K}$, liquid helium behaves as a normal fluid, called He I. But, if the temperature is lowered, at a temperature $T_\lambda = 2.17\,\mathrm{K}$, under saturated vapor pressure (SVP) conditions, liquid He undergoes a continuous transition, known as the λ-transition, whose signature is the very characteristic λ-shaped anomaly of the specific heat (Wilks, 1967).

Below T_λ, the most intriguing and astonishing property of liquid ^{4}He shows up, namely, *superfluidity.* Helium in this phase is called He II (Wilks, 1967). No other liquids are known to show a superfluid phase, except the lighter isotope ^{3}He at much lower temperatures (for $T < 3\,\mathrm{mK}$), if one does not consider as probable candidates the electron gas in a superconducting metal or the nucleonic fluid in a neutron star (Langlois *et al.*, 1998; Andersson and Comer, 2001; Lattimer and Prakash, 2001). For these systems the data are inevitably scant and they will not be considered here. We also neglect here the recent issue of Bose–Einstein condensation in gaseous assemblies of ultracold atoms.

At $T = 0\,\mathrm{K}$, helium is considered to be completely superfluid, with zero entropy and viscosity. Superfluidity means that a non-dissipative, irrotational flow can develop when liquid He is flowing in tubes or pipes. At the same time, this means that an object could move through the liquid without being retarded by the viscous drag that any normal fluids would be exerting on it because of viscosity.

At any other temperatures, $0 < T < T_\lambda$, the properties of liquid He II, in particular those related to heat flow and viscous effects, are described by a mathematical picture called the *two-fluid model.* Liquid He II is envisioned as a mixture of two completely miscible, non-interacting liquids, with their own densities. These two liquids are named the *normal fluid* and the *superfluid.* The total density is made up of the contribution of the two fluids:

$$\rho = \rho_n + \rho_s\,. \tag{1.1}$$

The density of the normal component vanishes at absolute zero, when the whole mass of He II consists of superfluid, and increases toward 100% for $T \to T_\lambda$.

This two-fluid model is a convenient tool to describe the flow properties of superfluid He II, but it does not imply that there are two real fluids physically mixed together. However, this model is based on a very clever microscopic picture of the atomic motion in a condensed phase at low temperature due to Landau (Landau, 1941).

A necessary, though not sufficient, condition for superfluidity to appear is that the substance forms a condensate, i.e., it must be found in a single coherent quantum state of the whole system that contains a macroscopic fraction of the substance (Guénault, 2003). The many-body quantum wave function of this ground state can be written as $\Phi(\mathbf{r}, t) = f(\mathbf{r}, t) \exp[i\phi(\mathbf{r}, t)]$ and it is easy to show, by taking local averages of density and momentum density operators (Wilks, 1967), that the local superfluid density and velocity fields are given by

$$\rho_s(\mathbf{r}, t) = f^2(\mathbf{r}, t) \tag{1.2}$$

and

$$\mathbf{v}_s = \frac{\hbar}{m} \boldsymbol{\nabla} \phi(\mathbf{r}, t) , \tag{1.3}$$

where $\hbar$ is Planck's constant divided by 2π.

From eqn (1.3) it immediately follows that the superfluid flow is irrotational:

$$\boldsymbol{\nabla} \times \mathbf{v}_s = \frac{\hbar}{m} \boldsymbol{\nabla} \times \boldsymbol{\nabla} \phi(\mathbf{r}, t) \equiv 0 . \tag{1.4}$$

Every weakly-excited state of this system can be considered as a collection of single *elementary excitations*, which form a dilute gas of non- or weakly-interacting quasiparticles (Landau, 1941; Donnelly, 1991; Guénault, 2003).

The superfluid is considered as the background fluid because at $T = 0\,\mathrm{K}$ He II is completely superfluid. The normal fluid is believed to be composed of the dilute gas of the thermal excitations of the superfluid, i.e., phonons, rotons, and ^{3}He atoms. The nature of these elementary excitations, together with the requirement of the quantum coherence of the ground state, is very important to determine why superfluidity can appear. In fact, if superfluidity has to show up, a deformed ground state must not immediately dissipate itself by forming excitations of the fluid above the ground state. In other words, there must be an energy gap between the ground state and the excited states of the substance and the nature of the excitations is essential to determine if superfluidity can appear.

The nature and shape of the quantal wave function describing the low-lying excited states of He II and its relation with the spectrum of the elementary excitations have been investigated and discussed by Feynman (Feynman, 1953, 1954; Feynman and Cohen, 1956), who has been able to show that the lowest-lying excited states are quantized longitudinal acoustic vibrations, i.e., *phonons*. He has also shown, in accordance with the suggestions based on quantum hydrodynamics (Landau, 1941), that a minimum energy is required to create excitations other than phonons. These higher-lying excitations are termed rotons, a name originally introduced by Tamm, although they do not possess any intrinsic angular momentum.

Rotons are more difficult to envisage than phonons. They can be thought of as a local compression of the medium (Careri *et al.*, 1959), or, following the suggestions of Feynman and Cohen (1956), they can be considered to be analogous to a small irreducible vortex ring of such a radius that only one atom at a time can pass through it. In any case, rotons are localized excitations with 3–4 Å size, of given momentum, with an associated irrotational backflow corresponding to the motion of the superfluid.

1.1 Superfluidity and the Landau critical velocity

The central issue of the research about superfluid helium is obviously the explanation of why helium can be superfluid. A quite simple argument, based on energy and momentum conservation, has been given by Landau in order to explain superfluidity in terms of the shape of the dispersion curve of the elementary excitations (Landau, 1941, 1947).

The theory predicts that the onset of dissipative behavior and the breakdown of superfluidity occur if the relative velocity between the fluid and its surroundings exceeds a well-defined critical velocity v_L. The fluid might either move with stationary velocity v with respect to the pipe walls which have effectively an infinite mass, or we could envision a massive object being dragged through the stationary fluid at the given velocity v. The Landau argument refers to this second case just because the physical picture is easier to grasp.

By looking at Fig. 1.1, let us consider a massive object moving through the stationary fluid with initial velocity v_i. Let us further assume that its mass m is much larger than the masses of the bosons constituting the condensate of the substance. The fluid can exert a drag on the moving object only if suitable excited states can exist in the liquid and dissipation sets in if excitations of the liquid are created. It is evident that the spectrum of the excited states must be such as to allow the conservation of energy and momentum in the process.

The kinetic energy of the object (in our case, a positive ion or a negative electron bubble) is reduced if an elementary excitation is created with energy ϵ and momentum $\hbar\mathbf{k}$. After the emission of the excitation the ion moves at a reduced speed v_f.

The requirements of energy and momentum conservation are enforced so as to yield

$$\frac{1}{2}mv_i^2 = \frac{1}{2}mv_f^2 + \epsilon \tag{1.5}$$

and

$$m\mathbf{v}_i = m\mathbf{v}_f + \hbar\mathbf{k}\,. \tag{1.6}$$

The final velocity $\mathbf{v}_f$ can be eliminated from these two equations so as to give

$$v_i \cos\theta = \frac{\epsilon}{\hbar k} + \frac{\hbar k}{2m}\,, \tag{1.7}$$

where θ is the angle between the momenta of the ion and of the quasiparticle. As $|\cos\theta| \leqslant 1$,

$$v_i \geqslant \frac{\epsilon}{\hbar k} + \frac{\hbar k}{2m}\,. \tag{1.8}$$

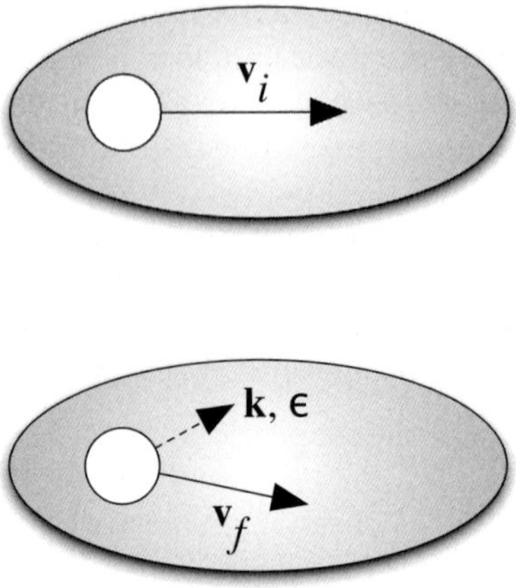

FIG. 1.1. Simplified scheme of a process of excitation creation. The moving ion with initial velocity $\mathbf{v}_i$ emits one quasiparticle with energy ϵ and momentum $\hbar\mathbf{k}$, and its velocity changes to the smaller value $\mathbf{v}_f$.

Equation (1.8) is the necessary condition to simultaneously enforce energy and momentum conservation. If the excitation dispersion curve is such that the right-hand side of eqn (1.8) has a positive minimum, then a threshold initial velocity of the moving object for the emission of a quasiparticle is defined. Below this threshold velocity the moving object has no chance to dissipate its kinetic energy to the fluid and therefore it must move without drag. In other words, the fluid shows superfluidity.

As it has been assumed that the moving ion has a very large mass, the second term in the right-hand side of eqn (1.8) is negligible. Thus, the so-called Landau *critical velocity* v_L can be defined as

$$v_L = \left(\frac{\epsilon}{\hbar k}\right)_{\min} . \tag{1.9}$$

The minimum value of $(\epsilon/\hbar k)$ is found by setting $\mathrm{d}(\epsilon/\hbar k)/\mathrm{d}k = 0$, yielding

$$\frac{\mathrm{d}\epsilon}{\mathrm{d}k} = \frac{\epsilon}{k} . \tag{1.10}$$

So, the existence of a threshold velocity is geometrically related to the coincidence of the tangent to the energy–momentum dispersion curve at a given point with the line joining that particular point to the origin. A liquid behaves as a superfluid if $v_L > 0$ depending on the shape of the dispersion curve of the elementary excitations of the liquid, $\epsilon \equiv \epsilon(k)$.

1.2 Spectrum of elementary excitations in He II

Consider, for instance, an ordinary liquid, such as liquid Ar. Its thermal energy is related to the thermal average of the kinetic energy of the motion of individual *real* particles, such as atoms or molecules. In this case, $\epsilon = (\hbar k)^2/2m$, where m is the particle mass. Because the particle spectrum extends down to $k = 0$, there is not any minimum value $(\epsilon/\hbar k)_{\rm min} > 0$. Thermal excitation of translational degrees of freedom of the constituent atoms is always possible, even in the limit of vanishing temperature, and the liquid is not superfluid.

The situation in superfluid He II is completely different. The thermal energy is carried by the elementary excitations, whose energy–momentum relationship has the peculiar features shown in Fig. 1.2. In Fig. 1.3 the dispersion curves measured by neutron scattering experiments at SVP and at a pressure of 2.56 MPa are shown (Henshaw and Woods, 1961*a*, 1961*b*).

For small values of the momentum, the energy is proportional to it, $\epsilon = ck$. This is the so-called *phonon* region of the spectrum. The only possible collective motions in the limit of long wavelength are quantized acoustic vibrations of the liquid. In this case, the critical velocity coincides with the sound velocity c, which, at SVP, varies in the range $c \approx 219$ m/s at the λ-point and $c \approx 238$ m/s at 0.1 K and increases up to $c \approx 370$ m/s close to the solid–liquid transition at a pressure $P \approx 2.5$ MPa (Findlay *et al.*, 1938, 1939; Atkins and Stasior 1953; Chase 1959; Whitney and Chase 1962; Rudnick and Shapiro 1965; Barmatz and Rudnick 1968).

For higher values of momentum, $\epsilon(k)$ shows a minimum in a range of momenta known as the *roton* region. Here, the dispersion curve can be approximated by

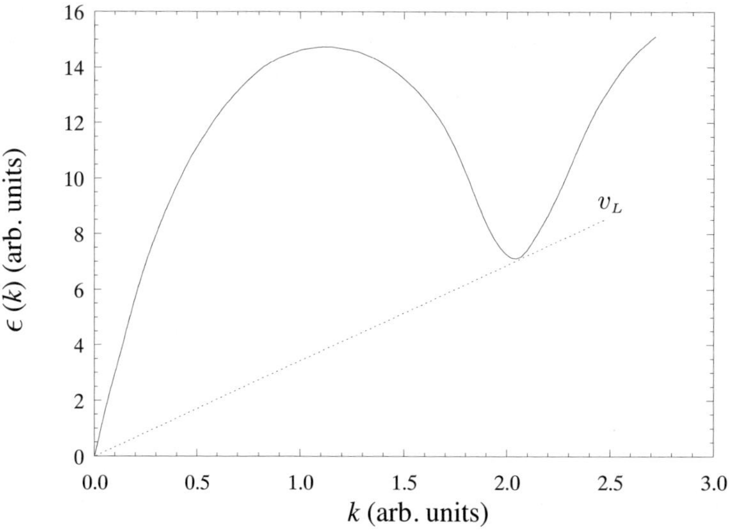

FIG. 1.2. Qualitative sketch of the energy spectrum of excitations in liquid He II. Excitations in the linear part at small k are phonons. Excitations close to the relative minimum are called rotons. The dashed line drawn from the origin is tangent to the dispersion curve near the roton minimum and satisfies eqn (1.10).

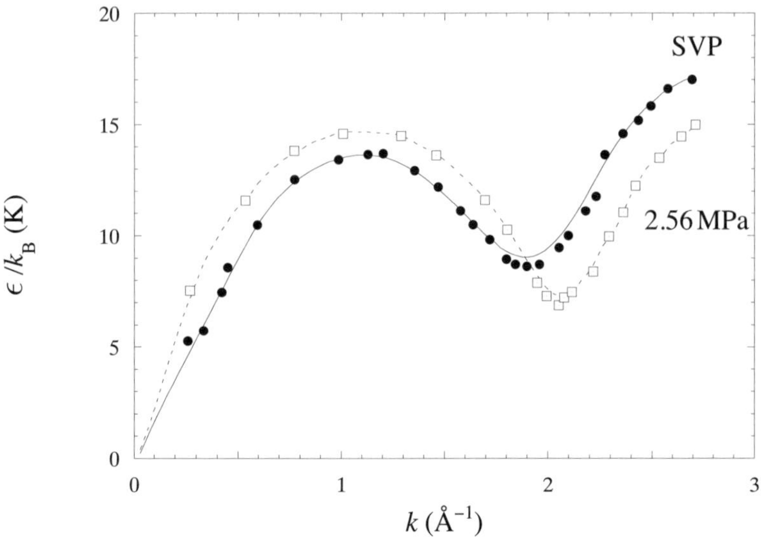

FIG. 1.3. Experimental energy spectrum of excitations in liquid He II at SVP for $T = 1.12\,\mathrm{K}$ (circles) and at pressure $P = 2.56\,\mathrm{MPa}$ (squares), as measured in neutron scattering experiments. (Henshaw and Woods, 1961*a*, 1961*b*.) The lines are only eyeguides.

a shifted parabola

$$\epsilon = \Delta + \frac{\hbar^2}{2m_r}(k - k_0)^2 \,, \tag{1.11}$$

where the energy gap Δ, the roton mass m_r, and the roton momentum $\hbar k_0$ are known as the Landau roton parameters that have been determined by means of neutron scattering experiments. For $T = 1.1\,\mathrm{K}$ at SVP their values are (Henshaw and Woods, 1961*a*)

$$\begin{aligned} \Delta/k_\mathrm{B} &= (8.65 \pm 0.04)\,\mathrm{K}, \\ k_0 &= (1.91 \pm 0.01)\,\mathrm{\mathring{A}}^{-1}, \\ m_r &= 0.16 m_4, \end{aligned}$$

where m_4 is the atomic mass of ^{4}He. Experimentally, small changes of Δ, k_0, and m_r have been observed as a function of temperature (Yarnell *et al.*, 1959) and pressure (Henshaw and Woods, 1961*b*).

The smallest value of $\epsilon/\hbar k$ is located near the minimum of the dispersion curve in the roton region, as shown in Fig. 1.2. From eqn (1.10), v_L is determined by the slope of the line drawn from the origin and tangent to the dispersion curve near the roton minimum. The point of contact is very close to the minimum itself and, within an accuracy of 1%, the critical velocity can be approximated by

$$v_L \approx \frac{\Delta}{k_0}. \tag{1.12}$$

Using the experimental values of the roton parameters (Donnelly, 1972), the values $v_L \approx 58\,\mathrm{m/s}$ at SVP and $v_L \approx 45.6\,\mathrm{m/s}$ close to the solidification pressure $P_m \approx 2.5\,\mathrm{MPa}$ are obtained.

1.3 Other localized "excitations"

The values of the critical velocity determined from the Landau criterion are, however, much larger than any observed critical velocities in experiments of hydrodynamic type as, for instance, the flow of superfluid helium through narrow channels or capillaries (Vinen, 1963, 1968; Wilks, 1967; Chrètien *et al.*, 1968; Keller, 1969; Fetter, 1976).

This difficulty was removed as soon as it was clear from the experiments that quantized vortices exist in the superfluid. They act as different types of localized excitations, for which much smaller $(\epsilon/p)_{\min}$ threshold velocities exist. Thermally-excited vortex lines are present in the bulk liquid but they can also be created when the superfluid flows in a narrow tube or when it is set into rotation. The transition to turbulence is believed to happen when the density of the vortex lines is increased enough to produce a tangled mass of them. For a complete review on quantized vortices, see Donnelly (1991).

Owing to the previous reasons, it is evident that the study of the breakdown of superfluidity is of paramount importance both from the theoretical as well as from the experimental point of view because it may give information on the microscopic dynamics and structure of the superfluid and on the nature of superfluidity itself.

1.4 Why ions and electrons as probes?

Although it is clear that quantized vortices are created by the flow of the liquid in tubes or by the motion of a macroscopic object in the liquid, the details of the breakdown of the superfluidity measured in such "macroscopic" hydrodynamical experiments are not well understood. In particular, complicated end effects are involved in flow experiments that are difficult to investigate and clarify. Microscopic roughness of the inner surfaces of capillaries or of the surface of moving objects may alter the boundary conditions of the flow, introducing vorticity at the same level. Some remnant vorticity, that depends on the conditions of sample preparation, might be present before the macroscopic flow is established and induce the initial superfluidity breakdown. Therefore, the process of fundamental interest, i.e., the process of intrinsic vorticity nucleation, is neither directly investigated nor accessible in this kind of experiment.

It was soon clear that the fundamental processes of superfluidity breakdown can be investigated by studying the motion of charge carriers (ions and electrons) through the liquid. Already starting with the pioneering work of Williams (1957), who measured the ionic mobility at high fields, and of Careri *et al.* (1959), who showed that ions participate in the normal fluid flow, experimental studies proved that negative and positive ions can be easily injected into the liquid by means of various techniques. They can be drawn through the liquid by applying small electric fields in a highly controllable fashion by means of suitably arranged assemblies of electrodes and their motion can be easily detected, even when their concentration is low, by measuring the time evolution of the current they induce in the external circuitry.

The so-called *ions* are in reality localized objects of semi-macroscopic size, with radii of the order of 10^{-9} m, and of large effective masses, $\approx 100\, m_4$. The associated structure is much larger than the bare ion.

The positive ion is rather made of a permanent cluster of He atoms strongly bound by electrostriction (Atkins, 1959) to the central He^+ or He_2^+ ion (Phelps and Brown, 1952). The negative ion, on the contrary, consists of an electron localized within a spherical, empty cavity, called a *bubble*, which is originated by the strong short-range repulsive exchange forces between the excess electron and the electron clouds of the atoms of the medium (Kuper, 1961).

Such entities and their associated structures, simply referred to as ions, are quite well defined on the atomic scale and can be regarded as practical tools to investigate the interaction between the superfluid and a boundary. They are of small enough size not to be spoiled by any nucleating remnant of vortices but they are large enough to give origin to a flow around them that, at least for not too large velocities, can be thought of as taking on a simple dipolar form.

The ions, when drawn through the liquid by the application of an external electric field, are subject to two different types of interaction. On one hand, they interact quantum-hydrodynamically with the liquid giving origin, owing to the well-defined boundary, to non-uniform superfluid velocity and density fields $\mathbf{v}_s$ and ρ_s in close proximity (Schwarz, 1975).

On the other hand, they are subjected to collisions with the elementary excitations present in the liquid. The velocity of the ion, therefore, changes randomly on a microscopic scale as a consequence of the scattering events but the ions move with an average drift velocity in the direction of the applied electric field. This kind of experiment is very closely related to the usual swarm experiments in dense gases (Huxley and Crompton, 1974; Borghesani and Santini, 1994*a*).

Only at a very low temperature, when the density of the excitation gas is low, can the two kinds of interaction be considered independent and distinct physical phenomena.

The experiments on ions in liquid He II can thus be roughly classified into two different types. On one side, there are experiments aimed at investigating the processes of scattering of elementary excitations off the moving ion, taking into account the fact that ions are endowed with a well-defined boundary. Below the critical velocity for quasiparticle creation ions can only scatter off pre-existing, thermally-excited quasiparticles. The outcome of these experiments is represented by a suitable average of the phenomenon under investigation weighted by the statistical effects arising from the velocity distribution function of the ions and from thermal fluctuations.

On the other side, the charged carriers with well-defined structures are used to investigate the quantum-hydrodynamic properties of the superfluid and its vortex structure. As clearly reported in Wilks (1967), the macroscopic appearance of rotational flow in the superfluid is related to the existence of vortex filaments, in the core of which the irrotational flow condition $\nabla \times \mathbf{v} = 0$ breaks down, whereas the superfluidity restriction applied outside the vortex core implies that the circulation around the vortex line $\kappa = \oint \mathbf{v} \cdot \mathbf{dl}$ is quantized in units of h/m_4. The non-superfluid core of vortices is estimated to be of size $\approx 1\,\text{Å}$ (Donnelly, 1991), i.e., its size is comparable to that of the ionic probe. It is thus assumed that vortex lines are sources of vorticity at the size scale of the ions so that the interaction between vortices and ions can be well approximated by the hydrodynamic interaction between a small sphere and a vortex string of given circulation.

At the same time, it has been recognized since the early experiments of Rayfield and Reif (1964) that ions moving in the liquid can create quantized vorticity and emit rotons (Rayfield, 1966). In any case, both processes are characterized by critical velocities, above which the probability of such processes occuring increases rapidly.

Over the years, a number of different experimental techniques have been designed, a great deal of experiments have been carried out, and several theoretical models have been devised to interpret the experimental data. Beyond the primary issues described above, ions, positive as well as negative, have been used to investigate vortex lines, vortex rings, phonons and rotons, the nucleation of vortices, and quasiparticle- and ^{3}He-impurity-scattering processes in the most widespread drift velocity experiments. In addition to mobility-related experiments, among other experiments, ions have been exploited to investigate the nature of steady-state turbulence in superfluid helium, the properties of normal liquid ^{4}He, of normal and superfluid ^{3}He at very low temperatures, and the nature of the electrostatically-bound states trapped just below the free surface of liquid helium. The structure of the ions themselves has been the subject of interesting experimental and theoretical studies. More complex ions have also been introduced into the liquid matrix in order to test the simple models of the ion structure.

In any case, ions have proven to be the best microscopic probe for investigating superfluidity in bulk liquid helium (Donnelly, 1967; Kestner, 1973; Schwarz, 1975; Fetter, 1976; Donnelly, 1991).

The goal of this book is to give a review of the large wealth of scientific results that the investigation of superfluid helium by means of ions has produced.

2

PRODUCTION OF IONS IN THE LIQUID

Several different techniques have been employed in order to inject ions and electrons into liquid helium. The characteristic features of the most widespread ones will be briefly described.

2.1 Direct liquid ionization

Probably, the most widespread technique to inject electrons into liquid helium is direct ionization and is shown in Fig. 2.1. A radioactive material is deposited on a metallic electrode that acts as an ion source. Typically, ^{210}Po (5.4 MeV) or ^{241}Am (5.6 MeV) alpha-particle emitters are used. The range of the alpha particles in

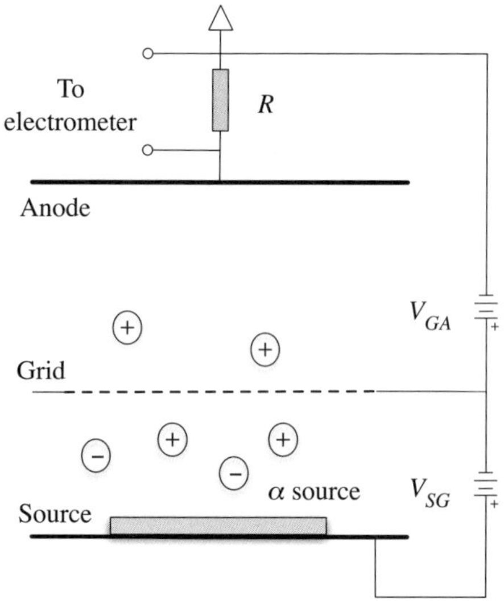

FIG. 2.1. Qualitative sketch of the simplest ion injection technique. A coating containing a radioactive source is deposited on the source electrode. Ionization is produced in the liquid in front of the source. A potential difference V_{SG} between the source and the grid selects the sign of the charge carriers injected into the drift volume between the grid and the anode. The potential difference V_{GA} determines the electric field pulling the ion toward the anode. The ion current is detected by amplifying the voltage drop across the high-impedance resistor R. More complicated grid assemblies and voltage set-ups, including gating grids and a.c. voltages, are used in real experiments.

liquid He is $\approx 0.3\,\mathrm{mm}$. A large amount of ionized atoms are produced in the liquid and a suitable arrangement of electric potentials applied to grids allows the extraction of ions of the desired polarity.

The strength of the alpha source varies in different experiments, but typically it is found to be in the range 10–200 μCi, yielding ion densities up to $10^{13}\,\mathrm{ions/m^3}$ and ionic currents in the range 10^{-13}–10^{-11} A (Meyer and Reif, 1958; Reif and Meyer, 1960; Careri *et al.*, 1964*a*; Cope and Gribbon, 1970*a*).

Alternatively, 18-keV β particles from tritium adsorbed on Ti evaporated on a stainless steel backing are used. In this case the source strength is in the range 1–8 mCi (Bruschi *et al.*, 1968*a*; Packard and Sanders, 1972; Sitton and Moss, 1972), yielding similar ion densities and currents as alpha-sources.

2.2 Field emission and field ionization

When a large negative potential is applied to a sharp metallic needle in the liquid, electrons may tunnel directly from the metal into the liquid if the electric field is strong enough to significantly reduce the barrier height. In *vacuo* the extracted electrons accelerate rapidly toward the collecting anode and the electric field is not modified by them (Gomer, 1961). In liquids, the charges, which are not necessarily free electrons, move much more slowly and spacecharge builds up that exerts an influence on the emitted current itself by reducing the field strength at the emitter. Moreover, secondary emission can take place in the proximity of the negative tip, thereupon complicating the situation (Halpern and Gomer, 1969*a*).

Field ionization is obtained by reversing the sign of the potential with respect to the emission case (Halpern and Gomer, 1969*b*). Now, an electron from an atom can tunnel inside the metallic tip, leaving an ion behind. In liquids, again, in contrast to rarefied gases, spacecharge phenomena and secondary emission are present.

A typical set-up for gated ion injection by means of spacecharge-limited discharges triggered by field ionization or emission is shown in Fig. 2.2. The ionic currents that can be obtained with these techniques are in the range 10^{-9}–10^{-6} A, depending on several factors such as the strength of the voltage applied to the tip, temperature, and the pressure of the liquid helium, among others. A drawback of these techniques is the ageing of the tip, whose emission characteristics degrade in the course of time owing to erosion phenomena.

A great deal of work has been devoted to characterize the properties of field ionization and emission in liquid He (McClintock, 1969, 1971, 1973*a*, 1973*c*; Gavin and McClintock, 1973; McClintock and Read-Forrest, 1973; McClintock and Webster, 1974; Phillips and McClintock, 1975; Bhatti and McClintock, 1979).

For currents $i > 10^{-9}$ A, it is proved that the characteristic features of the field emission or ionization yield information about the physical behavior of ions under a field of some tens of kV/m (Phillips and McClintock, 1975). In Figs 2.3 and 2.4 some typical current–voltage (I–V) characteristics of field ionization and emission in liquid He are shown.

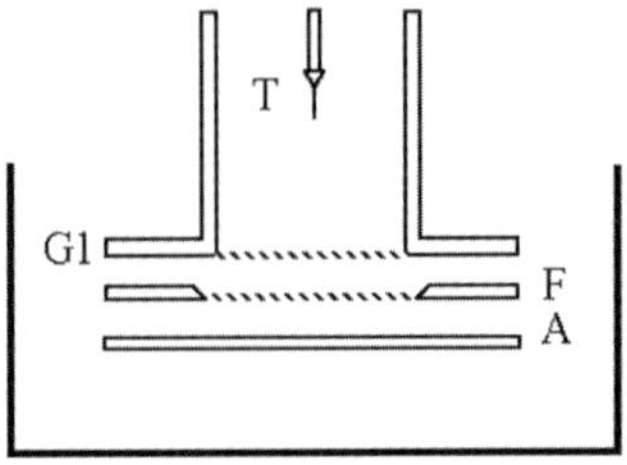

FIG. 2.2. Scheme of the electrical arrangement for gated injection of ions by means of field ionization. A voltage of several kV is applied between the tip T and the electrode G1. The region between T and G1 is the ion source region. A voltage between the grids G1 and F extracts the ions. The region between the two grids is the gating region. The Frisch grid F screens the collector A from the induction current due to the motion of the ions in the source and gating regions. The region between F and A is the drift region, where the electric field is determined by the potential difference between F and A and their relative distance. (Henson, 1964.)

By referring to Fig. 2.5, the current is obtained by solving the Poisson equation

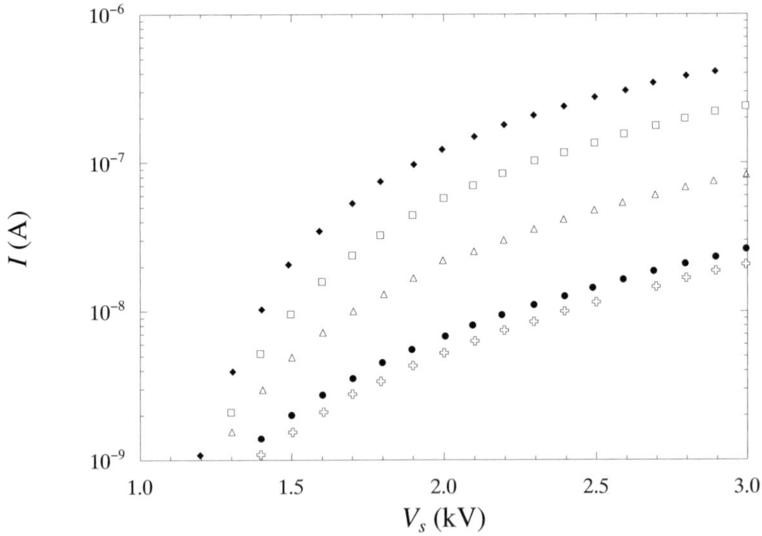

FIG. 2.3. I vs V_s for field ionization in superfluid He II for $P = 2.4\,\mathrm{MPa}$ and $T\,(\mathrm{K}) =$ 1.15 (diamonds), 1.44 (squares), 4.21 (triangles), 0.29 (circles), and 0.62 (crosses). (Phillips and McClintock, 1975.)

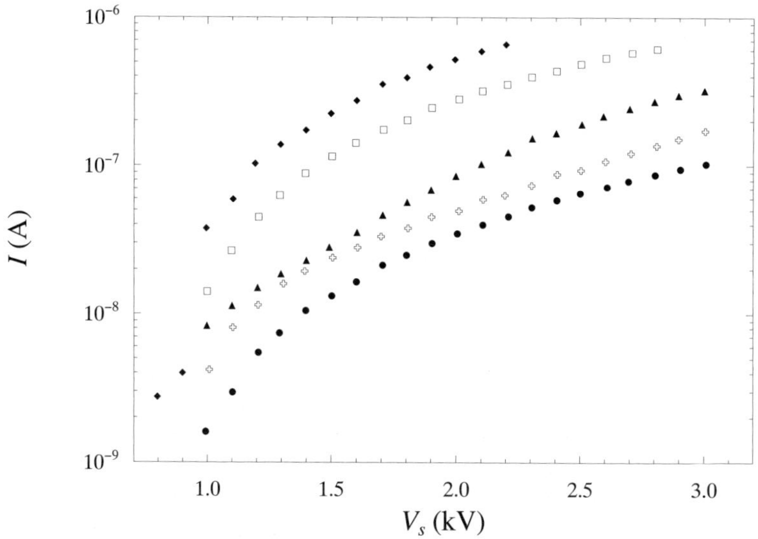

FIG. 2.4. I vs V_s for field emission in superfluid He II for $P = 2.5\,\text{MPa}$ and $T\,(\text{K}) = 1.21$ (diamonds), 1.46 (squares), 0.72 (triangles), 0.29 (crosses), and 2.52 (circles). (Phillips and McClintock, 1975.) The voltage polarity is the opposite of that in Fig. 2.3.

$$\nabla^2 V = -\frac{\rho}{\epsilon_r \epsilon_0}, \tag{2.1}$$

where ρ is the charge density, V is the potential between the tip and the collector, ϵ_r is the dielectric constant, and ϵ_0 is the vacuum permittivity . It is assumed that the problem has spherical symmetry and it is postulated that emission takes place only within a cone of solid angle $\alpha\pi$ sterad. The boundary conditions are $V = V_s$ for $r = r_s$ and $V = 0$ for $r = R$ (Phillips and McClintock, 1975). Moreover, $\mathrm{d}V/\mathrm{d}r = F_s$ for $r = r_s$. This last condition is enforced by the observation that the current is a very rapidly varying function of the electric field and that the field at the source can be effectively considered to be constant. The charge density is obviously related to the total current I by

$$\rho = \frac{I}{\alpha\pi r^2 v(r)}, \tag{2.2}$$

where $v(r)$ is the ion drift velocity at a distance r from the center.

In order to solve the Poisson equation, some assumptions about v must be made. The first one is that v is proportional to the local electric field $v = \mu \mathrm{d}V/\mathrm{d}r$ and the proportionality constant is the mobility μ. Equation (2.1) is then integrated, yielding

$$\frac{\mathrm{d}V}{\mathrm{d}r} = \left(\frac{2I}{3\alpha\pi\epsilon_r\epsilon_0\mu r}\right)^{1/2} \tag{2.3}$$

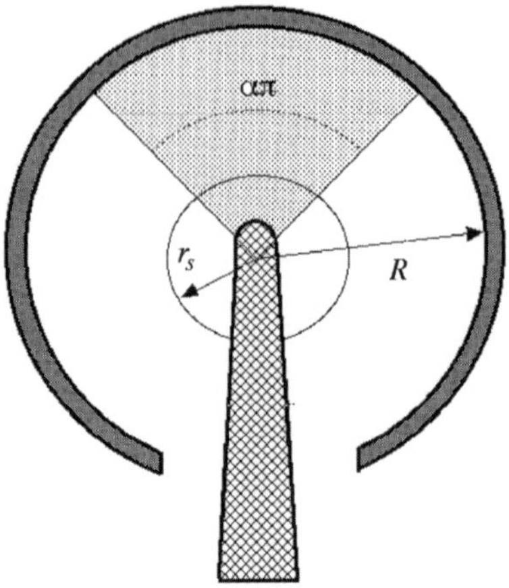

FIG. 2.5. Schematic geometry (not to scale) for describing field emission and ionization in liquid helium. r_s is the radius of curvature of the source region, which may be larger than the geometrical radius of curvature of the metallic needle. The size of r_s has been greatly exaggerated for the sake of clarity. $R \gg r_s$ is the radius of the collector. The emission is assumed to take place within a cone of solid angle $\alpha\pi$ steradians. (Phillips and McClintock, 1975.)

for $r \gg r_s$ and

$$V_s = V_0 + \left(\frac{8RI}{3\alpha\pi\epsilon_r\epsilon_0\mu} \right)^{1/2}. \tag{2.4}$$

V_s is the tip voltage and $V_0 = F_s r_s$ in the fully-spacecharge-dominated region.

In the opposite case, when the ion drift velocity is a constant independent of the electric field, the integration of the Poisson equation yields

$$\frac{\mathrm{d}V}{\mathrm{d}r} = \frac{I}{\alpha\pi\epsilon_r\epsilon_0 v r} \tag{2.5}$$

for $r \gg r_s$ and

$$V_s = V_0 + \frac{I}{\alpha\pi\epsilon_r\epsilon_0 v} \left[\ln\left(\frac{R}{r_s}\right) - 1 \right] \tag{2.6}$$

again in the fully-spacecharge-dominated regime.

These equations describe the two limiting cases. The details are fully accounted for in Phillips and McClintock (1975). The features of the (I–V) characteristics shown in Figs 2.3 and 2.4 depend on the physical behavior of the charge carriers through their mobility or drift velocity, as expressed by eqns (2.4) and (2.6). A general discussion of their properties as a function of temperature and pressure in liquid He will thus be postponed.

2.3 Glow discharge

D.c. glow discharges have been used as ion sources in experiments where the surface states of electrons are studied (Sommer, 1964). Typically, the glow discharge is initiated in the vapor above the liquid. A suitable electrode assembly then draws electrons toward the liquid–vapor interface. The field-assisted thermionic effect allows electrons to overcome the surface barrier and to be injected into the liquid.

2.4 Photoelectric effect

Ultraviolet (UV) light impinging on a photocathode easily extracts electrons from the metal because of the very well-known photoelectric effect. Although very often used in gas-phase electron swarm experiments (Grűnberg, 1969; Bartels, 1972; Christophorou, 1984; Borghesani *et al.*, 1986*a*; Borghesani and Santini, 1990*b*, 1992; Lamp and Buschhorn, 1994), the photoelectric effect has been seldom used in liquid helium, probably because the barrier to be overcome is quite high, $\approx 1\,\mathrm{eV}$, and drastically reduces the number of electrons that can be injected in comparison with gas-based experiments. In fact, the photoelectric effect has been practically exploited only in the experiment aimed at measuring the injection barrier and the electron conduction band energy in the liquid (Woolf and Rayfield, 1965).

2.5 Tunnel junctions

A solid-state device called an MIM has sometimes been used to inject electrons into helium and other insulating liquids (Silver *et al.*, 1967, 1970; Smejtek *et al.*, 1973). The cathode consists of a three-layer, solid-state structure, as shown schematically in Fig. 2.6, realizing a tunnel diode that is immersed in the liquid.

A thin Al_2O_3 film, whose thickness is of $\approx$ 100–150 Å, is grown on the top of a metallic Al film. A very thin Au metallization acts as the cathode of the structure. The acronym MIM stands for metal–insulator–metal. A voltage of a few volts (typically, 10 V) biases the tunnel diode so that electrons tunnel from the Al film into the Au film through the oxide layer. Some of the tunnel electrons have large enough energy to escape from the Au film into the liquid, where they become available for the experiment. These cathodes are excellent electron emitters. The *I–V* characteristics are obtained by taking into account the fact that the current is attenuated by back diffusion, which depends on the electron mean free path. Under suitable assumptions, the current density is given by

$$j_c \approx j_0 e^{-x_m/x_0}\,, \tag{2.7}$$

where $x_m = (e/16\pi\epsilon_r\epsilon_0 E)^{1/2}$ is the coordinate of the potential maximum in the liquid due to the combined action of the image potential and of the electric field E applied between the emitter and the anode and x_0 is the electron thermalization length in the liquid.

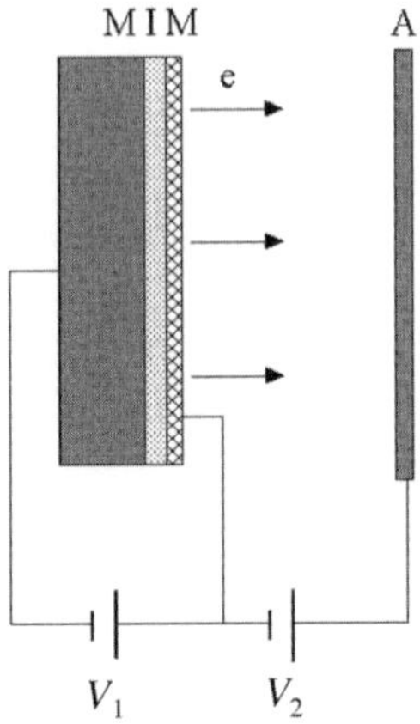

FIG. 2.6. Sketch of the metal–insulator–metal (MIM) tunnel diode electron emitter. A sandwich structure of Al–Al_2O_3–Au immersed in the liquid is polarized so that electrons tunnel from the Al film to the Au film through the ≈ 100 Å thick Al_2O_3 layer. Electrons entering the thin Au film (a few tens of Å thick) have enough energy to escape into the liquid, where they drift toward the collector A.

With an emitter area of $0.1\,\mathrm{cm}^2$, currents in the range 1–100 pA can be obtained, depending on the bias voltage. A typical MIM current–voltage characteristics is shown in Fig. 2.7 (Silver *et al.*, 1967).

This kind of tunnel diode has been successfully employed in experiments aimed at measuring the energy of the electron surface barrier in liquid ^{3}He and ^{4}He (Broomall *et al.*, 1976).

2.6 Thermionic emission

Traditional thermionic electron emission from a heated tungsten filament has been seldom used (Spangler and Hereford, 1968). A thin wire of tungsten (with diameter of some tenths of a μm) is immersed in liquid helium and heated up to 2000 K. A stable vapor film develops around the heated filament. Electrons are emitted by means of the thermionic effect into the vapor and then enter the liquid. Currents much larger than those produced by ionizing radiation or a tunnel cathode are achieved, up to 10^{-6} A.

The electron source is mounted in a diode configuration. The wire is stretched along the axis of a cylindrical collecting electrode. A potential difference is applied between the filament and the collector. A typical I–V characteristics is shown in Fig. 2.8.

For low voltages, the current is proportional to the applied voltage. At around 1 kV there is an increase of the slope, due to electron multiplication in the va-

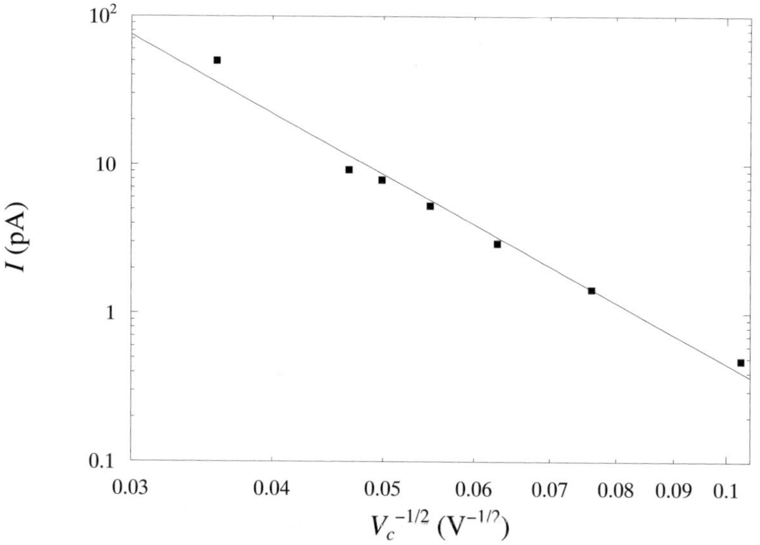

FIG. 2.7. Tunnel emission current in He at 1.8 K vs the inverse square root of the voltage applied across the liquid. (Silver *et al.*, 1967.) The line is a fit to eqn (2.7).

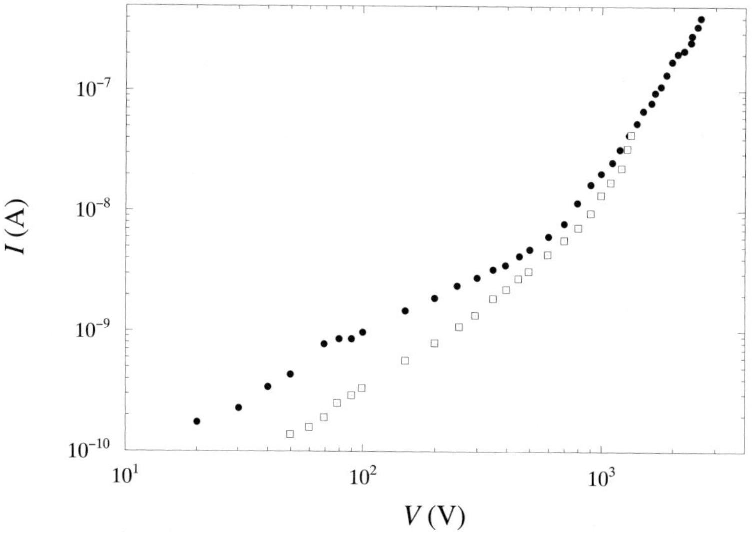

FIG. 2.8. I–V characteristics of a superfluid diode with a 5 μm diameter filament operated at the following values of power and temperature: 130 mW, 2110 K (circles); 88 mW, 2000 K (squares). Immersion depth: 3 cm. (Spangler and Hereford, 1968.)

por film surrounding the wire. The amount of emitted current also depends on the immersion depth of the filament below the helium surface. The cylindrical geometry is a limiting factor for ion mobility studies because of the field non-uniformity. This technique has been used in studies on turbulence in superfluid helium (Sitton and Moss, 1969).

2.7 Ion implantation by glow discharge or laser ablation

Positive alkali or alkaline–earth ions can be implanted into the liquid by adopting hot-filament techniques and glow-discharge or laser-ablation techniques. The simplified schemes of two apparatuses implementing such techniques are presented in Figs 2.9 and 2.10.

A slightly modified hot-filament technique has been adoptedby Ihas and Sanders (1970) and Glaberson and Johnson (1975). In this case there is a combination of the hot-filament and glow-discharge techniques. A tungsten wire is suspended above the free liquid helium surface and is biased with 0.5–1 kV with respect to a set of electrodes so as to act as a cathode. A glow-discharge is ini-

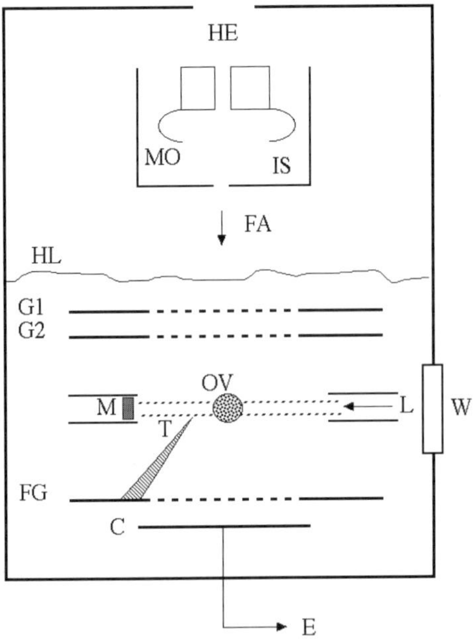

FIG. 2.9. Sketch of the drift mobility and laser spectroscopy cell with ion implantation by the flowing afterglow technique. HE: flow input of He, MO: micro-oven, IS: ion source, FA: flowing afterglow, HL: liquid helium level, G1 and G2: gating grids for ion pull and injection, M: mirror, OV: optical volume, L: excitation laser, W: optical window, T: electron-emitting tip for ion neutralization, FG: Frisch grid, and C: ion collector, connected to the electrometer amplifier via E. (Bauer *et al.*, 1989.)

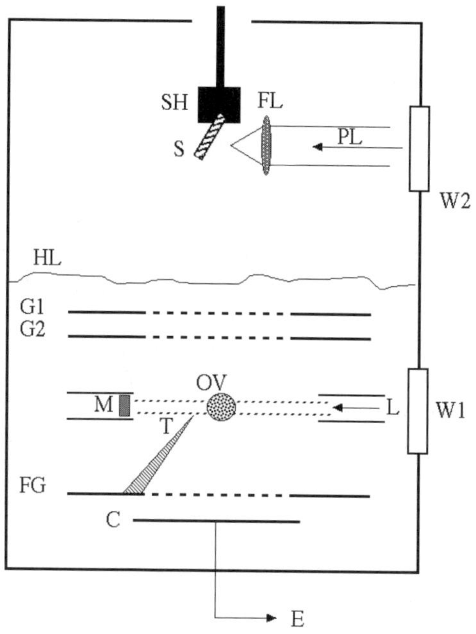

FIG. 2.10. Sketch of the drift mobility and laser spectroscopy cell with ion implantation by the laser-ablation technique. SH: sample holder, S: alkali or alkaline–earth sample to be evaporated, FL: focusing lens, PL: power nitrogen laser, HL: liquid helium level, G1 and G2: gating grids for ion injection, M: mirror, OV: optical volume, L: excitation laser, W1 and W2: optical windows, T: electron-emitting tip for ion neutralization, FG: Frisch grid, and C: ion collector, connected to the electrometer amplifier via E. (Tabbert *et al.*, 1995.)

tiated in the vapor and positive He ions are produced. The wire, coated with a solution of the element of interest, is also heated to a high temperature. Upon heating, the ions of the element are evaporated and positively ionized in the glow discharge. A suitable assembly of electrodes then pulls the ions toward and into the liquid. The exotic ions were observed using this technique (Ihas and Sanders, 1971).

In Fig. 2.9 the apparatus for ion implantation in the liquid by means of the glowing afterglow technique is shown. This technique, and its variation of ion implantation by laser ablation shown in Fig. 2.10, has also been used to investigate the spectroscopic properties of alkali and alkaline–earth atoms and ions embedded in liquid helium (Bauer *et al.*, 1985). These studies aim at investigating the structure of ions or atoms interacting with the superfluid (Tabbert *et al.*, 1995, 1997; Toennies and Vilesov, 1998).

The ion source consists of two micro-ovens filled with the metal under investigation. The He vapor flows through a region where a weak discharge is glowing and through the ovens. Metastable He atoms produced in the discharge and the evaporated metal atoms travel together in the He flow. Ions are produced by Penning ionization during the flow.

The resulting flowing afterglow reaches the free surface of the liquid. The ions enter the liquid because of the action of a system of gating grids and are drifted through a mobility cell, so that their mobility can also be measured. Two perpendicular sets of optical windows define two optical axes. One of them makes a path for laser excitation, and the other one leads to the detection system to observe the induced fluorescence of the laser-excited ions. If neutral atoms are to be investigated, a negatively-biased sharp tip injects electrons into the optical volume within the drift space and electron–ion recombination takes place. The atoms can then be further excited by means of the laser.

More recently, a modification to the previous implantation technique has been devised (Bauer *et al.*, 1990; Tabbert *et al.*, 1995). Laser ablation, instead of micro-ovens and glow discharge, is used to evaporate the alkali or alkaline–earth atoms, as shown in Fig. 2.10. Subsequent light absorption from the intense laser beam ($\approx 10^{13}\,\mathrm{W/m^2}$ inside the cryostat), in part also due to inverse Brehmsstrahlung, leads to excitation and ionization of the plasma constituents and to further heating of the atomic and ionic cloud up into the eV range. The implantation of the ions and the electron–ion recombination in the liquid to study the spectroscopic properties of the neutrals follow the procedure described previously.

Finally, we recall the recent introduction of the new "helium-jet" technique to inject microparticle impurities into liquid helium. A jet of helium containing traces of the microparticles to be investigated is directed towards the surface of the liquid (Boltnev *et al.*, 2002). This technique removes the need for neutralizing the ions to produce neutral species for spectroscopic investigation, as required in the laser-ablation or in the afterglow-implantation techniques.

3

THE STRUCTURE OF IONS

When the nature and structure of ions in liquid helium are discussed, a clear distinction must be made between positive and negative charge carriers. The positive $(^4\mathrm{He})_2^+$ ions are stable species with a binding energy of $\approx 2.4\,\mathrm{eV}$ (Pauling, 1967) and are detected in low-pressure discharges (Phelps and Brown, 1952), whereas the He^- ion is unstable.

Early measurements (Meyer and Reif, 1958; Careri *et al.*, 1959) have shown that the drift mobility of ions is much lower than that expected on the basis of the Nernst–Einstein relationship between the diffusion coefficient D and the ion mobility μ:

$$D = \frac{k_\mathrm{B}T}{e}\mu\,, \tag{3.1}$$

where e is the electron charge, k_B is the Boltzmann constant, and μ is the drift mobility. Actually, the diffusion coefficient of the positive ion at $T = 1.2\,\mathrm{K}$, calculated from the experimental mobility by means of eqn (3.1), is approximately two orders of magnitude smaller than the diffusion coefficient of $^3\mathrm{He}$ in dilute $^3\mathrm{He}$–$^4\mathrm{He}$ mixtures, although an ion and a $^3\mathrm{He}$ atom in solution should be quite similar objects.

This experimental observation has suggested that ions in liquid helium are complex structures. In particular, positive ions consist of a stable cluster of He atoms solidified around the ion, whereas the negative ion is an electron localized inside a fluid dilation. In any case, these structures endow the ion with a *static* effective mass, in addition to the usual *dynamical* hydrodynamic mass that amounts to several tens of $^4\mathrm{He}$ mass units.

3.1 Positive ions

Atkins (1959) devised the basic model for the description of the structure surrounding the positive ions. A positive ion produces a strong non-uniform electric field that polarizes the nearby fluid. In order to minimize the energy, the fluid is attracted toward the ion, where the field is strongest. This electrostriction effect thus produces a local enhancement of the fluid density around the ion.

By treating the fluid as a continuum and neglecting quantum corrections because atoms are well separated in the liquid with little overlap of the wave functions, electrostriction can be readily calculated. The relative dielectric constant ϵ_r of the fluid is related to the number density N and atomic polarizability α of the fluid by the Lorentz–Lorenz formula

$$\frac{\epsilon_r - 1}{\epsilon_r + 2} = \frac{\alpha}{3\epsilon_0}N\,, \tag{3.2}$$

where α is in SI units. Because the dielectric constant is close to unity, $\epsilon_r = 1.057$, eqn (3.2) can be inverted so as to yield

$$\epsilon_r \approx 1 + \frac{\alpha}{\epsilon_0} N . \tag{3.3}$$

In thermodynamic equilibrium in a non-uniform field the chemical potential must be constant throughout the fluid:

$$g_0 = \tilde{g}\left[P\left(r\right)\right] + V(r) , \tag{3.4}$$

where g_0 is the chemical potential of the unperturbed fluid at a large distance from the ion and $\tilde{g}[P(r)]$ is the local chemical potential that depends on the local pressure $P(r)$. $V(r)$ is the potential generated by the ion at a distance r.

The equilibrium condition is obtained by setting $\mathrm{d}g_0 = 0$, yielding, at constant T,

$$N^{-1}\mathrm{d}P = -\mathrm{d}V(r) , \tag{3.5}$$

where the thermodynamic relation $N\left(\partial g/\partial P\right)_T = 1$ has been used (Guggenheim, 1977).

Equation (3.5) is integrated between $r \to \infty$, where $V = 0$ and where the unperturbed fluid density and pressure are N_0 and P_0, respectively, and r, where the local density and pressure are N and P:

$$\int_{N_0}^{N} N^{-1} \left(\frac{\partial P}{\partial N}\right)_T \mathrm{d}N = -V(r) . \tag{3.6}$$

The long-range contribution to $V(r)$ comes from the interaction between the ion and the dipole induced on the atoms:

$$V(r) = -\frac{\alpha e^2}{2\left(4\pi\epsilon_0\epsilon_r\right)^2 r^4} . \tag{3.7}$$

So, by rearranging terms and integrating, one obtains

$$\mathcal{Z}\left(N, N_0\right) \equiv \left[\epsilon_r\left(N\right)\right]^2 \int_{N_0}^{N} N^{-1} \left(\frac{\partial P}{\partial N}\right)_T \mathrm{d}N = \frac{\alpha e^2}{2\left(4\pi\epsilon_0\right)^2 r^4} \tag{3.8}$$

or, equivalently,

$$r = \left[\frac{\alpha e^2}{2\left(4\pi\epsilon_0\right)^2 \mathcal{Z}\left(N, N_0\right)}\right]^{1/4} , \tag{3.9}$$

where the approximated expression given by eqn (3.3) can be used for $\epsilon_r(N)$. The density profile $N(r, N_0)$ is then calculated by inverting eqn (3.9) once the equation of state of the fluid is known. Below $T \approx 1.2\,\text{K}$, this can be approximated very well by a cubic (Abraham *et al.*, 1970), so that the integration in eqn (3.8) can be readily carried out analytically.

In Fig. 3.1 a typical density profile is shown. In close proximity to the ion, the pressure can be so large as to exceed the melting pressure, $P_m \approx 2.5\,\text{MPa}$ at low temperature (Grilly, 1966). It is therefore necessary to assume that He is solid very close to the ion, giving origin to a structure known as a *snowball.* In the figure the liquid–solid boundary is clearly shown. Owing to the spherical symmetry of the interaction potential $V(r)$, the snowball is a sphere with radius $R_+ \approx 6\,\text{Å}$.

This picture of a solid core around the positive charge is consistent with the results of trapping-time measurements of radioactive $^{12}B^+$ and $^{8}Li^+$ ions in liquid He (Takahashi *et al.*, 1995, 1996). The observed preservation of nuclear spin polarization of the core ions in the snowball suggests that the nuclear spins interact very weakly with the environment, thus supporting the model of a highly-symmetrical structure surrounding the ions.

A refinement of the calculations includes the possibility that a surface energy density σ_{ls} is associated with the existence of the liquid–solid boundary. In this case, the location of the boundary is implicitly defined by the condition

$$P_l\,(R_+) = P_m + \frac{2\sigma_{ls}}{R_+}\frac{v_s}{v_l - v_s}\,, \tag{3.10}$$

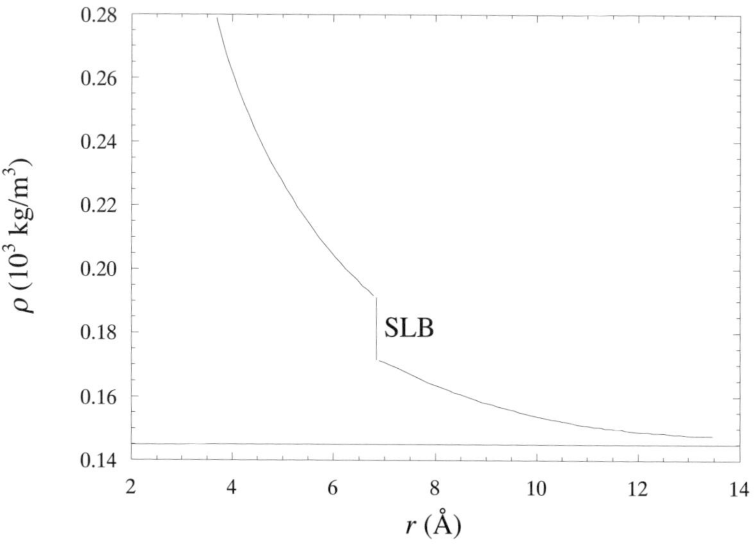

FIG. 3.1. Density profile $\rho(r)$ near a singly-charged positive ion in liquid He at $T = 1.25\,\text{K}$. The density of the unperturbed fluid at a large distance from the ion is $\rho \approx 145\,\text{Kg}\,\text{m}^{-3}$. The solid–liquid boundary (SLB) is clear. (Atkins, (1959).)

where $P_l(R_+)$ is the pressure in the liquid at the snowball boundary, v_l and v_s are the molar volumes of the liquid and of the solid, respectively, and R_+ is the ion radius.

In Fig. 3.2 several pressure profiles are shown for different values of the unperturbed fluid pressure $P_l(r \to \infty)$. The intersection of the right-hand side of eqn (3.10) with the electrostrictive curves $P(r)$ yields the core radius R_+. Unfortunately, the value of the liquid–solid surface tension σ_{ls} is not known, but experiments favor a value $\sigma_{ls} = (0.10 \pm 0.05) \times 10^{-3}\,\mathrm{N/m}$ (Schwarz, 1975).

For $T \geqslant 1.2\,\mathrm{K}$, the positive ion radius becomes temperature dependent, essentially because P_m increases rapidly with T and R_+ shrinks correspondingly, as shown in Fig. 3.3 (Ostermeier and Schwarz, 1972).

The ion drags around with itself in its motion an extra mass that is easily obtained by integrating the excess density over all the space:

$$M_e = 4\pi \int_0^\infty (\rho - \rho_0)\, r^2 \mathrm{d}r\,. \tag{3.11}$$

M_e is estimated to be of the order of approximately forty ^{4}He atoms for $T = 0$

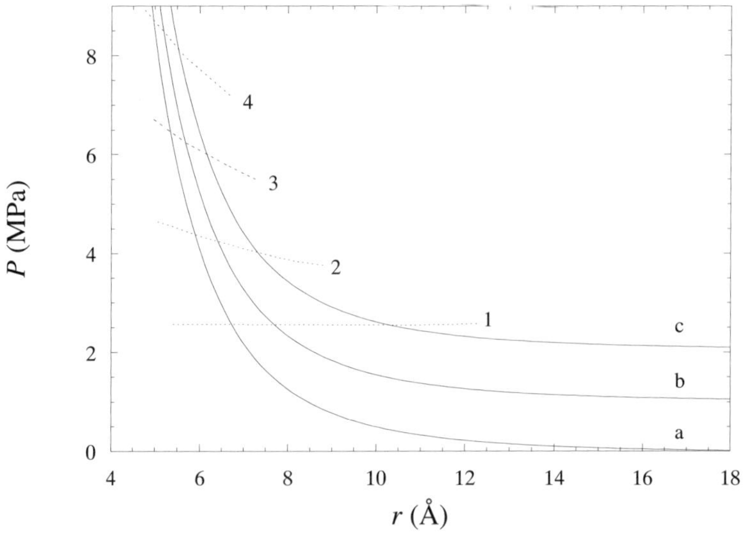

FIG. 3.2. Electrostriction-induced variation of the liquid pressure as a function of the distance from a positive ion in liquid He below 1 K. The unperturbed fluid pressure at an infinite distance from the ion for the different curves is $P\,(\mathrm{MPa}) = 0$ (a), 1.01 (b), and 2.02 (c). The dashed lines are the right-hand side of eqn (3.10) calculated for different values of the liquid–solid surface tension $\sigma_{ls}\,(10^{-3}\,\mathrm{N/m}\,) = 0$ (1), 0.05 (2), 0.10 (3), and 0.15 (4). The intersection of the dashed and solid lines determines the solidification radius R_+. (Schwarz, 1972*c*.)

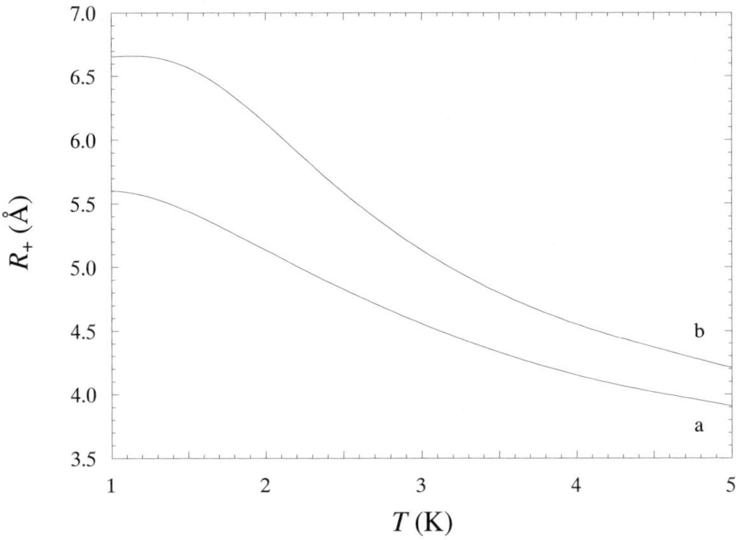

FIG. 3.3. A plot of R_+ vs T for σ_{ls} (10^{-3} N/m) = 0.1 (a) and 0 (b). (Ostermeier and Schwarz, 1972.)

Kand increases up to nearly sixty atoms at the normal boiling point ($T = 4.2\,\mathrm{K}$).

This extra mass remains attached to the ion as long as the ion velocity does not exceed the sound velocity. It is not clear, however, if this static extra mass also represents the hydrodynamic effective mass associated with the moving positive ion. Actually, it is known that a small sphere of radius R_+ and density equal to the mass density ρ_0 of the unperturbed fluid has an effective mass $2\pi\rho_0 R_+^3$ because the hydrodynamic flow contributes one half of the mass of the fluid displaced (Landau and Lifsits, 2000).

The success of the Atkins model shows that it correctly takes into account the relevant physical phenomena, i.e., the local compression due to the attractive polarization potential. However, the continuum approximation is unsatisfactory, in particular in the core region, where the density and pressure profiles vary appreciably on the atomic scale. The same kind of logical inconsistency has been encountered in a different context, namely, when describing the drift mobility of O_2^- ions in near-critical Ne and Ar gases (Borghesani *et al.*, 1993, 1997), of O_2^- ions in liquid Xe, and of positive ions in tetramethylsilane (Hilt *et al.*, 1994). Probably, the good results of the continuum approximation rely on the fact that thermal fluctuations on the scale of the ionic structure are sufficiently small and rapid so as to ensure that the thermodynamic relationships describe the long-time behavior of the density. In any case, the snowball model has to be considered phenomenologically sound.

According to the Atkins model, the properties of the He atoms clustering around the positive ion are independent of the nature of the core. There is, however, experimental evidence that there is a dependence on the ion core. Actually, the mobility of positive alkali ions is smaller than that of positive He ions and decreases with increasing atomic number, whereas for the positive alkaline–earth ions the trend is reversed (Ihas and Sanders, 1970; Glaberson and Johnson, 1975).

A refinement of the snowball model has been introduced by explicitly including three new kinds of ion–environment interaction (Cole and Bachman, 1977; Cole and Toigo, 1978). These are the direct interaction with the valence electrons of the central ion, an interaction with the excess or defect of He atoms near the ion, and a van der Waals interaction with the core ion. Again, this treatment requires the assumption of local thermodynamic equilibrium, i.e., a slow variation of the density near the ion. Thus, this assumption must be accepted insofar as the predictions of the model are in agreement with the experimental results.

For the alkali ions, two modifications to the Atkins model are introduced. The first interaction is an additional attractive potential of the van der Waals type:

$$V_I = -C_I r^{-6}\,, \tag{3.12}$$

associated with the interaction between the fluctuating dipole moments of the core ion and an He atom at a distance r.

The second interaction is an additional attraction exerted on an He atom of the environment due to the larger He density around the core ion with respect to the unperturbed fluid. Each atom is subjected to a potential energy of the form

$$U(\mathbf{r}-\mathbf{r}') = -C|\mathbf{r}-\mathbf{r}'|^{-6} \tag{3.13}$$

if $|\mathbf{r}-\mathbf{r}'| > 3.5$ Å. The interaction between an atom at r and the excess atoms in the solid cluster surrounding the core ion is then given by summing over all contributions:

$$V_C(r) = \int_0^{R_+} (N_s - N_l)\, U(\mathbf{r}-\mathbf{r}')\, \mathrm{d}\mathbf{r}'\,, \qquad r > R_+ + d, \tag{3.14}$$

$$V_C(r) = \text{constant}, \qquad R_+ < r < R_+ + d\,, \tag{3.15}$$

where $d = 2.56$ Å is the hard-core He–He interaction diameter and $C = 1.59 \times 10^{-60}\,\mathrm{J\,m^6}$. N_s and N_l are the solid and liquid number densities, respectively. These two contributions give an additional positive contribution to the alkali ion radius predicted by the snowball model that compares favorably with the experimental observations (Cole and Bachman, 1977).

For the positive alkaline–earth ions the interaction of the remaining electron outside of the filled shell cannot be neglected. Owing to the fairly large spatial extension of the valence-electron wave function ψ, a repulsive exchange interaction with the nearby atoms appears. This interaction is very similar to the case of electrons (Fetter, 1976) and excited He states in liquid helium (Hickman and Lane, 1971; Wisdom *et al.*, 1976). In these cases, a cavity is created within the liquid because of this short-range exchange repulsion. This additional repulsion can be cast in the following form (Shih and Woo, 1973; Padmore and Cole, 1974):

$$V_e(r) = \frac{2\pi\hbar^2 a}{m} |\psi(r)|^2 , \tag{3.16}$$

where a is the electron–atom scattering length, which, for He, is positive, $a \approx 0.63$ Å (O'Malley, 1963). The repulsive energy is practically infinite relative to the other energies in the problem. The remaining valence electron thus originates a hard-core potential of radius $R_1 \approx 4$ Å, which locally reduces to a great extent the He density so that the ion is practically localized within a nearly-hollow cavity up to ≈ 4.7 Å (Cole and Bachman, 1977). The effect of this cavity on the surrounding density is accounted for in approximately the same way as for the alkali ions through eqns (3.14) and (3.15) by noting that a depletion of atoms produces repulsion in this case.

The resulting density profile for the alkaline–earth positive ions is qualitatively shown in Fig. 3.4. Curve 1 is obtained by including the repulsive exchange potential and hole repulsion; curve 2 is calculated without hole repulsion. The solid–liquid boundary (SLB) is indicated by an arrow, as well as the Hartree–Fock hard-core interaction potential due to the positive Sr^+ ion.

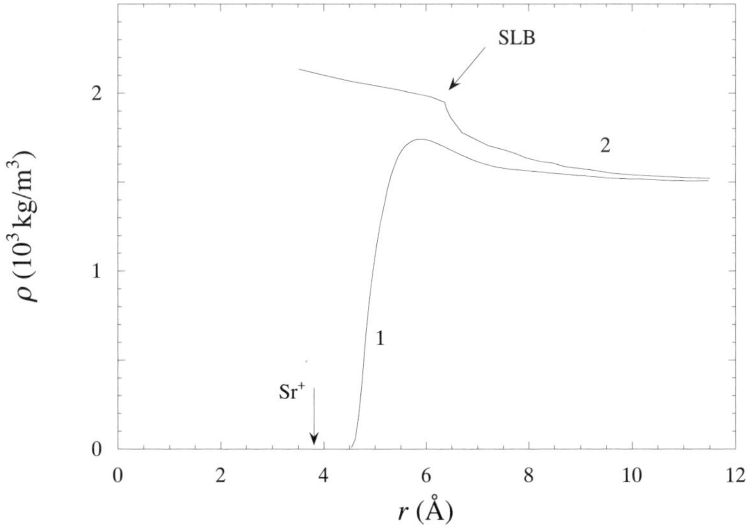

FIG. 3.4. Qualitative density profile near a singly-charged, positive alkaline–earth ion in liquid He. The exchange repulsion between the electron left outside the filled shell and the He atoms creates a nearly-empty cavity around the ion. The arrow shows the hard-core potential diameter due to the Sr^+ ion's valence electron (curve 1). The density profile obtained without this contribution is also shown for comparison (curve 2). SLB indicates the solid–liquid boundary. (Cole and Bachman, 1977.)

From Fig. 3.4 it can be clearly noted that the effect of the cavity produced by the exchange repulsion is to slightly reduce the ionic radius with respect to the simple snowball model of Atkins. Moreover, the absence of He atoms very close to the ionic core leads to an effective repulsion for $r > 4.7\,\text{Å}$, so that the density does not reach the high value necessary to produce a solid shell.

Also in the case of alkaline–earth ions, there is a great improvement of the predictions of this refined model toward the results of the experiments. We must, however, recall that even this model is based on the assumption of local thermodynamic equilibrium and is subject to the same kind of criticism as the snowball model.

3.2 Negative ions

The electron–atom interaction in He is dominated by the short-range repulsive exchange forces because of the small atomic polarizability of He that leads to a positive scattering length $a \approx 0.63\,\text{Å}$ (O'Malley, 1963). Therefore, the e–He interaction can be modeled as a hard-sphere repulsion (Hiroike *et al.*, 1965; Jortner *et al.*, 1965).

A very naive calculation of the mobility of an excess electron drifting through the liquid would yield $\mu = e\tau/m \approx (e/\pi m a^2 N)(m/3k_{\mathrm{B}}T)^{1/2} \approx 5 \times 10^{-2}\,\mathrm{m^2/V\,s}$, to be compared with the experimental value $\mu \approx 2 \times 10^{-6}\,\mathrm{m^2/V\,s}$ (Meyer and Reif, 1958; Careri *et al.*, 1959). Moreover, the predicted density and temperature dependencies are wrong. Even worse, the negative charge mobility is smaller than the corresponding positive ion mobility, thereby confirming that the effective mass of the negative ion is much larger than that of the positive one. This observation rules out the possibility that the negative ion is a quasi-free electron.

A very simple Wigner–Seitz calculation of the energy of an excess electron at the bottom of the conduction band yields an energy of $\approx 1\,\mathrm{eV}$ (Hernandez, 1991). Let us assume that the overlap of the electron–atom potentials and the density fluctuations in the liquid can be ignored. In such a model a unit Wigner–Seitz cell is assigned to each atom. An average, ordered model for the liquid has electron wave functions of the Bloch type (Ashcroft and Mermin, 1976; Hernandez and Martin, 1991) $\psi(\mathbf{r}) = \exp{(i\mathbf{k}\cdot\mathbf{r})}u_k(\mathbf{r})$, where u_k has the translational symmetry of the lattice. At the bottom of the band it is sufficient to take $k \approx 0$. Thus, if an atom is at the origin, the wave function u_0 is given by

$$u_0 = 0\,, \qquad r < a\,, \tag{3.17}$$

$$u_0 = \left(\frac{A}{r}\right)\sin\left[k_0\,(r-a)\right], \qquad r > a\,, \tag{3.18}$$

where k_0 is to be determined by the boundary conditions. To satisfy the average translational symmetry of the liquid, the logarithmic derivative of u_0 must vanish at the Wigner–Seitz cell boundary r_s, defined by the condition $4\pi r_s^3 N = 1$. The boundary condition yields

$$\tan\left[k_0\,(r_s - a)\right] = k_0 r_s\,. \tag{3.19}$$

The energy at the band bottom is then

$$V_0 = \frac{\hbar^2 k_0^2}{2m} \approx \frac{4\pi\hbar^2 a}{2m} N\,, \tag{3.20}$$

where the last expression, known as the *optical potential*, is obtained by a power series expansion of the tangent in the limit of low densities. With $a \approx 0.63\,\text{Å}$, eqn (3.20) yields $V_0 \approx 0.98\,\text{eV}$ (Jortner *et al.*, 1965; Springett *et al.*, 1967). This prediction is confirmed by experimental results of electron injection from the vapor into the liquid or extraction from the liquid (Careri *et al.*, 1960; Sommer, 1964; Woolf and Rayfield, 1965; Bruschi *et al.*, 1966*b*; Broomall *et al.*, 1976).

Because of this large energy, and owing to the natural compliance of a fluid, a lower-energy state can be accomplished if the excess electron gets trapped in an unstable density fluctuation, where the local density is lower than the average one. Once trapped, the electron stabilizes the density inhomogeneity by pushing away atoms from its neighborhood and becomes self-trapped.

The conditions for electron self-trapping, as the most probable state, are a trade-off between a large density of the average fluid that makes the energy of the delocalized excess electron state very high and a compliant medium that can undergo distortions without too large an energy cost.

The dynamics of electron self-trapping is not completely clarified yet (Dexter and Fowler, 1969; Hernandez and Silver, 1970; Hernandez and Silver, 1971; Schwarz and Prasad, 1976; Artem'ev and Khrapak, 1986; Sakai *et al.*, 1992; Rosenblit and Jortner, 1995; Schmidt *et al.*, 2001), in particular as far as the origin of the incipient bubble is concerned. However, it is known that electrons injected into liquid helium thermalize in a few picoseconds (Onn and Silver, 1969) by creating the charged bubble.

The calculation of the equilibrium cavity structure is a well-assessed topic (Hernandez, 1991). A very detailed *bubble* model was first suggested by Kuper (1961). Since then, several different and more refined models have been proposed (Miyakawa and Dexter, 1970; Hernandez, 1973, 1975; Khrapak and Iakubov, 1979). This issue is discussed in detail in another chapter of this book. I therefore discuss here only the simplest model of the spherically-symmetric square well in order to grasp the physics of self-trapping.

It is assumed that the electron resides in an empty cavity that can be modeled as the spherically-symmetric, infinite square well of basic quantum mechanics. This assumption is justified by the large barrier $V_0 \gg k_\mathrm{B}T$ at the temperatures of liquid He. The surrounding fluid exerts a net inward pressure on the bubble boundary, which is compensated by the outward pressure exerted by the electron owing to its zero-point energy. If the cavity boundary is assumed to be a step function, the effect of the neighboring atoms can be described by an effective surface tension σ.

The equilibrium radius R_- is obtained by minimizing the electron excess free energy with respect to the radius. The excess free energy can be written as follows (Miyakawa and Dexter, 1970):

$$\Delta\mathcal{A} = \mathcal{E}_e + 4\pi\sigma R_-^2 + \frac{4}{3}\pi P R_-^3 - \frac{1}{2}\left(\frac{\epsilon_r - 1}{\epsilon_r}\right)\frac{e^2}{4\pi\epsilon_0 R_-} - V_0\,, \tag{3.21}$$

where $\mathcal{E}_e$ is the energy eigenvalue of the Hamiltonian for the square well. The second contribution on the right-hand side is the surface energy. The third term represents the pressure–volume work at constant temperature necessary to expand the cavity. It can be cast in this very simple form just because the cavity is empty. The fourth contribution is the Born electron solvation energy. It is added here only for the sake of completeness, but it can be safely neglected with respect to all other energies involved. Finally, V_0 is given by eqn (3.20) or by the experiment.

The electron energy $\mathcal{E}_e$ is obtained by solving the Schrődinger equation for the radial wave function. If the s-wave ground state is sought, the electron energy is obtained as

$$\mathcal{E}_e = \frac{\hbar^2}{2mR_-^2}X^2\,, \tag{3.22}$$

where X is obtained by solving the usual eigenvalue equation

$$(\tan X)^{-1} = -\frac{\sqrt{\mathcal{P} - X^2}}{X} \tag{3.23}$$

and $\mathcal{P} = (2m/\hbar^2)V_0R_-^2$ is the well strength (Gasiorowicz, 1974).

The radial wave function of the electron ground state and the radial probability density are shown in Fig. 3.5. In the case shown, the barrier height is 1 eV, the cavity radius is 17 Å, and the first energy eigenvalue is $\approx$ 100 meV. It can be clearly seen that the electron wave function is practically confined completely within the cavity and it spills very little over the barrier because of its large height.

If the leakage of the wave function over the barrier can be neglected, i.e., if the barrier is assumed to be indefinitely high, then $X = \pi$. The equilibrium radius is obtained by minimizing eqn (3.21), thus yielding the condition

$$P = \left(\frac{\hbar^2\pi}{4mR_-^5} - \frac{2\sigma}{R_-}\right)\,. \tag{3.24}$$

Using the bulk helium surface tension $\sigma \approx 0.37 \times 10^{-3}$ N/m (Atkins and Narahara, 1965), the zero-pressure value of the equilibrium radius is readily obtained as $R_-(P = 0) \approx 19$ Å, corresponding to an excess free energy $\Delta\mathcal{A} \approx -0.81$ eV and to an effective mass of several hundreds of helium atom units (Springett *et al.*, 1967). The relative probability of an electron to be still free, $\exp \Delta\mathcal{A}/k_BT$, is thus completely negligible.

The value of the radius thus obtained is in reasonable agreement with the experimental results of electron injection (Sommer, 1964; Woolf and Rayfield, 1965) in spite of the coarse assumptions to neglect the bubble curvature and the leakage of the wave function into the liquid, and to assume a step-function density profile of the bubble.

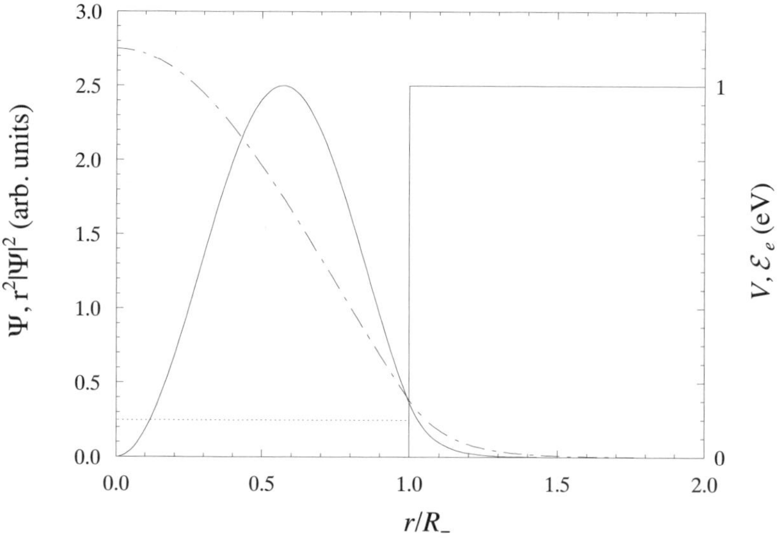

FIG. 3.5. Radial wave function (dash–dotted line) and radial probability density (solid line) of the s-wave solution of the Schrődinger equation of the electron trapped in the hollow cavity. The step-function shape of the potential energy is also shown. A horizontal line represents the first energy eigenvalue of the electron. The scale on the right refers to the energies. The distance is expressed in units of the bubble radius R_-.

For $P \neq 0$, the surface tension is not known directly and must therefore be inferred from theoretical estimates (Amit and Gross, 1966), which predict that σ scales as $c(P)\rho_\infty(P)$, where c is the sound velocity and $\rho_\infty(P)$ is the density of the unperturbed fluid at the given pressure. With these assumptions, it is easy to minimize eqn (3.21) with respect to the radius in order to obtain its equilibrium value as a function of P.

Upon increasing P up to 2.5 MPa, there is a decrease of nearly 30% in the equilibrium radius. This decrease is the result of two opposing effects. On one hand, there is an increase in the electron kinetic energy due to the reduction of the localization volume, which acts so as to withstand the radius contraction.

On the other hand, the decrease in the radius produces a stronger diminution of the excess free energy due to the volume- and surface-work terms. The net effect is that the excess-free-energy minimum is reached for radii values smaller than for $P = 0$. In Fig. 3.6 the pressure dependence of the equilibrium radius of the electron bubble is reported for two different values of the surface tension. The model is compared with some experimental data showing the nice agreement between theory and experiment.

In a similar way, the temperature dependence of the equilibrium radius can be easily calculated and in Fig. 3.7 this dependence is shown at vapor pressure.

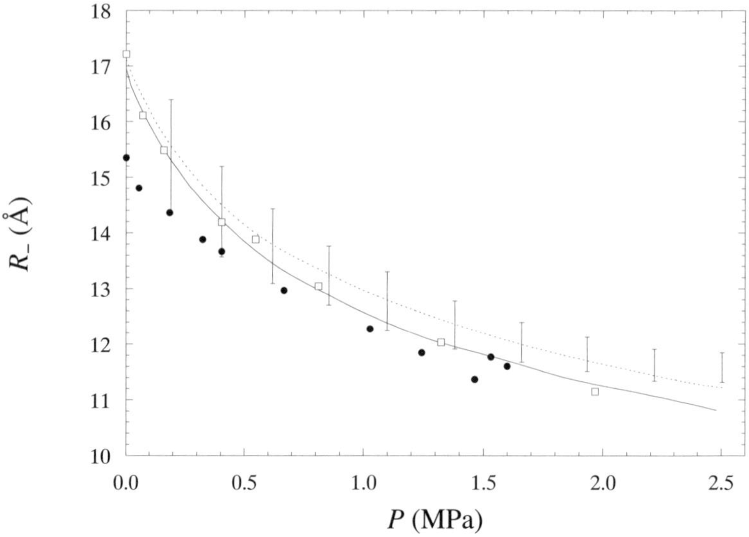

FIG. 3.6. R vs P. Dashed line: eqn (3.24) with the bulk liquid surface tension. Solid line: eqn (3.24) with $\sigma \propto c(P)\rho_\infty(P)$ (Amit and Gross, 1966). Squares: vortex line trapping experiments (Springett, 1967). Dots: photoejection experiments (Zipfel, 1969). Bars: phonon-limited mobility measurements (Ostermeier, 1973).

The radius increases slightly with T because the surface tension decrease with T overcompensates for the pressure increase.

This simplified square-well model allows the calculations of other interesting properties of the electron bubble such as its optical (Fowler and Dexter, 1968; Miyakawa and Dexter, 1970) and vibrational properties (Celli *et al.*, 1968; Gross and Tung-Li, 1968).

Further refinements of the theoretical model have been introduced by minimizing the free energy as a functional of the He density $\rho(r)$ and of the electron wave function $\psi_e(r)$. The minimization of the free-energy functional gives origin to a set of nonlinear coupled equations for both the local density profile and the electron wave function. In fact, this is an envelope function, since the electron–atom interaction details have been hidden by a contact potential (an average pseudopotential approach). The shape of the density profile and of the wave function are calculated in a self-consistent manner simultaneously (Khrapak and Iakubov, 1973; Miller and Reese, 1989; Hernandez, 1991).

These density-functional models (Padmore and Cole, 1974; Ancilotto and Toigo, 1994) have the advantage of describing in much better detail the features of the bubble surface than the simple step-function model. The density profile turns out to be smoother than in the square-well model. As a consequence, the radius of the bubble is not precisely defined but has an intrinsic uncertainty of ≈ 2 Å. In any case, these density-functional models have proven very successful

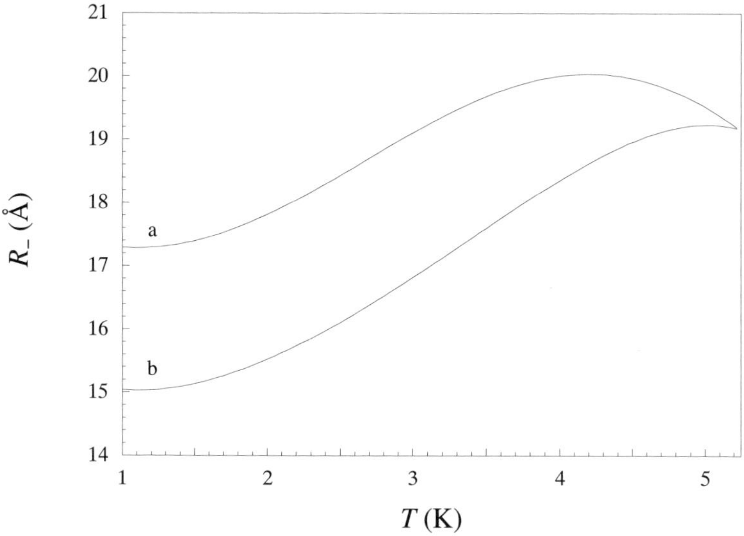

FIG. 3.7. R_- vs T for two different assumptions about the surface tension σ_{lv}. Curve a: experimental σ_{lv}. Curve b: experimental σ_{lv} enhanced by a factor of 1.7. (Ostermeier and Schwarz, 1972.)

in tackling several problems in liquid helium, such as the spectroscopic properties of the alkali ions (De Toffol *et al.*, 1996), or the approach of the electron bubble to the liquid helium surface (Ancilotto and Toigo, 1994).

It has to be noted that the conditions of electron self-trapping in a bubble may fail if the helium density is too low, so that V_0 cannot support bound states or the binding energy is less than the fluid distortion energy. Indeed, a continuous transition from high mobility values in the dilute gas to low mobility, typical of the electron bubble, at higher gas density has been reported for helium gas (Levine and Sanders, 1962, 1967; Harrison *et al.*, 1973; Borghesani and Santini, 2002).

Moreover, the conditions for self-trapping are not peculiar to helium only. In fact, self-trapping has to be expected in a fluid, whose electron–atom interaction is sufficiently strong and repulsive to produce a large energy of the conduction band, whose density is large enough to make the binding energy greater than the distortion energy, and whose temperature is low enough not to thermally excite the electron outside the cavity. Actually, localized states have been detected by investigating the excess electron mobility in liquid neon (Bruschi *et al.*, 1972; Loveland *et al.*, 1972; Sakai *et al.*, 1992; Storchak *et al.*, 1996), in dense hydrogen gas (Harrison and Springett, 1971), and in dense neon gas (Borghesani *et al.*, 1988; Borghesani and Santini, 1990*a*, 1992). The square-well model or the more refined density-functional, self-consistent models or even computer simulations have also been adopted in these cases to explain the observed mobility behavior of the excess electrons (Hernandez, 1991; Hernandez and Martin, 1991; Ancilotto and Toigo, 1992).

As a final remark it has to be recalled that localized states of light particles in liquid helium have also been detected for positron or positronium. In fact, the term "bubble" was introduced for the first time by Ferrell (1957), when he aimed to explain why positronium has a longer lifetime in liquid helium. Details about this issue can be found elsewhere (Khrapak and Iakubov, 1979; Nieminen *et al.*, 1980; Iakubov and Khrapak, 1982; Hernandez, 1991).

4

IONS AND ELEMENTARY EXCITATIONS

Ions slowly moving in liquid He at zero temperature are expected not to encounter any resistance owing to the superfluid nature of the liquid. At any higher temperatures, the ions suffer collisions with the thermal excitations that are present in the fluid. Under the influence of an applied electric field, E, the charge carriers accelerate rapidly until the mean drag force $\mathcal{F}$ exerted by the liquid balances the applied electric force eE. The ions then move with an average drift velocity v_D in the direction of the field. For small enough electric field strengths, v_D is proportional to E and the proportionality constant is the *mobility* (Huxley and Crompton, 1974):

$$v_D = \mu E \,. \tag{4.1}$$

In this restricted velocity regime, the mobility at low T depends on the ion–elementary excitation scattering processes. Thus, a study of μ or v_D as a function of E, T, and P gives a great deal of information about the interaction of the charge carriers with the liquid and on their structure.

The spectrum of elementary excitations in liquid helium is shown in Fig 1.3. Only the regions around $k = 0$ and around the roton minimum for $k = k_0$ are thermally populated. At high temperatures, say $T \geqslant 1\,\mathrm{K}$, the normal fluid properties are dominated by rotons owing to their large momenta. The roton density decreases exponentially with decreasing temperature (Wilks, 1967) and, eventually, phonons dominate.

In addition, ^{3}He impurities are to be considered as elementary excitations characterized by the quasiparticle spectrum

$$\epsilon = -\epsilon_{03} + \frac{\hbar^2 k^2}{2m_3^\star} \,, \tag{4.2}$$

with $\epsilon_{03}/k_\mathrm{B}T \approx 2.8\,\mathrm{K}$ and effective mass $m_3^\star \approx 2.4\,m_3$ (Ebner and Edwards, 1970), where m_3 is the mass of the ^{3}He atom. ^{3}He occupies a larger specific volume than ^{4}He because of its smaller mass. The ^{3}He quasiparticles can thus be envisaged as a low-density region with a radius of $\approx 4\,\mathrm{Å}$ and a fractional volume excess ≈ 0.29. The impurity-limited mobility of the ions can be studied experimentally by adding small amounts of the lighter He isotope for $T \leqslant 1\,\mathrm{K}$.

5

TRANSPORT PROPERTIES OF CHARGE CARRIERS AT LOW TEMPERATURE

In the limit of low temperature the number of excitations is so low that the quasiparticles can be treated as a mixture of non- or weakly-interacting gases and the contribution of the different scattering processes can be singled out. At a higher temperature the excitations cannot be treated as non-interacting any longer and, for instance, close to the λ-transition the situation is far more complicated than at low temperature.

5.1 Transport at small fields

In typical experiments the charge carrier density is sufficiently low to consider them as a weakly-ionized plasma. Under the combined action of a small electric field and of excitation scattering, the ions reach a steady-state velocity. Although there are interesting data on the field dependence in the various scattering regimes, the present discussion is limited here to the case of vanishingly-small electric fields. This is the case in which the energy imparted to an ion from the electric field in a mean free path ℓ is smaller than its thermal energy: $eE\ell \ll k_{\mathrm{B}}T$. Henceforth, μ will denote the zero-field mobility.

To determine the steady-state properties of the ion gas, the momentum distribution function $f_{\mathbf{p}}$ of the carriers must be known. Under the assumptions of steady-state conditions and of spatial uniformity, $f_{\mathbf{p}}$ is obtained by solving the Boltzmann equation (Huang, 1963; Chapman and Cowling, 1970; Huxley and Crompton, 1974)

$$e\mathbf{E}\cdot\frac{\partial f_{\mathbf{p}}}{\partial \mathbf{p}} = \sum_{\mathbf{p}'}\left[f_{\mathbf{p}'}\Gamma\left(\mathbf{p}'\rightarrow\mathbf{p}\right) - f_{\mathbf{p}}\Gamma\left(\mathbf{p}\rightarrow\mathbf{p}'\right)\right] , \qquad (5.1)$$

where $\Gamma\left(\mathbf{p}\rightarrow\mathbf{p}'\right)$ is the scattering rate, i.e., the probability per unit time that a carrier of momentum $\mathbf{p}$ is scattered to a different state $\mathbf{p}'$. Obviously, Γ depends on the number and type of the elementary excitations responsible for scattering, as well as on the nature of the carrier itself.

Equation (5.1) implicitly assumes that the time interval during which the charged carrier interacts with excitations is a small fraction of its time. This condition is fulfilled (Schwarz, 1975) for $T < 1.7\,\mathrm{K}$ for the positive ions and for $T < 1.1\,\mathrm{K}$ for the negative ones, and limits the ^{3}He impurity concentration to a maximum of 2.5% in the case of positive ions and 0.1% for the negative ions. At a higher temperature or impurity concentration, multiple-scattering effects must be taken into account.

Generally speaking, the ion perturbs the distribution function and, consequently, also the scattering rate Γ. However, if the excitation mean free path ℓ_{exc} is much larger than the ion size, the perturbation due to the charge carrier can be safely neglected (Chapman and Cowling, 1970). The mean free path can be estimated from the kinetic theory formula for the viscosity:

$$\eta = \frac{1}{3}\rho_n \bar{v} \ell_{\mathrm{exc}}\,, \tag{5.2}$$

where $\bar{v}$ is an average thermal velocity.

By using experimental viscosity data (Staas *et al.*, 1960; Woods and Hollis-Hallett, 1963) and taking into account the fact that phonon scattering is an inefficient momentum relaxation process, the roton mean free path is estimated to be $\ell_r \approx 10^{-9}\exp(\Delta/k_{\mathrm{B}}T)\,\mathrm{m}$. The requirement $\ell_r \geqslant 10R$, where R is the radius of the charged probe, leads to the restriction $T < 1.4\,\mathrm{K}$ for the positive ions and $T \leqslant 1.2\,\mathrm{K}$ for the negative ones (Schwarz, 1975). Moreover, the phonon mean free path is very large anyway ($\ell_{\mathrm{ph}} \approx 200\,\text{Å}$ at $T = 1.4\,\mathrm{K}$) and does not represent a limiting factor. Also, the addition of ^{3}He impurities does not alter the situation significantly.

Owing to these considerations, the Boltzmann transport equation can be safely used for $T \leqslant 1.4\,\mathrm{K}$ for the positive carriers and $T \leqslant 1.1\,\mathrm{K}$ for the negative carriers because, in these conditions, the mean free path of the relevant excitations is very large compared to the radius of the ions and the presence of ions does not perturb the distribution function of the excitations.

5.2 Experimental techniques

The drift velocity measurements have been carried out mainly with time-of-flight methods. There are several different implementations of these techniques. All of them are based on a precise timing of the time interval spent by the ions to cross a well-defined distance. Here a brief description of the commonly used techniques is given.

5.2.1 *Double-gate velocity spectrometer*

Probably the most widespread technique, originally developed by Meyer and Reif (1958, 1961), and subsequently improved by, for instance, Kuchnir *et al.* (1972), is the double-gate drift velocity spectrometer. A schematic drawing of the geometry of the electrode assembly is shown in Fig. 5.1.

The alpha particles emitted by the ^{210}Po-plated source electrode S ionize the He atoms just in front of S. A d.c. field of suitable polarity drives the ions of the right sign toward the collector C. An a.c. field of frequency ν is superimposed between the closely-spaced pair of grids g_1-g_2. The pair of grids acts like an electrical shutter allowing ions to pass through predominantly during one particular part of each cycle only. Similarly, the same a.c. voltage is applied between the pair of grids g_3-g_4, which also act as an identical shutter. The ionic current reaching the collecting electrode C is therefore a maximum whenever the drift

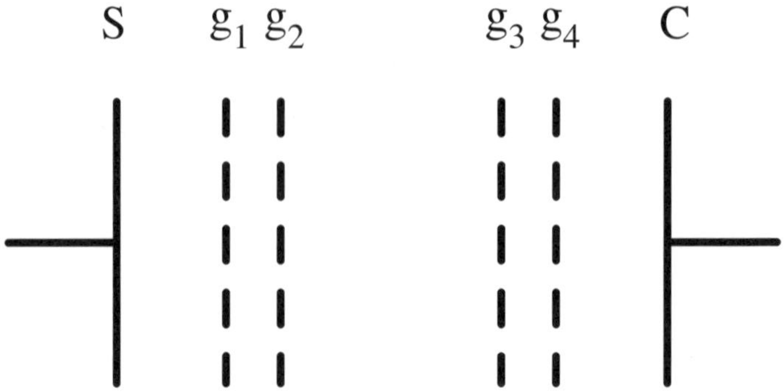

FIG. 5.1. Schematic drawing of the electrode assembly in the drift velocity spectrometer used by Meyer and Reif (1958). Ions are produced by the α-particles emitted by the Po-plated source electrode S and drift toward the collector C under the action of a small d.c. electric field. The pairs of grids g_1-g_2 and g_3-g_4 are driven by an a.c. voltage and act as electrical shutters.

time, i.e., the time spent by the ions traveling from g_2 to g_3, is equal to an integral number of the periods ν^{-1} of the a.c. voltage.

5.2.2 *Single-gate velocity spectrometer*

The single-gate drift velocity spectrometer was originally devised by Cunsolo (1961) and later improved by Bruschi and Santini (1970). This technique was exploited by several researchers (Careri *et al.*, 1964*a*; Bruschi *et al.*, 1966*c*, 1970; Tanner, 1966; Neeper and Meyer, 1969).

A schematic diagram of the geometry of the electrode assembly is shown in Fig. 5.2. The charges produced by alpha-particle ionization are extracted and directed toward the grid by an adjustable d.c. voltage of either polarity, depending on the sign of the carriers under investigation. Between the grid and the collector an a.c. square-wave voltage is applied.

During one half-period the a.c. electric field drives the charges toward the collector, while during the other half-period the field is reversed and the charges are swept back. If the spreading of the swarm due to diffusion and end-effects can be neglected, the working equation of the velocity spectrometer is

$$I = \frac{I_0}{2}\left[1 - \left(2d^2\nu/\mu V_{sw}\right)\right] \qquad \text{for } \nu^{-1} \geqslant 2d^2/\mu V_{sw}\,, \tag{5.3}$$

$$I = 0 \qquad \text{for } \nu^{-1} \leqslant 2d^2/\mu V_{sw}\,, \tag{5.4}$$

where I_0 is the current collected if a d.c. potential is applied between the grid and the collector, d is the grid-to-collector distance, ν and V_{sw} are the frequency and amplitude of the square wave, respectively, and μ is the mobility. μ is determined by zeroing the current as a function of the square-wave frequency.

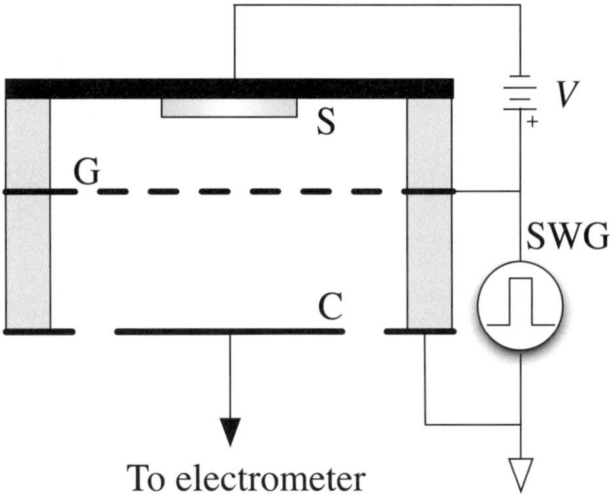

FIG. 5.2. Schematic drawing of the single-gate velocity spectrometer developed by Cunsolo (1961). S: Po-plated source electrode, G: grid, GR: guard ring, C: collector, V: d.c. injection voltage, and SWG: square-wave generator.

5.2.3 *Charge pulse time-of-flight*

A very accurate technique for the measurement of the ionic drift time has been devised by Schwarz (1972*a*). A schematic of the electrode assembly is shown in Fig. 5.3.

In this method, charges created by ionization close to the source electrode S are injected into the drift space by applying a suitable d.c. voltage across the gap between the source and ths grid G1. The drift distance is the long distance ($\approx$ 27 cm) between the grid G1 and the Frisch grid FG. Such a long distance makes the drift times quite large, so that they can be measured with great accuracy. The drift voltage V_D is equally divided between a network of twenty-five equally-spaced guard rings, GR, so as to ensure a great uniformity of the drift field.

The ion current is gated on or off by applying an additional voltage pulse across the S–G1 region. The current pulse propagates across the drift space with a characteristic drift velocity. The collector is screened from the current pulse by the Frisch grid FG, so that the collector records the current pulse only as soon as it appears just beyond FG. The signal is averaged by a signal averager in order to improve the signal-to-noise ratio. The drift time is defined by the time delay between the S–G1 gating pulse and the appearance of the current pulse at the collector. Details about the accuracy of the method, including the signal distortion introduced by the rise time of the amplifier, by space–charge effects at very low drift velocities and at high signal levels, or by focusing fields are found in the original paper by Schwarz (1972*a*).

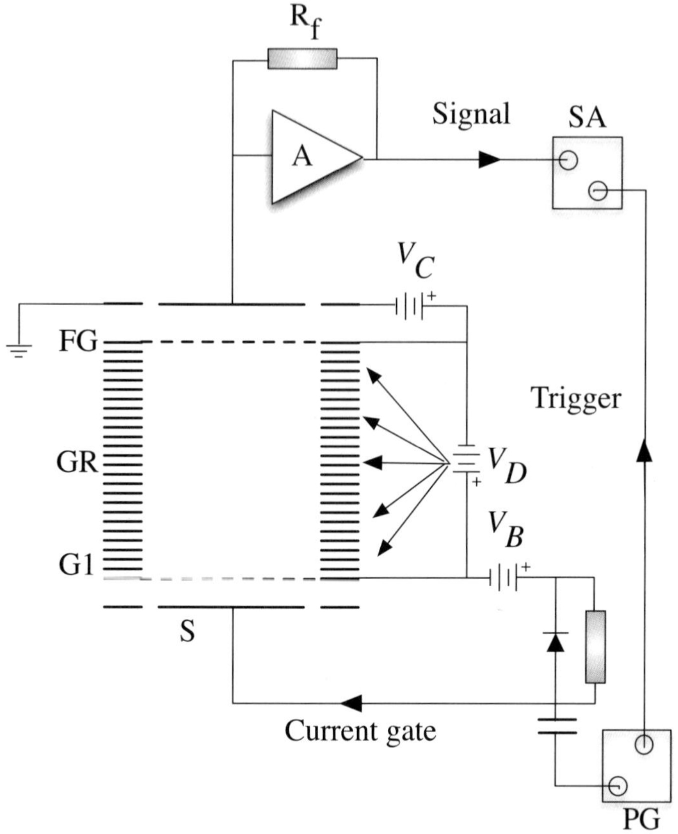

FIG. 5.3. Sketch of the long drift tube. S: Am-plated source electrode, G1: extraction grid, GR: guard rings, FG: Frisch grid, C: collector, A: amplifier, R_f: feedback resistor, SA: signal averager, PG: pulse generator, and V_B, V_C, and V_D: d.c. bias. The gap between the grids G1 and FG is the drift distance. V_D is the drift voltage, equally divided between twenty-five equally-spaced guard rings GR. (Schwarz, (1972*a*).)

The nucleation of quantized vortex rings by the source deserves, however, particular care, especially with positive ions at low temperatures (Rayfield, 1968*a*). This problem will be discussed in detail later in this review. At low T, the vortex rings generated by the energetic ionization close to the source can propagate without significant energy loss to G1 and penetrate into the drift region G1–FG, thus altering significantly the measured overall drift time t_D. In order to prevent this problem a weak bias is applied to the region S–G1, sufficiently large to make the vortex rings decay in this region but not strong enough to produce a rapid collection of the resulting normal charge carriers.

The accuracy of this method is the highest among the drift velocity spectrometers. The absolute error in the mobilities is conservatively estimated to be ±2%. Many high-accuracy data have been obtained by using this method (Schwarz and Stark, 1969; Schwarz, 1972*b*, 1978.)

5.3 Early measurements at saturated vapor pressure

The first low-field mobility data under saturated vapor pressure (SVP) conditions are due to Meyer and Reif (Meyer and Reif, 1958, 1960, 1961; Reif and Meyer, 1960). Their results show that the drift velocity in the low-field limit is indeed proportional to the applied electric field.

In Fig. 5.4 the measured (zero-field) mobility is plotted as a function of the inverse temperature T^{-1}. The mobility of the negative ions, μ_-, turns out to be smaller than that of the positive ions, μ_+.

The most striking feature of the experimental results is the exponential increase of the mobility with decreasing temperature that suggests a correlation with the roton component of the normal fluid density: $\mu \propto \rho_n^{-1}$. This is what is expected from a simple model in which the ion mean free path varies inversely with the number of scattering excitations (predominantly rotons for $T \geqslant 1\,\mathrm{K}$), in a situation where their number is not too large, i.e., below the λ-temperature.

If it is assumed that $\rho_n \propto \exp\,[-\Delta/k_\mathrm{B}T]$, an exponential fit to the mobility data for $T < 2\,\mathrm{K}$ in the form

$$\mu \propto \exp\,(\Delta/k_\mathrm{B}T) \tag{5.5}$$

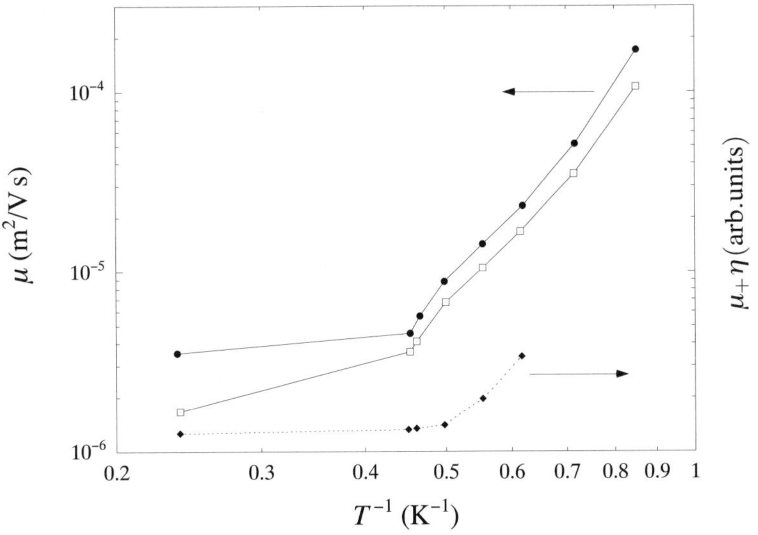

FIG. 5.4. μ vs T^{-1} for positive (circles) and negative (squares) ions (Meyer and Reif, 1958). The diamonds are the mobility–viscosity product, $\mu_+\eta$, for positive ions. The lines are only eyeguides.

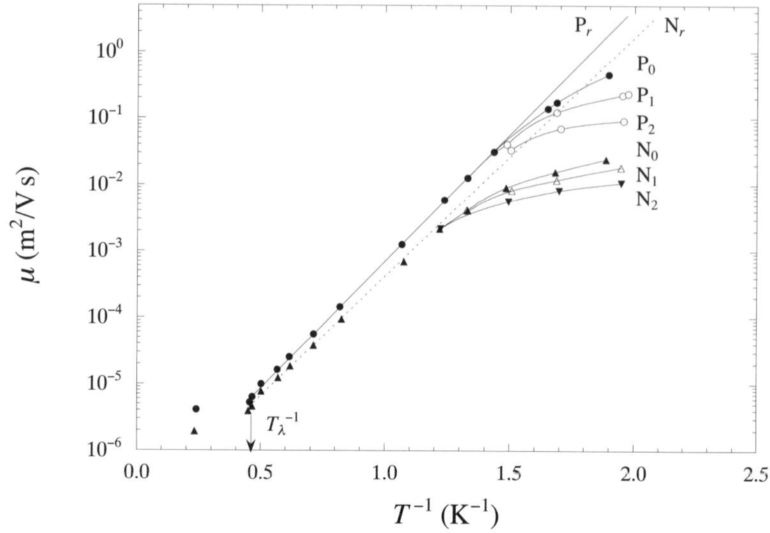

FIG. 5.5. μ_+ (P) and μ_- (N) vs T^{-1} in liquid He for several ^{3}He concentrations x_3. Solid (P_r) and dotted (N_r) lines: eqn (5.5). x_3 (ppm) $\leqslant 0.14$ (P_0 and N_0), $x_3 = 13$ (P_1 and N_1), and $x_3 = 41$ (P_2 and N_2). (Meyer and Reif, 1960.)

yields $\Delta/k_B = 8.8\,\text{K}$ for positive ions and $\Delta/k_B = 8.1\,\text{K}$ for negative ions (Reif and Meyer, 1960), in very nice agreement with the experimental value of the roton energy gap (Yarnell *et al.*, 1959; Wilks, 1967).

It shoud also be noted that in the vicinity of the λ-point the experimental data lie slightly above the straight lines (eqn (5.5)). This small deviation is consistent with the determination of Δ by neutron-scattering experiments (Yarnell *et al.*, 1959), which shows that the effective value of Δ decreases upon approaching T_λ. The physical reason for this behavior is that the number of excitations greatly increases as T increases, so that their mutual interaction is no longer negligible.

In contrast, close to the λ-point, where the elementary excitation gas is very dense, or in He I, helium should be considered as an ordinary liquid. In this case, the ion mobility should be related to the liquid viscosity η by means of the usual Stokes formula for a small rigid sphere moving through the liquid, yielding $\mu\eta = \text{const}$. This high-T behavior of $\mu_+\eta$ is displayed in Fig. 5.4.

Extended measurements at lower temperatures (Meyer and Reif, 1960) show that the data deviate from the fit (eqn (5.5)) starting from $T \approx 0.65\,\text{K}$ for the positive ions and $T \approx 0.8\,\text{K}$ for the negative ions, as shown in Fig. 5.5. These data clearly indicate that other different scattering mechanisms become dominant at lower temperatures, where the roton density is quite small. These are scattering of ions by phonons and by ^{3}He atoms. It is evident from Fig. 5.5 that the impurity ^{3}He atoms act as additional scattering centers, although their isotopic abundance in natural ^{4}He is too small to account for deviations from the straight lines described by eqn (5.5). Therefore, the observed deviations in natural helium are essentially due to phonon scattering.

If it can be assumed that the different types of scatterers act independently of each other, then the scattering rate τ^{-1} is the sum of the scattering rates due to the different processes. As the Drude law states that $\mu = (e/M)\tau$, where M is the effective mass of the ion, the following relationship is obtained:

$$\mu^{-1} = \mu_r^{-1} + \mu_{\rm ph}^{-1} + \mu_3^{-1}, \tag{5.6}$$

where μ_j refers to the mobility in the presence of a single type of scatterer and $j = r$, ph, 3 refers to rotons, phonons, and ^{3}He impurities, respectively. The values of μ_j are summarized in Table 5.1.

A subtraction procedure can, thus, yield the contribution to the mobility due to all types of scatterer. If μ_0 denotes the mobility in ordinary or purified He,then the roton contribution μ_r is obtained by the fit using eqn (5.5), whereas the phonon contribution is obtained as $\mu_{\rm ph}^{-1} = \mu_0^{-1} - \mu_r^{-1}$. Moreover, if μ_x denotes the mobility in ^{3}He-enriched liquid, then the impurity dependent contribution is obtained as $\mu_3^{-1} = \mu_x^{-1} - \mu_0^{-1}$. From each contribution the effective ion–scatterer cross-section $\sigma_{i,j}$ can be deduced for an assumed effective mass M of the ion because the scatterer number density is known (Meyer and Reif, 1960).

From these measurements it is evident that the negative ions are much larger objects than the positive ones. As the effective ion mass is much greater than that of the He atom, the effective collision diameter scales approximately as $D_{is} \propto M^{-1/2}$. Comparison with data at $T = 0.9\,$K suggests that rotons are more effective scattering centers than ^{3}He atoms (Meyer and Reif, 1960).

It should be noted that the temperature dependence of the different contributions to the mobilities is of the form $\mu_j \propto T^{-k}$. From Table 5.1 it can be seen

Table 5.1 *Contribution to the ion mobilities due to the different scatterer types for $T = 0.55$ K. The ^{3}He impurity contribution is reported for two different concentrations. $D_{is} = (\sigma_{is}/\pi)^{1/2}$ is the ion–scatterer collision diameter estimated for $M = m_4$. k in the last column is the index of the power law $\mu_j \propto T^{-k}$. (Meyer and Reif, 1960.)*

Ion	Scatterer	μ_j (cm^2/V s)	D_{is} (Å)	k
+	Phonon	5900	1.3	3.3 ± 0.3
+	^{3}He: $c_3 = 1.3 \times 10^{-5}$	4350	...	≈ 0
+	^{3}He: $c_3 = 5.1 \times 10^{-5}$	1200	8.8	≈ 0
−	Phonon	240	6.2	2.4 ± 0.4
−	^{3}He: $c_3 = 1.3 \times 10^{-5}$	560	...	$\leqslant 1$
−	^{3}He: $c_3 = 5.1 \times 10^{-5}$	180	22.9	$\leqslant 1$

that the impurity contribution is practically temperature independent, as one would expect for hard-sphere scattering. However, the exponent $k \approx 3$ for the phonon contribution suggests that the ion–phonon scattering cross-section $\sigma_{i,\mathrm{ph}}$ is nearly temperature independent, as the phonon number density is $n_{\mathrm{ph}} \propto T^3$, in contrast with the prediction T^{-9} for the scattering of a rigid sphere by a long-wavelength sound wave (Khalatnikov and Zharkov, 1957) and with the T^{-4} behavior predicted by the solution of the Boltzmann equation by means of a variational principle (Abe and Aizu, 1961).

5.4 Early low-field measurements at high pressure

It is known that the dispersion relation of the elementary excitations is appreciably affected if helium is pressurized (Henshaw, 1959; Yarnell *et al.*, 1959; Wilks, 1967). This fact significantly alters the ion mobilities according to eqn (5.5).

Cunsolo and Mazzoldi (1961) and Meyer and Reif (1961) carried out the first measurements at high pressures up to melting ($\approx 2.5\,\mathrm{MPa}$ at $T = 0\,\mathrm{K}$). In Fig. 5.6 the positive ion mobility is shown as a function of T^{-1} for three values of P.

The bending of the mobility at low temperature for $P = 0$ is due to the relative decrease of the roton-scattering contribution and to the rise of phonon scattering. At higher pressure there is no such bending. This means that phonon scattering does not become important, at least down to $T \approx 0.5\,\mathrm{K}$. The physical reason is that the roton gap energy Δ decreases with increasing pressure (Wilks,

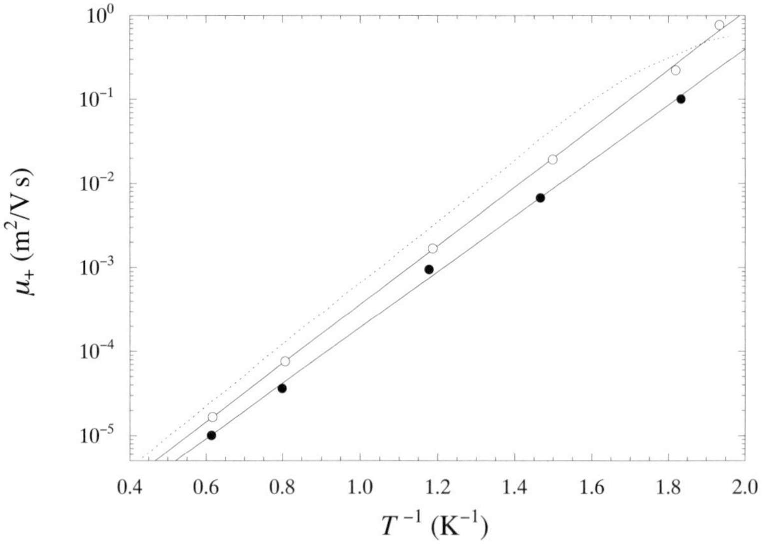

FIG. 5.6. μ_+ vs T^{-1} in liquid He for $P\,(\mathrm{MPa}) = 1.08$ (open circles) and 2.17 (closed circles). The solid lines are fits to eqn (5.5). The dashed line represents the result for $P = 0$ (Meyer and Reif, 1958, 1960; Reif and Meyer, 1960). At these pressures, $\mu_- = \mu_+$ (Meyer and Reif, 1961).

1967). In Table 5.2 the values of the roton parameters are reported. This fact produces an increase of the roton density, which therefore remains the dominant scattering process at much lower temperatures than at $P = 0$.

In Fig. 5.7 the pressure dependence of μ at several temperatures is shown. The behavior of the negative ions as a function of P is quite remarkable. At $P = 0$, for all temperatures, μ_- is always less than μ_+. However, for $P \geqslant 0.$ MPa, irrespective of T, μ_- becomes equal to μ_+. As P is increased at constant T, μ_- increases at first so as to approach μ_+, and then decreases at the same rate with a further increase of P.

The behavior of negative ions is explained by assuming that rotons limit their mobility in the kinetic regime. The mobility can thus be expressed as (Springett and Donnelly, 1966)

$$\mu_- = \frac{e}{Mv\sigma N_r}, \tag{5.7}$$

Table 5.2 *Pressure dependence of the roton parameters.*

P (MPa)	Δ/k_B (K)	$p_0/\hbar$ (Å^{-1})	m_r/m_4
0	8.65	1.91	0.16
2.56	7.0	2.05	(0.09)

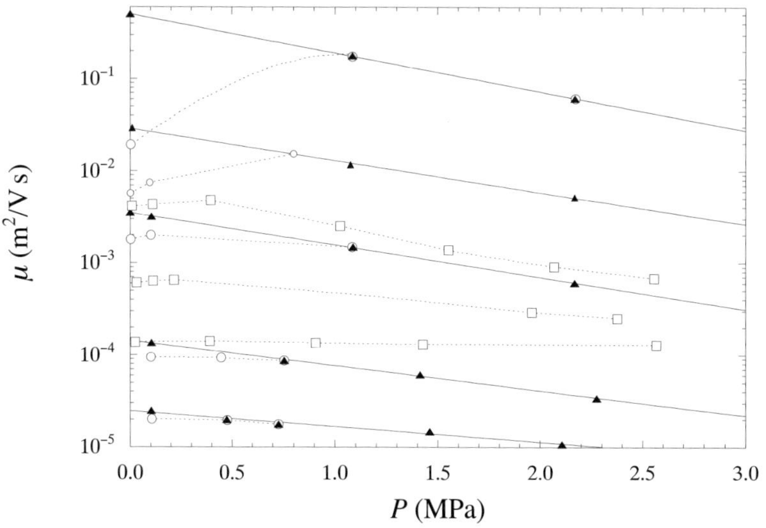

FIG. 5.7. μ_+ [triangles (Meyer and Reif, 1961)] and μ_- [open circles (Meyer and Reif, 1961) and open squares (Cunsolo and Mazzoldi, 1961)] for several values of T. From top: $T = 0.57, 0.70, 0.85, 1.23$, and $1.64\,$K (Meyer and Reif, 1961), $T = 1.13, 1.22$, and $1.66\,$K (Cunsolo and Mazzoldi, 1961). Solid lines: fits to eqn (5.5) for positive ions. Dashed lines: eyeguides.

where M is the reduced mass, v is the mean relative velocity of ions and rotons, N_r is the roton number density, and σ is an effective cross-section. e/Mv is expected not to depend very much on pressure. $(N_r\sigma)^{-1}$ can be considered as the ion–roton mean free path. N_r depends on P through Δ. The cross-section is taken to be $\sigma = \pi(R_- + a_r)^2$, where R_- is the radius of the electron bubble and $a_r \approx 4\,\text{Å}$ is the roton radius (Landau and Khalatnikov, 1949). By using the ion radius determined as a function of P from experiments of ion trapping on vortex lines (Springett and Donnelly, 1966; Springett, 1967) and shown in Fig. 3.6, the position of the mobility maximum of the negative ions for $T \approx 1\,\text{K}$ is reproduced quite well (Springett and Donnelly, 1966).

The behavior of the positive ion mobilities under pressure shown in Figs 5.6 and 5.7, and, later, in Fig. 5.8, can be explained if the pressure dependence of the roton parameters listed in Table 5.2 is accounted for. To leading order in pressure, both the roton energy and momentum can be interpolated linearly as a function of pressure:

$$\Delta = \Delta^\star \left(1 - aP\right) , \tag{5.8}$$

$$p_0 = p_0^\star \left(1 + bP\right) , \tag{5.9}$$

where the starred quantities refer to $P = 0$ and the interpolation coefficients are $a = 7.44 \times 10^{-2}\,\text{MPa}^{-1}$ and $b = 2.86 \times 10^{-2}\,\text{MPa}^{-1}$. Assuming, to a first approximation, that the ion effective mass and the ion–roton cross-section do not depend on P, a fit to the data shown in Fig. 5.6 yields the values $\Delta/k_\text{B} = 8.03\,\text{K}$

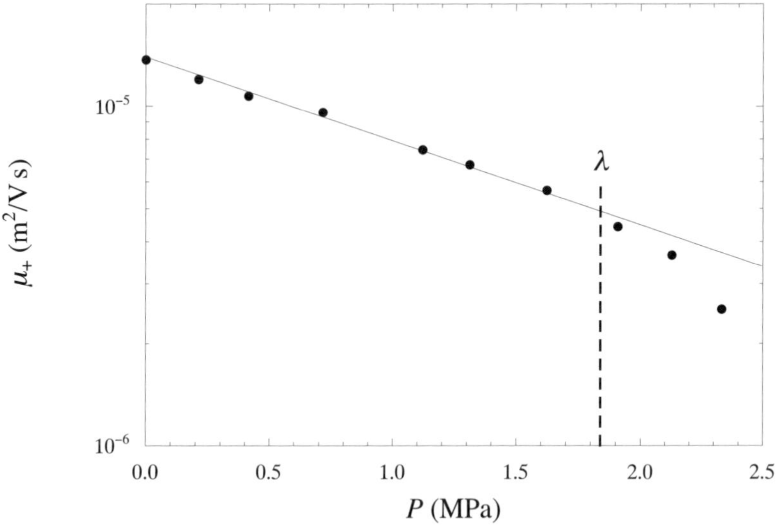

FIG. 5.8. μ_+ vs P for $T = 1.86\,\text{K}$ (Cunsolo and Mazzoldi, 1961). Solid line: eqn (5.11), valid up to the λ-transition (dotted line).

for $P = 1.08\,\text{MPa}$ and $\Delta/k_B = 7.6\,\text{K}$ for $P = 2.17\,\text{MPa}$, which agree fairly well with the values 7.95 and 7.3, respectively, calculated by means of eqn (5.8).

At constant temperature, assuming that the roton mass is much smaller than the ion effective mass, the mobility in the roton-dominated region can be written as (Meyer and Reif, 1961)

$$\mu = \frac{3}{4}(2\pi)^{3/2}h^3\left(\frac{e}{M}\right)\frac{1}{k_B T \sigma p_0^2}e^{[\Delta/k_B T]}, \tag{5.10}$$

where M is the ion mass and σ is an effective cross-section. Using the interpolation formulas (5.8) and (5.9) in eqn (5.10), a good approximation is obtained:

$$\ln\frac{\mu}{\mu^\star} = -\left[\frac{\Delta^\star}{k_B T} + 2b\right]P, \tag{5.11}$$

which yields a straight line in Fig. 5.7, as observed experimentally. For instance, for $T = 1.64$ and $0.5\,\text{K}$, the slopes calculated by eqn (5.11) are -0.45 and $-1.18\,\text{MPa}^{-1}$, respectively, to be compared favorably with the experimental values -0.41 and $-1.04\,\text{MPa}^{-1}$. This behavior holds true until the λ-transition is approached very closely (Cunsolo and Mazzoldi, 1961), as shown in Fig. 5.8.

5.5 Impurity, phonon, and roton contributions to the ion mobility

More accurate velocity measurements in a wider temperature range, such as those shown in Fig. 5.9, have clarified the contributions of the three ion-excitation

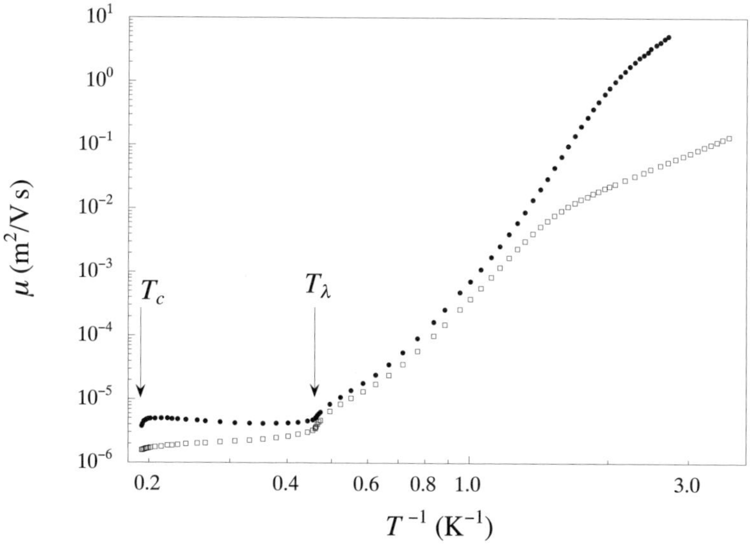

FIG. 5.9. μ vs T in superfluid He II at SVP (Schwarz, 1972*a*). Closed symbols: positive ions and open symbols: negative ions. T_c is the critical point. T_λ is the λ-point.

scattering processes described by eqn (5.6) by means of the subtraction procedure mentioned in Section 5.3. In the following sections, the mechanisms of scattering off different quasiparticles are described.

5.6 ^{3}He impurity-limited mobility

The phonon and roton contributions to the drag on ions decrease rapidly with T below 1 K, as shown in Fig. 5.10. For this reason, ^{3}He scattering can easily be made the dominant momentum-transfer process at low T just by adding a small amount of impurities. The dependence of the impurity contribution to the drag is obtained as a function of the ^{3}He concentration by first measuring the drag e/μ in pure ^{4}He and subtracting it from the values of e/μ measured in the mixture.

The early measurements of Meyer and Reif (1960) were first extended by Neeper and Meyer (1969), and then by Kuchnir *et al.* (1972) and Schwarz (1972*b*) to much lower temperatures. For $0.3\,\mathrm{K} \leqslant T \leqslant 1\,\mathrm{K}$ and for impurity concentrations from 10 up to 1800 ppm the impurity drag contribution e/μ_3 turns out to be directly proportional to the ^{3}He number density n_3, as reported in Fig. 5.11 (Schwarz, 1972*b*). This behavior confirms the hypothesis that, at low temperature, the excitations act as a dilute gas and demonstrates that there are no changes of the ion structure due to interactions with the impurities. The data can be interpreted in terms of single ^{3}He scattering of the complex charge carriers.

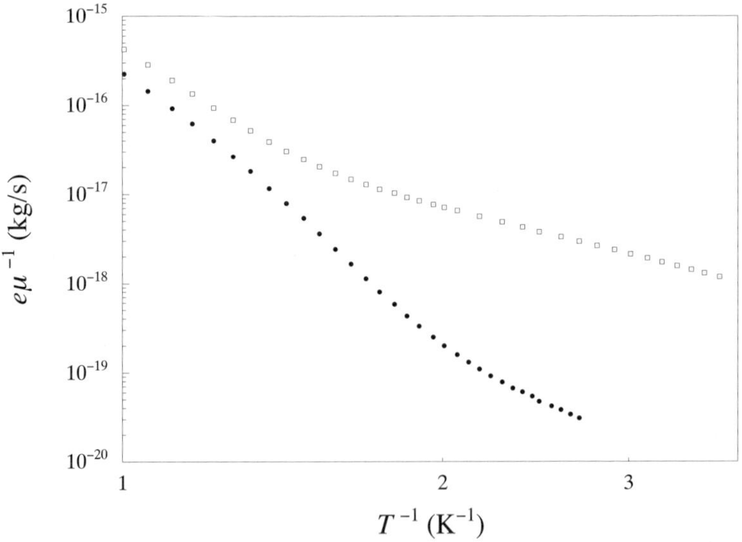

FIG. 5.10. Drag coefficient e/μ for positive (closed circles) and negative ions (open squares) (Schwarz, 1972*a*) in pure He for $0.27\,\mathrm{K} \leqslant T \leqslant 1\,\mathrm{K}$ at SVP, showing the rapid decrease of the roton- and phonon-drag contributions.

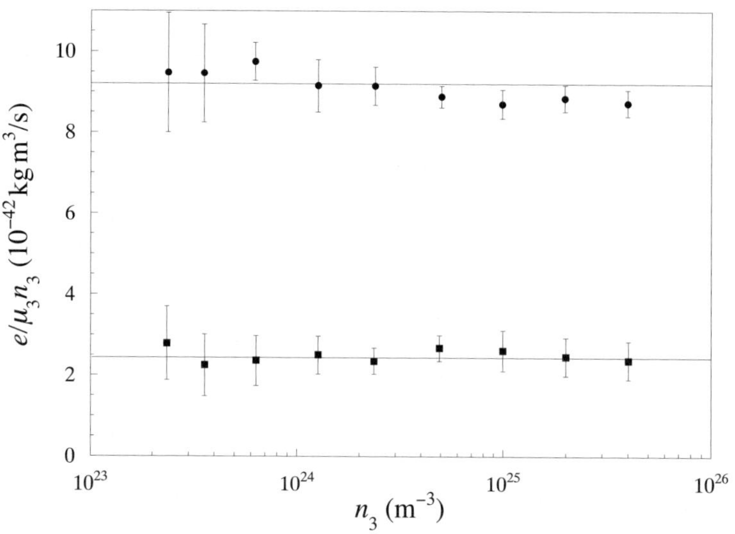

FIG. 5.11. $e/\mu_3 n_3$ vs n_3 in superfluid He II for $T = 0.508$ K (Schwarz, 1972*b*). Circles: negative ions, squares: positive ions. The phonon and roton contributions have been subtracted out.

The proportionality constant between e/μ_3 and n_3 is temperature dependent and is shown in Figs 5.12 and 5.13. This temperature dependence is quite weak, though non-negligible also for the positive species. It varies as $T^{1/2}$ for the negative ions, whereas it becomes constant below $T \approx 0.6$ K for the positive ones.

Data at much lower temperatures, down to 17 mK, confirm that μ_+ remains independent of T (Kuchnir *et al.*, 1972), as shown in Fig. 5.14.

In contrast, the negative ion mobility shows a different, more interesting behavior, displayed in Fig. 5.15, indicating that the ^{3}He–negative ion interaction changes, in particular in the $0.1 - 0.3$ K region, and also that ion structural changes of the negative charge carriers take place (Kuchnir *et al.*, 1972).

Because ions are much more massive than ^{3}He impurities and as the impurity atoms can be treated as a gas of quasiparticles moving through the inert superfluid background, the mobility can be calculated in the frame of the kinetic theory of transport by introducing a scattering cross-section that depends on the details of the interaction between the charge carrier and the quasiparticles. As usual in kinetic theory, measurements of the mobility are a sensitive test of this interaction.

If recoil is neglected (Bowley, 1971*c*; Fetter, 1976), the contribution to the ion drag due to ^{3}He scattering can be written as (Bowley and Lekner, 1970; Kramer, 1970*b*)

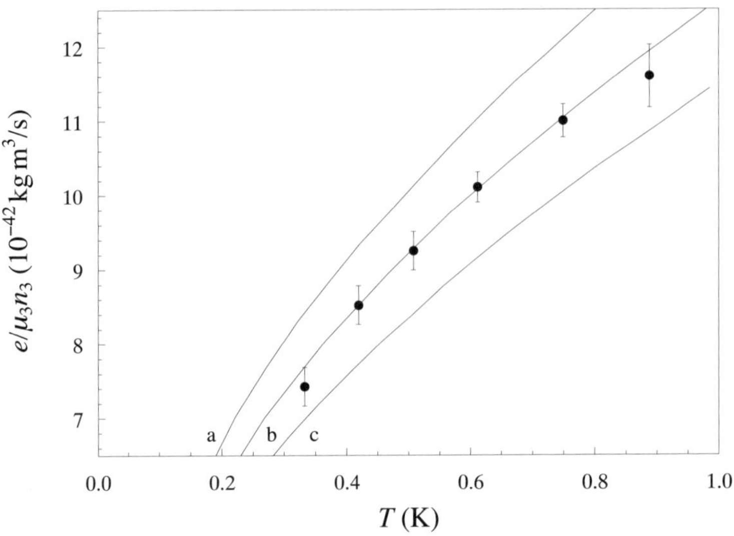

FIG. 5.12. $e/n_3\mu_3$ vs T for negative ions (Schwarz, 1972*b*). The curves are described in the text.

$$\frac{e}{\mu_3} = -\frac{\hbar}{3\pi^2}\int_0^\infty k^4 \frac{\partial n(k)}{\partial k}\sigma_{mt}(k)\,\mathrm{d}k\,, \tag{5.12}$$

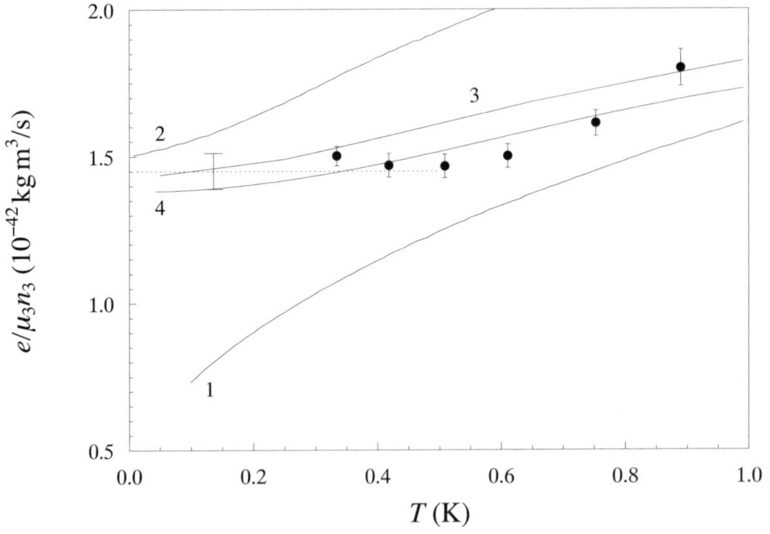

FIG. 5.13. $e/n_3\mu_3$ vs T for positive ions (Schwarz, 1972*b*). The curves are described in the text.

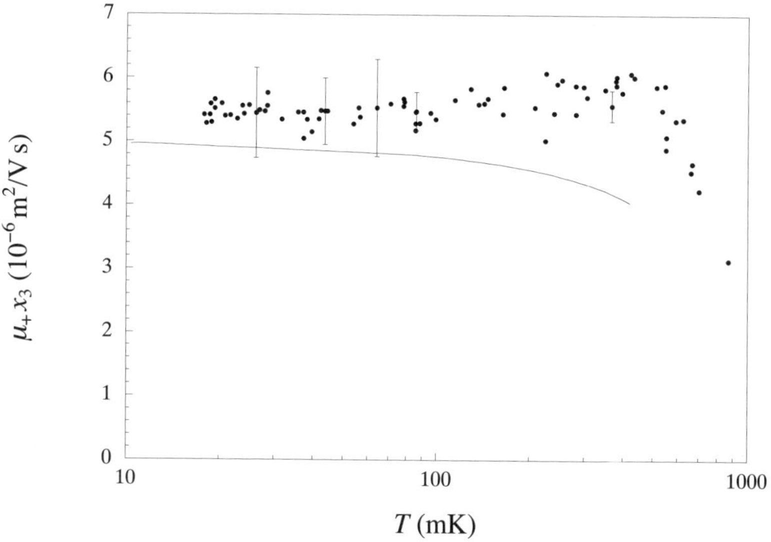

FIG. 5.14. $\mu_+ x_3$ vs T (Kuchnir *et al.*, 1972). The data are a compilation of several concentrations in the range 100 – 3300 ppm. Line: theory (Bowley and Lekner, 1970).

where $n(k)$ is the momentum distribution of the ^{3}He atoms and $\sigma_T(k)$ is the momentum-transfer scattering cross-section. The distribution function is the Maxwell–Boltzmann distribution for quasiparticles of effective mass $m_3^\star$:

$$n(k) = n_3 \left(\frac{2\pi\hbar^2 k_B T}{m_3^\star} \right)^{-3/2} \exp\left(-\frac{\hbar^2 k^2}{2m_3^\star k_B T} \right), \tag{5.13}$$

and the momentum-transfer cross-section is given in terms of the asymptotic values of the phase shifts $\delta_l(k) = \lim_{r\to\infty} \eta_l(r,k)$ as

$$\sigma_{mt}(k) = \frac{4\pi}{k^2} \sum_{l=0}^{\infty} (l+1) \sin^2(\delta_l - \delta_{l+1}) \,. \tag{5.14}$$

The phase shifts are obtained from the scattering potential $V(r)$ by integrating the differential equation (Levy and Keller, 1963; Calogero, 1967; Atrazhev *et al.*, 2002)

$$\frac{\mathrm{d}\eta_l(r)}{\mathrm{d}r} = -kr^2 \frac{2m_3^\star}{\hbar^2} V(r) \left[j_l(kr) \cos\eta_l(r,k) - n_l(kr) \sin\eta_l(r,k) \right]^2 \tag{5.15}$$

with initial condition $\tan\eta_l(k, r = R) = j_l(k,R)/n_l(k,R)$ out to infinity. j_l and n_l are the Bessel and von Neumann functions, respectively.

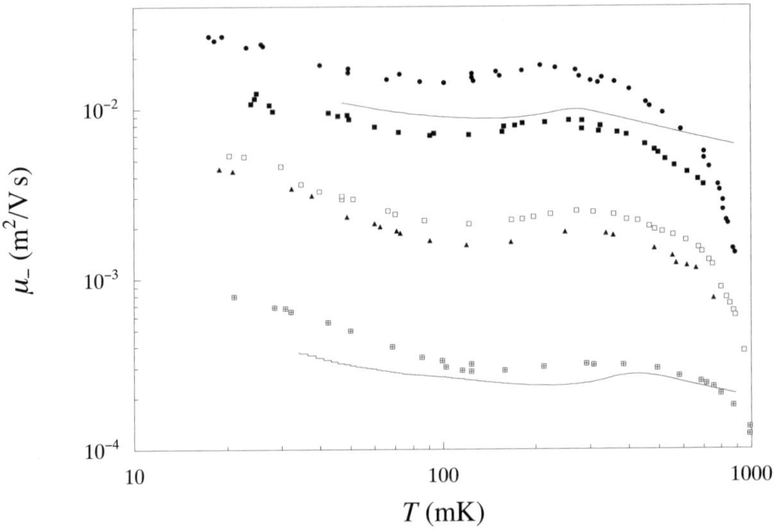

FIG. 5.15. μ_- vs T for several ^{3}He concentrations x_3 = 77, 170, 500, 655, and 3300 ppm from top (Kuchnir *et al.*, 1972). Solid lines: theory of ^{3}He ion scattering that accounts for ^{3}He condensation on the electron bubble surface and for the Knudsen limit correction (Kramer, 1970*b*). The calculations are for $x_3 = 180$ and 2900 ppm, respectively (from top).

If a hard-sphere interaction is assumed to describe the ^{3}He–ion interaction, with hard-core radius R,then the cross-section takes on the values $\sigma_{mt} = 4\pi R^2$ in the low-momentum limit ($kR \ll 1$) and $\sigma_{mt} = \pi R^2$ in the large-momentum limit ($kR \gg 1$). If the root mean square thermal wave vector $(3k_{\mathrm{B}}Tm_3^{\star}/\hbar^2)^{1/2}$ is in one of these limits, a $T^{-1/2}$ dependence of the mobility results (Kuchnir *et al.*, 1972).

Consider first the case of negative ions. As the electron bubble has a quite definite interface and exerts negligible electrostriction on the surrounding liquid, the hard-core model describes fairly well the interaction with the impurities. An effective hard-core, or collision, radius R_-^{eff} conglomerates the details of the interface region of ≈ 2 Å. In the temperature range $0.3\,\mathrm{K} < T < 1\,\mathrm{K}$, this approximation is in good agreement with the data if the value $R_-^{\mathrm{eff}} = 21$ Å is chosen. In Fig. 5.12, the curves labeled a, b, and c correspond to $R_-^{\mathrm{eff}} = 22$, 21, and 20 Å, respectively (Schwarz, 1972*b*). The best curve is for $R_-^{\mathrm{eff}} = 21$ Å, in reasonable agreement with other determinations of the electron bubble radius (see Section 3.2).

The behavior of the electron bubble mobility in the range below 0.3 K, as shown in Fig. 5.15, indicates that there are structural changes of the ion, especially between 0.1 and 0.3 K. This anomaly can be qualitatively understood if it is considered that ^{3}He atoms lower their zero-point energy relative to a ^{4}He atom if they stay on a free surface (Dahm, 1969). Thus, there are bound states for ^{3}He atoms at the bubble surface that get occupied around $T \approx 0.25\,\mathrm{K}$. This occupation changes the surface tension (Andreev, 1966; Guo *et al.*, 1971) and, consequently, the bubble radius.

Kramer (1970*b*) calculated the temperature dependence of μ_- including the temperature-dependent radius in the hard-sphere scattering model and also taking into account the distortion of the momentum distribution of the quasiparticles by the moving ion due to the fact that the ^{3}He mean free path is not much larger than the ion size. His results for two different concentrations are shown as solid lines in Fig. 5.15. Although the agreement is not perfect, the experimental results are fairly accurately described by scattering off hard spheres whose radius depends on temperature because the surface tension is modified by the condensation of ^{3}He atoms on the bubble surface (Kramer, 1970*b*).

The same approach for the positive ions, by assuming a collision radius $R_+^{\mathrm{eff}} \approx 7\,\mathrm{Å}$, gives incorrect results, as shown by curve 1 in Fig. 5.13. Nonetheless, the mobility of positive ions can still be understood in terms of a simple potential scattering if the ^{3}He atom–ion interaction is modeled as a hard-core repulsion plus a long-range repulsive polarization potential of the form

$$V_P(r) = \frac{\delta_{34}\alpha e^2}{(4\pi\epsilon_0)^2\, r^4} \tag{5.16}$$

for r greater than the hard-core radius. $\delta_{34} = (V_3 - V_4)/V_4$ is the excess specific volume of ^{3}He atoms, and α is the atomic polarizability. This repulsive contribution arises from the fact that the lighter ^{3}He atom occupies a larger volume than a ^{4}He atom. The energy of the system is then lowered if the ^{3}He atom is moved away from the ion, thus getting more ^{4}He atoms in the attractive polarization field of the ion. This phenomenon finally leads to an effective ion–^{3}He atom repulsion (Bowley and Lekner, 1970; Bowley, 1971*a*).

The combined effect of the two terms in the potential gives a strongly k-dependent momentum-transfer cross-section, whose effect is to practically nullify the temperature dependence of μ_+ (Bowley and Lekner, 1970) (Fig. 5.16). The results of the calculation for an effective collision radius of 8.4 Å and an effective ^{3}He mass $m_3^\star = 2.34\, m_3$ (Brubaker *et al.*, 1970) are shown as curve 2 in Fig. 5.13. Although in disagreement with the data over the entire $0.3\,\mathrm{K} < T < 1\,\mathrm{K}$ range, nonetheless the low-temperature limit is correctly reproduced.

Curve 3 is computed by letting $R_+^{\mathrm{eff}} = 7\,\mathrm{Å}$, not in disagreement with the value 5.1 Å derived from phonon-limited mobility data (Schwarz and Stark, 1968). Its behavior supports the essential correctness of the Bowley–Lekner approach.

A further, though tiny, improvement can be obtained by taking into account the rapid variation of the fractional excess volume δ_{34} with the distance from the ion because of the steep density gradient induced by electrostriction. The repulsive polarization potential thus takes the form

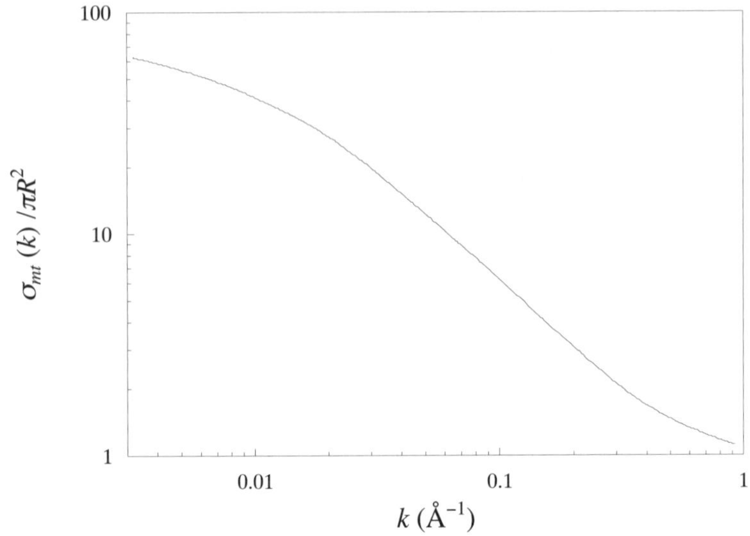

FIG. 5.16. $\sigma_{mt}/\pi R^2$ for the ion–^{3}He atom interaction potential given by a short-range, hard-core repulsive term plus the long-range repulsive polarization contribution (eqn (5.16)) with $R = 8.4$ Å. (Bowley and Lekner, 1970.)

$$V(r) = \frac{\delta_{34}(r)\alpha e^2}{2\,(4\pi\epsilon_0)^2\,r^4\,[1 + 4\pi\beta\rho(r)]}, \qquad (5.17)$$

where β is the polarizability per unit mass and $\delta_{34}(r)$ is calculated from the local pressure in the electrostriction model. Equation (5.17) includes a small correction for dielectric screening. Using this scattering potential, curve 4 in Fig. 5.13 is obtained with little, if any, improvement.

The results of the calculation of Bowley *et al.* with a collision radius $R_+^{\text{eff}} = 8.4$ Å in the low-temperature domain are shown in Fig. 5.14, thus confirming the essential correctness of the model.

5.6.1 *Electric field dependence of the drift velocity in mixtures*

At low temperature, $T < 0.3$ K, the ion mobility in ^{3}He–^{4}He solutions is determined essentially by collisions with the impurities. Whereas the mobility of negative ions decreases weakly with increasing field and can be qualitatively interpreted as being determined by collisions between impurities and hard spheres of suitable size, the mobility of positive ions increases superlinearly with the applied field even in the range of low field (Kuchnir *et al.*, 1971, 1972; Esel'son *et al.*, 1971, 1972).

Esel'son *et al.* (1972) have suggested that ^{3}He impurities might be sucked into a region of higher superfluid velocity that develops around the ion drifting through the solution (Reut and Fizher, 1969).

The electric field dependence of the drift velocity of positive ions has been thus calculated by Bondarev and Fisher (1974) using classical mechanics for the prediction of trajectories and cross-sections for particles moving in a potential that takes into account this suction effect. It is assumed that the ion–impurity scattering potential consists of three parts:

1. V_0, a hard-sphere potential of radius a equal to the sum of the radii of the positive ion and of the ^{3}He atom;
2. a repulsive polarization potential given by eqn (5.17), stemming from the different specific volumes of the two helium isotopes and decreasing as r^{-4} for $r > a$;
3. an attractive potential of the form

$$V_2(r) = -Bv_s^2(r), \qquad B > 0\,, r > a\,, \tag{5.18}$$

where v_s is the velocity of the superfluid component of the liquid flowing around the ion because of its motion in a solution at rest.

The square of the superfluid velocity is given by classical hydrodynamics as

$$v_s^2\,(r) = \frac{a^6 v_D^2}{4r^6}\left(1 + 3\cos^2\gamma\right)\,, \qquad r > a\,, \tag{5.19}$$

where v_D is the ion drift velocity, and γ is the angle between $\mathbf{r}$ and $\mathbf{v}_D$. This *suction* potential introduces an additional r^{-6} contribution to the ion–impurity interaction.

In order to simplify the problem, eqn (5.19) is replaced by its angular average and the complete interaction potential becomes

$$V(r) = V_0 + \frac{\delta_{34} e^2}{\left(4\pi\epsilon_0\right)^2 r^4} - \frac{B a^6 v_D^2}{2r^6}\varphi(y) \tag{5.20}$$

with

$$\varphi\,(y) = \begin{cases} 1 + \frac{2}{5}y^2 - \frac{4}{35}y^4 + \frac{8}{315}y^6 - \frac{16}{3465}y^8 + \ldots & \text{for } y \leqslant 1\,, \\ \to 2 & \text{for } y \to \infty\,, \end{cases} \tag{5.21}$$

where $k_{\mathrm{B}}Ty^2 = mv_D^2/2$ and m is the reduced mass of the ion–impurity system.

This approximation for the potential is valid for ion velocities up to $\approx 30\,\mathrm{m/s}$. Above this value the potential is no longer monotonic (Bondarev and Fisher, 1974).

The cross-section for this potential is calculated according to standard classical mechanics methods (Maitland *et al.*, 1981) and the mobility according to standard kinetic theory (Huxley and Crompton, 1974). A typical result is plotted as a dashed line in Fig. 9.13. In all cases, the results of the calculations show an upward inflection occurring for $v_D \approx 30\,\mathrm{m/s}$ because direct scattering off the hard-core part of the ion potential is expected at high relative particle velocities (Bondarev and Fisher, 1974).

5.7 Phonon-limited mobility

For $T \leqslant 0.6\,\mathrm{K}$, the mobility of the charge carriers is essentially dominated by scattering with thermally-excited phonons (Meyer and Reif, 1960, 1961; Reif and Meyer, 1960; Schwarz and Stark, 1968, 1969, 1972*a*), as can be seen in Fig. 5.10. Again, at very low electric fields the drift velocity v_D is proportional to the strength of the electric field E, as shown in Fig. 5.17.

The zero-field mobility of the negative ions for $T \leqslant 0.5\,\mathrm{K}$ varies as T^{-3} (Fig. 5.18) and is described by the relation (Schwarz and Stark, 1968)

$$\mu_- = 3.247 \times 10^{-3} T^{-3}\,. \tag{5.22}$$

In contrast, the positive carriers display a much stronger temperature dependence than the electron bubbles in the same temperature range, as shown in Fig. 5.19, where the average momentum loss per meter (or drag) e/μ_+ of positive ions is plotted.

The interpretation of the ion mobility in terms of the microscopic phonon-scattering process was given by Baym *et al.* (1969) for the negative bubble and has been extended to positive ions by Schwarz (1972*c*). At the low temperature of the experiment, the phonons relevant for the transport properties of the carriers have wavelengths several atomic spacings long and can therefore be described reasonably well by continuum sound waves.

The effective mass of ions, in particular of negative ions, is very large. Thus, the ion velocity is negligibly affected by collisions with phonons of long wave-

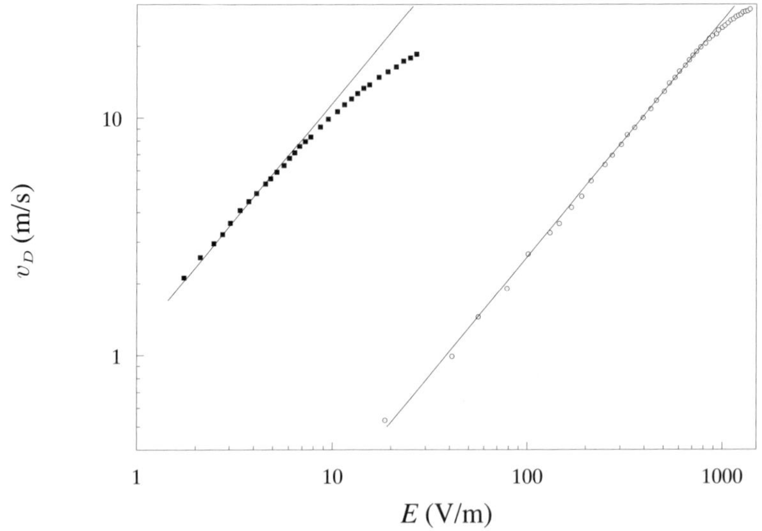

FIG. 5.17. v_D vs E for positive (squares) (Schwarz and Stark, 1969) and negative (circles) ions (Schwarz and Stark, 1968) at $T = 0.5\,\mathrm{K}$. The lines help to detect the range, over which $v_D \propto E$.

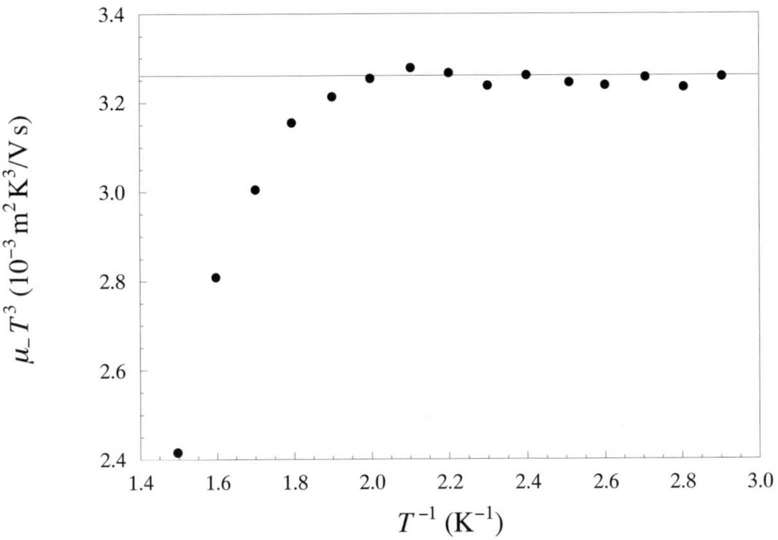

FIG. 5.18. μ_-T^3 vs T. The line is eqn (5.22). (Schwarz and Stark, (1968).)

length and recoil effects can be neglected.

The rate at which phonons are scattered from wave vector $\mathbf{k}$ to wave vector $\mathbf{k}'$ by a heavy bubble moving with steady average velocity $\mathbf{v}_D$ is given by (Baym *et al.*, 1969)

$$\nu_s = n_{\mathbf{k}} (1 + n_{\mathbf{k}'}) \Gamma_{\mathbf{v}_D} (\mathbf{k} - \mathbf{k}') , \tag{5.23}$$

where $\Gamma_{\mathbf{v}_D}(\mathbf{k}-\mathbf{k}')$ is the transition rate. $n_{\mathbf{k}}$ is the equilibrium phonon distribution function given by

$$n_{\mathbf{k}} = [\exp(\hbar sk/k_{\mathrm{B}}T) - 1]^{-1} , \tag{5.24}$$

where s is the sound velocity. It is assumed that the presence of the bubble does not perturb $n_{\mathbf{k}}$.

The rate of change of phonon momentum by collisions with the bubble is given by

$$\frac{\mathrm{d}\mathbf{P}_{\mathrm{ph}}}{\mathrm{d}t} = \sum_{\mathbf{k},\mathbf{k}'} \hbar (\mathbf{k}' - \mathbf{k}) n_{\mathbf{k}} (1 + n_{\mathbf{k}'}) \Gamma_{\mathbf{v}_D}(\mathbf{k} - \mathbf{k}') . \tag{5.25}$$

In order for $\mathrm{d}\mathbf{P}_{\mathrm{ph}}/\mathrm{d}t$ to be related to the drift velocity, terms linear in v_D must be retained in the right-hand side of eqn (5.25). These terms stem from an intrinsic dependence of $\Gamma_{\mathbf{v}_D}$ on $\mathbf{v}_D$ and from the values of $\mathbf{k}'$ contributing to eqn (5.25), which are Doppler shifted in magnitude from k.

If phonons were in equilibrium with the moving bubble, the distribution function $n_{\mathbf{k}}$ would be replaced by (Wilks, 1967)

$$\bar{n}_{\mathbf{k}} = \{\exp[\hbar(sk - \mathbf{k}\cdot\mathbf{v}_D)/k_{\mathrm{B}}T] - 1\}^{-1} \tag{5.26}$$

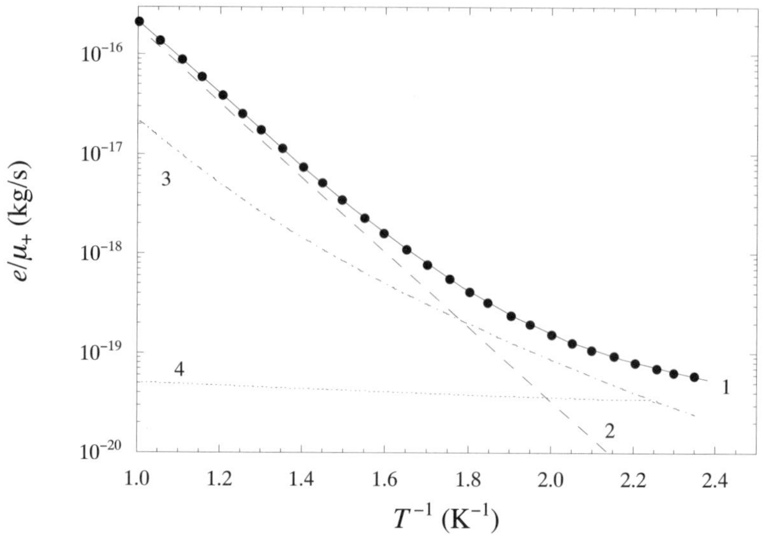

FIG. 5.19. e/μ_+ vs T^{-1} in the range $0.426\,\mathrm{K} < T < 1.001\,\mathrm{K}$. Curve 1 is a best fit to the data including the roton- (curve 2), the phonon- (curve 3), and the ^{3}He-scattering (curve 4) contributions. (Schwarz and Stark, 1969.)

and $\mathrm{d}\mathbf{P}_{\mathrm{ph}}/\mathrm{d}t = 0$. Thus, eqn (5.25) must be replaced by

$$\frac{\mathrm{d}\mathbf{P}_{\mathrm{ph}}}{\mathrm{d}t} = \sum_{\mathbf{k},\mathbf{k}'} \hbar\left(\mathbf{k}' - \mathbf{k}\right)\left[n_{\mathbf{k}}\left(1 + n_{\mathbf{k}'}\right) - \bar{n}_{\mathbf{k}}\left(1 + \bar{n}_{\mathbf{k}'}\right)\right]\Gamma_{\mathbf{v}_D}\left(\mathbf{k} - \mathbf{k}'\right). \qquad (5.27)$$

The right-hand side of eqn (5.27) is now explicitly of order $v_D \equiv |\mathbf{v}_D|$. Thus, $\Gamma_{\mathbf{v}_D}$ can be replaced by its value Γ_0 for $v_D = 0$ and the Doppler shift of k' can be neglected.

Owing to the time-reversal symmetry of the collisions with a stationary object, $\Gamma_0(\mathbf{k} - \mathbf{k}') = \Gamma_0(\mathbf{k}' - \mathbf{k})$, the term in square brackets in eqn (5.27) reduces to

$$n_{\mathbf{k}} - \bar{n}_{\mathbf{k}} = \left(\frac{\partial n_{\mathbf{k}}}{\partial sk}\right)\mathbf{k}\cdot\mathbf{v}_D \qquad (5.28)$$

to first order in v_D. By inserting eqn (5.28) into eqn (5.27), and by taking into account the isotropy of the problem, one obtains

$$\frac{\mathrm{d}\mathbf{P}_{\mathrm{ph}}}{\mathrm{d}t} = \frac{1}{3}\hbar\mathbf{v}_D \sum_{\mathbf{k}} k^2 \frac{\partial n_{\mathbf{k}}}{\partial k}\sigma_{mt}\left(k\right), \qquad (5.29)$$

where $\sigma_{mt}(k)$ is the momentum-transfer scattering cross-section defined, as usual, in terms of the differential cross-section $\sigma(k,\theta)$ as

$$\sigma_{mt}(k) = \int \left(1 - \cos\theta\right)\sigma\left(k,\theta\right)\,\mathrm{d}\Omega \qquad (5.30)$$

with the differential cross-section given by

$$\sigma(k,\theta) = \frac{1}{s}\sum_{\mathbf{k}'}\Gamma_0(\mathbf{k}-\mathbf{k}')\,\delta\left(\cos\theta - \hat{k}\cdot\hat{k}'\right). \tag{5.31}$$

Equating the left-hand side of eqn (5.29) to the force due to the electric field, $e\mathbf{E}$, the following expression for the drag is obtained:

$$\frac{e}{\mu} = -\frac{\hbar}{6\pi^2}\int_0^\infty k^4 \frac{\partial n_{\mathbf{k}}}{\partial k}\sigma_{mt}(k)\mathrm{d}k\,. \tag{5.32}$$

If σ_{mt} were a constant, eqn (5.32) would yield

$$\frac{e}{\mu} = \rho_n s \sigma_{mt}\,, \tag{5.33}$$

where $\rho_n \propto T^4$ is the phonon normal mass density (Wilks, 1967).

Moreover, if it is assumed that σ_{mt} varies as k^j, then eqn (5.32) gives $e/\mu \propto T^{4+j}$. This means that the T^{-3} dependence of the negative bubble mobility implies a k^{-1} dependence for phonon scattering of the bubble, while the nearly T^{-8} variation of the positive ion mobility implies a k^4 dependence for the scattering by positive ions. As the typical phonon wavelengths in this temperature range are of the order of 15 Å (Schwarz, 1972*c*), this k^4 dependence for the positive species can be interpreted as the Rayleigh limit for the scattering of a wave from an object small compared to the wavelength.

In contrast, the k^{-1} dependence of σ_{mt} arises from the existence of normal modes of vibration of the electron bubble with characteristic frequencies comparable with those of the thermally-excited phonons.

The central problem is now to calculate the differential cross-section and, hence, the momentum-transfer cross-section. The problem is tackled by calculating the scattering of classical sound waves from an elastic sphere. Electrostrictive effects and the response of the solid–liquid boundary to the phonon field are to be taken into account for the case of positive ions (Schwarz, 1972*c*, 1975). The negative ion is treated by taking into account that it is an empty bubble with a well-defined boundary.

The basic equation to be solved is the Navier–Stokes equation (Landau and Lifsits, 2000), supplemented with the equation of mass conservation in the presence of electrostrictive effects (Schwarz, 1972*c*):

$$\rho\left[\frac{\partial\mathbf{v}}{\partial t} + (\mathbf{v}\cdot\boldsymbol{\nabla})\,\mathbf{v}\right] = -\boldsymbol{\nabla}p + (\mathbf{P}\cdot\boldsymbol{\nabla})\,\mathbf{E}\,, \tag{5.34}$$

$$\frac{\partial\rho}{\partial t} + \boldsymbol{\nabla}\cdot(\rho\mathbf{v}) = 0,. \tag{5.35}$$

Here p denotes the pressure and $\mathbf{P}$ the polarization per unit volume. With sufficient approximation $\mathbf{P} = (\rho N_A\alpha/4\pi\epsilon_0 M)\mathbf{E} = \beta\rho\mathbf{E}$, where N_A is Avogadro's number, α is the atomic polarizability, and M is the atomic weight. β is thus the polarizability per unit mass. With this approximation, eqn (5.34) becomes

$$\frac{\partial \mathbf{v}}{\partial t} + (\mathbf{v} \cdot \boldsymbol{\nabla})\,\mathbf{v} = -\frac{1}{\rho}\boldsymbol{\nabla} p + \frac{\beta}{2}\boldsymbol{\nabla}\left[\frac{D^2}{(1+4\pi\beta\rho)^2}\right], \tag{5.36}$$

where $\mathbf{D} = (e/4\pi)\,[\mathbf{r} - \mathbf{r}_0(t)]\,/|\mathbf{r} - \mathbf{r}_0(t)|^3$ is the dielectric displacement vector that depends on the instantaneous position $\mathbf{r}_0(t)$ of the charge.

In the absence of the phonon field, the electrostriction problem has been solved for a stationary positive ion (Atkins, 1959), yielding the well-known density and pressure profiles.

In order to treat the scattering of sound, the equilibrium solution in the presence of electrostriction is indicated by a "0" subscript and small deviations about this solution are sought. The Navier–Stokes and mass conservation equations are linearized, also taking into account the spatial non-uniformity of the density. This problem is similar to sound propagation through an inhomogeneous atmosphere in mechanical equilibrium under the gravitational field (Lamb, 1945). Writing $\mathbf{v} = \mathbf{v}_s$, $p = p_0 + p_s$, and $\rho = \rho_0 + \rho_s$, eqns (5.35) and (5.36) are linearized so as to give

$$c_0^{-2}\frac{\partial p_s}{\partial t} + \boldsymbol{\nabla}\cdot(\rho_0 \mathbf{v}_s) = 0\,, \tag{5.37}$$

$$\frac{\partial \mathbf{v}_s}{\partial t} = -\boldsymbol{\nabla}\left(\frac{p_s}{\rho_0}\right), \tag{5.38}$$

where c_0 is the sound velocity. The relation $\rho_s = (\partial\rho/\partial p)_0 p_s = p_s c_0^{-2}$ is used to eliminate ρ_s in eqn (5.37). Small terms arising from the dependence of the local dielectric constant on ρ_s have been neglected without introducing significant errors (Schwarz, 1972*c*).

By the introduction of the scalar velocity potential Φ, defined by $\mathbf{v} = \boldsymbol{\nabla}\Phi$, the previous equations are transformed into the following ones:

$$p_s = -\rho_0 \frac{\partial \Phi}{\partial t}\,, \tag{5.39}$$

$$\boldsymbol{\nabla}^2\Phi + \boldsymbol{\nabla}(\ln \rho_0)\;\cdot\;\boldsymbol{\nabla}\Phi - \frac{1}{c_0^2}\frac{\partial^2\Phi}{\partial t^2} = 0\,. \tag{5.40}$$

In these equations the local density ρ_0 and the local sound velocity c_0 depend strongly on the radial distance from the charge because of electrostriction. At large distances, the density and sound velocity approach their values in the unperturbed uniform liquid: $\rho_0 \to \rho_\infty$ and $c_0 \to c_\infty$, respectively.

The scattering equations (5.39) and (5.40), are solved by means of the partial-wave analysis, seeking a solution for Φ in the form

$$\Phi = \sum_{l=0}^{\infty} A_l \frac{u_{l(r)}}{kr} P_l(\cos\theta)\, e^{-i\omega t}\,, \tag{5.41}$$

where $P_l(\cos\theta)$ are the Legendre polynomials. The associated radial equation is then

$$\frac{\mathrm{d}^2 u_l}{\mathrm{d}r^2} + \frac{\mathrm{d}\,(\ln\rho_0)}{\mathrm{d}r}\frac{\mathrm{d}u_l}{\mathrm{d}r} + \left[\frac{k^2 c_\infty^2}{c_0^2} - \frac{1}{r}\frac{\mathrm{d}\,(\ln\rho_0)}{\mathrm{d}r} - \frac{l\,(l+1)}{r^2}\right] u_l = 0 \qquad (5.42)$$

with $k = \omega / c_\infty$.

The boundary conditions at the radius R of the ion complex, either the snowball or the electron bubble, are treated by describing the distortion of the bubble or core surface of radius R as (Celli *et al.*, 1968)

$$R\,(\theta) = R_0 + \sum_{l=0}^{\infty} \delta R_l P_l\,(\cos\theta) \qquad (5.43)$$

in response to the excess-pressure field

$$\delta p\,(\theta) = \sum_{l=0}^{\infty} \delta p_l P_l\,(\cos\theta)\,. \qquad (5.44)$$

The response can also be described in terms of the coefficients $\lambda_l = \delta R_l / \delta p_l$ (Schwarz, 1972*c*). The same approach has also been used to describe the parametric resonance of the O_2^- ion bubble in near-critical argon gas that influences the ion drift by causing the emission of sound waves (Borghesani and Tamburini, 1999).

The boundary condition at $r = R$ is then obtained from eqn (5.39)

$$\frac{\mathrm{d}}{\mathrm{d}r}\left(\frac{u_l}{r}\right) = \lambda_l \rho_0(R) c_\infty^2 k^2 \left(\frac{u_l}{r}\right)\,. \qquad (5.45)$$

Equation (5.42) is transformed again into an equation for the phase shifts $\eta_l(k, r)$ (Levy and Keller, 1963; Calogero, 1967):

$$\begin{aligned}\frac{\mathrm{d}\eta_l}{\mathrm{d}r} &= (kr)^2 \frac{\mathrm{d}\,(\ln\rho_0)}{\mathrm{d}r}\left[j_l'\,(kr)\cos\eta_l - n_l'\,(kr)\sin\eta_l\right]\left[j_l\,(kr)\cos\eta_l - n_l\,(kr)\sin\eta_l\right] \\ &\quad - k^3 r^2 \left(1 - \frac{c_\infty^2}{c_0^2}\right)\left[j_l\,(kr)\cos\eta_l - n_l\,(kr)\sin\eta_l\right]^2\,, \end{aligned} \qquad (5.46)$$

where primes indicate differentiation with respect to the argument. The phase shifts $\eta_l(k, r)$ must be interpreted as the phase shifts that would result by setting $\rho_0 = \rho_\infty$ and $c_0 = c_\infty$ for radii greater than r, i.e., if the scattering terms were cut off at r (Schwarz, 1972*c*). The boundary condition at the radius of the ion structure becomes

$$\tan\eta_l\,(k, R) = \frac{j_l'\,(kR) + \gamma_l kR j_l\,(kR)}{n_l'\,(kR) + \gamma_l kR n_l\,(kR)}\,, \qquad (5.47)$$

where the coefficients γ_l are defined as (Baym *et al.*, 1969)

$$\gamma_l = -\lambda_l \left(\frac{\rho(R) c_\infty^2}{R} \right). \tag{5.48}$$

The initial phase shifts given by eqn (5.47) are those arising from the central-core structure alone and therefore describe the scattering when electrostriction can be neglected.

Once the phase shifts are calculated by numerically integrating eqn (5.46) with initial condition eqn (5.47), the differential cross-section is given by

$$\sigma(k, \theta) = k^{-2} \left| \sum_{l=0}^{\infty} (2l+1) P_l(\cos\theta) \sin\delta_l \exp(i\delta_l) \right|^2, \tag{5.49}$$

where $\delta_l(k) = \lim_{r \to \infty} \eta_l(k, r)$.

As an example of the influence of electrostriction, the first three phase shifts are shown in Fig. 5.20. It is clear that for an electron bubble, for which electrostriction can be safely neglected, the boundary conditions at the bubble surface dominate the behavior of the cross-section, while for the positive ions σ_{mt} strongly depends on the fluid environment of the core.

The other significant difference between the electron bubble and the positive ion is related to the calculation of the coefficients γ_l. In the case of the electron bubble, because the excitation energies of the electron inside the bubble are much

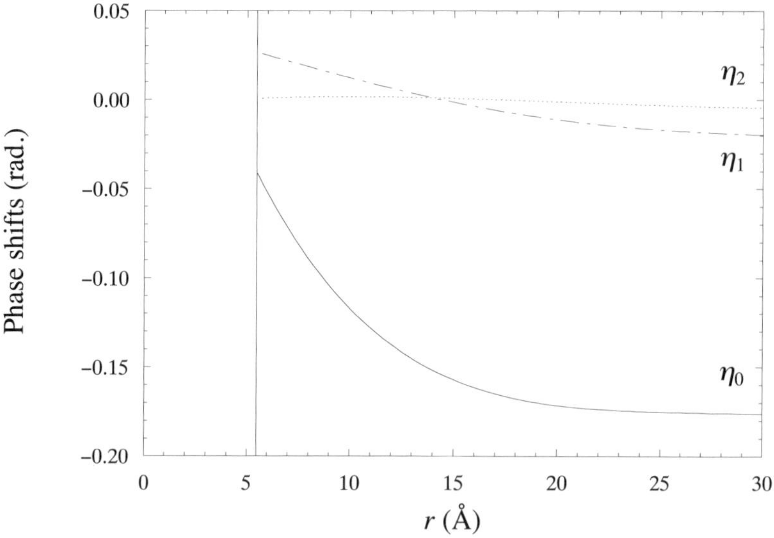

FIG. 5.20. The first three phase shifts calculated by integrating eqn (5.46), including the effects of electrostriction in the fluid surrounding an immovable hard core of 5.5 Å diameter. In this case the phonon wave vector is $k = 0.1\,\text{Å}^{-1}$. (Schwarz, 1972*c*.)

larger than the phonon energies, all the γ_l can be taken as static derivatives (Baym *et al.*, 1969). For instance, for the s-wave, $\delta R_0/\delta p_0$ is just $\partial R/\partial p$, i.e., the variation of the equilibrium bubble radius with the applied pressure.

A special case is the $l = 1$ partial-wave contribution. In fact, an infinitesimal $l = 1$ displacement of the surface of the sphere corresponds to a uniform translation, for which there is no restoring force. In this case the Newton equation is $M_B \delta \dot{a}_1 = -(4\pi/3)R^2 \delta p_1$, where M_B is the mass within the bubble, and $\gamma_1 = -(4\pi^3 \rho/3M_B k^2 R^2)$ (Baym *et al.*, 1969).

In Fig. 5.21 the momentum-transfer cross-section and the thermal factor $-k^4(\partial n_k/\partial k)$ are shown. The strong s- and d-wave resonances can be clearly seen. The s-wave resonance corresponds to a breathing mode of the bubble (Gross and Tung-Li, 1968), whereas the d-wave resonance corresponds to a quadrupole oscillation of the bubble. Resonances of higher angular momentum are not shown because the thermal factor limits the range of k values significantly contributing to the mobility (see eqn (5.32)). However, the higher l-partial waves at large kR ensure that σ_{mt} approaches its geometrical limit πR^2.

The results for the mobility of the electron bubble are reported in Fig. 5.22. The constancy of the product $\mu_- T^3$ is accidental and is related to the fact that the thermal factor samples the cross-section just beyond the s-wave resonance, where the cross-section behaves approximately as k^{-1}.

For $T \geqslant 0.5\,\mathrm{K}$, both experiment and theory deviate from the $\mu_- \propto T^{-3}$ law.

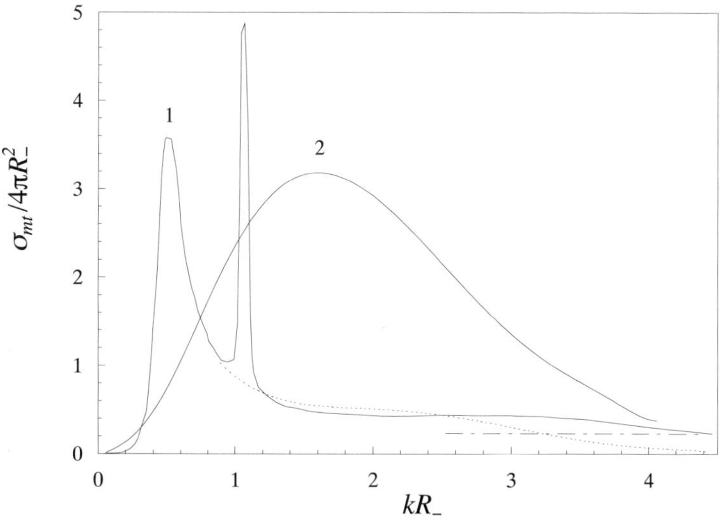

FIG. 5.21. $\sigma_{mt}/4\pi R_-^2$ for scattering on sound waves (curve 1). Curve 2 is the thermal factor $-k^4(\partial n_k/\partial k)$ for an electron bubble. σ_{mt} is calculated by including the s-, p-, and d-waves. The dotted line is obtained if the d-wave contribution is neglected. The dash–dotted line is the geometrical cross-section. (Baym *et al.*, 1969.)

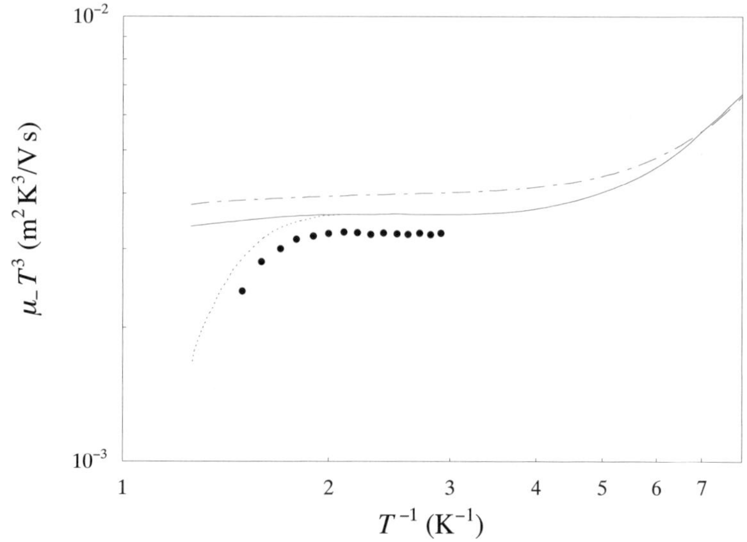

FIG. 5.22. Calculated phonon-limited mobilities for the electron bubble. Dash–dotted line: cross-section with only s- and p-waves included and solid line: cross-section also including d-waves and higher partial waves (Baym *et al.*, 1969). The dotted line takes into account phenomenologically the roton contribution. The points are the measurements of Schwarz and Stark (1968).

The experimental data fall off very rapidly because of the roton contribution to the scattering.

In any case, the agreement between theory and experiment is reasonably good in the region where $\mu_- T^3$ is constant. The residual 10–15% discrepancy might be attributed to the crude approximation of too steep a transition region at the bubble surface (Schwarz, 1972*c*).

In the case of the positive ions the response coefficient γ_l must be calculated by taking into account the special structure of the complex, including the liquid–solid boundary due to electrostriction. This phase transition occurs at an equilibrium radius R_+ given by eqn (3.10). Although the liquid–solid surface tension σ_{ls} is not known accurately, there is strong evidence that it is of the order of 1×10^{-4} N/m (Keshishev *et al.*, 1969) and eqn (3.10) allows us to determine the equilibrium radius as a function of σ_{ls}. The response of the core to the sound pressure field, i.e., the response coefficients λ_l, is calculated by assuming that the liquid–solid boundary reacts instantaneously to local pressure changes. It is interesting to note that a local increase of the pressure due to the sound field causes the phase boundary to move outwards (Schwarz, 1972*c*). The response coefficients are given by

$$\lambda_l = -\left(\left.\frac{\partial p_0}{\partial r}\right|_{r=R_+} + \frac{2\sigma_{ls}}{R_+^2}\frac{v_s}{v_l - v_s}\right)^{-1}, \qquad l = 0, 2, 3, \ldots . \tag{5.50}$$

The $l = 1$ term is also explicitly excluded in this case because the $l = 1$ partial-pressure wave exerts a net force on the core and the resulting motion is accounted for by adding a term of the form (Lamb, 1945)

$$\Delta\lambda_1 = \frac{4\pi R_+^4}{3c_\infty^2 M_+ \left(kR_+\right)^2}, \tag{5.51}$$

where M_+ is the mass of the core calculated within the electrostriction model. Once the λ_ls are calculated, the previous formalism (eqns (5.46) and (5.47)) is applied and the cross-section can be calculated by means of eqn (5.49).

In Fig. 5.23 the calculated momentum-transfer cross-section for the positive ion is shown in three different cases. Curve 1 is calculated for an immovable hard core without electrostrictive effects in the surrounding fluid. Curve 2 is calculated including electrostriction. Finally, curve 3 is obtained for an immovable responsive core with electrostriction included.

The resulting phonon-limited mobility of positive ions is shown in Fig. 5.24 along with the experimental data (Schwarz, 1972*a*) for the core parameters corresponding to $\sigma_{ls} = 0.15 \times 10^{-3}$ and $0.10 \times 10^{-3}\,\mathrm{N/m}$. The agreement with the experimental data is fairly good. An improvement of the fit can be obtained by

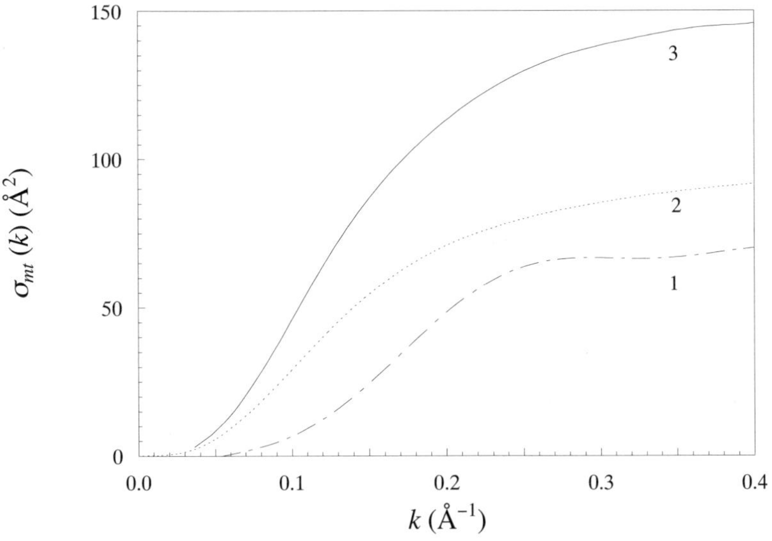

FIG. 5.23. $\sigma_{mt}(k)$ for phonons scattering off the snowball. Curve 1: no electrostriction, curve 2: with electrostriction, and curve 3: with electrostriction and core responsivity. The hard core radius is 5.37 Å. (Schwarz, 1972*c*.)

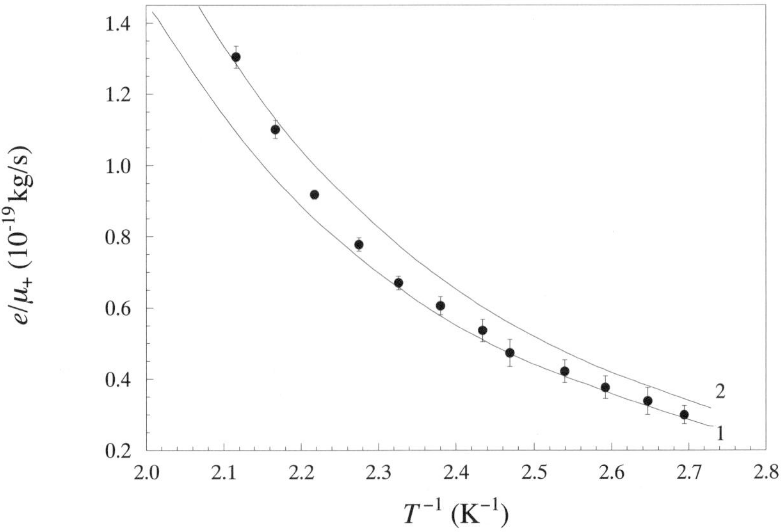

FIG. 5.24. e/μ_+ in the phonon-dominated region. Curves 1 and 2 are calculated with core parameters corresponding to $\sigma_{ls} = 0.15 \times 10^{-3}$ and 0.10×10^{-3} N/m, respectively. The deviations for $T^{-1} < 2.25\,\mathrm{K}^{-1}$ are due to the influence of rotons. (Schwarz, 1972*c*.)

using the value $\sigma_{ls} = 0.135 \times 10^{-3}$ N/m, but this is not significant, owing to the uncertainty in the experimental value of the surface tension.

As the mobility in the phonon-limited regime gives information on the radius and structure of the charge carriers, mobility measurements at higher pressures are aimed at investigating how these properties change.

At high P, the importance of phonon relative to roton scattering decreases (Ostermeier, 1973). This can be understood qualitatively by looking at the normal fluid densities for the two types of excitation (Wilks, 1967). The ratio of the two densities is given by

$$\frac{\rho_{\mathrm{ph}}}{\rho_r} \propto \frac{T^{9/2} e^{\Delta(P)/k_{\mathrm{B}}T}}{c^5\,(P)}, \tag{5.52}$$

where $\Delta(P)$ and $c(P)$ are the pressure-dependent roton energy gap and sound velocity, respectively. At vapor pressure, rotons and phonons contribute equally to the normal fluid density for $T \approx 0.5$ K. This value is consistent with the approximate temperature at which μ switches from the roton- to the phonon-dominated regime ($T \approx 0.5$ K for the positive ions and $T \approx 0.7$ K for the negative ones) (Schwarz and Stark, 1968, 1969).

From eqn (5.52) it is clear that phonons become more important as T is lowered. However, as $\Delta(P)$ decreases and $c(P)$ becomes larger with increasing P, the roton contribution dominates the scattering toward lower T at higher P. This situation is depicted in Figs 5.25 and 5.26. As P is increased, the temperature of the transition between the roton- and the phonon-dominated regions shifts to lower values.

A fit of the phonon-limited mobility for the negative species using the phonon-scattering models outlined above enables the determination of the radius $R_-(P)$ of the electron bubble. The results are shown in Fig. 3.6. This determination of the radius is in good agreement with other experimental and theoretical predictions (Amit and Gross, 1966; Springett, 1967). The bubble is *squeezed* as P increases.

Actually, it is interesting to note that this specific determination of the radius, which basically probes the outside of the electron bubble via phonon scattering, gives a radius value somewhat larger than photoejection experiments, which probe its inner radius (Zipfel, 1969). The difference between the two determinations is of approximately 1 Å, i.e., comparable to the width of the bubble interface.

For the positive carriers the experimental data cannot give a detailed test of the pressure dependence of the theory (Schwarz, 1972*c*) because of the too-large roton contribution, even though the calculated phonon–limited mobilities shown in Figs 5.25 and 5.26 appear to be of the correct order of magnitude.

Measurements at an even higher temperature in the range $1.3\,\mathrm{K} < T < T_\lambda$ up to the melting pressure (Brody, 1975) are not significant for the determination of the phonon contribution to the scattering because rotons dominate the scattering

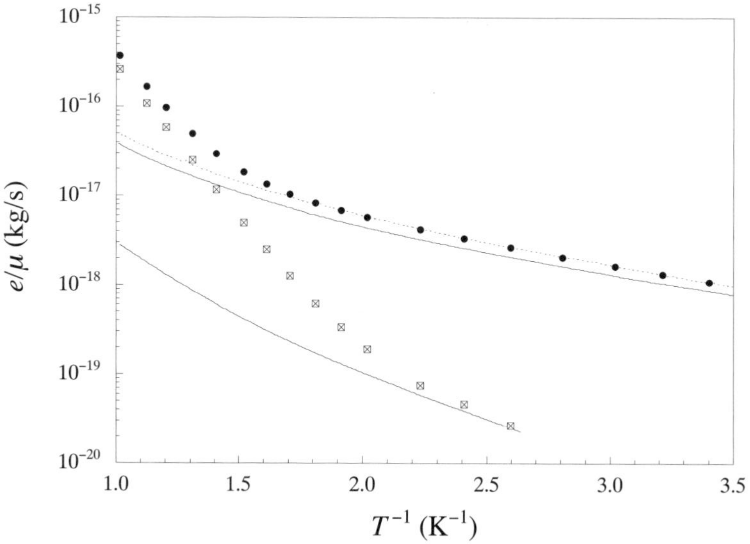

FIG. 5.25. e/μ vs T^{-1} for $P = 0.214\,\mathrm{MPa}$. Circles: negative ions, and squares: positive ions (Ostermeier, 1973). Solid lines: phonon contribution (Baym *et al.*, 1969). Dashed line: phonon-contribution for the electron bubble (Schwarz, 1975).

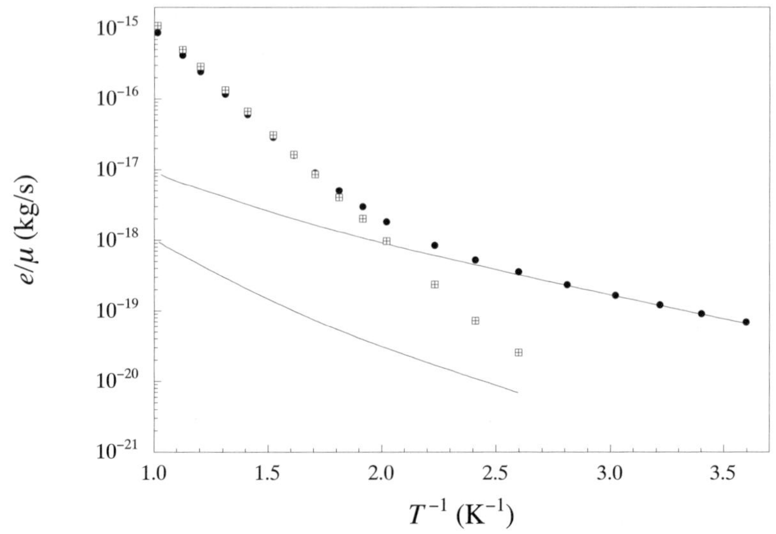

FIG. 5.26. e/μ vs T^{-1} for $P = 2.224$ MPa. Circles: negative ions, and squares: positive ions (Ostermeier, 1973). Lines: phonon contribution (Baym *et al.*, 1969).

at these temperatures and phonons do not contribute more than a few per cent at most.

5.8 Roton-limited mobility

For temperatures above 0.6–0.8 K, the ion mobility is basically limited by collisions with rotons (Barrera and Baym, 1972). In the kinetic regime for $0.8\,\mathrm{K} < T < 1.7\,\mathrm{K}$, roton–roton scattering can be neglected, whereas for $T \geqslant 2\,\mathrm{K}$ the roton–roton collisions are so frequent that rotons appear to the ions as a viscous fluid.

Several experiments have been carried out in the kinetic regime at SVP as well as at higher pressures (Reif and Meyer, 1960; Meyer and Reif, 1961; Schwarz and Stark, 1968, 1969; Doake and Gribbon, 1971; Strayer *et al.*, 1971; Schwarz, 1972*c*; Ostermeier, 1973; Allum *et al.*, 1975; Brody, 1975). Experimental data on alkali and alkaline–earth ions (Ihas and Sanders, 1970; Főrste *et al.*, 1997) and on exotic ions (Ihas and Sanders, 1971) will be discussed later.

The roton-limited drag coefficient e/μ_r can be extracted from the experimental data by subtracting out the phonon contribution discussed previously. In any models considering collisions between ions and rotons the scattering rate is proportional to the density of thermal rotons, ρ_r, given by

$$\rho_r = \frac{1}{3}\hbar k_0^4\,(P,T)\left[\frac{g\,(P,T)}{2\pi^3 k_{\mathrm{B}} T}\right]^{1/2} e^{-\Delta(P,T)/k_{\mathrm{B}}T}\,, \qquad (5.53)$$

where g is the chemical potential, and k_0 and Δ are the roton momentum and energy gap at the roton minimum, respectively (Ostermeier, 1973). As a result, the drag coefficient is dominated by the exponential factor $\exp(-\Delta/k_BT)$, as is easily observed, e.g., in Fig. 5.4.

In order to remove the dominating exponential, it is customary to plot the data in the form (Yarnell *et al.*, 1959; Dietrich *et al.*, 1972; Donnelly, 1972)

$$f_\pm(P,T) = \frac{e}{\mu_r}\exp\left[\Delta(P,T)/k_BT\right]. \tag{5.54}$$

The relevant physics of scattering is therefore contained in the prefactor $f_\pm(P,T)$, where the symbol $\pm$ refers to the positive and negative species.

In Fig. 5.27 the prefactors $f_\pm$ at vapor pressure are shown as a function of the inverse temperature, whereas in Figs 5.28 and 5.29 their temperature dependence is shown for higher pressures.

Below 1 K both $f_\pm$ increase monotonically with decreasing T, but for $T \geqslant$ 1 K their behavior is much more complicated. It is suggested that the smooth behavior for $T < 1$ K is related to the temperature invariance of the carrier structure and roton parameters in this temperature range (Ostermeier, 1973).

Because the roton wavelengths are shorter than the carrier size, a simple kinetic approach would yield the mobility in the form

$$\frac{e}{\mu_r} \propto \rho_n\sigma v_{\rm rel} = \frac{\hbar k_0^4}{3\pi^2}\sigma\exp\left(-\Delta/T\right), \tag{5.55}$$

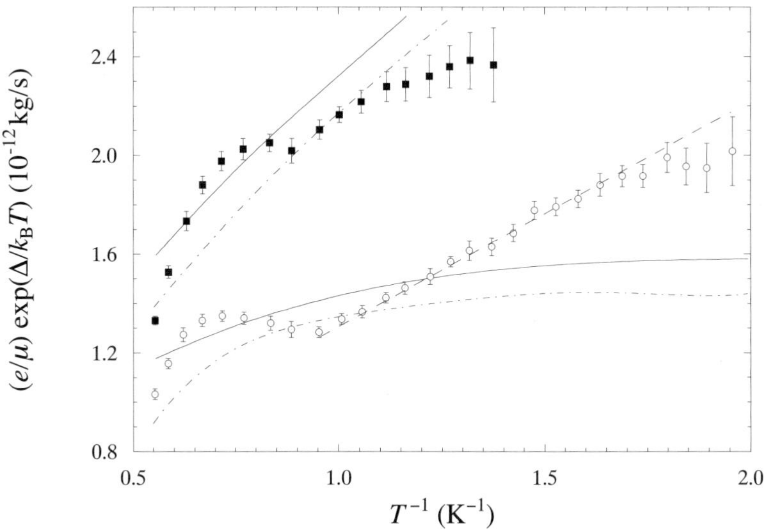

FIG. 5.27. f_+ (circles) and f_- (squares) vs T^{-1} at vapor pressure (Ostermeier, 1973). The meaning of the curves is explained in the text.

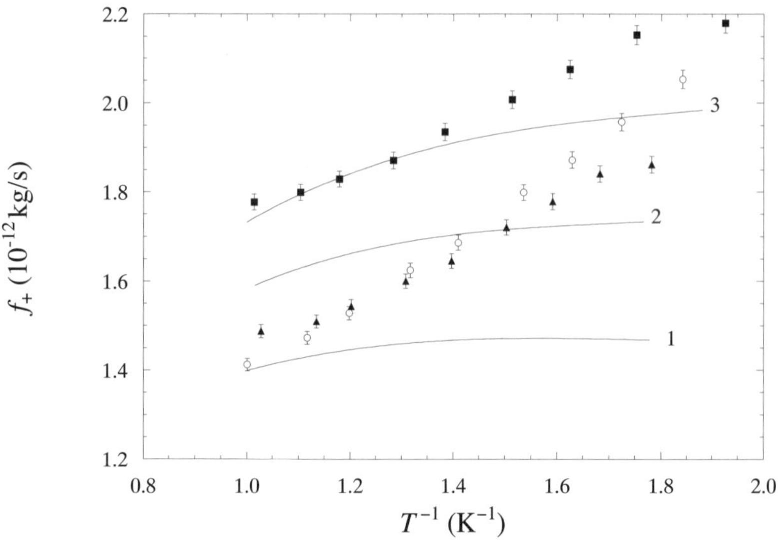

FIG. 5.28. f_+ vs T^{-1} for P (MPa) = 0.214 (circles), 1.38 (triangles), and 2.51 (squares). (Ostermeier, 1973.) Curves 1, 2, and 3 refer to the same pressures, respectively, and their meaning is explained in the text.

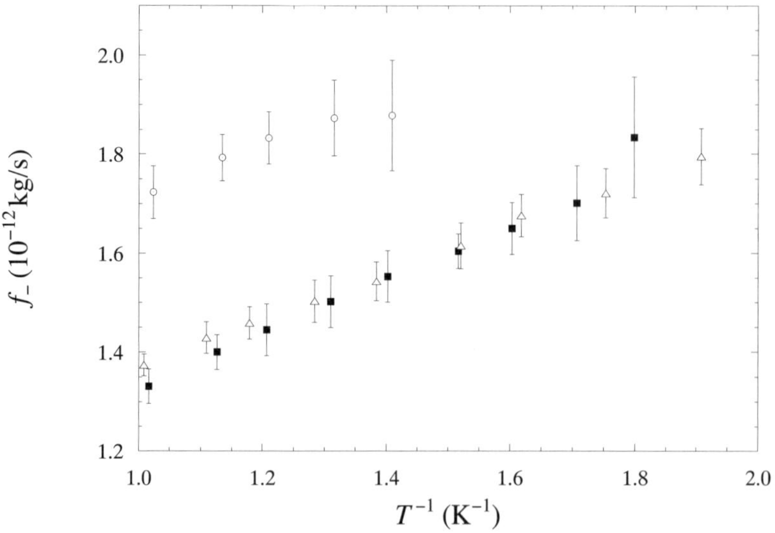

FIG. 5.29. f_- vs T^{-1} for P (MPa) = 0.214 (circles), 1.38 (squares), and 2.51 (triangles). (Ostermeier, 1973.)

where ρ_n is the roton normal mass density, $v_{\rm rel}$ is the average relative velocity between rotons and ions, and σ is a mean roton–ion cross-section of the order of the geometrical cross-section (Schwarz, 1975).

Unfortunately, this approach fails to account for the relative magnitude of f_-/f_+, 1.5 at vapor pressure, whereas the ratio of the geometrical cross-sections is of order 10 (Schwarz, 1975). This fact suggests the possibility that the electrostriction-induced density variation around the positive ion greatly influences the scattering of rotons (Schwarz, 1972*c*).

Moreover, the kinetic model fails to explain the differences in the temperature dependence observed for the positive and negative carriers because it does not take into account the detailed ion–roton kinematics allowing for recoil of the ions due to their finite mass (Barrera and Baym, 1972).

As $f_\pm$ include the physics of scattering they should be related in some way to the effective radii $R_\pm$ and effective masses $M^\star_\pm$ of the charge carriers. However, the correlation is not definitive but only qualitative, as shown in Fig. 5.30.

Bowley (1971*a*, 1971*c*) and Barrera and Baym (1972) have treated the effect of ion recoil in terms of the van Hove scattering functions of the impurities and of the ion–roton scattering amplitudes. The main problem for the explicit calculation of the ion mobility is the lack of any knowledge about the ion-scattering matrix elements. Although a microscopic theory for the scattering of rotons by

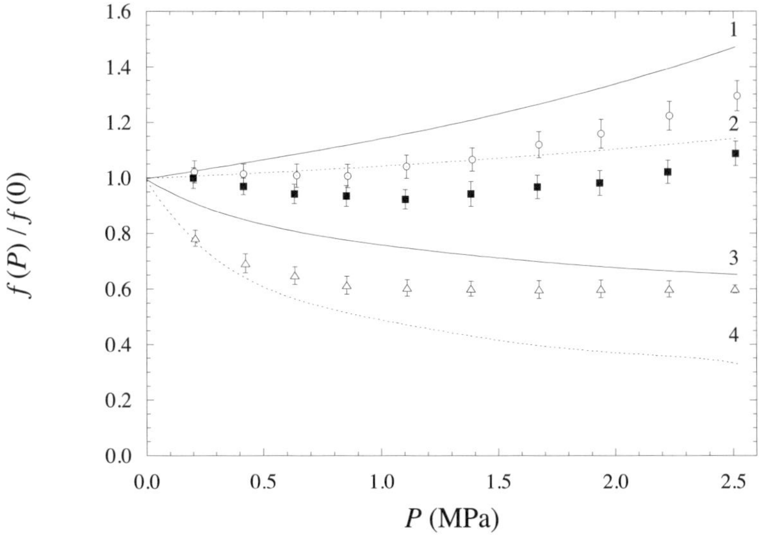

FIG. 5.30. $f_\pm(P)/f_\pm(0)$, ion radius, and effective mass vs P (Ostermeier, 1973). Triangles: $f_-(P)/f_-(0)$ at $T = 0.985$ K, circles: $f_+(P)/f_+(0)$ at $T = 0.985$ K, and squares: $f_+(P)/f_+(0)$ at $T = 0.586$ K. Curve 1: $R_+(P)/R_+(0)$, curve 2: $M^\star_+(P)/M^\star_+(0)$, curve 3: $R_-(P)/R_-(0)$, and curve 4: $M^\star_-(P)/M^\star_-(0)$. $R_\pm$ and $M^\star_\pm$ are calculated with surface tension $\sigma_{ls} = 0.135 \times 10^{-3}$ N/m (Schwarz, 1972*c*).

fixed non-recoiling ions via a two-body potential has provided a model for these matrix elements (Iguchi, 1971), nonetheless a phenomenological approach for their calculation has still to be followed.

As usual, the calculation of the roton-limited mobility starts with the Boltzmann transport equation (Barrera and Baym, 1972)

$$e\mathbf{E} \cdot \nabla_{\mathbf{p}} f_{\mathbf{p}}^0 = - \sum_{\mathbf{p}'\mathbf{k}\mathbf{k}'} \left[f_{\mathbf{p}} n_{\mathbf{k}} \left(1 + n_{\mathbf{k}'}\right) - f_{\mathbf{p}'} n_{\mathbf{k}'} \left(1 + n_{\mathbf{k}}\right) \right] \Gamma \left(\{\mathbf{p}, \mathbf{k}\} \to \{\mathbf{p}', \mathbf{k}'\}\right) , \tag{5.56}$$

where e is the ionic charge, $\mathbf{E}$ is the electric field, and $\nabla_{\mathbf{p}}$ is the gradient in the momentum space. Here

$$f_{\mathbf{p}}^0 = n_i \left(\frac{2\pi\beta\hbar^2}{M^\star} \right)^{3/2} e^{-\beta p^2/2M^\star} \tag{5.57}$$

is the equilibrium distribution function of the ion, $M^\star$ is its effective mass, $\beta = 1/k_{\mathrm{B}}T$, and n_i is the ion density. $n_{\mathbf{k}}$ is the distribution function of the elementary excitations. Also,

$$\Gamma \left(\{\mathbf{p}, \mathbf{k}\} \to \{\mathbf{p}', \mathbf{k}'\}\right) = \frac{2\pi}{\hbar} \delta \left(\frac{p^2}{2M^\star} + \hbar\omega_k - \frac{p'^2}{2M^\star} - \hbar\omega_{k'} \right) \left| \langle \mathbf{p}', \mathbf{k}' | T | \mathbf{p}, \mathbf{k} \rangle \right|^2 , \tag{5.58}$$

where Γ is the transition rate at which an ion of momentum $\mathbf{p}$ scatters into a state of momentum $\mathbf{p}'$ by collision with an excitation of initial energy $\hbar\omega_k$ and momentum $\hbar\mathbf{k}$, and final energy and momentum $\hbar\omega_{k'}$ and $\hbar\mathbf{k}'$, respectively. $\langle \mathbf{p}', \mathbf{k}' | T | \mathbf{p}, \mathbf{k} \rangle$ is the scattering amplitude. The Dirac delta ensures energy conservation.

If the temperature is sufficiently low, the mean free path for roton–roton scattering is much larger than the ion–roton effective interaction radius R. Therefore, the distribution function is negligibly modified near the ion in this kinetic regime. Moreover, because the ion density is quite small, roton–roton collisions are much more frequent than the ion–roton ones. This means that the equilibrium distribution function of the excitations is quickly re-established by roton–roton collisions after an ion–roton collision long before the next ion–roton collision takes place, and the net momentum gained by the elementary excitations is eventually dissipated at the container walls (Barrera and Baym, 1972).

Because of these conditions, the excitation distribution function n_k can be replaced by the equilibrium one:

$$n_{\mathbf{k}} \approx n_{\mathbf{k}}^0 = \left(e^{\beta\hbar\omega_k} - 1 \right)^{-1} . \tag{5.59}$$

By defining $\beta\Phi_{\mathbf{p}} \equiv \left(f_{\mathbf{p}} - f_{\mathbf{p}}^0 \right) / f_{\mathbf{p}}^0$ and taking into account the detailed balance

$$f_{\mathbf{p}}^0 n_{\mathbf{k}}^0 (1 + n_{\mathbf{k}}^0) = f_{\mathbf{p}'}^0 n_{\mathbf{k}'}^0 (1 + n_{\mathbf{k}}^0) , \tag{5.60}$$

eqn (5.56) can be written as

$$\left(e\mathbf{E}\cdot\frac{\mathbf{p}}{M^\star}\right)f^0_{\mathbf{p}} = -\sum_{\mathbf{p}'\mathbf{k}\mathbf{k}'}\left(\Phi_{\mathbf{p}}-\Phi_{\mathbf{p}'}\right)f^0_{\mathbf{p}}n^0_{\mathbf{k}}\left(1+n^0_{\mathbf{k}'}\right)\Gamma\,. \tag{5.61}$$

The drift velocity is expressed in terms of Φ_p as

$$\mathbf{v}_D = \frac{\beta}{N_i M^\star}\sum_{\mathbf{p}}\mathbf{p}\,\Phi_{\mathbf{p}}f^0_{\mathbf{p}}\,, \tag{5.62}$$

where N_i is the total number of ions.

Owing to the lack of an explicit solution of eqn (5.61), the mobility is calculated by means of a variational principle. It has been shown (Ziman, 1960) that the kernel of the right-hand side of eqn (5.61) is positive and that, for this reason, the following inequality holds true for any given trial function Ψ_p:

$$\left(e\mathbf{E}\cdot\mathbf{p}\frac{\Phi_{\mathbf{p}}f^0_{\mathbf{p}}}{M^\star}\right)^{-1} = \frac{\beta}{\mu N_i e E^2} \leqslant \frac{1}{2}\frac{\sum\limits_{\mathbf{p}\mathbf{p}'\mathbf{k}\mathbf{k}'}\left(\Psi_{\mathbf{p}}-\Psi_{\mathbf{p}'}\right)f^0_{\mathbf{p}}n^0_{\mathbf{k}}\left(1+n^0_{\mathbf{k}'}\right)\Gamma}{\left(\sum\limits_{\mathbf{p}}\Psi_{\mathbf{p}}e\mathbf{E}\cdot\mathbf{p}f^0_{\mathbf{p}}/M^\star\right)^2}\,. \tag{5.63}$$

If the simplest trial function is chosen:

$$\Psi_{\mathbf{p}} = \mathbf{p}\cdot\mathbf{v}_D\,, \tag{5.64}$$

which represents an ion in drifting equilibrium, then a lower bound for the mobility is obtained:

$$\frac{e}{\mu} = \frac{\beta}{6N_i}\sum_{\mathbf{p}\mathbf{p}'\mathbf{k}\mathbf{k}'}\left(\mathbf{p}-\mathbf{p}'\right)^2 f^0_{\mathbf{p}}n^0_{\mathbf{k}}\left(1+n^0_{\mathbf{k}'}\right)\Gamma\,. \tag{5.65}$$

Only some educated guesses can be put forward to calculate the matrix elements (or scattering amplitudes) $t(k) = \langle\mathbf{p},\mathbf{k}|T|\mathbf{p}',\mathbf{k}'\rangle$ that determine Γ. Some pieces of information on the structure of the scattering amplitudes can be obtained by calculating the cross-section for an ion initially at rest ($\mathbf{p}=0$). Assuming that the scattering amplitude $t(k)$ depends only on the initial roton momentum that is related to the initial roton group velocity $\partial\omega/\partial k = |\mathbf{k}-\mathbf{k}_0|/m_r$, where $\mathbf{k}_0$ is the wave vector at the roton minimum in the dispersion curve and m_r is the roton mass, the total cross-section is given by

$$\sigma_{mt} = \left(\frac{m_r}{|\mathbf{k}-\mathbf{k}_0|}\right)\left|t(k)\right|^2\rho(\omega)\,, \tag{5.66}$$

where $\omega = \hbar(\mathbf{k}-\mathbf{k}_0)^2/2m_r$ and the density of states $\rho(\omega)$ is defined as

$$\rho(\omega) = \frac{2\pi}{\hbar}\int\frac{\mathrm{d}^3k'}{(2\pi\hbar)^3}\delta\left[\frac{\hbar^2(\mathbf{k}-\mathbf{k}')^2}{2M^\star}+\frac{\hbar^2(\mathbf{k}'-\mathbf{k}_0)^2}{2m_r}-\frac{\hbar^2(\mathbf{k}-\mathbf{k}_0)^2}{2m_r}\right]. \tag{5.67}$$

The integration is carried out easily, yielding

$$\rho(\omega) = \begin{cases} \frac{M^\star|\mathbf{k}-\mathbf{k}_0|}{\pi\hbar^4}, & (\mathbf{k}-\mathbf{k}_0)^2 \leqslant \frac{4m_r k_0^2}{M^\star}, \\ \frac{M^\star|\mathbf{k}-\mathbf{k}_0|}{\pi\hbar^4}\left\{1-\left[1-\frac{4m_r k_0^2}{M^\star(\mathbf{k}-\mathbf{k}_0)^2}\right]^{1/2}\right\}, & (\mathbf{k}-\mathbf{k}_0)^2 \geqslant \frac{4m_r k_0^2}{M^\star} \end{cases} \tag{5.68}$$

if terms of order $m_r/M^\star$ are neglected and if the initial roton momentum is limited to $|\mathbf{k}-\mathbf{k}_0| \ll k_0$.

In order to introduce a k-dependence, it is assumed that the cross-section is approximately equal to the geometrical cross-section for all values of k. An interpolating formula then gives

$$|t(k)|^2 = \frac{\pi^2\hbar^4}{M^\star m_r}\left(a^2 + a_1^2\frac{m_r}{M^\star}\frac{|\mathbf{k}-\mathbf{k}_0|^2}{2k_0^2}\right), \tag{5.69}$$

where a is the scattering length, of the order of the ionic radius R, and a_1 is another effective diameter. The first term accounts for processes in which $|\mathbf{k}-\mathbf{k}_0| \to 0$, while the second term describes the situation for a very massive ion and agrees with the result obtained following a microscopic approach (Iguchi, 1971).

This choice of $t(k)$ has given origin to severe criticism. In particular, it is not justified to assume that, in the limit as $|\mathbf{k}-\mathbf{k}_0| \to 0$, a geometrical cross-section should be arrived at. Moreover, if only the second term in eqn (5.69) was retained, i.e., in the limit of very massive ions, the kinetic formula (5.55) with a temperature-independent prefactor would again be obtained. On the other hand, the inclusion of the first term in eqn (5.69) gives a wrong pressure dependence of the electron bubble mobility (Schwarz, 1975). In any case, it is instructive to carry out the calculations of the model explicitly.

The scattering length a turns out to be an effective ion radius and has now to be fitted to the experimental results. With this scattering amplitude the mobility is readily obtained from eqn (5.65). By introducing the auxiliary variables $q = |\mathbf{k}-\mathbf{k}'|$, $\omega = (\hbar/2m_r)[(\mathbf{k}-\mathbf{k}_0)^2 - (\mathbf{k}'-\mathbf{k}_0)^2]$, and $\epsilon = (1/2)(\hbar^2/2m_r)[(\mathbf{k}-\mathbf{k}_0)^2 + (\mathbf{k}'-\mathbf{k}_0)^2]$, $\epsilon_i = \Delta + (\hbar^2/2m_{r)}(k-k_0)^2$, $\epsilon_f = \Delta + (\hbar^2/2m_{r)}(\mathbf{k}'-\mathbf{k}_0)^2$, and indicating by n the roton distribution function, the mobility turns out to be (Bowley, 1971*c*; Ostermeier, 1973)

$$\frac{e}{\mu} = \frac{\beta}{3(2\pi\hbar)^3}\int_0^\infty \mathrm{d}\epsilon \int_{-2\epsilon}^{2\epsilon}\mathrm{d}\omega \int_{k-k'}^{k+k'} \mathrm{d}q\, q^3 S(q,\omega) n(\epsilon+\hbar\omega/2)[1+n(\epsilon-\hbar\omega/2)]kk'$$
$$\times\left(\sum_j \frac{\left|\langle k|T|k'\rangle\right|^2}{|\mathrm{d}\epsilon_i/\mathrm{d}k|\,|\mathrm{d}\epsilon_f/\mathrm{d}k'|}\right), \tag{5.70}$$

where the sum over the transition matrix elements is for the three possible channels of scattering available to the incident roton (Bowley, 1971*c*).

The van Hove scattering function $S(q,\omega)$ is defined as

$$S(q,\omega) = \left(\frac{M^\star\beta}{2\pi\hbar^2q^2}\right)^{1/2} \exp\left[-\frac{M^\star\beta}{2\hbar^2q^2}\left(\hbar\omega - \frac{\hbar^2q^2}{2M^\star}\right)^2\right] \tag{5.71}$$

and is obtained by summing the δ-function in eqn (5.58) over all possible values of $\mathbf{p}$ (Josephson and Lekner, 1969; Bowley, 1971*a*, 1971*c*).

Owing to the formidable theoretical difficulty in deriving the matrix elements from first principles, the mobility is explicitly calculated by assuming that the matrix elements are constant (Bowley, 1971*c*; Barrera and Baym, 1972), i.e., that $t(k)$ consists only of the first term in eqn (5.69), or equivalently

$$\langle \mathbf{p},\mathbf{k}|T|\mathbf{p}',\mathbf{k}'\rangle = \frac{\pi\hbar^2 R}{(m_r M^\star)^{1/2}\,\Omega}\delta_{\mathbf{p}+\mathbf{k},\mathbf{p}'+\mathbf{k}'}\,, \tag{5.72}$$

where Ω is the volume of the system and R is the effective ion radius. The final result is in the form of eqn (5.55):

$$\frac{e}{\mu_r} = \frac{4}{3}(2\pi)^{1/2}\rho_r v_r R^2 \mathcal{G}\left(\frac{\beta\hbar^2k_0^2}{4M^\star}\right), \tag{5.73}$$

where the roton normal mass density is

$$\rho_r = \frac{\hbar k_0^4}{3\pi}\left(\frac{\beta m_r}{2\pi}\right)^{1/2} e^{\beta\Delta}\,. \tag{5.74}$$

The roton thermal velocity is $v_r = (\beta m_r)^{-1/2}$ and

$$\mathcal{G}(x) = x^{-1} + e^{-x}\left[xK_0(x) - (1+x)K_1(x)\right]\,, \tag{5.75}$$

where the K_ns are the modified Bessel functions of the second kind.

Within this model the prefactors $f_\pm$ can be written as

$$f_\pm = \frac{4}{9\pi}\hbar k_0^4 R_\pm^2 \mathcal{G}\,. \tag{5.76}$$

Using this equation with $M_-^\star = 170m_4$ and $R_- = 18.65$ Å for the electron bubbles and $M_+^\star = 40m_4$ and $R_+ = 9.81$ Å for the positive ions, and accounting for the asymmetry of the roton dispersion curve (Ostermeier, 1973), the solid lines in Fig. 5.27 are obtained at vapor pressure (Barrera and Baym, 1972).

It is interesting to note that, in spite of its limitations, the model predicts prefactors that generally increase with decreasing temperature. Moreover, by taking into account recoil explicitly, the relative magnitude of f_+/f_- is accounted for quite satisfactorily. No improvement of this result is obtained by neglecting the purported asymmetry of the roton spectrum (Roach *et al.*, 1972) and including the T and P variations in the effective ionic radius R in a reasonable way (Ostermeier, 1973), as indicated by the dash–dotted curves in Fig. 5.27.

Following this approach, the temperature dependence of the prefactors as a function of the pressure can also be calculated. The results are shown in Fig. 5.28 for the positive ions and in Fig. 5.29 for the electron bubbles. For the positive ion the predicted prefactor qualitatively describes both the temperature and pressure behavior of the experimental data, as shown by the solid curves in Fig. 5.28. In contrast, the prefactors calculated for the electron bubble, though they correctly increase with decreasing temperature, show a behavior as a function of pressure that is opposite to that of the experimental data (Ostermeier, 1973) and is not therefore shown in the figure.

In summary, in spite of the limitations of the model (Ostermeier, 1973; Schwarz, 1975), a variational solution of the Boltzmann transport equation for the roton contribution to the ion mobility gives a semi-quantitative agreement with the experimental data.

Some improvements in the prefactors at vapor pressures have been obtained by considering the effect of localized roton states in the proximity of the ion as a consequence of the density variation around the ion due to electrostriction (Bondarev, 1973) and of the change of the roton spectrum in the region of potential flow of the superfluid around the ion at large drift velocities (Doake and Gribbon, 1971; Strayer *et al.*, 1971). The discussion of these effects will be postponed to Chapter 6, which is devoted to transport at intermediate fields at low pressures, because they have been derived from observations at high drift velocities.

Here, we observe further that at fairly high temperatures and pressures, at which roton scattering is the dominant mechanism determining the ion mobility, the drift velocity of negative ions shows relevant superlinearities. Results for $T = 1.05\,\mathrm{K}$ at $P = 2.5\,\mathrm{MPa}$ (Allum *et al.*, 1975) are shown in Fig. 5.31. It is believed that the decrease in the ratio E/v_D for non-vanishingly-small electric fields as v_D increases is due to a decrease in the density of fthe inal states, into which rotons can scatter when the ion moves at higher speed (Bowley, 1976*c*).

If the Doppler shift $\hbar\mathbf{q}\cdot\mathbf{v}_D$ of the ion energy due to the relative motion of ions and rotons is introduced into the van Hove scattering function (5.71), the theory for roton scattering (Barrera and Baym, 1972) can be extended to higher ion velocities (Bowley, 1976*c*). This extended theory predicts that the drag on the ion is given by a power expansion in terms of v_D^2:

$$\frac{eE}{v_D} = \frac{1}{\mu_0}\left[1 + Av_D^2 + Bv_D^4 + \ldots\right], \tag{5.77}$$

where the expansion coefficients can be calculated in terms of the ion effective mass, of the momentum at the roton minimum, and of temperature.

In Fig. 5.32 the variation of the drag E/v_D for negative ions for $T = 1.05\,\mathrm{K}$ and $P = 2.5\,\mathrm{MPa}$ is shown as a function of v_D^2. The data for $v_D \leqslant 3\,\mathrm{m/s}$ actually depend linearly on v_D^2. The coefficient $A = (1.04 \pm 0.02) \times 10^{-4}\,\mathrm{m^2/V\,s}$ (Allum *et al.*, 1975) yields an estimate $M^\star = (73 \pm 9)m_4$, which is consistent with the estimate of the ion radius that yields $M^\star = 68m_4$ (Bowley, 1976*c*).

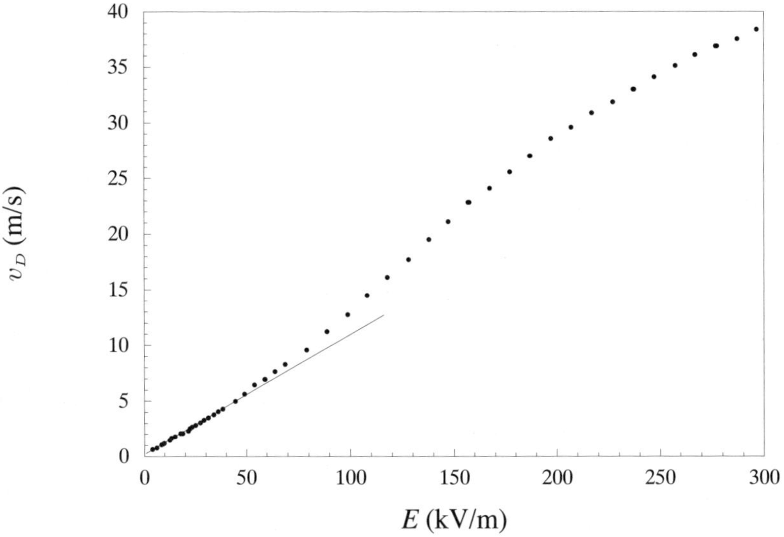

FIG. 5.31. v_D vs E for negative ions for $T = 1.05\,\mathrm{K}$ at $P = 2.5\,\mathrm{MPa}$, showing large nonlinearities. (Allum *et al.*, 1975.)

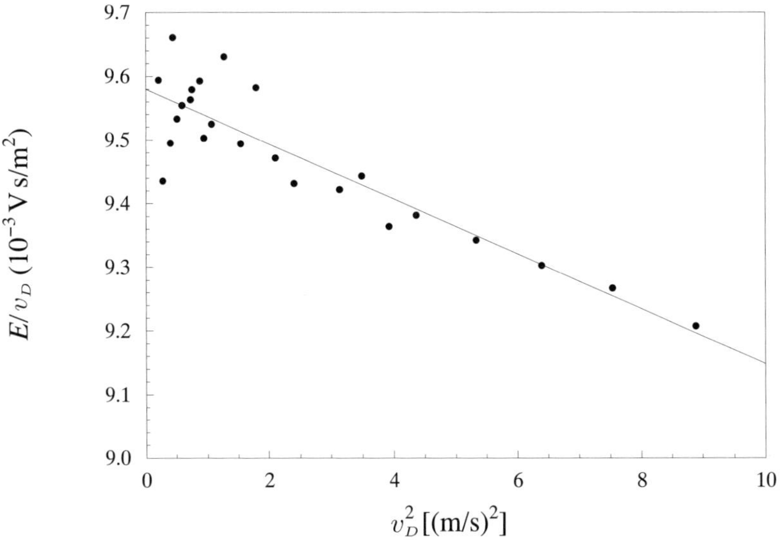

FIG. 5.32. E/v_D vs v_D^2 for negative ions for $T = 1.05\,\mathrm{K}$ at $P = 2.5\,\mathrm{MPa}$ (Allum *et al.*, 1975). The straight line is a fit to eqn (5.77) (Bowley, 1976*c*).

Quite interestingly, nonlinearities of the drift velocity that are qualitatively very similar to those observed in liquid He II at high pressure (Allum *et al.*, 1975) have also been detected in normal liquid He for $T = 3.86\,\mathrm{K}$ (Sakai *et al.*, 1982; Schmidt *et al.*, 2003), and in dense He gas (Schwarz, 1980; Borghesani and Santini, 2002) and Ne gas (Borghesani and Santini, 1990*a*). In these latter cases, the nonlinearities have been interpreted as being due to the relative increase in the population of high-mobility quasi-free electrons with respect to that of negative ions (Schwarz and Prasad, 1976; Borghesani and Santini, 1990*a*; Schmidt *et al.*, 2001).

6

ION TRANSPORT AT INTERMEDIATE FIELDS AT LOW PRESSURE

At higher electric fields E the drift velocity v_D is no longer proportional to E. The way in which v_D varies as E deviates from small-field linear behavior depends on which momentum-transfer process is dominant and could in principle be calculated from the knowledge of the probe scattering rate $\Gamma(\mathbf{p} \rightarrow \mathbf{p}')$ for a non-stationary probe.

Unfortunately, the microscopic models discussed previously have not been extended to carriers moving with a finite velocity (Schwarz and Jang, 1973). In any case, it is reasonable to assume that the deviations from linearity can be interpreted in terms of the simple dilute-gas theory.

At higher field strengths the ions may gain enough energy from the field to no longer be in thermal equilibrium with the gas of the scatterers; if ℓ is the ionic mean free path, the energy gain from the field is $eE\ell$. If $\varphi = eE\ell/k_{\mathrm{B}}T \ll 1$, the ions are in thermal equilibrium with the scatterers. If $\varphi \geqslant 1$, they are not. In the latter case, the ions scatter more frequently and the average drift velocity increases less than proportionally with increasing E. An example of such a behavior for positive ions at SVP is shown in Fig. 6.1.

It is well known that the drift velocity of bare ions depends on the ion energy in excess of the thermal one (Huxley and Crompton, 1974). According to the Drude model $\mu = e\tau/M$, where μ is the zero-field mobility, τ is the mean time between collisions, and M is the ion mass. The mean free path is $\ell = v_{\mathrm{th}}\tau$, where v_{th} is the average thermal velocity. Therefore, the ratio of the energy gained from the field to the thermal energy is $\varphi \propto E/\mu T^{1/2}$. If v_D is plotted as a function of the reduced field $E' = (2/3) \times 10^{-5}(E/\mu T^{1/2}) \propto \varphi$, a universal curve should be obtained from the data taken along different isotherms (Reif and Meyer, 1960; Rayfield, 1968*a*). This is indeed true, e.g., for negative ions in the range $0.49\,\mathrm{K} < T < 0.65\,\mathrm{K}$ in Fig. 6.2 (Rayfield, 1968*a*).

Eventually, if the electric field is sufficiently large, it becomes possible for ions to create rotons in the background fluid, and/or quantized vortex rings (Reif and Meyer, 1960; Rayfield and Reif, 1963; Careri *et al.*, 1965).

If the electric field becomes sufficiently large so that the condition $\varphi \ll 1$ is no longer satisfied, but rather the condition $\varphi \geqslant 1$ is fulfilled, then charge-carrying quantized vortex rings are created in the superfluid (Rayfield and Reif, 1963; Rayfield, 1968*a*).

At low temperatures, $T \ll 1\,\mathrm{K}$, where the gas of elementary excitations is quite dilute, quantized vortex ring creation can be accomplished by means of

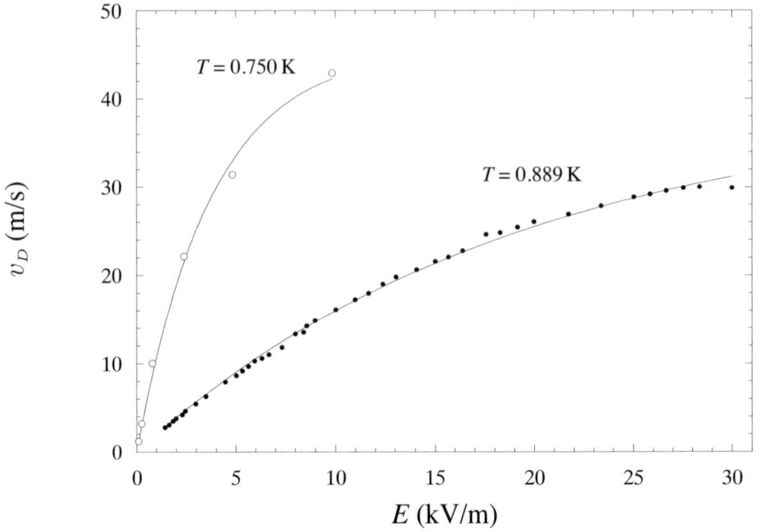

FIG. 6.1. v_D vs E for positive ions in liquid He II at SVP for $T = 0.75$ K (Reif and Meyer, 1960) and $T = 0.889$ K (Careri *et al.*, 1964*a*). The lines are eyeguides.

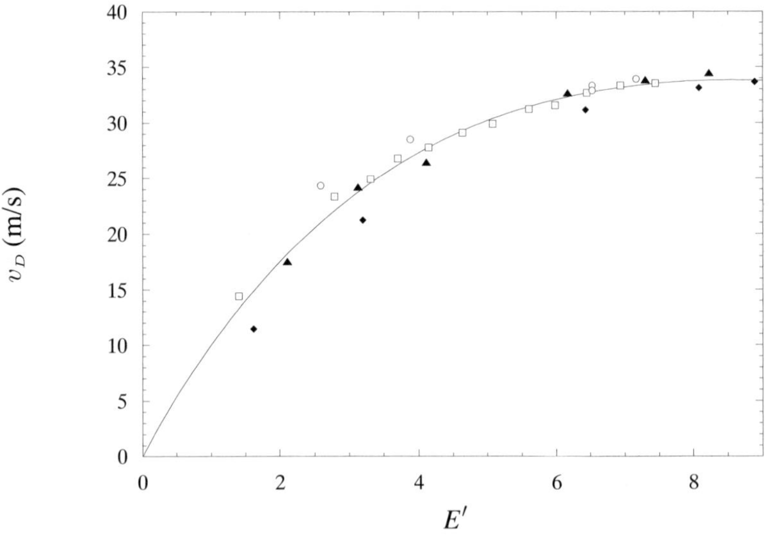

FIG. 6.2. v_D vs $E' = (2/3) \times 10^{-5}(\mu_- E/T^{1/2})$ for negative ions in liquid He II at SVP for T (K) $= 0.493$ (diamonds), 0.545 (triangles), 0.602 (circles), and 0.645 (squares). (Rayfield, 1968*a*.) The line is an eyeguide.

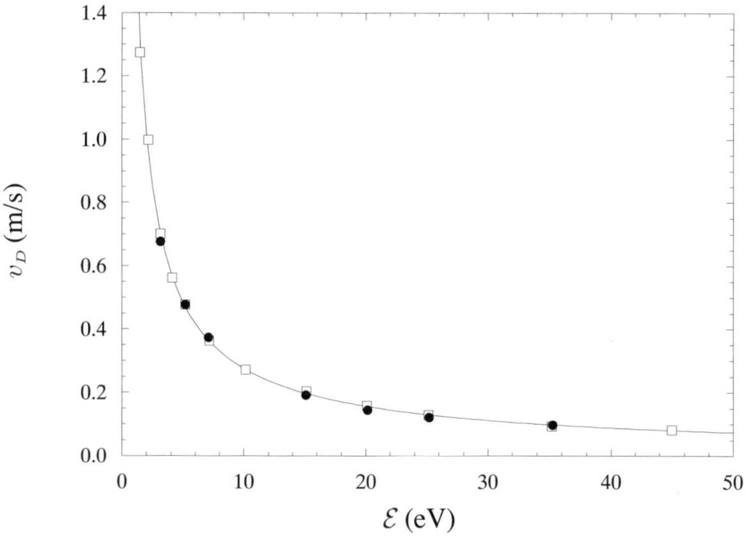

FIG. 6.3. v_D vs $\mathcal{E}$ for positive (squares) and negative (circles) ions in liquid He for $0.28\,\mathrm{K} < T < 0.6\,\mathrm{K}$. Line: theoretical prediction for a vortex ring of one quantum of circulation. (Rayfield and Reif, 1963.)

fairly low fields (Rayfield and Reif, 1963, 1964). In Fig. 6.3 the relationship between the drift velocity v_D and energy $\mathcal{E}$ of these quantized charged vortex rings is shown and will be discussed later.

At higher temperatures, $T \geqslant 1\,\mathrm{K}$, for which the density of elementary excitations, predominantly rotons, is quite large, the ion mean free path is much shorter than at lower T and the condition $\varphi \gg 1$ is fulfilled for much larger field strengths only (Careri *et al.*, 1965). In this case the drift velocity-field data of Figs 6.4 and 6.5 show an abrupt transition, known as the *giant fall* or *giant discontinuity*, in correspondence with a threshold electric field (Careri *et al.*, 1964*a*, 1965). For electric fields larger than the threshold, the observed drift velocity corresponds to the motion of a charged vortex ring drifting through the excitation gas.

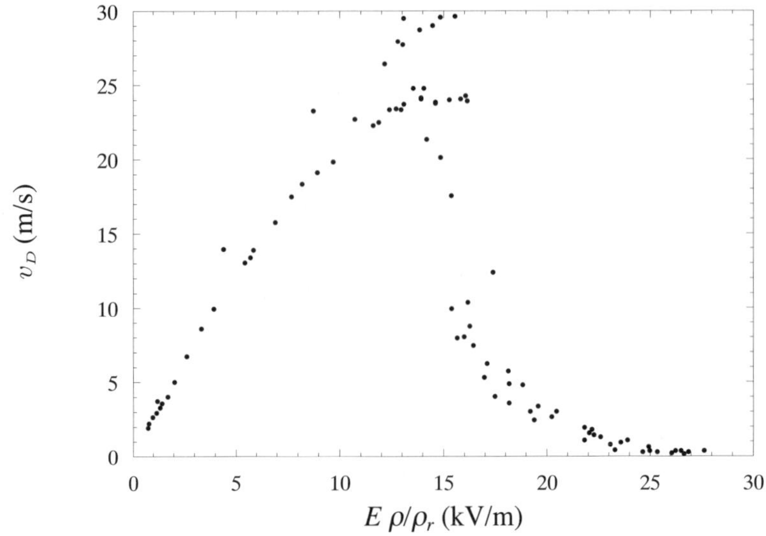

FIG. 6.4. v_D vs $E\rho/\rho_r$ for negative ions. ρ_r and ρ are the roton and liquid density, respectively. A compilation of data for $T = 0.895$, 0.899, 0.902, 0.915, and 0.954 K at SVP is shown. (Careri *et al.*, 1965.)

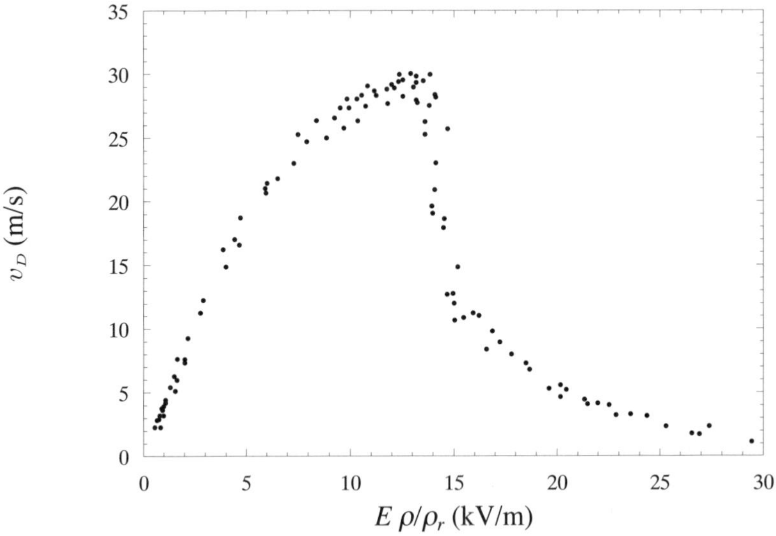

FIG. 6.5. v_D vs $E\rho/\rho_r$ for positive ions. Data for $T = 0.883$, 0.910, 0.920, and 0.975 K at SVP, $P \approx 0.5$ and 0.6 MPa are shown. (Careri *et al.*, 1965.)

6.1 The localized roton model

In the electric field range up to the threshold field for the appearance of quantized vortex rings there is no such description of the ion motion at finite velocities as in the mobility regime at very low fields. Nonetheless, a semi-hydrodynamic approach to the roton-limited mobility of ions in liquid helium provides quite an accurate description of the dependence of the drift velocity on the applied electric field strength (Doake and Gribbon, 1971; Strayer *et al.*, 1971). This is a heuristic approach based on a scheme of localized rotons around the ion, first proposed by Glaberson *et al.* (1968*a*) and by Glaberson (1969) for calculating the properties of the vortex core.

In the usual kinetic theory the scattering rate is proportional to the scatterer density and, hence, the drift velocity is inversely proportional to it. For not too low temperatures, the dominant elementary excitations available for ion scattering in pure superfluid helium are rotons. Their equilibrium density is easily obtained by integrating the distribution function:

$$n_r(T,0) = h^{-3} \int_0^\infty \{\exp[\epsilon(p)/k_B T] - 1\} \, d^3 p \,. \tag{6.1}$$

In eqn (6.1) $\epsilon(p)$ and p are the roton energy and momentum, respectively. $n_r(T,0)$ is the roton density in a reference frame moving with the superfluid.

However, in order to calculate the roton drag on the ion, the roton density in the immediate vicinity of the ion has to be known. This can be done by changing the reference frame from the one moving with the superfluid to another one moving with the ion and its associated rotons. This is necessary because ions interact with the roton density modified by the potential flow of the rotons past the ion surface (Doake and Gribbon, 1971).

As a result of the Galilean transformation to a reference frame moving with some velocity $\mathbf{w}$ with respect to the superfluid, the roton dispersion relation is changed to

$$\epsilon^\star(p) = \Delta + \frac{(p-p_0)^2}{2m_r} - pw\cos\phi \,, \tag{6.2}$$

where ϕ is the angle between $\mathbf{p}$ and $\mathbf{w}$. The new roton density in the presence of relative motion of the normal fluid and superfluid is obtained by inserting $\epsilon^\star$ in place of ϵ into eqn (6.1).

The effect of w is to depress the roton minimum to $\Delta_m = \Delta - m_r w^2/2 - p_0 w$ for $\phi = 0$ and to change the corresponding roton momentum to $p_m = p_0 + m_r w$ (Strayer *et al.*, 1971).

For the usual experimental conditions, $\Delta_m \gg k_B T$ and $p_0 \gg m_r w$, and, to a good approximation, the new local roton number density is now

$$n_r(T,w) = n_r(T,0) \frac{\sinh p_0 w/k_B T}{p_0 w/k_B T} \,. \tag{6.3}$$

The meaning of eqn (6.3) is that, at higher velocities, rotons begin to localize near the ion, thus increasing the drag on it. Higher field strengths are required to maintain a given velocity (Strayer and Donnelly, 1971). As the velocity increases, rotons appear at the ion equator, predominantly polarized in the direction of motion owing to the $\mathbf{p} \cdot \mathbf{w}$ term (Strayer *et al.*, 1971).

The number N_r of excess rotons around the ion can be obtained by integrating $n_r(T, w) - n_r(T, 0)$ over all the space outside the ion. As this quantity diverges, it is necessary to introduce a cut-off (Strayer *et al.*, 1971). This can be done with some arbitrariness by introducing a new length into the problem, namely, the distance d over which trapping occurs. In particular, the cut-off has been chosen by integrating over the space outside the ion up to the distance where $|\mathbf{w}| = v_D/3$, where v_D is the ion drift velocity. This produces a finite estimate for N_r. The distance d turns out to be of the order of the ion radius.

Further quantum corrections based on the uncertainty principle are required, owing to the fact that the roton wavelength is nearly 4 Å, comparable to the size of both negative and positive ions. To first order, localization of rotons in a volume of order d^3 produces a spread of momenta of the order of $\Delta p \sim \hbar/d$, with a consequent increase in the roton energy by the quantity $\Delta\epsilon = (\Delta p)^2/2m_r$. This correction must be added to $\epsilon^\star$.

The excess roton number is obtained by numerical integration and can be approximated in the temperature range $0.6\,\mathrm{K} \leqslant T \leqslant 1.0\,\mathrm{K}$, and in the drift velocity range $3 \leqslant \mathcal{V} = p_0 v_D/k_\mathrm{B}T \leqslant 8$, by the following equation (Strayer *et al.*, 1971):

$$N_r = n_r\,(T, 0)\, R_i^3 \mathcal{E}_r\,, \tag{6.4}$$

where R_i is the ion radius and

$$\mathcal{E}_r = c_\pm \left(\frac{\sinh m_\pm \mathcal{V}}{m_\pm \mathcal{V}} - 1 \right). \tag{6.5}$$

Here $m_\pm$ and $c_\pm$ are best-fit parameters. The values $m_- = 0.9$ for negative ions and $m_+ = 1.0$ for positive ions are temperature independent, while c_- and c_+ have a linear temperature dependence (Strayer *et al.*, 1971).

As $N_r \ll 1$, the ion is bare for most of the time. When a roton approaches the ion, it is localized for a time that can be estimated to be $\sim 3 \times 10^{-5}\,\mathrm{s}$ at $T = 0.6\,\mathrm{K}$ for $v_D = 30\,\mathrm{m/s}$ (Donnelly and Roberts, 1969*a*). The time between collisions in the roton bath is, in contrast, of order $2 \times 10^{-7}\,\mathrm{s}$. The trapped roton has a thermal velocity $v_\mathrm{th} \sim \sqrt{k_\mathrm{B}T/m_r} \approx 90\,\mathrm{m/s}$. So, the roton fluctuates back and forth in the well around the ion with frequency $v_\mathrm{th}/R_i \approx 6 \times 10^{10}\,\mathrm{s}^{-1}$. During the time of localization, the ion undergoes about 150 collisions with the rotons of the bath and the trapped roton revolves in its well approximately 10^3 times between these collisions. It is understood that collisions with phonons are also present at the same time.

Each time the trapped roton fluctuates in the well away from the ion, it comes into contact with the surrounding normal fluid, which is approximately at rest, and exchanges momentum with it. The amount of momentum loss can be written as $(p_0^2/3k_\mathrm{B}T)v_D f$, where $p_0^2/3k_\mathrm{B}T$ is the effective mass of the roton in the well (Donnelly, 1967) and f is an adjustable parameter that takes into account the fact that the surrounding normal fluid consists of rotons, phonons, and ^{3}He atoms, and that the exact amount of momentum loss depends on the type of scatterer. In principle, f is expected to depend on T, P, ^{3}He concentration, and on the ion type.

The resulting drag on the ion $\mathcal{F}_r$ turns out to be the product of the number of trapped rotons, of the fluctuation rate, and of the momentum loss per fluctuation (Strayer *et al.*, 1971):

$$\mathcal{F}_r = f\left(\frac{p_0^2}{3k_\mathrm{B}T}\right) v_D \left(\frac{k_\mathrm{B}T}{m_r R_i^2}\right)^{1/2} N_r = f\rho_r \mathcal{E}_r v_D \left(\frac{k_\mathrm{B}T}{m_r}\right)^{1/2} R_i^2\,, \tag{6.6}$$

where the roton contribution to the normal density far from the ion is $\rho_r = p_0^2 n_r(T,0)/3k_\mathrm{B}T$ (Donnelly, 1967).

Equation (6.6) can also be obtained by calculating the viscous stress on the surface of the ion by means of elementary kinetic theory arguments, in which the fluctuating roton acts as the momentum-transfer agent between the moving ion and the stationary fluid outside. In this sense, this process is called *quasi-viscous* (Strayer *et al.*, 1971).

The quasi-viscous drag (6.6) is to be added to the contributions due to phonon, roton, and ^{3}He atom scattering at small drift velocity. By recalling the definition of the drag coefficient μ^{-1} at low fields, eqn (5.6), one finally obtains (Strayer *et al.*, 1971)

$$E = v_D\left[\frac{1}{\mu} + \left(\frac{f}{e}\right)\left(\frac{k_\mathrm{B}T}{m_r}\right)^{1/2} \rho_r R_i^2 \mathcal{E}_r\right]. \tag{6.7}$$

In Figs 6.6 and 6.7 experimental high-field drift velocity data for positive and negative ions in the range $0.6\ \mathrm{K} < T < 0.75\,\mathrm{K}$ are shown. The solid curves are eqn (6.7) with $f_+ = 0.065$ for the positive ions and $f_- = 0.01$ for the negative ones. The adjustable parameters $f_\pm$ turn out to be practically temperature independent. The localized roton model, though in a simplified form, has been successfully used to describe v_D up $T = 1.14$ K (Schwarz, 1970; Doake and Gribbon, 1971).

In the case of positive ions, however, a further effect related to electrostriction must be considered, which again leads to an increase in the roton density around the ion (Bondarev, 1973). Namely, as the local density and pressure around a positive ion differ very much from the values in the unperturbed liquid, the energy spectrum of a roton depends on its distance from the ion through the local density. If $\delta\rho$ is the density deviation from the unperturbed value ρ, the roton energy spectrum becomes, to first order in $\delta\rho$,

$$\epsilon = \Delta + \frac{(p-p_0)^2}{2m_r} + \left(\frac{\partial\Delta}{\partial\rho}\right)_0 \delta\rho\,. \tag{6.8}$$

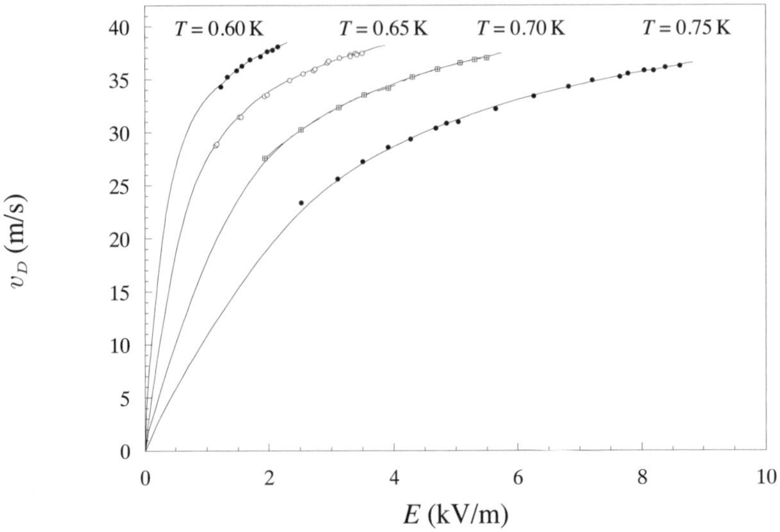

FIG. 6.6. v_D vs E for positive ions at SVP for $T = 0.6$, 0.65, 0.7, and 0.75 K. (Strayer and Donnelly, 1971.) Curves: eqn (6.7) with $f_+ = 0.065$.

By recalling that $\partial\Delta/\partial\rho < 0$ (Wilks, 1967), and taking into account the equation of electrostriction (Atkins, 1959), one obtains

$$\epsilon\,(p, r) = \Delta + \frac{(p - p_0)^2}{2m_r} - \frac{A}{r^4}, \tag{6.9}$$

where $A = |(\partial \ln \Delta/\partial \ln \rho)_0|(\Delta\alpha e^2/(4\pi\epsilon_0)^2 m_4 c^2)$. Here, α is the atomic polarizability of helium and c is the ordinary sound velocity.

The last term in eqn (6.9) is a potential energy of the roton in the attractive field of the positive ion (Bondarev, 1973) and it leads to a dependence of the roton density on the distance from the ion of the form

$$n_r(r) = n_\infty \exp\left(\frac{A}{2k_B T r^4}\right), \tag{6.10}$$

where n_∞ is the equilibrium roton density at a large distance from the ion. The number of excess rotons near the ion is calculated, as usual, as

$$N_r = 4\pi n_\infty \int_{R_i}^{\infty} \left(\frac{n_r(r)}{n_\infty} - 1\right) r^2 \mathrm{d}r = \frac{4}{3}\pi R_i^3 n_\infty \left[4B\Phi\left(\frac{1}{4}, \frac{5}{4}; B\right) - e^B + 1\right], \tag{6.11}$$

where $B = A/(2k_B T R_i^4)$ and $\Phi\,(1/4, 5/4; B)$ is a hypergeometric confluent series.

The calculation of the additional drag contribution due to this number of excess localized rotons follows essentially the arguments of Strayer *et al.* (1971)

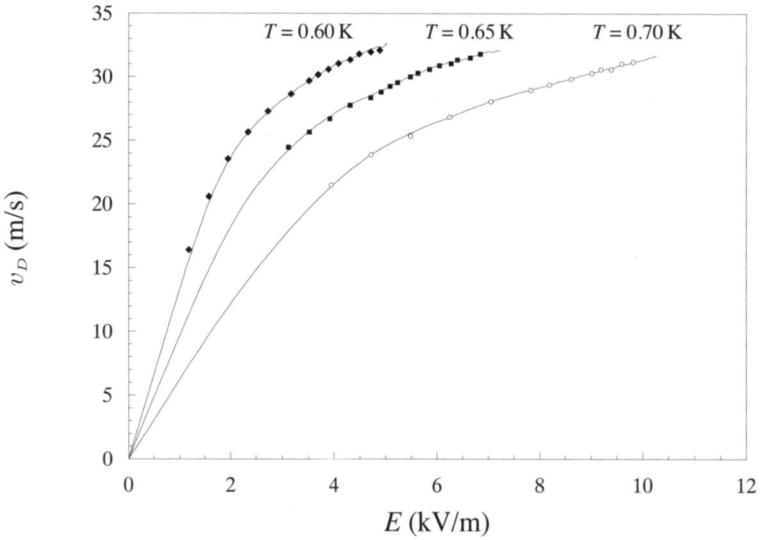

FIG. 6.7. v_D vs E for negative ions at vapor pressure for $T = 0.6$, 0.65, and 0.70 K. (Strayer and Donnelly, 1971.) Curves: eqn (6.7) with $f_- = 0.01$.

and the complete expression for the roton-limited positive ion mobility becomes (Bondarev, 1973)

$$\frac{e}{\mu_+} = \frac{4}{9}R_+^2 \left\{ \frac{1}{\pi}\hbar k_0^4 \mathcal{G}\left(\frac{\hbar^2 k_0^2}{4M_+^\star k_\mathrm{B}T}\right) + f_+ \frac{1}{\sqrt{2\pi}}\hbar k_0^4 R_+^2 \left[4B\Phi\left(\frac{1}{4},\frac{5}{4};B\right) - e^B + 1\right]\right\} e^{-\Delta/k_\mathrm{B}T}, \quad (6.12)$$

where $M_+^\star \approx 40m_4$ is the effective positive ion mass and R_+ is its radius.

The first term on the right-hand side of eqn (6.12) is the roton-limited mobility for small fields derived by Barrera and Baym (1972). The factor f_+ in the second term is the adjustable parameter introduced by Strayer *et al.* (1971). Using an effective ion radius $R_+ = 7$ Å and an adjustable parameter $f_+ = 0.096$, a good agreement is obtained for the temperature dependence of the roton-limited mobility for temperatures in the range $0.6\,\mathrm{K} < T < 1\,\mathrm{K}$ at vapor pressure, as shown by the dashed line in Fig. 5.27 (Bondarev, 1973). This localization of rotons as a consequence of electrostriction is negligible for the electron bubbles owing to their much larger radius.

Although the agreement of the predictions of the localized roton model with the high-electric field drift velocity data is quite good, it is, however, necessary to remember that it has been strongly criticized (Schwarz and Jang, 1973) by noting that the concept of a localized roton is meaningless unless the roton–roton mean free path is short compared to the ion size. If this is not the case, the excitation distribution function cannot be modified to a large extent by the presence of the ion.

6.2 The elusive drift velocity discontinuities

Before switching to the topic of quantum hydrodynamics where the interaction of ions with quantized vortices is discussed, it is interesting to briefly mention the long-debated question of the drift velocity discontinuities first discovered by Careri's group (Careri *et al.*, 1961, 1964*a*, 1964*b*) and then also observed by different groups (Cope and Gribbon, 1970*a*, 1970*b*).

These steps appear as a decrease of the ion mobility by small discrete amounts at approximately integral values of a critical velocity, $v_c \approx 2.4\,\mathrm{m/s}$ for the negative ions and $v_c \approx 5.2\,\mathrm{m/s}$ for the positive ones. Typical results for positive ions are shown in Figs 6.8 and 6.9.

This critical velocity shows a very complicated temperature dependence (Bruschi *et al.*, 1966*c*, 1968*a*), as shown in Fig. 6.10. After the initial plateau, v_c decreases sharply with increasing T, approximately linearly in terms of ρ_n/ρ, where ρ_n is the density of the normal fluid. v_c shows a great dip at $T \approx 1.5\,\mathrm{K}$ with a subsequent increase up to the low-temperature value. Finally, for $T > 1.6\,\mathrm{K}$, v_c decreases in a way that is directly proportional to ρ_s/ρ_n, the ratio of the superfluid to the normal fluid density. The positive ions have a similar behavior with the great dip shifted to a lower temperature (Bruschi *et al.*, 1968*a*).

Although a large number of papers, both experimental (Careri *et al.*, 1961, 1964*a*, 1964*b*; Gaeta, 1962; Bruschi *et al.*, 1966*c*, 1968*a*; Cope and Gribbon,

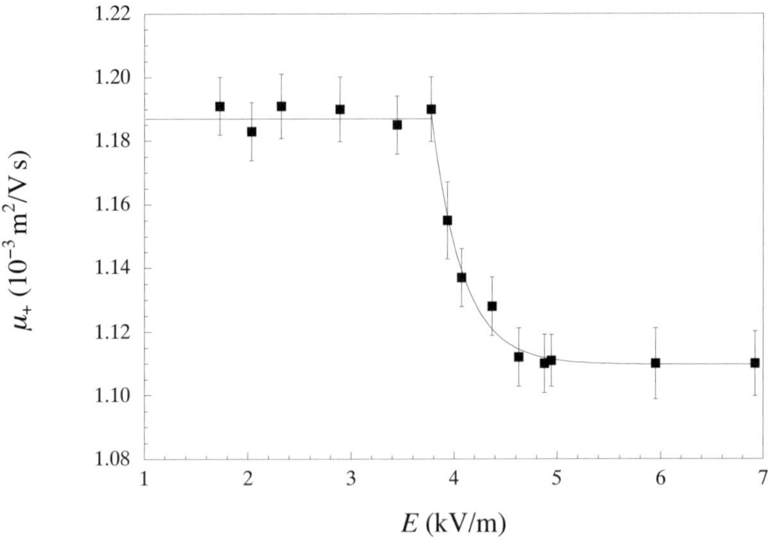

FIG. 6.8. Discontinuity of μ_+ vs E in liquid He II at $T = 0.927\,\mathrm{K}$. (Careri *et al.*, 1961.) The line is an eyeguide.

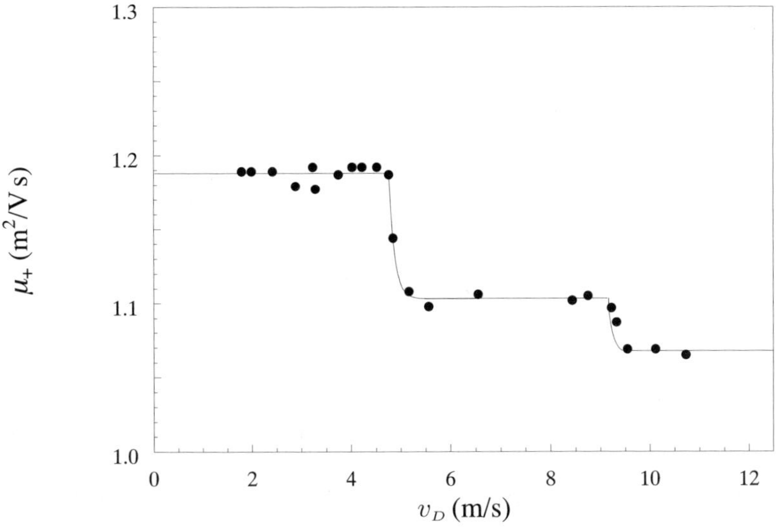

FIG. 6.9. Discontinuities of μ_+ vs v_D for $T = 0.928\,\mathrm{K}$. The steps appear approximately at integral multiples of a critical velocity $v_c \approx 5\,\mathrm{m/s}$. (Careri *et al.*, 1961.) The line is an eyguide.

1970*a*, 1970*b*) as well as theoretical (Cope and Gribbon, 1965; Huang and Olinto, 1965; Di Castro, 1966; Jones, 1969) are concerned with this phenomenon, none of the previously described behaviors have been satisfactorily explained, and all the theoretical attempts at explaining the discontinuities have been entirely phenomenological. Furthermore, some experimental facts cast serious doubts on the real existence of these steps. First of all, the steps disappear for $T > T_\lambda$ (Bruschi *et al.*, 1966*c*, 1968*a*). Moreover, inexplicably, they have also been detected in non-quantum liquids, such as Ar or CCl_4 (Henson, 1964; Bruschi *et al.*, 1970), in which vorticity is not expected to be quantized.

The strongest objection to the steps, however, is that they have only been observed in cells with electrodes in a triode configuration using the single-gate method developed by Cunsolo (1961). Some non-convincing results were obtained with the heat-flush method, where ions are dragged perpendicularly to the lines of the electric field by a convective flow of the normal fluid driven by a controlled thermal gradient (Careri *et al.*, 1959, 1964*b*; Cope and Gribbon, 1970*b*). Measurements carried out with different techniques, and, more specifically, with a charge–pulse time-of-flight technique (Schwarz, 1970; Steingart and Glaberson, 1970), or in a double, symmetric differential triode cell (Goodstein *et al.*, 1968), or even in an improved triode cell (Doake and Gribbon, 1971), do not show any steps.

In particular, in the charge–pulse time-of-flight method (Schwarz, 1970), the cell is always in a conducting state because of a forward bias between the source

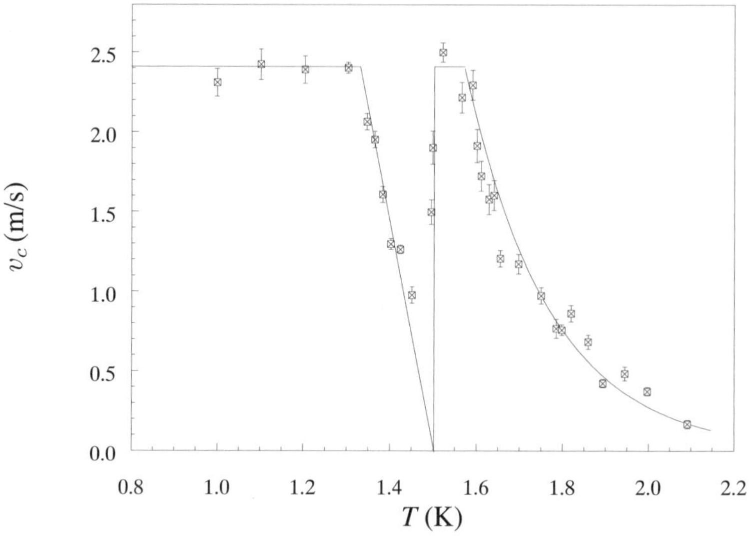

FIG. 6.10. v_c vs T for the first step of negative ions at SVP. (Bruschi *et al.*, 1968*a*.)

and the gating grid. The ion current is periodically switched off by a reverse voltage pulse applied to this region. The off pulse then propagates across the long drift space and the collector region, so that the drift time is easily measured. Digital signal averaging is implemented to improve the signal-to-noise ratio. The use of an "off" pulse eliminates any source effects due to the fact that the ionizing source is generally strong enough so that the ions are emitted with sufficient speed to nucleate a vortex ring with no additional electric field (Rayfield, 1968*a*; Neeper and Meyer, 1969). The drift mobility values are thus independent of the voltages used in the source region, and do not depend on the pulse height, duration, repetition rate, or signal strength.

On the other hand, the single-gate technique is subject to a number of possible errors, including large build-ups of space charge, generation of second sound resonances because of eddy currents in the drift space, and source effects that cannot be reliably controlled (Schwarz, 1970). Among the latter, an alteration of the transmission coefficient of the gating grid has been suggested because of an accumulation, at the source side of the grid, of charged vortex rings, which are not able to decay rapidly and release the trapped charge to the grid (Doake and Gribbon, 1972). Such source effects might be responsible for the (lower) critical velocity observed at low temperature (Cunsolo *et al.*, 1968*b*; Cunsolo and Maraviglia, 1969).

Even though the elusive steps might not be a real physical phenomenon, nonetheless, they have, at least, historical importance because they stimulated the production of theoretical, though phenomenological, models (Huang and Olinto, 1965; Di Castro, 1966; Jones, 1969) that have contributed to shed light on the physical phenomena leading to the less-than-proportional increase of the drift velocity with increasing field and to the giant discontinuity, at which ions get trapped to quantized vortex rings.

Among those models, a special place is occupied by the so-called Huang–Olinto (HO) model (Huang and Olinto, 1965). Briefly, the model assumes that quantized vortex rings are produced and shed in the low-field region. During the nucleation process, as soon as the bare ion reaches the critical velocity v_c, a vortex ring is created.

After creation, the ion is captured by the vortex ring by a kind of hydrodynamic suction, because the fluid velocity near the vortex core is larger than in the unperturbed fluid. The ion transfers energy from the electric field to the nucleating vortex ring, whose size grows suitably. As soon as a quantized vortex ring is fully developed, but the electric field is below the critical field of the giant discontinuity, collisions of the ring and of the ion with the elementary excitations present in the liquid make the ring detach from the ion. The fate of the ring is to decay into smaller and smaller rings, dissipating its energy to the fluid, until it practically disappears.

Once freed from the ring, the ion can accelerate again in the electric field and reach the right velocity to produce another ring. The whole process is repeated until a value of the field is reached for which the ion–ring complex is stable and the mobility shows the great decrease for the critical field E_c, see Figs 6.4 and 6.5. In the low-to-intermediate-field region, ions dissipate energy from the field to create the vortex rings. Thus, their velocity increases less than proportionally with increasing field.

In the intermediate-field region just above the critical field for the creation of stable ion–vortex ring complexes, the HO model describes semi-quantitatively the drift velocity of the charged vortex ring as a function of the electric field by phenomenologically introducing a plausible form for the viscous reaction on the trapped ion in addition to the viscous drag on the ring.

Finally, by assuming that vortex rings are created during the ion motion and that a stability condition for the ion–vortex ring complex is reached at the electric field of the giant discontinuity, the HO model qualitatively explains why the Landau critical velocity for roton creation ($\approx 58\,\mathrm{m/s}$) is never reached in experiments at vapor pressure (Reif and Meyer, 1960; Careri *et al.*, 1965; Rayfield, 1968*a*).

In summary, although the HO model was stimulated by experimental results that might have proven false and although it is only a phenomenological description of the physical situation, nonetheless, it has drawn the attention of researchers to the fact that ions dissipate energy by creating quantized vortex rings. Subsequent measurements have elucidated in great detail the process of vortex nucleation and these phenomena will be treated in the next chapters.

7

VORTEX HYDRODYNAMICS

The study of the ionic motion in a range of intermediate to high electric fields has provided the experimental evidence that ions create quantized vortex rings and remain stuck on them. More generally, ions interact with the vorticity present in superfluid helium and can thus be used to investigate the fascinating realm of quantum hydrodynamics. A great deal of our present knowledge about vortices is obtained from experiments based on ions that are used as a probe to microscopically examine the properties of the superfluid.

The existence of quantized circulation seems to require long-range order, whereas the ion structure basically depends on the interaction between the ion and the liquid at an atomic scale. Nonetheless, in spite of the different scale range, ions and vortices are closely related and interact mutually. This interaction is investigated by experiments based on the capture and escape of ions from vortex lines or on vortex ring creation.

In order to analyze and discuss the experiments involving ions and vortices, a brief introduction to (quantized) vortices is necessary. A very complete survey on vortices can be found in the book of Donnelly (1991).

Many experiments in superfluid He have confirmed the existence of vortices, whose properties closely resemble the properties of vortices in classical hydrodynamics (Wilks, 1967). Vortices in superfluid He, however, are of quantum-mechanical nature (Fetter, 1967). Experiments (Vinen, 1958, 1961; Rayfield and Reif, 1963, 1964) have proved that the circulation around vortices is quantized:

$$\oint \mathbf{v} \cdot \mathrm{d}\mathbf{l} = n \frac{h}{m_4} = n\kappa \,, \tag{7.1}$$

where h is the Planck constant and n an integer. The unit of circulation is $\kappa = 9.97 \times 10^{-8}\,\mathrm{m}^2/\mathrm{s}$.

This quantization can be intuitively understood if one considers a Bose gas near $T = 0\,\mathrm{K}$. In this case, the particles of the system condense into the same quantum state. This is known for a stationary Bose gas, but it also happens in more general situations. Suppose, for instance, that all particles have the same z-component $n\hbar$ of angular momentum. In this case, their azimuthal velocity $n\hbar/m_4 r$ varies as $1/r$, which is typical of a rectilinear vortex (Lamb, 1945).

The vortex strength can be defined simply in terms of the circulation $\oint \mathbf{v} \cdot \mathrm{d}\mathbf{l}$. By symmetry, the line integral gives exactly $n\kappa$. This result was suggested for the first time by Onsager (1949) and can also be deduced from the usual Bohr–Sommerfeld quantization rules $\oint \mathbf{p} \cdot \mathrm{d}\mathbf{l} = nh$. Although it is not clear if this rule can be immediately applied to an N-body system, experiments apparently confirm this hypothesis.

7.1 Semi-classical vortex hydrodynamics

In order to clarify the nature and origin of quantized circulation, a quantum-mechanical description of the vortices is required. However, for the purpose of analyzing their interaction with ions in superfluid helium, it is sufficient to describe them in a phenomenological way, in which they satisfy the equations of classical hydrodynamics for an incompressible fluid but are endowed with the property of quantized circulation (Fetter, 1976). Here, the treatment of the semi-classical quantum hydrodynamics of vortices due to Fetter (1976) will be followed.

7.1.1 *Flow patterns and energy of vortices*

In a classical incompressible fluid the velocity field is subject to the condition that it must be solenoidal over the entire fluid: $\nabla \cdot \mathbf{v} = 0$ (Landau and Lifsits, 2000). If, in addition, $\nabla \times \mathbf{v} = 0$, the flow is called *irrotational.*

A fluid is said to contain vortices if it is everywhere irrotational, except along certain singular lines that constitute the core of the vortices. In this case, the vorticity

$$\boldsymbol{\zeta} \equiv \nabla \times \mathbf{v} \tag{7.2}$$

is confined to the core regions and is zero everywhere else.

As $\nabla \cdot \boldsymbol{\zeta} \equiv 0$ identically, $\kappa = \oint \mathbf{v} \cdot \mathrm{d}\mathbf{l}$ is independent of the integration path provided that it does not cross the core. Moreover, the lines of $\boldsymbol{\zeta}$ are closed, i.e., the vortex axis must close on itself, thus forming an object topologically equivalent to a ring, or it must end on the fluid boundary.

It is easy to find the velocity field generated by vortices by noting the mathematical analogy with the magnetic field generated by current-carrying wires. The equivalent of the Biot–Savart law is thus obtained immediately:

$$\mathbf{v}(\mathbf{r}) = \frac{\kappa}{4\pi} \oint_C \frac{\mathrm{d}\mathbf{s}' \times (\mathbf{r} - \mathbf{r}')}{|\mathbf{r} - \mathbf{r}'|^3} \,. \tag{7.3}$$

The line integral is calculated by following the vortex axis in a positive sense.

The solution of Euler's equation for an incompressible, inviscid fluid (Landau and Lifsits, 2000),

$$\frac{\mathrm{d}\mathbf{v}}{\mathrm{d}t} \equiv \frac{\partial \mathbf{v}}{\partial t} + (\mathbf{v} \cdot \nabla)\,\mathbf{v} = -\frac{1}{\rho}\nabla P \,, \tag{7.4}$$

describes the dynamics of the system. Here P and ρ are the pressure and mass density of the liquid, respectively.

Taking the curl of eqn (7.4) one gets

$$\frac{\partial \boldsymbol{\zeta}}{\partial t} = \nabla \times (\mathbf{v} \times \boldsymbol{\zeta}) \,. \tag{7.5}$$

A further simplification is obtained by recalling that $\mathbf{v}$ and $\boldsymbol{\zeta}$ are of solenoidal nature, thus obtaining

$$\frac{\mathrm{d}\boldsymbol{\zeta}}{\mathrm{d}t} \equiv \frac{\partial \boldsymbol{\zeta}}{\partial t} + (\mathbf{v} \cdot \boldsymbol{\nabla})\, \boldsymbol{\zeta} = (\boldsymbol{\zeta} \cdot \boldsymbol{\nabla})\, \mathbf{v}\,. \tag{7.6}$$

This last equation, which can be obtained from the Kelvin circulation theorem, states that the vortex core contains the same real particles and that each element of the core moves with the actual fluid velocity at that point, provided that the flow is isoentropic (Landau and Lifsits, 2000).

7.1.1.1 *Rectilinear vortices* This is by far the most simple vortex configuration, whose cross-sectional view is shown in Fig. 7.1. Suppose that the vortex axis is aligned along the z-axis. Equation (7.3) then gives

$$\mathbf{v} = \frac{\kappa}{2\pi r}\hat{\theta}\,, \tag{7.7}$$

where cylindrical polar coordinates are used and $\hat{\theta}$ is the unit azimuthal vector. It should be noted that if r becomes less than a cut-off parameter a, known as the *core radius*, the velocity must deviate from the law expressed by eqn (7.7) because it would otherwise become infinite (Rayfield and Reif, 1964).

Apart from the divergence of the velocity field for $r \to 0$, the symmetry of this problem implies that the core of a rectilinear vortex remains stationary. The rectilinear vortex shows no self-induced motion.

If the fluid is incompressible, the vortex energy is simply obtained by integrating the kinetic energy density over the volume: $\int (1/2)\rho v^2 \mathrm{d}^3 r$. Owing to the divergence of eqn (7.7) at the origin, a specific model of the core is required. One simple choice is to assume a hollow core for $r < a$. Such a vortex is called a

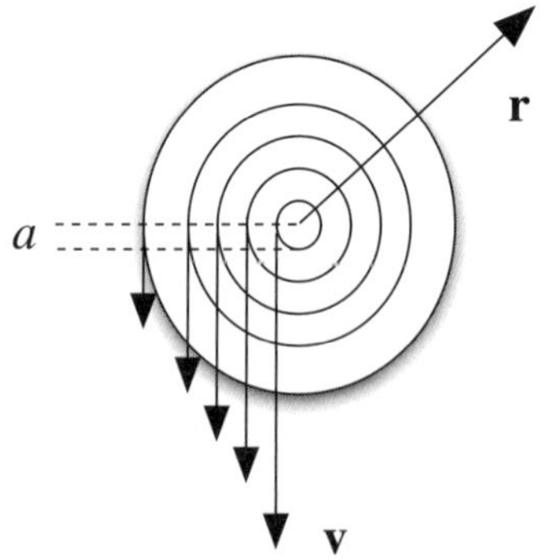

FIG. 7.1. Schematic cross-sectional view of a rectilinear vortex. a: core radius, $\mathbf{v}$: tangential velocity, and $\mathbf{r}$: distance from the vortex axis in the plane.

potential vortex (Donnelly, 1991). In this case, the kinetic energy per unit length of the vortex is

$$\mathcal{E} = \pi\rho \int_a^b rv^2 \mathrm{d}r = \rho \left(\frac{\kappa^2}{4\pi}\right) \ln\left(\frac{b}{a}\right), \tag{7.8}$$

where a is the radius of the vortex core and b is some characteristic distance, such as the distance between vortices or the container size (Fetter, 1976; Donnelly, 1991).

For different structures of the core, the kinetic energy per unit vortex length can be cast in the form

$$\mathcal{E} = \pi\rho \int_0^b rv^2 \mathrm{d}r = \rho \left(\frac{\kappa^2}{4\pi}\right) \left[\ln\left(\frac{b}{a}\right) + \delta\right], \tag{7.9}$$

where δ is a dimensionless constant whose value depends on the explicit model of the core. For instance, if the core executes a solid-body rotation (a so-called *Rankine* vortex), with velocity

$$\mathbf{v} = \left(\frac{\kappa r}{2\pi a^2}\right) \hat{\theta}, \qquad r \leqslant a, \tag{7.10}$$

then the core contribution to $\mathcal{E}$ is $\mathcal{E}_{\text{core}} = \rho\kappa^2/16\pi$, so that $\delta = 1/4$.

The previous arguments can be generalized to a system of parallel rectilinear vortices in an unbounded liquid. If the lth vortex, with circulation κ_l, intersects the x–y plane at $\mathbf{r}_l$, the integration in eqn (7.3) runs along each vortex axis and the resulting fluid velocity outside the core region is

$$\mathbf{v}(\mathbf{r}) = \sum_l \left(\frac{\kappa_l}{2\pi}\right) \frac{\hat{z} \times (\mathbf{r} - \mathbf{r}_l)}{|\mathbf{r} - \mathbf{r}_l|^2}, \tag{7.11}$$

where $\hat{z}$ is a unit vector along the z-axis.

It is interesting to note that each vortex moves under the influence of all other vortices, owing to the absence of self-induced motion. So, the velocity of every vortex core is simply given by

$$\dot{\mathbf{r}}_m = \sum_l \left(\frac{\kappa_l}{2\pi}\right) \frac{\hat{z} \times (\mathbf{r}_m - \mathbf{r}_l)}{|\mathbf{r}_m - \mathbf{r}_l|^2}, \tag{7.12}$$

where the dot means differentiation with respect to time.

Equations (7.11) and (7.12) mean that the velocity field of the fluid and the translational motion of the vortices are completely determined by the instantaneous configuration of the vortices and that the dynamics is described by first-order time derivatives.

Two rectilinear vortices with opposite circulation $\pm\kappa$ and separated by a distance $2R$ are called a *vortex pair*, shown in Fig. 7.2. The pair moves with uniform translational speed

$$v = \frac{\kappa}{4\pi R} \tag{7.13}$$

perpendicular to the plane joining their axes, as is readily seen from eqn (7.12). It is also easy to see the close relationship between this situation and that of a vortex ring.

Another configuration of interest is realized when a large number N_v of identical vortices are uniformly distributed over a circle of radius R. A useful quantity is then the vortex density per unit area, $n_v = N_v/\pi R^2$. If n_v is large, the summation in eqn (7.12) can be replaced by an integral:

$$\begin{aligned}
\mathbf{v}(\mathbf{r}) &\approx \frac{n_v\kappa}{2\pi}\hat{z}\times\int\frac{\mathbf{r}-\mathbf{r}'}{|\mathbf{r}-\mathbf{r}'|^2}\mathrm{d}^2r' \\
&= \frac{n_v\kappa}{2\pi}\left(\hat{z}\times\hat{r}\right)\int_0^R r'\mathrm{d}r'\int_0^{2\pi}\frac{r-r'\cos\theta}{r^2+r'^2-2rr'\cos\theta}\mathrm{d}\theta \\
&= n_v\frac{\kappa}{r}\left(\hat{z}\times\hat{r}\right)\int_0^r r'\mathrm{d}r' = \frac{1}{2}n_v\kappa\left(\hat{z}\times\mathbf{r}\right),
\end{aligned} \tag{7.14}$$

where $\hat{r}$ is the radial unit vector. Equation (7.14) describes a uniform rotation with tangential velocity $\mathbf{v} = \boldsymbol{\Omega}\times\mathbf{r}$, where the angular velocity $\boldsymbol{\Omega}$ is defined by

$$\boldsymbol{\Omega} = \frac{1}{2}n_v\kappa\hat{z}\,. \tag{7.15}$$

The velocity field at a given position $\mathbf{r}_0$ depends on what is occurring in the region for $r < r_0$.

Equation (7.15) is important for the analysis of the experiments of ion trapping by vortex lines in a rotating bucket (Careri *et al.*, 1962; Packard and Sanders, 1969), in which the angular velocity of the bucket provides a direct measure of the areal density of vortex lines. Equation (7.15) has also been shown to hold true by second–sound attenuation experiments (Hall and Vinen, 1956).

The uniform distribution of vortex lines with areal density n_v is the configuration that minimizes the free energy of the system. However, when boundaries are included, the theoretical vortex distribution is slightly modified (Hall, 1960; Stauffer and Fetter, 1968) and (at least) one row of vortices is missing near the boundary, thus giving origin to a vortex-free region of width proportional to the inter-vortex separation $n_v^{-1/2}$, as has been experimentally observed (Northby and Donnelly, 1970).

7.1.1.2 *Vortex rings* The condition that the vorticity is a solenoidal vector, $\boldsymbol{\nabla}\cdot\boldsymbol{\zeta} = 0$, implies that the vortex lines either terminate on the boundary or on themselves. In the latter case, the topology of a ring is obtained; hence the

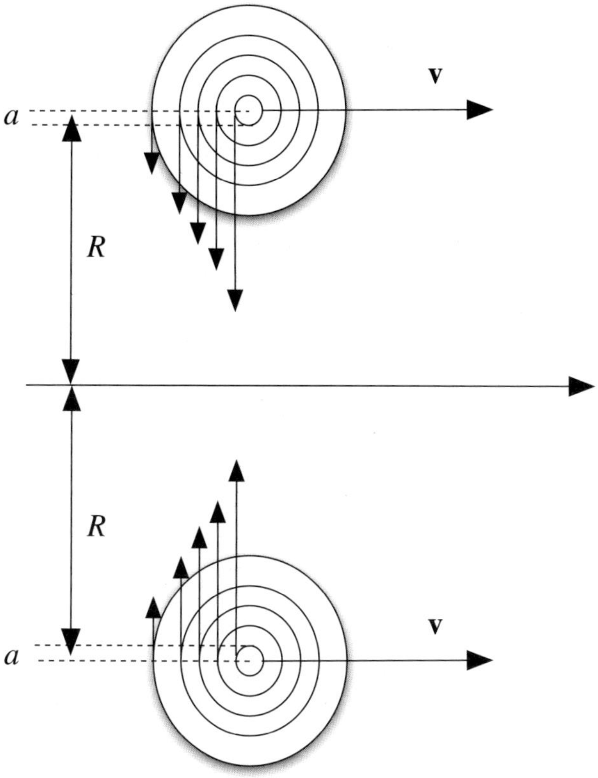

FIG. 7.2. Cross-sectional view of a pair of counter-rotating rectilinear vortices separated by a distance $2R$. It also represents a cross-sectional view of a vortex ring of radius R.

name of *vortex rings*. Figure 7.2 can also represent a cross-sectional view of such a vortex ring.

A suitable description of the ring is given in cylindrical polar coordinates. It is also convenient to introduce, by analogy with the vector potential of magnetostatics, a vector $\mathbf{A}(\mathbf{r})$ such that the velocity field can be obtained by taking its curl:

$$\mathbf{v}(\mathbf{r}) = \boldsymbol{\nabla} \times \mathbf{A}(\mathbf{r}) , \tag{7.16}$$

with

$$\mathbf{A}(\mathbf{r}) = \frac{\kappa}{4\pi} \oint \frac{\mathrm{d}\mathbf{l}'}{|\mathbf{r} - \mathbf{r}'|} , \tag{7.17}$$

If a vortex ring of radius R in the x–y plane is considered, the vector potential has only an azimuthal component (Walraven, 1970):

$$\mathbf{A}(\mathbf{r}) = A_\theta(r,z)\hat{\theta} = \frac{\kappa R}{4\pi}\hat{\theta}\int_0^{2\pi} \frac{\cos\theta}{(r^2 - 2rR\cos\theta + R^2 + z^2)^{1/2}} \mathrm{d}\theta\,. \tag{7.18}$$

The vector potential can be expressed in terms of the complete elliptic integrals $K(y)$ and $E(y)$ (Arfken, 1985) as (Walraven, 1970)

$$A_\theta(r,z) = -\frac{1}{2}\left(\frac{\kappa}{\pi y}\right)\left(\frac{R}{r}\right)^{1/2}\left[\left(2-y^2\right)K(y) - 2E(y)\right], \tag{7.19}$$

with

$$y^2 = 4rR\left[(r+R)^2 + z^2\right]^{-1}. \tag{7.20}$$

The velocity field is obtained by direct differentiation. The ring energy is completely kinetic, as for the rectilinear vortex, and is given by

$$\begin{aligned}\mathcal{E} &= \frac{1}{2}\rho\int v^2 \mathrm{d}^3r \frac{1}{2}\rho\int \mathbf{v}\cdot(\boldsymbol{\nabla}\times\mathbf{A})\,\mathrm{d}^3r \\ &= \frac{1}{2}\rho\oint \mathrm{d}\mathbf{S}\cdot(\mathbf{A}\times\mathbf{v}) - \frac{1}{2}\rho\int(\mathbf{A}\cdot\boldsymbol{\zeta})\,\mathrm{d}^3r\,.\end{aligned} \tag{7.21}$$

The surface integral must be extended over the fluid boundaries. In an unbounded fluid, as $r\to\infty$, the surface integral gives no contribution because $A\sim r^{-2}$ and $v\sim r^{-3}$. If the ring has a hollow core (potential vortex), whose size is $a \ll R$, then $\boldsymbol{\zeta}$ is zero everywhere. As a consequence, eqn (7.21) reduces to a surface integral over the core surface, where both $r-R$ and z are of order a. In this approximation, the azimuthal vector potential is simplified to

$$A_\theta(r,z) \approx \frac{\kappa}{2\pi}\left\{\ln\left[\frac{8R}{\left[(r-R)^2+z^2\right]^{1/2}}\right] - 2\right\} \qquad \text{for } (r-R)^2 + z^2 \ll R^2\,. \tag{7.22}$$

The resulting streamlines are approximately concentric circles encircling the core, as shown in Fig. 7.2, with flow speed

$$v(r,z) \approx \frac{\kappa}{2\pi\left[(r-R)^2+z^2\right]^{1/2}} \qquad \text{for } (r-R)^2 + z^2 \ll R^2\,. \tag{7.23}$$

The kinetic energy is then easily calculated, yielding

$$\mathcal{E} = \frac{1}{2}\rho\kappa^2 R\left[\ln\left(\frac{8R}{a}\right) - 2\right] \tag{7.24}$$

if the core is hollow. If the core executes a solid-body rotation with constant vorticity $\kappa/\pi a^2$, then

$$\mathcal{E} = \frac{1}{2}\rho\kappa^2 R\left[\ln\left(\frac{8R}{a}\right) - 2 + \delta\right], \qquad \delta = \frac{1}{4}. \tag{7.25}$$

The derivation of the motion of the vortex ring is not as simple as in the case of a pair of linear vortices because of the curvature of the ring axis (Hicks, 1884, 1885) and will not be repeated here. For more details, see Fetter (1976). For the present purpose, it suffices to quote the final result for the vortex ring speed:

$$v = \left(\frac{\kappa}{4\pi R}\right)\left[\ln\left(\frac{8R}{a}\right) - \frac{1}{2} + \delta\right]. \tag{7.26}$$

If the core has uniform vorticity, again $\delta = 1/4$, as obtained for the first time by von Helmholtz (1867). Similar relationships have been derived by following a quantum-mechanical variational approach (Amit and Gross, 1966). The only relevant difference is the value of the constant δ. However, as the correct value of the core radius is inaccessible to direct experimental determination, its exact value is irrelevant.

In contrast, a direct experimental determination of the energy–radius relationship, eqn (7.25), for quantized charged vortex rings has been obtained by using a kind of vortex ring sieve consisting of a two-grids drift cell (Gamota and Sanders, 1965, 1971). Direct ionization produces ions that are accelerated toward a coarse-mesh grid by a large electric field. Such energetic ions generate charged vortex rings. After passing the grid, the rings drift toward a second grid under the action of a constant electric field. When the charged vortex rings reach the grid, their energy is given by the total voltage drop and their radius is determined by eqn (7.25). The experiment is carried out at low temperature ($0.29\,\mathrm{K} < T < 0.40\,\mathrm{K}$) so as to neglect energy losses due to collisions with the elementary excitations of the liquid.

Grids of different openings of square shape with side $4\,\mu\mathrm{m} < l < 10\,\mu\mathrm{m}$ select the maximum size of the vortex rings that can pass through the grid and reach the collector, at which they are detected. For square openings of side l, the grid transmission is proportional to $(l/2-R)^2$, where R is the ring radius, for $R < l/2$, and zero for $R \geqslant l/2$. The cut-off energy is defined as the energy for which the grid transmittance is zero, yielding $R = l/2$. Thus, the analysis of the current transmitted through the second grid as a function of the total voltage drop gives a direct determination of the ring radius.

In Fig. 7.3 the cut-off energy is plotted as a function of the grid size. The solid line is calculated from eqn (7.25) with the cut-off condition $R(\mathcal{E}_{\mathrm{co}}) = l/2$, assuming $\delta = 1/4$, $a = 1.28$ Å, and one quantum of circulation as determined from experiments (Rayfield and Reif, 1963, 1964). The agreement between experiment and theory strongly supports the hypothesis that the quantized vortex rings are described by the equations of classical hydrodynamics supplemented with the condition of quantized circulation $\kappa = h/m_4$.

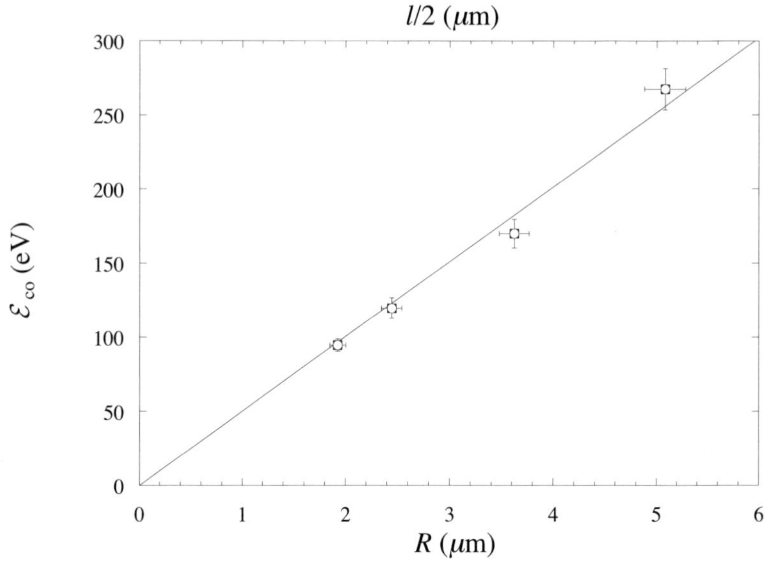

FIG. 7.3. $\mathcal{E}_{co}$ vs R (lower scale) or $l/2$ (upper scale). The line is eqn (7.25) in which $R(\mathcal{E}_{co}) = l/2$ is used. (Gamota and Sanders, 1965.)

7.1.2 *Momentum and impulse of vortices*

As a vortex moves at a definite speed of order $(\kappa/4\pi R)\ln(R/a)$ and has an effective mass $\sim \rho R^3$, the corresponding momentum would be $\sim \rho\kappa R^2$. This is not true, though appealing, owing to the rapid decrease of the speed with the distance from the vortex.

The rigorous way to proceed is to define a quantity called the *impulse*, whose time variation equals the changes in time of the total momentum of the fluid. By so doing, the forces acting on the vortex can be readily calculated.

In the simply connected region outside the vortex core, the fluid flow is irrotational everywhere. Only in the vortex core the flow has nonzero vorticity. Moreover, the liquid can be assumed to be incompressible and to have (nearly) zero viscosity. For these reasons, the flow patterns outside the cre can be obtained by solving the equations of change for *potential flow*. This means that the velocity field of the fluid flow outside the vortex core can be derived from a potential (Byron Bird *et al.*, 1960) as

$$\mathbf{v}(\mathbf{r}) = -\boldsymbol{\nabla}\Phi(\mathbf{r}) \; . \tag{7.27}$$

Equation (7.17) for the vector potential can be rewritten with the aid of Stokes's theorem as a surface integral:

$$\mathbf{A}(\mathbf{r}) = \frac{\kappa}{4\pi}\int \mathrm{d}\mathbf{S}' \times \boldsymbol{\nabla}'\frac{1}{|\mathbf{r}-\mathbf{r}'|} \, , \tag{7.28}$$

where $\boldsymbol{\nabla}'$ means that the differentiation is to be carried out with respect to the coordinates $\mathbf{r}'$. The surface integral consists of the circular region of radius $\approx R$ in the plane of the ring enclosed by the vortex core. The orientation of the surface $\mathbf{S}'$ depends on the vorticity in the core and is the same as the direction of the flow through the ring center. The velocity field is obtained by taking the curl of eqn (7.28)

$$\begin{aligned}\mathbf{v}(\mathbf{r}) &= \frac{\kappa}{4\pi}\boldsymbol{\nabla}\times\int \mathrm{d}\mathbf{S}'\times\boldsymbol{\nabla}'\frac{1}{|\mathbf{r}-\mathbf{r}'|}\\ &= \frac{\kappa}{4\pi}\int(\mathrm{d}\mathbf{S}'\cdot\nabla')\,\boldsymbol{\nabla}'\frac{1}{|\mathbf{r}-\mathbf{r}'|} - \frac{\kappa}{4\pi}\int \mathrm{d}\mathbf{S}'\left(\boldsymbol{\nabla}'\right)^2\frac{1}{|\mathbf{r}-\mathbf{r}'|}\\ &= -\frac{\kappa}{4\pi}\boldsymbol{\nabla}\int \mathrm{d}\mathbf{S}'\cdot\frac{\mathbf{r}-\mathbf{r}'}{|\mathbf{r}-\mathbf{r}'|^3}\,,\end{aligned} \tag{7.29}$$

where the relations $\boldsymbol{\nabla}'|\mathbf{r}-\mathbf{r}'|^{-1} = -\boldsymbol{\nabla}|\mathbf{r}-\mathbf{r}'|^{-1}$ and $\nabla^2|\mathbf{r}-\mathbf{r}'|^{-1} = 0$ for $\mathbf{r}\neq\mathbf{r}'$ have been used.

By a direct comparison of eqn (7.29) with eqn (7.27), one immediately obtains

$$\Phi(\mathbf{r}) = \frac{\kappa}{4\pi}\int \mathrm{d}\mathbf{S}'\cdot\frac{\mathbf{r}-\mathbf{r}'}{|\mathbf{r}-\mathbf{r}'|^3}\,. \tag{7.30}$$

Clearly, Φ is $\kappa/4\pi$ times the solid angle subtended at $\mathbf{r}$ by the oriented surface $\mathbf{S}'$ enclosed by the ring core and is continuous everywhere, except in the ring plane for $z = z_0$, where it has a discontinuity of κ, namely,

$$\lim_{\epsilon\to 0}\Phi(r,\theta,z\pm\epsilon) = \begin{cases}\pm\kappa/2\,, & r<R\,,\\ 0\,, & r>R\,.\end{cases} \tag{7.31}$$

Let S_0 be an imaginary surface around the instantaneous location of the vortex ring, enclosing a disc-shaped region of radius just greater than R and negligible thickness. V is the volume outside S_0 and bounded by a spherical surface S_1 of radius $R_1\to\infty$. The balance of momentum in the region bounded by the two surfaces is given by

$$\frac{\partial}{\partial t}\int_V \rho\mathbf{v}\mathrm{d}^3r = -\int P\mathrm{d}\mathbf{S} - \int \rho\mathbf{v}\left(\mathbf{v}\cdot\mathrm{d}\mathbf{S}\right)\,, \tag{7.32}$$

where P is the hydrostatic pressure. The last integral represents the convection of momentum density $\rho\mathbf{v}$. The surface integrals are extended to both S_0 and S_1 with the normal pointing outwards from V. The motion is irrotational and solenoidal in V. Thus, one can use Bernoulli's theorem $P - \rho\partial\Phi/\partial t + (1/2)\rho v^2 = \text{constant}$ to obtain

$$\frac{\partial}{\partial t}\int_V \rho\mathbf{v}\mathrm{d}^3r = -\rho\frac{\partial}{\partial t}\int_{S_0}\rho\Phi\mathrm{d}\mathbf{S} + \frac{1}{2}\rho\int_{S_0} v^2\mathrm{d}\mathbf{S} - \rho\int_{S_0}\mathbf{v}\left(\mathbf{v}\cdot\mathrm{d}\mathbf{S}\right)\,, \tag{7.33}$$

where the surface integrals only include S_0 since the surface S_1 makes negligible contribution as $R_1\to\infty$, because both $\mathbf{v}$ and $\partial\Phi/\partial t$ are of order R_1^{-3}. In eqn

(7.33) the last two integrals are identically zero, because v is continuous through the ring plane. So, one finally gets

$$\frac{\partial}{\partial t}\int \rho \mathbf{v} \mathrm{d}^3 r = -\frac{\partial}{\partial t}\mathbf{I}\,. \qquad (7.34)$$

The quantity

$$\mathbf{I} = -\int_{S_0} \rho \Phi \mathrm{d}\mathbf{S} \qquad (7.35)$$

is named the *impulse* of a vortex ring. Equation (7.35) states that the rate of change of the impulse is equal to the rate of change of the total momentum of the fluid.

For a single vortex ring in an unbounded fluid, the impulse is

$$\mathbf{I} = \pi\rho\kappa R^2 \hat{v}\,, \qquad (7.36)$$

where $\hat{v}$ is a unit vector parallel to the translational velocity of the ring. If no external forces are exerted on a ring in an infinite fluid, i.e., if it is free, then its impulse is constant and, as a consequence of eqn (7.36), both its radius and velocity do not vary with time.

Whereas the linear momentum of the vortex, defined as $\int \rho \mathbf{v} \mathrm{d}^3 r$, is indeterminate because its value depends on the shape of the bounding surface at infinity (London, 1964; van Vijfeijken *et al.*, 1969), the impulse $\mathbf{I}$ is always well defined and has customarily been used as a measure of the corresponding momentum. This identification, although generally wrong, is sometimes justified. For instance, if an infinitesimally-small charged vortex ring of initial radius R_i grows into a large ring of final radius $R_f \gg R_i$ under the influence of an externally-applied electric field, then the impulse $\pi\rho\kappa(R_f^2 - R_i^2) \approx \pi\rho\kappa R_f^2$ represents the change of total momentum of the fluid and can be considered as the momentum associated with the presence of the final vortex ring of radius R_f.

A clever experiment by Gamota and Barmatz (1969) has directly shown the equivalence between changes in impulse and momentum. The experimental set-up is schematically shown in Fig. 7.4. A radioactive source produces the necessary ionization to create vortices in the liquid. The charged vortex rings gain energy by means of the voltages applied to the grids (G_1 through G_4). A square-wave voltage is used to chop the vortex beam. The gated beam impinges on the flexible Kapton membrane (K) that forms a parallel-plate capacitor together with grid G_4. The deformation of this membrane due to the impulse supplied by the impinging vortex rings produces a variation of the capacitance, which can be accurately detected by a sensitive capacitor bridge. The method is highly sensitive because it can detect capacitance changes as small as a few 10^{-18} F.

The measurements are carried out at low temperature ($T = 0.3$ K) to neglect the momentum loss of the vortex rings due to the interaction with the elementary excitations (Rayfield and Reif, 1964). The impulse p of a singly-charged ring is

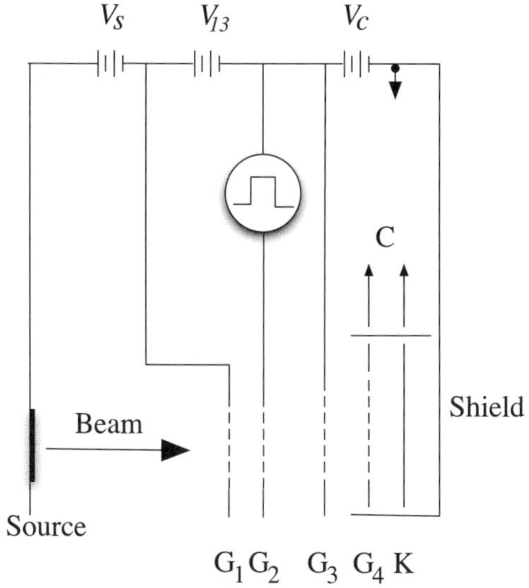

FIG. 7.4. Experimental set-up for the measurement of the impulse of charged quantized vortex rings. (Gamota and Barmatz, 1969.)

related to the capacitance change ΔC by $p = \eta(e/i)\Delta C$, where i is the electrical current detected and η is a proportionality factor that depends on the physical properties of the Kapton-based capacitor. The total energy of the charged vortex rings, assuming that they are singly-charged and that friction is negligible, is simply given by eV_T, where V_T is the total voltage drop.

The impulse of the charged ring is given by eqn (7.36), and its energy $\mathcal{E}$ is given by eqn (7.25). By eliminating R between these two equations, assuming a solid-body rotation of the core with $\delta = 1/4$, the relationship between energy and momentum is obtained as

$$\mathcal{E} = \left(\frac{\rho\kappa^3}{4\pi}\right)^{1/2}\left[\ln\left(\frac{8}{a\sqrt{\pi\rho\kappa}}\mathcal{E}p^{1/2}\right) - \frac{7}{4}\right], \tag{7.37}$$

that can be solved for the impulse p.

The experimental relationship between $\mathcal{E}$ and p is shown in Fig. 7.5. The excellent agreement between theory and experiment shown by the solid line in the figure is obtained by inserting into eqn (7.37) the experimental values for $\kappa \approx 1 \times 10^{-7}\,\mathrm{m}^2/\mathrm{s}$ and for the vortex core $a \approx 1.3\,\text{Å}$ (Rayfield and Reif, 1964). It should be recalled, however, that the interpretation of the experiment is not as straightforward as it may appear at first glance (Meyer and Soda, 1965; Huggins, 1970, 1972; Fetter, 1972; Cross, 1974).

It has been pointed out (Roberts and Donnelly, 1970) that the energy $\mathcal{E}$ of the vortex ring given by eqn (7.25) serves as an effective Hamiltonian, from which

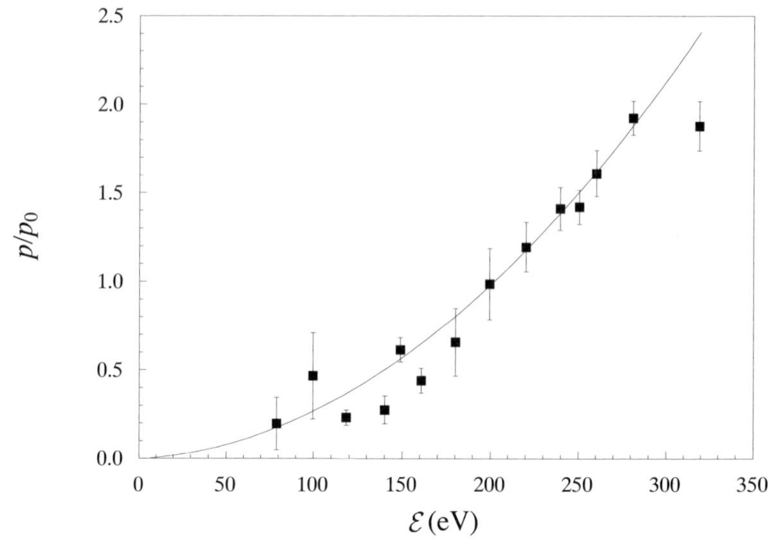

FIG. 7.5. p/p_0 vs $\mathcal{E}$ for quantized charged vortex rings. (Gamota and Barmatz, 1969.) The experimental data are normalized to the impulse p_0 at an energy of 200 eV. Line: eqn (7.37).

the ring velocity can be obtained by differentiation with respect to the impulse I or the momentum p:

$$v = \frac{\mathrm{d}\mathcal{E}}{\mathrm{d}I} = \frac{\mathrm{d}\mathcal{E}}{\mathrm{d}p}\,. \tag{7.38}$$

This identification has been verified experimentally to also hold in three dimensions by analyzing the motion of vortex rings deflected by either electric (Rayfield and Reif, 1963, 1964; Hess, 1972) or magnetic fields (Meyer, 1966).

In this context it is interesting to briefly describe the experiment of Meyer (1966). In this case, a split collector is used and the amount of charges reaching each part of the collector is a function of the strength of a transverse deflecting magnetic field. The experiment is carried out at 0.3 K to make the energy losses due to friction negligible. Small voltage drops are used to produce small vortex rings of radii in the range 50–200 Å. The results confirm the validity of eqn (7.38) leading from eqn (7.25) to eqn (7.26). In Fig. 7.6 the magnetic field strength B necessary to deflect charged vortex rings by 0.41 rad. as a function of the ring energy $\mathcal{E}$ is shown. The solid line is eqn (7.25) with $\delta = 1/4$ and $a = 1$ Å, while the dashed line is the slightly modified version

$$\mathcal{E} = \frac{1}{2}\rho\kappa^2 R\left(\ln\frac{8R}{a_p} - 1.67\right) \tag{7.39}$$

calculated by a variational quantum-mechanical approach, with 1.67 in place of the standard classical value 7/4 for a solid-core ring and $a_p \approx 0.5$ Å (Amit

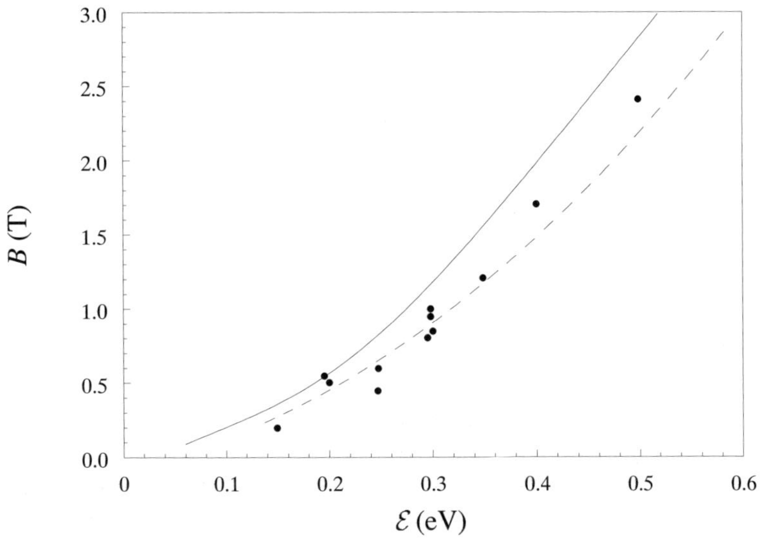

FIG. 7.6. B vs $\mathcal{E}$ (Meyer, 1966). Solid line: eqn (7.25) with $\delta = 1/4$ and $a = 1\,\text{Å}$. Dashed line: eqn (7.39) (Amit and Gross, 1966).

and Gross, 1966). Once more, the agreement between theory and experiment confirms the validity of the identification of momentum and impulse for a vortex ring and the correctness of eqn (7.38) that relates the energy and velocity of the vortex rings.

8

MOTION OF CHARGED QUANTIZED VORTEX RINGS IN SUPERFLUID HE II

In this chapter the motion of charged vortex rings is analyzed in order to gather important pieces of information on the interaction of the rings with the superfluid environment. In the limit of zero temperature no elementary excitations would be present and there would not be any possibility for the rings to interact with the fluid. At the same time, the type and density of elementary excitations can be controlled, to some extent, by changing the temperature and/or by modifying the content of ^{3}He impurities. For these reasons, the knowledge about the interaction of charged vortex rings and the liquid has been derived from experiments carried out at low temperature.

8.1 Frictionless measurements at low temperature

If the velocity of ions moving through the superfluid under the influence of a suitably-large electric field exceeds a threshold velocity of the order of 30–40 m/s, quantized vortex rings are created. The ions get stuck on them and the ion–ring complex moves through the liquid as a single entity. The nucleation process as well as the nature of the ion capture process are quite complicated and will be discussed in a separate chapter. In the present chapter, the features of the motion of the ion–vortex ring complex, once established, are described.

Critical parameters for the creation of vortex rings by drifting ions are low temperatures and large drift velocities. Low temperatures are required to keep the number of elementary excitations low enough that rings are not destroyed upon collisions. Large velocities mean high electric drift fields, in which the ions gain sufficient energy during a mean free path to largely exceed their thermal energy. High velocities also ensure that the ion motion is not adiabatic and that it is possible to create macroscopic excitations that involve the flow of large amounts of liquids and are characterized by higher energies than phonons or rotons (Rayfield and Reif, 1964).

The first experimental evidence of the creation of vortex rings in superfluid helium at low temperature ($0.3\,\mathrm{K} < T < 0.7\,\mathrm{K}$) was obtained using a drift velocity spectrometer with several different arrangements of retarding and gating grids (Rayfield and Reif, 1963). The simplest electrical set-up is similar to that shown in Fig. 2.1. An accelerating potential is applied between the Am-coated source electrode and the first grid. Charged vortex rings are created in this region if the field is strong enough and they may reach an equilibrium distribution if the source-to-grid distance is large enough (Gamota *et al.*, 1971).

If the temperature is sufficiently low, the charges passing through the grid are able to reach the collector if the voltage applied between the grid and collector is zero or even retarding. A retarding potential nearly equal to the accelerating one must be applied to restrain the charged rings from reaching the collector. The rings, which move much more slowly than bare ions, behave almost like free particles with well-defined kinetic energy. The low-temperature condition ensures that the frictional effects due to collisions with the elementary excitations of the liquid are negligible or small. In the investigated temperature range the fractional energy loss is of the order of 5–10% per cm (Rayfield and Reif, 1963).

The drift velocities of the ion–ring complex are measured with the double-gate drift velocity spectrometer described earlier. The energy of the vortex rings is set by the total voltage drop across the drift space. The experimental results are shown in Fig. 6.3, where the drift velocity of charged vortex rings is shown as a function of their energy.

The striking feature of these data is that the velocity, which is itself small, decreases with increasing energy. Its small value suggests that the charges are strongly coupled to a large amount of the surrounding liquid and the peculiar energy $\mathcal{E}$–velocity v relationship can be described accurately by using eqns (7.25) and (7.26), which relate energy and velocity for a vortex ring of given strength κ and radius R much larger than its core a. If R is eliminated between these two equations, and if it is assumed that the core parameter a is so small that $\eta \equiv \ln(8R/a) \gg 1$, the desired v–$\mathcal{E}$ relationship is obtained (Rayfield and Reif, 1963, 1964):

$$\left(\frac{v\mathcal{E}}{C}\right)^{1/2} = \left\{\ln \mathcal{E} - \ln\left[\left(\frac{v\mathcal{E}}{C}\right)^{1/2} - \frac{3}{4}\right]\right\} + \left[\ln\left(\frac{16}{\rho\kappa^2 a}\right) - 1\right], \tag{8.1}$$

where $C = \rho\kappa^3/8\pi$. By fitting this equation to the data shown in Fig. 6.3, the solid line in the figure is obtained if the fit parameters a and κ take on the values

$$\kappa = (1.00 \pm 0.03) \times 10^{-7}\,\mathrm{m^2/s}\,,$$
$$a = (1.28 \pm 0.13)\,\text{Å}\,.$$

Within the limit of the experimental accuracy the value of the circulation is equal to one quantum: $\kappa = h/m_4 = 0.997 \times 10^{-7}\,\mathrm{m^2/s}$.

In the energy range under investigation, $1.5\,\mathrm{eV} < \mathcal{E} < 45\,\mathrm{eV}$, the vortex ring radius is very large, $500\,\text{Å} < R < 10^4\,\text{Å}$, i.e., much larger than the radius of the bare[1] ions (either positive, $R_+ \approx 6\,\text{Å}$, or negative, $R_- \approx 17\,\text{Å}$) (Rayfield and Reif, 1963, 1964).

The velocity of the ion–ring complex depends neither on T nor on the sign of the associated charge, in agreement with what one would expect for a vortex

[1]It should be noted that in the context of quantum hydrodynamics the expression *bare ions* indicates ions that are not stuck on vortex lines or rings, although they are still surrounded by either the electrostrictive He atoms cluster, if positive, or by the empty cavity, if negative. In the following this terminology will be steadily adopted.

ring, i.e., $\mathcal{E}$ and v are determined by the hydrodynamic properties of the large amount of fluid involved in the ring rather than on the properties of the small individual charges bound to the ring itself.

Some questions arise naturally. The first one is related to the creation of vortex rings by the drifting ions and will be addressed in the following chapters. Other questions are how the motion of the charge in the ring can be envisioned and what physical mechanism keeps the ion stuck on the ring. These last two questions can be answered on a phenomenological basis by keeping in mind that large semi-macroscopic rings are considered here (Rayfield and Reif, 1964).

The core region acts as a potential well for the ion (either electron bubble or positive cluster). Therefore, the ion wave function is localized around the core. As far as the ion motion around the ring circumference is concerned, the potential well can be considered as a one-dimensional box with length $l \approx 2R \geqslant 1000\,\text{Å}$. The spacing of the levels associated with this degree of freedom is consequently much less than $k_B T$ and the ion moves as a classical particle on a circle. The thermal velocity of the ion, though its effective mass is several tens of helium mass units large (Atkins, 1959), is still high enough that the ring velocity does not vary appreciably in the time taken by the charge to complete one circle. Moreover, the motion of the charge along the ring is stopped by random collisions with the elementary excitations of the liquid (mainly phonons at this low temperature). As the ion mean free path, determined from mobility measurements, is of the same order as the ring circumference, the charge can be regarded as being effectively distributed uniformly around the vortex ring core.

This naive picture has been somewhat modified by recent calculations based on modern vortex dynamics (Donnelly, 1991; Samuels and Donnelly, 1991). In fact, measurements of the mobility of ions along the vortex core have shown that for $T \geqslant 0.4\,\text{K}$ the ion must be considered to be fixed in place on one side of the ring (Ostermeier and Glaberson, 1976). Nevertheless, the presence of localized forces on vortex rings, due to the combined effects of the electric force acting on the ion and the Magnus force acting on the ring, induce a superposition of waves on any real vortex. The combination of a point force and a specific series of vortex waves results in a uniform growth of the entire vortex ring with sideways velocity (Samuels and Donnelly, 1991), thus preserving the basic idea that vortex rings remain circular when an electric field is acting upon them (Rayfield and Reif, 1963).

The answer to the second question is quite simple. The physical process that maintains the ion stuck on the ring is simply hydrodynamic suction. The fluid velocity field around the core has a $1/r$ dependence. According to Bernoulli's principle (Landau and Lifsits, 2000), the ion (which is rather to be thought of as a small sphere) is subject to a restoring force that pulls it toward the core. Once localized on the core, the ion replaces a small volume of fast-rotating fluid, thereby reducing the kinetic energy associated with the vortex ring (Rayfield and Reif, 1964). Because this binding energy is much larger than $k_B T$, the ion–ring complex is stable against thermal fluctuations.

8.2 Interaction of quantized vortex rings with elementary excitations

At low enough temperatures the number of elementary excitations is so small that a vortex ring can negotiate an appreciable distance with negligible energy loss. At higher temperatures, however, the collisions between the vortex ring and the quasiparticles produce significant energy losses. If the fractional energy loss is negligible for $T = 0.3\,\mathrm{K}$, at $T = 0.65\,\mathrm{K}$ it is as large as $15\,\mathrm{eV/cm}$ (Rayfield and Reif, 1964).

At the lowest T, the elementary excitations are predominantly phonons and ^{3}He atoms, whereas rotons dominate at higher temperatures. The systematic analysis of the energy loss as a function of temperature gives information on the cross-sections describing the scattering processes of the quasiparticles off the vortex rings.

The frictional force on the vortex rings, which is in principle a function of the temperature and of the ring energy, can be easily related to the ring velocity on the basis of a very simple argument (Rayfield and Reif, 1964).

First consider a long, straight vortex moving with velocity v with respect to the quasiparticles. v is much smaller than their mean speed. The frictional force per unit length $\mathcal{F}'$ due to the interaction of the vortex with the elementary excitations must vanish as soon as $v \to 0$. For small v, the leading term in the expansion of $\mathcal{F}'$ as a function of v is retained, yielding $\mathcal{F}' \propto v$.

Moreover, the mean free path of the quasiparticles, especially at low temperatures, is larger than the size of the vortex core, so that kinetic arguments can be used to calculate the effect of individual quasiparticles scattering off the vortex line.

Finally, the typical radius of fairly energetic vortex rings is quite large ($R \geqslant 500\,\text{Å}$), larger than the typical distance over which a vortex line interacts appreciably with a quasiparticle, and also larger than the quasiparticle wavelength λ. This last condition is fulfilled exactly for rotons with $\lambda \approx 3\,\text{Å}$ but only approximately for phonons with $\lambda \leqslant 400\,\text{Å}$ at $T \approx 0.3\,\mathrm{K}$ (Rayfield and Reif, 1964).

For these reasons, the frictional force on a vortex ring can be approximated by the force on a vortex line bent so as to form a circle of radius R, i.e., $\mathcal{F} = 2\pi R\mathcal{F}' \propto Rv$, and can thus be cast in the form (Rayfield and Reif, 1964)

$$\mathcal{F} = \alpha(T)\left(\eta - \frac{1}{4}\right) \equiv \alpha(T)\chi(\mathcal{E}) , \tag{8.2}$$

with $\chi = \eta - 1/4$, where $\eta = \ln(8R/a)$ and R is the ring radius. $\alpha(T)$ is the friction coefficient that, for a ring of given circulation, must be a function only of T. The definition (8.2) explicitly separates the dependence of $\mathcal{F}$ on temperature and ring energy. The explicit value of η for a given ring energy can be calculated using eqn (7.25).

The coefficient $\alpha(T)$ at low temperature, $0.3\,\mathrm{K} < T < 0.7\,\mathrm{K}$, has been measured by Rayfield and Reif (1964) using two different techniques, both based on drift velocity spectrometry. In the first, the *constant-velocity* method, the velocity of the traveling vortex rings is maintained constant by applying a constant electric field that counterbalances their energy loss.

In the second, the *stopping-potential* method, the energy loss is directly determined by measuring the stopping potential required to arrest the vortex rings after they have drifted though a field-free region. The results of the two methods agree very well.

In Fig. 8.1 the frictional force on a vortex ring as a function of its energy is shown for $T = 0.615\,\mathrm{K}$. The solid line represents eqn (8.2) with a choice of suitable parameters $\mathcal{F} = 1.04\chi(\mathcal{E}) = 1.04[\ln(8R/a) - 1/4]$. It is calculated using the relationship between $\mathcal{E}$ and R given by eqn (7.25), with the value $a \approx 1.28\,\text{Å}$ determined from the fit to the velocity data shown in Fig. 6.3 (Rayfield and Reif, 1963, 1964).

Careri *et al.* (1965) carried out energy-loss measurements at much higher temperatures. They investigated the drift velocity of vortex rings as a function of the electric field above the giant mobility discontinuity.

In this situation, the density of thermally-activated excitations is much higher than in the conditions of the experiment of Rayfield and Reif (1963). Thus, vortex rings move at constant speed as the result of the balance between the driving force due to the electric field E and the retarding force due to friction, yielding

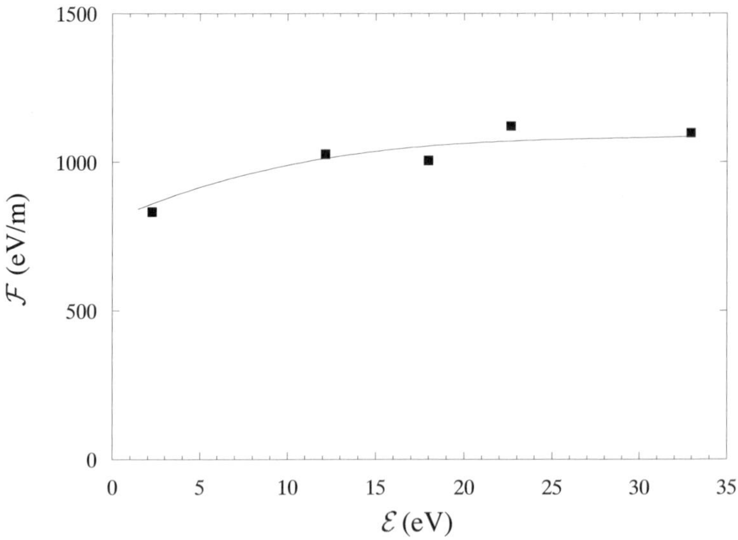

FIG. 8.1. $\mathcal{F} = \alpha(T)\chi(\mathcal{E})$ vs $\mathcal{E}$ for $T = 0.615\,\mathrm{K}$. (Rayfield and Reif, 1964.) Line: $\mathcal{F} = 1.04\chi(E)$, with $\chi(\mathcal{E})$ given by eqn (7.25).

$$eE = \alpha(T)\left(\eta - \frac{1}{4}\right). \tag{8.3}$$

By using eqn (7.26) for the drift velocity of the charged rings in order to eliminate R, eqn (8.3) gives

$$\ln \mu \equiv \ln \frac{v}{E} = \ln \frac{2e\kappa}{\pi a \alpha(T)} + \frac{1}{4} - \frac{eE}{\alpha(T)} \tag{8.4}$$

and the friction coefficient α can be obtained from the slope of the semi-logarithmic plot of the mobility as a function of the electric field E.

In Fig. 8.2 mobility data for fields beyond the giant discontinuity (see, for instance, Figs 6.4 and 6.5) are shown (Careri *et al.*, 1965). Similar results have also been obtained in the range $0.6\,\mathrm{K} < T < 0.8\,\mathrm{K}$ and for pressures up to 2.4 MPa (Rayfield, 1968*a*).

In Fig. 8.3 the values of α are shown in the extended temperature range $0.4\,\mathrm{K} < T < 1\,\mathrm{K}$ (Careri *et al.*, 1965; Rayfield, 1968*a*). The high-temperature data at vapor pressure have been subsequently confirmed in the range $0.8\,\mathrm{K} < T < 1.3\,\mathrm{K}$ with a different experimental set-up based on field emission or field ionization to inject charges of the required sign into the liquid (van Dijk *et al.*, 1977). At higher pressures the friction coefficient turns out to be larger than at $P = 0$, but it retains its exponential increase with increasing T (van Dijk *et al.*, 1977).

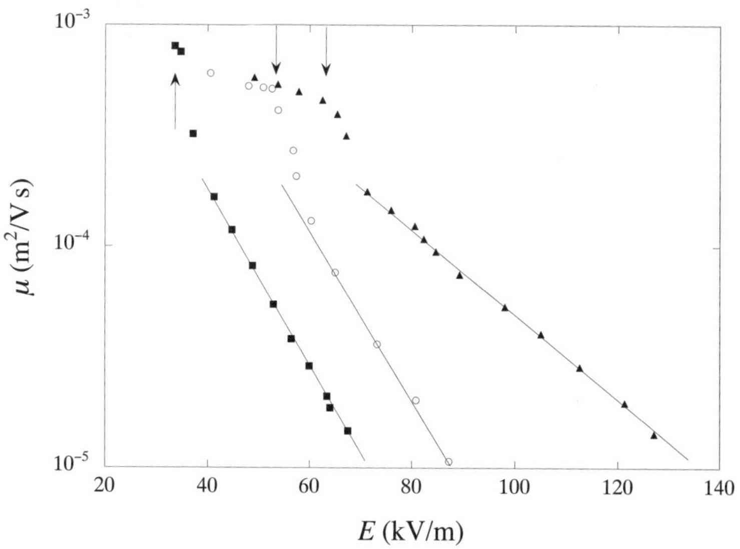

FIG. 8.2. μ vs E above the discontinuity (indicated by arrows). (Careri *et al.*, 1965.) Lines: fits to eqn (8.4). $T\,(\mathrm{K}) = 0.892$ (positive ions at SVP, squares), 0.938 (negative ions at SVP, circles), and 0.920 (positive ions for $P = 0.61\,\mathrm{MPa}$, triangles).

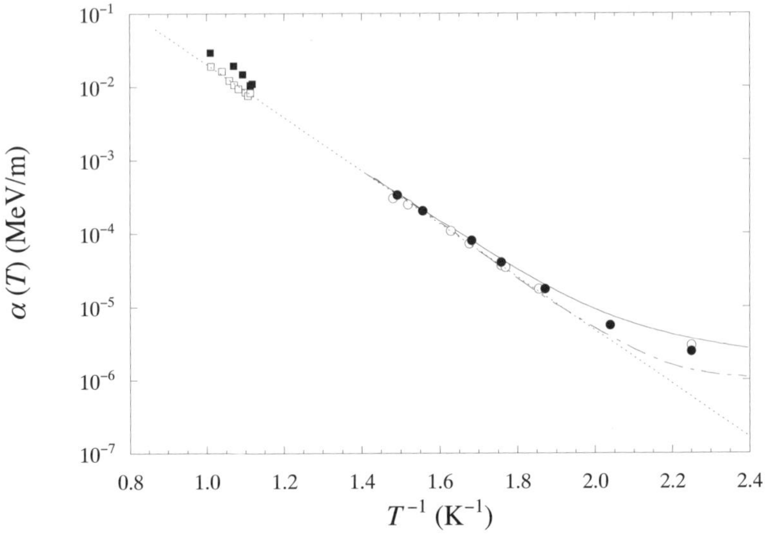

FIG. 8.3. α vs T^{-1} in an extended range. Closed squares: positive ions, and open squares: negative ions (Careri *et al.*, 1965). Closed circles: positive ions, and open circles: negative ions (Rayfield and Reif, 1964). The lines are explained in the text.

The calculation of the friction exerted on the vortex ring by collisions with the elementary excitations of the superfluid is carried out in the frame of kinetic theory by assuming that the scattering of elementary excitations off the ring is elastic and that the ring radius and the excitation wavelength satisfy the previously-described conditions (Rayfield and Reif, 1964).

Because of the symmetry due to the motion of the charged rings along the z-direction of the electric field, the force acting on them is related only to the mean component of momentum transferred in the $-z$-direction by virtue of collisions with the quasiparticles. The result for the friction force per unit length due to each type of quasiparticle is given by

$$\mathcal{F}' = -\frac{3\pi^2}{8} v \int_0^\infty \frac{\partial f_0}{\partial \epsilon} u p^4 \sigma_{mt}(p) \mathrm{d}p \,, \tag{8.5}$$

where f_0 is the distribution function of excitations, the solution of the two-term expansion of the Boltzmann transport equation (Huxley and Crompton, 1974). Here $u = |\partial \epsilon / \partial p|$ is the group velocity of the excitation under investigation, ϵ and p are its energy and momentum, respectively, v is the drift velocity of the vortex ring, and σ_{mt} is the momentum-transfer cross-section related to the particular excitation considered. It is worth noting that eqn (8.5) is just the same as, for instance, eqn (5.12). The only difference is that the dynamics specific to the present case is contained in the cross-section. For scattering off vortex lines, the cross section has the dimension of length, not of area as usual, and represents a sort of effective width of the vortex line responsible for scattering.

By introducing a suitable average cross-section $\bar{\sigma}_{mt}$, defined as

$$\bar{\sigma}_{mt} = \int_0^\infty \frac{\partial f_0}{\partial \epsilon} up^4 \sigma_{mt} \mathrm{d}p \Big/ \int_0^\infty \frac{\partial f_0}{\partial \epsilon} up^4 \mathrm{d}p\,, \tag{8.6}$$

the friction coefficient α_i turns out to be

$$\alpha_i = \frac{3\pi^2}{16} \kappa \bar{\sigma}_{mt,i} \int_0^\infty \left(\frac{\partial f_0}{\partial \epsilon} \right) up^4 \mathrm{d}p\,. \tag{8.7}$$

The index $i = r, 3, \mathrm{ph}$ indicates if the scattering is due to rotons, ^{3}He atoms, or phonons, respectively. The total drag coefficient is the sum of the three contributions because the inverse mobility of bare ions is the sum of the contributions due to the three types of quasiparticle, as described by eqn (5.6).

8.2.1 *Roton scattering*

At higher temperatures, vortex rings move essentially through a roton gas, in the same way as bare ions at vanishingly-small fields. The energy loss of vortex rings due to roton scattering is proportional to the roton number, which depends exponentially on T as described by eqn (5.53).

Rotons obey Bose–Einstein statistics with distribution function is $f_0(\epsilon) = h^{-3}[\exp(\epsilon/k_{\mathrm{B}}T) - 1]^{-1}$. As $\epsilon/k_{\mathrm{B}}T \gg 1$, only rotons with momentum $p \approx p_0$ contribute to the integral in eqn (8.7), yielding

$$\alpha_r = \frac{3\pi^2}{8} \frac{\kappa}{h^3} p_0^4 e^{-\Delta/k_{\mathrm{B}}T} \bar{\sigma}_{mt,r}\,, \tag{8.8}$$

where p_0 is the momentum at the roton minimum.

In this high-temperature regime, by analogy with the case of bare ions, friction is dominated by rotons. The exponential temperature dependence of the friction coefficient masks any possible temperature dependence of the effective cross-section. The dotted line in Fig. 8.3 represents this exponential factor with $\Delta \approx 8.65\,\mathrm{K}$, in agreement with the known values of the roton energy gap (Henshaw and Woods, 1961*a*). From the experimental data the cross-section for roton scattering is found to be $\bar{\sigma}_{mt,r} = (9.5 \pm 0.7)\,\mathrm{Å}$, in fairly good agreement with experiments of second-sound attenuation (Hall and Vinen, 1956) and of vortex waves in rotating He II (Hall, 1960). In spite of several theoretical attempts (Hall and Vinen, 1956; Lifsits and Pitaevskii, 1957; Hall, 1960), no microscopic models are able to predict such a value.

At lower temperatures, the experimental values of the friction coefficient increasingly deviate from the simple exponential law because the importance of roton scattering decreases with respect to scattering off phonons and isotopic impurities.

8.2.2 ^{3}*He-impurity scattering*

Isotopic-impurity scattering is responsible for the limiting value of the friction coefficient at very low temperature, where the number of rotons and phonons is negligible. The expression for the ^{3}He-scattering contribution can be obtained in the same way as the roton one by recalling that ^{3}He is so dilute as to obey Maxwell–Boltzmann statistics. α_3 is thus given by

$$\alpha_3 = \frac{3}{8}\kappa \left(2\pi m_3^\star k_B T\right)^{1/2} n_3 \bar{\sigma}_{mt,3}\,, \tag{8.9}$$

where $m_3^\star$ and n_3 are the effective mass and the number density of the light He isotope, respectively. As the natural isotope abundance is quite low, α_3 is determined by adding a known amount of impurities. The experiment yields $\sigma_{mt,3} = (18.3 \pm 0.7)\,\text{Å}$ if $m_3^\star = 2.5 m_3$ (Rayfield and Reif, 1964).

The dash–dotted line in Fig. 8.3 represents the sum of the roton and ^{3}He contributions to the friction coefficient, $\alpha_r + \alpha_3$, where α_3 is calculated by using the value 1.4×10^{-7} for the natural isotopic abundance (Rayfield and Reif, 1964).

8.2.3 *Phonon scattering*

The interaction of the vortex rings with phonons gives an attenuation coefficient α_{ph}, which is calculated by means of eqn (8.5) by recalling the dispersion relation for phonons, $\epsilon = cp$, where c is the velocity of first sound:

$$\alpha_{\text{ph}} = \frac{\pi^2}{20}\frac{\kappa}{h^3}\left(\frac{k_B T}{c}\right)^4 \bar{\sigma}_{mt,\text{ph}}\,. \tag{8.10}$$

Theoretical calculations (Pitaevskii, 1959) give the following prediction for the cross-section: $\sigma_{mt,\text{ph}} = (\pi/2)(\kappa^2 p/c^2 h)$, that, in turn, yields an effective cross-section $\bar{\sigma}_{mt,\text{ph}} \approx 0.3\,\text{Å}$. If this contribution is added to the first two, the solid line in Fig. 8.3 is obtained, showing a fairly good agreement with the experimental data.

8.3 Determination of the vortex core parameter

Once the friction coefficient α is known, the equation of motion of the charged rings can be explicitly integrated, allowing the experimental determination of the vortex core parameter a (Glaberson and Steingart, 1971).

Accurate measurements of the vortex ring velocity in the range $0.35\,\text{K} < T < 0.65\,\text{K}$ at vapor pressure (Glaberson and Steingart, 1971) have, indeed, shown that the vortex core parameter a has a slight temperature dependence (see Fig. 8.4) that is well described by the vortex core model of Glaberson *et al.* (1968*a*), in which the vortex core radius is associated with a strongly-increasing excitation density as the center of the vortex is approached, although quasi-thermodynamic calculations (Pollock, 1971) cannot definitely be ruled out.

The pressure dependence of the vortex core parameter has also been measured (Steingart and Glaberson, 1972) and the results are reported in Fig. 8.5. The lines

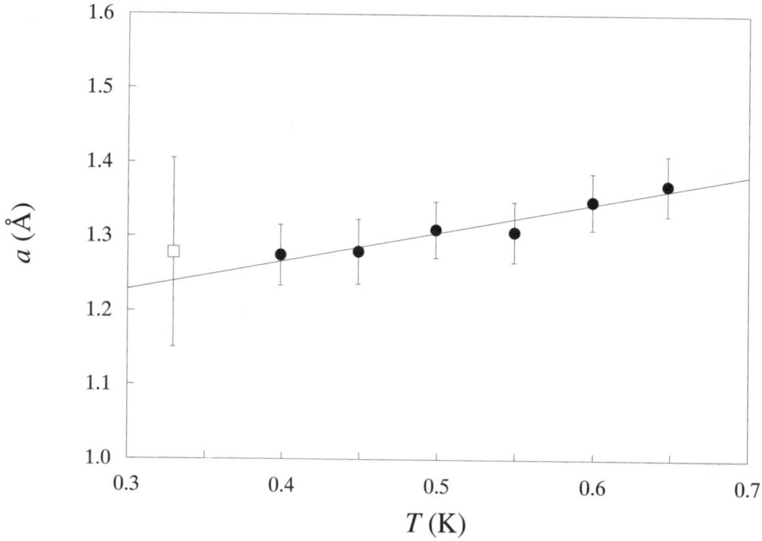

FIG. 8.4. a vs T (Glaberson and Steingart, 1971). Square: result of Rayfield and Reif (1964). Line: vortex core model (Glaberson *et al.*, 1968*a*).

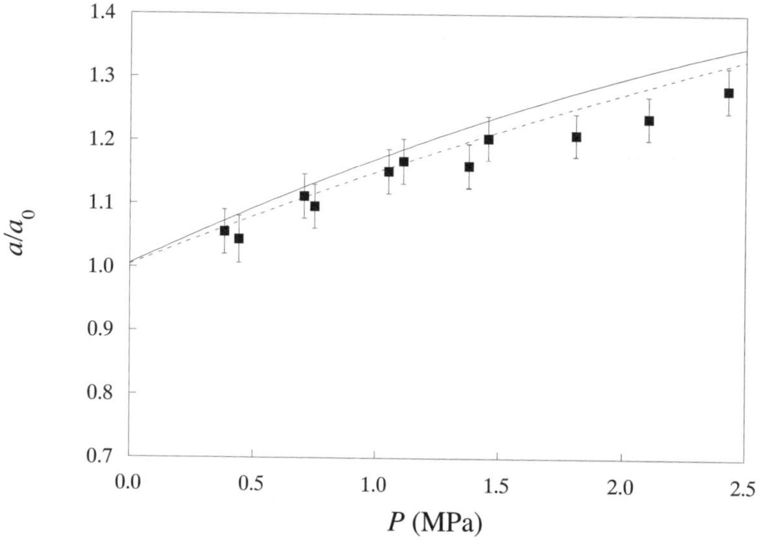

FIG. 8.5. a vs P for $T = 0.36\,\text{K}$ (Steingart and Glaberson, 1972). Vortex core model with (solid line) or without (dashed line) localization energy (Glaberson *et al.*, 1968*a*).

are the prediction of the vortex core model (Glaberson *et al.*, 1968*a*), already exploited in the localized roton model. The dashed and the solid lines correspond, respectively, to the addition or exclusion of the localization energy (Glaberson *et al.*, 1968*a*).

9

NUCLEATION OF VORTEX RINGS

The creation of quantized vortex rings in superfluid helium is one of the ways in which superfluidity breaks down and is therefore a subject of paramount importance to understand the nature of superfluidity. The discovery of the existence of remnant vorticity (Awschalom and Schwarz, 1984), apparently independent of the past history of the liquid, has vindicated the suspicion that the threshold critical velocities for vortex formation in flow and thermal counterflow experiments (Tough, 1982) refer to the physical situations needed to expand pre-existing vorticity to form dissipative tangles, not to those necessary to *create* a vortex anew (Bowley *et al.*, 1984; McClintock, 1999).

However, the experiments in which quantized vortex rings are generated by moving ions are the only ones where the intrinsic nucleation mechanisms can be directly investigated because the small size of the probes makes them insensitive to remnant vorticity (Bowley *et al.*, 1982). It is also expected that the fluid flow around the ions is laminar (Donnelly and Roberts, 1971), thereby reducing the complexity of the problem. In addition, owing to the spherical shape of the ions, problems of variable surface characteristics and impurities do not occur and it appears that the nucleation of vorticity by drifting ions is not related to vorticity already present in the liquid (Zoll, 1976).

The transition between ion- and ring-like behavior has been investigated essentially by measuring the dependence of the ion velocity on the applied electric field over a drift space of a few centimeters (Careri *et al.*, 1965; Rayfield, 1968*a*; Cunsolo and Maraviglia, 1969). In "pure" liquid He, the transition to a charged, quantized vortex ring appears as a sudden discontinuity (Careri *et al.*, 1965), thus indicating that the ring grows to a large size as soon as it is formed before coming to an equilibrium situation, in which the drag acting on it balances the external force due to the applied electric field. Depending on several physical parameters such as T, or ^{3}He-impurity content, or P, the field and mobility values at the transition may vary over several orders of magnitude, while the drift velocity is limited to a range between 25 and 50 m/s.

Experiments have shown that the transition region around the discontinuity has a stochastic nature (Donnelly and Roberts, 1969*a*; Strayer and Donnelly, 1971). In fact, if a pulse of bare ions is injected into the drift space, their number decreases exponentially because of the generation of charged vortex rings. The vortex ring generation rate increases dramatically as the electric field is increased above the threshold at which the giant discontinuity occurs.

When interpreting these observations, it must be recalled that the ion drift velocity v_D is a statistical concept in the sense that it is an average over a distribution of individual velocities and that the actual velocity of a single charge carrier fluctuates within a narrow range around it. Thus, it is commonly accepted that an individual charge carrier suddenly exchanges energy and momentum with the superfluid by creating or growing a quantized vortex ring as soon as it reaches a certain critical velocity with respect to the superfluid rest frame (Schwarz and Jang, 1973). In spite of this common view, there is no general consensus on the details of the microscopic processes leading to a critical velocity.

Apart from some early speculations about the nucleation of vortices (Huang and Olinto, 1965), there are essentially three more or less contradictory models that attempt to explain the experimental results: the *peeling*, *fluctuation*, and *girdling* models.

9.1 The peeling model

In pure liquid He, at $T \approx 0.6\,\mathrm{K}$, in the region where roton scattering already dominates the vortex ring–liquid and the ion–liquid interactions, the curves of drift velocity versus electric field do not allow one to ascertain whether the transition is continuous or discontinuous because, when the field is large enough to push the ion to the critical velocity, it is also great enough to support quite large rings. Any small ring that can be originated grows rapidly to the large equilibrium radius pertaining to such large field strengths.

In Fig. 9.1 such a curve is shown for positive ions. These data are obtained by using a double-gate velocity spectrometer. Rings are created in the source region and are brought to equilibrium at a lower value of the field before their

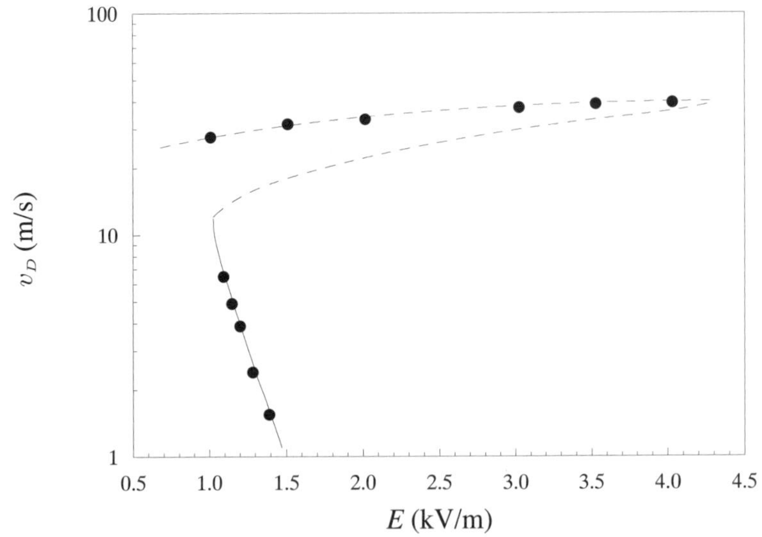

FIG. 9.1. v_D vs E for positive ions for $T = 0.643\,\mathrm{K}$. (Rayfield, 1967.) Solid line: eqn (8.4). The dashed lines are eyeguides.

velocity is measured (Rayfield, 1968*a*). These results clearly state that quantized rings are created when the bare-ion velocity exceeds the critical value but, once created, the rings persist and are stable even at much lower fields, thus ruling out the Huang–Olinto model for vortex creation, ion capture, and ion–vortex ring-complex stability (Huang and Olinto, 1965).

If the transition was continuous, the low-velocity branch of the quantized vortex rings would then smoothly join the upper, bare-ion branch, as suggested by the dashed line in Fig. 9.1. In order to elucidate this point, and trace out the vortex ring spectrum, measurements were carried out by lowering the temperature down to $T = 0.3\,\mathrm{K}$, in order to significantly reduce the drag on both the ion and ring exerted by the roton gas, and adding $\approx 187\,\mathrm{ppm}$ of ^{3}He impurities. In fact, ^{3}He atoms have a larger vortex line-scattering cross-section (Rayfield and Reif, 1964) but a smaller ion-scattering cross-section (Rayfield, 1966) than thermally-excited quasiparticles. In this way, by selectively increasing the drag on the quantized vortex rings, the vortex ring curve should shift to higher fields with respect to the bare-ion one.

This is, indeed, the case as shown in Fig. 9.2. Thus, it appears that the ion–vortex ring transition is smooth, without discontinuities in the drift velocity when the critical velocity for ring formation is reached. According to Rayfield (1967), these results prove that the vortex line associated with the formation of the vortex ring is slowly *peeled* away from the ion in the form of a steadily-growing loop as the electric field is increased. In this physical picture the ion

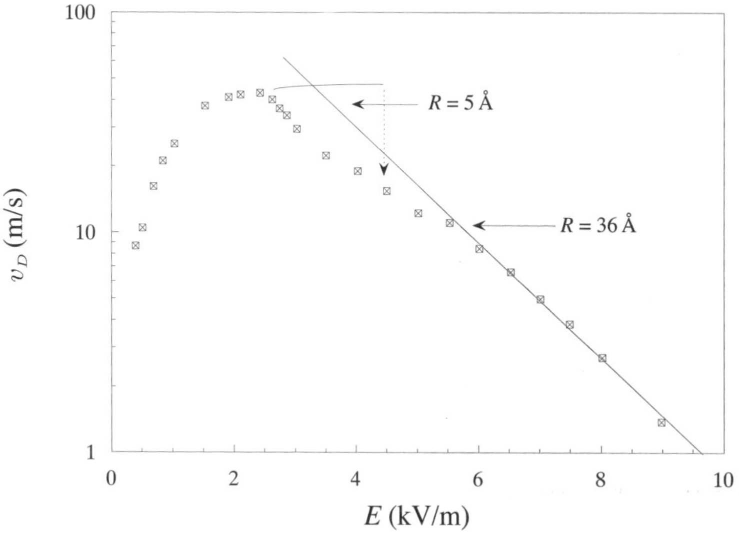

FIG. 9.2. v_D vs E for the positive ion complex at $T = 0.3\,\mathrm{K}$ in a dilute ($\approx 190\,\mathrm{ppm}$) ^{3}He–^{4}He mixture. (Rayfield, 1967.) The theoretical ring radius at each of two different velocities is indicated. Straight line: eqn (8.4).

is not required to hop or jump onto the quantized ring because it is always connected or trapped on the vortex core. Apparently, the nucleation of vortex rings is not related to the creation of rotons (Rayfield, 1966, 1968*a*).

It is interesting to note that for $v_D > 1\,\mathrm{m/s}$ (i.e., for rings of radius $R < 36\,\text{Å}$) the hydrodynamical prediction, given by eqn (8.4) and shown as the line in Fig. 9.2, deviates from the data in such a way as to suggest that the presence of the ion reduces the total drag on the ring.

Measurements of the negative ion drift velocity at low temperature under pressure apparently confirm the picture of the peeling model (Rayfield, 1966, 1968*a*; Neeper, 1968). For $P \geqslant 1.2\,\mathrm{MPa}$, the maximum velocity of negative ions is limited by roton emission, not by vortex ring nucleation.

In Fig. 9.3 the limiting drift velocity of negative ions is plotted as a function of pressure. For $P \geqslant 1.2$–$1.4\,\mathrm{MPa}$, the maximum velocity of negative ions, v_m, is equal, within the experimental accuracy, to the Landau critical velocity for roton emission and no vortex rings are detected.

At lower pressures, however, the plateau velocity v_m is determined by the point at which the creation of vortex rings occurs. In this region v_m increases with P. This can be understood by recalling that the radius of the electron bubble, i.e., of the negative ion, is a decreasing function of pressure. A decreasing bubble

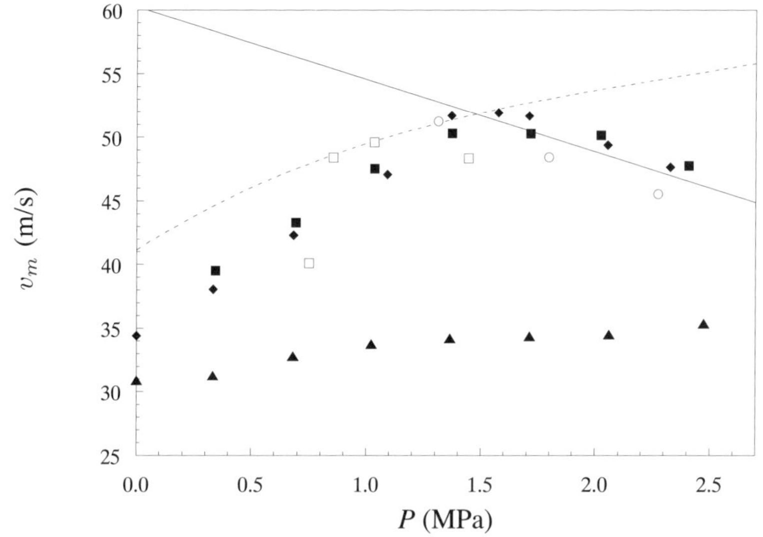

FIG. 9.3. Limiting drift velocity v_m of negative ions as a function of P in the range $0.4\,\mathrm{K} < T < 0.7\,\mathrm{K}$ (Rayfield, 1966, 1968*a*). $T\,(\mathrm{K}) = 0.601$ (closed squares and diamonds), 0.48 (open squares), and 0.36 (circles). Triangles: $T = 0.3\,\mathrm{K}$ in a dilute ^3He–^4He mixture (Rayfield, 1968*b*). Solid line: Landau critical velocity $v_L = \Delta/p_0$ from neutron-scattering data (Henshaw and Woods, 1961*b*). Dashed line: prediction of the girdling model (Schwarz and Jang, 1973).

size produces an increase in the critical velocity $v_c = v_m$ for ring production because circulation is quantized. For a spherical object of radius R, the condition $\oint \mathbf{v}_s \cdot \mathrm{d}\mathbf{l} = h/m_4$ requires $Rv_c =$ const; hence, the increase of v_m with P. This conclusion is of heuristic validity since there is no theory for the quantitative dependence of v_c on P.

These measurements also show that vortex ring creation and roton emission are different and competing processes. However, the pressure dependence of the critical velocity of positive and negative ions in a dilute ^{3}He–^{4}He mixture is quite different. While the critical velocity of positive ions does not change in the mixture with respect to the pure liquid, the presence of isotopic impurities greatly depresses v_c for negative ions as pressure is increased. This is shown in Fig. 9.3. This behavior is easily explained within the peeling model (Rayfield, 1968*b*) by recalling that ^{3}He impurities have a smaller specific volume than ^{4}He atoms and therefore tend to condense at the electron bubble surface (Andreev, 1966; Rayfield, 1968*a*; Dahm, 1969). This phenomenon increases the size of the negative ion, thereby reducing the critical velocity required to create one quantum of vorticity.

9.2 The fluctuation model

In the peeling model it is supposed that the ion is slowed down at the instant of nucleation and that the vortex ring is formed by the impulse imparted to the surrounding liquid by this braking action. The ring energy comes from the decrease in the ion kinetic energy.

A simple calculation enforcing energy and momentum conservation shows that the model predicts that ions, in order to produce vortex rings, must be accelerated to a speed several times larger than the speed of the fastest moving ring. Suppose for simplicity, in fact, that the ion is a point particle of mass m_i and initial velocity v_i. Let v_f be its velocity after nucleation. Momentum conservation yields

$$m_i(v_i - v_f) = \rho_s \pi \kappa R^2 \,, \tag{9.1}$$

where R is the ring radius and κ is the quantum of circulation. Conservation of energy yields the condition

$$\frac{1}{2} m_i (v_i^2 - v_f^2) \geqslant \frac{1}{2} \rho_s \kappa^2 \eta R \,, \tag{9.2}$$

with $\eta = \ln(8R/a)$. Dividing eqn (9.2) by eqn (9.1) one obtains

$$v_i + v_f \geqslant \frac{\kappa \eta}{\pi R} \,. \tag{9.3}$$

Requiring that the ion at the moment of nucleation moves with the self–induced velocity of the ring, i.e., $v_f \approx \kappa\eta/4\pi R$, the following condition is obtained: (Donnelly and Roberts, 1971)

$$v_i \geqslant 3 v_f \,. \tag{9.4}$$

Measurements, however, show that the ion–ring complexes move at about the same velocity as the ions that produce them (Rayfield, 1967, 1968*a*; Strayer and

Donnelly, 1971). As v_f is in the range 30–40 m/s (Bruschi *et al.*, 1968*b*; Cunsolo *et al.*, 1968*b*; Cunsolo and Maraviglia, 1969), $v_i \approx$ 90–120 m/s is expected. Because this condition is not verified experimentally, doubt is cast on the validity of the peeling model.

The hypothesis thus has been suggested that thermal fluctuations can supply sufficient energy and momentum to produce vortex rings (Donnelly and Roberts, 1969*a*, 1971; Strayer and Donnelly, 1971).

In this *fluctuation* model, based on previous suggestions of Iordanskii (1965) and of Langer and Fisher (1967), it is proposed that the initial loop of the vortex core grows from a protoring as a consequence of a thermal fluctuation. It is assumed, just for making quantitative predictions, that the protoring is a roton localized around the equator of the moving ion, as discussed in Section 6.1. Collisions of the protoring–ion complex with the thermal excitations of the liquid will occasionally make one of these rings grow to finite size. The natural attraction of the ion near the vortex core makes the ion attach to the ring. If the ring grows to a size such that its self-induced velocity equals that of the ion, a critical vortex ring is formed.

Within this model, the probability that a vortex is nucleated, or, in other words, that a localized roton grows to its critical size, is assumed to be $\mathcal{P} \propto \exp(-\Delta\mathcal{A}/k_{\mathrm{B}}T)$, where $\Delta\mathcal{A}$ is the free-energy barrier for this process. Once this barrier is surmounted, the nascent vortex ring peels away continuously in much the same way as envisaged by Rayfield (1967) and expands spontaneously by draining energy from the applied electric field.

The nucleation probability is calculated by using the formalism of Brownian motion theory for describing the vortex rings (Chandrasekhar, 1943). The stochastic effects of collisions with quasiparticles are divided into two parts: the systematic dynamical friction created by the preferential direction of quasiparticle impact because of the persistent ring motion relative to the quasiparticle gas and the remaining stochastic force, which is random in direction. The ring is assumed to be sufficiently large that upon each collision its total momentum does not change very much.

Under these assumptions, the probability for the occurrence of a vortex ring with momentum within a given range satisfies a Fokker–Planck equation that introduces a diffusion constant defined as (Donnelly and Roberts, 1969*a*, 1971)

$$\Lambda_c = 4\alpha \left(\frac{\pi}{\rho_s \kappa^3}\right)^{1/2} p^{1/2}, \tag{9.5}$$

where p is the ring impulse, and α is the friction coefficient introduced by Rayfield and Reif (1964). $\alpha = \alpha_r + \alpha_{\mathrm{ph}} + \alpha_3$ contains contributions from all of the quasiparticles, rotons, phonons, and ^{3}He atoms and is determined from experiments.

The evolution of the protoring to a critical fluctuation or critical vortex ring is envisaged as a diffusion from a well in momentum space over a saddle point lying $\Delta\mathcal{A}$ higher in free energy than the bottom of the well. The probability per unit time of this diffusion is (Chandrasekhar, 1943; Donnelly and Roberts, 1969*b*)

$$P = \frac{\omega_A \omega_C}{2\pi} \Lambda_c e^{-\Delta\mathcal{A}/k_\mathrm{B}T}\,, \tag{9.6}$$

where ω_A and ω_C are the curvatures of the well and barrier in momentum space. If rotons are assumed to be the protorings, then $\omega_A \sim \omega_C \approx m_r^{-1/2}$, where m_r is the roton mass. The free-energy barrier is obtained as a Legendre transform of the ring energy in terms of its impulse:

$$\Delta\mathcal{A} = \mathcal{E} - \frac{\partial \mathcal{E}}{\partial p} p\,. \tag{9.7}$$

The energy and impulse of the nascent ring are assumed to be given by eqns (7.25) and (7.36), after a slight modification to take into account the fact that the nascent vortex may be represented as a circular segment of radius of curvature R and length of arc $R_i\varphi\left(\pi + 2\tan^{-1}\varphi\right)$, where R_i is the ion radius and a is the radius of the vortex core (Donnelly and Roberts, 1969*a*). The following equation for the energy is obtained:

$$\mathcal{E} = \frac{1}{4\pi}\rho_s \kappa^3 R_i \left[\ln\left(\frac{8R_i\varphi}{a}\right) - \frac{3}{2}\right]\varphi\left(\pi + 2\tan^{-1}\varphi\right)\,. \tag{9.8}$$

The momentum is taken as the hydrodynamic impulse corresponding to the area between the vortex and the ion, namely,

$$p = \rho_s \kappa R_i^2 \left[\frac{1}{2}\varphi^2\left(\pi + 2\tan^{-1}\varphi\right) + \varphi - \tan^{-1}\varphi\right]. \tag{9.9}$$

The self-induced velocity

$$v = \frac{\partial \mathcal{E}}{\partial p} = \frac{\kappa}{4\pi R_i \varphi}\left[\ln\left(\frac{8R_i\varphi}{a}\right) - \frac{1}{2}\right] + \frac{2\varphi}{\left(1+\varphi^2\right)\left(\pi + 2\tan^{-1}\varphi\right) + 2\varphi} \tag{9.10}$$

is equated to the superfluid velocity at the outer rim of the nascent vortex,

$$v_i \left\{1 + \frac{1}{2\left[\left(1+\varphi^2\right)^{1/2} + \varphi\right]^3}\right\}. \tag{9.11}$$

In a typical experiment, N bare ions enter the drift space. A fraction

$$N_i/N = \exp\left(-\nu\tau\right) \tag{9.12}$$

survive for τ seconds without nucleating a ring. If the drift distance is L and the average ion drift velocity is v_i, a fraction

$$N_i/N = \exp\left(-\nu L/v_i\right) \tag{9.13}$$

of the ions arrives at the collector without nucleating rings.

The validity of this steady-state theory has been experimentally tested for time intervals τ down into the microsecond region (Titus and Rosenshein, 1973). The nucleation rate can be written as $\nu = n_r\mathcal{P}$, where n_r is the number of protorings localized near the ion and $\mathcal{P}$ is the probability per unit time of one of them becoming a ring. In the localized roton model, discussed in Section 6.1, the protorings are rotons and n_r is the number of those localized near the ion with the correct polarization with respect to the superfluid velocity.[2]

The critical velocity for nucleation, $v_{i,c}$, is obtained by setting $\nu\tau = 1$. In Fig. 9.4 the temperature dependence of the predicted critical velocity for positive ions is quite favorably compared with experimental results for $T \leqslant 1\,\mathrm{K}$ (Cunsolo and Maraviglia, 1969) and for $1\,\mathrm{K} < T < 1.6\,\mathrm{K}$ (Bruschi *et al.*, 1968*b*). A similar agreement is obtained for negative ions for $T < 1\,\mathrm{K}$ (Cunsolo and Maraviglia, 1969). The discrepancy at higher temperature is due to the fact that, among others, a large field is necessary to maintain the ring velocity of this order owing to the high energy loss at high temperatures. Moreover, as will be discussed later, at high temperatures the positive ion ceases to be permanently attached to the vortex (Bruschi *et al.*, 1968*b*).

In this context, it should be observed that the second branch of rings, observed by Rayfield (1968*a*) and by Cunsolo and Maraviglia (1969) at low tem-

[2]Strictly speaking, the assumption of rotons as protorings is only necessary to make quantitative predictions because the theory does not depend on the true nature of protorings.

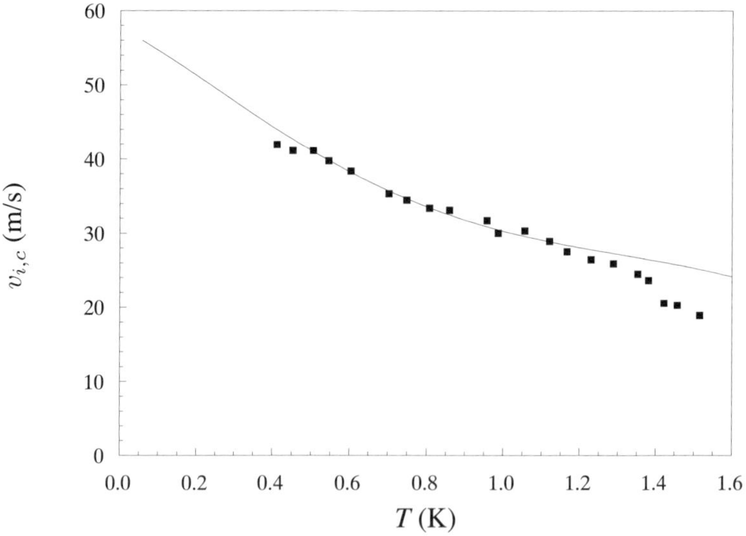

FIG. 9.4. Critical velocity for vortex nucleation $v_{i,c}$ vs T for positive ions. Line: fluctuation model. Points: collection of data of different authors: $T \leqslant 1\,\mathrm{K}$ (Cunsolo and Maraviglia, 1969) and $T > 1\,\mathrm{K}$ (Bruschi *et al.*, 1968*a*).

peratures, is due to the fact that alpha-particle ionization at the source produces fast enough ions with velocity above the critical value that vortex rings are produced directly. These may survive and be injected into the drift space if the field in the source region is large enough to overcome the energy loss.

The quantity N_i/N is also proportional to the bare-ion current detected at the collector. Thus, a direct measure of the current can be compared to the prediction of the model. In Fig. 9.5 the amplitude of the negative ion current as a function of the voltage across the 5.1 cm-long drift space is shown (Strayer and Donnelly, 1971). As soon as the average drift velocity reaches the critical value, N_i/N decreases so rapidly with increasing field beyond a given threshold that the region around this value of the field is called the *lifetime-edge* region (Donnelly and Roberts, 1969*a*).

The solid line is eqn (6.7) and is used to relate the velocity that the bare ions would have at fields above the lifetime edge to the field strength if they were not nucleating rings. By combining the data shown in Fig. 9.5 and the data of Figs 6.6 and 6.7, together with eqn (6.7), N_i/N can be calculated as a function of the ion velocity and compared with the prediction of the model.

This goal is accomplished in Fig. 9.6 for positive and negative ions at vapor pressure. Similar results are also obtained for $T = 0.6$ and $0.75\,\mathrm{K}$ for the positive ions and for $T = 0.6$ and $0.65\,\mathrm{K}$ for the negative ions. For the calculations, the values of the positive and negative ion radii have been chosen to be $R_i^+ = 6.5\,\text{Å}$ and $R_i^- = 16\,\text{Å}$ (Strayer and Donnelly, 1971). Similar results are obtained at the

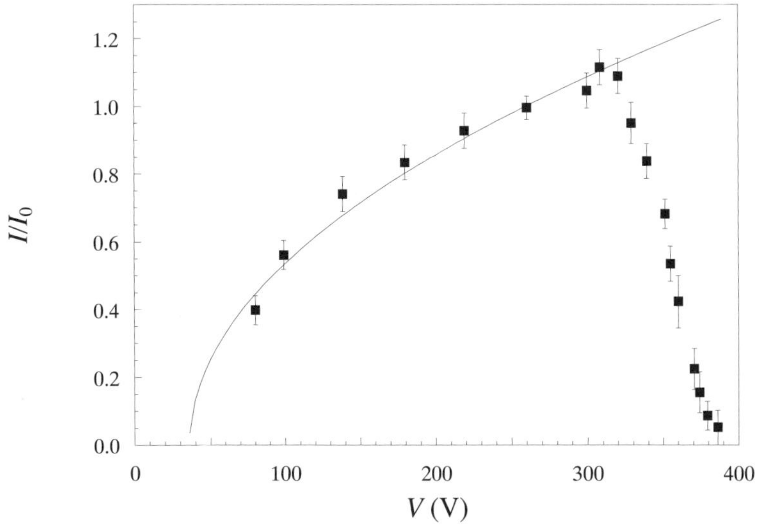

FIG. 9.5. Amplitude of the negative ion current in liquid He at $T = 0.645\,\mathrm{K}$ at SVP. (Strayer and Donnelly, 1971.) The current is normalized at unity for $V = 260\,\mathrm{V}$. Line: eqn (6.7).

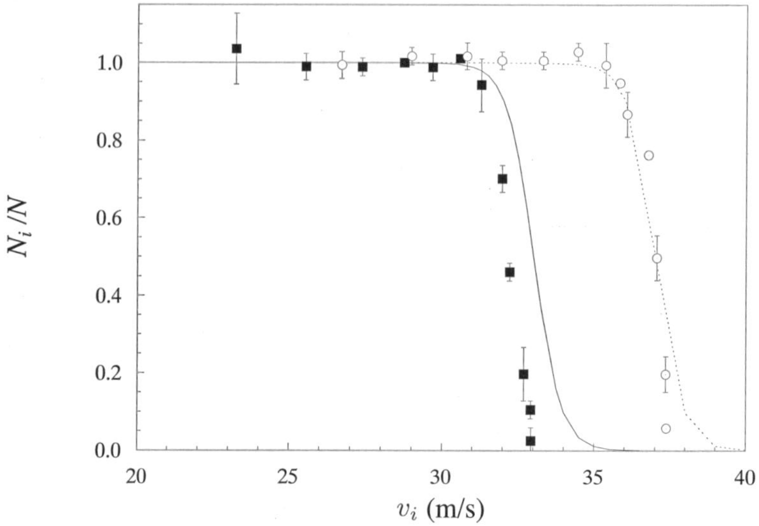

FIG. 9.6. N_i/N vs v_i for positive ions at $T = 0.70\,\mathrm{K}$ (circles) and for negative ions at $T = 0.60\,\mathrm{K}$ (squares). (Donnelly and Roberts, 1971; Strayer and Donnelly, 1971.) Lines: fluctuation model.

other temperatures explored. The agreement between theory and experiment is quite satisfactory and the direct observation of the lifetime edge is claimed to strongly support the stochastic nucleation theory.

However, some authors (Schwarz and Jang, 1973) have cast considerable doubt on the concept of localized rotons, as discussed previously (see Section 6.1). Moreover, the results of the model for dilute ^{3}He–^{4}He mixtures only qualitatively predict the isotopic effect on the observed pressure dependence of the critical velocity of negative ions (Rayfield, 1968*b*).

9.3 The girdling model

The statistical nature of the nucleation process in the fluctuation model, eqn (9.12), does not necessarily show up because the vortex ring creation process is stochastic by itself. It could manifest just because the charge carrier velocity distribution has a spread of $\approx \sqrt{T}$ around the mean velocity. Thus, as the field is increased toward the lifetime-edge region, the rate at which the bare carriers fluctuate toward the critical velocity for vortex nucleation and make the ion–ring transition increases very rapidly (Zoll and Schwarz, 1973). As a consequence, the velocity fluctuation of the bare carriers gives rise to the stochastic features observed in real experiments, independently of the microscopic nature of the nucleation process.

A completely different approach to understand the microscopic nature of vortex ring nucleation has been followed by other authors (Schwarz and Jang, 1973; Blount and Varma, 1976).

In this new *girdling* model the ring is supposed to be created suddenly so that energy and momentum conservation arguments apply. It is assumed that the probability per unit time at which an ion produces a vortex ring is only a function of the ion structure and of its instantaneous microscopic velocity (Zoll, 1976). The conservation laws also imply that the state of the superfluid must change and excited states of it must be taken into account.

The model's goal is to find a condition on the ion velocity, $u = u(M_i, R_i, \xi)$, at which the moving probe, of radius R_i and mass M_i, can generate an excited state ξ of the superfluid. The determination of the critical value u_{cr} of the velocity, at which the ion can first interact with the superfluid, is obtained by examining the allowed excited states of it and finding the minimum of $u(M_i, R_i, \xi) \equiv u_{\mathrm{cr}}$. This is why this model is also sometimes referred to as the *quantum transition model* (Bowley *et al.*, 1982).

The detailed hydrodynamical calculations of the model are very intricate (Schwarz and Jang, 1973). It suffices to say that the lowest-lying state corresponds to a configuration in which a ring just *girdles* the ion at its equator. A very schematic picture of the differences between the peeling and the girdling model is shown in Fig. 9.7. In Fig. 9.3 the prediction of the girdling model for the critical velocity of negative ions as a function of pressure is shown. The agreement is quite good. For the positive ions, the prediction is approximately 40% too large.

The final fate of the just-created ion–ring complex depends on the balance between the friction force due to excitation scattering and that of the applied electric field (Zoll, 1976). Qualitative considerations show that presumably the ring and the carrier may also separate (Padmore, 1972*a*).

The analysis of the experimental results about the behavior of the nucleation rate of charged vortex rings suggests two possible situations (Zoll, 1976). On one hand, if the field at which bare ions produce rings is larger than the drag force on the ion–ring complex, the rings still gain energy from the field and slow down further. They eventually reach the large radius for which the drag force and the electric force are equal. In this case, a conversion of fast, bare ions into very slow ion–ring complexes is observed.

On the other hand, if the drag on the newly-generated ion–ring complexes is larger than the electric force at creation, the charged ring loses energy to the field and becomes smaller with increasingly larger velocity, turning back into a bare ion, and this cycle is repeated continuously. In such a situation, the charge spends a greater fraction of its time in the slow-ring state as the field is increased and the macroscopically-observed drift velocity appears to be a smoothly-decreasing function of the field (Zoll and Schwarz, 1973; Zoll, 1976).

This model has also been criticized, particularly because it implies a large discontinuous change in the wave function of the system at the instant of nucleation (Bowley *et al.*, 1982).

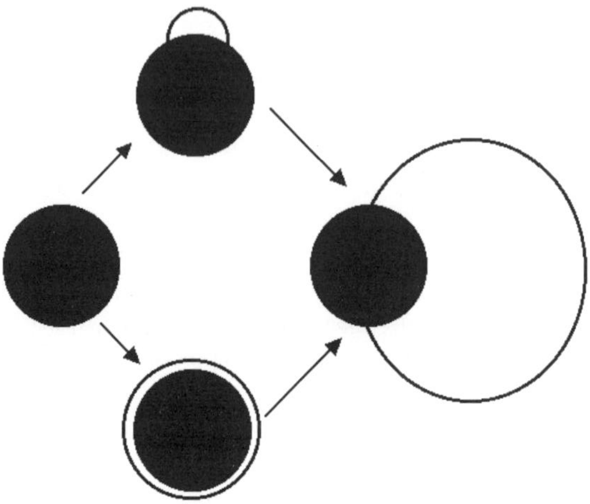

FIG. 9.7. Sketch of two alternative models for the microscopic vortex creation process. The ring may either grow out from a small loop of vortex core attached at both ends to the ion (top) or it may appear spontaneously (bottom), girdling the ion as a consequence of a quantum transition. Finally, in both cases, the ion moves sideways and gets trapped on the vortex core. The ion velocity is assumed to be directed toward the reader and the time sequence is from left to right. (Bowley *et al.*, 1982.)

9.4 Vortex ring nucleation at intermediate electric fields

Very accurate measurements of the nucleation rate of vortex rings are carried out by stepping the drift field across the lifetime edge during the ion flight (Titus and Rosenshein, 1973; Zoll and Schwarz, 1973; Zoll, 1976).

A pulse of bare ions is gated into the drift space (see, for instance, Fig. 5.3). The electric field in the drift space is kept at low enough values so that no rings are produced but it is sufficiently high that any rings that are formed do not decay back to bare ions.

After the pulse is in the drift space, the field is stepped up into the transition region for a time Δt and then it is brought back to the initial value before the carriers completely cross the drift space. The current is then composed of two pulses: the fast one due to the residual bare ions, and the slower one due to the rings that have been created. As expected for random ring creation, the bare-ion current amplitude decreases as $I = I_0 \exp(-\Delta t/\tau)$. Here $\nu \equiv \tau^{-1}$ is interpreted as the probability per unit time that an ion goes into a ring-coupled state (Zoll, 1976); ν is also known as the *vortex ring nucleation rate.*

9.4.1 *Pure ^{4}He at vapor pressure*

In Fig. 9.8 the drift velocity and nucleation rate for positive ions are shown as a function of the applied electric field (Zoll, 1976). Qualitatively similar results are obtained at all temperatures in the range $0.4\,\mathrm{K} < T < 0.7\,\mathrm{K}$ for both types of ion (except for the different behavior of v_D for negative ions as a function of the field, as previously discussed) (Zoll, 1976).

The critical field E_c, defined as the field strength at which the nucleation rate is $\nu(E_c) = 10^4\,\mathrm{s}^{-1}$, increases sharply with increasing temperature, as shown in Fig. 9.9.

The interesting result is that the nucleation rate curves, once appropriately normalized, appear basically the same for both types of charge carrier at all temperatures. For $E/E_c \approx 0.9$, the nucleation rate is already high ($\nu \approx 10^3\,\mathrm{s}^{-1}$) and the current due to bare ions becomes difficult to detect. Experimentally, this situation corresponds to the discontinuous drift velocity transition shown in Figs 6.4 and 6.5.

Within the girdling model, the small decrease in the average drift velocity at the discontinuity shown in Fig. 9.4 is probably due to the larger ion velocity fluctuations as the temperature is increased (Zoll, 1976).

9.4.2 *Influence of ^{3}He impurities*

As observed by Rayfield (1967, 1968*a*), the addition of a fairly large amount of ^{3}He impurities at low T makes the bare ion–vortex ring transition continuous

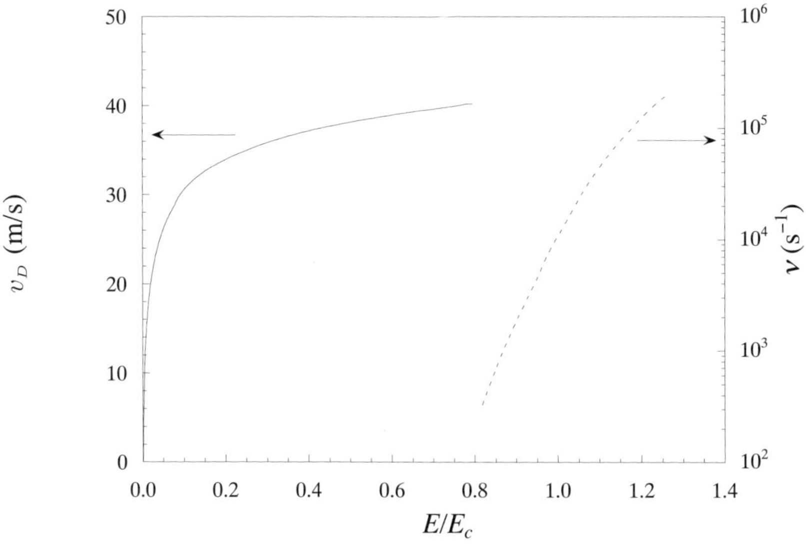

FIG. 9.8. v_D (left scale) and ν (right scale) vs E/E_c for positive ions in pure ^{4}He at SVP for $T = 0.395\,\mathrm{K}$. (Zoll, 1976.) Abscissae are normalized to the field E_c, at which $\nu(E_c) = 10^4\,\mathrm{s}^{-1}$.

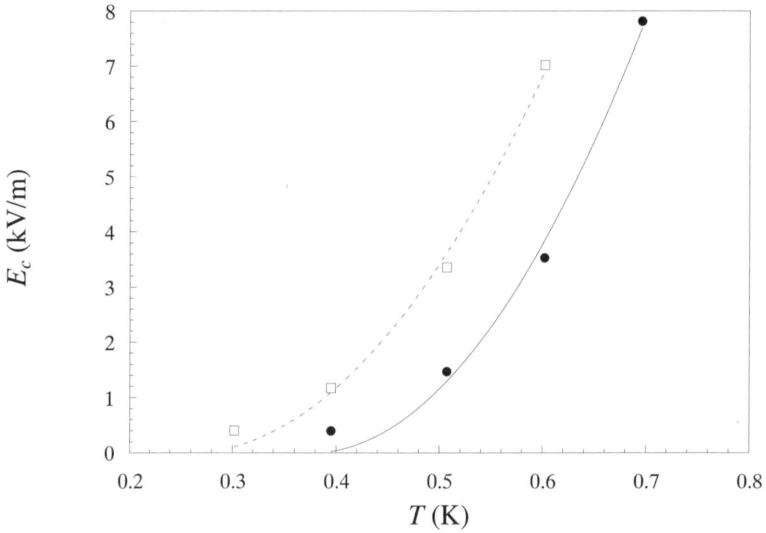

FIG. 9.9. E_c vs T. (Zoll. 1976.) Circles: negative ions, and squares: positive ions. The curves are only eyeguides.

(Fig. 9.2). However, the shape of the drift velocity-field curves depends on the concentration x_3 of impurities. This behavior is shown in Fig. 9.10 for positive ions at low temperature.

By increasing the impurity concentration, the drag on the vortex ring is increased significantly and the drag curves shift to larger fields. The stability field E_m, below which stable rings do not exist, also increases. For low x_3 values, $E_m < E_c$ and the bare ion–vortex ring transition is discontinuous. For larger x_3, $E_m > E_c$ and the transition becomes continuous (Zoll and Schwarz, 1973). Also, the field E_c increases as the concentration of ^{3}He atoms is increased (Zoll and Schwarz, 1973; Zoll, 1976), as is shown in Fig. 9.11, but less rapidly than E_m.

In the case of negative ions, the addition of isotopic impurities is not so effective as for positive ions, because the shift of the stability field E_m is such to keep it below E_c at all concentrations. Therefore, the velocity curve of negative ions is always discontinuous (Zoll, 1976).

This heuristic approach of Zoll (1976) can explain the spectacular second discontinuity observed for positive ions in dilute ^{3}He–^{4}He solutions at very low temperature (Kuchnir *et al.*, 1971, 1972). In Fig. 9.12 the experimental data of drift velocity as a function of the electric field strength for a ^{3}He concentration of 106 ppm at several temperatures are shown.

At high $T = 550$ mK, the giant discontinuity is very evident. As T is decreased, the discontinuity disappears, at first. By continuining to lower T, a bump in the curve develops at intermediate fields and, eventually, at still lower temperatures, the discontinuity appears again at a much lower field strength

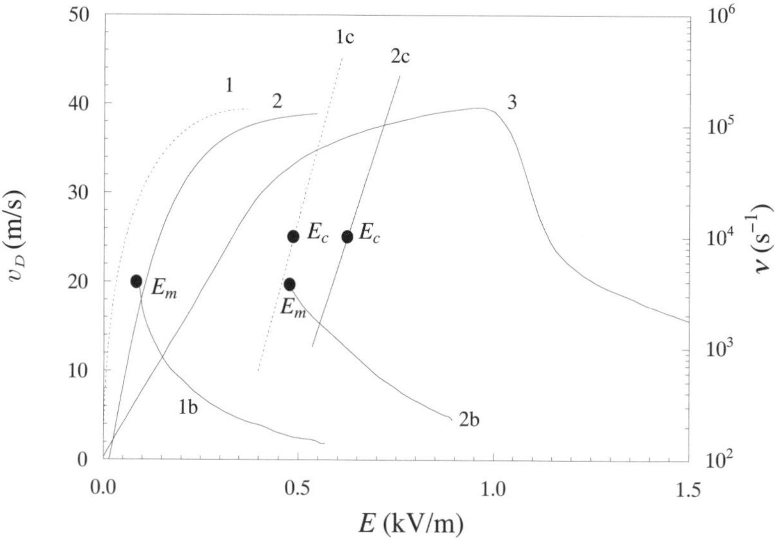

FIG. 9.10. Electric field dependence of the vortex ring transition for positive ions at $T = 0.395$ K for different ^{3}He concentrations (x_3). Curves 1, 1b, and 1c refer to $x_3 = 6.9$ ppm. Curves 2, 2b, and 2c refer to $x_3 = 29$ ppm. The v_D curve 3 (left scale) refers to $x_3 = 230$ ppm and shows the continuous transition. Bare-ion drift velocities: curves 1 and 2 (left scale). Vortex ring velocities: curves 1b and 2b (left scale). Nucleation rates: curves 1c and 2c (right scale). E_m is the vortex ring stability field. E_c is the critical field. (Zoll and Schwarz, 1973.)

(Kuchnir *et al.*, 1971, 1972). Similar results are also obtained at higher concentrations, as shown in Fig. 9.13 for a concentration $x_3 = 170$ ppm (Kuchnir *et al.*, 1971, 1972) and in Fig. 9.14 for $x_3 = 502$ ppm (Kuchnir *et al.*, 1972).

The thermal wavelength of ^{3}He atoms increases as T is decreased. This fact gradually changes the total drag curve for positive ions at high concentrations in such a way that the stability field moves far away above E_c. When the field is approximately equal to E_m, the second discontinuity appears.

Negative ions, in contrast, do always show the discontinuity. The addition of ^{3}He atoms, however, decreases the maximum bare-ion velocity at the discontinuity. This fact is considered a vindication of the girdling model (Zoll, 1976), according to which the observed statistical transition rates reflect the fluctuations of the ion velocities to values above some fundamental critical velocity u_{cr}. Actually, the addition of more isotopic impurities changes the number density n_3 of the excitation gas without affecting its momentum distribution. Thus, the distribution function of the ion velocities relative to a given average velocity is independent of n_3, whereas the field necessary to produce that average velocity depends on it. Thus, one expects ν/n_3 to be a universal function of E_c, as is indeed shown in Fig. 9.15.

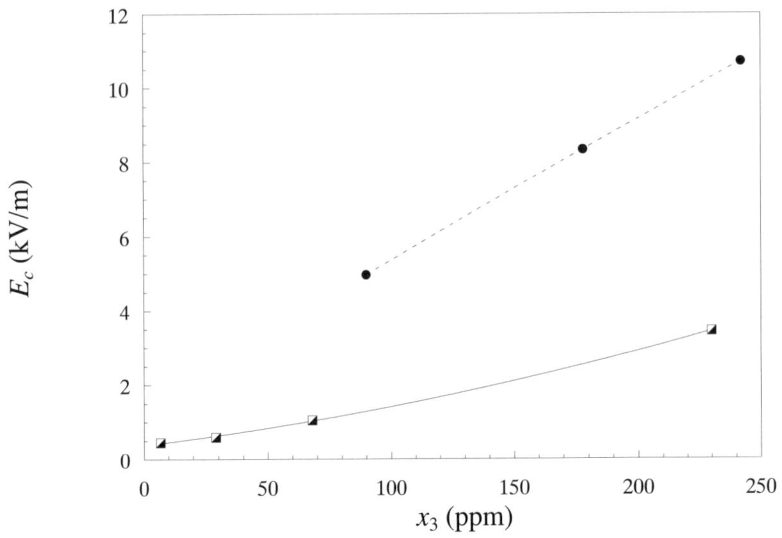

FIG. 9.11. E_c vs x_3 for $T = 0.395\,\mathrm{K}$ for positive (squares) and negative (circles) ions. (Zoll and Schwarz, 1973.)

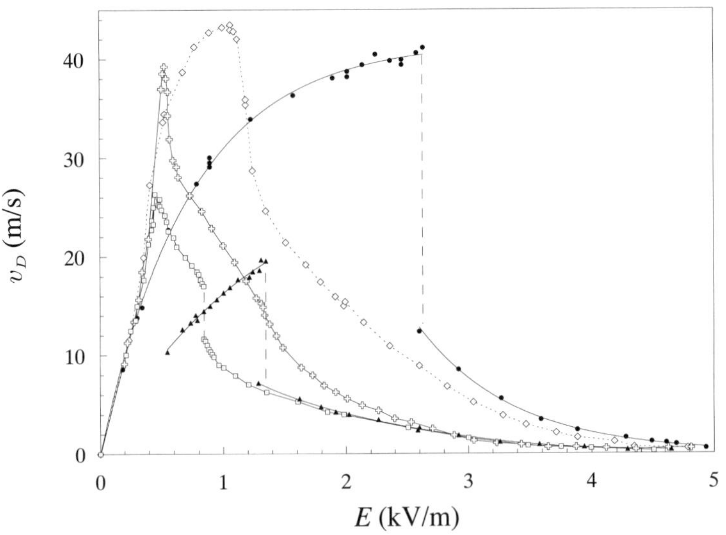

FIG. 9.12. v_D vs E for positive ions in a 106 ppm ^{3}He–^{4}He solution. T (mK) = 550 (circles), 370 (diamonds), 85 (crosses), and 19 (squares). Triangles: negative ions at $T = 18\,\mathrm{mK}$. (Kuchnir *et al.*, 1972.)

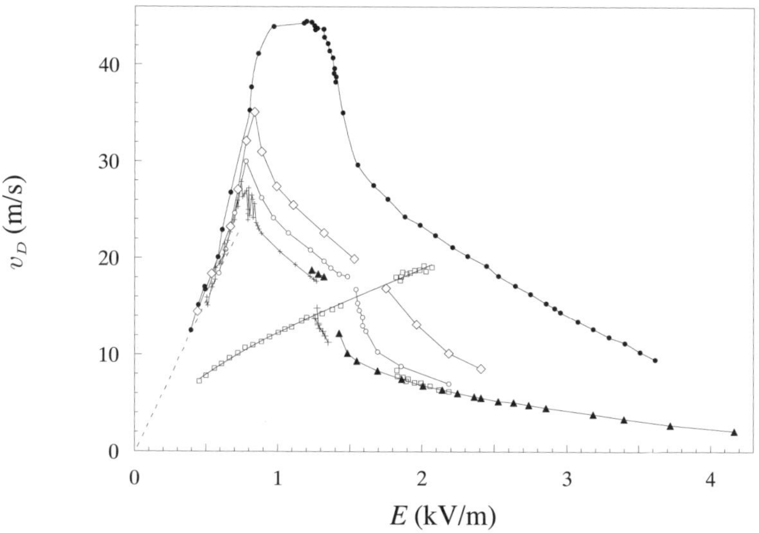

FIG. 9.13. v_D vs E for positive ions in a 170 ppm ^{3}He–^{4}He solution. T (mK) = 286 (closed circles), 76.8 (diamonds), 42.7 (open circles), 28.6 (triangles), and 18.5 (crosses). Squares: negative ions at $T = 20$ mK. (Kuchnir *et al.*, 1971.)

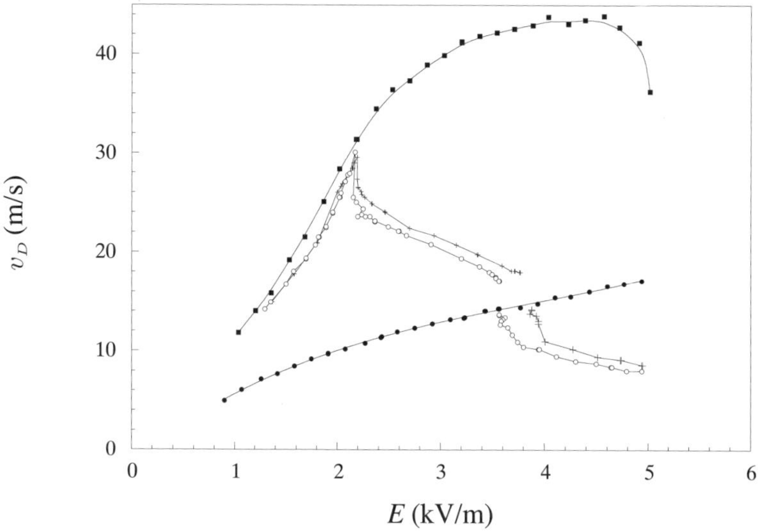

FIG. 9.14. v_D vs E for positive ions in a 502 ppm ^{3}He–^{4}He solution. T (mK) = 300 (squares), 28 (crosses), and 18 (open circles). Closed circles: negative ions at $T = 20$ mK. (Kuchnir *et al.*, 1972.)

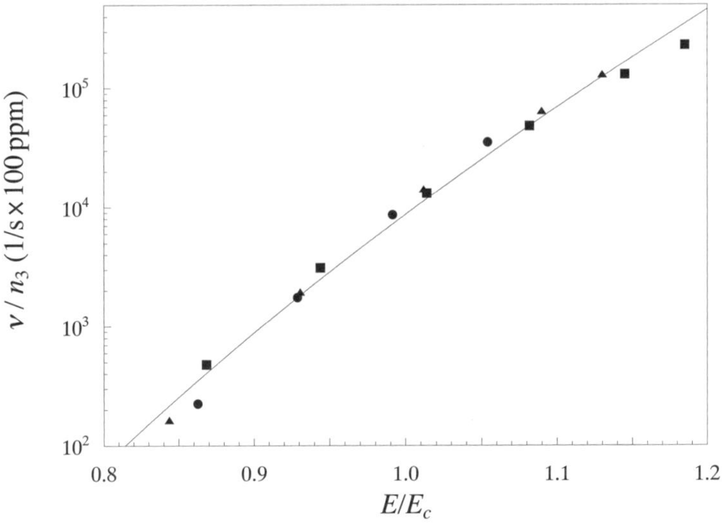

FIG. 9.15. ν/n_3 vs E/E_c for negative ions at $T = 0.395\,\mathrm{K}$. (Zoll, 1976.) x_3 (ppm) = 90 (circles), 178 (squares), and 242 (diamonds).

9.4.3 *Influence of pressure*

Pressure applied to the liquid influences the structure of the ions in a known way and changes in the vortex nucleation rate are expected. Because the predictions of the girdling model depend on the ion structure, pressure measurements should help in falsifying the theory. Moreover, in the case of negative ions, the Landau critical velocity for roton emission drops below the critical velocity for vortex ring nucleation at pressures in excess of $\approx 1.2\,\mathrm{MPa}$. Thus, pressure measurements clarify the issue if roton emission and vortex ring nucleation are competing or mutually-excluding processes.

Measurements at low $T = 0.395\,\mathrm{K}$, show that the nucleation rate for positive ions is practically unaffected by P up to $2.5\,\mathrm{MPa}$ (Zoll, 1976) and the drift velocity and nucleation curves are similar to those shown in Fig. 9.8, except that, with increasing pressure, the maximum positive ion velocity decreases slightly. The positive bare ion–vortex ring transition is thus of discontinuous type.

In contrast, the nucleation rate for the negative ion depends strongly on pressure, as shown in Fig. 9.16 for $P \leqslant 1.0\,\mathrm{MPa}$. The increase in the limiting velocity as P increases is the same as shown in Fig. 9.3. The striking feature is the reduction of the nucleation rate with increasing P. The nucleation curves greatly flatten out. Eventually, for $P > 1\,\mathrm{MPa}$ the nucleation rate hardly reaches a value in excess of $\nu = 10^4\,\mathrm{s}^{-1}$. By increasing the temperature from $T = 0.332\,\mathrm{K}$ up to $T = 0.507\,\mathrm{K}$, the flattening of the nucleation rate is even more pronounced and ν is always less than $10^4\,\mathrm{s}^{-1}$.

The conclusion is drawn that vortex rings are also generated at high pressures

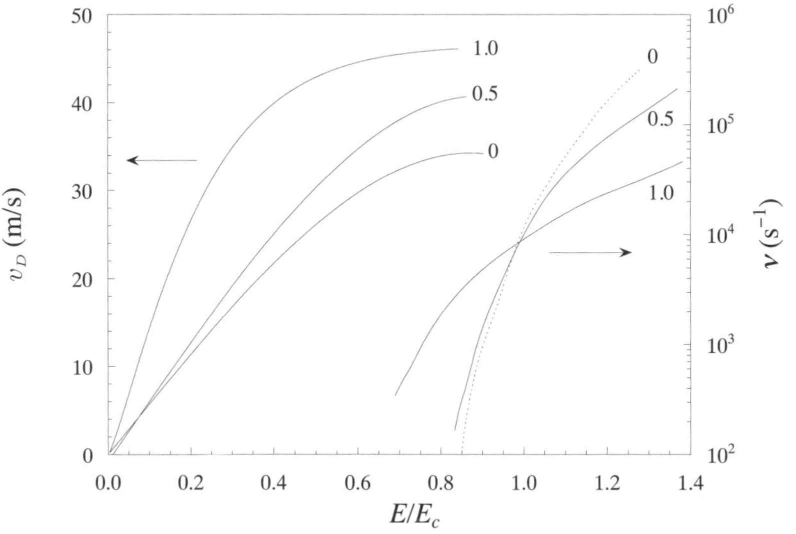

FIG. 9.16. v_D (left scale) and ν (right scale) vs E/E_c for negative ions at $T = 0.395$ K. (Zoll, 1976.) The numbers near each curve are the values of P in MPa.

but in a quantity that, upon increasing P, may fall below the detection limit of experiments. Inspection of Fig. 9.3, in which the prediction of the critical velocity for vortex nucleation is compared with the experimental data and with the Landau critical velocity for roton emission, therefore supports the idea that vortex rings are generated or rotons are emitted as soon as an individual ion reaches the appropriate critical velocity determined by energy and momentum conservation for either process (Zoll, 1976).

The peculiar behavior of ν as P is increased can be explained by assuming that ring creation or roton emission are competing processes. If the critical velocity for nucleation is much less than the Landau critical velocity for roton emission, $u_{\mathrm{cr}} \ll v_L$, then rings are readily created. But, as soon as v_L becomes smaller than u_{cr}, ions fluctuating toward the critical velocity quickly lose energy by roton emission and their velocity distribution is limited at the upper end by v_L. As most of the ion energy loss is accounted for by roton emission, an increase in E increases the rate of roton emission rather than modifying the velocity distribution. Therefore, v_D and ν become rather insensitive to increases of field for velocities close to v_L (Zoll, 1976).

9.5 Vortex ring nucleation by negative ions at high P and E

Interesting features of the vortex ring nucleation process are discovered by investigating the behavior of negative ions at high pressure for strong electric fields in isotopically-pure ^{4}He. Under such conditions, the negative ions move at speeds near to the Landau critical velocity for roton emission (Phillips and McClintock, 1974*b*). As discussed previously, the nucleation rate of vortex rings is much smaller at high pressure than at low pressure, but it is still large enough for there to be a significant probability that rings are created when ions transit across the short distance separating the source from the collector.

A large volume of the multi-dimensional parameter space in terms of T, P, E, and ^{3}He concentration x_3 has been investigated thoroughly by the Lancaster group. For a quick review see McClintock (1999).

The technique devised by this group to investigate the nucleation process is based on electric induction. A scheme of the electrode structure is shown in Fig. 9.17. Electrons are injected by field ionization into the liquid by the point source S. Electron bubbles are readily formed in front of S and are gated into the drift space as a thin charge disk by a voltage pulse applied to grids G1 and G2. Electrodes H1 through H3 ensure field uniformity in the drift space. The Frisch grid FG shields the collector C from the moving charge. The signal is induced in the collector while the charge disk is traveling between FG and C under the action of a strong electric field.

The shape of the charge signal induced at the collector depends in a very complicated way on the cell geometry, on the size of the disk, on the amount of the drifting charge, and on the electric field. Details about the signal shape are found in the literature (Bowley *et al.*, 1982).

Under ideal conditions (i.e., injection pulse much shorter than drift time, cell large in comparison with the transverse size of the charge disk, and negligible space charge (Borghesani *et al.*, 1986*b*; Borghesani and Santini, 1990*b*)), if the ion disk of total charge Q is moving at constant speed v_D, the induced current, in the absence of vortex nucleation, is $i_c = Qv_D/L$, where L is the drift distance between FG and G2 in Fig. 9.17, if the collector is kept at a constant potential. If vortex rings are nucleated, they very quickly expand and slow down, practically

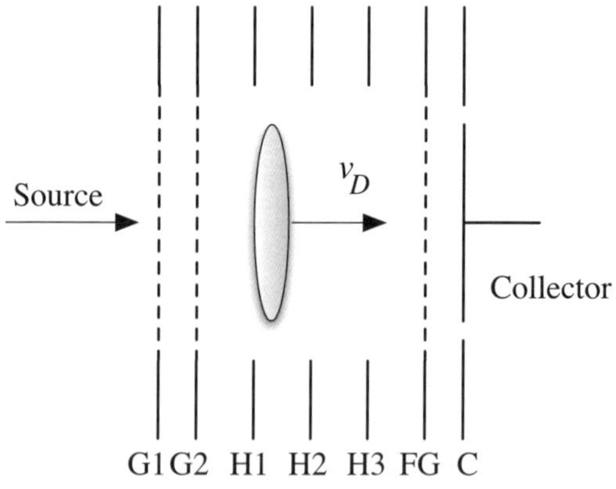

FIG. 9.17. Electrode structure of the electric induction cell for nucleation rate measurements. (Phillips and McClintock, 1974*a*.)

giving no contribution at all to the measured current, whose time evolution is thus given by

$$I = \frac{Qv_D}{L} \exp(-\nu t), \tag{9.14}$$

where ν is the nucleation rate. It must be stressed that the technique is limited to rather high fields, for which the newly-formed rings grow very rapidly and slow down quickly, thus no longer contributing to the observed current. This condition limits the minimum field to $\approx 50\,\mathrm{kV/m}$ (Bowley *et al.*, 1982).

The magnitude of the field-emission current in the liquid is controlled by the formation of spacecharge that, for $T < 1\,\mathrm{K}$ and low P, appears to take the form of a dense tangle of charged vortex lines (McClintock, 1973*c*).

The maximum amplitude I_m of the induction signal depends in an interesting way on P and T, as shown in Figs 9.18 and 9.19. The decrease in the maximum signal amplitude at constant P with increasing T is mainly due to the increased rate of temperature-dependent vortex nucleation in the drift space before ions reach the Frisch grid and are detected. On the other hand, it increases with increasing P because the nucleation rate of vortex rings decreases.

Whereas in earlier experiments the control of the ion speed was achieved by means of balancing the electric force on the charges against the drag force arising from excitations scattering at low T in the range K $0.3 < T < 1\,\mathrm{K}$, in the Lancaster experiments ions are drifted in pressurized liquid helium above $1.1\,\mathrm{MPa}$ so that they can move close to the Landau critical velocity v_L. By so

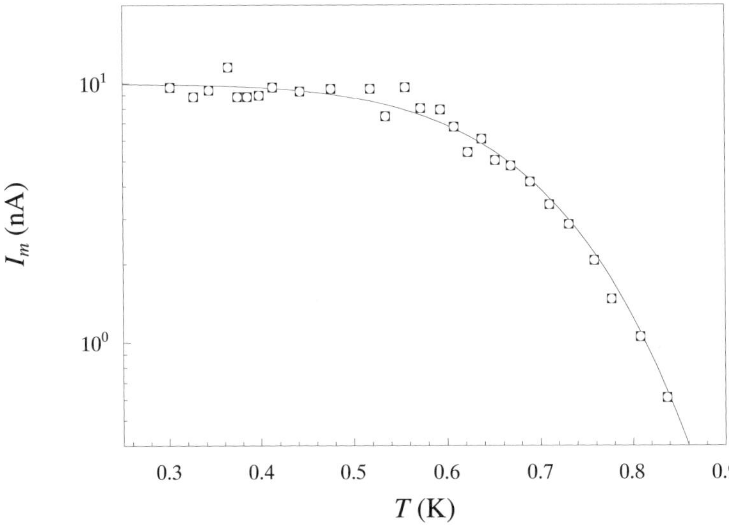

FIG. 9.18. I_m vs T for $P = 2.3\,\mathrm{MPa}$ with emitter potential $V_t = 1.2\,\mathrm{kV}$ and field $E = 0.45\,\mathrm{MV/m}$. (Bowley *et al.*,1982.) The solid line is only an eyeguide.

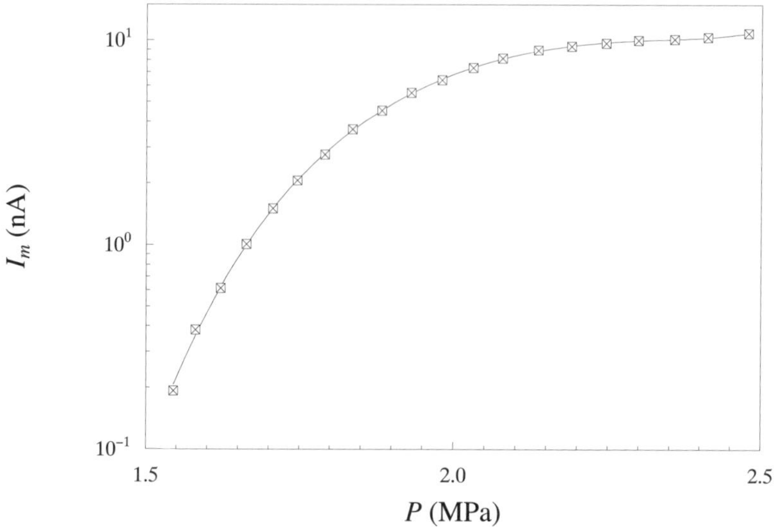

FIG. 9.19. I_m vs P for $T = 0.32$ K with emitter potential $V_t = 1.2$ kV and field $E = 0.43$ MV/m. (Bowley *et al.*, 1982.) The solid line is an eyeguide.

doing, the speed of the ion is controlled by balancing the electric force against the rate of momentum loss caused by roton emission above v_L.

This method can be used because experiments have shown that vortex nucleation and roton emission are independent processes and that negative ions above 1.1 MPa can reach velocities near v_L without immediately nucleating and being trapped on vortex rings (Meyer and Reif, 1961; Rayfield, 1966, 1968*a*).

In Fig. 9.20 the drag on negative ions moving through superfluid He at $T = 0.35$ K and $P = 2.5$ MPa is displayed. Under pressure, the drag remains quite small until the Landau critical velocity is reached. Now, the emission of rotons so dramatically increases the drag that the ion speed can be accurately controlled by adjusting the drifting field (Allum *et al.*, 1976*b*). For comparison purposes, the drag on a negative ion in the normal liquid at $T = 4.0$ K is also shown.

In a typical field-emission experiment, the current increases rapidly with P for $P \geqslant 1.0$ MPa and for $T \leqslant 0.7$ K, as shown in Fig. 9.21, because the fraction of bare ions in the cell increases. The fraction α of ions reaching the collector without nucleating vortex rings is given, as a function of P, by $\alpha(P) \approx 1 - I(P)/I(1)$, where $I(1)$ is the current at $P = 0.1$ MPa. Thus, the data show that, at $T = 0.7$ K and $P = 2.5$ MPa, more than 90% of emitted ions reach the collector without nucleating vortex rings (Phillips and McClintock, 1973).

A detailed analysis of the data in the figure allows the determination of the vortex ring nucleation rate. It is assumed that ions, upon emerging from the field-ionization region around the source, travel at the Landau velocity for roton emission v_L because of the large field strength. They negotiate a distance that

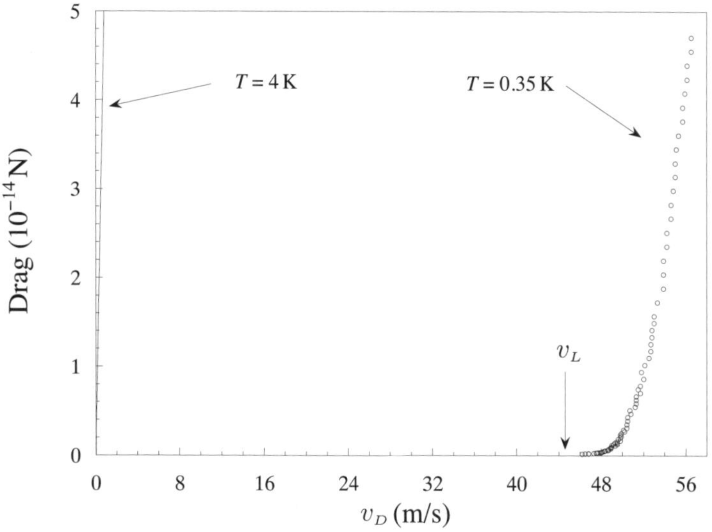

FIG. 9.20. Drag on negative ions vs v_D above v_L in He II for $T = 0.35\,\mathrm{K}$ at $P = 2.5\,\mathrm{MPa}$. (Allum *et al.*, (1976*b*).) Solid line: drag at $T = 4.0\,\mathrm{K}$.

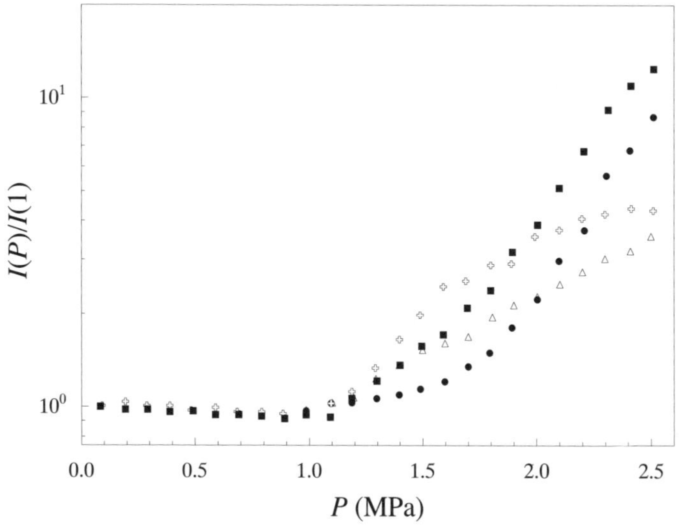

FIG. 9.21. Field-emission current I vs P in He II for different temperatures and emission voltages. [T (K), V (kV)] = [0.7, 3] (squares), [0.7, 24] (circles), [0.3, 3] (crosses), and [0.3, 2] (triangles). (Phillips and McClintock, 1973.) The data are normalized to unity at $P = 0\,\mathrm{MPa}$.

depends on ν and then nucleate vortex rings that rapidly grow and tangle with the vorticity left by previous ions (Phillips and McClintock, 1974*b*).

The average ionic transit time $\bar{\tau}$ between the emitter (considered as a point source) and the anode of radius R_a is calculated under the assumption that after a time t, at a distance $r = v_L t$, only a fraction $\exp(-\nu t)$ of ions is still untrapped. The sum of all their transit times consists of three contributions from the following: times spent as bare ions before nucleation and trapping, times spent trapped on vortex rings, and the transit time $t_f = R_a/v_L$ of those ions that are not trapped at all.

One thus obtains

$$\bar{\tau} = \int_0^{t_f} \nu t e^{-\nu t} \mathrm{d}t + \int_0^{t_f} \left(\frac{2\nu}{A\mu_v}\right) \left[R_a^{3/2} - (v_L t)^{3/2}\right] e^{-\nu t} \mathrm{d}t + t_f e^{-\nu t_f} , \qquad (9.15)$$

where $\mu_v \approx 8 \times 10^{-7}\,\mathrm{m^2/V\,s}$ is the effective vortex-limited mobility in the space-charge-dominated region (Gavin and McClintock, 1973; Phillips and McClintock, 1973). Here A is the constant relating the electric field strength E to the radial distance r from the emitter: $E = Ar^{-1/2}$. It depends on the ionic mobility and on the emitter potential (Halpern and Gomer, 1969*a*; Phillips and McClintock, 1974*b*).

By carrying out the integrations and defining $t_v = 2R_a^{3/2}/3A\mu_v$ as the transit time for an ion nucleating a vortex ring at the source, one gets

$$\bar{\tau} = \left(t_v + \nu^{-1}\right)\left[1 - e^{-\nu t_f}\right] - \left[\frac{t_v}{(\nu t_f)^{3/2}}\right] \mathcal{I}(\nu t_f) , \qquad (9.16)$$

where

$$\mathcal{I} = \int_0^{t_f} (\nu t)^{3/2} e^{-\nu t} \mathrm{d}t . \qquad (9.17)$$

As $\nu^{-1} \ll t_v$, eqn (9.16) can be approximated as

$$\bar{\tau} = t_v \left[1 - e^{-\nu t_f} - \frac{\mathcal{I}(\nu t_f)}{(\nu t_f)^{3/2}}\right] , \qquad (9.18)$$

where $\mathcal{I}(x) = \Gamma(5/2, x)$ is the incomplete Euler Γ function (Arfken, 1985). Under spacecharge conditions, the measured current is $I \propto \bar{\tau}$ and one finally gets

$$\varphi = \frac{I}{I_v} = \left[1 - e^{-\nu t_f} - \frac{\Gamma(5/2, \nu t_f)}{(\nu t_f)^{3/2}}\right] . \qquad (9.19)$$

Here I is the measured emission current and I_v is the vortex-borne component, obtained from the data for $P < 1\,\mathrm{MPa}$ (see Fig. 9.21).

From eqn (9.19) the nucleation rate ν can be obtained. In Fig. 9.22 the nucleation rate is shown as a function of P for several temperatures. These data confirm and extend to higher electric fields the results of Zoll (1976), according to which the nucleation rate of vortex rings due to negative ions decreases with increasing pressure.

The stochastic nature of the transition, as intuitively elucidated by Zoll and Schwarz (1973) and Zoll (1976), is such that the average ion velocity increases with increasing electric field strength E, thus making it easier for an ion to reach and overcome the critical velocity for nucleation.

In order to calculate the nucleation rate as a function of E, it is necessary to determine the distribution function of the ionic velocities, $f(v, E)$. This can be done by solving the Boltzmann equation for the steady-state distribution when roton-emission processes dominate the ionic motion (Bowley, 1976d):

$$\frac{\partial f(v,E)}{\partial t} + \frac{\partial f(v,E)}{\partial v}\frac{eE}{M_i} - R_2(v)\, f(v,E) + R_2(u)\, f(u,E)\frac{\mathrm{d}u}{\mathrm{d}v} - R(v)\, f(v,E) = 0\,, \tag{9.20}$$

where R_2 is the rate of roton emission for an ion traveling with speed v, $u - v$ is the average decrease in velocity of an ion traveling at velocity v after roton emission, and $R(v)$ is the vortex ring nucleation rate.

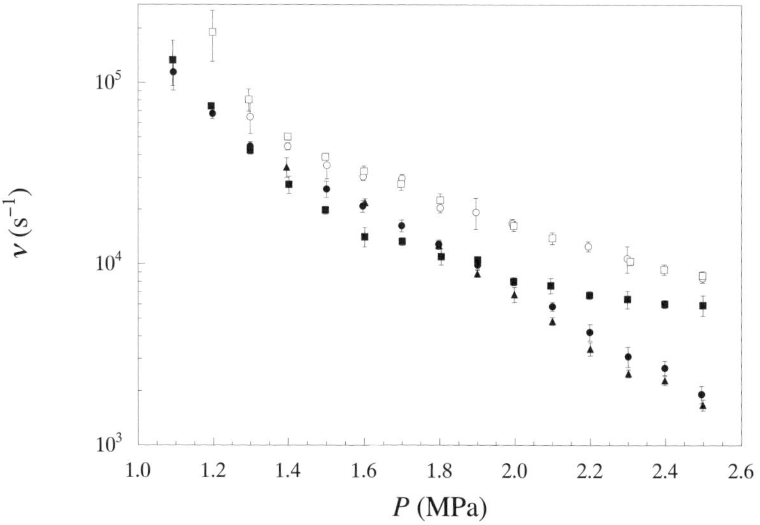

FIG. 9.22. ν vs P for negative ions in He II for several T and emitter potentials V. (Phillips and McClintock, 1974b.) [T (K), V (kV)] = [0.3, 3] (closed squares), [0.3, 2] (open circles), [0.35, 2] (dotted squares), [0.6, 3] (closed circles), and [0.71, 3] (triangles).

In the steady state, $f(v, E)$ decreases exponentially with time because bare ions are lost to ions bound to vortex rings (Titus and Rosenshein, 1973; Zoll and Schwarz, 1973; Zoll, 1976). Thus, the distribution function can be written as $f(v, E) = f_0(v, E)\exp(-\nu t)$. Moreover, the experimentally-observed values of $R_2 \approx 10^{11} - 10^{13}\,\mathrm{s}^{-1} \gg R \approx 10^5 - 10^6\,\mathrm{s}^{-1}$ are such that eqn (9.20) can be solved by letting $R = 0$ and the equilibrium distribution function f_0 is obtained. In other words, it is assumed that the shape of the distribution function is determined only by the process of roton emission and it is not affected by vortex nucleation.

The nucleation frequency is then given by

$$\nu = \int_0^\infty R(v)\, f_0(v, E)\,\mathrm{d}v\,. \tag{9.21}$$

Assuming that the nucleation rate has the analytical form $R(v) = Ru_{-1}(v - v_{\mathrm{cr}})$, where u_{-1} is the unit step function, i.e., the nucleation rate is zero below the critical velocity for vortex nucleation (see Fig. 9.3) and takes on the constant value R for $v \geqslant v_{\mathrm{cr}}$, then the Boltzmann equation (9.20) can be solved analytically for small fields and the nucleation frequency can be calculated, yielding

$$\nu = \frac{eER}{2\hbar k_0 \alpha\,(v_{\mathrm{cr}} - v')^2}\exp\left[-\left(\frac{M_i\alpha\,(v_{\mathrm{cr}} - v')^3}{3eE}\right)\right], \tag{9.22}$$

where M_i is the ion mass. Here α and v' are two constants describing the roton-emission rate $R_2(v) = \alpha(v - v')^2$, where $\alpha \approx 4.4 \times 10^8\,\mathrm{s/m^2}$ and $v' \approx v_L + \hbar k_0/M_i \approx 50.2\,\mathrm{m/s}$ is the threshold velocity for the emission of two rotons (Bowley and Sheard, 1977). For higher fields, the Boltzmann equation must be solved numerically.

In Fig. 9.23 the dependence of ν/R on E^{-1} is shown for several values of the ratio v_{cr}/v_L at $T = 0.85\,\mathrm{K}$, a temperature at which it is unlikely that ^{3}He atoms condense on the surface of the electron bubble (Allum and McClintock, 1976*a*).

The data are consistent with the theoretical prediction and from them the critical velocity v_{cr} can be determined. As the data lie between the curves corresponding to $v_{\mathrm{cr}}/v_L = 1.28$ and 1.30, the critical velocity for nucleation is estimated to be $v_{\mathrm{cr}} = (58.9 \pm 0.5)\,\mathrm{m/s}$, if $v_L = 45.6\,\mathrm{m/s}$. A Lagrangian approach (Bowley, 1984) predicts the critical velocities as a function of P at high pressures and compares quite favorably with experiment (Bowley *et al.*, 1982).

As already noted, pressure strongly influences the nucleation rate of vortex rings. The most accurate experiments, carried out with the electrostatic induction technique, show that the nucleation rate decreases dramatically with P at low T, as shown in Fig. 9.24.

The nucleation rate also depends strongly on the electric field, under which the bare ions are drifted above the Landau critical velocity for roton emission. All $\nu(E)$ curves increase with E, but most of them exhibit a maximum at large E that depends on P. This behavior is shown in Fig. 9.25. The same behavior is observed for $P > 1.5\,\mathrm{MPa}$ (Bowley *et al.*, 1982).

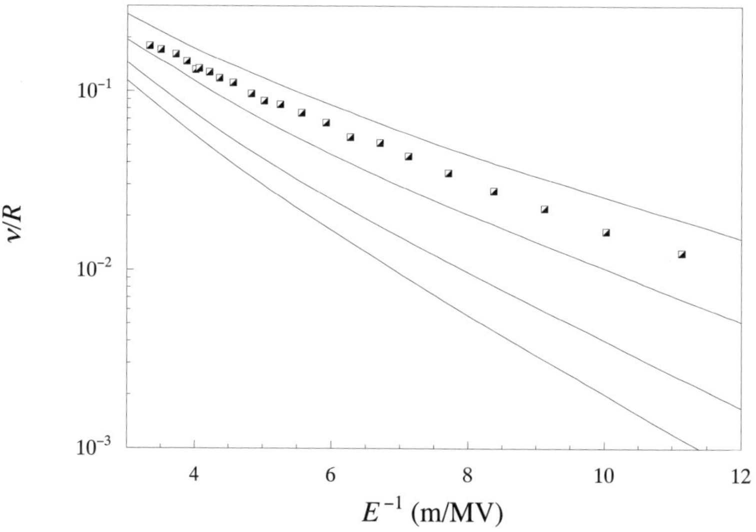

FIG. 9.23. ν/R vs E^{-1} for $P = 2.5\,\text{MPa}$ and $T = 0.85\,\text{K}$ for different values of v_{cr}. The curves correspond to $v_{\text{cr}}/v_L = 1.28$, 1.30, 1.32, and 1.34 (from top) with $R = 9.3 \times 10^4\,\text{s}^{-1}$. Experimental data: $T = 0.85\,\text{K}$. (Bowley, 1976*d*.)

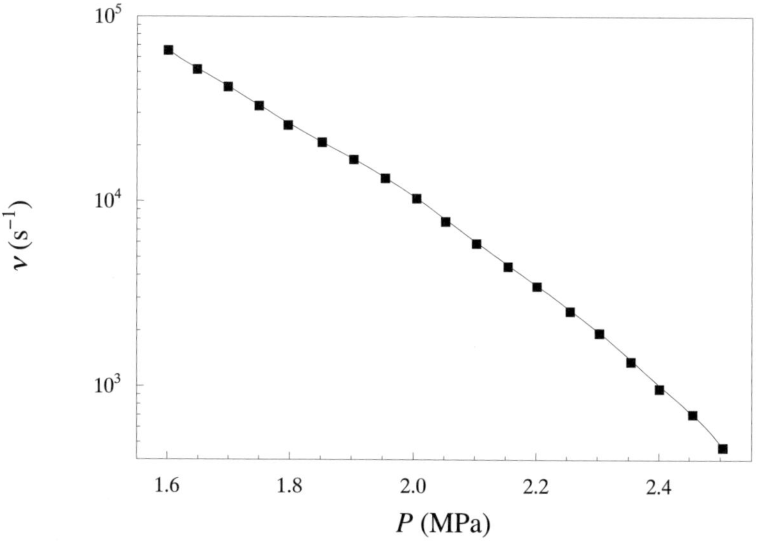

FIG. 9.24. ν vs P for negative ions in He II for $T = 0.322\,\text{K}$ and $E = 0.43\,\text{MV/m}$. (Bowley *et al.*, 1982.)

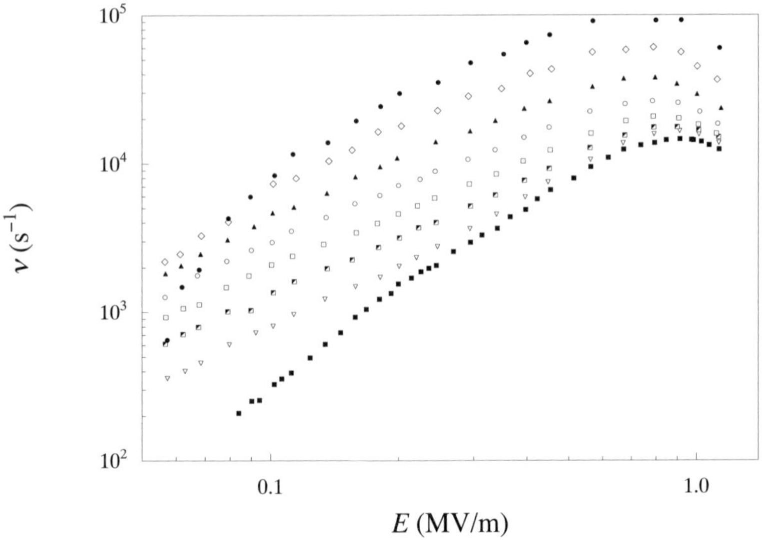

FIG. 9.25. ν vs E for negative ions at $P = 2.1\,\mathrm{MPa}$. (Bowley *et al.*, 1982.) $T\,(\mathrm{K}) =$ 0.784, 0.734, 0.692, 0.655, 0.623, 0.595, 0.555, and 0.324 (from top).

If a specific analytical form for the rate R is assumed, namely, $R = R_1 u_{-1}(v - v_1) + R_2 u_{-1}(v - v_2) - (R_1 + R_2) u_{-1}(v_3 - v)$, with R_1, v_1, R_2, v_2, and v_3 fitting parameters, and if the distribution function derived by Bowley and Sheard (1977) is used, then the general features of the electric field dependence can be reproduced (McClintock *et al.*, 1980), as shown in Fig. 9.26.

At still higher fields, the nucleation rate shows a great decline with increasing E (McClintock *et al.*, 1980; Nancolas and McClintock, 1982). This inhibition of vortex nucleation by strong electric fields has been attributed to the fact that the ion may evade capture by the nascent vortex ring at the instant of creation (McClintock *et al.*, 1980; Nancolas and McClintock, 1982). This is an instability of the initial ion–ring complex that quenches the ion–ring transition.

The quenching process fits easily into the quantum transition model. The initial configuration of the nascent ring girdling the ion is intrinsically unstable. As soon as the ring grows the ion moves sideways and eventually becomes trapped on the core. When an electric field is applied, the trapping potential is tilted down in the direction of the field. For strong enough fields, the ion may escape over the reduced barrier, is accelerated away from the generated ring, and the nucleation rate decreases (Nancolas and McClintock, 1982).

A very interesting feature observed in experiments for $0.3\,\mathrm{K} < T < 1\,\mathrm{K}$ under high pressure is that the nucleation rate, ν, is highly temperature dependent for relatively weak electric fields, as shown in Fig. 9.27, whereas the drift velocity under the same conditions is practically temperature independent (Allum and McClintock, 1978*a*, 1978*b*; Stamp *et al.*, 1979; Bowley *et al.*, 1982). For rela-

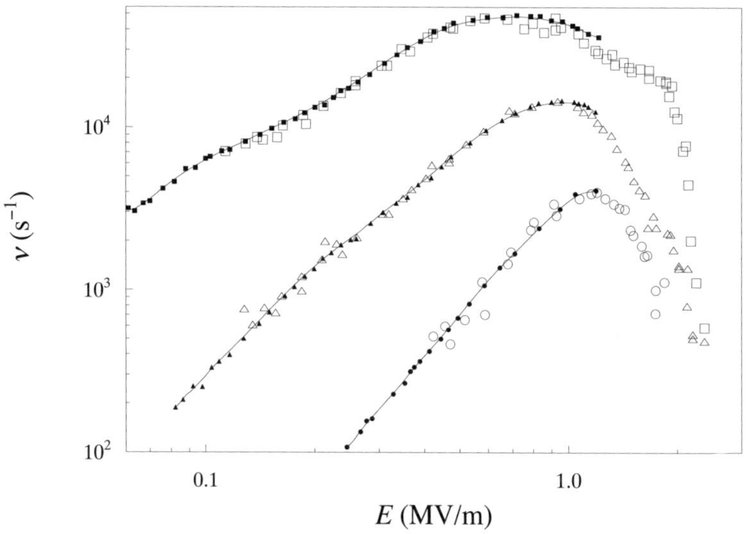

FIG. 9.26. ν vs E for negative ions in He II at $T = 0.3241$, K for $P = 1.7$, 2.1, and 2.5 MPa (from top). Small symbols: McClintock *et al.* (1980), and large symbols: Nancolas and McClintock (1982). The meaning of the lines is described in the text (Bowley, 1976*d*; Bowley and Sheard, 1977; Bowley *et al.*, 1982).

tively stronger fields, the nucleation rate becomes temperature independent, as expected theoretically (Bowley, 1976*d*).

For $P > 1.7$ MPa, at a given E, the nucleation rate can be described by a temperature-independent contribution plus a contribution of the Arrhenius type (Stamp *et al.*, 1979), as shown in Fig. 9.28:

$$\nu = \nu(0) + Ae^{-B/T}\,. \tag{9.23}$$

The activation energy, $B \approx 6.2$ K at $P = 2.5$ MPa, is very close to the roton energy gap $\Delta \approx 7$ K. This fact inevitably suggests that the nucleation of vortex rings is thermally assisted by mechanisms involving rotons (Stamp *et al.*, 1979).

It is believed (Bowley *et al.*, 1982) that the low-T (temperature-independent) contribution $\nu(0)$ describes a spontaneous nucleation process, which could be related to the quantum transition (or girdling) model (Schwarz and Jang, 1973). However, the rapid increase of ν, which depends on T with the same exponential law for the roton density, may lend some credibility to the peeling model and could exclude the girdling model.

This mutual exclusion of the models can be overcome with the following assumptions (Bowley *et al.*, 1982). Suppose that the instantaneous rate of vortex nucleation R, for a given P, is independent of E, and depends only on temperature and ion velocity. The measured rate is an average over the ionic velocity distribution function:

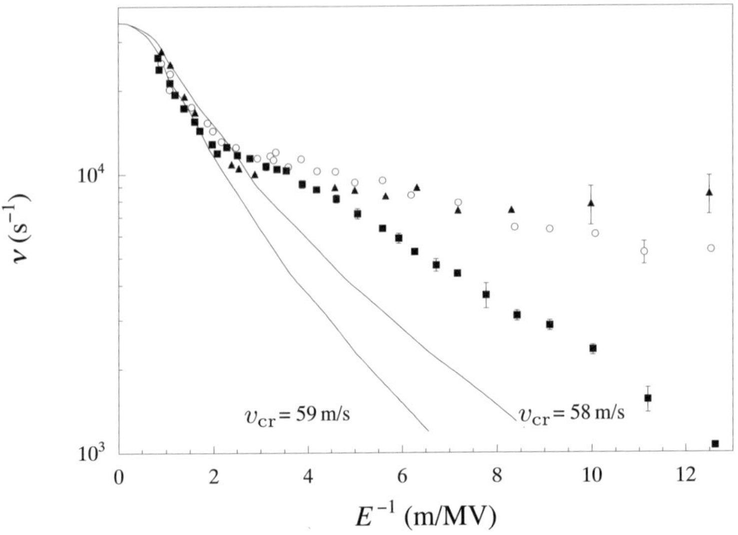

FIG. 9.27. ν vs E^{-1} for negative ions at $P = 2.5\,\mathrm{MPa}$ (Allum and McClintock, 1978*a*). $T\,(\mathrm{K}) = 0.4$ (triangles), 0.5 (circles), and 0.6 (squares). The lines are theoretical fits to the T-independent part of the data, with $v_{cr} = 58$ and $59\,\mathrm{m/s}$ (Bowley, 1976*d*).

$$\nu\,(E,T) = \int_0^\infty f\,(v,E,T)\,R\,(v,T)\,\mathrm{d}v\,. \tag{9.24}$$

For each T, the drift velocity v_D is a unique function of E and eqn (9.24) can be rewritten as

$$\nu\,(v_D,T) = \int_0^\infty f\,(v,v_D,T)\,R\,(v,T)\,\mathrm{d}v\,, \tag{9.25}$$

with $v_D = \int_0^\infty v f\,(v,E,T)\,\mathrm{d}v$.

It is further assumed that $f(v,v_D,T)$ is independent of T, apart from the temperature dependence of v_D. This is equivalent to assuming that processes involving excitation scattering do not alter the shape of the distribution function f but only the mean value v_D (Bowley *et al.*, 1982). If it is finally assumed that R can be formally written as

$$R\,(v,T) = R_s\,(v) + n_r R_r\,(v)\;, \tag{9.26}$$

where n_r is the roton number density, and R_s and R_r are two functions of v, then one gets

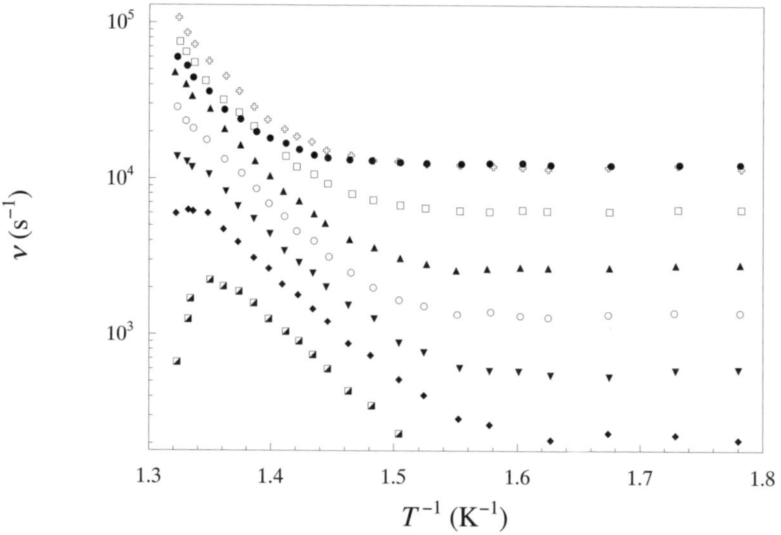

FIG. 9.28. ν vs T^{-1} for negative ions at $P = 2.1\,\mathrm{MPa}$. $E\,(\mathrm{MV/m}) = 1.13$ (closed circles), 0.675 (crosses), 0.45 (open squares), 0.293 (triangles), 0.203 (open circles), 0.135 (inverted triangles), 0.09 (diamonds), and 0.0562 (half-filled squares). (Bowley *et al.*, 1982.)

$$\nu(v_D, T) = \int_0^\infty f(v, v_D)\, R_s(v)\, \mathrm{d}v + n_r \int_0^\infty f(v, v_D)\, R_r(v)\, \mathrm{d}v$$
$$\equiv \nu_s(v_D) + n_r \nu_r(v_D)\,. \tag{9.27}$$

This formulation leads to a nucleation rate that is constant for $T < 0.5\,\mathrm{K}$ and increases rapidly with T because the roton density increases, in agreement with the experimental results (Bowley *et al.*, 1982).

The nucleation rate at high pressure in isotopically-pure ^{4}He at higher temperatures is dominated by the exponential term in eqn (9.23), which is related to a roton-assisted vortex nucleation process. The temperature-dependent contribution ν_r does indeed scale with the number density of thermal rotons $n_r = (k_0^2/2\pi^2\hbar)(2\pi m_r k_B T)^{1/2} \exp(-\Delta/k_B T)$ and a universal plot for all pressures is obtained by plotting $(\nu(v_D, T) - \nu_s(v_D))T^{-1/2} \exp(\Delta/k_B T)$ as a function of v_D (Nancolas *et al.*, 1981; Bowley *et al.*, 1982; Nancolas and McClintock, 1982), as shown in Fig. 9.29. Similar results are obtained for all $P > 1.6\,\mathrm{MPa}$ (Bowley *et al.*, 1982).

It is interesting to note that the maximum of the temperature-dependent contribution to the nucleation rate, ν_r, occurs for $v_D \approx 63\,\mathrm{m/s}$ (Fig. 9.29). On the other hand, if the data of Fig. 9.26 are plotted as a function of v_D instead of E, their maximum is found at the slighlty higher velocity $v_D \approx 66$–$67\,\mathrm{m/s}$

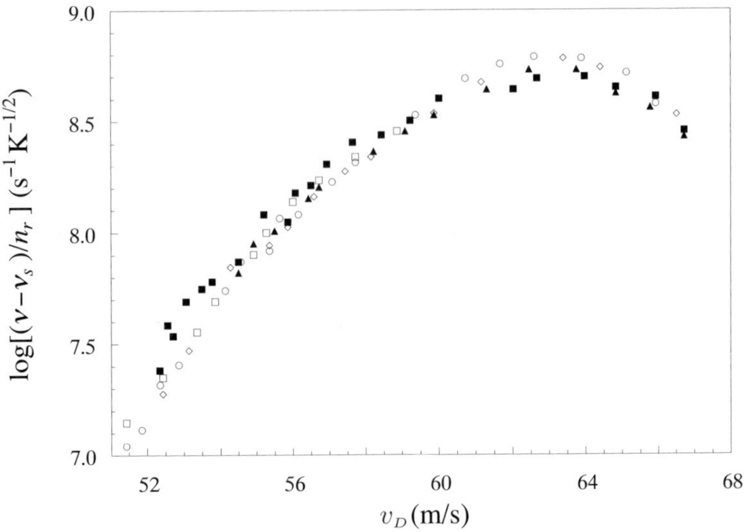

FIG. 9.29. $(\nu - \nu_s)/n_r$ vs v_D in isotopically-pure He II for $P = 2.3$ MPa. (Nancolas and McClintock, 1982.) T (K) $= 0.606$ (closed squares), 0.652 (triangles), 0.708 (diamonds), 0.753 (circles), and 0.806 (open squares).

(Bowley *et al.*, 1982).

The data shown in Fig. 9.26 are at such a low temperature that $\nu = \nu_s$. They thus describe the temperature-independent, intrinsic nucleation mechanism. The small difference in the position of the maximum nucleation rate for the two different mechanisms can be easily rationalized by recalling that in the girdling model the nucleation of a vortex ring occurs as soon as the conditions of energy and momentum conservation are satisfied.

In the case of the roton-assisted mechanism, the momentum and energy of the roton must be taken into account, leading to a small decrease in the ion velocity required for the nucleation process to take place. A proper treatment of the roton contribution leads to a correction of approximately 4 m/s for the critical velocity, in agreement with the experimental results (Bowley *et al.*, 1982).

It has to be recalled, however, that the analysis of measurements of the drift velocity of ions as a function of P and T for fields strong enough to quench the creation of charged vortex rings has led to the conclusion that the inference that the intrinsic and thermally-activated nucleation processes are very similar, except for the absorption of a roton in the latter case, might be an oversimplification (Nancolas *et al.*, 1986).

9.5.1 *The quantum-tunneling process*

The extension of the measurements of vortex ring nucleation by negative ions in isotopically-pure He to lower pressure has provided evidence of further different physical processes. On one hand, below approximately 1 MPa, but for E strong enough to suppress the creation of charged vortex rings, the main ion-energy dissipation mechanism at $T = 0.3\,\text{K}$ is apparently related to the formation of a succession of microscopic, uncharged vortex rings (Nancolas *et al.*, 1985*b*).

On the other hand, measurements of the nucleation rate at approximately 1.2 MPa at very low T (down to 50 mK) have shown that the creation of vortices in He II involves macroscopic quantum tunneling through, or thermal activation over, a potential-energy barrier (Hendry *et al.*, 1988*a*). These data are reported in Fig. 9.30.

In spite of the similarity with the data at higher pressure shown in Fig. 9.28 (Stamp *et al.*, 1979; Bowley *et al.*, 1980, 1982), the rapid rise in ν with increasing T of the present data takes place at $T \approx 0.2\,\text{K}$, when the roton density is negligible. In fact, the data are accurately fitted by eqn (9.23) with an activation energy $B/k_\text{B} = (3.1 \pm 0.1)\,\text{K}$, much smaller than the roton energy gap at the same pressure.

This kind of behavior is usually associated with the presence of an energy barrier; the temperature-independent term in eqn (9.23) represents quantum-mechanical tunneling through the barrier and the Arrhenius-type term represents thermal excitation over it.

Muirhead *et al.* (1984) have proposed a theory, according to which the creation of vortex rings by ions at low T occurs through a sort of macroscopic quantum-mechanical tunneling. Basically, they calculate the total energy change ΔE that occurs when a vortex loop of radius R_0 is formed at constant impulse.

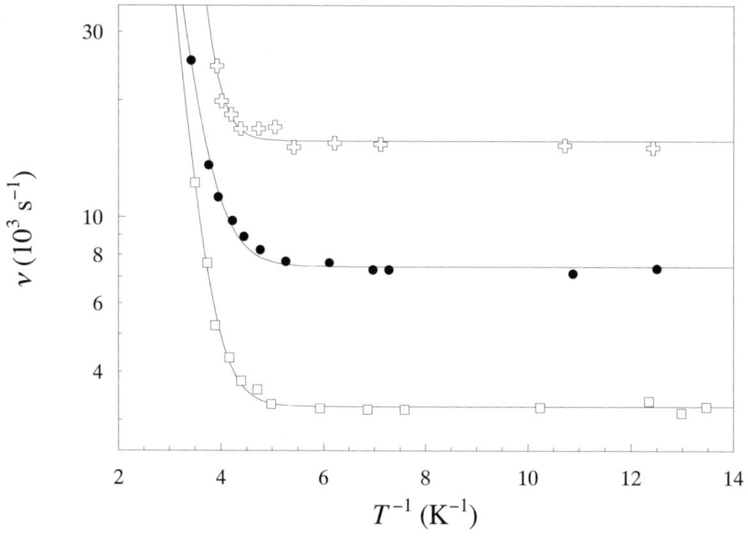

FIG. 9.30. ν vs T^{-1} for negative ions in pure He II at $P = 1.2\,\text{MPa}$. (Hendry *et al.*, 1988*a*.) $E\,(\text{kV/m}) = 44.0$, 17.7, and 8.85 (from top).

The vortex loop may then tunnel through the barrier, thus giving origin to a ring.

In Fig. 9.31 the results of a sample calculation of ΔE for three different ion velocities are shown. The calculations are of hydrodynamic nature and cannot be extended down to R_0, though it can be expected that $\Delta E \to 0$ for $R_0 \to 0$.

The resulting barrier, for an ion with a velocity slightly exceeding the critical one, is the shaded area in the figure. Its height favorably compares with the experimental 3.1 K-high barrier. Moreover, the minimum critical velocity ($\approx 58\,\mathrm{m/s}$), for which ΔE shows a minimum as a function of R_0, is in remarkably good agreement with the value $v_c \approx 59.5\,\mathrm{m/s}$ determined by fitting the theory (Allum *et al.*, 1976*b*; Bowley, 1976*d*) to the electric field dependence of the nucleation rate (Hendry *et al.*, 1988*a*).

This theoretical approach is very similar to the fluctuation and peeling models (Rayfield, 1967; Donnelly and Roberts, 1969*a*, 1969*b*) and, in some sense, is contrary to the girdling model (Schwarz and Jang, 1973; Bowley, 1984). In fact, by using energy barrier arguments, Muirhead *et al.* (1984) conclude that loop nucleation is favored over complete ring nucleation.

However, detailed numerical investigations of the two scenarios (nascent vortex loop or complete vortex ring) have been carried out by using the nonlinear

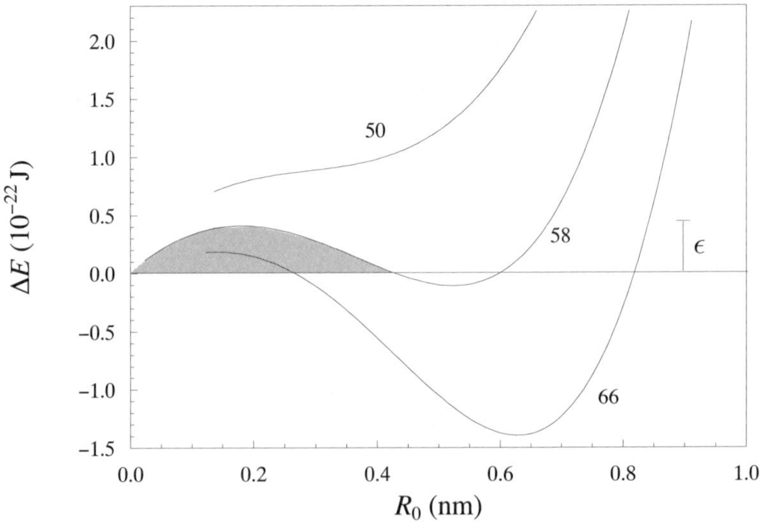

FIG. 9.31. Energy change of the vortex ring–ion complex ΔE, at constant impulse, when a vortex loop of radius R_0 is formed in the equatorial plane of a negative ion for three ion velocities $v_D = 50$, 58, and $66\,\mathrm{m/s}$ (from top) at $P = 1.7\,\mathrm{MPa}$. (Hendry *et al.*, 1988*a*.) The energy barrier for the case of an ion slightly exceeding the critical velocity (middle curve) is the shaded area. The experimental barrier of $\approx 3.1\,\mathrm{K}$ is shown as a vertical bar on the right.

Schrődinger equation in order to analyze the superfluid flow around the ion without revealing any marked tendency for vortex loops to nucleate in preference to vortex rings at criticality (Berloff, 2000; Berloff and Roberts, 2000*a*, 2000*b*, 2001). It is argued there that vortices are nucleated when the liquid velocity around the ion exceeds the velocity of sound.

The facts that, at higher pressure, the temperature-dependent part of the nucleation rate is proportional to the number of thermally-excited rotons, and that the temperature at which the crossover from the intrinsic to the thermally-activated nucleation process occurs shifts to higher values, can be explained by assuming that the barrier height responsible for the quantum tunneling of vortices increases with increasing pressure (Hendry *et al.*, 1988*a*).

The nature of the entity that undergoes tunneling is not very clear yet, though it appears that vortex nucleation is an example of macroscopic tunneling (Nancolas *et al.*, 1985*a*).

9.5.2 *Effect of ^{3}He impurities*

Tiny proportions of ^{3}He impurities dramatically change the creation of vortex rings by negative ions, even at natural isotopic abundance (Bowley *et al.*, 1980; McClintock *et al.*, 1981; Nancolas *et al.*, 1985*a*).

In Fig. 9.32 the effect of impurities on the nucleation rate of vortex rings is shown. Only at high fields does the nucleation rate in natural helium approach the value of isotopically-pure ^{4}He (Bowley *et al.*, 1980). The peak observed for natural He is an artifact due to the presence of ^{3}He atoms condensed on the surface of the electron bubble (Allum and McClintock, 1976*a*).

However, the effect of the isotopic impurity is strong even at such a low concentration that the average number of adsorbed ^{3}He atoms per ion is less than unity (Bowley *et al.*, 1984).

Typical results showing the effect of the concentration x_3 of isotopic impurities on the vortex ring nucleation rates are shown in Figs 9.33 and 9.34, in which either x_3 or T are kept fixed.

The nucleation rate shows a peak at a given field strength because of the adsorption of the impurity. There is no such peak in isotopically-pure He. If the concentration x_3 is kept fixed, the peak position shifts to smaller fields as T is increased, whereas it is insensitive to x_3 at constant T (Nancolas *et al.*, 1985*a*).

The dependence of the nucleation rate on x_3, at constant T and P, shows two different regimes, as depicted in Figs 9.35 and 9.36. For a small electric field, ν increases superlinearly with x_3, suggesting the onset of a condition in which a significant fraction of negative ions possesses two (or more) trapped ^{3}He atoms. On the other hand, for a larger electric field, ν increases linearly with x_3 because the probability of there being more than one trapped ^{3}He atom on an ion is negligible (Bowley, 1984).

In order to rationalize the experimental observations and treat the case in which at most one ^{3}He atom is trapped on the ion, ν is fitted, at constant T, P, and E, by a power-law expansion in x_3 truncated to the second term:

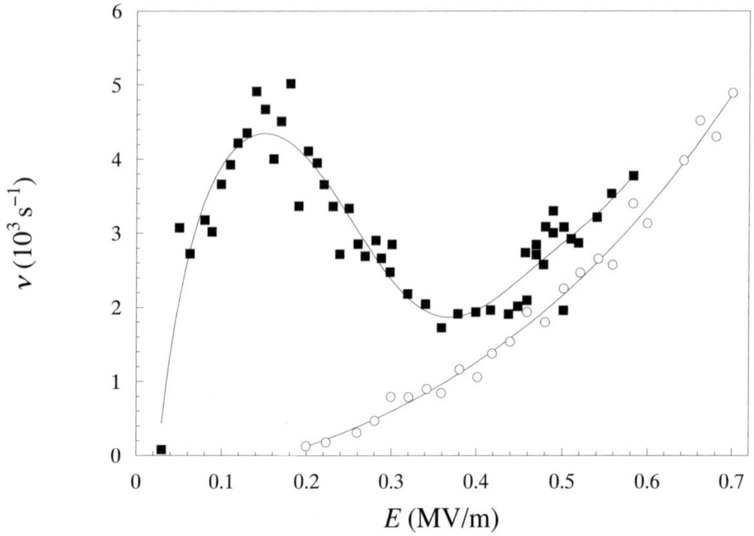

FIG. 9.32. ν vs E for negative ions for $P = 2.4\,\text{MPa}$ and $T = 0.52\,\text{K}$. (Bowley *et al.*, 1980.) Closed symbols: natural He. Open symbols: isotopically-pure ^{4}He. The lines are eyeguides.

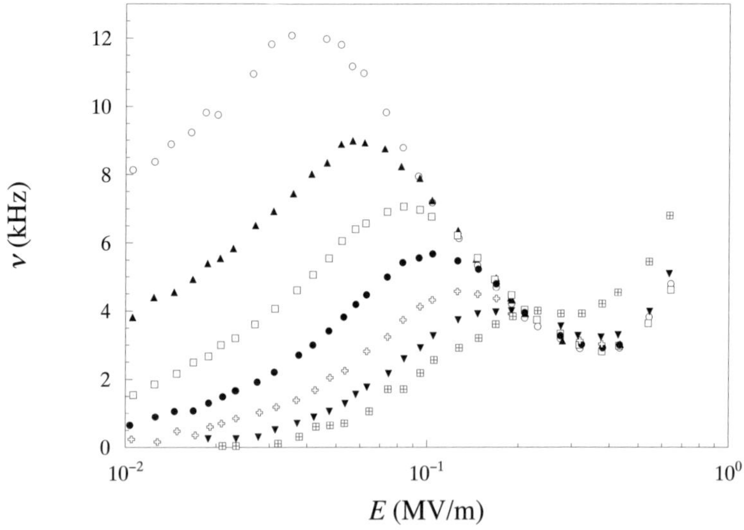

FIG. 9.33. ν vs E in a solution with $x_3 = 0.172\,\text{ppm}$ at $P = 2.3\,\text{MPa}$. $T\,(\text{K}) = 0.329$, 0.369, 0.414, 0.458, 0.507, 0.549, and 0.608 (from top). (Nancolas *et al.*, 1985*a*.)

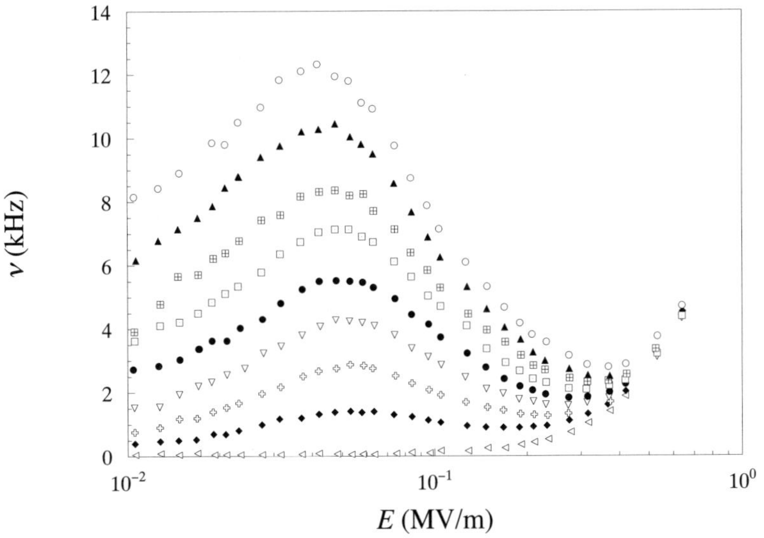

FIG. 9.34. ν vs E at $P = 2.3\,\text{MPa}$ and $T = 0.329\,\text{K}$ in solutions with x_3 (ppm) = 0.172, 0.150, 0.129, 0.107, 0.0858, 0.0644, 0.0429, 0.0214, and 0.19×10^{-3} (from top). (Nancolas *et al.*, 1985*a*.)

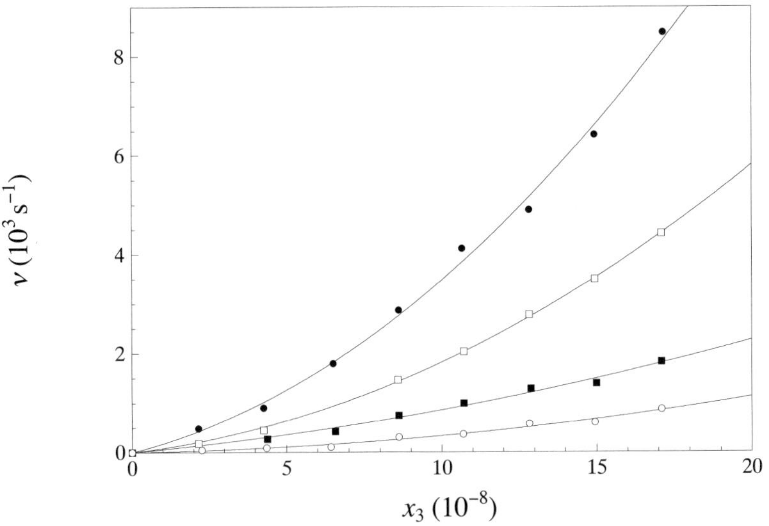

FIG. 9.35. ν vs x_3 for $P = 2.3\,\text{MPa}$ at $E = 95\,\text{kV/m}$. T (K) = 0.329, 0.414, 0.458, and 0.507 (from top). ((Nancolas *et al.*, 1985*a*), 1985*a*.)

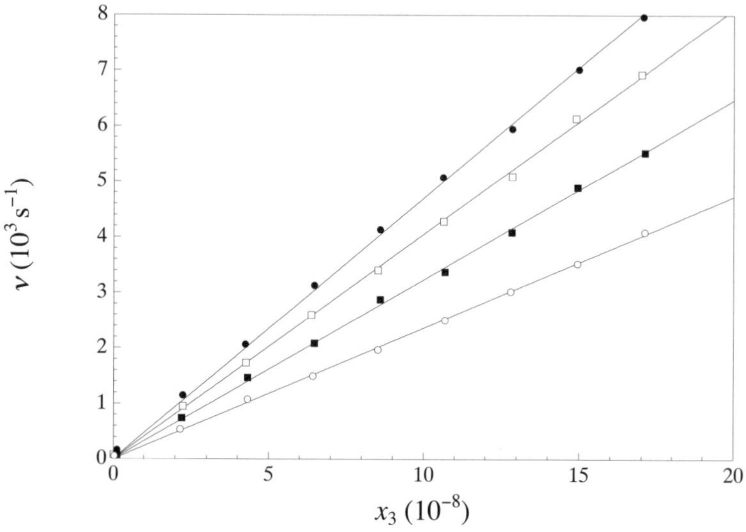

FIG. 9.36. ν vs x_3 for $P - 2.3$ MPa at $E = 12.7$ kV/m. T (K) $= 0.329$, 0.369, 0.414, and 0.458 (from top). ((Nancolas *et al.*, 1985*a*), 1985*a*.)

$$\nu(x_3) = \nu_0 + \nu' x_3 + \nu'' x_3^2 \,, \tag{9.28}$$

where ν_0 is the nucleation rate in pure ^{4}He at the same T, P, and E, and ν' represents the ^{3}He contribution to the nucleation rate (per unit concentration) that would be measured in the limit of very low concentration. The nonlinear term is usually very small. Thus, the statistical analysis of the coefficient ν'' does not give significant information because of the low accuracy of the data (Nancolas *et al.*, 1985*a*).

It is assumed that, for given E, T, and P, two different nucleation rates ν_0 and ν_1 exist. The former is the nucleation rate for bare ions (Bowley *et al.*, 1982), whereas the latter is the one appropriate for an ion with one ^{3}He atom trapped on it. In a dilute solution, such that $\nu'' x_3^2 \approx 0$, the measured rate can be written as

$$\nu = (1 - n_B)\, \nu_0 + n_B \nu_1 \,, \tag{9.29}$$

where n_B is the average number of ^{3}He atoms adsorbed on the ion. As $n_B < 1$, it can be considered as the probability of finding one ^{3}He atom bound on the ion.

For small concentrations, eqn (9.29) can be rewritten as

$$\Delta\nu \equiv \nu - \nu_0 = n_B\,(\nu_1 - \nu_0) \simeq \nu' x_3 \,, \tag{9.30}$$

where $\Delta\nu$ is measured directly.

In Fig. 9.37 ν' is plotted as a function of E for several T. There are three

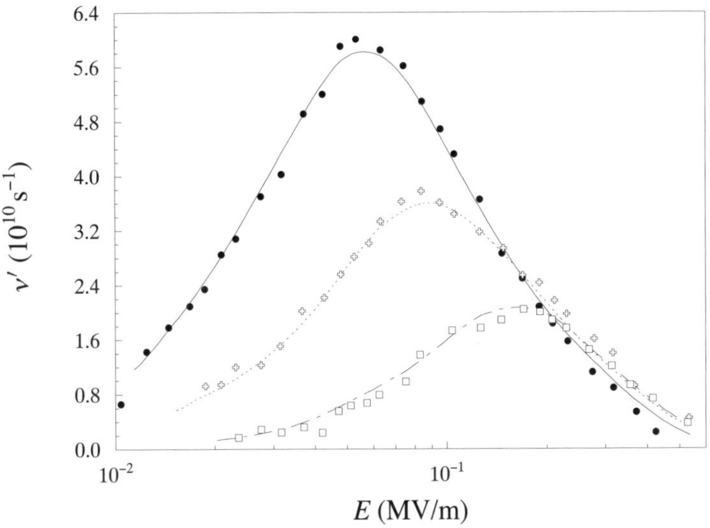

FIG. 9.37. ν' vs E for T (K) $= 0.329$, 0.414, and 0.549 (from top). (Bowley *et al.*, (1984).) Lines: fits of the model described in the text.

main features displayed by ν': it has a nice temperature-dependent maximum; it apparently goes to zero for small and large E; and it is almost independent of T at high E.

Owing to the lack of understanding of the microscopic mechanisms of nucleation, ν_1 is left as an adjustable parameter to be determined by fitting the data (Nancolas *et al.*, 1985*a*), while the variations of ν with T and E are then basically due to the behavior of n_B (Bowley, 1984).

Shikin (1973) has shown that there are several bound states of ^{3}He atoms on the surface of the electron bubble, which can be labeled by the quantum number l of their angular momentum. The energy eigenvalues are

$$\epsilon_l = E_B + gl\,(l+1)\,, \qquad 0 \leqslant l \leqslant L\,, \tag{9.31}$$

where $g = \hbar^2/2m_{3,s}R_i^2$. Here $m_{3,s}$ is the effective mass of a ^{3}He atom sitting on the surface, and R_i is the ion radius. E_B is the binding energy, i.e., the energy of the lowest-lying level, and L is the largest value for which $\epsilon_L < 0$, i.e., the quantum number of the least-bound state.

The determination of the behavior of n_B as a function of E and T is achieved by considering the rates of absorption and desorption of ^{3}He atoms onto or from an ion. Two main processes are supposed to be active (Bowley, 1984; Nancolas *et al.*, 1985*a*):

1. one ^{3}He atom is absorbed and the binding energy, ≈ 2 K, is carried away by one or more phonons; the inverse process, namely the absorption of one or more phonons, produces the desorption of one impurity atom;

2. the emission of an impurity atom accompanied by the emission of two rotons.

At small fields, n_B is determined by normal thermal equilibrium processes. Thus, owing to the scarcity of thermal rotons at the low temperatures of the experiments, process 1 is dominant and ν' should depend strongly on T, as experimentally observed.

On the other hand, at high fields E, process 2 becomes energetically favorable (Bowley, 1976*d*) and n_B falls below its thermal equilibrium value. For large enough fields, process 1 can be neglected with respect to process 2 and ν' becomes temperature independent, again in agreement with experiment (Bowley, 1984).

According to Shikin (1973),

$$n_B = \sum_{l=0}^{L} 2(2l+1)n_l \,, \tag{9.32}$$

with occupation numbers n_l given by

$$n_l = \frac{n_l^{\mathrm{eq}}}{1 + n_l^{\mathrm{eq}} R'/n_3 K} \,, \tag{9.33}$$

where the occupation numbers at thermal equilibrium are

$$n_l^{\mathrm{eq}} = \left[1 + \exp\left(\epsilon_l - \mu\right)\right]^{-1} \,. \tag{9.34}$$

Moreover, $R' = R_e + \nu_1 - \nu_0$ and $R_e(E)$ is the average rate of two-roton emission together with a ^{3}He atom (Nancolas *et al.*, 1985*a*). The chemical potential μ is given by

$$\mu = -k_{\mathrm{B}} T \ln \left[\frac{2}{n_3} \left(\frac{m_3^\star k_{\mathrm{B}} T}{2\pi\hbar^2} \right)^{3/2} \right], \tag{9.35}$$

where n_3 is the ^{3}He number density and $m_3^\star$ its effective mass.

The absorption rate K is written as

$$K = \alpha \bar{v}_{\mathrm{rel}} \mathcal{P}\,(T)\, \sigma_g \,, \tag{9.36}$$

where σ_g is the geometrical cross-section of the ion, α is a constant representing the probability that an impurity atom reaching the electron bubble surface is eventually trapped, and $\mathcal{P}$ is the fraction of ^{3}He quasiparticles with enough energy to surmount the small potential barrier (Bowley and Lekner, 1970) that exists near the ion (Nancolas *et al.*, 1985*a*).

The model can be fitted to the data by leaving R'/α, ν_1, and E_B free as adjustable parameters. The lines in Fig. 9.37 show an example of the quality of the fit. Also, the temperature dependence of ν_1 at constant E is fitted by the model to the same accuracy (Bowley *et al.*, 1984).

It is found that $E_B/k_{\mathrm{B}} = (-2.83 \pm 0.17)\,\mathrm{K}$, in fair agreement with the value $E_B = (-2.22 \pm 0.03)\,\mathrm{K}$ deduced for ^{3}He atoms binding to a planar He II surface at zero pressure (Edwards and Saam, 1978; Sadd *et al.*, 1999).

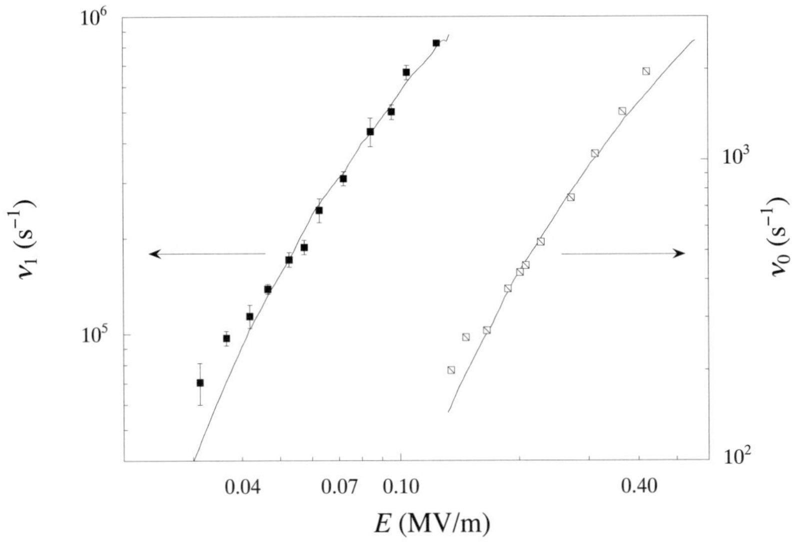

FIG. 9.38. Electric field dependence of the vortex ring nucleation rate, ν_1, for ions with one bound ^{3}He atom as a function of E at $P = 2.3$ MPa (left scale). ν_0, the rate for bare ions, is plotted for comparison (right scale) (Bowley *et al.*, 1984). The lines represent fits to ν_1 and ν_0 according to the model proposed by Bowley (1976*d*) and Bowley *et al.* (1982).

The dependence of $\nu_1(E)$ on E is shown in Fig. 9.38. It is interesting to note that the nucleation rate for an ion with a trapped impurity is two orders of magnitude larger than that of a bare ion, although the functional dependence on E is rather similar.

This difference is qualitatively consistent with the model of vortex nucleation as a sort of macroscopic, quantum-mechanical tunneling process (Muirhead *et al.*, 1984). If the trapped impurity has to be incorporated in the core of the nascent vortex, the height of the nucleation-inhibiting barrier is expected to be lowered, thereby reducing the critical velocity for nucleation and correspondingly increasing the nucleation rate (Bowley *et al.*, 1984).

This explanation is consistent with the behavior of the field-emission current in ^{3}He–^{4}He mixtures at high pressure $P = 2.5$ MPa (Allum and McClintock, 1976*b*), in which the presence of ^{3}He impurities suppresses the fraction of current carried by bare ions, leaving only the vortex-borne component.

However, different alternatives are possible and are discussed exhaustively in thr literature (Nancolas *et al.*, 1985*a*).

10

BARE-ION TRANSPORT AT HIGH FIELDS

The discovery of quantized charged vortex rings and the thorough investigation of their dynamics and nucleation processes have shed light on the behavior of ions in a range of temperatures, electric fields, and pressures, in which both free ions and charged vortex rings coexist.

In this context, it is customary to term *bare* those ionsthat are not stuck on vortices while being dragged along under the action of an externally-applied electric field, although the charge carriers in liquid helium are not bare at all because they are either endowed with a solvation cluster, if they are positive, or with an empty cavity, if electrons.

Deep physical insight into the interaction between ions and fluid excitations can be gathered by investigating the transport properties of those ions that have not been captured by vortex lines or rings. For this reason, the present discussion of the transport properties of bare ions was necessarily postponed until after the chapter devoted to the analysis of the nucleation and transport of quantized vortex rings generated by ions drifting under quite large electric fields.

10.1 Escape of bare ions from vortex rings

Early evidence that ions can move without being stuck on vortices is produced in Fig. 10.1 (Bruschi *et al.*, 1966*a*). At the electric field E_{c1}, the drift velocity v_D shows the giant discontinuity associated with the formation of quantized charged vortex rings whose dynamics is well known and accurately describes the decrease of the drift velocity of the ion–ring complex with increasing field.

There exists, however, another critical field E_{c2}, above which v_D starts increasing again with E and can no longer be described by ion–vortex ring dynamics. The increase in v_D for $E > E_{c2}$ is associated with an increase in the second-sound attenuation in the same field range (Bruschi *et al.*, 1966*a*).

It is clear that the increase in v_D with $E > E_{c2}$ is related to the possibility that the ion escapes from the slow ring and spends a fraction of its time as a fast free ion before getting retrapped. If the time interval spent as a free ion increases with E, so does the observed average drift velocity (Bruschi *et al.*, 1968*b*).

In particular conditions of temperature and field, the drifting ions may nucleate vortex rings without being trapped on them. Exploiting this fact, the transport properties of bare ions can be investigated in a much wider electric field range. Measurements of the lifetime for positive ions in vortex rings (Cade, 1965) and of negative ions in vortex lines (Douglass, 1964) have shown that the negative ion is strongly bound to vortex rings for temperatures as high as 1.8 K, whereas the escape probability of positive ions is large for T as low as

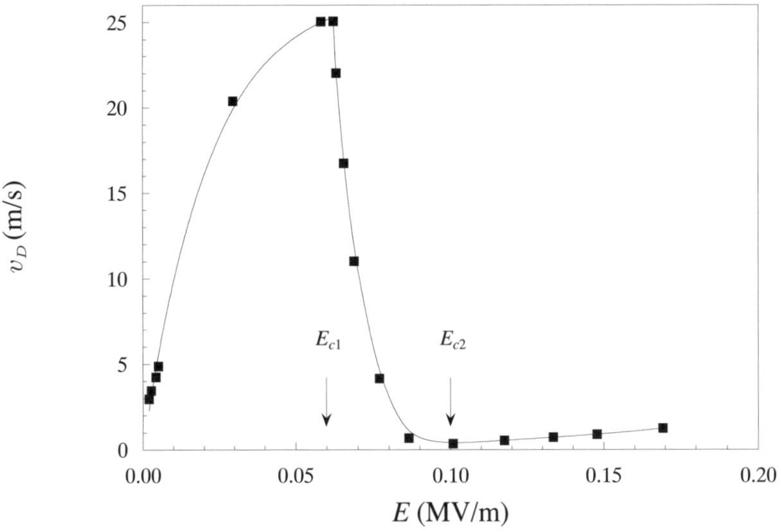

FIG. 10.1. v_D vs E for negative ions for $T = 0.905$ K at SVP. (Bruschi *et al.*, 1966*a*.) The line is an eyeguide.

≈ 1 K. Therefore, it is possible to attain a situation in which ions create vortex rings without being trapped on them and their average drift velocity should be independent of the field over a broad interval.

In Fig. 10.2 the drift velocity of ions, both positive and negative, is plotted as a function of the electric field at high temperatures ($T = 1.41$ and 1.50 K) at vapor pressure (Bruschi *et al.*, 1968*b*). At such high temperatures, the giant discontinuity for positive ions has gradually changed into a plateau, thus suggesting that the positive ion is continuously nucleating vortex rings without getting trapped on them and that its escape probability is quite large.

At high fields, the increase in dissipation due to the increase of the vortex nucleation rate leads to a steady-state condition, in which there is a balance between the energy gained by the ion from the field and the energy dissipated by nucleating vortices, so that v_D becomes independent of E. In contrast, negative ions remain stuck on the vortex rings because of their much smaller escape probability.

At lower T, the escape probability of positive ions is not so large and their drift velocity shows an intermediate behavior (Bruschi *et al.*, 1968*b*), as plotted in Fig. 10.3 at SVP.

The qualitative explanation of the observed behavior is quite simple: for E beyond the minimum of v_D, the ion is able to escape from the ring it has generated (Padmore, 1971) and the ion–ring complex breaks up under the opposing action of the electric field and the normal fluid drag (Jones, 1969).

After escape, the ion accelerates under the action of the field once more

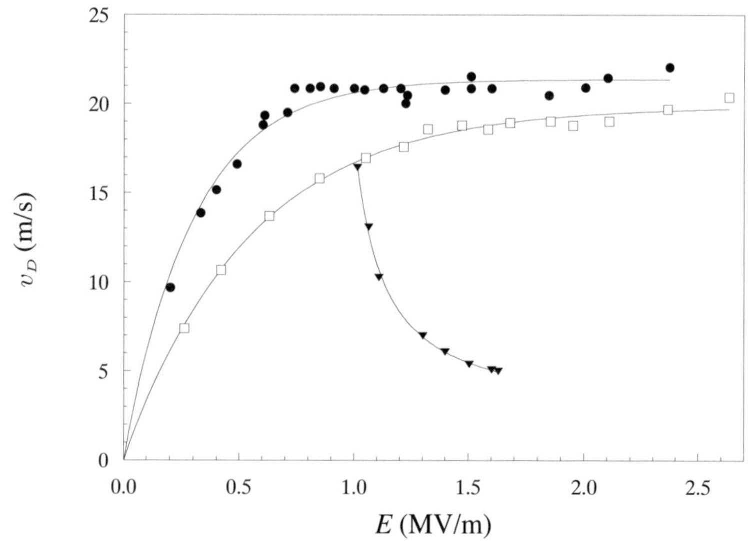

FIG. 10.2. v_D vs E for positive ions for T (K) = 1.41 (circles), and 1.50 (squares) and for negative ions for T = 1.41 K (triangles) at SVP. (Bruschi *et al.*, 1968*b*.) The lines through the positive ion data are eyeguides. The line through the negative ion data is the hydrodynamic calculation for the ion–vortex ring complex.

toward the critical velocity for vortex ring nucleation; it again nucleates a ring, on which it is captured. The ring starts growing toward the equilibrium radius for the given E and decelerates.

This cycle is repeated again and again. The observed velocity, which is an average over the time fraction that the ion spends as free or trapped, happens to be larger than it would have been in the case when the ion had always remained trapped on the ring. Escape thus provides the explanation of the drift velocity minimum and its subsequent increase for larger fields (Padmore, 1972*a*).

The description of the drift velocity for fields above the minimum in Fig. 10.3 (Padmore, 1971, 1972*a*) is based on the assumption that the nucleation of a ring by a bare ion is a thermally-activated process (Donnelly and Roberts, 1969*a*) with probability per unit time $\mathcal{P}_n(v)$, where v is the ion velocity. It is assumed that the ion will escape from a ring with probability $\mathcal{P}_e$ per unit time.

In an external field E, the ion is bound to the vortex with an energy $W(E)$ that is calculated from classical hydrodynamics (Donnelly and Roberts, 1969*b*). Thus, one expects $\mathcal{P}_e \propto \exp[-W(E)/k_B T]$. $W(E)$ has been directly measured for positive ions at $T = 0.5$ K for $E = 1$ MV/m, yielding $W = (\rho_s/\rho)(15.9 - 1.44E + 0.04E^2)$ (Cade, 1965).

For the description of the drift velocity around the minimum it is further assumed that

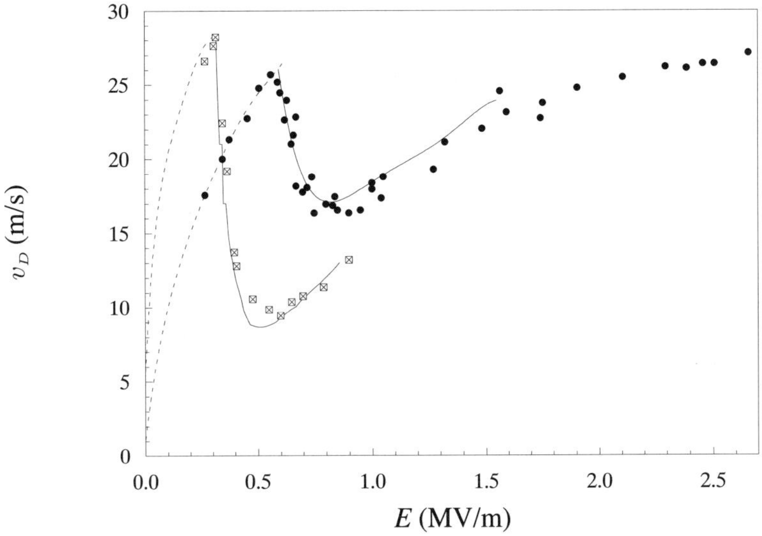

FIG. 10.3. v_D vs E for positive ions for $T\,(\mathrm{K}) = 1.18$ (circles) and 1.30 (squares). (Bruschi *et al.*, (1968*b*).) Dashed lines: eyeguides. Solid lines: theory (Padmore, 1972*a*).

$$\mathcal{P}_n\,(v_D) \gg \mathcal{P}_e\,, \tag{10.1}$$

$$\mathcal{P}_e \gg \tau_d^{-1}\,, \tag{10.2}$$

where v_D is the average bare-ion velocity and τ_d is the macroscopic drift time. These two conditions imply that the ion spends most of its time trapped on the ring and that a large number of escape–retrapping events occur during its time-of-flight.

The motion of the ring between nucleation and escape is determined by the following equation (Lamb, 1945; Rayfield and Reif, 1963):

$$M_i \dot{v} + \dot{p}_r = eE - \mathcal{F}_{ir} - \zeta \mathcal{F}_i\,, \tag{10.3}$$

where M_i is the bare-ion mass, p_r is the impulse of the ring, $\mathcal{F}_{ir}$ is the drag acting on the ion–ring complex, and $\mathcal{F}_i$ is the drag acting on the bare ion. $\mathcal{F}_{ir}$ is given by (Huang and Olinto, 1965)

$$\mathcal{F}_{ir} = \alpha \left[1 - \frac{R_i}{\pi R}\right] \ln\left(\frac{R}{\xi}\right), \qquad \xi = a e^{1/4}/8\,, \tag{10.4}$$

where R_i is the ion radius, R the ring radius, a the core radius, and α is the friction coefficient introduced by Rayfield and Reif (1964). $\mathcal{F}_i = ev/\mu$, where μ is the ion mobility, v its velocity, and $\zeta \approx 1/3$ (Huang and Olinto, 1965).

For large rings $\zeta = 0$, and $M_i\dot{v}$ can be neglected. Moreover, $\mathcal{F}_{ir} = \alpha \ln (R/\xi)$. In this case, the solution of eqn (10.3) for the steady state is the Careri formula (8.4) for the drift velocity of charged rings:

$$v_D = \frac{\kappa}{4\pi\xi} \tilde{E} \exp(-\tilde{E}), \qquad \tilde{E} = eE/\alpha\,, \tag{10.5}$$

which describes the rapid decrease of v_D after the critical field E_{c1} (see Fig. 10.1).

The minimum of v_D observed for $E = E_{c2}$ is a consequence of the dynamics of the rings. If the approximation for large rings is used, eqn (10.3) is integrated so as to obtain the time $t(R)$ that a just-nucleated ring needs to grow to its equilibrium radius R:

$$t(R) = 2\pi\rho\kappa\xi^2\alpha^{-1} \exp\left(\tilde{E}\right) \left[E_1\left(2\tilde{D}\right) - E_1\left(2\tilde{E}\right)\right], \tag{10.6}$$

with initial condition $R = \xi$ for $t = 0$. E_1 is the exponential integral (Arfken, 1985) and $\tilde{D} = \tilde{E} - \ln(R/\xi)$.

Using the relation $\mathrm{d}x = v\mathrm{d}t$ and the relation between v and R for a ring, eqn (7.26), the distance $x(R)$ negotiated by the ring during time $t(R)$ is obtained as

$$x(R) = \frac{1}{2}\kappa^2\rho\alpha^{-1} \left\{\xi\tilde{E} \exp\left(\tilde{E}\right) \left[E_1\left(\tilde{D}\right) - E_1\left(\tilde{E}\right)\right] - R + \xi\right\}. \tag{10.7}$$

$x(t)$ is obtained parametrically from these two equations.

These expressions are valid for large R. For smaller rings the complete equation of motion, eqn (10.3), is to be used, instead. However, the time spent as a small ring is very short and can be neglected (Padmore, 1971).

The ring growth is interrupted by the stochastically-determined escape of the ion at a rate $\mathcal{P}_e$, whose inverse is the lifetime of the ion–ring complex (van Dijk *et al.*, 1977). If the escape probability $\mathcal{P}_e$ is assumed independent of time (Padmore, 1971), the average distance $\langle x \rangle$ a ring survives is approximately given by

$$\langle x \rangle = \int_0^\infty \mathcal{P}_e e^{-\mathcal{P}_e t} x(t)\,\mathrm{d}t\,. \tag{10.8}$$

The approximation arises because of the finite length of the drift space and because condition (10.2) ensures that the fraction of very long-lived rings is negligible (Padmore, 1972*a*).

The mean duration of a ring is similarly given by

$$\langle t \rangle = \int_0^\infty \mathcal{P}_e e^{-\mathcal{P}_e t}\, t\mathrm{d}t = \mathcal{P}_e^{-1}\,. \tag{10.9}$$

The average drift velocity v_D measured in a time-of-flight experiment is then calculated as

$$v_D = \frac{\langle x \rangle}{\langle t \rangle} = \mathcal{P}_e \langle x \rangle \,. \tag{10.10}$$

In this way one can also calculate the persistence current observed by Bruschi *et al.* (1966*a*), which is a consequence of the possibility that large rings can propagate a macroscopic distance in a field-free region.

The above results can be extended to the case when the fraction of time spent as a free ion is relevant, therefore relaxing condition (10.1) to the less stringent one $\mathcal{P}_n \geqslant \mathcal{P}_e$. The bare-ion equation of motion $M_i \dot{v} = eE - \mathcal{F}_i$ is integrated numerically so as to give the time $\mathcal{T}(v)$ necessary for the ion to reach the velocity v, and the distance $\mathcal{X}(v)$ it travels during this time interval. The probability that a ring is nucleated between $\mathcal{T}$ and $\mathcal{T} + \mathrm{d}\mathcal{T}$ is

$$\mathcal{P}_n(v) \exp\left[-\int_0^{\mathcal{T}} \mathcal{P}_n\left(v\left(t'\right)\right) \mathrm{d}t' \right] \mathrm{d}\mathcal{T} \,. \tag{10.11}$$

The mean duration $\langle \mathcal{T} \rangle$ and extent $\langle \mathcal{X} \rangle$ of the bare-ion state are given by

$$\langle \mathcal{T} \rangle = \int_0^{\infty} \mathcal{P}_n(v) \exp\left[-\int_0^{\mathcal{T}} \mathcal{P}_n\left(v\left(t'\right)\right) \mathrm{d}t' \right] \mathcal{T} \mathrm{d}\mathcal{T} \,, \tag{10.12}$$

$$\langle \mathcal{X} \rangle = \int_0^{\infty} \mathcal{P}_n(v) \exp\left[-\int_0^{\mathcal{T}} \mathcal{P}_n\left(v\left(t'\right)\right) \mathrm{d}t' \right] \mathcal{X}\left(\mathcal{T}\right) \mathrm{d}\mathcal{T} \,. \tag{10.13}$$

The mean distance between escapes is then $\langle x \rangle + \langle \mathcal{X} \rangle$ and the mean time is $\langle t \rangle + \langle \mathcal{T} \rangle$, so that the drift velocity is now

$$v_D = \frac{\langle x \rangle + \langle \mathcal{X} \rangle}{\mathcal{P}_e^{-1} + \langle \mathcal{T} \rangle} \,. \tag{10.14}$$

The combined results in eqn (10.5) for $E_{c1} < E < E_{c2}$ and eqn (10.14) are shown as solid lines in Fig. 10.3.

It is particularly interesting to note that, whereas for $T = 1.18\,\mathrm{K}$ the velocities are small enough to justify the large-ring approximation, at $T = 1.30\,\mathrm{K}$ the velocities are sufficiently large to ensure that rings are always small. Thus, the phenomenological approach represented by eqn (10.3) is quite accurate. This behavior of the drift velocity of positive ions occurs over an extended range of T and P (van Dijk *et al.*, 1977), as reported in Fig. 10.4 for several temperatures at $P \approx 0.01\,\mathrm{MPa}$ and in Fig. 10.5 for $P = 1.5\,\mathrm{MPa}$.

A second, less-pronounced change in the velocity curve is observed at a field E_m just above the critical field of the giant discontinuity (van Dijk *et al.*, 1977). Both values of these fields are marked by arrows in Fig. 10.4 for the curve at $T = 0.939\,\mathrm{K}$. This change progressively disappears with increasing T. An analysis

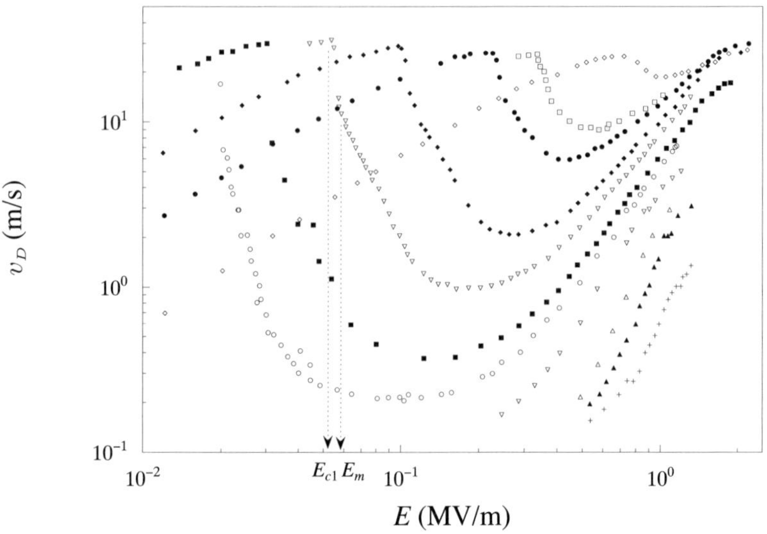

FIG. 10.4. v_D vs E for positively-charged ion–vortex ring complexes at $P = 0.01$ MPa. T (K) $= 0.821$ (open circles), 0.866 (closed squares), 0.939 (inverted open triangles), 1.008 (closed diamonds), 1.119 (closed circles), 1.185 (dotted squares), 1.338 (open diamonds), 0.752 (inverted closed triangles), 0.645 (open triangles), 0.568 (closed triangles), and 0.458 (crosses). (van Dijk *et al.*, 1977.)

of the shape of the ionic signal suggests that E_m is the minimum field required to sustain a charge carrier in the ring-coupled state. For fields $E_{c1} < E < E_m$ the ring can decay back into the bare-carrier state, as suggested by Zoll and Schwarz (1973).

The escape rate can be deduced from the drift velocity measurements following the previously-described procedure (Padmore, 1972*a*; van Dijk *et al.*, 1977). Escape rates for positive ions at $P \approx 0.01$ MPa are shown in Fig. 10.6 as a function of T^{-1} for several electric fields. At higher P, $\mathcal{P}_e$ decreases in magnitude but its dependence on E and on T remain nearly the same (van Dijk *et al.*, 1977).

The presence of a second critical field E_{c2} for the negative ions is observed only at high pressures in natural He II. The first evidence of it has been produced by Allum and McClintock (1976*c*) at $T = 1.00$ K and $P = 2.5$ MPa (Fig. 10.7). The presence of the minimum drift velocity is very clear and the increase with the field of the average drift velocity as a consequence of the succession of escape–nucleation–retrapping events is observed. However, for $E \gg E_{c2}$, the negative ion drift velocity reaches the roton-emission-limited value observed at lower T (Allum *et al.*, 1976*a*).

The same behavior of the drift velocity of negative ions is also shown for all temperatures around 1 K, as reported in Fig. 10.8, and is also observed at pressures as low as 1.5 MPa for $T > 0.9$ K (van Dijk *et al.*, 1977). Results at

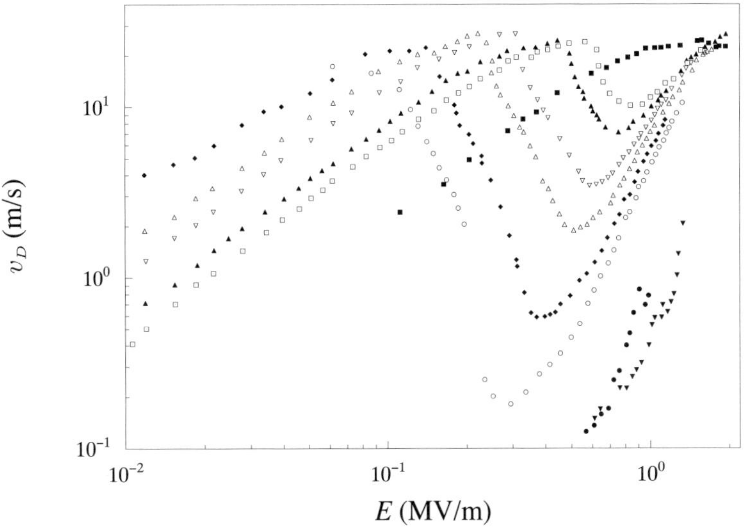

FIG. 10.5. v_D vs E for positively-charged ion–vortex ring complexes at $P = 1.5\,\text{MPa}$. (van Dijk *et al.*, 1977.) $T\,(\text{K}) = 1.303$ (closed squares), 1.119 (dotted squares), 1.076 (closed triangles), 1.008 (inverted open triangles), 0.970 (open triangles), 0.909 (diamonds), 0.866 (open circles), 0.650 (closed circles), and 0.487 (inverted closed triangles).

$P = 1.5\,\text{MPa}$ are reported in Fig. 10.9. Thus, thermally-activated, field-assisted escape processes also influence the drift velocity of negative ions.

For negative ions, however, the determination of the escape rate is not as straightforward as for the positive ions, because the nucleation rate in pressurized He II is greatly depressed at high P (Phillips and McClintock, 1974*a*; Zoll, 1976; Bowley *et al.*, 1982). Therefore, after escaping from a ring, the negative ion does not immediately produce a new ring, as assumed by Padmore (1971, 1972*a*). Rather, it travels for a significant time interval with a drift velocity that is mainly controlled by roton emission, before the next ring is nucleated. For this reason, v_D depends on $\mathcal{P}_e$ as well as on ν, and $\mathcal{P}_e$ cannot be determined without a prior knowledge of ν, that can, fortunately, be gathered from the analysis of the shape of the measured current pulse (Allum and McClintock, 1976*c*). v_D is then given, according to eqn (10.14), by

$$v_D = \frac{\bar{x}_r + \bar{x}_i}{\mathcal{P}_e^{-1} + \nu^{-1}}, \tag{10.15}$$

where $\bar{x}_r$ and $\bar{x}_i$ are the average survival distances for rings and bare ions, defined, respectively, as

$$\bar{x}_r = \int_0^\infty v_r(t) e^{-\mathcal{P}_e t} \mathrm{d}t \tag{10.16}$$

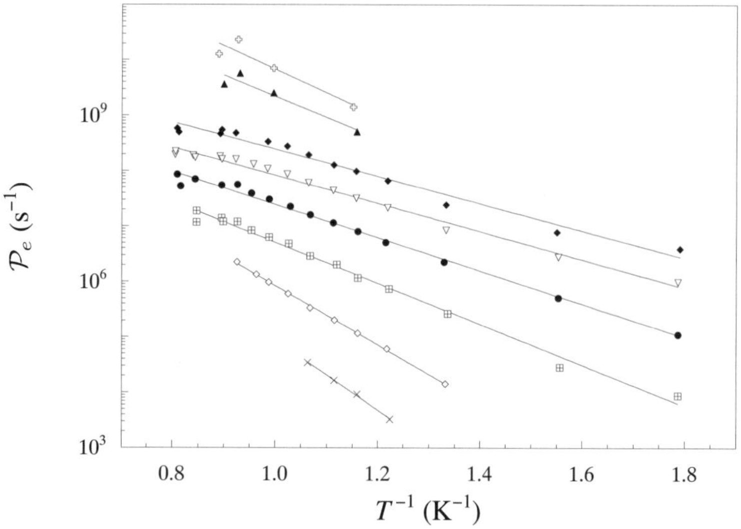

FIG. 10.6. $\mathcal{P}_e$ vs T^{-1} for positive ions at $P \approx 0.01$ MPa. E (MV/m) = 2.0, 1.6, 1.2, 1.0, 0.8, 0.6, 0.4, and 0.2 (from top). Lines: exponential fits. (van Dijk *et al.*, 1977.)

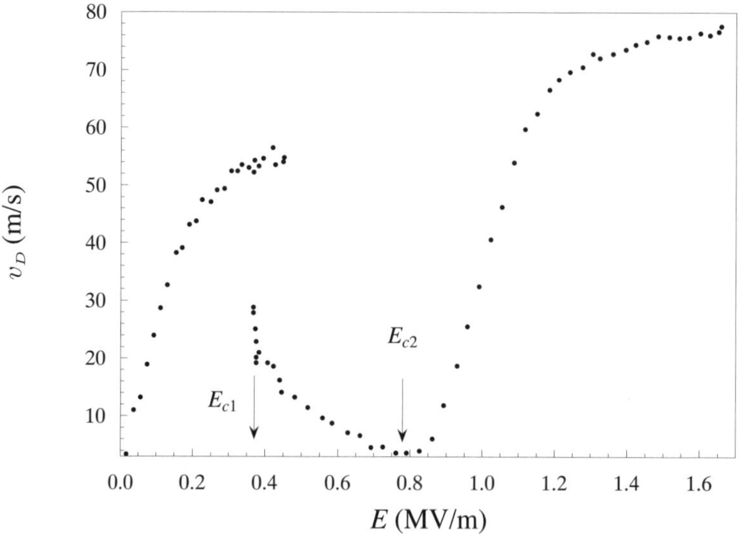

FIG. 10.7. v_D vs E for negative ions at $T = 1$ K and $P = 2.5$ MPa. (Allum and McClintock, 1976*c*.)

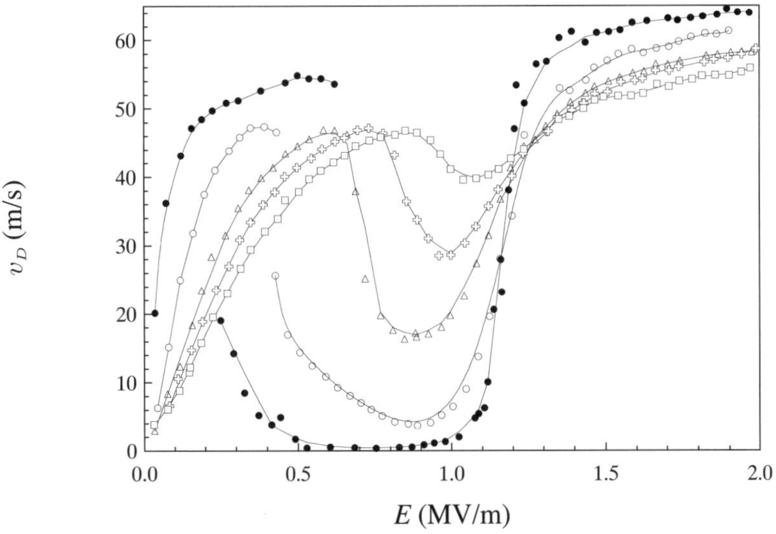

FIG. 10.8. v_D vs E for negative ions at $P = 2.5$ MPa in the range $0.90\,\mathrm{K} < T < 1.13\,\mathrm{K}$. T (K) = 0.90 (closed circles), 1.00 (open circles), 1.07 (triangles), 1.10 (crosses), and 1.13 (squares). (Allum and McClintock, 1976*c*.)

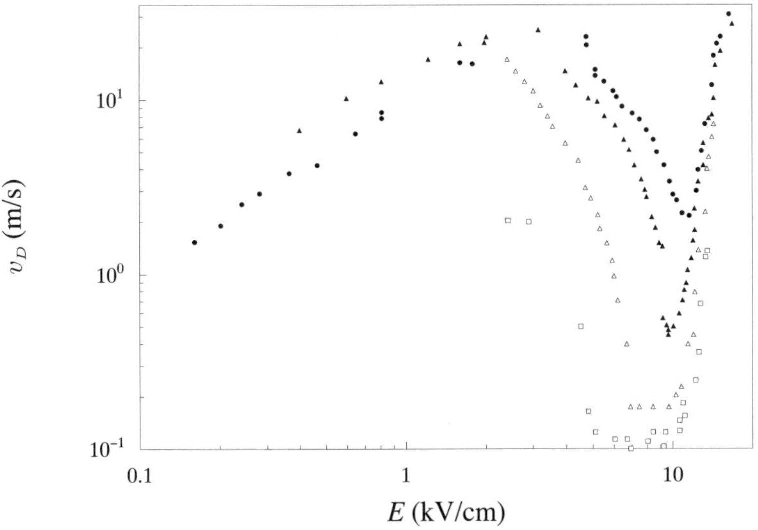

FIG. 10.9. v_D vs E for negative ions at $P = 1.5$ MPa. (van Dijk *et al.*, 1977.) T (K) = 0.970 (squares), 1.007 (open triangles), 1.076 (closed triangles), and 1.119 (circles).

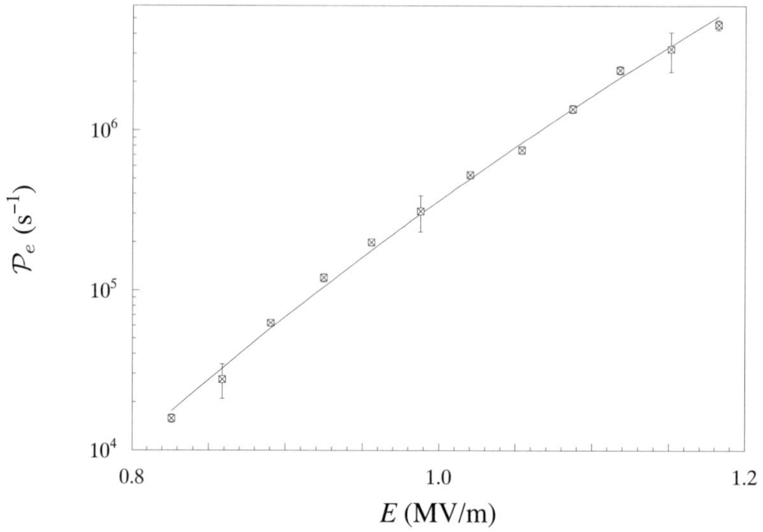

FIG. 10.10. $\mathcal{P}_e$ vs E for negative ions at $T = 1.00$ K and $P = 2.5$ MPa. (Allum and McClintock, 1976*c*.)

and

$$\bar{x}_i = \int_0^\infty v_i(t)e^{-\nu t}\mathrm{d}t\,. \tag{10.17}$$

Here v_r and v_i are the ring and ion velocities, respectively. v_r is calculated according to the procedure given earlier (eqns (10.3)–(10.14)) (Padmore, 1972*a*). The instantaneous bare-ion velocity $v_i(t)$ can be replaced by its time-averaged value $\bar{v}_i$ because the roton-emission rate, $\approx 10^{10}\,\mathrm{s}^{-1}$, is much greater than the vortex ring nucleation rate ν (Allum and McClintock, 1976*c*), thus yielding

$$\bar{x}_i = \bar{v}_i \nu^{-1}\,. \tag{10.18}$$

The nucleation rate is determined by the analysis of the signal waveform and it can be extrapolated to high E by using the theory of Bowley (1976*d*). In Fig. 10.10 the values of the escape rate determined in this way are shown.

It should be noted that the escape rate for the negative ions is much smaller (nearly three orders of magnitude) than that of positive ions (van Dijk *et al.*, 1977) (Fig. 10.6), as is to be expected in view of its much larger radius.

10.2 Roton-emission-limited mobility of bare ions

The studies of vortex ring generation by means of ions have led to the discovery that, under special circumstances, i.e., in pressurized He II at low temperatures, negative ions can be drifted at high electric field strengths with speed close to and in excess of the Landau critical velocity for roton emission without nucleating any vortex rings (Phillips and McClintock, 1973). This observation has given researchers the opportunity to investigate the phenomenon of supercritical energy dissipation leading to the breakdown of superfluidity via roton-emission processes.

The feasibility of such an investigation is basically due to the fact that the nucleation of vortex rings by negative ions is quenched at low T and high P in isotopically-pure ^{4}He if the applied electric field is strong enough, as discussed earlier with reference to Fig. 9.26.

The roton barrier was broken for the first time by Phillips and McClintock (1974*a*) and subsequent measurements have elucidated the issue of supercritical dissipation (Allum and McClintock, 1976*c*, 1977; Allum *et al.*, 1976*b*; Ellis *et al.*, 1980*a*, 1980*b*). A more recent review of this issue can be found in the literature (McClintock, 1995).

Typical experimental results are shown in Fig. 10.11 (Allum *et al.*, 1976*b*). In the range $2\,\mathrm{kV/m} < E < 200\,\mathrm{kV/m}$, the velocity data are well described by the following equation:

$$v_D = v_L + AE^{1/3}\,. \tag{10.19}$$

The behavior of v_D appears to be very similar to what would be expected on the basis of Landau theory. For $v_D < v_L$ no drag would be exerted by the superfluid on the moving object. For larger velocities, superfluidity breaks down because of the creation of elementary excitations and the drag would increase enormously. This picture is immediately grasped if the data are looked at in the way presented in Fig. 9.20.

The straight line in Fig. 10.11 is a fit to eqn (10.19). The drift velocity v_D in the linear region does not depend on T, as shown in Fig. 10.12. This observation supports the idea that the drift velocity is limited by roton emission, which is a temperature-independent process. The quite large values of the difference $v_D - v_L$ rule out the Takken wave radiation model, in which a conical wave of coherent roton radiation is generated (Takken, 1970; Phillips and McClintock, 1974*a*; Allum *et al.*, 1976*b*).

The interpretation of the electric field dependence of v_D in this regime requires a dynamical theory of supercritical dissipation and a statistical theory of the ionic motion (Allum *et al.*, 1976*b*; Bowley and Sheard, 1977).

By using Fermi's golden rule, it is shown that a process in which a single roton is emitted leads to $v_D - v_L \propto E^{2/3}$ (Sheard and Bowley, 1978), in contrast with experiment (Allum *et al.*, 1976*a*), whereas the observed $E^{1/3}$ dependence is consistent with a process in which two rotons are emitted.

The steady-state kinetics of the process is quite simple (see Fig. 10.13). An ion with initial velocity $v_0 < v_L$ is accelerated by the field past the threshold velocity for two-roton emission $v' = v_L + p_0/M$, where p_0 is the roton momentum at the minimum of the liquid-helium dispersion curve and M is the effective ion mass. The ion reaches a velocity $v_e > v'$, at which the rotons are emitted with total momentum $2p_0$, and, after recoil, comes back to its initial velocity.

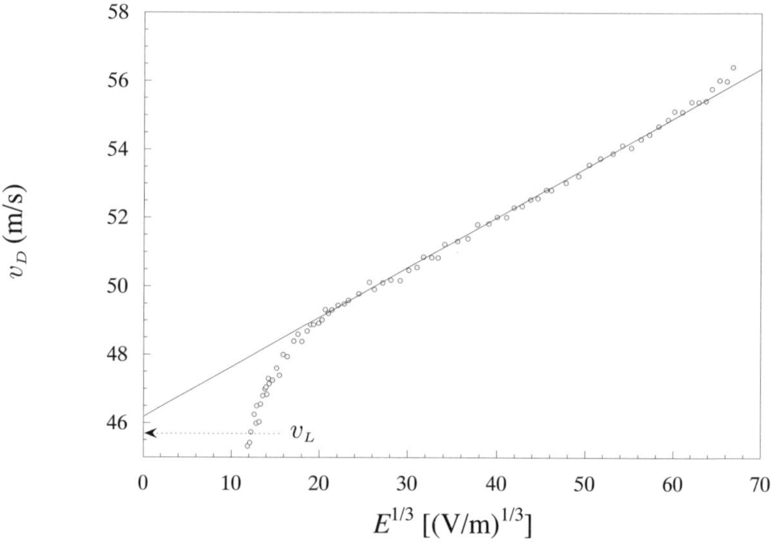

FIG. 10.11. v_D vs $E^{1/3}$ for negative ions in isotopically-pure ^{4}He at $T = 0.45$ K for $P = 2.5$ MPa. (Allum *et al.*, 1976*b*.)

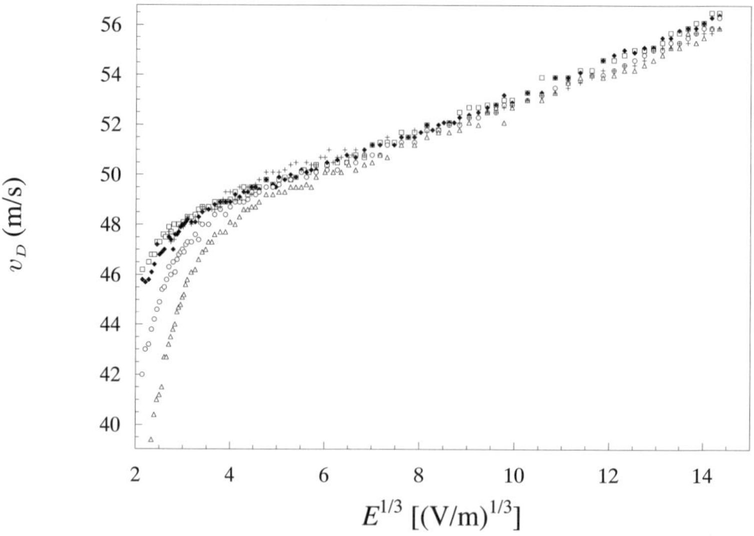

FIG. 10.12. v_D vs $E^{1/3}$ in isotopically-pure ^{4}He at $P = 2.5$ MPa. (Allum *et al.*, 1976*b*.) T (K) = 0.30 (crosses), 0.35 (squares), 0.40 (diamonds), 0.45 (circles), and 0.50 (triangles). At high E, the roton-limited drift velocity is independent of T.

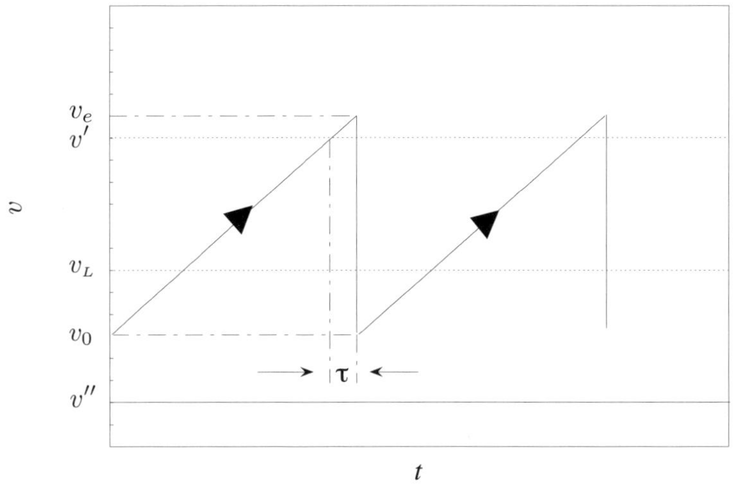

FIG. 10.13. Velocity trajectory of an ion showing the acceleration of the ion by the field and the instantaneous recoil after roton emission. (Bowley and Sheard, 1977.)

The average time τ for which the ion velocity exceeds v' is calculated from the probability of roton emission. An approximate formula for the transition rate R for two-roton emission for ions moving with $v > v'$ is given as

$$R(v) = \alpha (v - v')^2 . \tag{10.20}$$

After exceeding the velocity v' the ion has the probability $\mathcal{P}(t)$ of surviving for a time interval t before emitting a roton pair. $\mathcal{P}$ satisfies the differential equation

$$\frac{\mathrm{d}\mathcal{P}}{\mathrm{d}t} = -R(v)\,\mathcal{P}(t) . \tag{10.21}$$

As $v = v' + eEt/M$, eqn (10.21) is integrated to give

$$\mathcal{P}(t) = \exp\left[-\frac{\alpha M}{3eE}\left(\frac{eEt}{M}\right)^3\right]. \tag{10.22}$$

By defining, quite arbitrarily, τ as the time for which $\mathcal{P}(\tau) = 1/e$, one gets

$$\frac{eE\tau}{M} \approx \left(\frac{3eE}{\alpha M}\right)^{1/3} . \tag{10.23}$$

The average velocity, i.e., the ion drift velocity, is easily found to be

$$v_D = v_L + \left(\frac{3eE}{\alpha M}\right)^{1/3} , \tag{10.24}$$

which accurately describes the field dependence of the drift velocity data, provided that $v_D - v_L < 2p_0/M$.

Departures from eqn (10.24) are expected when the initial velocity v_0 lies entirely above the threshold velocity, i.e., when $v_D - v_L > 2\hbar k_0/M$. In this case, a similar calculation yields

$$(v_D - v_L)^3 - \left(v_D - v_L - \frac{2\hbar k_0}{M}\right)^3 \approx \frac{3eE}{\alpha M}, \tag{10.25}$$

which gives a field dependence closer to $E^{1/2}$ than to $E^{1/3}$ (Bowley and Sheard, 1977). Such a behavior is, indeed, observed at even higher electric fields (Allum *et al.*, 1976*a*; Ellis *et al.*, 1980*b*), as shown in Fig. 10.14.

In order to quantitatively compare the data with the two-roton-emission theory, the rate of two-roton emission and the ion velocity distribution function must be calculated. The golden rule allows us to calculate the rate of two-roton emission for an ion of velocity v in the form

$$R(v) = \sum_{\mathbf{k}} \sum_{\mathbf{q}} \frac{|V_{\mathbf{k},\mathbf{q}}|^2}{\Omega^2} \delta\left(\epsilon_{\mathbf{k}} + \epsilon_{\mathbf{q}} - \hbar(\mathbf{k}+\mathbf{q})\cdot\mathbf{v} + \frac{\hbar^2(\mathbf{k}+\mathbf{q})^2}{2M}\right), \tag{10.26}$$

where Ω is the system volume (Allum *et al.*, 1976*a*; Bowley and Sheard, 1977). The matrix element $V_{\mathbf{k},\mathbf{q}}$ is unknown but it is assumed that it can be approximated, in the momentum range of interest, by the constant value $V_{\mathbf{k}_0,\mathbf{k}_0}$ that is to be adjusted by fitting the experimental data.

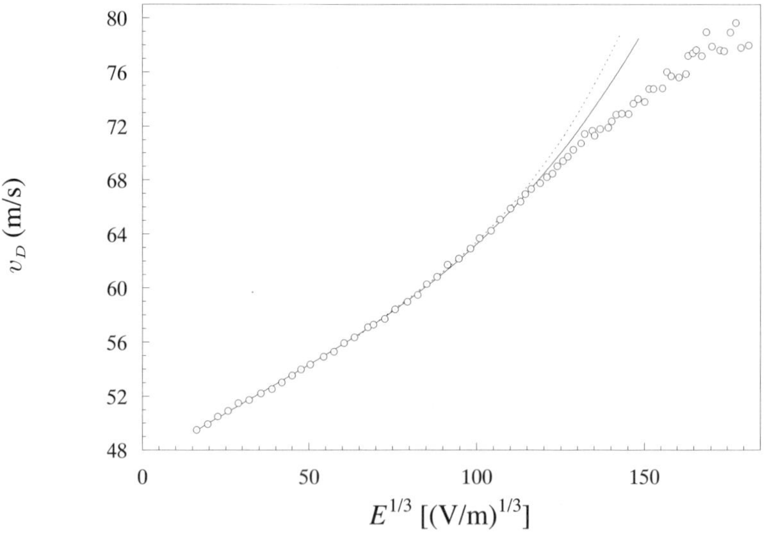

FIG. 10.14. v_D vs $E^{1/3}$ for $P = 2.5$ MPa and $T = 0.34$ K in isotopically-pure ^{4}He. Ellis *et al.*. Dashed line: theory (Bowley and Sheard, 1977). Solid line: extension of this theory that accounts for the non-parabolicity of the real dispersion curve.

With this approximation the two-roton-emission rate becomes

$$R(v) = \frac{m_r k_0^4 |V_{\mathbf{k}_0,\mathbf{k}_0}|^2}{2\pi^2 \hbar^3 (v - \hbar k_0/M)^2} (v - v')^2 \approx \frac{m_r k_0^4 |V_{\mathbf{k}_0,\mathbf{k}_0}|^2}{2\pi^2 \hbar^3 v_L^2} (v - v')^2 , \qquad (10.27)$$

where, again, $v' = v_L + \hbar k_0/M$. The last expression is obtained for $v \approx v'$, and in this case eqn (10.27) reduces to eqn (10.20).

The velocity v_D is calculated as an average of the ion velocity over the distribution function: $v_D = \int v f(v) \mathrm{d}v / \int f(v) \mathrm{d}v$. The distribution function itself is obtained, at steady state, by solving the Boltzmann transport equation in the form quoted previously, eqn (9.20), with $R_2 = R$ and with the recoil velocity given, for isotropy reasons, by

$$u - v = \frac{2\hbar k_0}{M} \left[1 - \frac{2}{3} \frac{(u - v')}{u} \right]. \qquad (10.28)$$

The dashed line in Fig. 10.14 is the prediction of the model of two-roton emission. It gives an excellent fit to the data for ion velocities up to $\approx 70\,\mathrm{m/s}$.

A slight improvement is obtained by including corrections due to the anharmonicity of the dispersion curve near the roton minimum, to the momentum dependence of the pole strength for excitations with k far away from k_0, and to the fact that the average momentum of the emitted excitations is slightly larger than k_0 for large v_D. The residual discrepancy occurring for even larger v_D is yet to be understood (Ellis *et al.*, 1980*b*).

A reduction of the residual scattering by thermal excitations, accomplished by lowering T, allows an accurate determination of the pressure dependence of the Landau critical velocity and of the matrix element $V_{\mathbf{k}_0,\mathbf{k}_0}$. Actually, although $v_L = \Delta/\hbar k_0$ is a very good approximation, the exact expression for v_L is (Ellis *et al.*, 1980*a*)

$$v_L = \frac{\left(2m_r \Delta + \hbar^2 k_0^2\right)^{1/2} - \hbar k_0}{m_r} , \qquad (10.29)$$

and $V_{\mathbf{k}_0,\mathbf{k}_0}$ is related to the coefficient A of eqn (10.19) as follows (Nancolas *et al.*, 1985*b*):

$$A = \left(\frac{5.36 e \pi^2 \hbar^3 v_L^2}{M k_0^4 m_r |V_{\mathbf{k}_0,\mathbf{k}_0}|^2} \right)^{1/3} . \qquad (10.30)$$

In Fig. 10.15 the drift velocities of negative ions in pressurized, isotopically-pure ^{4}He at $T = 90\,\mathrm{mK}$ are reported (Ellis *et al.*, 1980*a*; Ellis and McClintock, 1985). The field range investigated is quite limited in order to ensure that eqn (10.19) is accurately followed. The lines are a linear fit to the data and the intercept is the value of $v_L(P)$ for the given pressure.

In Fig. 10.16 the values of $v_L(P)$ are shown as a function of P. The solid line is calculated by using eqn (10.29) with the roton parameters $\Delta(P)$, $k_0(P)$, and $m_r(P)$ found in the literature (Donnelly, 1972). The dashed line is the usual approximation $v_L = \Delta/\hbar k_0$. Owing to the lack of accurate knowledge of the

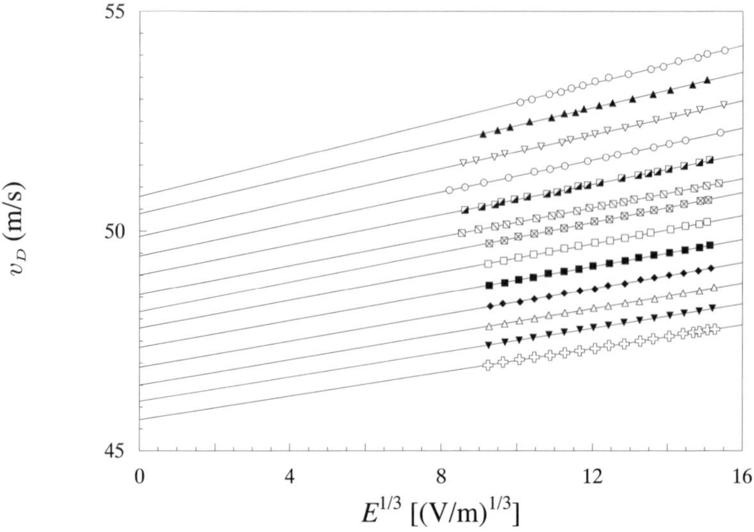

FIG. 10.15. v_D vs E for negative ions in isotopically-pure He II at $T = 90$ mK. (Ellis *et al.*, 1980*a*.) From top: P (MPa) = 1.3 , 1.4, 1.5, 1.6, 1.7, 1.8, 1.9, 2.0, 2.1, 2.2, 2.3, 2.4, and 2.5.

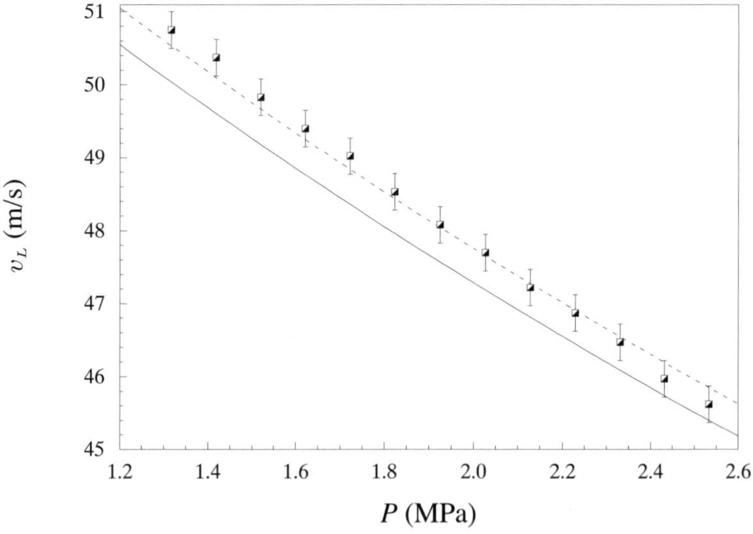

FIG. 10.16. $v_L(P)$ at $T = 90$ mK. (Ellis *et al.*, 1980*a*; Ellis and McClintock, 1985.) Solid curve: eqn (10.29). Dashed curve: $v_L = \Delta/\hbar k_0$.

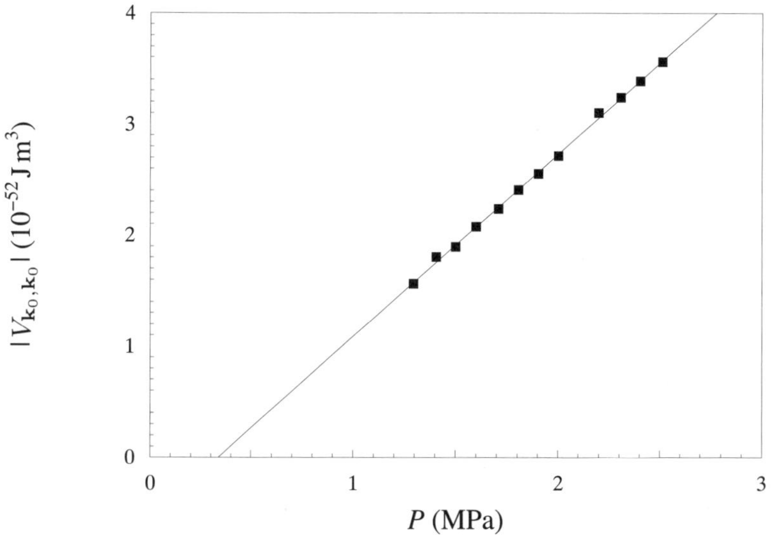

FIG. 10.17. The matrix element $V_{\mathbf{k}_0,\mathbf{k}_0}(P)$ for roton-pair emission at $T \approx 80\,\mathrm{mK}$. (Ellis and McClintock, 1985; Nancolas *et al.*, 1985*b*.)

roton parameters in this range (Donnelly and Roberts, 1977), the agreement between experiment and theory is to be considered quite good.

Finally, in Fig. 10.17 the absolute value of the matrix element $V_{\mathbf{k}_0,\mathbf{k}_0}$ for roton-pair emission as a function of P at $T \approx 80\,\mathrm{mK}$ is shown. The experimental determination of $M(P)$ has been used (Ellis *et al.*, 1983).

Although not justified by any theories, the behavior of $V_{\mathbf{k}_0,\mathbf{k}_0}$ as a function of P indicates that the emission of roton pairs becomes less important as P is decreased. However, it must be recalled that vortex ring nucleation prevents this sort of experiment below $P \approx 1.3\,\mathrm{MPa}$ (Ellis and McClintock, 1985; Nancolas *et al.*, 1985*b*).

A physical reason for the decrease of the matrix element with decreasing P might be associated with the expansion of the ionic radius: it might be conceived that there could be a critical radius at which the nascent rotons might interfere destructively, effectively reducing the probability of the emission event (Nancolas *et al.*, 1985*b*).

11

INTERACTION OF IONS WITH VORTEX LINES

Rotating helium II produces a strong anisotropy to the passage of ions. The earliest evidence of it is the observation by Careri *et al.* (1962) that in He II, at $T = 1.801$,K, the fully-spacecharge-limited negative ion current in a cylindrical geometry was attenuated by as much as 80% upon rotation at 6 rad/s if the current was flowing perpendicular to the rotation axis, whereas it was independent on the rotation frequency if flowing along it. The authors explained this attenuation as being due to the capture and retention of negative ions by quantized vortex lines.

The ion trapping on vortex lines is interpreted in terms of the bubble model of the negative ion and the vortex line model of the rotating superfluid (Donnelly, 1965; Parks and Donnelly, 1966). A generalized hydrodynamic formalism for the description of the motion of a sphere interacting with a vortex line can be found elsewhere (Schwarz, 1974).

The superfluid is considered to rotate, resembling a solid body by, developing a uniform array of quantized rectilinear vortices lying parallel to the rotation axis with an areal density n_v proportional to the angular frequency Ω, as expressed by eqn (7.15). Hydrodynamics provides an attractive interaction between the ion and the vortex lines. As a consequence, the vortices provide an array of potentials wells, in which ions may become trapped.

Trapping is believed to be limited by the thermally-activated escape of the ion over the barrier in the radial direction and by migration of the ions along the vortices to their ends. In this way, experiments give information on both ions and vortex lines (Pratt and Zimmermann, 1969).

The experiments are typically carried out by using traditional drift cells that are set into rotation either parallel or perpendicular to their cylindrical axis. Suitable sets of crossed electrodes allows the investigation of the transport of ions in both directions (Douglass, 1964; Springett *et al.*, 1965; Tanner, 1966).

A very schematic electrode set-up is shown in Fig. 11.1. Ions produced at the source usually drift toward the collector C if the cell is at rest. If the cell is set into rotation with angular frequency Ω, vortex lines develop parallel to the rotation axis and capture some of the ions. If, after a loading period, the rotation is stopped, the trapped charge is released and reaches the collector again. The amount of charge trapped Q is proportional to the areal density of the vortex lines and, experimentally, to Ω (Pratt and Zimmermann, 1969). The proportionality is shown in Fig. 11.2, confirming the validity of eqn (7.15).

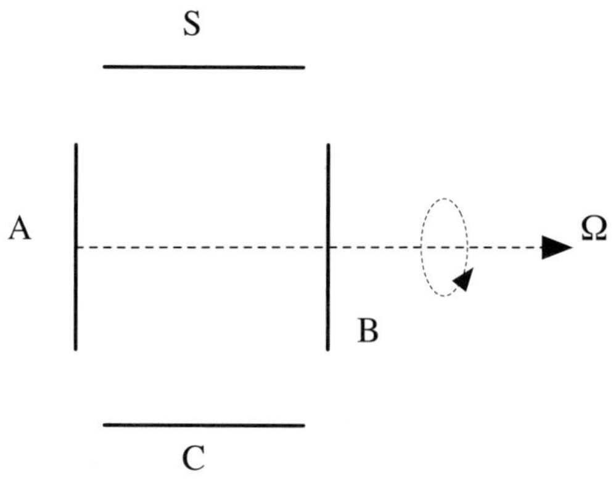

FIG. 11.1. Schematic electrode arrangement for measuring the interaction of ions and vortex lines. S is the ion source, A, B, and C are the collectors, and Ω is the rotation frequency.

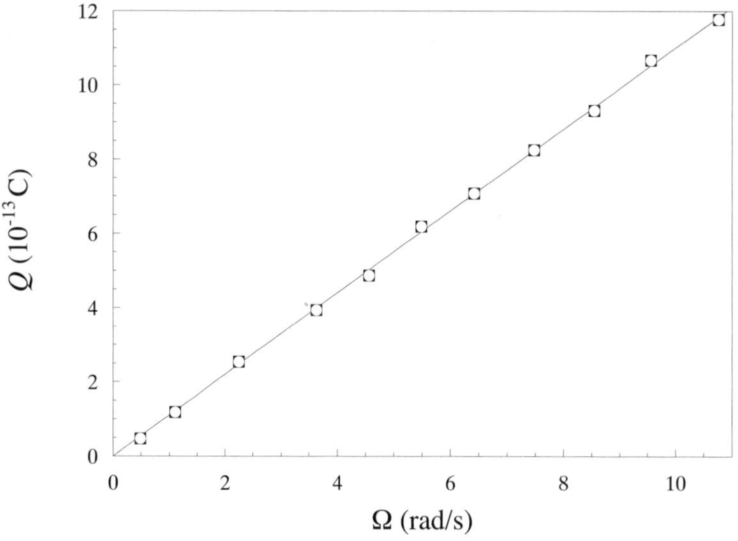

FIG. 11.2. Charge Q released from the vortex lines after rotation was stopped vs Ω. (Pratt and Zimmermann, 1969.)

11.1 Basic phenomenology of ion capture on vortex lines

Under suitable field conditions, ions are trapped on vortex lines for considerable periods of time, if they are prevented from leaking out of the ends of the lines. Both negative and positive ions are trapped below 0.6 K, but only negative ions are trapped above 1.1 K (Douglass, 1966).

The trapping time is typically measured by first loading the vortex lines by injecting ions into the liquid during rotation. Then, injection is cut off by suitably reversing the polarity of the injection electrodes. After a given time t, the rotation is stopped and the remaining ions trapped on vortices are released and collected (Pratt and Zimmermann, 1969). The collected charge depends exponentially on t:

$$Q = Q_0 \exp(-t/\tau), \tag{11.1}$$

where τ is the lifetime or trapping time. A configuration with a second collector, perpendicular to the rotation axis (Douglass, 1964), allows us to perform the measurements without stopping rotation.

The lifetime for negative ions in the range $1.62\,\mathrm{K} < T < 1.70\,\mathrm{K}$ at SVP is shown in Fig. 11.3. In the given range, the trapping time τ is described very well by an Arrhenius-type law:

$$\tau = \tau_0 \exp(\mathcal{E}_0/k_\mathrm{B}T), \tag{11.2}$$

with $\mathcal{E}_0 = 0.012\,\mathrm{eV}$ (Douglass, 1964), suggesting that $\mathcal{E}_0$ is associated with the depth of a potential well in which ions are trapped.

The magnitude of the trapping time τ depends on the electric field, especially at low strength, as shown in Fig. 11.4. Here τ is reported as a function of the average electric field $E_\mathrm{av} = 2V/[(b+a)\ln(b/a)]$, where V is the potential difference between the source and collector, a is the outer radius of the source, and b

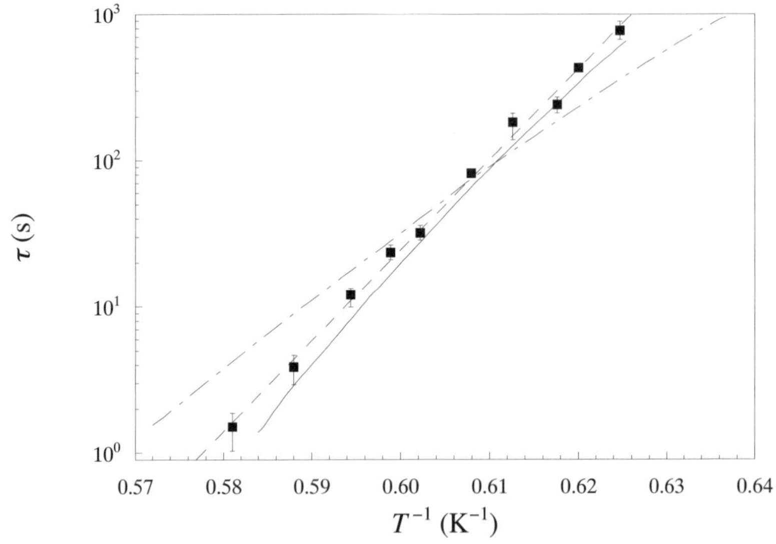

FIG. 11.3. τ vs T^{-1} for negative ions for $E = 2.5\,\mathrm{kV/m}$ (Douglass, 1964). Dash–dotted line: model of Parks and Donnelly (1966). Solid line: model of Padmore (1972b). Dashed line: fit to the data using eqn (11.2).

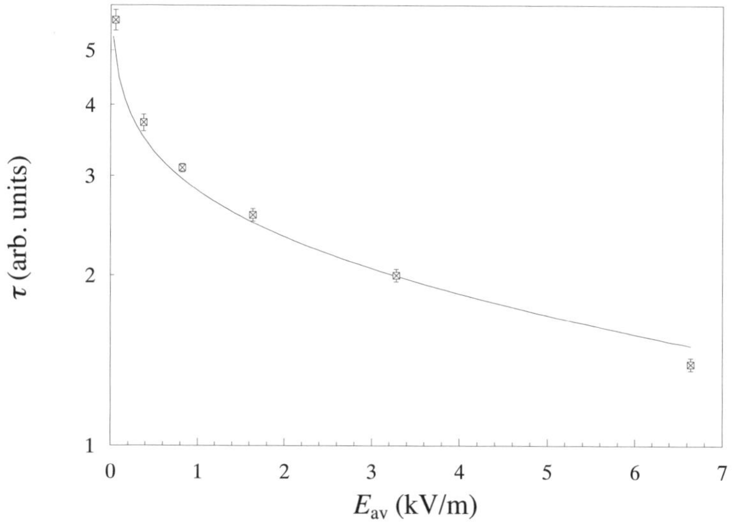

FIG. 11.4. τ vs E_{av} for $T = 1.65\,\mathrm{K}$ (Pratt and Zimmermann, 1969). Line: low-field Brownian motion theory (McCauley and Onsager, 1975*b*).

is the inner radius of the collector. τ increases with decreasing E, although the activation energy $\mathcal{E}_0$ does not depend on E (Pratt and Zimmermann, 1969).

11.1.1 *Capture cross-section or capture width*

The trapped ions can also move along the vortex lines under the influence of the electric field applied across the electrodes A and B.

In Fig. 11.5 the way in which the current is divided between the two mutually-perpendicular electrodes is shown as a function of Ω. The total current, i.e., the sum of the contributions of both electrodes, is approximately constant. However, the fraction of the charge that is collected at C after crossing the array of vortex lines decreases exponentially with increasing Ω (Tanner, 1966).

When crossing a region filled with quantized vortex lines, the ions are continuously captured by them, and their number, and hence the current, decreases as the distance they have to negotiate increases, and the concept of a cross-section is naturally introduced.

As the vortex lines extend to the boundaries in one dimension, the relevant physical quantity is the width a line presents to the ions. The ion–vortex cross-section thus has dimensions of length (Springett *et al.*, 1965; Tanner, 1966).

If the density of lines is large enough to use the continuum approximation, the ion current flowing in the y-direction perpendicular to the axis of rotation is given by

$$I = I_0 \exp\left[-\left(2\Omega\sigma y/\kappa\right)\right], \tag{11.3}$$

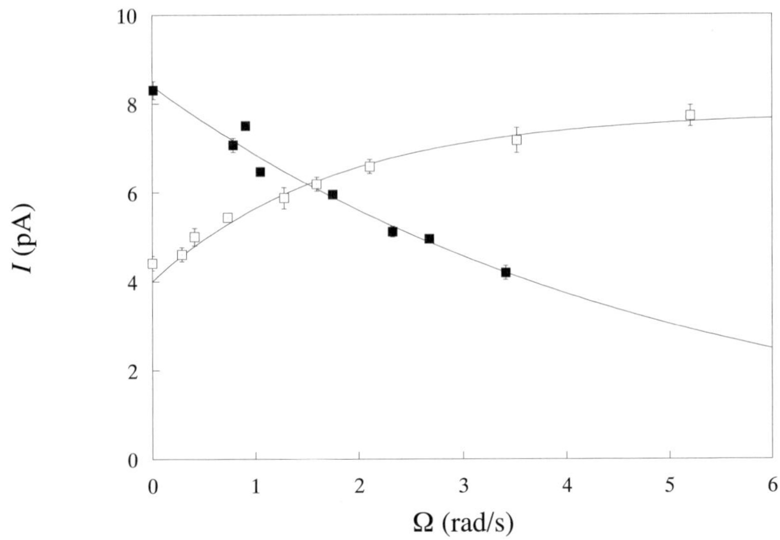

FIG. 11.5. I vs Ω for negative ions arriving at two mutually-perpendicular collectors. (Tanner, 1966.) Closed squares: collector C, and open squares: collector B.

where $2\Omega/\kappa = n_v$ is the areal density of lines, and σ is the ion–vortex line cross-section or capture width. Henceforth, capture cross-section and capture width will be used synonymously.

Equation (11.3) is the basis for any beam experiments aimed at measuring a cross-section. In Fig. 11.6 the exponential dependence of I on Ω is shown. A fit to the data yields the capture width σ. The measurements are carried out in a triode cell (see Fig. 5.2) that is rotated perpendicular to the cylindrical cell axis joining the source, grid, and collector (Springett *et al.*, 1965; Tanner, 1966).

In Fig. 11.7 the cross-section σ, measured for $E \approx 4.35\,\text{kV/m}$, is shown as a function of T. The solid line has no theoretical meaning. The dashed line is the prediction of the model of Donnelly (1965). The sharp cut-off at high T is due to the rapid decrease of the trapping time.

Similar to the trapping times, the capture cross-section also depends on E, as shown in Fig. 11.8.

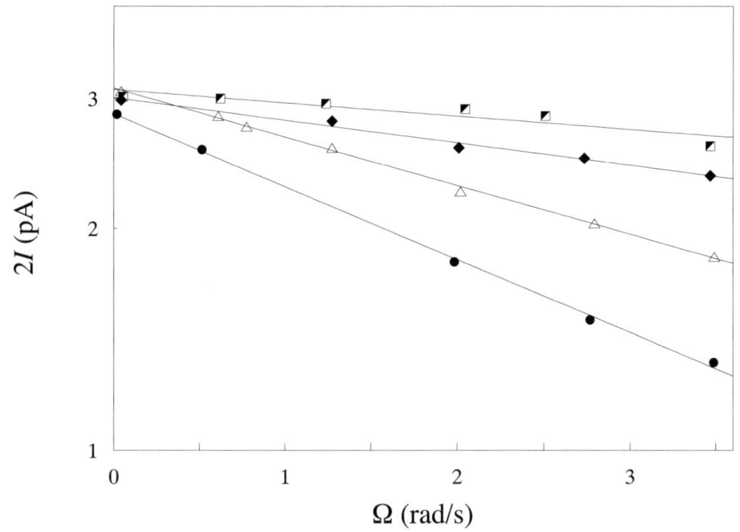

FIG. 11.6. I vs Ω for negative ions in He II at SVP. (Springett *et al.*, 1965.) $T\,(\mathrm{K}) =$ 1.845, 1.148, 1.352, and 1.579 (from top).

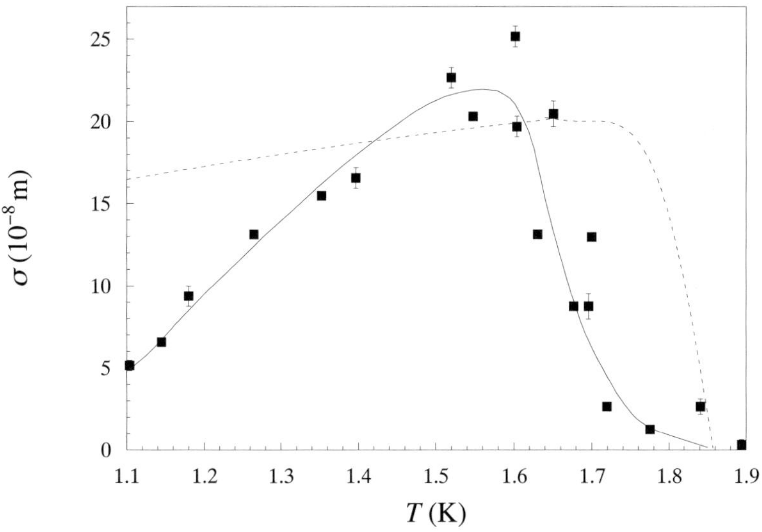

FIG. 11.7. σ vs T for negative ions at $E \approx 4.35\,\mathrm{kV/m}$ (Springett *et al.*, 1965). Solid line: eyeguide. Dashed line: theory for a negative ion of radius of 12.1 Å and mass $M_{-} = 100 m_4$ (Donnelly, 1965).

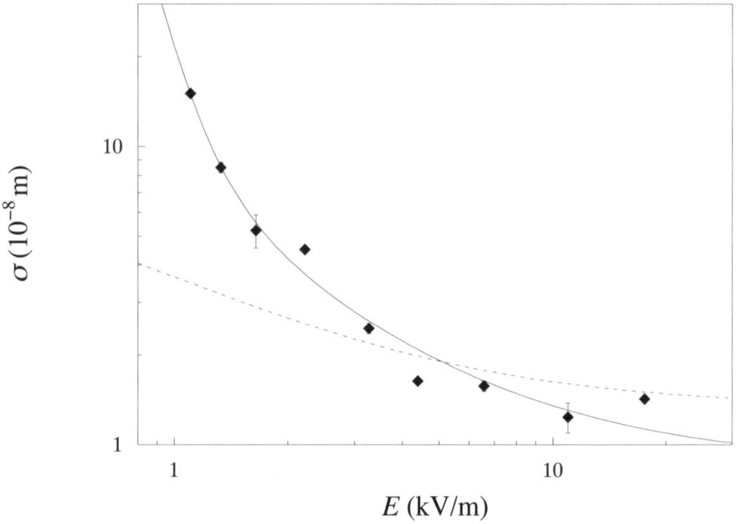

FIG. 11.8. σ vs E for negative ions at $T = 1.46\,\mathrm{K}$ (Springett *et al.*, 1965). Solid line: eyeguide. Dashed line: theory (Donnelly, 1965).

11.2 The model of Brownian diffusion

The mechanism of ion trapping on vortex lines has been elucidated by a hydrodynamic model (Donnelly, 1965, 1967; Parks and Donnelly, 1966).

The ion is assumed to be a Brownian particle immersed in the gas of elementary excitations of He II. It moves under the action of the field E. In the neighborhood of a rectilinear vortex the ion also experiences a Bernoulli force directed toward the center of the vortex and eventually becomes localized in the potential well at the vortex core (Williams and Packard, 1978). The motion of the ion toward the center of the vortex is assumed to be two-dimensional, with r being the distance between the ion and the center of the vortex.

The rotating superfluid in the vortex gives rise to a radial pressure gradient

$$\frac{\mathrm{d}P}{\mathrm{d}r} = \rho_s \frac{v_s^2}{r} \tag{11.4}$$

that attracts the ion toward the center of the vortex itself. ρ_s and v_s are the density and velocity of the superfluid, respectively. For $r \gg R_i$, where R_i is the ion radius, this hydrodynamic attraction can be derived from the hydrodynamic potential $U(r)$, first obtained long ago by Thomson (1873), on applying the quantization restriction to the circulation (Pratt and Zimmermann, 1969):

$$U(r) \approx -\frac{3}{4}\rho_s \left(\frac{\hbar}{m_4}\right)^2 \left(\frac{4\pi R_i^3}{3}\right) r^{-2} \equiv -\frac{C_v}{r^2}\,. \tag{11.5}$$

When the ion is close to a vortex, the energy of the system is reduced by an amount equal to the kinetic energy of the rotating superfluid replaced by the ion. By assuming that the ion does not perturb the superfluid density outside its radius R_i, the substitution energy $U(r)$ is obtained as

$$U(r) = \int_{\text{ion}} \frac{1}{2}\rho_s' v_s^2 \mathrm{d}V \,, \tag{11.6}$$

where $v_s = \kappa/2\pi r$. The superfluid density close to the vortex core ρ_s' is given by Fetter's approximate relation (Fetter, 1963, 1971) that takes into account the healing length. The integration is extended to the volume occupied by the ion.

The substitution energy can be calculated for two cases in a simple way. For the ion on the vortex, i.e., $r = 0$, one gets

$$U(0) = -2\pi\rho_s \left(\frac{\hbar}{m_4}\right)^2 R_i \left[1 - \left(1 + \frac{a^2}{R_i^2}\right)^{1/2} \sinh^{-1}\left(\frac{R_i}{a}\right)\right], \tag{11.7}$$

where a is the healing length. This is simply the negative of the kinetic energy of a spherical volume of fluid centered on the vortex in the absence of the ion.

The second case is for $r \gg a$ and one obtains

$$U(r) = -2\pi\rho_s \left(\frac{\hbar}{m_4}\right)^2 R_i \left[1 - \left(\frac{r^2}{R_i^2} - 1\right)^{1/2} \sin^{-1}\left(\frac{R_i}{r}\right)\right]. \tag{11.8}$$

It is easily shown that eqn (11.8) gives eqn (11.5) for $r \gg R_i$, that represents the substitution energy for a point ion (Parks and Donnelly, 1966).

If, in addition, there is the presence of an electric field in the x-direction, the total energy of the ion is

$$U_T(r) = U(r) - qEx \,, \tag{11.9}$$

where q is the ionic charge. In Figs 11.9 and 11.10, $U_T(r)$ is shown for positive and negative ions, respectively. Once the ion–vortex interaction potential is known, the two problems of the calculation of the escape rate and of the capture cross-section can be addressed separately.

11.2.1 *Escape rate*

The problem of the escape of the ion from the potential well is treated as the thermally-activated escape of a Brownian particle over a potential barrier (Parks and Donnelly, 1966), following Chandrasekhar (1943). The potential sketched in Figs 11.9 and 11.10 has a minimum a A and a saddle point at C. At C the electric field lowers the height of the well by an amount proportional to E.

The well depth is $\Delta u = U_C - U_A \approx U_C - U(0)$. At C the total potential energy is given by (Springett, 1967)

$$U_C = -2\pi\rho_s \left(\frac{\hbar}{m_4}\right)^2 \frac{R_i^3}{3x_c} - qEx_c \,, \tag{11.10}$$

where x_c is the coordinate of the maximum of the potential.

If the potential is expanded in series around A and C, four characteristic frequencies can be defined:

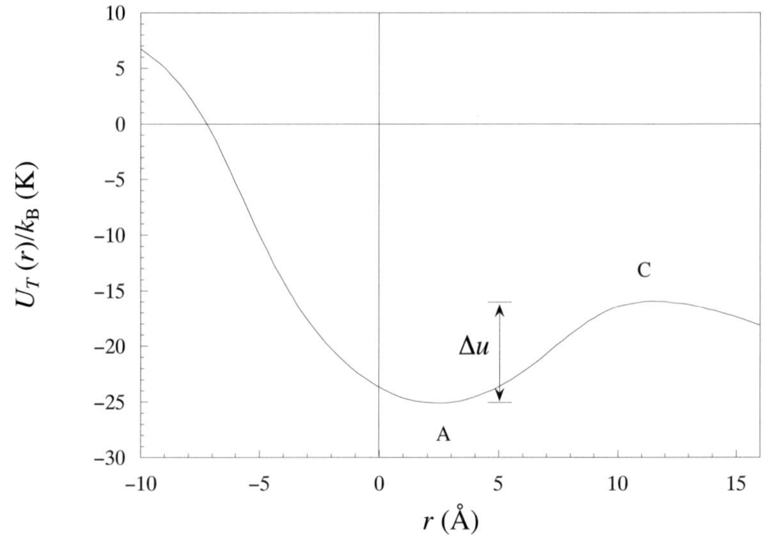

FIG. 11.9. U_T (in the $y = 0$ plane) vs r for a positive ion in the combined vortex and electric fields. (Parks and Donnelly, 1966.) U_T is calculated for $R_i^+ = 7.9$ Å, $E = 0.7$ MV/m, and $T < 1$ K.

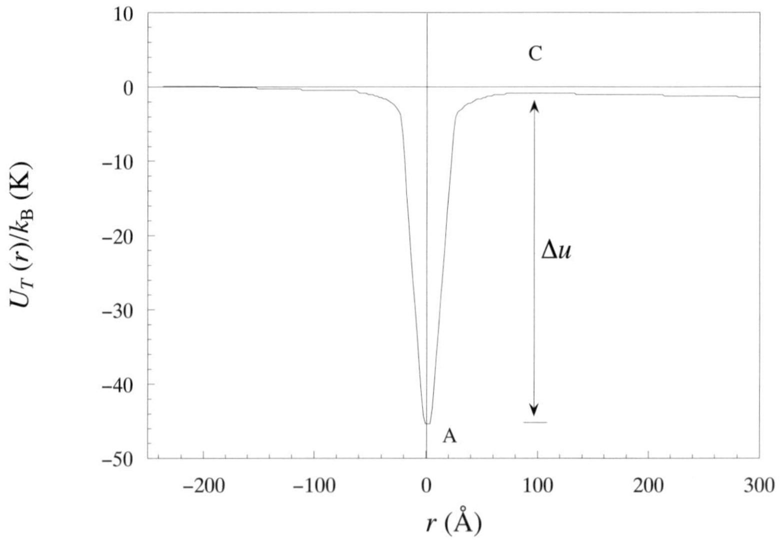

FIG. 11.10. U_T (in the $y = 0$ plane) vs r for a negative ion in the combined vortex and electric fields. (Parks and Donnelly, 1966.) U_T is calculated for $R_i^- = 15.96$ Å, $E = 2.5$ kV/m, and $T = 1.64$ K.

$$\omega_{Ax}^2 = \frac{1}{M_i}\left(\frac{\partial^2 U_T}{\partial x^2}\right)_A, \quad \omega_{Ay}^2 = \frac{1}{M_i}\left(\frac{\partial^2 U_T}{\partial y^2}\right)_A, \tag{11.11}$$

$$\omega_{Cx}^2 = -\frac{1}{M_i}\left(\frac{\partial^2 U_T}{\partial x^2}\right)_C, \quad \omega_{Cy}^2 = \frac{1}{M_i}\left(\frac{\partial^2 U_T}{\partial y^2}\right)_C, \tag{11.12}$$

where M_i is the ion mass.

The escape probability per unit time, $\mathcal{P}$, is obtained by integrating the Fokker–Planck equation, yielding

$$\mathcal{P} = \frac{1}{2\pi}\frac{\omega_{Ax}\omega_{Ay}}{\omega_{Cx}\omega_{Cy}}\left[\left(\omega_{Ax}^2 + \frac{1}{4}\beta^2\right)^{1/2} - \frac{\beta}{2}\right] e^{-\Delta u/k_B T}, \tag{11.13}$$

where $\beta = e/M_i\mu$ is the friction coefficient and μ the ion mobility. Physically, ω_{Ax} and ω_{Ay} represent the frequencies at which the ion comes up against the barrier, whereas ω_{Cx} and ω_{Cy} represent tunneling probabilities per unit time (Springett, 1967). The trapping time is simply $\tau = \mathcal{P}^{-1}$.

Some results of the model are shown in Fig. 11.11. The escape rate for positive ions is much larger than for negative ions because the well is much shallower.

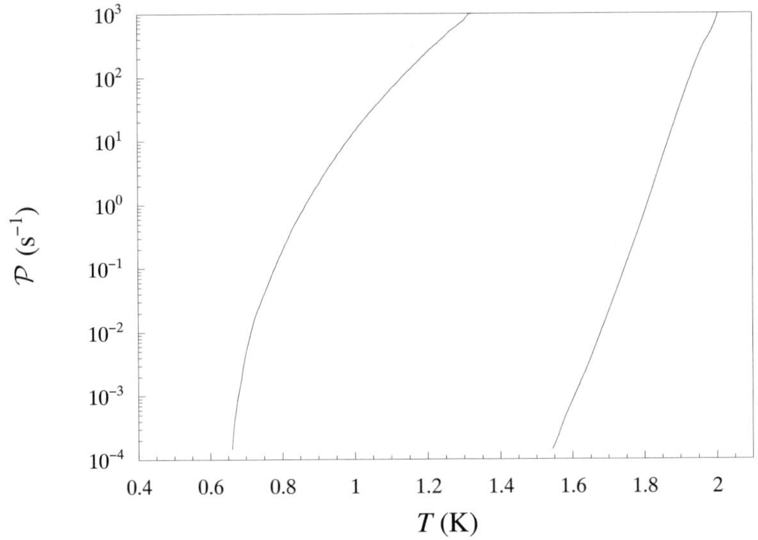

FIG. 11.11. $\mathcal{P}$ vs T for trapped ions in an exernal field $E = 2.5\,\text{kV/m}$. (Donnelly, 1965.) Positive ions: left curve, and negative ions: right curve. The curves are calculated by assuming $R_i^+ = 6.3$ Å and $R_i^- = 12.1$ Å, respectively, and $M_i = 100m_4$. For the positive ion $\Delta u/k_B = 39.4\,\text{K}$ at $T = 0.5\,\text{K}$, and $\Delta u/k_B = 19.6\,\text{K}$ at $T = 1.0\,\text{K}$. For the negative ion $\Delta u/k_B = 50.7\,\text{K}$ at $T = 1.0\,\text{K}$, and $\Delta u/k_B = 19.3$ K at $T = 1.8\,\text{K}$.

This explains why the trapping of positive ions on vortices is observed only at much lower temperatures than for the negative ions.

The previous results can also be applied to large vortex rings (Donnelly, 1965). The mean lifetime of positive ions in vortex rings has been determined experimentally at low T in a field-assisted escape experiment (Cade, 1965). Under these conditions, $\beta \ll \omega_{Cx}$ and $\mathcal{P}$ is inversely proportional to the ion mass.

In Fig. 11.12 the experimental results are compared with the present theory. An ionic radius $R_i^+ = 7.9$ Å has been used for the computation of $\mathcal{P}$. The theory accounts quantitatively for the experimental results.

The mean lifetime of negative ions in vortex lines measured by Douglass (1964) is also quite well described by this theory, as shown in Fig. 11.3. At the much higher temperature of this experiment the opposite condition $\beta \gg \omega_{Cx}$ is fulfilled and M_i influences $\mathcal{P}$ only through second-order terms.

An improvement of the theory considers the reaction force exerted on the vortex line by the approaching ion (Padmore, 1972*b*). The vortex line is not fixed in space and it deforms in response to this force. The vortex core will move so as to generate a Magnus force that cancels the external forces (Rayfield and Reif, 1964). Because the force depends on the vortex configuration, and hence on its past history, it realizes a "non-Markoffian" process. An educated guess leads to the introduction of a relative correction to the acceleration imparted to the ion by the vortex given by $1 - \gamma e/(\mu \rho_s \kappa R_i)$ (Padmore, 1972*b*), where μ is the ion mobility and $\gamma < 1$ is an adjustable parameter. This produces the same correction in the potential well, which is now $\Delta u_{\text{eff}} = \Delta u(1 - \gamma e/\mu \rho_s \kappa R_i)$. For $0 < \gamma < 1$,

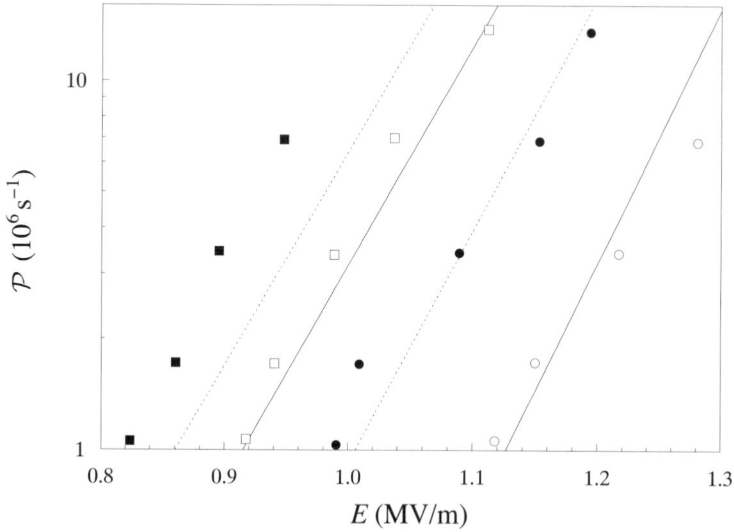

FIG. 11.12. $\mathcal{P}$ vs E for positive ions. (Cade, 1965; Parks and Donnelly, 1966.) T (K) = 0.571, 0.532, 0.472, and 0.397 (from left). Lines: theory.

the barrier height is lowered, but a strong temperature dependence is introduced through the mobility. This effect brings the theory closer to the experiment, as shown by the dashed line in Fig. 11.3 that is obtained with $R_i^- = (17.75 \pm 0.1)$ Å and $\gamma = 0.330$.

The theory of Parks and Donnelly (1966) also describes quantitatively the trapping of negative ions observed in turbulent superfluid helium (Sitton and Moss, 1969). This experiment shows that negative ions are indeed captured by vortex lines produced by a supercritical heat current up to above 1.8 K and that the turbulent structure can be described as an irregular, or *tangled* mass of vorticity, as suggested by Vinen (1957) and Hall (1960), with vortex lines identical to those produced by steady rotation of the liquid except for the spatial configuration.

The solution to the problem of the diffusion of a Brownian particle given by Parks and Donnelly (1966) is valid for fairly large fields. This condition is not always met in the experiments that are instead performed at quite small fields. A more rigorous treatment of the low-field situation has been presented by McCauley and Onsager (1975*a*, 1975*b*) in order to better explain the increase of the mean trapping time at low fields observed in the experiments (Pratt and Zimmermann, 1969) and shown in Fig. 11.4.

11.2.2 *Capture width*

In order to calculate the capture width, ions are still assumed to diffuse as classical Brownian particles through the gas of elementary excitations of He II. This hypothesis, of course, limits the validity of this approach to rather high temperatures, at which the mean free path for ion–roton interaction is short enough.

The driving forces for ion diffusion are their concentration gradient, the applied electric field, and the hydrodynamic suction force. The appropriate equation for the problem at hand is the Smoluchowski equation for the probability density w of finding an ion at a specified position (Chandrasekhar, 1943):

$$\frac{\partial w}{\partial t} = \boldsymbol{\nabla} \cdot \left(D \boldsymbol{\nabla} w - \frac{\mathbf{K}}{\beta} w \right) , \tag{11.14}$$

where $\mathbf{K} = -\boldsymbol{\nabla}\varphi$ is the ion acceleration due to all forces, $\beta = e/\mu M_i$ is the friction coefficient, and μ is the ion mobility.

The potential φ is given by the Bernoulli potential eqn (11.5) and by that due to the applied electric field:

$$\varphi = -\frac{C_v}{r^2} - q\mathbf{E} \cdot \mathbf{r} \, . \tag{11.15}$$

In steady state, eqn (11.14) reduces to

$$\nabla^2 w = -\boldsymbol{\nabla} \cdot \left[(\boldsymbol{\nabla} u) \, w \right] , \tag{11.16}$$

where the Nernst–Einstein relation $D/\mu = k_{\mathrm{B}}T$ has been used and $u = \varphi/k_{\mathrm{B}}T$.

By introducing the explicit analytic form for the potentials, one gets (Donnelly and Roberts, 1969*b*)

$$\nabla^2 w = 2\boldsymbol{\nabla} \cdot \left(\frac{1}{r_E} \hat{u}_x - \frac{r_v^2}{r^3} \hat{u}_r \right) w\,, \tag{11.17}$$

where $\hat{u}_r$ and $\hat{u}_x$ are unit vectors. Three length scales are naturally introduced:

$$r_E = \frac{2eE}{k_\mathrm{B} T}\,, \tag{11.18}$$

$$r_v = \left(\frac{C_v}{k_\mathrm{B} T} \right)^{1/2}\,, \tag{11.19}$$

$$r_c = \left(r_E r_v^2 \right)^{1/3}\,, \tag{11.20}$$

where r_E is the characteristic length of ion diffusion in an electric field, r_v is the length at which hydrodynamic attraction dominates thermal fluctuations in the absence of the field (Donnelly, 1965), and r_c defines the boundary at which the diffusion and attraction to the vortex core balance the electric force. For $T \approx 1\,\mathrm{K}$, $r_v \ll r_c \ll r_E$.

If the vortex is aligned along the z-axis of a polar coordinate system, the boundary conditions for w are $\lim_{r\to 0} w(r,\theta) = 0$ and $\lim_{r\to\infty} w(r,\theta) = 1$. The ion capture cross-section σ is the ratio of the ion flux into the vortex divided by the flux at infinity and is given by (Ostermeier and Glaberson, 1975*c*)

$$\sigma = r_E \lim_{r\to 0} \int_0^\pi \left(r\frac{\partial w}{\partial r} + rw\frac{\partial u}{\partial r} \right) \mathrm{d}\theta\,. \tag{11.21}$$

The solution of eqn (11.21) with the given boundary conditions is quite difficult because of the mixed symmetry of the problem. An exact solution exists for some limiting cases. In the absence of an electric field, kinematic arguments yield $\sigma = 2r_v$ (Tanner, 1966; Donnelly and Roberts, 1969*b*). Assuming that the radii of positive and negative ions are $R_+ = 7.9\,\text{Å}$ and $R_- = 16\,\text{Å}$, respectively, and that their mass is of order 100 atomic masses, then one gets $r_v \approx 12\,\text{Å}$ at $T = 1.6\,\mathrm{K}$.

In the opposite case of such strong a drag that inertial terms can be neglected, the integration of the equation of motion gives capture for an impact parameter $b \leqslant b^\star = [(3/2)\pi]^{1/3} r_c$, and $\sigma = 2b^\star = (12\pi)^{1/3} r_c$ is obtained.

In order to solve eqn (11.21), the vortex is approximated by a *soft* square well of radius d with capture strength a defined by the condition $w(r = d, \theta) = a$, where $0 < a < 1$ is an adjustable parameter (Tanner, 1966). This condition replaces the condition $\lim_{r\to\infty} w(r,\theta) = 1$.

The cross-section therefore becomes (Donnelly and Roberts, 1969*b*)

$$\sigma = \pi r_E \left(1 - a\right) \left[\frac{I_0\left(d/r_E\right)}{K_0\left(d/r_E\right)} + 2\sum_{n=1}^{\infty} \left(-1\right)^n \frac{I_n\left(d/r_E\right)}{K_n\left(d/r_E\right)} \right], \tag{11.22}$$

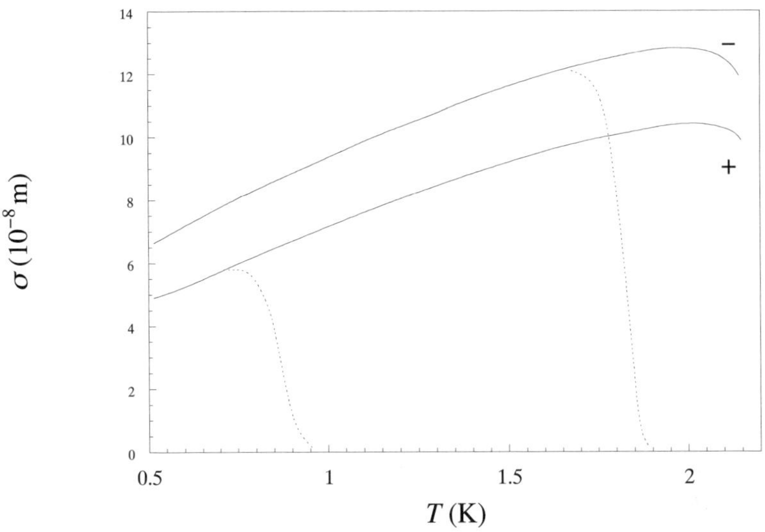

FIG. 11.13. σ vs T for positive (+) and negative (−) ions at $E = 2.5\,\mathrm{kV/m}$. (Donnelly, 1965.) Solid line: eqn (11.22). Dashed lines: effective capture width $\sigma \exp(-\mathcal{P}t^{\star})$ for $t^{\star} = 1\,\mathrm{s}$. $\mathcal{P}$ is the escape probability per unit time.

where I_n and K_n are the modified Bessel functions of first and second kind.

Donnelly and Roberts (1969*b*) assume that d is of the order of the radius at which the electric and vortex fields are nearly equal, so that $d = r_c$. It is further conjectured (Tanner, 1966) that $a = 1/2$ at $r = d$, because ions have nearly equal probabilities of being captured by the vortex or to move away from it in the direction of the electric field, owing to the nearly equal magnitudes of the two competing fields.

In Fig. 11.13 the capture width for $E = 2.5\,\mathrm{kV/m}$ is shown as a function of T for both types of ion. The solid lines are eqn (11.22). The effective cross-section, however, shows a cut-off (or *lifetime edge*) at high T owing to the finite trapping time, and can be written as $\sigma_{\mathrm{eff}} = \sigma \exp(-\mathcal{P}t^{\star})$, where $\mathcal{P}$ is the ion escape probability per unit time and $t^{\star}$ is some characteristic time of the experiment (such as the length of the collector divided by the ion velocity in the vortex lines). For $t^{\star} = 1\,\mathrm{s}$, σ_{eff} for the positive ions is cut off rapidly at $T \approx 0.85\,\mathrm{K}$, whereas the cut-off for the negative ions occurs for $T \approx 1.8\,\mathrm{K}$. This is the reason why no trapping of positive ions is observed at high T, while at low T both ions are expected to be trapped. The reasonable agreement of this model with experiment is presented in Fig. 11.7.

11.3 The effect of pressure on the radius of negative ions

Important pieces of information about the pressure dependence of the radius of the electron bubble are gathered from measurements of trapping time and capture width of negative ions on vortex lines (Springett and Donnelly, 1966; Springett, 1967; Springett *et al.*, 1967; Pratt and Zimmermann, 1969).

It can be shown (Pratt and Zimmermann, 1969) that, for $\beta \gg \omega_{Cx}$ and for not too low temperatures, the trapping time can be written as

$$\tau \approx \tau_0 \exp\left\{-\ln\frac{\rho_s}{\rho_{s,0}} + \frac{1}{k_{\mathrm{B}}T}\left[-\Delta(P) + \left(\frac{\rho_s}{\rho_{s,0}}\right)^{1/3} U_{C,0} - \frac{\rho_s}{\rho_{s,0}} U_{A,0}\right]\right\}, \tag{11.23}$$

where U_A and U_C are given by eqns (11.7) and (11.10), respectively. The subscript "0" denotes some convenient common fixed reference T and P at which the quantities so labeled are to be evaluated. Using the approximate expression for the ion mobility $\mu(T,P) \approx \mu_\infty \exp \Delta(P)/k_{\mathrm{B}}T$, with $\mu_\infty = 1.2 \times 10^{-7}\,\mathrm{m^2/V\,s}$, where Δ is the roton energy gap, the constant τ_0 becomes

$$\tau_0(R_-) = (2\pi e/\sqrt{3})\mu_\infty M_i(R_-)\omega^2_{A,x}(\rho_{s,0}/\rho_s)\,. \tag{11.24}$$

Owing to the lack of accurate knowledge of ω_{Ax}, the pressure dependence of R_- is obtained by fitting the slope of the trapping time (Pratt and Zimmermann, 1969) and the results are shown in Fig. 11.14.

The data are compared with the prediction of a simplified bubble model, in which the well depth is assumed to be infinite. The radius is then obtained by minimizing for each pressure P the electron energy in the form

$$\mathcal{E} = \frac{\pi^2\hbar^2}{2m_e R_-^2} + 4\pi\sigma_{lv}R_-^2 + \frac{4\pi}{3}R_-^3 P \tag{11.25}$$

with respect to variations of R_-. The value of the bulk surface tension of the liquid is adjusted so as to give $R_- = 19.5\,\text{Å}$ at $P = 0$, in agreement with previous results (Pratt and Zimmermann, 1969). The data agree well with the prediction of the model. Although the absolute value of the ion radius might be slightly different from values obtained in other experiments (see Fig. 3.6), these measurements of the lifetime of negative ions trapped on vortex lines confirm the idea that the electron bubble is *squeezed* by an increase in the applied pressure.

Measurements of the pressure dependence of the capture width for negative ions on vortex lines give similar results (Springett and Donnelly, 1966; Springett, 1967). In Fig. 11.15 the capture cross-section is reported as a function of T for several P, while in Fig. 11.16 the cross-section is shown as a function of P for two temperatures. The cut-off (*lifetime edge*), observed for $T \approx 1.7\,\mathrm{K}$ at SVP (Springett *et al.*, 1965) and interpreted by Parks and Donnelly (1966) as due to the finite trapping time of ions on vortex lines, is also observed at all other pressures. The lifetime edge shifts to lower T as P is increased (see Fig. 11.15). This can easily be explained by the fact that the ion radius decreases with increasing P (Springett and Donnelly, 1966; Springett, 1967). In fact, the capture width and lifetime edge are well described by

$$\sigma = \sigma_0 e^{-\mathcal{P}t^\star}\,. \tag{11.26}$$

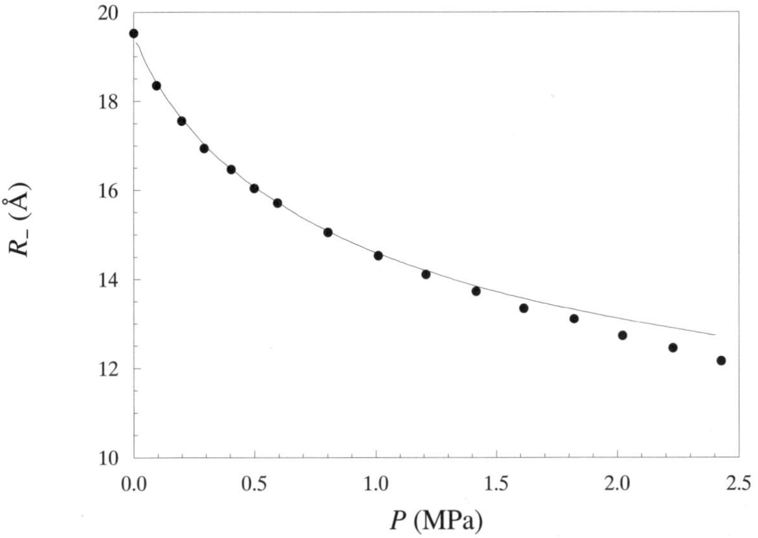

FIG. 11.14. R_{-} vs P from measurements of trapping time on vortex lines. (Pratt and Zimmermann, 1969.) Line: simple bubble model with surface tension adjusted so as to give $R_{-} = 19.5$ Å at $P = 0$.

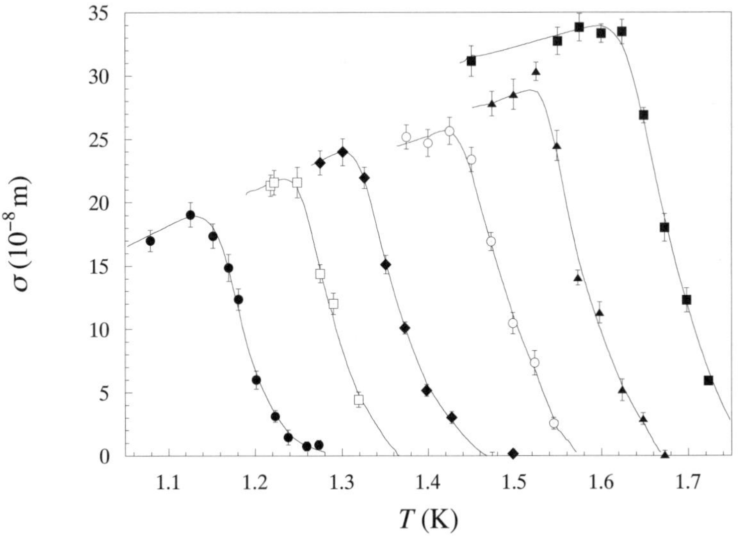

FIG. 11.15. σ vs T for negative ions at $E = 2.0$ kV/m. P (MPa) = 1.956, 1.327, 0.838, 0.419, 0.172, and vapor pressure (from the left). (Springett, 1967.) Lines: eyeguides.

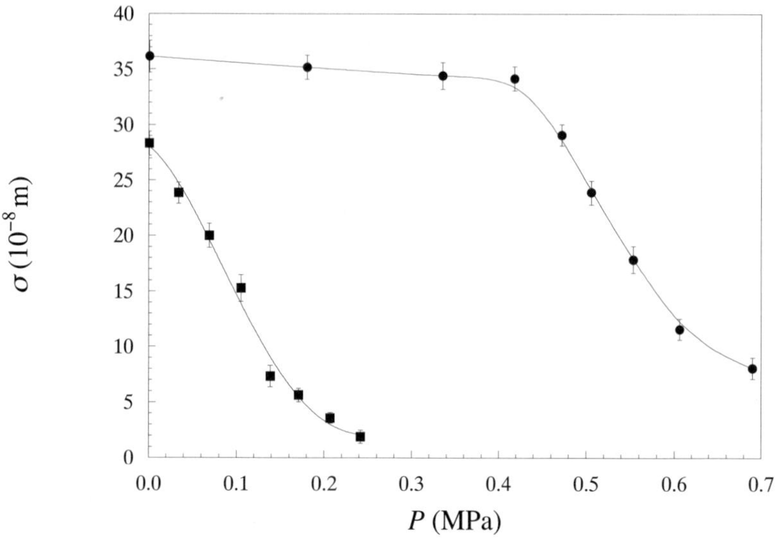

FIG. 11.16. σ vs P at $E = 2.0\,\mathrm{kV/m}$ for $T\,(\mathrm{K}) = 1.635$ (squares) and 1.448 (circles). (Springett, 1967.) Lines: eyeguides.

σ_0 is the cross-section well below the cut-off and is calculated theoretically (Donnelly, 1965; Donnelly and Roberts, 1969*b*; Ostermeier and Glaberson, 1975*c*).

$\mathcal{P}$ is the ion escape rate from vortices, which strongly depends on the ion radius, as stated by eqn (11.13). $t^\star$ is a characteristic time for the ion to remain within the experimental region. It is determined by recording the (P, T) coordinates at which $\sigma = \sigma_0/2$. At vapor pressure $R_- = 15.96$ Å (Parks and Donnelly, 1966), $\mathcal{P}$ is calculated by means of eqn (11.13), and, hence, $t^\star$ is determined at vapor pressure and is assumed to remain constant throughout the whole (P, T) range. At any other pressures and temperatures, $\mathcal{P}$ is calculated and inverted so as to give R_-. The results are shown in Fig. 3.6, in reasonable agreement with the prediction of the bubble model.

The cross-section also shows a cut-off as a function of P at constant T (Fig. 11.16). Again, this fact is consistent with both the model of thermally-activated escape over the barrier and with the electron bubble model of the negative ion. If P is increased, the radius of the negative ion decreases. The depth of the potential well also decreases and the increasingly larger escape rate consequently lowers the capture width.

11.4 Ion trapping on vortex lines at low temperature

The phenomenon of ion trapping on vortex lines shows new and unexpected features at low temperature. First of all, for $T < 0.7\,\mathrm{K}$ (Ostermeier and Glaberson, 1975*d*; Williams *et al.*, 1975), the trapping and escape of positive ions are found to be in agreement with the prediction of the Brownian diffusion model (Donnelly, 1965; Parks and Donnelly, 1966).

As $R_+ < R_-$, the well depth in the vortex for the positive ion is much shallower (see Fig. 11.9), and a much lower temperature is necessary to trap it. Moreover, since the capture cross-section of bare positive ions is very small (Donnelly, 1965), charging of the vortex lines is sometimes achieved by crossing the region containing the vortex lines with a beam of positively-charged vortex rings (Ostermeier and Glaberson, 1975*d*).

The interaction cross-section of vortex rings incident on vortex lines is fairly large (Schwarz and Donnelly, 1966; Schwarz, 1968), as shown in Fig. 11.17. Upon hitting a vortex line, the vortex ring is destroyed and the positive ion has a finite probability of remaining trapped on the line. Similar results are obtained for ring–ring interaction, showing that their cross-section is nearly geometrical because their interaction is hydrodynamic rather than electrostatic (Gamota and Sanders, 1968*a*, 1968*b*).

In Fig. 11.18 the trapping time τ of positive ions on vortex lines is shown. There is an *intrinsic* regime at higher temperatures, where $\tau = \tau_0 \exp(A/T)$, in agreement with eqn (11.13). A fit of the data yields $A \approx 17.7\,\mathrm{K}$. If the radius of the positive ion is calculated from this experimental value of the activation energy by using the equations of the Donnelly model, one gets $R_+ \approx (7.7\pm0.1)\,\text{Å}$. This value agrees with the value of 7.9 Å obtained from the analysis of high-field escape from vortex rings (Cade, 1965).

This determination of the radius includes the length over which the superfluid

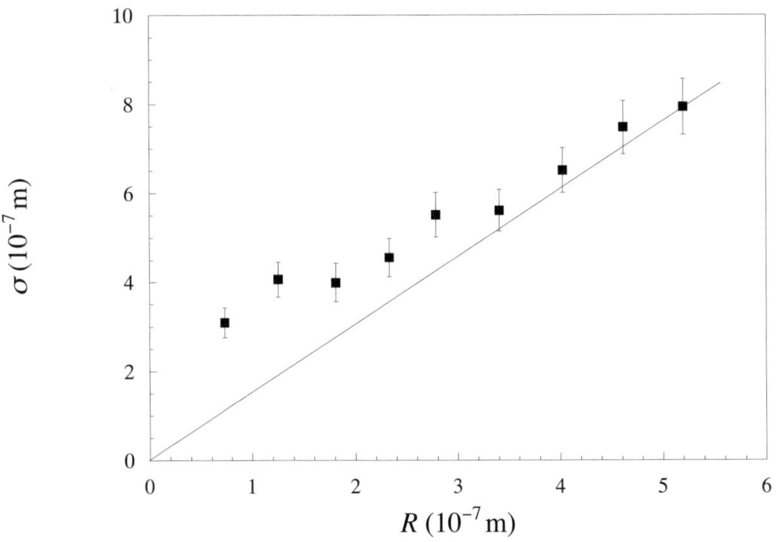

FIG. 11.17. Effective capture cross-section for vortex rings incident on vortex lines as a function of the ring radius. (Schwarz, 1968.) Line: hydrodynamic theory.

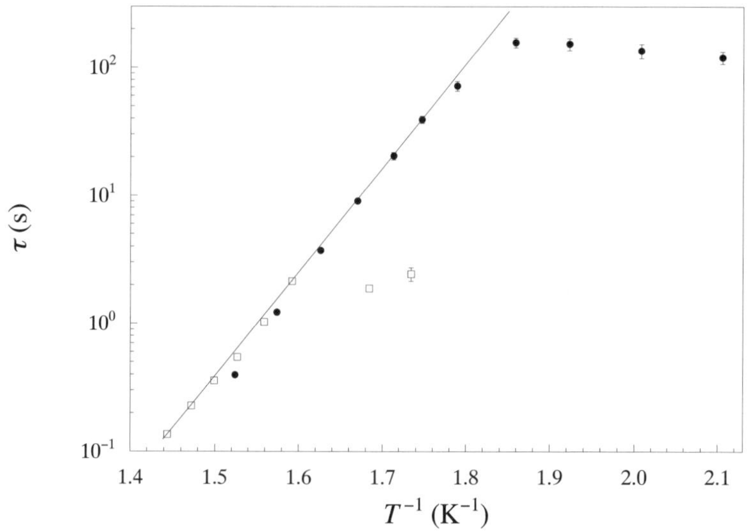

FIG. 11.18. τ vs T^{-1} for positive ions. Squares: Williams *et al.* (1975), and circles: Ostermeier and Glaberson (1975*d*). Line: equation $\tau = \tau_0 \exp(A/T)$, with $A \approx 17.7\,\mathrm{K}$.

density falls to zero at the ion surface. By subtracting a healing length of 1.5 Å, a hard-core ion radius of 6.2 Å is obtained, in agreement with the value 6.1 Å determined in resonance experiments (Poitrenaud and Williams, 1972, 1974).

It is interesting to note that there are deviations from the intrinsic behavior below a temperature that depends on the cell size. The trapping time no longer increases with decreasing T, but instead it saturates or even decreases. This behavior is also shown by negative ions, as first observed by Douglass (1969), subsequently confirmed by DeConde *et al.* (1974), and reported in Fig. 11.19. The data of DeConde *et al.* (1974) are taken with cells of different size, the longer lifetime being observed in the larger-sized container. For higher T, the intrinsic regime is followed closely with an activation energy $A \approx 130\,\mathrm{K}$, in good agreement with other authors (Douglass, 1964; Pratt and Zimmermann, 1969; Sitton and Moss, 1969). For lower T, τ decreases with T and then saturates.

The obvious conclusion is that there are two different escape mechanisms with their own rates, and that the overall rate is the sum of the two. In the high-T region, the thermally-activated intrinsic escape dominates. In the low-T region the extrinsic mechanism is dominant.

It is clear, owing to the dependence on the container size, that the extrinsic process is related to the motion of the charged vortex lines. It is assumed that a vortex line in an array of vortex lines moves toward the walls of the container.

Eventually, upon hitting the wall, the vortex annihilates and releases the trapped charge, which is collected at the wall. In order to maintain the equilibrium line density another vortex line will soon grow to replace the destroyed

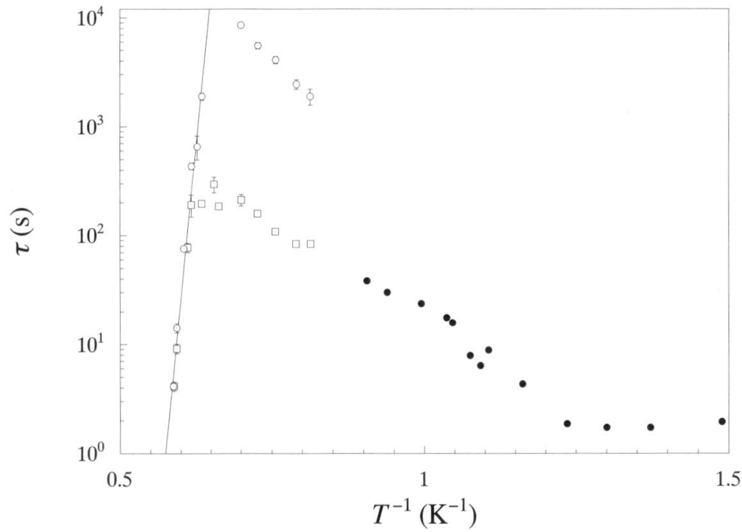

FIG. 11.19. τ vs T^{-1} for negative ions. Open symbols: DeConde *et al.* (1974). Closed symbols: Douglass (1969).

one, but this new line is not charged. If the characteristic time for a vortex line to cross the container is shorter than the intrinsic lifetime, vortex motion will be the dominant charge-loss mechanism (Williams *et al.*, 1975).

The lifetime in the low-T region is thus believed to be representative of the time a vortex lines takes to cross the container. However, this time depends on so many variables, including the past history of the liquid, angular velocity of the container, and so on, that it is nearly impossible to make any quantitative predictions.

It can be, nonetheless, conjectured that the decrease of the lifetime in the extrinsic region with decreasing T might be related to the decrease of the normal fluid density ρ_n that is damping the vortex motion. At sufficiently low temperature, ρ_n becomes negligible and the vortex line motion and lifetime are determined by the undamped vortex line hydrodynamics, and τ should eventually become independent of T, as experimentally observed (Williams *et al.*, 1975). This interpretation is consistent with the observation that electron bubble trapping by vortex lines appears to be reduced in the presence of an axial heat current (Cheng *et al.*, 1973).

11.4.1 *Effect of* ^{3}*He impurities*

If the behavior of the trapping time in the extrinsic region depends on how the motion of vortex lines is damped by collisions with the quasiparticles of the normal fluid, then the addition of ^{3}He atoms should increase the lifetime. Actually, this behavior has been observed in earlier measurements (Douglass,

1969; Ostermeier and Glaberson, 1975*b*).

Moreover, ^{3}He profoundly affects the trapping lifetime because it condenses on the electron bubble surface (Shikin, 1973; Bowley, 1984; Nancolas *et al.*, 1985*a*), thereby changing its radius, and consequently both the potential well Δu and τ_0. If ^{3}He further condenses on the vortex core (Ohmi *et al.*, 1969), the potential well and τ_0 are expected to change (Williams and Packard, 1975). Thus, the measurements of the trapping time provide a sensitive insight into the influence of ^{3}He on these parameters.

The expression for the trapping time is still given by $\tau = \tau_0 \exp{(\Delta u)}/k_{\mathrm{B}}T$, with τ_0 keeping its expression in terms of ion radius R_i and healing length a_0:

$$\tau_0^{-1} = 1.25 \left(\frac{\mu}{e}\right) \frac{\rho_s \left(\hbar/m_4\right)^2}{R_i \left[1 + \left(a_0/R_i\right)^2\right]} \left[-1 + \frac{2 + \left(a_0/R_i\right)^2}{\left[1 + \left(a_0/R_i\right)^2\right]^{1/2}} \sinh^{-1} \frac{R_i}{a_0}\right]. \tag{11.27}$$

This expression remains valid for the mixtures because, at low T, the collision rate of ions with ^{3}He quasiparticles is high in comparison to the case of pure ^{4}He, in which the thermal excitations available for collisions are very few (Williams and Packard, 1978).

The addition of ^{3}He produces several effects on the trapping well and on the ion radius. The well is altered because of changes in the superfluid density ρ_s near the vortex core and in the healing length a_0. It is predicted that the impurity tends to be concentrated near the center of the vortex (Ohmi *et al.*, 1969; Reut and Fizher, 1969; Kuchnir *et al.*, 1972). This phenomenon reduces the trapping potential because the excess ^{3}He atoms lower the superfluid density around the core, thereby reducing the integrand in eqn (11.6).

The ion binding energy would be further decreased if the isotopic-impurity concentration were high enough to produce phase separation. In this case, the vortex core would be practically pure ^{3}He and the core parameter a_0 should be increased from its value in pure ^{4}He. To quite a good approximation, eqn (11.7) still remains valid (Williams and Packard, 1978).

The analysis of the positive ion data is complicated further by the repulsive ion–^{3}He atom interaction that arises from the fact that the electrostrictive interaction due to the positive ion repels the ^{3}He atoms, which have a larger atomic volume (Bowley and Lekner, 1970). This repulsive potential should contribute 1 K, in temperature units, and is comparable to the binding energy of the ^{3}He atoms on the vortex (Williams and Packard, 1978).

Theoretical models give a core radius increasing from 2 Å at 0.6 K to 4 Å at 20 mK for ^{3}He concentrations up to 0.05% (Ohmi *et al.*, 1969; Kuchnir *et al.*, 1972). Moreover, the binding energy of a ^{3}He atom replacing the ion trapped on the vortex is calculated to be $\approx$ 3 K. Therefore, a further reduction of this order of magnitude of the well depth for ion trapping on vortex lines in a dilute ^{3}He–^{4}He mixture is expected (Williams and Packard, 1978).

In Fig. 11.20 the trapping lifetime of positive ions for three different ^{3}He

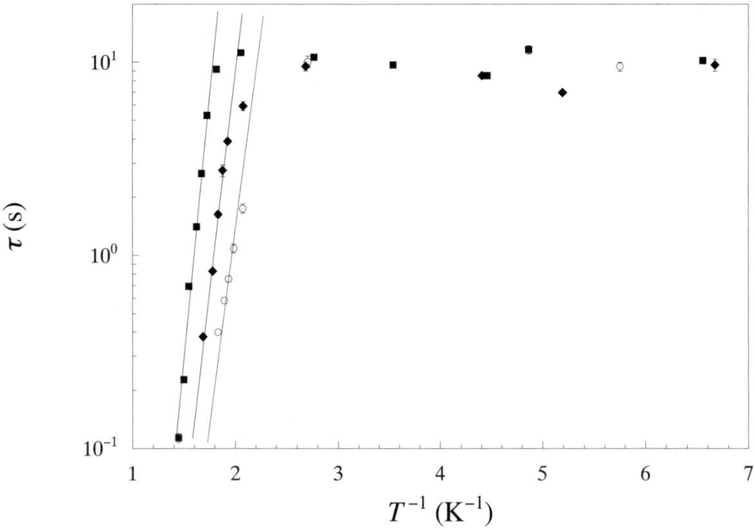

FIG. 11.20. τ vs T^{-1} for positive ions in ^{3}He–^{4}He solutions. (Williams and Packard, 1975.) ^{3}He concentration x_3 (%) = 0.02 (squares), 0.24 (diamonds), and 0.34 (circles). Lines: exponential fits to the data in the intrinsic region.

concentrations is shown. For $T < 0.5$ K, the lifetime becomes constant (and larger than in pure ^{4}He). For $0.5\,\mathrm{K} < T < 0.7$ K, τ shows exponential dependence on T^{-1}, typical of the intrinsic regime, indicating the thermal excitation of the positive ions out of the Bernoulli potential well. The increase of the impurity concentration shifts the transition between the intrinsic and extrinsic regions to lower temperatures.

If the data in the intrinsic region are fitted to the usual exponential form, the binding energy and τ_0 can be determined for each concentration and are shown in Fig. 11.21. Δu decreases linearly with increasing x_3, as expected on the basis of the lowered superfluid density near the core due to ^{3}He condensation. One can assume the ion radius to be 7.7 Å and treat the healing length a_0 as an adjustable parameter, which then varies from 1.9 Å for $x_3 = 0.02\%$ to 4.4 Å for $x_3 = 0.34\%$. However, if these values are used in eqn (11.27), the results are several orders of magnitude smaller than the experimental data. Such a discrepancy was also noted in the measurements of the trapping time of negative ions as a function of pressure (Pratt and Zimmermann, 1969).

The dependence of the lifetime of negative ions at low temperature in dilute ^{3}He–^{4}He mixtures in a wide concentration range below the phase separation is quite strange (Ostermeier and Glaberson, 1975*b*; Williams and Packard, 1975, 1978), as illustrated in Fig. 11.22.

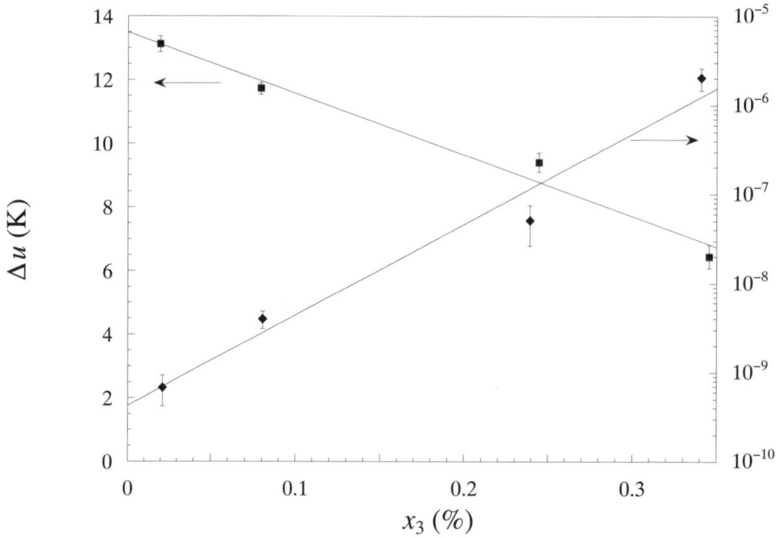

FIG. 11.21. Δu (squares, left scale) and τ_0 (diamonds, right scale) vs x_3 for positive ions. (Williams and Packard, 1975.) Lines: fits to the data.

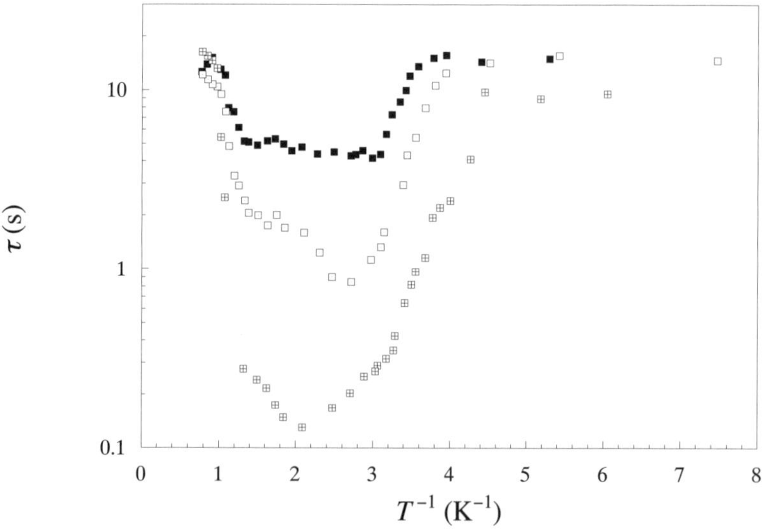

FIG. 11.22. τ vs T^{-1} for negative ions at low T. (Williams and Packard, 1975, 1978.) $x_3\,(\%) = 0.03$ (crossed squares), 0.8 (open squares), and 4.8 (closed squares).

The lifetime in the range $0.3\,\mathrm{K} < T < 0.9\,\mathrm{K}$ shows a marked dependence on the impurity concentration, increasing with increasing x_3, as already observed (Douglass, 1969). This increase may be related to the enhancement of the damping of the vortex line motion due to the increasingly large number of collisions with ^{3}He quasiparticles (Williams and Packard, 1978).

The rapid increase of τ for $T < 0.3\,\mathrm{K}$ is possibly associated with a further enhancement of the damping of vortex lines (Ostermeier and Glaberson, 1975*b*), due to an increase in the bubble radius because of the condensation of impurity atoms at the bubble interface, which is assumed to occur exactly in this temperature range (Zinov'eva and Peshkov, 1960; Dahm, 1969; Kramer, 1970*b*; Shikin, 1973).

Alternatively, it can be attributed to the condensation of ^{3}He atoms on the vortex core, whose larger size would give origin to an increased mutual friction drag force on the vortex as a consequence of an increased scattering with the bulk ^{3}He quasiparticles (Williams and Packard, 1978). In any case, why the lifetime for $T < 0.3\,\mathrm{K}$ is so short still remains a puzzling issue.

11.4.2 *Lifetime effects*

An interesting behavior at low T is also shown by the capture cross-section of negative ions (Ostermeier and Glaberson, 1974, 1975*c*), reported in Fig. 11.23. The upper line is the prediction of the Brownian diffusion model (Donnelly, 1965; Parks and Donnelly, 1966; Donnelly and Roberts, 1969*b*). The agreement between theory and experiment occurs for $T > 1.3\,\mathrm{K}$. For lower T there is an evident discrepancy. It is known that the results of Donnelly are obtained in the approximation of large fields. McCauley and Onsager (1975*a*, 1975*b*) performed a more rigorous calculation of the Brownian motion of the ion in the limit of low fields that gives a nice description of the field dependence of the ion trapping time (Pratt and Zimmermann, 1969). However, their results are even 20% higher than those of Donnelly *et al.*

The discrepancy at low T may be due to several reasons. It has been pointed out (Donnelly and Roberts, 1969*b*) that the stochastic theory is valid only above 1 K at best, and that the conditions for the use of the Smoluchowski equation are even more restrictive. The basic assumptions of the theory are that the ion velocity distribution function is Maxwellian and that the forces on the ion do not change appreciably over a diffusion length l_D.

In the case of the ion–vortex interaction, the latter requirements are valid as long as the ion remains at a distance

$$r \gg l_D = \left(\frac{M_i \mu}{e}\right)\left(\frac{2k_\mathrm{B}T}{M_i}\right)^{1/2} . \tag{11.28}$$

If the ion is now thermalized, i.e., if the first condition is met, and if it is within a distance r_v from the vortex, where r_v (eqn (11.19)) is the radius at which the vortex potential is less than $-k_\mathrm{B}T$, then the ion is effectively trapped. Then, the validity of the Smoluchowski equation is ensured by the condition

$$r_v \gg l_D \tag{11.29}$$

or $\mu T \ll 1.5\times10^{-4}\,\mathrm{m^2\,K/V\,s}$. The inequality (11.29) starts to break down around 1.2–1.3 K, where the anomalous behavior shown in Fig. 11.23 sets in.

Because the stochastic theory cannot be applied below this temperature, Monte Carlo (MC) calculations have been performed (Ostermeier and Glaberson, 1974, 1975*c*), in which a negative ion interacts with a fixed, straight vortex line and individual ion–roton scattering events are taken into account by assuming isotropic scattering. The collision frequency is adjusted so as to give the correct mobility. Phonon scattering can be neglected because T is still high enough.

Better results are obtained from the calculations if it is assumed that an ion is captured when it approaches the vortex core within one ionic radius. The results of the MC simulations are in excellent agreement with the experimental data, as shown by the solid line through the low-temperature points (Ostermeier and Glaberson, 1974, 1975*c*).

The electric field dependence of the capture cross-section is also reproduced extremely well by the MC calculations, whereas the agreement with the Smoluchowski approach is far less satisfactory, as reported in Fig. 11.24.

Another reason for the discrepancy between the Smoluchowski theory and the data is the existence of lifetime effects, as first suggested by Douglass (1968). These effects lower the effective cross-section because the escape rate of the ions

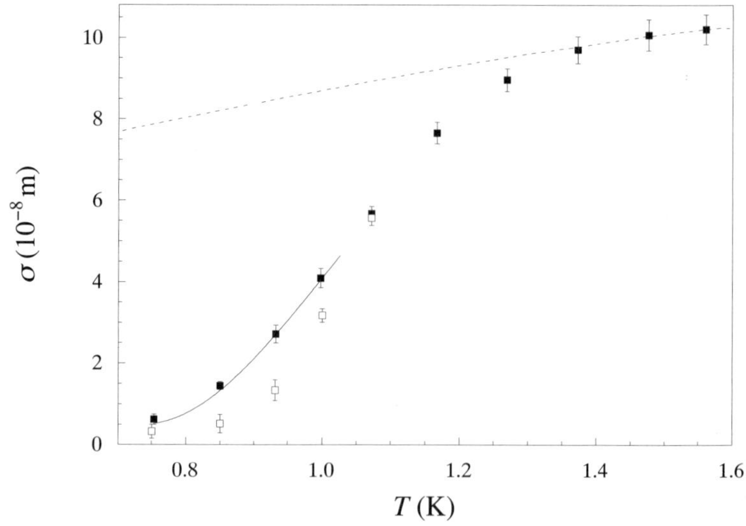

FIG. 11.23. σ vs T for negative ions in pure ^{4}He for $E = 3.3\,\mathrm{kV/m}$ (Ostermeier and Glaberson, 1975*c*). Dashed line: Brownian model (Donnelly, 1965). Solid line: Monte Carlo simulations. Open symbols: raw data. Closed symbols: data obtained with a parallel clearing field to limit the effects due to the finite trapping time of ions.

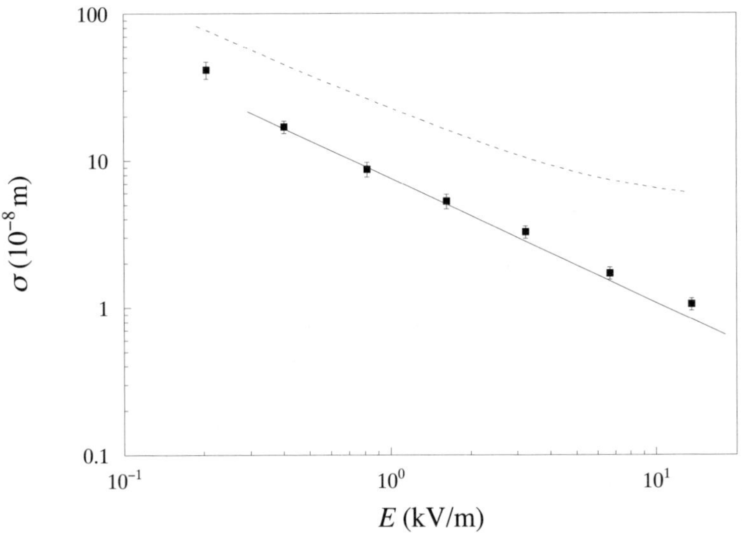

FIG. 11.24. σ vs E for negative ions in pure ^{4}He at $T = 0.996\,\mathrm{K}$. (Ostermeier and Glaberson, 1974, 1975*c*.) Dashed line: Brownian diffusion model. Solid line: Monte Carlo simulations.

from the vortex lines is too large, leading to an effective cross-section $\sigma_{\mathrm{eff}} = \sigma \exp\left[-(\mathcal{P}t^{\star})\right]$ (Donnelly, 1965).

If the transit time of the ions along the vortex lines is made smaller by applying a parallel clearing field, the measured cross-section should increase. This is indeed the case (Ostermeier and Glaberson, 1975*c*). The open points in Fig. 11.25 are the raw data and the closed symbols are obtained by applying a parallel clearing field to reduce $t^{\star}$. The effect produces a small correction.

At lower temperatures the Smoluchowski equation loses its validity, mainly because of the reduced density of scatterers (mainly rotons) in pure ^{4}He. The range of validity of this approach can be extended by adding small amounts of ^{3}He, thereby increasing the scatterer density.

On the other hand, the addition of the isotopic impurity in a temperature range in which the escape rate is dominated by the intrinsic mechanism (Williams and Packard, 1975) lowers τ, thus leading to an increase of the lifetime effects on the capture cross-section.

This effect is evident in Fig. 11.25, where the cross-section in a 0.94% ^{3}He–^{4}He mixture is shown as a function of T. The deviation of the prediction of the Smoluchowski equation starts at a slightly lower T than in pure ^{4}He, as is to be expected on the basis of the artificially-enhanced scatterer number. Moreover, if a clearing field, parallel to the vortex lines, is applied to reduce the characteristic transit time $t^{\star}$, then a substantial increase of the cross-section is obtained, bringing the data more in line with the theoretical prediction that now

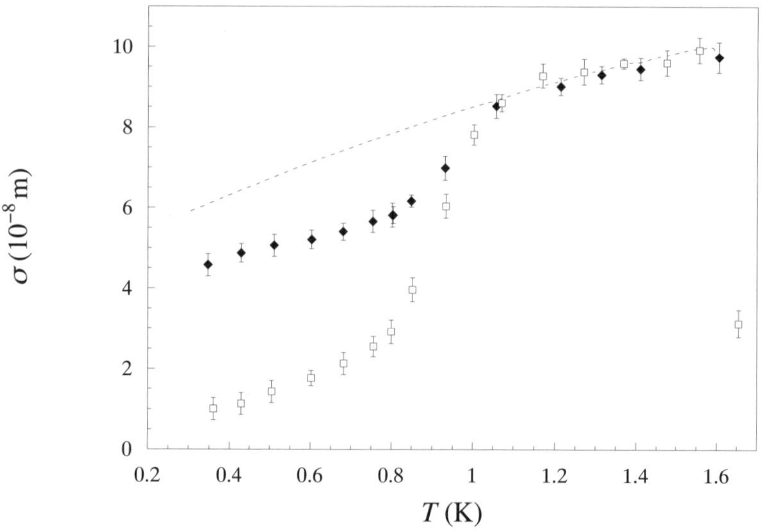

FIG. 11.25. σ vs T for negative ions in an $x_3 = 0.94\%$ mixture. (Ostermeier and Glaberson, 1975*c*.) Line: Brownian diffusion model. Open symbols: raw data. Closed symbols: data taken with a parallel clearing field $E = 3.3\,\mathrm{kV/m}$.

also describes the field dependence of σ well (Ostermeier and Glaberson, 1975*c*).

It thus appears that lifetime effects play a primary role in producing the observed low-T anomaly of the capture cross-section. These results are consistent with the measured temperature behavior of the trapping lifetime in mixtures (Ostermeier and Glaberson, 1975*b*; Williams and Packard, 1975).

12

MOTION OF IONS ALONG LINEAR VORTICES

Ions trapped on vortex lines can be drifted along them by applying an electric field parallel to the rotation axis of the container, as discovered quite early by Douglass (1964).

This particular phenomenon, on one hand, has allowed the possibility of investigating the appearance of individual quantized vortex lines in a rotating bucket of narrow diameter in very elegant experiments (Packard and Sanders, 1972; Williams and Packard, 1974; Yarmchuk and Packard, 1982). On the other hand, the study of ion mobility along the vortex lines has given information about the structure of the vortical lines (Glaberson *et al.*, 1968*a*, 1968*b*).

12.1 Detection of individual vortex lines

For superfluid ^{4}He in a rotating vessel there is a well-defined distribution of quantized vortices characteristic of the speed and size of the vessel. Their number is zero below a critical angular velocity, at which the first vortex appears, and becomes proportional to the angular velocity Ω for rapid enough rotation (Fetter, 1976). The fluid comes into rotation discontinuously in a series of quantized jumps. If the number of vortices is large, the macroscopic appearance of the superfluid is practically indistinguishable from a normal fluid.

Usually, quantized vortex lines have been detected in several kinds of experiments, in which their number is large (Andronikashvili and Mamaladze, 1966). The situation in which individual vortices are present has hardly been investigated, with some notable exceptions (Hess and Fairbank, 1967).

Consider superfluid He contained in a right cylinder of radius R, that rotates at constant uniform angular velocity Ω about its axis. The equilibrium state is obtained by minimizing the free energy

$$\mathcal{E}' = \mathcal{E} - \Omega L\,, \tag{12.1}$$

where $\mathcal{E}$ and L are the kinetic energy and angular momentum of the fluid (Landau and Lifsits, 1958).

The superfluid flow is assumed to be two-dimensional and irrotational except for the occurrence of quantized vortices. The most favorable state contains only singly-quantized vortices. When a single vortex is present at a distance r from the rotation axis, the free energy is

$$\mathcal{E}' = \frac{\rho\kappa^2}{4\pi}\left[\ln\left(\frac{R}{a}\right) + \ln\left(1 - \frac{r^2}{R^2}\right) - \tilde{\Omega}\ln\left(1 - \frac{r^2}{R^2}\right)\right]. \tag{12.2}$$

Here a is the vortex core parameter, $\tilde{\Omega} = 2\pi R^2\Omega/\kappa$ is the dimensionless angular frequency, and κ is the circulation quantum (Packard and Sanders, 1972). For $\tilde{\Omega} > \ln R/a$, the free energy has a minimum in the central position, $r = 0$, whose value is lower than that of the vortex-free state. Thus, the one-vortex state becomes preferred to the vortex-free state for $\Omega \geqslant \Omega_{c1} = (\kappa^2/2\pi R^2)\ln(R/a)$. For $\Omega < \Omega_{c1}$, the fluid is vortex free. One vortex is stable for a frequency range $\Omega_{c1} < \Omega < \Omega_{c2}$.

A second vortex is generated for $\Omega > \Omega_{c2}$, and so on. For higher angular velocities, additional vortices appear at well-defined values of the angular frequency (Hess, 1967). When many vortices are present, a new line appears whenever $\Delta\tilde{\Omega} = 1$, i.e., whenever the angular frequency changes by the amount $\hbar/m_4R^2$.

The technique developed by Packard and Sanders (1972) for detecting the appearance of individual vortex lines as a function of Ω is to measure the variation in the amount of charge trapped on the vortex lines, which is proportional to their number. In order to have $\Omega_{c1} \approx 1\,\mathrm{rad/s}$, the bucket radius must be as small as $R \approx 0.5\,\mathrm{mm}$. Thus, the active volume is small and the amount of charge to be detected is also small.

To obtain a great sensitivity, the negative charges are extracted through the meniscus (Cunsolo *et al.*, 1968*a*) and are detected in a proportional counter in the vapor.

Negative charges are first injected into the measuring volume by using an ionization source with a suitable electrode arrangement, and the existing vortex lines are charged. After a loading period, the source is shut off and the charges trapped on the vortex lines are drifted vertically by means of a suitable electric field along the bucket axis. The trapped charges are therefore pulled up to the meniscus, where they easily leave the liquid (Cunsolo *et al.*, 1968*a*; Rayfield and Schoepe, 1971*b*) and enter the vapor-filled, top section of the cell, in which a stretched wire kept at high voltage acts as the collecting anode of a proportional counter, allowing the detection of very small amounts of charge.

Although the results of the experiment depend on several factors, including the past history of the liquid, the angular acceleration, and the appearance of hysteresis between acceleration and deceleration, nonetheless this experiment has clearly shown that the trapped charge varies in a discontinuous way and that the liquid comes into "rotation" in a discontinuous fashion.

In Fig. 12.1 a typical result is shown. As the apparatus is accelerated from rest, no signal is detected below a characteristic angular frequency, above which it appears in a step-like way. This frequency is close to the expected one. The increments of the signal strength at higher angular frequency similarly appear in a step-like manner. This behavior is what is expected if each step represents the contribution from a single vortex line.

In a further experiment the proportional counter on the top of a rotating bucket of larger radius was replaced by a phosphor screen and an image intensifier, and a direct image of the two-dimensional array of vortex lines was photographed (Williams and Packard, 1974; Yarmchuk and Packard, 1982).

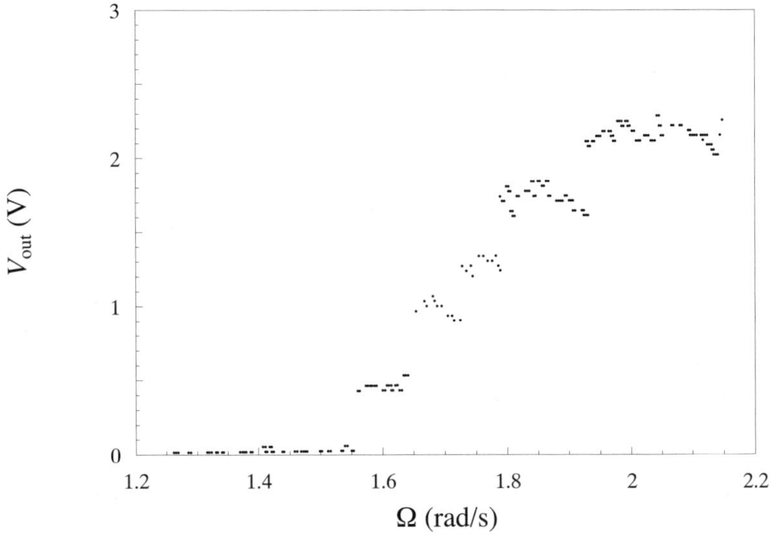

FIG. 12.1. Electrometer output (proportional to the trapped charge on the vortex lines) as a function of the angular velocity Ω for a bucket with radius $R = 0.5$ mm. (Packard and Sanders, 1972.) The angular acceleration rate is $1.15 \times 10^{-4}\,\mathrm{s}^{-2}$.

12.2 Mobility of ions trapped on linear vortices

From the experiments involving the motion of quantized vortex rings, the capture and escape of ions from vortex lines, and the interaction between charged vortex rings and vortex lines, a picture emerges that is consistent with the idea that vortices behave classically, except for the quantization of circulation. The interesting region near the core, however, where many-body effects may become important within a few Å of the vortex center, can be investigated only by detecting the motion of trapped ions along the vortex lines, because the probe is located directly on the site of interest (Glaberson *et al.*, 1968*b*).

The ion represents a large discontinuity in a vortex line. Its motion under the action of a field parallel to the vortices will be affected by collisions related to those that the ion undergoes when it is free, by additional collisions from extra excitations that might concentrate near the vortex core (Glaberson *et al.*, 1968*b*), and also by thermal excitations of the vortex itself (Douglass, 1968).

Trapped ions moving along vortex lines are subjected to a larger drag than in the free state, as observed in earlier experiments (Douglass, 1964). The results of the drift velocity measurements for negative ions along vortex lines in pure ^{4}He in a fairly wide temperature range, $0.7\,\mathrm{K} < T < 1.6\,\mathrm{K}$, show that the mobility is field independent in the explored range. It does not depend on the angular velocity of the vessel and is actually lower than that of free ions, as shown in Fig. 12.2.

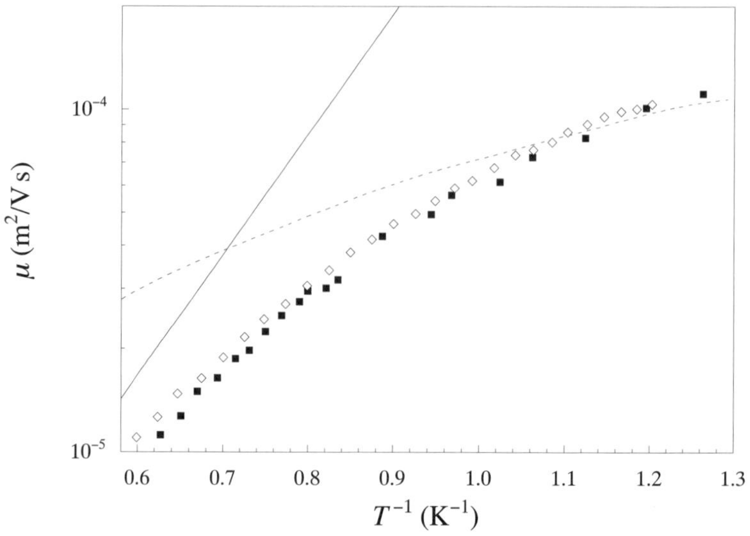

FIG. 12.2. μ vs T^{-1} for negative ions trapped on vortex lines at SVP. Closed symbols: Douglass (1968), and open symbols: Glaberson *et al.* (1968*b*). Solid line: free-ion mobility. Dashed line: vortex wave-limited mobility (Douglass, 1968; Fetter and Iguchi, 1970).

Three basic processes contribute to the drag on trapped ions: (i) the scattering of quanta of vortex waves; (ii) the creation of additional vortex waves; and (iii) the frictional drag arising from excess rotons trapped near the core. Additional drag from ^{3}He impurity atoms is absent in pure ^{4}He (Glaberson *et al.*, 1968*b*).

12.2.1 *Scattering on vortex waves*

It was first proposed that the enhanced drag experienced by trapped ions is due to scattering of the atoms off thermally-excited vortex waves (Douglass, 1968). An approximate calculation yields a contribution independent of the ion effective mass. A more refined model, in which vortex waves are reflected by a movable impurity located in the core, gives results that depend on the ion effective mass but that agree with the simplified calculations if an effective mass $M_i \approx 100 m_4$ is assumed (Fetter and Iguchi, 1970). The predictions for the vortex wave-limited mobility are shown as a dashed line in Fig. 12.2.

A vortex wave is assumed to be a boson of energy $\hbar\omega$ and momentum $\hbar k$ in the direction of propagation. It is further assumed that the classical dispersion relation for waves on a vortex with a hollow core is given by

$$\omega = \left(\frac{\kappa}{2\pi a^2}\right)\left\{\left[1 + ka\frac{K_0\,(ka)}{K_1\,(ka)}\right]^{1/2} - 1\right\}, \tag{12.3}$$

where a is the core radius and K_0 and K_1 are modified Bessel functions (Thomson, 1880; Douglass, 1968). In the range of interest, $0.3 \leqslant ka \leqslant 2$, eqn (12.3) can be linearized so as to give $T' \equiv \hbar\omega/k_B = 4.10ka - 0.49$, with $a \approx 1\,\text{Å}$.

If a trapped ion with momentum p along the vortex reflects a vortex wave of momentum $\hbar k$, the change of p is quite accurately given by

$$\Delta p = 2\hbar k\left(\frac{\mathrm{d}\omega}{\mathrm{d}k} - \frac{p}{M_i}\right)\left(\frac{\mathrm{d}\omega}{\mathrm{d}k}\right)^{-1} \tag{12.4}$$

because the thermal velocity of the ion is much smaller than that of the vortex wave (Douglass, 1968).

It is further assumed that ions have an equilibrium Boltzmann distribution corresponding to a small drift velocity v_D along the core and that there is one ion per unit vortex length. The collision rate per unit vortex length between ions in the range $\mathrm{d}p$ and vortex waves in the range $\mathrm{d}k$ is

$$\begin{aligned}\mathrm{d}N = \; & \frac{\mathcal{R}}{2\pi}\left|\frac{\mathrm{d}\omega}{\mathrm{d}k} - \frac{p}{M_i}\right|\left(e^{T'/T} - 1\right)^{-1} \\ & \times \left[(2\pi M_i k_B T)^{-1/2}\, e^{-(p^2/2M_i + pv_D)/k_B T}\mathrm{d}p\right].\end{aligned} \tag{12.5}$$

The reflection coefficient $\mathcal{R}$ can be linearly approximated in the range of interest (Douglass, 1968).

The rate of momentum loss per ion due to collisions is the integral of the product $\mathrm{d}N \times \Delta p$. By equating the external force eE to the rate of momentum loss, one finally gets the vortex wave-limited mobility as

$$\mu_{vw}^{-1} = \left(\frac{4\hbar}{\pi e}\right)\int_0^\infty \frac{\mathcal{R}k}{\exp(T'/T) - 1}\mathrm{d}k\,, \tag{12.6}$$

which is shown as a dashed line in Fig. 12.2. At fairly low T, where the roton contribution to scattering should be small, μ_{vw} agrees with the data.

The roton contribution to the mobility can be obtained by subtracting out the vortex wave contribution from the total mobility: $\mu_r^{-1} = \mu^{-1} - \mu_{vw}^{-1}$, if it is assumed that roton and vortex wave scattering are independent processes. It turns out that $\mu_r^{-1} \propto \exp(A/T)$ with $A \approx 6.2\,\mathrm{K}$ (Douglass, 1968; Glaberson *et al.*, 1968*b*; Glaberson, 1969).

The fact that $A < \Delta$, where Δ is the roton energy gap, suggests that an excess roton density develops near the core due to a $\mathbf{p}\cdot(\mathbf{v}_n - \mathbf{v}_s)$ Doppler shift effect, which is described by the localized roton model (Douglass, 1968; Glaberson *et al.*, 1968*a*, 1968*b*; Iguchi, 1972).

Though already described in Section 6.1, it is useful to recall here the main features of this model (Glaberson *et al.*, 1968*a*). The distribution of rotons near the vortex line described by the superfluid velocity field $\mathbf{v}_s = \mathbf{V}(r) = (\kappa/2\pi r)\hat{u}_\theta$ is given by

$$h^3 N_r(T,V,P) = \int \left\{\exp\left[\left[\Delta + (p-p_0)^2/2m_r + pV\cos\theta\right]/k_\mathrm{B}T\right]\right\}^{-1} \mathrm{d}^3p$$
$$= \frac{2\pi k_\mathrm{B}T}{V}\int_0^\infty p\ln\left\{\frac{1-\exp\left\{-\left[\Delta+(p-p_0)^2/2m_r+pV\right]/k_\mathrm{B}T\right\}}{1-\exp\left\{-\left[\Delta+(p-p_0)^2/2m_r-pV\right]/k_\mathrm{B}T\right\}}\right\}dp \quad (12.7)$$

because the rotons are fixed and $\mathbf{v}_n = 0$. As a consequence, there is a Landau critical velocity at a given radius R_c where rotons exist with zero energy, i.e., where

$$\left[\Delta + \frac{(p-p_0)^2}{2m_r} - pV\right]_{\min} = 0\,. \quad (12.8)$$

Within this radius, the excitations can no longer be considered as rotons and form a central stagnant core. For $p_0V \ll \Delta$, eqn (6.3) is obtained.

In eqn (12.7) one has to take into account the fact that Δ decreases with increasing T (Yarnell *et al.*, 1959). This behavior is considered as a manifestation of a roton–roton attractive interaction. Thus, Δ can be approximated as $\Delta(N) = \Delta - 0.7383 \times 10^{-37}N$ (Bendt *et al.*, 1959). Moreover, the roton parameters vary nearly linearly with the pressure P (Springett, 1967). Finally, the quantum localization energy for rotons, $\hbar^2/8m_r r^2$, is added, yielding

$$\left(\frac{\beta h^3 V}{2\pi}\right) N_r(T,V,P,r)$$
$$= \int_0^\infty p\ln\left\{e^{-\beta pV}\frac{\sinh\left\{\beta\left[\Delta+(p-p_0)^2/2m_r+pV+\hbar^2/8m_r r^2\right]/2\right\}}{\sinh\left\{\beta\left[\Delta+(p-p_0)^2/2m_r-pV+\hbar^2/8m_r r^2\right]/2\right\}}\right\}dp\,, \quad (12.9)$$

where $\beta = 1/k_\mathrm{B}T$. The normal fluid fraction ρ_n is calculated in a similar way.

In Fig. 12.3 the results for $T = 1.6\,\mathrm{K}$ at SVP are shown. $N_r(r)$ increases very rapidly for $r < R_c \approx 4.25\,\text{Å}$ and the superfluid fraction plunges down rapidly. These results suggest that the quantized vortex has a central core of normal fluid, whose radius depends on T and P, and a tail in which the roton density increases dramatically as R_c is approached. The rotons are highly polarized because most of them have momenta (though not group velocity) aligned along a direction antiparallel to the superfluid velocity. There is no net rotation of either the core or tail (Glaberson *et al.*, 1968*a*).

The transit time τ of negative ions can be calculated as a function of T and P by assuming an ion radius $R_- \approx 16\,\text{Å}$ at $P = 0$ with a known pressure dependence (Springett, 1967; Glaberson *et al.*, 1968*b*). The extra density of excitations

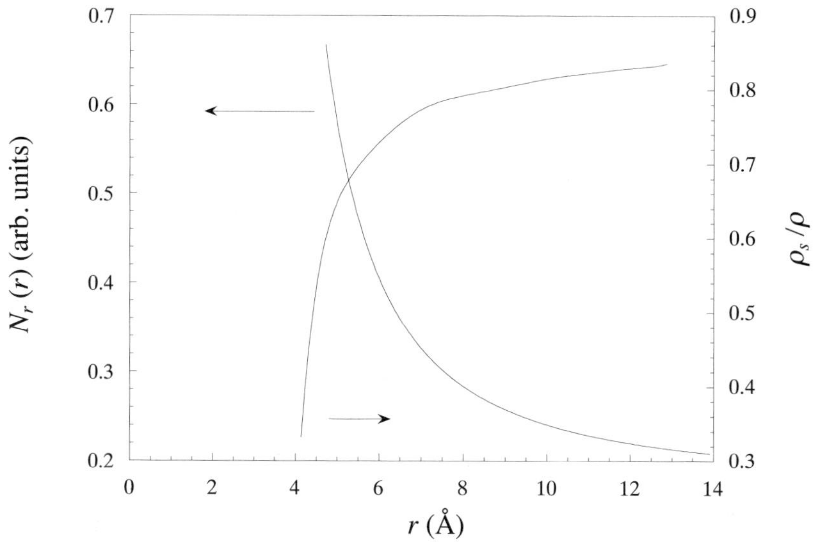

FIG. 12.3. $N_r(r)$ and $\rho_s(r)/\rho$ vs r, the distance from the vortex core, for $T = 1.6$ K and $P = 0$ MPa. (Glaberson *et al.*, 1968*a*.)

contributes an additional drag on the ion. The transit time thus has two contributions: τ_t due to the rotons in the tail and τ_c due to the core (Glaberson *et al.*, 1968*a*).

The rotons in the tail, except very close to R_c, form a dilute gas and their contribution to the drag is proportional to their number density. Thus, one writes

$$\frac{\tau_t\,(P,T)\,[\pi(R_-)^2(P=0)]N_r(T,V,P=0,r\to\infty)}{\tau_{\text{free}}\,(P=0,T)} = 2\pi \int_{R_c(P,T)}^{R_-(P)} rN_r(T,V,P,r)\mathrm{d}r\,, \tag{12.10}$$

where $R_-(P)$ is known from experiments and the transit time τ_{free} of the free ions is measured.

The core contribution to the transit time is calculated by realizing that the medium in the core is so dense that it behaves *viscously* (Glaberson, 1969). It is assumed that the velocity distribution of the core medium near the ion surface in the forward core region is the same as would occur for a sphere moving through an infinite viscous medium. The drag is calculated in a manner similar to that of Stokes (Landau and Lifsits, 2000) but integrating the force only over the forward core region, yielding

$$\mathcal{F}_{\text{core}} = \frac{3}{2}\pi\eta u \frac{R_c^2}{R_-}\,, \tag{12.11}$$

where u is the ion velocity and η the fluid viscosity. Since the drag force is proportional to the ion velocity, the total transit time of trapped ions is

$$\tau_T = \tau_t + \tau_c\,, \tag{12.12}$$

where

$$\tau_c(P,T) = 1.5\pi \frac{L}{eE}\frac{R_c^2(P,T)}{R_-(P,T)}\eta(P,T) \tag{12.13}$$

and L is the drift length (Glaberson *et al.*, 1968*a*; Glaberson, 1969). It is also assumed that the thermodynamic parameters of the core medium are those of He I. The fluid viscosity η is taken from experimental data (Tjerkstra, 1952) and extrapolated to the desired (P,T) range. The contribution of rotons in the tail, τ_t, dominates the drag at high T and low P, while τ_c is more important at high P and low T (Glaberson, 1969).

In Fig. 12.4 the computed transit time of trapped negative ions at SVP is shown as a function of T^{-1}, whereas in Fig. 12.5 the pressure dependence of τ_T is reported for $T = 1.2\,\text{K}$. The agreement with the data is quite good and it is evident that Δ (and hence rotons) dominates the scattering of trapped ions. Similar results are also obtained for all other T (Glaberson, 1969).

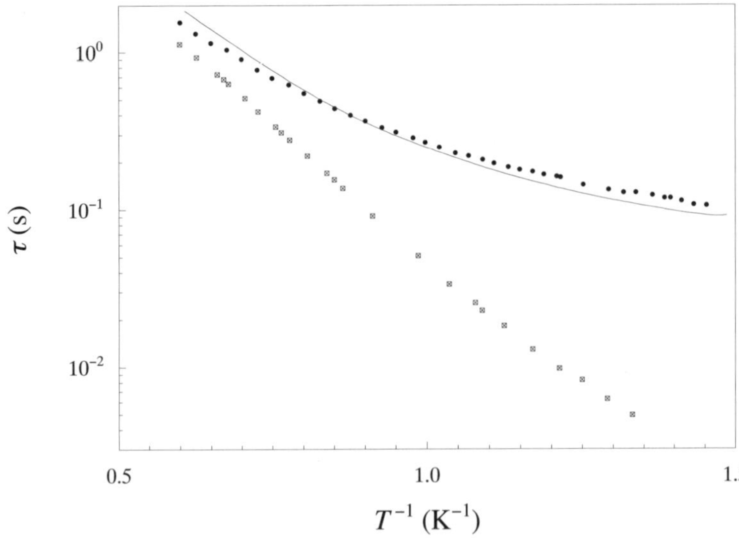

FIG. 12.4. Transit time τ vs T^{-1} for trapped (squares) and free (circles) negative ions along vortex lines over a distance $L = 4.93\,\text{cm}$ at SVP. (Glaberson *et al.*, 1968*a*.) Solid line: eqn (12.12).

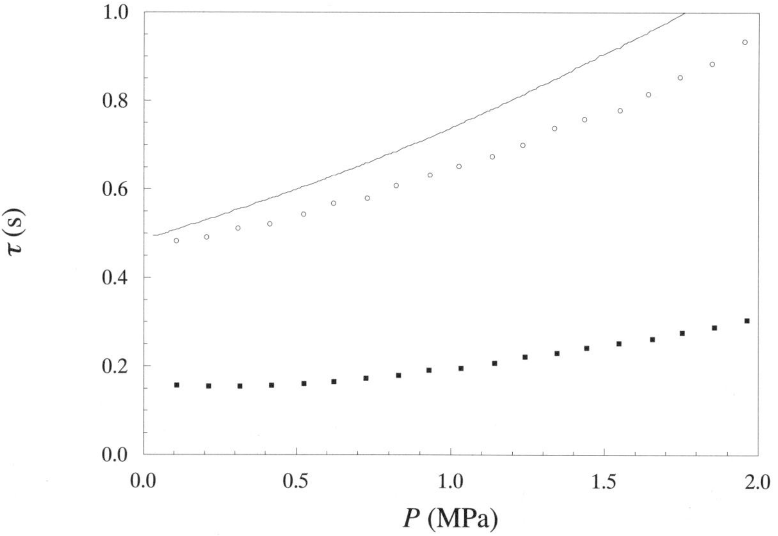

FIG. 12.5. Transit time τ vs P for trapped (open symbols) and free (closed symbols) negative ions along vortex lines over a distance $L = 4.93\,\mathrm{cm}$ for $T = 1.2\,\mathrm{K}$. (Glaberson *et al.*, 1968*a*.)Solid line: eqn (12.12).

It appears that the data can be explained without the need to introduce vortex wave scattering (Douglass, 1968), which, if present, would contribute only at the lowest temperatures (see Fig. 12.2). Similar conclusions are also drawn by using a quantum-mechanical approach for the calculation of the bound roton states (Iguchi, 1972).

Vortex waves, however, do exist and have been detected directly (Ashton and Glaberson, 1979). Vortex waves are helicoidal deformations of the line (Thomson, 1880; Raja Gopal, 1964). These waves are so polarized that an element of the line executes a circular motion in a sense opposite to the circulation sense of the vortex velocity field.

A radio frequency (r.f.) field, transverse to a vortex line charged with ions moving along it, will strongly couple to vortex waves under suitable conditions (Halley and Cheung, 1968; Halley and Ostermeier, 1977). The resonant generation of vortex waves is expected when the following criteria are met:

$$\omega_{rf} = \omega(k) - kv_i \tag{12.14}$$

$$v_i = \frac{\partial \omega(k)}{\partial k} \tag{12.15}$$

where k is the wave vector and v_i is the ion velocity.

The first condition simply states that the frequency and rotation sense of the vortex wave are the same as those of the r.f. field, in the frame of reference of the moving ion.

The second condition states that the energy pumped by the r.f. field into the vortex wave remains in close proximity to the ions.

These two conditions determine a characteristic ion velocity that depends on ω_{rf}. If the ion velocity along the line is measured as a function of the d.c. longitudinal field, an anomaly is expected at the characteristic velocity v_i. The vortex waves produced near resonance have phase velocities in the same direction as the ions. Therefore, the ions experience an enhanced drag (Ashton and Glaberson, 1979).

Because the group velocity of the vortex waves is larger than their phase velocity, the sign of ω_{rf} is opposite to that of $\omega(k)$. Thus, conditions (12.14) and (12.15) can be satisfied simultaneously only if the r.f. electric field is circularly polarized in the same sense as the vortex circulation.

The ion velocity anomaly is expected to appear only for one particular sense of rotation of the experimental apparatus with respect to the polarization of the r.f. field.

A schematic of the experimental electrode set-up for investigating the interaction of ions and vortex waves is shown in Fig. 12.6. Ions are injected into the region L filled with vortex lines generated by the rotation of the cell. The gating grids GG pull ions into the drift velocity cell, where an r.f. field is applied by means of the r.f. electrodes RF. The ions moving along the lines are finally collected at C.

In Fig. 12.7 the velocity of negative ions for $T = 0.377\,\mathrm{K}$ at SVP is shown

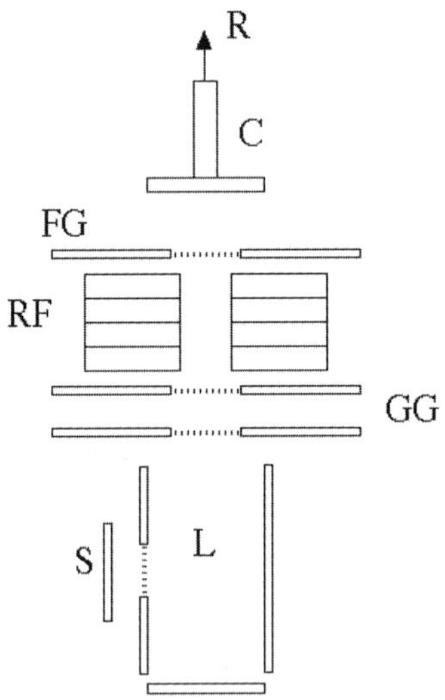

FIG. 12.6. Schematic side view of the cell for investigating vortex waves. (Ashton and Glaberson, 1979.) S: ion source, L: line-charging region, GG: gating grids, RF: r.f. electrodes, FG: Frisch grid, C: collector, and R: rotation axis.

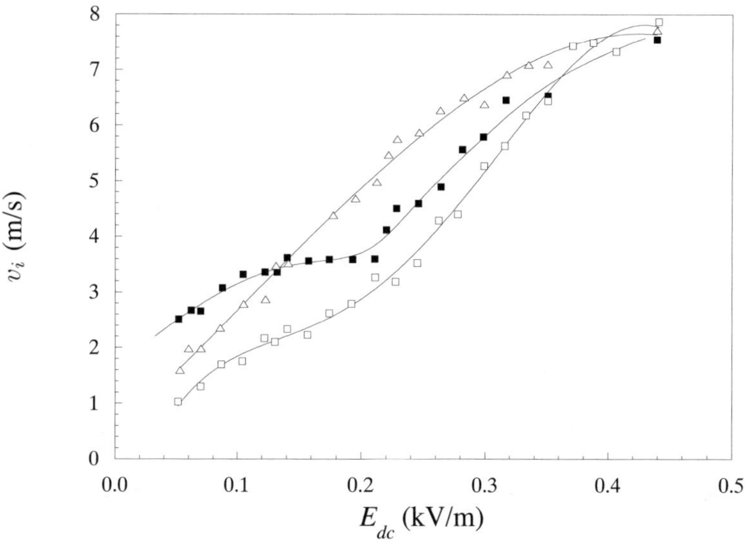

FIG. 12.7. v_i vs E_{dc} for negative ions at $T = 0.377\,\mathrm{K}$ at SVP. R.f. field amplitude: 6.6 kV/m. R.f. angular frequency: $\omega_{rf} = 2\pi \times 10^7$ rad/s. (Ashton and Glaberson, 1979.) Triangles: no r.f. field, open squares: clockwise-polarized r.f. field, and closed squares: counterclockwise-polarized r.f. field. Lines: eyeguides.

as a function of the longitudinal d.c. field. The ion velocity is anisotropic with respect to the field polarization. An anomalous kink and a plateau of the ion velocity as a function of the d.c. field for a counterclockwise polarization of the r.f. field appear where the ion velocity does not increase with increasing E_{dc}. That means that the energy gained by the ions from the r.f. field is resonantly transferred to vortex waves, leading to an increase of the drag on the ions.

The plateau is very close to the predicted value of 3 m/s (Ashton and Glaberson, 1979). It is not very sharp because of residual field inhomogeneities and of finite vortex wave damping. At the characteristic velocity the vortex waves have a wavelength of ≈ 2000 Å, much larger than both the ion radius and the vortex core. These results confirm the most important theoretical prediction (Halley and Ostermeier, 1977).

It is interesting to note that, except for the case of counterclockwise polarization and velocities smaller than the characteristic one, the net effect of the generation of vortex waves is to always increase the drag on ions, as deduced from the fact that the velocities in the presence of an r.f. field are always smaller than those in the absence of it for a given d.c. field (Ashton and Glaberson, 1979).

12.2.2 *Emission of vortex waves*

Measurements of positive ion mobility along the vortex lines at lower T have revealed a number of peculiar features that cannot be obtained by a simple extrapolation from the high-temperature data (Ostermeier and Glaberson, 1975*a*).

At low T, the time width of the ion pulse traveling along the vortex lines becomes very broad and the measured time-of-flight depends on the time delay between the charging of the lines and the gating of the ions into the drift region. This apparently indicates that the passage of ions produces some kind of disturbance in the array of vortex lines, which slows down the ion motion if it is not allowed to dissipate.

The first ions travel along unperturbed lines and their drift time is the shortest. The following ions travel along (probably) deformed lines that hinder their motion (Ostermeier and Glaberson, 1975*a*). As the ion motion is very supersonic with respect to the speed of propagation of vortex waves, perturbations of the lines are likely. Probably, the vortex lines are unstable with respect to deformation (Gamota, 1972).

This generation of vortex deformation might also be the origin of the high-field saturation of the drift velocity v_D, shown in Fig. 12.8. Attempts at predicting this limiting velocity from eqn (12.15) and from the assumed binding potential have proven unfruitful. In any case, the most striking feature of the low-T experiments on the motion of trapped ions is the enormous rise in the mobility as the temperature is lowered, compared with the high-T data shown in

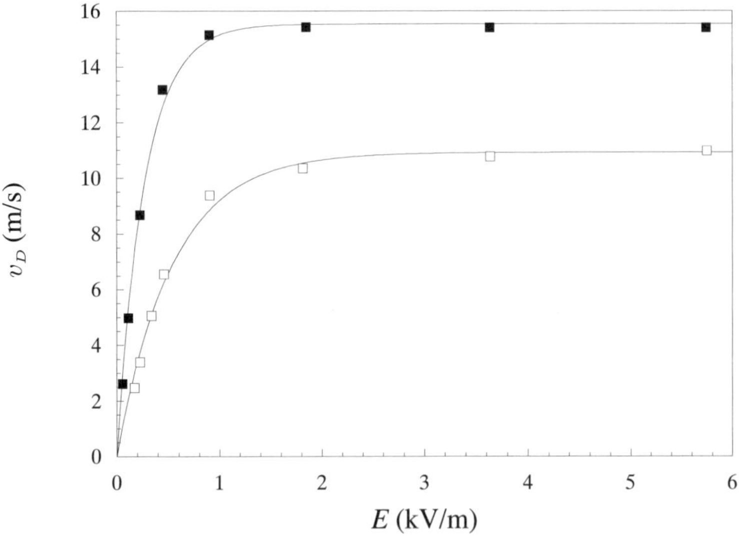

FIG. 12.8. v_D vs E for $T = 0.41$ K. v_D is defined in terms of the leading edge of the ion pulse. (Ostermeier and Glaberson, 1975*a*.) Open squares: negative ions, and closed squares: positive ions. Lines: eyeguides.

Fig. 12.2. In Fig. 12.9 the temperature dependence of the positive and negative ion mobility along the vortex line is shown.

The low-T data smoothly join the high-T data, but μ^{-1} decreases by three orders of magnitude when T is lowered to $\approx 0.3\,\mathrm{K}$ and becomes comparable with that of free negative ions (Schwarz and Jang, 1973). The lines in the figure represent the prediction of the vortex wave–ion-scattering theory for two different values of the ion effective mass (Fetter and Iguchi, 1970). It is clear that the addition of any other drag mechanism would increase the disagreement with the theory.

12.2.3 *Scattering on ^{3}He impurities*

One possible explanation for the large discrepancy in μ^{-1} at low T with the prediction of vortex wave theory (Fetter and Iguchi, 1970) or of the bound rotons model (Glaberson *et al.*, 1968*a*, 1968*b*; Iguchi, 1972), which give reasonable agreement with the data for $T > 0.7\,\mathrm{K}$, is that the ion might not be located exactly on the center of the vortex line because of the interaction between the flow fields produced by its motion and the vortex line, which is moving in response to off-axis motion (Ostermeier and Glaberson, 1975*a*). Once off axis, energy would be pumped into the resultant vortex deformations by the electric field. At higher T the motion of the vortex lines motion is heavily damped and this off-axis pro-

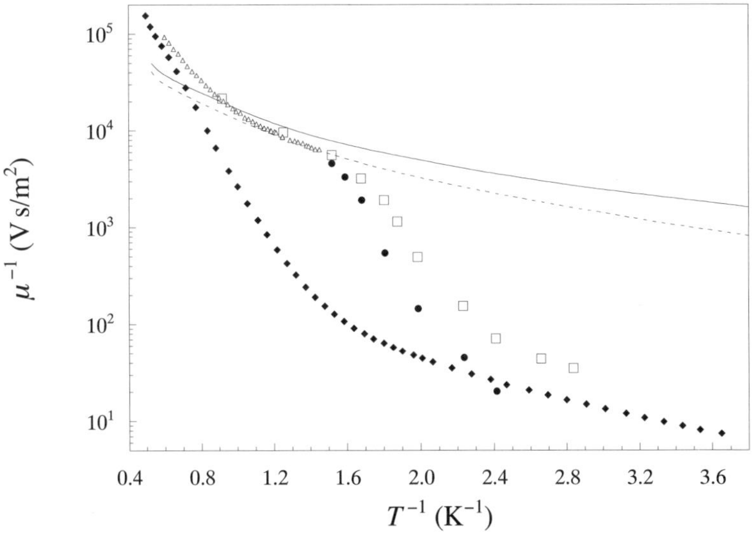

FIG. 12.9. μ^{-1} vs T^{-1} for negative and positive ions along vortex lines (Ostermeier and Glaberson, 1975*a*). Open squares: negative ions, closed circles: positive ions, open triangles: high-T negative ions (Glaberson *et al.*, 1968*b*), and closed diamonds: free negative ions (Schwarz, 1972*a*). Lines: vortex wave-scattering theory for two different values of the effective ion mass (Fetter and Iguchi, 1970).

cess would not occur. An off-axis ion would experience less drag and its mobility should approach that of free ions.

The addition of a small amount of ^{3}He, which increases the damping of the vortex line motion, in fact restores the agreement with vortex wave theories at low T (Ostermeier *et al.*, 1975; Ostermeier and Glaberson, 1976). In Fig. 12.10 the inverse mobility of trapped and free negative ions for several ^{3}He–^{4}He mixtures are shown as a function of T^{-1}. The addition of ^{3}He reduces the increase of μ observed in pure ^{4}He. Moreover, none of the anomalous features observed in pure ^{4}He below 0.7 K, such as, for instance, the broadening of the ion pulse, occur in the mixtures.

If the various scattering processes were acting independently, the observed zero-field mobility could be written as

$$\mu_{\text{trap}}^{-1} = \alpha\mu_{\text{free}}^{-1} + \mu_{\text{vort}}^{-1}, \tag{12.16}$$

where α is an unknown constant. The last term includes contributions due to the presence of the vortex, i.e., drag associated with vortex waves, enhancement of local thermal excitations, and ^{3}He atoms.

It is not clear, however, if it is justified to separate the drag on trapped ions into two distinct parts; for instance, vortex wave scattering might be modulated

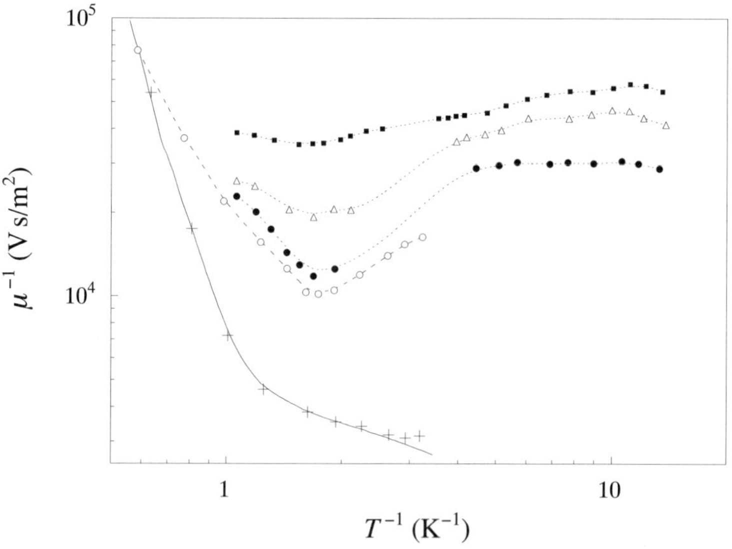

FIG. 12.10. Trapped negative ion mobility μ^{-1} vs T^{-1} in ^{3}He–^{4}He mixtures of concentration x_3 (%) = 1.6 (closed squares), 0.8 (open triangles), 0.4 (closed circles) (Williams and Packard, 1978), and 0.27 (open circles), and for free ions at $x_3 = 0.27\%$ (crosses) (Ostermeier *et al.*, 1975; Ostermeier and Glaberson, 1976). Solid line: extrapolation of the low-concentration free-ion data (Schwarz, 1972*a*). Dashed lines: eyeguides.

by excitations of the bulk and there could also be screening of the vortex core from ambient excitations (Ostermeier and Glaberson, 1976). Similar effects are difficult to assess but there is experimental evidence that screening by the vortex is negligible (Steingart and Glaberson, 1972) and, for positive ions, it is always much larger than the first term. For these reasons α is set equal to unity.

The experimental results for the vortex contribution μ_{vort} to the mobility of trapped negative ions are shown in Fig. 12.11. At high T, most of the data lie on the same curve, independent of the concentration, i.e., ^{3}He plays only a small role in determining the trapped-ion drag at high T.

The common curve is well described by the vortex wave-scattering theory of Fetter and Iguchi (1970), shown as the solid line. This fact suggests that at high temperature the drag on the trapped ions is mainly due to thermal activation of vortex lines. Deviations from the solid line at the highest temperatures and concentrations might be due to changes in the core parameter in these conditions (Ostermeier *et al.*, 1975). Similar results are obtained with positive ions, as shown in Fig. 12.12.

The most striking feature of the data is that the vortex contribution to the drag on trapped ions, both negative and positive, increases below a particular temperature T_p for each concentration (Ostermeier *et al.*, 1975; Ostermeier and Glaberson, 1976). This behavior is consistent with the expectation that ^{3}He

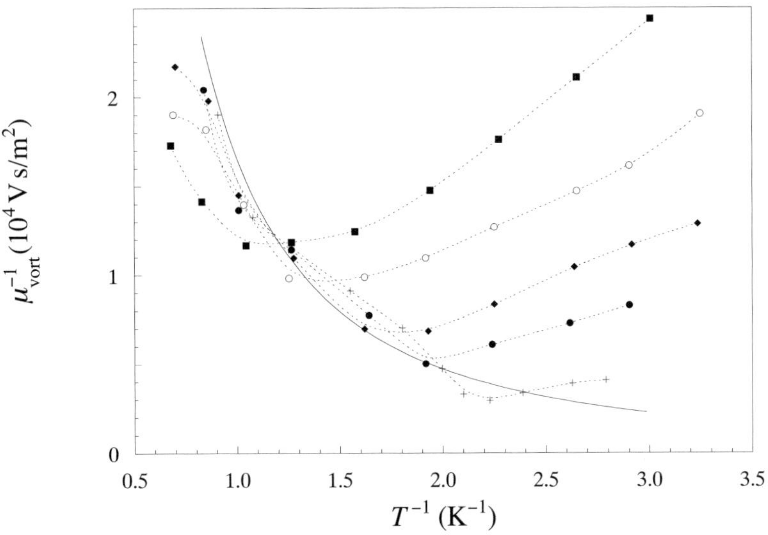

FIG. 12.11. μ_{vort}^{-1} vs T^{-1} for negative ions trapped on vortex lines in different isotopic mixtures (Ostermeier *et al.*, 1975; Ostermeier and Glaberson, 1976). x_3 (%) = 2.90 (closed squares), 1.04 (open circles), 0.27 (closed diamonds), 0.072 (closed circles), and 0.017 (crosses). Solid line: vortex wave theory (Fetter and Iguchi, 1970). Dashed lines: eyeguides.

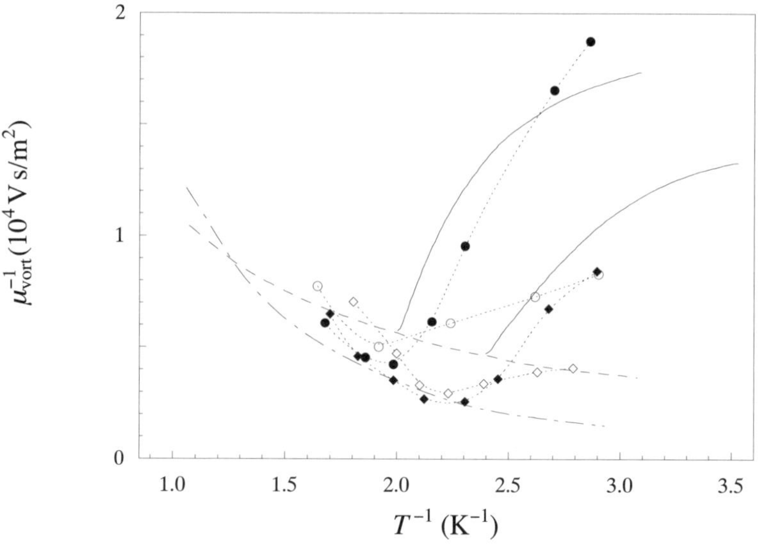

FIG. 12.12. $\mu_{\rm vort}^{-1}$ vs T^{-1} for trapped positive ions in different isotopic mixtures (Ostermeier *et al.*, 1975; Ostermeier and Glaberson, 1976). $x_3\,(\%) = 0.072$ (closed circles), and 0.017 (closed diamonds). The same, though open, symbols refer to negative ions in the same mixtures (Ostermeier *et al.*, 1975; Ostermeier and Glaberson, 1976). Dotted lines: eyeguides. Dash–dotted line: vortex wave theory for negative ions (Fetter and Iguchi, 1970). The solid lines and the dashed line are the prediction, for positive ions at the two given concentrations, of the model that takes into account ^{3}He condensation on the vortex lines (Huang and Dahm, 1976).

condenses onto the vortex cores.

In contrast, the enhancement of the drag is not believed to be associated with the changes of the ion structure due to ^{3}He condensation on the surface of the electron bubble (Kuchnir *et al.*, 1972), which have not been observed for positive ions, while the minimum vortex drag occurs at approximately the same temperature for both positive and negative ions.

Strong evidence that ^{3}He condensation on vortex cores is responsible for the enhancement of the ion drag is produced by plotting in Fig. 12.13 the ambient ^{3}He concentration as a function of the temperature T_p at which the vortex drag deviates from the universal behavior for high T.

In the presence of a vortex line, the ^{3}He number density at a distance r from the core, by assuming local thermodynamic equilibrium, is given by

$$n_3(r) = n_\infty e^{-U(r)/k_{\rm B}T}\,, \tag{12.17}$$

where n_∞ is the ^{3}He number density far away from the vortex. $U(r)$ is an effective hydrodynamic potential related to the kinetic energy of the superfluid displaced by a ^{3}He atom.

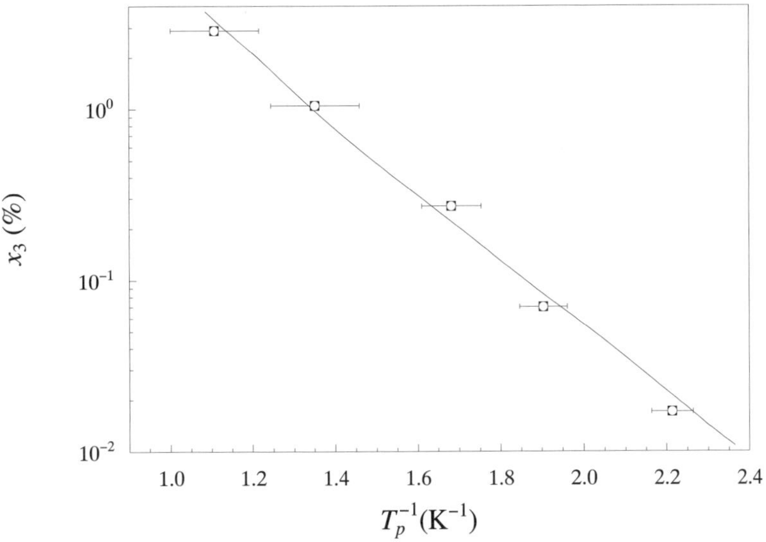

FIG. 12.13. Bulk ^{3}He concentration x_3 vs T_p^{-1} (Ostermeier *et al.*, 1975; Ostermeier and Glaberson, 1976). Line: ^{3}He condensation model (Ostermeier and Glaberson, 1975*d*).

The idea is that ^{3}He condenses on the vortex core as soon as $n_3(0) = n_c(T_p)$, where $n_c(T_p)$ is the critical ^{3}He density at phase separation in bulk solutions. The curve in Fig. 12.13 is given by

$$n_\infty = n_c(T_p)e^{U(0)/k_\mathrm{B}T_p} \tag{12.18}$$

provided that the number density is converted to a concentration.

In eqn (12.18) data for $n_c(T)$ from the literature are used (Zinov'eva and Peshkov, 1960). For $U(0)$ the hydrodynamic potential (11.7) developed by Donnelly and Roberts (1969*b*) is adopted with the obvious substitution of the effective ^{3}He atom radius a_3 in place of the ion radius:

$$\begin{aligned} U(0) = &-2\pi\rho_s\left(\frac{\hbar}{m_4}\right)^2 a_3 \\ &\times\left\{\left[1+\left(\frac{a}{a_3}\right)^2\right]^{1/2}\ln\left(\frac{a_3}{a}+\left[1+\left(\frac{a_3}{a}\right)^2\right]^{1/2}\right)-1\right\}. \end{aligned} \tag{12.19}$$

The effective ^{3}He radius is obtained from knowledge of the effective mass $m_3^\star$ (Brubaker *et al.*, 1970) by

$$m_3^\star = m_3 + \frac{2\pi}{3}\rho_s a_3^3\,, \tag{12.20}$$

and a is the vortex core parameter found in the literature (Steingart and Glaberson, 1972).

This model fits the experimental points with great accuracy, varying smoothly between -2.7 and $-3.1\,\mathrm{K}$ in the experimental (T, x_3) range. It is important to note the agreement with the prediction of a binding energy of 3.5 K by Ohmi and Usui (1969), Muirhead *et al.* (1985), and Sadd *et al.* (1999).

The present model is highly questionable because the properties of bulk solutions and continuum mechanics are used in order to explain a phenomenon that is localized over a few Å about the vortex core, taking into account the fact that the de Broglie wavelength of a ^{3}He atom is of the order of 15–20 Å. However, the quite good agreement with the data is an indication of the true physics underlying this phenomenon.

The increase of the drag for temperatures below the various critical temperatures T_p for the different concentrations of ^{3}He suggests that the drag at lower T is dominated by a different mechanism than at higher T (Ostermeier and Glaberson, 1976). In particular, if the vortex drag is assumed to be written, by analogy with eqn (12.16), as

$$\mu_{\mathrm{vort}}^{-1} = \mu_{\mathrm{th}}^{-1} + \mu_3^{-1}, \tag{12.21}$$

where the first term on the right-hand side is the contribution from scattering by thermal vortex waves and bound rotons, and the last term gives the contribution associated with condensed ^{3}He and any other excess ^{3}He above the ambient concentration.

It is further assumed that μ_{th}^{-1} is represented by the nearly concentration-independent vortex drag at higher temperatures and that this contribution can be extrapolated at low temperatures. In this way μ_3^{-1} can be determined as a function of T.

In Fig. 12.14 the resulting ^{3}He contribution μ_3^{-1} to the vortex drag for both ionic species is shown as a function of the reduced temperature $T_p/T - 1$, where T_p is the temperature at which the drag starts increasing for each concentration.

Some interesting features can be noticed. The first one is that, although in a very limited range, μ_3^{-1} is linearly proportional to $T_p/T - 1$ and all data for each ion, within experimental accuracy, fall on a universal curve, thus suggesting a relationship with the usual critical phenomena (Stanley, 1971).

Moreover, the drag for the positive ion is approximately two to three times larger than that for negative ions. This fact is unique among the known drag mechanisms in liquid helium.

Two more or less different approaches have been pursued to explain the experimental observations (Huang and Dahm, 1976; Ostermeier and Glaberson, 1976). Both are based on the condensation of ^{3}He atoms on the vortex core.

The first approach (Huang and Dahm, 1976) assumes that the vortex core consists only of condensed ^{3}He atoms, which are localized with respect to motion perpendicular to the vortex line. The ^{3}He density $n(r)$ in the core is set equal to

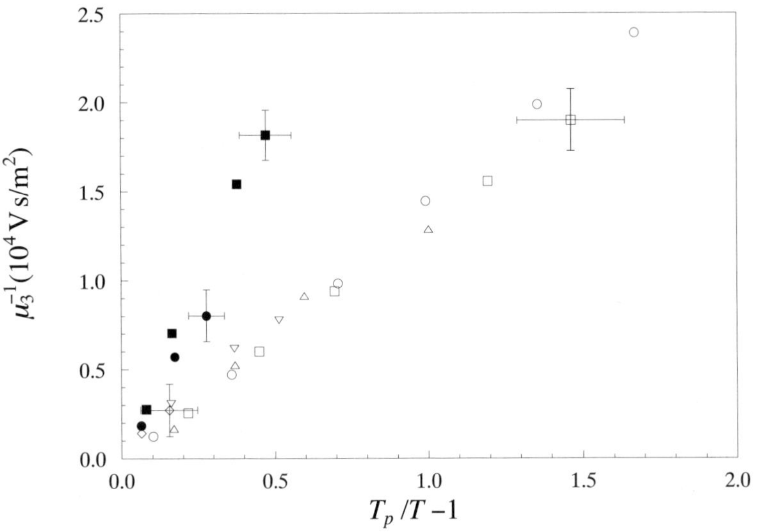

FIG. 12.14. μ_3^{-1} vs $T_p/T-1$ for positive and negative ions. (Ostermeier and Glaberson, 1976.) Closed symbols: positive ions, with $x_3\,(\%) = 0.072$ (squares), and 0.017 (circles). Open symbols: negative ions, with $x_3\,(\%) = 2.90$ (circles), 1.04 (squares), 0.27 (triangles), 0.072 (inverted triangles), and 0.017 (diamonds).

the bulk density n_3 of pure ^{3}He for $r < a$, where a is the vortex core parameter. For $r > a$, $n = 0$.

The system is thus treated as a one-dimensional, degenerate Fermi gas with linear density n_L. The ion, of radius R_i, separates the two halves of the vortex core, acting like a rigid wall. The two portions of the vortex core communicate with each other through an indirect equilibrium process between the core and the bulk. Since the Fermi gas is degenerate, only those atoms within $k_B T$ of the Fermi surface can scatter.

As the motion in this problem is uni-dimensional, no vector notation is required, and the information on the direction of motion is indicated by the sign of the velocities. The ^{3}He atoms are considered to undergo elastic collisions with a massive ion moving to the right with velocity V. The ^{3}He atom incident with velocity v_i rebounds, upon collision with the ion, with velocity $v_f = -v_i + 2V$. The total force on the right-hand side of the ion is

$$\mathcal{F}_R = -\sum_i{}' m_3^\star (v_f - v_i)\,|v_i - V| n_L f(v_i)\,[1 - f(v_f)] \Big/ \sum_j f(v_j)\,, \qquad (12.22)$$

where $m_3^\star$ is the effective mass of the ^{3}He atoms, f is the Fermi–Dirac distribution function, and $n_L f(v_i)/\sum_j f(v_j)$ is the linear density of ^{3}He atoms with velocity v_i. The prime means that the summation is restricted to those values of v_i such that $v_i - V < 0$. However, as $V \ll v_F$, the Fermi velocity, the summation is carried out over all $v_i < 0$.

The force $\mathcal{F}_L$ on the left-hand side of the ion is identical to eqn (12.22), except that the summation is over all $v_i > 0$.

Using the usual symmetry properties of the Fermi distribution, one obtains the total force on the ion as

$$\mathcal{F} = \mathcal{F}_R + \mathcal{F}_L = -8m_3^\star n_L V \sum_i^{v_i>0} v_i f(v_i)\left[1 - f(v_f)\right] \Big/ \sum_j f(v_j)\,, \qquad (12.23)$$

from which μ_3 is obtained as

$$\mu_3^{-1} = 8m_3^\star \frac{n_L}{e} \sum_i^{v_i>0} v_i f(v_i)\left[1 - f(v_f)\right] \Big/ \sum_j f(v_j)\,. \qquad (12.24)$$

For small ion velocities V, $|v_f| \approx |v_i|$ and $f(v_i)[1 - f(v_f)] \approx f(v_i)[1 - f(v_i)] = 0.25\,\mathrm{sech}^2[(\epsilon - \epsilon_F)/2k_\mathrm{B}T]$, where ϵ_F is the Fermi energy. The summation in the numerator of eqn (12.24) can be replaced by the integral

$$\int_0^\infty \frac{l}{8\pi\hbar}\,\mathrm{sech}^2\left[(\epsilon - \epsilon_F)/2k_\mathrm{B}T\right]\mathrm{d}\epsilon = \frac{lk_\mathrm{B}T}{4\pi\hbar}\left[1 + \tanh\left(\epsilon_F/2k_\mathrm{B}T\right)\right]\,, \qquad (12.25)$$

where l is the total vortex length.

The denominator in eqn (12.24) is given in terms of the Fermi momentum k_F as $\sum_j f(v_j) = l/\pi k_F$. Finally, the mobility is expressed as

$$\mu_3^{-1} = \left(\frac{2m_3^\star n_L k_\mathrm{B}T}{e\hbar k_F}\right)\left[1 + \tanh\left(\epsilon_F/2k_\mathrm{B}T\right)\right]\,. \qquad (12.26)$$

The linear density corresponding to a single line of ^{3}He atoms condensed onto the vortex core is taken as $n_L = n_3\pi a_1^2 = d_3^{-1}$, where d_3 is the diameter of the Wigner–Seitz cell in pure ^{3}He. The corresponding mobility μ_1 is calculated by substituting the one-dimensional Fermi momentum $k_F = 0.5\pi n_1$ into eqn (12.26) and setting $m_3^\star = m_3$, the bare ^{3}He atom mass, thus obtaining

$$\mu_1^{-1} = \frac{8m_3k_\mathrm{B}}{e\pi\hbar}T = 1.04 \times 10^4\,T \qquad (\mathrm{V\,s/m^2})\,. \qquad (12.27)$$

As T is lowered, both the core radius and the linear density, which are related by $n_L = n_3\pi a^2(n_\infty, T)$, increase because the phase-separation boundary moves away from the vortex center and occurs for a radius r_0 given by

$$n_\infty = n_c(T)\exp\left(U(r_0)/k_\mathrm{B}T\right)\,, \qquad (12.28)$$

where $n_c(T)$ is the phase-separation concentration of ^{3}He at temperature T (Zinov'eva and Peshkov, 1960; Edwards *et al.*, 1965).

The density of ^{3}He atoms at a distance r is calculated using eqn (12.17) with U given by eqn (11.8), the substitution energy difference of a ^{3}He atom (Donnelly and Roberts, 1969*b*), where $a_3 = 2.8$ Å is the ^{3}He radius from the effective mass formula (12.20), and a is the core parameter for pure ^{3}He. The core radius is now set to $a = r_0 + a_1$.

The results of these calculations for positive ions are compared with the experimental data in Fig. 12.12. The theoretical curves, eqn (12.26), are the solid lines. The dashed line is eqn (12.27). If one assumes that the onset of the increased ion drag occurs when a single line of ^{3}He atoms has condensed onto the vortex core, then this onset is given by the intersection of the theoretical lines with the dashed one, in quite good agreement with experiment. The use of the effective mass $m_3^\star = 2.88m_3$ and of $k_F = (3\pi^2 n_3)^{1/3}$ yields results that are approximately 15% higher.

Ions are bound to vortex lines as the result of the displacement of high-velocity superfluid near the core. As the normal core grows, the ions may be displaced from the center of the line, and the model is not expected to apply for core radii much larger than one half of the ion radius (Huang and Dahm, 1976).

The difference between positive and negative ions shown in Fig. 12.12 is explained by recalling the structural differences among the two kinds of ion. In addition to the attraction toward the center of the vortex, ^{3}He atoms also condense at the surface of the electron bubble. The binding energy on the surface, nearly 2 K, is comparable to the near 3 K of binding energy on the vortex core. The ^{3}He atoms of interest on the core have kinetic energy near the Fermi energy that is much greater than the difference in the ion and vortex core binding energies, and are not expected to undergo complete reflection. They might even be absorbed on the ions or, in the extreme case, slip along the ion surface to the opposite side, thus yielding a smaller drag. In any case, they transfer less momentum to negative than to positive ions.

The second approach by Ostermeier and Glaberson (1976), whose results are presented in Fig. 12.15, again assumes that, as soon as T is lowered below T_p for each concentration, the phase-separation boundary moves away from the center to a distance r_0 defined by eqn (12.28), yielding a new core parameter $a = a_1 + r_0$. The linear density of ^{3}He atoms onto the vortex core is the same $n_L = n_3 \pi a_1^2$ as calculated in the previous model.

The ion mobility is calculated as (Baym *et al.*, 1969)

$$\mu^{-1} = -\frac{2\hbar}{e\pi} \int_0^\infty k^2 \frac{\partial n}{\partial k} |\mathcal{R}(K)|^2 \mathrm{d}k\,, \tag{12.29}$$

where $\mathcal{R}(k)$ is the reflection coefficient, k is the axial wave number, and n is the distribution function.

As only states near the Fermi energy are able to scatter, as eqn (12.29) should include a summation over the various radial and azimuthal quantum numbers,

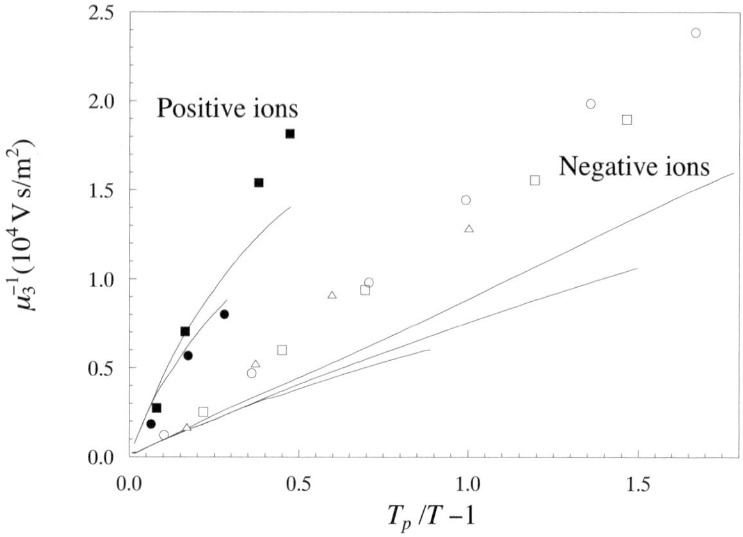

FIG. 12.15. μ_3^{-1} vs $T_p/T-1$ for positive and negative ions in mixtures. (Ostermeier and Glaberson, 1976.) THE ^{3}He concentrations and symbols are as in Fig. 12.14. The curves are the calculated ^{3}He contribution to the vortex drag.

and as n_L is quite large, it is assumed that the Fermi–Dirac distribution function can be replaced by the Maxwell–Boltzmann distribution

$$n = n_L \left(\frac{\pi\hbar^2}{2m_3^\star k_\mathrm{B}T}\right)^{1/2} \exp\left(-\frac{\hbar^2k^2}{2m_3^\star k_\mathrm{B}T}\right). \tag{12.30}$$

Thus, the mobility turns out to be

$$\mu_3^{-1} = \left(\frac{32m_3^\star k_\mathrm{B}T}{\pi}\right)^{1/2} \frac{n_L}{e} \int_0^\infty \frac{\epsilon e^{-\epsilon/k_\mathrm{B}T}}{(k_\mathrm{B}T)^2} \left|\mathcal{R}\left(\frac{(2m_3^\star/\epsilon)^{1/2}}{\hbar}\right)\right|^2 \mathrm{d}\epsilon\,. \tag{12.31}$$

The difference between the positive and negative ions lies in their reflection coefficients and in their different structure. In particular, the electrostrictive nature of the positive ion structure (Atkins, 1959) leads to a repulsive buoyant interaction with a ^{3}He atom (eqn (5.16)) (Bowley and Lekner, 1970), which forces the ^{3}He atom to pass around the positive ion at a distance of the order of 25 Å, therefore transferring more momentum to it than to the negative ion (Ostermeier and Glaberson, 1976). One thus expects the reflection coefficients $\mathcal{R}_+(k) \approx 1$ for wavelengths $\lambda \leqslant 25$ Å and $\mathcal{R}_+(k) = 0$ for longer λ. This cut-off leads to a value ≈ 1 for the integral in eqn (12.31), whereas the existence of a binding energy of ≈ 2 K for the ^{3}He atom on the electron bubble surface leads to a model in which the ^{3}He atom is scattered by a 2 K deep and πR_- wide square well (Guo *et al.*, 1971). This gives a value of 0.25 for the integral in eqn (12.31).

The results of this model are shown in Fig. 12.15. Though quite crude, it describes quite well the general temperature dependence of μ^{-1} and the differences between positive and negative ions, suggesting that the basic physics of the problem has been correctly accounted for, namely, that the most important contribution to the drag increase at low temperature comes from ion scattering off ^{3}He atoms condensed onto and about the vortex core.

13

TRANSPORT PROPERTIES OF DIFFERENT IONS

Different ionic species have been investigated, though in a much less thorough way than helium ions. Besides spectroscopic studies that are not of concern here (Bauer *et al.*, 1985, 1989, 1990; Beijersbergen *et al.*, 1993; Kanorsky *et al.*, 1994; Tabbert *et al.*, 1995; De Toffol *et al.*, 1996;Stienkmeier *et al.*, 1996), the usual transport measurements have been carried out with results that are in line with the measurements on He ions, though with some notable exceptions.

13.1 Positive impurity ions

Elements in the first two columns of the periodic table can be easily ionized so as to obtain singly-charged positive ions. Their mobility is studied in order to test the picture of the positive ion in superfluid He II. Actually, as previously discussed, the influence of the remaining electron in the valence shell produces a radius different from that expected on the basis of the simple electrostriction model (Cole and Bachman, 1977; Cole and Toigo, 1978).

The main experimental efforts have been directed at alkali and alkaline–earth ions (Ihas and Sanders, 1970; Johnson and Glaberson, 1972, 1974; Glaberson and Johnson, 1975; Főrste *et al.*, 1997, 1998). As for the positive He ions, there is a low-field region where the drift velocity v_D is proportional to the electric field, and so the mobility is constant. In Fig. 13.1 the inverse zero-field drift mobility of several positive ions in the range $1.1\,\mathrm{K} < T < 1.7\,\mathrm{K}$ is shown.

The interesting feature is that the alkaline–earth ions (Ca, Sr, Ba) have mobilities larger than those of He^+, whereas the alkali metal ions have lower μ than those of He^+. This difference cannot be explained in terms of the simple electrostriction model as in this model the ion-complex structure depends only on the ionic charge and on the thermodynamic parameters of the fluid (Atkins, 1959).

These observations have thus led researchers to consider the effect of the electron remaining in the valence shell of the alkaline–earth ions (Cole and Bachman, 1977). The wave function of this electron is quite extended and its repulsive exchange interaction with the electron clouds of the surrounding He leads to the formation of a cavity, whose radius is a bit smaller than that of the alkali ions, rather than that of a snowball.

Notable exceptions are the mobilities of Be^+ and Mg^+ (Főrste *et al.*, 1997, 1998), whose difference cannot be explained in terms of their slightly different polarizability and valence-electron wave function. Actually, the mobility of Be^+ ions is smaller than that of positive He ions, although it is an alkaline–earth

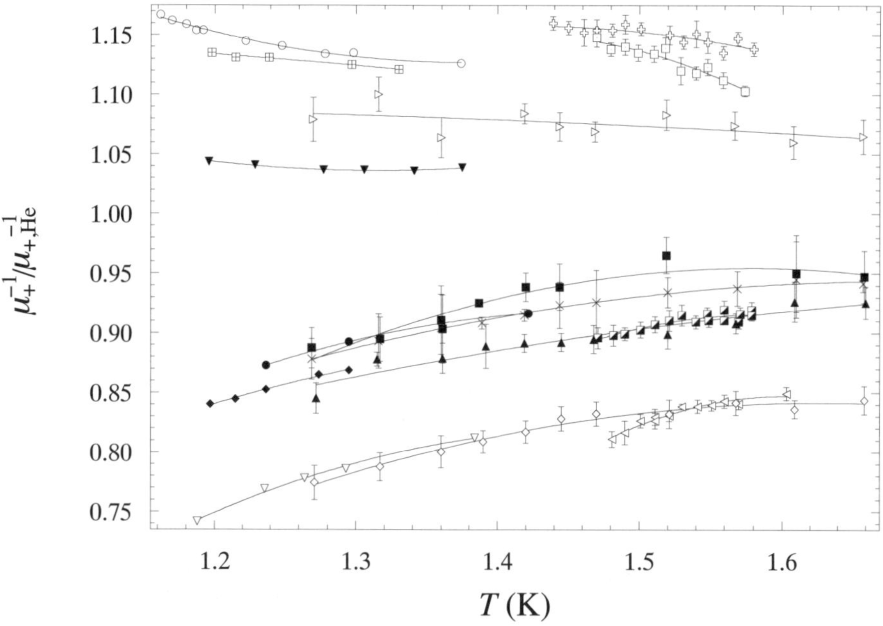

FIG. 13.1. $\mu_+^{-1}/\mu_{+,\mathrm{He}}^{-1}$ vs T. Rb^+ (open circles), Cs^+ (crossed squares), K^+ (inverted closed triangles), Ca^+ (closed circles), Sr^+ (closed diamonds), and Ba^+ (inverted open triangles) (Johnson and Glaberson, 1972; Glaberson and Johnson, 1975). Cd^+ (thick crosses), Zn^+ (dotted squares), Sr^+ (half-filled squares), and Ba^+ (left-pointing triangles) (Bauer *et al.*, 1985). Be^+ (right-pointing triangles), Mg^+ (closed squares), Ca^+ (oblique crosses), Sr^+ (closed triangles), and Ba^+ (open diamonds) (Gűnther *et al.*, 1996; Főrste *et al.*, 1997, 1998). Lines: eyeguides.

element, and has a temperature dependence closer to that of alkali ions than to that of the other alkaline–earth ions.

At the fairly high temperatures under investigation, scattering is dominated by rotons, and specific differences in the roton–ion interaction can be determined by factoring out the exponential roton density factor. In Fig. 13.2 the prefactor of the roton-limited mobility, $f_+/e = \mu_+^{-1}\exp(\Delta/T)$, is shown (Főrste *et al.*, 1997, 1998).

In this figure the similarity between the mobilities of Be^+ and of the positive helium atom is even more evident. Be^+ behaves much more like a snowball ion than like the other alkaline–earth ions, which are surrounded by a cavity.

If it is assumed that the temperature dependence of the prefactor f_+ is determined by the ion-complex structure, then this hypothesis suggests that Mg^+ has a bubble-like structure, whereas Be^+ has a snowball-like structure.

It is important to mention here the theoretical treatment due to Cole and Toigo (1978), who proposed that for Mg^+ a structural transition occurs as a

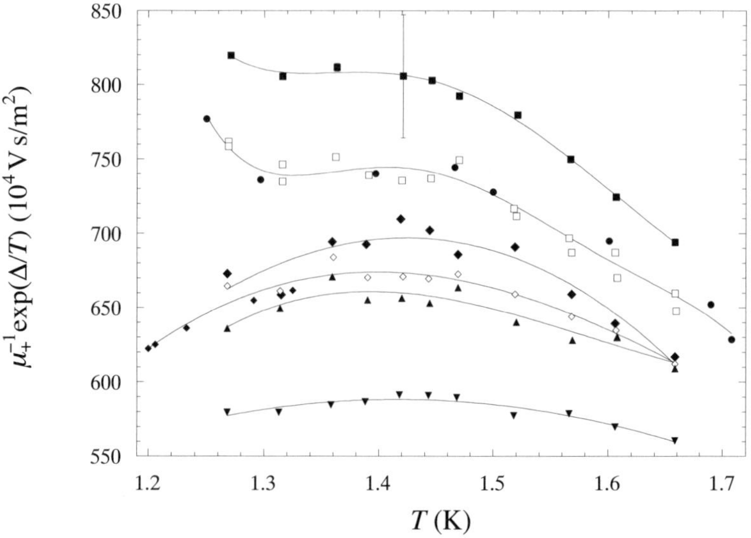

FIG. 13.2. $\mu_+^{-1}\exp(\Delta/T)$ vs T. Be^+ (closed circles), He_2^+ (closed squares), Mg^+ (closed triangles), Ca^+ (open diamonds), Sr^+ (open triangles), and Ba^+ (inverted closed triangles) (Főrste *et al.*,1997, 1998). He_2^+ (open squares) (Schwarz, 1972*a*). Ca^+ (closed diamonds) (Johnson and Glaberson, 1972).

function of the applied pressure. At a given distance from the ion, the local pressure increase induced by electrostriction could lead to the formation of a thin "crust" surrounding the cavity because of exchange forces. Probably, this may happen for Be^+, instead. This effect might have some similarity with the anomaly of the positive normal ion radius near T_λ, which is due to an electrostriction-induced λ-transition near the ion (Ahlers and Gamota, 1972; Goodstein *et al.*, 1974; Scaramuzzi *et al.*, 1977*a*; Cole *et al.*, 1978; Goodstein, 1978).

In any case, attempts at getting information on the ion–roton interaction along the lines indicated by Barrera and Baym (1972) and by Bowley (1971*c*) have led nowhere (Glaberson and Johnson,1975; Főrste *et al.*, 1997, 1998), also because of the lack of a better knowledge of the ion–roton transition matrix elements.

In contrast, good estimates of the ionic radii for some alkali and alkaline–earth species have been obtained by investigating the electric field dependence of the positive impurity ions in conditions in which they generate vortex rings and are trapped by them (Johnson and Glaberson, 1974). In Fig. 13.3 the drift velocity of positive impurity ions is reported as a function of the electric field E beyond the critical field of the giant discontinuity and close to the second critical field (Bruschi *et al.*, 1966*a*).

The low-field data, not shown here, are essentially in proportion to the low-field mobility and have approximately the same critical field.

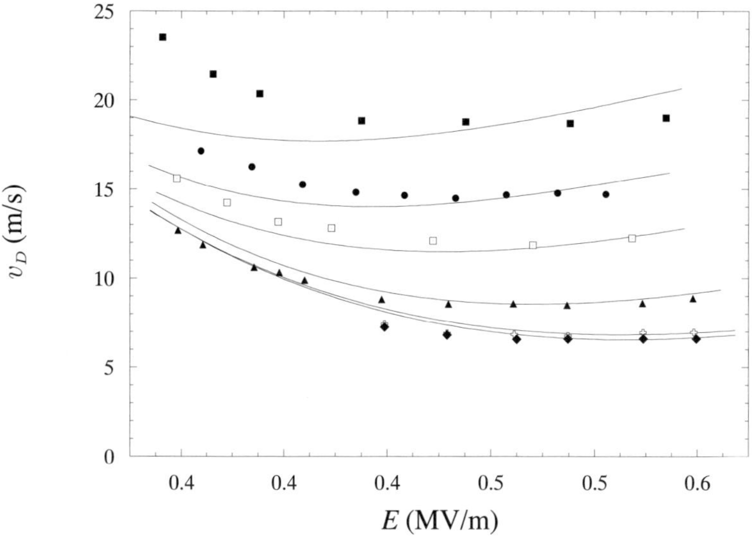

FIG. 13.3. v_D vs E for the positive impurity ions Ba^+, Sr^+, Ca^+, He^+, K^+, and Cs^+ (from top) for $T = 1.18\,K$ (Johnson and Glaberson, 1974). Solid curves: theoretical model (Padmore, 1971).

For larger E, the positive ions follow a dynamics similar to that of He ions: v_D depends on the sequence of capture–escape from vortex rings.

As positive ions are smaller than negative ions, they are less tightly bound to vortex rings and their thermally-activated escape rate is quite large and depends on the ion radius. The time spent as a bare ion is quite negligible. The major reason why positive ions have higher v_D at higher E than ions permanently bound to vortex rings is that a significant amount of time is spent by the vortex ring accelerating to its equilibrium velocity.

The positive ion data are analyzed in terms of the Padmore model (Padmore, 1971, 1972*a*), with the escape rate given by Donnelly and Roberts (1969*b*), and the ion radius left as an adjustable parameter. The corresponding radii are reported in Table 13.1 and the calculated v_D are shown in Fig. 13.3 as solid lines that are in good agreement with experiment. It should be recalled that the ion radii determined in this way are rather the radii of the regions of excluded superfluid, which may be larger by something like a healing length (Johnson and Glaberson, 1974).

Table 13.1 *Values of R_+ (Å) at $T = 1.18\,K$. (Johnson and Glaberson, 1974.)*

Ba	Sr	Ca	He	K	Cs
6.1	6.1	7.1	7.9	8.35	8.40

13.2 Exotic negative ions

A brief mention must be made of the still-elusive, fast negative ions (Doake and Gribbon, 1969; Ihas and Sanders, 1971; Eden and McClintock, 1984; Sanders and Ihas, 1987; Hendry *et al.*, 1988*b*; Maris, 2000).

Under very specific experimental conditions, related to the height of the liquid level above the ion source and the appearance of the glow discharge in the vapor (Ihas and Sanders, 1971), several types of fast negative ions appear, having much larger drift velocities than either the negative or positive helium ion.

The fastest ones apparently reach a saturation velocity limited by roton emission at high fields without nucleating vortex rings (Doake and Gribbon, 1969; Ihas and Sanders, 1971), while at low electric E their mobility is much larger than that of helium ions. In Fig. 13.4 the zero-field mobility of such ions is reported.

A plethora of ions with intermediate mobilities between the fast and normal ions were discovered later, but their nature is still elusive. A study of the behavior of these exotic ions at high E has shown that v_D flattens out close to $v_L \approx 60\,\mathrm{m/s}$ at SVP for the fastest ion, whereas the exotic ions of intermediate mobility show the giant discontinuity associated with the creation of charged quantized vortex rings (Eden and McClintock, 1984). Such a behavior is reported in Fig. 13.5.

Each exotic species exhibits the giant fall that normal negative ions do at $E = E_c$, though at its own peculiar value of the critical field, $E = E_{ci}$. The critical velocities are approximately 10 m/s higher than the corresponding critical velocity of the normal ions. Owing to the lack of sensitivity of the ion current

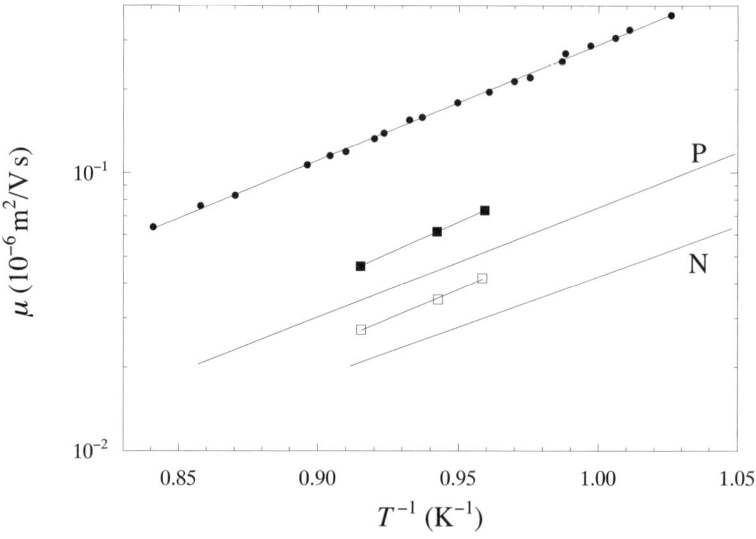

FIG. 13.4. μ vs T for fast negative ions (closed circles) and for two other types of exotic negative ions. (Ihas and Sanders, 1971.) Solid lines: exponential fits. Lines labeled "P" and "N" are the mobilities of normal positive and negative He ions.

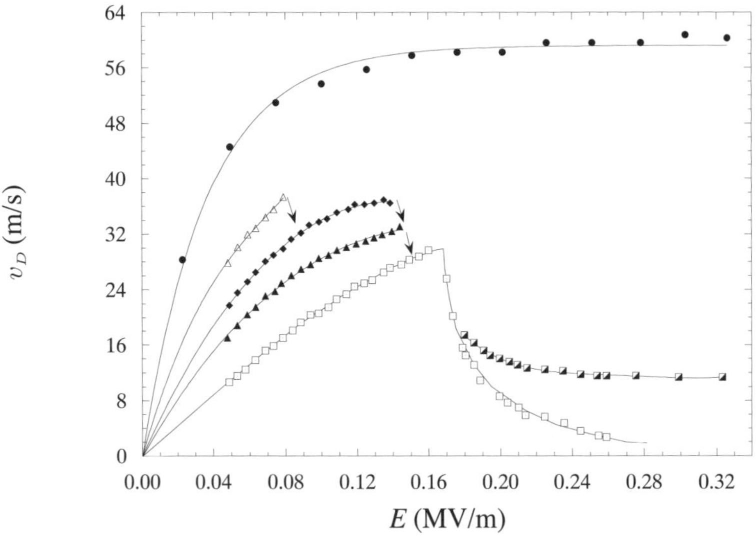

FIG. 13.5. v_D vs E for fast ions (circles) and for three exotic ions of intermediate mobility (open triangles, diamonds, and closed triangles) at $T = 1.03\,\text{K}$. Only the fastest of them is plotted for $E > E_c$ (half-filled squares). Open squares: normal negative ions. Lines: eyeguides. The arrows indicate the field at which the exotic ion signal abruptly disappears for each species. (Eden and McClintock 1984.)

measurements for the weak exotic ion signal, it is not possible to unambiguously associate individual species observed for $E < E_c$ with those for $E > E_c$ (Eden and McClintock, 1984).

A detailed plot for several exotic ions in the field range for which the dynamics of charged quantized vortex rings is expected to occur is shown in Fig. 13.6. The behavior in this region is even more puzzling because the velocity of each of the five most prominent exotic ions appears to tend to an individual field-independent value of several m/s at high field. If the exotic carriers in this regime are in fact charged vortex rings, then it appears very likely that their motion should be determined by a continuous sequence of escape–trapping events (Padmore, 1972*a*). However, this process should not necessarily lead to a field-independent limiting v_D.

On the other hand, owing to their higher zero-field mobility, exotic ions are expected to be smaller than normal ions, so that their binding energy to the vortex ring should be weaker and their thermally-activated escape rate should consequently be higher (Donnelly, 1965; Parks and Donnelly, 1966). One, thus, expects that the limiting v_D should decrease with increasing T, but this possibility has not been investigated.

Several hypotheses about the nature of these negative ions have been advanced, especially for the case of the fastest ions. In particular, it has been sug-

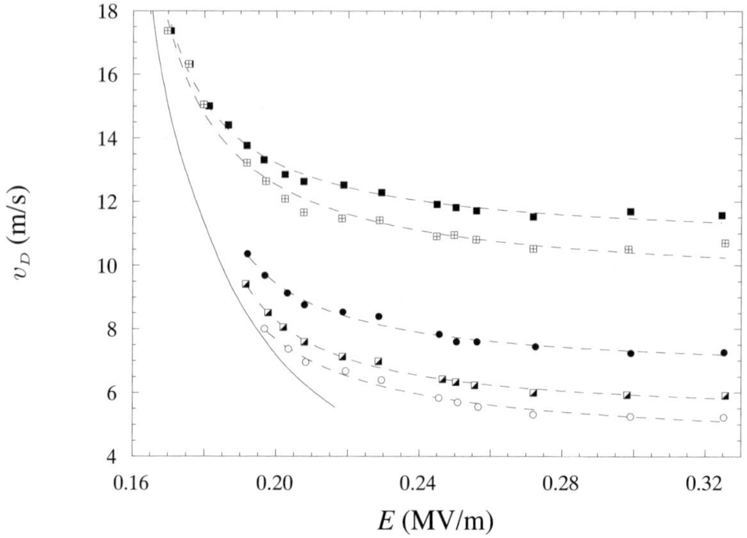

FIG. 13.6. High-E behavior for the five most prominent exotic ions. (Eden and McClintock, 1984.) Solid line: normal negative ion. Dashed lines: eyeguides.

gested that they might be multi-electron bubbles (Tempere *et al.*, 2001), whose mobility is determined by the unusual kinematics of rotons due to the existence of two momentum states available for scattering with antiparallel velocities corresponding to the same energy, located nearly symmetrically about p_0 (Sanders and Ihas, 1987). If the radius of the multi-electron bubble is left as an adjustable parameter, a good agreement with the fastest ion data (see the solid line in Fig. 13.4) is obtained.

This interpretation has been strongly criticized for several reasons (Hendry *et al.*, 1988*b*). In fact, it has been proved that two-electron bubbles are unstable because of a too strong Coulombian repulsion between electrons (Fowler and Dexter, 1968; Dexter and Fowler, 1969; Tempere *et al.*, 2003) and ESR experiments have confirmed that negative ions are singly-charged (Reichert and Dahm, 1974; Zimmermann and Reichert, 1977; Reichert and Jarosik, 1983).

Moreover, the drift velocity of the exotic ions in the region where they undergo the transition to charged vortex rings is approximately 10 m/s larger than the corresponding velocity of the normal negative ions, thus suggesting that they are smaller than normal negative ions.

Finally, the investigation of the influence of the nature and current density of the discharge in the vapor on the magnitudes of the relative fluxes of normal and exotic ions entering the liquid suggests that exotic ions arise from precursors in the vapor (Williams *et al.*, 1987).

However, the most intriguing, though unrealistic, hypothesis about the nature of the exotic ions ever advanced is the idea of fractional electron bubbles (Maris,

2000), based on the older suggestion of the possibility of electron fission: the *electrino* hypothesis (Ivanenko and Kolesnikov, 1952).

According to this hypothesis, the electron, localized in the bubble in its $1s$ ground state, can be optically excited by the light of the discharge into the $1p$ state. The existence of such excited states of the electron in the bubble has been directly confirmed by means of different techniques (Grimes and Adams, 1990, 1992; Konstantinov and Maris, 2003).

Upon excitation of the electron, the bubble walls are set in motion and the bubble profile is deformed so as to match the new shape of the electron wave function. The inertia of the liquid surrounding the bubble can be sufficient to lead to the break-up of the bubble into two pieces, with the electron wave function fractionally shared by the two of them. The resulting bubbles would contain only a fraction of the electron charge.

This proposal, though very appealing, has been strongly criticized, and it has been concluded that, if such a state were to form, it would quickly collapse into an incoherent quantum superposition of two separated ground-state bubbles and all the measurable properties of this state are identical to those of a single bubble (Rae and Vinen, 2001).

Just for the sake of completeness, it is important to mention the recent *grand unified theory of classical quantum mechanics* (Mills, 2002), according to which electrons would be a new entity called an *electron orbitsphere*. Exotic ions would be a manifestation of the existence of fractional principal quantum excited energy states of free electrons in superfluid helium (Mills, 2001). As this theory puts at stake the foundations of quantum mechanics, because it postulates that the notion of probability waves must be abandoned and that atomic theory must be based on reality (Mills, 2001), it will not be pursued further.

14

DIRECT DETERMINATION OF THE EFFECTIVE MASS OF IONS

A very important physical quantity for the analysis of the transport properties of ions is their effective mass. In transport experiments the effective mass is indirectly inferred from the transport data, but it is not directly measured, in spite of its importance. In fact, its measurements offer a potentially very precise method for the determination of the ion radius R_i through the simple hydrodynamic relation (Landau and Lifsits, 2000)

$$M_i^\star = M_i + \frac{2\pi}{3}\rho R_i^3 \,, \tag{14.1}$$

where M_i is the bare-ion mass and the second term is half the mass of fluid displaced by the ion.

Actually, the bare-ion mass takes into account the fact that the ions in liquid helium are complex structures, namely, the snowball and the electron bubble. For this reason, eqn (14.1) has to be modified for the two species of ions as follows:

$$M_+^\star = \frac{4\pi}{3}R_+^3\left[\rho_{\mathrm{sol}} + \frac{\rho}{2}\right] \,, \tag{14.2}$$

$$M_-^\star = m_e + \frac{2\pi}{3}\rho R_-^3 \approx \frac{2\pi}{3}R_-^3 \,, \tag{14.3}$$

where ρ_{sol} is the density of the solid cluster induced by electrostriction around the positive ion and m_e is the electron mass. In particular, as the negative ion is squeezed by pressurizing the liquid, a strong pressure dependence of the negative ion radius is expected.

Essentially, three different methods have been devised to directly measure $M_i^\star$. Two of them are based on microwave techniques (Dahm and Sanders, 1966; Poitrenaud and Williams, 1972), and the third one investigates the inertial effects on drifting ions subject to temporary reversals of the applied electric field (Ellis and McClintock, 1982). In spite of the differences in the methods, the results give a consistent picture of the large effective mass of the ions, and, in particular, of the differences in the effective masses of the two different species of ions.

14.1 The microwave loss technique

Ions are introduced into a microwave cavity, resonant at approximately 9.3 GHz (Dahm and Sanders, 1966). Their presence changes the complex reflection coefficient Γ of the cavity by an amount proportional to the complex ion mobility.

The analysis of the experimental results is most simply carried out in terms of a time-averaged Langevin equation: (Hall, 1974)

$$\mathbf{F} = e\mathbf{E} = M_i^{\star}\left(\frac{\mathrm{d}\mathbf{v}}{\mathrm{d}t} + \frac{\mathbf{v}}{\tau}\right), \tag{14.4}$$

where $\mathbf{F}$ is the force acting on the ions, τ is the momentum relaxation time of the ions, and $M_i^{\star}$ is the ion effective mass. In the case of a d.c. field, the usual Drude result for the ion mobility is obtained:

$$\mathbf{v} = \mu(0)\,\mathbf{E} = \left(\frac{e\tau}{M_i^{\star}}\right)\mathbf{E}\,. \tag{14.5}$$

If the applied field varies sinusoidally with time with angular frequency ω, the ion drift velocity becomes

$$\mathbf{v} = \mu(\omega)\,\mathbf{E} = \left\{\frac{\mu(0)\,\mathbf{E}}{1+(\omega\tau)^2}\right\}(1 - i\omega\tau)\,. \tag{14.6}$$

A measurement of the ratio of the imaginary to the real part of Γ, Γ_i/Γ_r, yields $\omega\tau$ directly, and the effective mass is then obtained from the d.c. mobility by using eqn (14.5).

An alternative way, useful in the case of high noise levels, is to measure only the variation of Γ_r. In this case, the change of Γ_r can be written as

$$\delta\Gamma_r = \left\{\frac{2\pi Q_0 e\mu(0)}{\omega\left[1+(\omega\tau)^2\right]}\right\}\frac{\int n(\mathbf{r})\,\mathbf{E}_m(\mathbf{r})\,\mathrm{d}^3r}{\int[\mathbf{E}_m(\mathbf{r})]^2\,\mathrm{d}^3r}\,. \tag{14.7}$$

Here, Q_0 is the quality factor of the unloaded cavity, and n and $\mathbf{E}_m$ are, respectively, the spatial distribution of the ion density and microwave electric field that are known by previous calibration of the apparatus. Thus, $\omega\tau$ can be determined.

The reflection cavity is contained in one arm of a two-bolometer bridge (homodyne) spectrometer, allowing the measurement of the reflection coefficient. The cavity is cylindrical and is used in the TE_{011} mode. Ions are generated by direct ionization outside the sensing volume and are injected into it by means of gating grids with electric fields of suitable polarity.

In Fig. 14.1 the values $M_i^{\star}$ of positive and negative ions measured at SVP are shown as a function of T. The value $\mu(0)$ is from the literature (Reif and Meyer, 1960). The positive ion data can be explained in terms of the electrostriction model (Atkins, 1959). The radius of the ion is assumed to be the distance from its center, where the pressure equals the melting pressure $P_m(T)$. An approximate expression for it is

$$R_+(P_m) = \left\{\frac{(\epsilon_r - 1)\,e^2}{8\pi{\epsilon_r}^2(P_m - P_0)}\right\}^{1/4}, \tag{14.8}$$

where ϵ_r is the relative dielectric constant and P_0 is the ambient pressure.

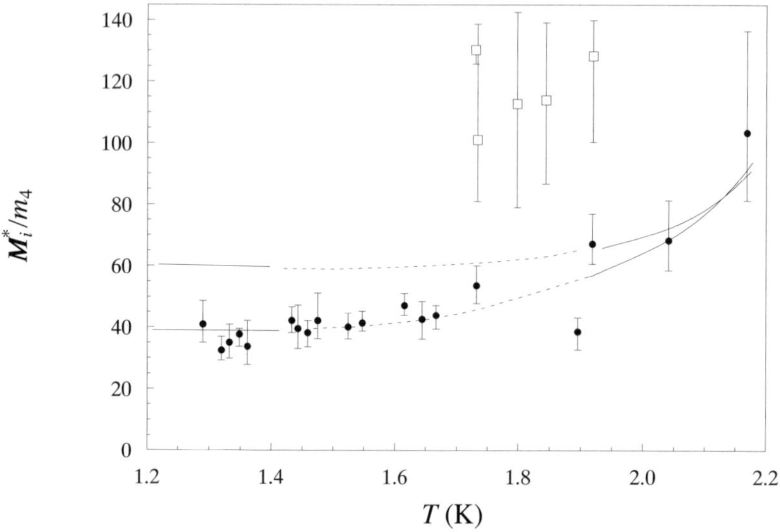

FIG. 14.1. $M_i^\star/m_4$ vs T. (Dahm and Sanders, 1966.) Circles: positive ions. Squares: negative ions. Theory with $R_+ = 5.8\,\text{Å}$ (lower curve) and with $R_+ = R_+(P_m)$, where P_m is the melting pressure (upper curve).

At low T, $\rho \approx \rho_s$ and the effective mass is given by

$$M_+^\star = M_+ + \frac{2\pi}{3}\rho_s R_+^3 = (47 + 13)m_4 \qquad (T \ll T_\lambda)\,, \tag{14.9}$$

where $M_+ = (4\pi/3)\rho_{\text{sol}} R_+^3$ is the mass of a solid sphere of the given radius.

For T near T_λ, the mean free path of excitations is small compared to R_+ and the contribution to the effective mass due to the viscous flow around the ion in the normal fluid must be taken into account (Landau and Lifsits, 2000). Adding together the contributions from the normal and superfluid components, one has the expression

$$M_+^\star = M_+ + \frac{2\pi}{3}\rho R_+^3 \left[1 + \frac{9}{2}\left(\frac{\rho_n \delta}{\rho R_+}\right)\right] \qquad (T < T_\lambda)\,, \tag{14.10}$$

where $\delta = \sqrt{2\eta/\omega\rho_n}$ is the viscous penetration depth and η is the viscosity of the normal fluid.

In this temperature region the viscous drag exerted by the normal fluid must be considered and eqn (14.4) has to be modified, yielding

$$M_+^\star \left(\frac{\mathrm{d}\mathbf{v}}{\mathrm{d}t}\right) + 6\pi\eta R_+ \left(1 + \frac{R_+}{\delta}\right)\mathbf{v} = e\mathbf{E}\,. \tag{14.11}$$

The real part of the microwave mobility is now

$$\mu_r(\omega) = \frac{\mu(0)}{1 + R_+/\delta}\left\{1 + \left[\frac{\omega\mu(0)\,M_+^\star}{e\,(1 + R_+/\delta)}\right]^2\right\}^{-1}. \tag{14.12}$$

The relationship between the effective mass $M_+^\star$ and the experimental value $(M_+^\star)_{\rm exp}$, plotted in Fig. 14.1, is

$$(M_+^\star)_{\rm exp} \equiv \frac{e}{\omega\mu(0)}\left[\frac{\mu(0)}{\mu_r(\omega)} - 1\right]^{1/2} = \left[\left(\frac{e}{\omega\mu(0)}\right)^2 \frac{R_+}{\delta} + \frac{\left(M_+^\star\right)^2}{1 + R_+/\delta}\right]^{1/2}. \tag{14.13}$$

The predictions of eqn (14.9) with R_+ evaluated from eqn (14.8) or with $R_+ = 5.8$ Å are shown as the solid lines for $T < 1.4$ K. The calculated value of eqn (14.13) with $M_+^\star$ given by eqn (14.10) is also shown for $T > 1.9$ K for the same radii. The calculated values of $(M_+^\star)_{\rm exp}$ differ from $M_+^\star$ by less than 20% for these radii.

In the intermediate temperature range, the quasiparticle mean free path is comparable with R_+ and the normal fluid contribution to the effective mass cannot be calculated.

Several other experiments can be explained in terms of the picture of the positive ion gathered in this experiment. The measured binding energy of pos-

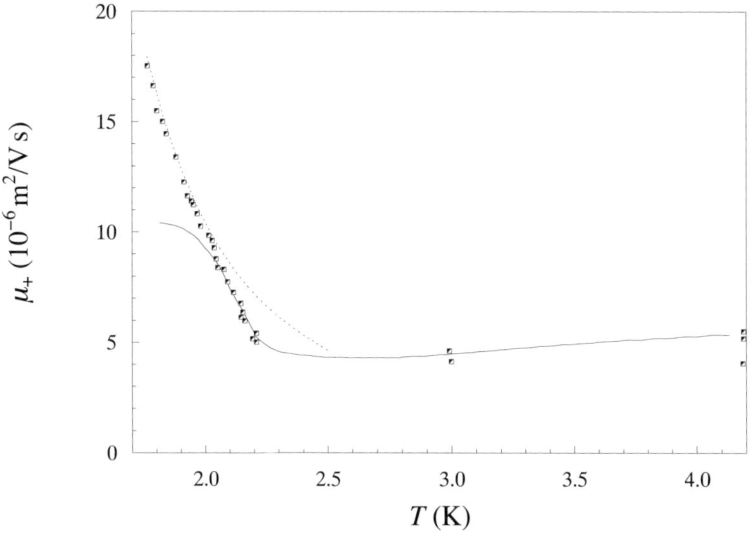

FIG. 14.2. μ_+ vs T in liquid He for $T < T_\lambda$ (Reif and Meyer, 1960) and for $T > T_\lambda$ (Meyer *et al.*, 1962; Dahm and Sanders, 1966). Dotted line: kinetic theory $\propto \exp(\Delta/T)$. Solid line: Stokes mobility with R_+ given by eqn (14.8) and η from the literature (Taylor and Dash, 1957; Tough *et al.*, 1963).

itive ions to vortex rings (Cade, 1965) is explained (Parks and Donnelly, 1966) by assuming a radius of 6.44 Å, in reasonable agreement with both the low-temperature value predicted by eqn (14.8), $R_+ = 6.7$ Å, and the best-fit value $R_+ = 5.8$ Å.

Moreover, the mobility of positive ions in He near T_λ is observed to deviate from the simple $\exp(\Delta/T)$ law observed at lower T (Reif and Meyer, 1960). This is the range in which viscous hydrodynamics is at work and the ion mobility should be described by the Stokes law:

$$\mu_+(0) = \frac{e}{6\pi\eta R_+} . \tag{14.14}$$

In Fig. 14.2 the values of the Stokes mobility calculated using the values of the radius given by eqn (14.8) and viscosity data from the literature (Taylor and Dash, 1957; Tough *et al.*, 1963) are compared to experimental mobility data (Reif and Meyer, 1960; Meyer *et al.*, 1962) for $T > 1.8$ K.

14.2 The microwave resonance technique

Ions can be trapped beneath the surface of the liquid by the combined effect of their image force and an electric field E applied perpendicular to the surface (Bruschi *et al.*, 1966*b*).

If z is the distance from the surface in the liquid and if the vapor density is negligible, the potential energy of an ion can be written as

$$V(z) = -\frac{A}{z} + eEz , \tag{14.15}$$

with A given by (Jackson, 1998)

$$A = \left(\frac{e^2}{16\pi\epsilon_0\epsilon_r}\right)\frac{\epsilon_r - 1}{\epsilon_r + 1} . \tag{14.16}$$

The minimum of the potential lies at a distance of approximately 300 Å beneath the liquid surface for a typical field strength of 10 kV/m.

At low T, the resulting potential well can be filled with ions with areal density up to $2\epsilon_0 E/e \approx 10^{12}\,\mathrm{m}^{-2}$, which form a two-dimensional plasma in which the average ion spacing greatly exceeds any thermally-induced vertical displacement. The pool of ions can be confined in the horizontal direction by a suitable fringing field.

At sufficiently low T all ions are thermalized at the bottom of the potential well, where they vibrate at the frequency

$$\omega_0^2 = \frac{2}{M_i^\star}\frac{(eE)^{3/2}}{A^{1/2}} , \tag{14.17}$$

where $M_i^\star$ is the ion effective mass. The sharpness of the resonance depends on the parameter $\omega_0\tau$, where τ is the characteristic time for momentum relaxation related to the mobility. For a frequency of 200 MHz, $\omega_0\tau > 100$ for $T < 0.6$ K.

The experiment consists of finding the electric field E at which the system resonates with an exciting z-directed r.f. field of 200 MHz (Poitrenaud and Williams, 1972). The absorption is measured as a function of E with a superheterodyne spectrometer (Poitrenaud, 1970). The results at $T = 0.7\,\mathrm{K}$, corrected for the anharmonicity of the potential, yield $M_-^\star = (243 \pm 5)m_4$ and $M_+^\star = (45 \pm 2)m_4$. By converting masses into radii by means of eqns (14.3) and (14.2), one gets $R_- = (11.4 \pm 0.1)\,\text{Å}$ and $R_+ = (6.0 \pm 0.1)\,\text{Å}$, respectively, in reasonable agreement with other experiments (Dahm and Sanders, 1966; Parks and Donnelly, 1966; Schwarz and Stark, 1969) and with theoretical estimates (Clark, 1965; Hiroike *et al.*, 1965).

Capacitance–conductance bridge techniques have been used to study the plasma resonances for ion motion in a direction parallel to the surface at much smaller frequencies (Theobald *et al.*, 1981; Ott-Rowland *et al.*, 1982; Barenghi *et al.*, 1986, 1991). In these experiments, the liquid surface is between the top and bottom electrodes in such a way that a vertical d.c. field confines the ions beneath the surface. Sideways electrodes generate the a.c. field necessary to induce the ion motion parallel to the surface.

The solution of the continuity, acceleration, and Poisson equations in a cylindrical geometry gives the dispersion relation

$$\omega^2 = \left(\frac{n_0 e^2 k}{2\epsilon_0 M_i^\star}\right) \frac{2\sinh(kd)\sinh[k(h-d)]}{\sinh(kh) + (\epsilon_r - 1)\sinh[k(h-d)]}, \tag{14.18}$$

where h is the distance between the top and bottom electrodes, d is the height of the ion layer above the bottom electrode, n_0 is the ion density, ϵ_r is the relative dielectric constant of the liquid, and $M_i^\star$ is the effective ion mass (Grimes and Adams, 1976; Barenghi *et al.*, 1986, 1991).

The plasma resonances have sufficiently high Q at low T to be easily observed. The measurement of the effective mass of negative ions at low T ($50\,\mathrm{mK} < T < 130\,\mathrm{mK}$) (Barenghi *et al.*, 1986, 1991) yields $M_-^\star = (237 \pm 7)m_4$, independent of T, in good agreement with data at $T = 0.7\,\mathrm{K}$ (Poitrenaud and Williams, 1972), as shown in Fig. 14.3.

The linewidths of the resonances of the negative ions are strongly temperature dependent (Barenghi *et al.*, 1986, 1991). If it is assumed that dissipation is entirely due to the finite ion mobility, the mobility itself can be deduced from the linewidths. Thus, it has been possible to extend the observations of Schwarz (1972*a*) to much lower T, as shown in Fig. 14.4.

The line in the figure is the theory of phonon scattering of Baym *et al.* (1969). As only s- and p-wave scattering has been included, the theory is expected to be reasonably accurate only below 100 mK (Barenghi *et al.*, 1986, 1991). Owing to the very good agreement of the bulk phonon scattering theory with the data, and owing to the fact that the results do not depend on the distance z_0 of the bottom of the well from the surface, at least for $z_0 \geqslant 30\,\text{Å}$, it appears that no significant ripplon scattering (Shikin, 1970) occurs.

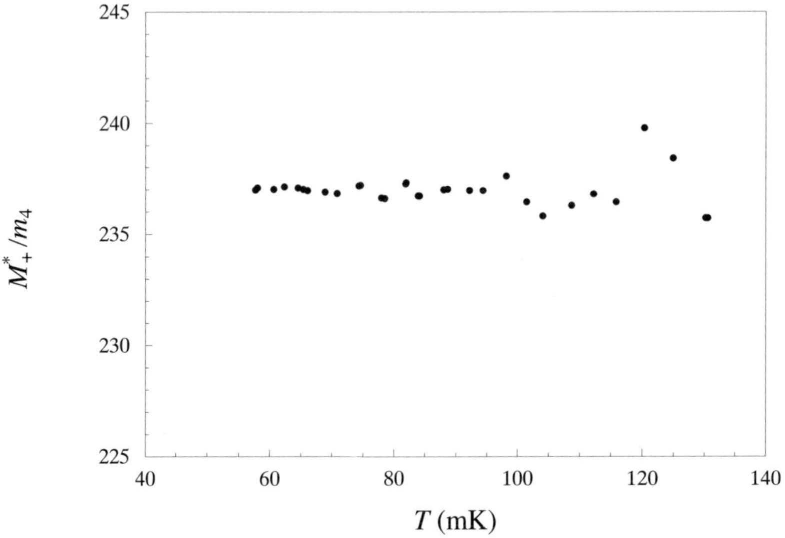

FIG. 14.3. $M_-^\star/m_4$ vs T at SVP and low temperature. (Barenghi *et al.*, 1986, 1991.)

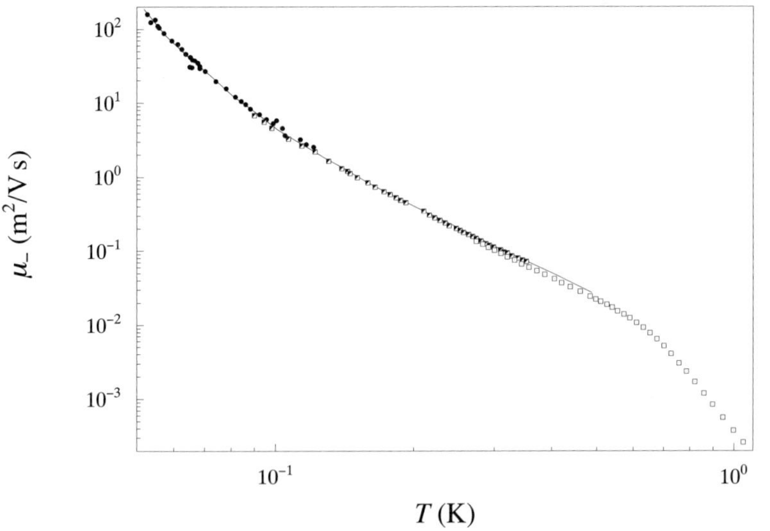

FIG. 14.4. μ_- vs T at low T from plasma resonance measurements of ions trapped beneath the liquid He II surface. (Barenghi *et al.*, 1986, 1991). Circles: data from linewidth measurements, and half-filled squares: data from plasma capacitance—conductance measurements (Barenghi *et al.*, 1986, 1991). Dotted squares: drift data (Schwarz, 1972*a*). Solid line: phonon-scattering theory (Baym *et al.*, 1969).

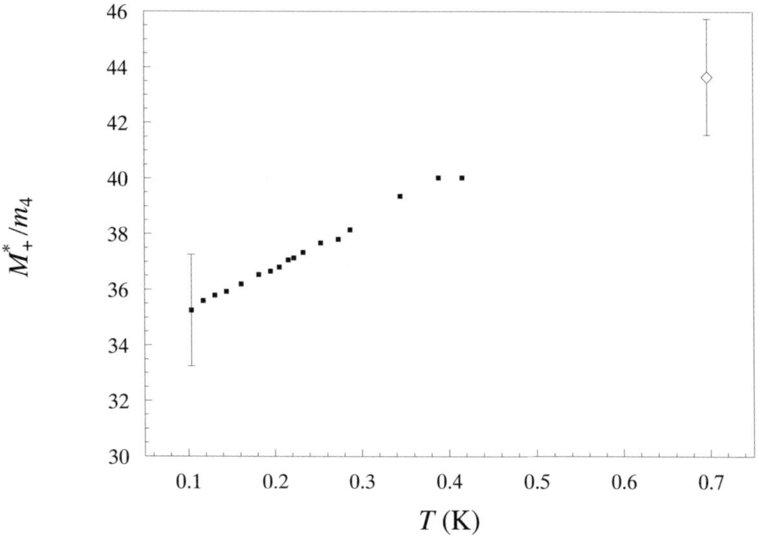

FIG. 14.5. $M_+^\star/m_4$ vs T (Ott-Rowland *et al.*, 1982). Diamond: high-T value reported by Poitrenaud and Williams (1972).

In contrast, $M_+^\star$, measured with a similar technique, shows an unexpected temperature dependence (Ott-Rowland *et al.*, 1982), as shown in Fig. 14.5, although the extrapolation to high T agrees with previous measurements (Poitrenaud and Williams, 1972). This weak temperature dependence has been attributed to a small temperature dependence of the liquid–solid surface tension of the cluster surrounding the positive ion, which might be different from the bulk value (Ott-Rowland *et al.*, 1982).

It is interesting to recall that in this experiment nonlinear effects of the two-dimensional plasma resonances due to the ponderomotive force (the $(\mathbf{v}\cdot\nabla)\mathbf{v}$ term in the equation of motion) have been observed for large-oscillation amplitudes (Ott-Rowland *et al.*, 1982).

14.3 The acceleration method

A straightforward acceleration method has been used to directly measure $M_-^\star$ at low $T \approx 70\,\text{mK}$ and at high $P = 2.5\,\text{MPa}$ in isotopically-pure ^{4}He by using a drift velocity-like technique (Ellis and McClintock, 1982; Ellis *et al.*, 1983).

Under these experimental conditions, the drag on the ions due to residual excitation scattering is negligible and the rate of vortex nucleation is also negligible (Bowley *et al.*, 1980, 1982). Thus, the ion moves through the *mechanical vacuum* provided by the isotopically-pure He II.

In response to a steady force from a d.c. electric field E, the ion undergoes an acceleration $eE/M_-^\star$ toward the Landau critical velocity v_L. Thereafter, it proceeds with an average drift velocity v_D, which is determined by the aperiodic emission of roton pairs (Allum *et al.*, 1976*a*, 1976*b*; Bowley and Sheard, 1977; Ellis *et al.*, 1980*a*, 1980*b*; McClintock *et al.*, 1981).

At $T = 90\,\text{mK}$, $v_D = v_L$ to within 1% for $E = 100\,\text{V/m}$. Provided that the instantaneous ion velocity lies within the dissipationless range $|v| < v_L$, it is expected that the ion dynamics is similar to those of a free particle. The acceleration is determined from suitable time-of-flight measurements, using a modified version of the charge–pulse technique shown in Fig. 5.3 (Schwarz, 1972*a*). Even for fields as low as 100 V/m, the ions reach the allowed velocity range of $\pm v_L$ in a few μs, so that they are already moving at the desired speed when tehy enter the drift space.

During their motion, ions are subjected to a repeated transient reversal of the field E for an adjustable period Δt in the range $0.5\,\mu\text{s} < \Delta t < 50\,\mu\text{s}$. This reversal produces a uniform deceleration of the ion, followed by a further acceleration as soon as the forward field is re-established. The ions' arrival is thus delayed by an amount that depends on the product $E\Delta t$. Depending on whether $E\Delta t > 2v_L M_-^\star/e$ or not, the ion may, or may not, decelerate to $-v_L$ before the forward field is re-established. In the former case, there is a delay in the ionic arrival time at the collector given by $2\Delta t$. In the latter case, the delay is $(eE/M_-^\star v_L)\Delta t^2$.

In order to introduce an easily measurable delay, the field is reversed N times during the ion flight. Thus, the final delay is $\tau_d = N\Delta t$. If a critical time interval is defined as $\Delta t_c = 2v_L M_-^\star/eE$, the expected arrival delay is

$$\tau_d = 2N\Delta t \qquad \text{for } \Delta t \geqslant \Delta t_c \tag{14.19}$$

$$\tau_d = \left(\frac{NeE}{M_-^\star v_L}\right)\Delta t^2 \qquad \text{for } \Delta t \leqslant \Delta t_c \tag{14.20}$$

In Fig. 14.6 the signal delay caused by N reversals of the electric field of duration Δt is shown. As expected, the data for different E fall on a universal curve. From eqns (14.19) and (14.20), a plot of $\tau_d/2N\Delta t$ should be a constant equal to 1 for $\Delta t \geqslant \Delta t_c$. For $\Delta t < \Delta t_c$, a straight line with slope $e/2M_-^\star v_L$ is expected, as experimentally observed (Ellis and McClintock, 1982).

The values of $M_-^\star$ determined from the slopes turn out to be independent of the electric field strength, as shown in Fig. 14.7.

The average value of $M_-^\star = (87 \pm 5)m_4$ is considerably less than the values $M_-^\star = (243 \pm 5)m_4$ and $M_-^\star = (237 \pm 7)m_4$ determined at SVP (Poitrenaud and Williams, 1972; Barenghi *et al.*, 1986, 1991). However, it is in excellent agreement with the value $M_-^\star = (83 \pm 6)m_4$ deduced from the hydrodynamic formula (14.3) using the ionic radius $R_- = (11.5 \pm 0.3)$ Å determined by Ostermeier (1973).

The pressure dependence of $M_-^\star$ has been investigated down to $P = 1.1\,\text{MPa}$, the lowest pressure at which vortex nucleation is still negligible (Ellis *et al.*, 1983). The results are shown in Figs 14.8 and 14.9. $M_-^\star$ increases with decreasing pressure because of the expansion of the ion radius.

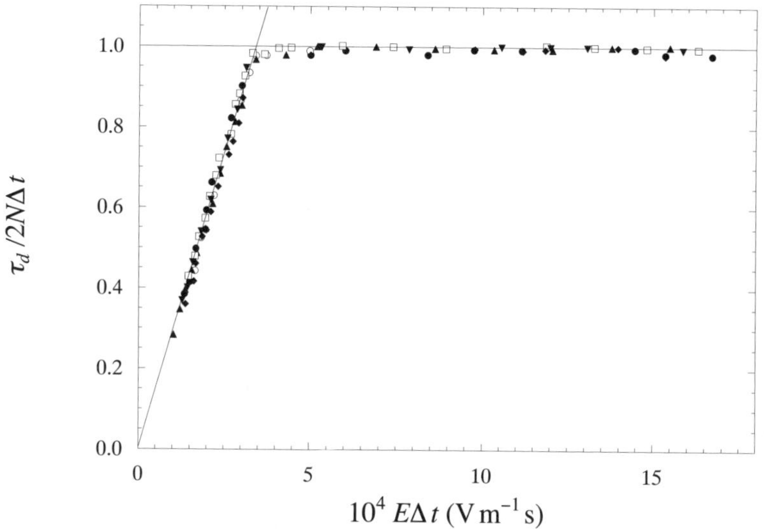

FIG. 14.6. $\tau_d/2N\Delta t$ vs $E\Delta t$. (Ellis and McClintock, 1982.) E (V/m) = 85.6 (closed triangles), 98.8 (open circles), 120 (closed circles), 132 (inverted close triangles), 139 (diamonds), and 147 (squares).

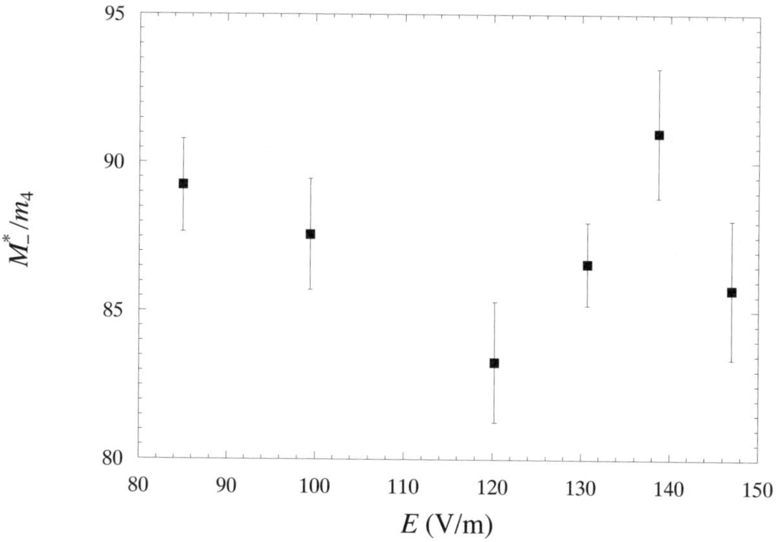

FIG. 14.7. $M^\star_-/m_4$ vs E determined by the acceleration method. (Ellis and McClintock, 1982.)

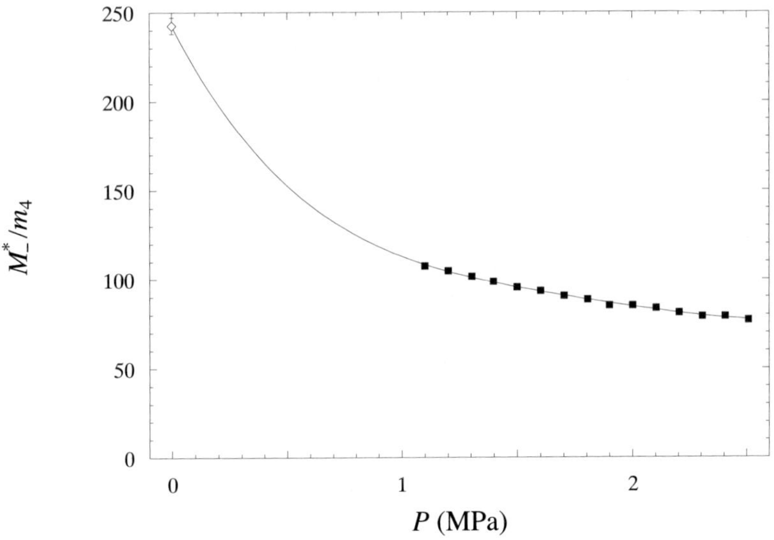

FIG. 14.8. $M_-^\star/m_4$ vs P measured by the inertial method (Ellis and McClintock, 1982). Square: SVP determination (Poitrenaud and Williams, 1972). Line: eyeguide.

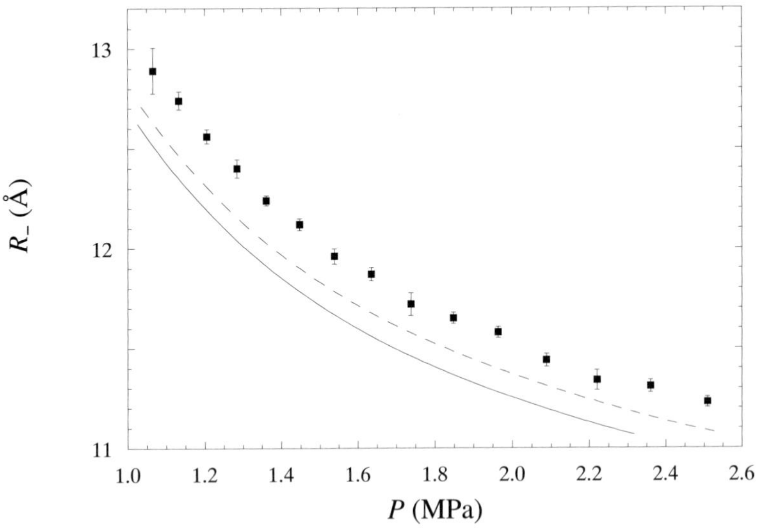

FIG. 14.9. R_- vs P from measurements of M_- with the inertial method. (Ellis and McClintock, 1982.) Lines: electron bubble model (explained in the text).

The corresponding ionic radii are plotted in Fig. 14.9. The lines are the prediction of the electron bubble model, obtained by minimizing eqn (3.21) with respect to the radius. The dashed curve is obtained by assuming that the surface energy of He II under pressure scales linearly with the bulk density, whereas the solid line is calculated by assuming that it scales with the square of the density (Chang and Cohen, 1973). The values of V_0 are taken from the literature (Springett *et al.*, 1967).

The agreement between experiment and theory is quite good. The residual discrepancy can be attributed to two causes. The first one is experimental. $M_-^\star$ has been determined by assuming that the electron bubble moves with the kinetic energy of a free particle $\epsilon = p^2/2M_-^\star$. A more detailed analysis of the experimental data suggests that the ionic acceleration is not completely independent of the ion velocity, leading to a dispersion relation $\epsilon = (p^2/2M_-^\star)(1 - \alpha p^2/2)$, where α is an adjustable parameter. This effect leads to an 8% reduction of the determined effective mass and to a 2% decrease of the ionic radius, improving the agreement with the theoretical model.

The second reason might be due to the crude approximation of a step-like density discontinuity at the bubble wall (Fetter, 1976).

15

OTHER RELEVANT EXPERIMENTS WITH NEGATIVE IONS

Several experiments, in which properties other than transport are investigated, are of great importance in the physics of the interaction of ions with superfluid helium. In particular, one relevant problem is the direct determination of the structure and properties of the electron bubble. More specifically, the difference in the energies of the localized electron and of the electron in the conduction band of the liquid determines, along with the thermodynamic parameters of the liquid, the cavity radius that minimizes the free energy of the system. These experiments involve spectroscopic investigations of the electron bubble and the transmission of the electrons through the liquid–vapor interface.

15.1 Spectroscopic investigation of the electron bubble

The direct determination of the energy of the electron in the conduction band of the liquid has been accomplished in several ways, either by looking at the electric field dependence of the current of electrons injected into the normal liquid He by an MIM structure (Broomall *et al.*, 1976), or by looking at the transmission through the liquid–vapor interface (Sommer, 1964; Bruschi *et al.*, 1966*b*; Schoepe and Probst, 1970; Rayfield and Schoepe, 1971*b*, 1971*c*; Schoepe and Rayfield, 1973), or, most notably in this context, by looking directly at the shift in the work function of a metal cathode when it is immersed in the liquid (Woolf and Rayfield, 1965).

In this latter experiment, a phototube with a Cs–Sb photocathode is immersed in an He bath at $T = 1.1\,\mathrm{K}$. Monochromated light impinges on the photocathode and the spectral response I of the phototube is measured as a function of the photon energy E_ν. The same procedure is repeated *in vacuo* and the shift of the spectral response of the phototube, reported in Fig. 15.1, gives a direct determination of the energy V_0 of the delocalized electron in the conduction band of the liquid.

The electrons emitted from the photocathode enter the liquid very rapidly so that the He atoms are initially unperturbed. The electrons are then initially untrapped and get localized a short while thereafter (Hernandez and Silver, 1970; Rosenblit and Jortner, 1995; Schmidt *et al.*, 2001). Therefore, this experiment should, indeed, probe the energy of the conduction band of the delocalized electrons, V_0. The shift of the spectral response amounts to $1.02\,\mathrm{eV}$, in line with other theoretical (Kuper, 1961; Burdick, 1965; Jortner *et al.*, 1965; Springett *et al.*, 1967; Miyakawa and Dexter, 1970) and experimental determinations (Bruschi *et al.*, 1966*b*; Schoepe and Probst, 1970; Rayfield and Schoepe, 1971*b*; Broomall *et al.*, 1976).

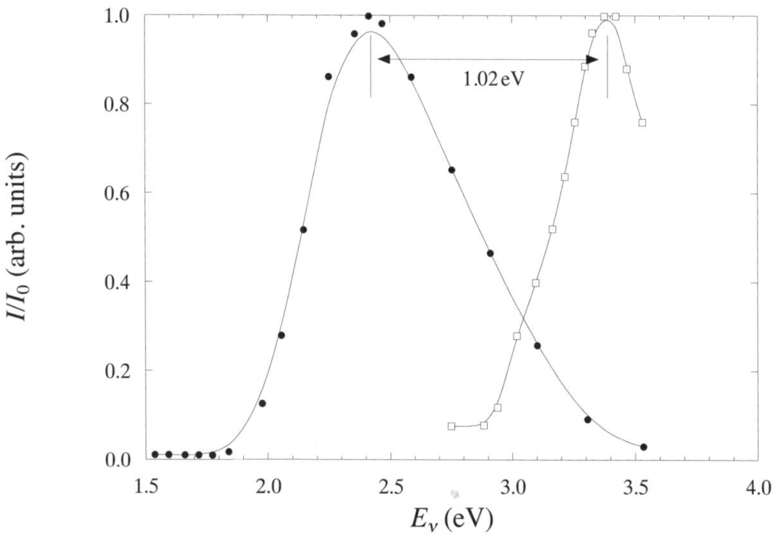

FIG. 15.1. I/I_0 vs E_ν *in vacuo* (circles) and in liquid He (squares) at $T = 1.1\,\mathrm{K}$. (Woolf and Rayfield, 1965.) The curves are normalized to unity at their maximum.

Another way to get direct information about the energy difference between the electron energy inside and outside the bubble is by a photo-modulated conductivity measurement. The experiment is carried out in the range $1.3\,\mathrm{K} < T < 1.7\,\mathrm{K}$ (Northby and Sanders, 1967).

A single-gate velocity spectrometer (Cunsolo, 1961) is operated at a gating frequency for which the ion transit time is slightly greater than one half-period. In this way no charges arrive at the collector and no current is detected unless fast electrons are ejected out of the bubble upon irradiating the sample with light of short enough wavelength. The minimum wavelength required to free electrons from the bubble corresponds to the height of the potential well. By irradiating the sample with a modulated light source, the current is also modulated and can be detected synchronously.

The detector output can be written as $S = \sigma_t(\lambda)P(\lambda)f(\lambda)$, where σ_t is the cross-section for the ejection of electrons from the bubbles (in $\mathrm{m^2/J}$) by radiation of wavelength λ, P is the light flux (in $\mathrm{W/m^2}$), and f is the detector efficiency, which is obtained by calibration.

The cross-section σ_t is plotted in Fig. 15.2 as a function of the light wavelength λ. The solid line corresponds to the theoretical calculation of the cross-section based on the assumptions that the electron is initially in a spherically-symmetric square-well potential of depth V_0 and radius R, that there is electric-dipole coupling, and that the final state is a continuum p-state in the same potential. The theoretical curve is characterized by an infrared threshold followed by diffraction-like peaks and zeros.

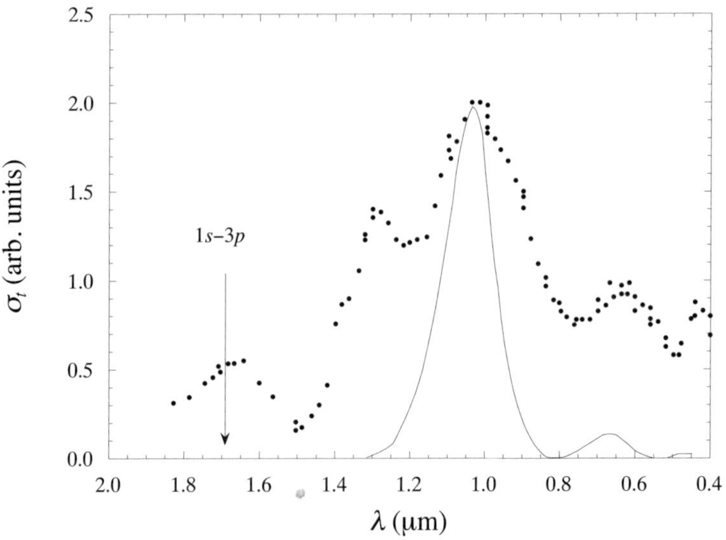

FIG. 15.2. Photoejection cross-section σ_t vs λ. (Northby and Sanders, 1967.) Solid line: σ_t predicted according to the simple square-well model.

The calculation is very sensitive to the ion radius, but less sensitive to the well depth. The good agreement with the highest peak is obtained with $V_0 = 1.02\,\mathrm{eV}$ and $R_- = 21.35\,\mathrm{\AA}$, in very nice agreement with the values determined by Woolf and Rayfield (1965). A very good agreement is also obtained for the position of the minima and maxima of higher energy (Northby and Sanders, 1967).

However, some discrepancies between theory and experiment still remain. In particular, the theory is neither able to explain the low-energy peak at $\lambda \approx 1.75\,\mu\mathrm{m}$, which is believed to be related to a transition from the $1s$ ground state to the highest bound $3p$ state in the well, nor the existence and temperature dependence of the peak at $\lambda \approx 1.28\,\mu\mathrm{m}$. Finally, the theory predicts zeros while the experiment detects minima. This fact may be due to several factors. Among them are a distribution of bubble sizes and shapes, contributions from higher multipole radiative processes, and instrumental factors.

It is also suggested that the signal-to-intensity ratio might not be the true optical absorption spectrum σ_t of the electron in the bubble because, when the bubble relaxes to the new size and shape after the electron has made the transition from the s-ground state to the excited p-state, phonons are radiated, which produce an acceleration of the bubble (Fletcher and Bowley, 1973). This acceleration might not be sufficient for all bubbles that have absorbed a photon to be detected and the resulting electron bubble signal/light intensity ratio would only reflect the main features of the optical absorption spectrum.

Photoconduction measurements have been extended to higher pressure in the range $1.2\,\mathrm{K} < T < 1.6\,\mathrm{K}$ (Zipfel, 1969; Grimes and Adams, 1990) because they give information on the pressure dependence of the electron bubble radius that can be compared with different determinations of it (Springett, 1967; Ostermeier, 1973; Ellis *et al.*, 1983) and with the results for the electron bubble model (Fowler and Dexter, 1968; Miyakawa and Dexter, 1970).

The equilibrium radius in the simplest bubble model is related to the pressure and surface tension σ by eqn (3.24): $P = \hbar^2\pi/4mR_-^5 - 2(\sigma/R_-)$. Taking $\sigma = 0.341 \times 10^{-3}\,\mathrm{N/m}$, the radius becomes 19.36 Å at $P = 0$ and decreases to ≈ 12.5 Å at the solidification pressure (Woolf and Rayfield, 1965; Iino *et al.*, 1985). For the infinite, spherical square well, the $1s$–$1p$ transition energy is $\Delta\mathcal{E}_{1s-1p} = 0.105\,\mathrm{eV}$ at $P = 0$ and increases to $\Delta\mathcal{E}_{1s-1p} = 0.232\,\mathrm{eV}$ at 2 MPa, basically because the well shrinks under pressure.

For a well of finite depth V_0, the electronic energy eigenvalue can be written as $\mathcal{E}_{nl} = C_{nl}(V_0)/R_-$, where the coefficients C_{nl} are obtained by solving simple transcendental equations (Gasiorowicz, 1974). The finite repulsive barrier allows the penetration of the electron wave function into the liquid. This phenomenon, for a given radius R_-, decreases $\mathcal{E}_{nl}$ relative to its value for the infinitely-deep potential well.

In Figs 15.3 and 15.4 the experimental values of the transition energies relative to the $1s$–$1p$ transition (Grimes and Adams, 1990) and to the $1s$–$2p$ transition (Zipfel, 1969) are shown along with the predictions of the theory. The lines in the figures are the prediction of the improved simple bubble model (3.21),

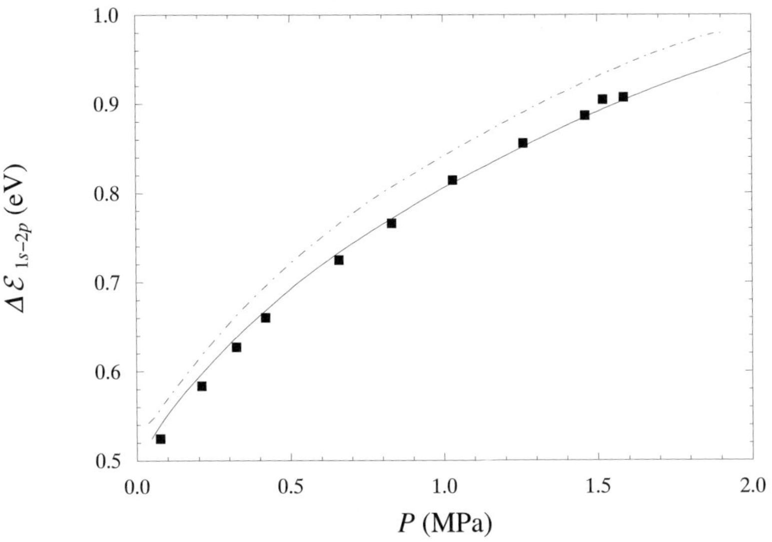

FIG. 15.3. $\Delta\mathcal{E}_{1s\text{-}2p}$ vs P in He II at $T = 1.3\,\mathrm{K}$ (Zipfel, 1969; Grimes and Adams, 1990). Dash–dotted line: electron bubble model (Miyakawa and Dexter, 1970) with a surface tension that increases with P. Solid line: same model with constant surface tension (Grimes and Adams, 1990).

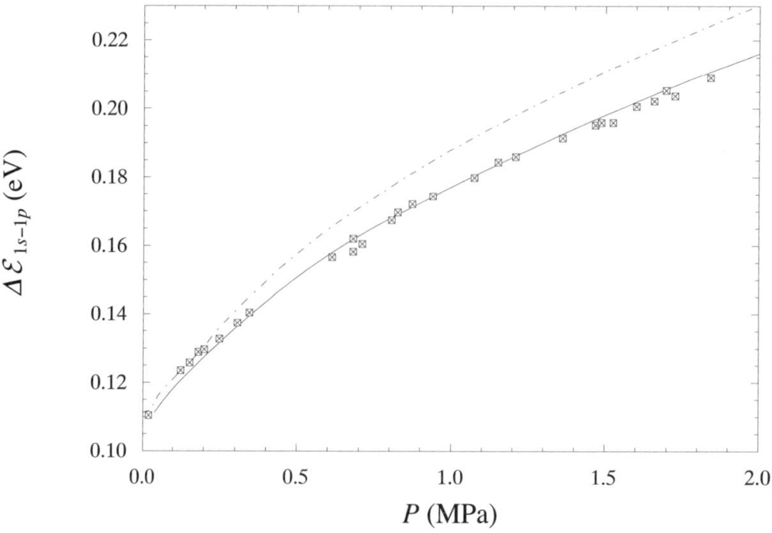

FIG. 15.4. $\Delta\mathcal{E}_{1s-1p}$ vs P at $T = 1.3\,\mathrm{K}$ (Grimes and Adams, 1990). Lines: same meaning as Fig. 15.3.

with the inclusion of the polarization contribution (Miyakawa and Dexter, 1970) and the use of the Wigner–Seitz model for calculating $V_0(P)$.

The pressure dependence of the surface tension predicted by the theory of Amit and Gross (1966) is used, i.e., σ varies from $0.36 \times 10^{-3}\,\mathrm{N/m}$ at $P = 0$ to $0.66 \times 10^{-3}\,\mathrm{N/m}$ at $P = 2.53\,\mathrm{MPa}$. However, if a constant value of the surface tension is used, $\sigma = 0.341 \times 10^{-3}\,\mathrm{N/m}$, independent of P, then a much better agreement with the data is obtained (Grimes and Adams, 1990), as indicated by the solid line in Figs 15.3 and 15.4. Since there are no direct measurements of σ, the issue is open to speculation.

It is interesting to note that the observed lines have widths that are attributed essentially to breathing and quadrupole oscillation modes of the bubble (Fowler and Dexter, 1968; Grimes and Adams, 1990).

A more fundamental issue that has not been elucidated yet is the mechanism of photoionization. It has been suggested by Northby and Sanders (1967) that direct photoionization in the continuum occurs for bubbles trapped on vorticity. Measurements at high pressure by Zipfel (1969) have shown that photoionization disappears near the temperature and pressure at which Springett (1967) found that vortices no longer trap electron bubbles.

Miyakawa and Dexter (1970) suggest that the electron bubble expands to a new larger radius in order to match the more-extended electron wave function in the excited state. The expansion to a new equilibrium radius occurs in a time of order $R_-/v_s \approx 10^{-11}\,\mathrm{s}$, where v_s is the speed of sound. The sudden expansion raises the local temperature and the electron bubble is thermally-excited off of the vortex line, thus contributing to the photoconductivity. However, experiments have revealed that the photoconductive response takes place with a characteristic time of $\approx 0.1\,\mathrm{s}$ (Grimes and Adams, 1990).

The fact that the photoconductivity signal appears in the (P, T) region where bubbles are known to spend most of their time trapped on vortices (McClintock, 1973*c*; Grimes and Adams, 1990) and the fact that it disappears along the line in the (P, T) plane for which the capture of the bubble by vortex lines becomes negligible (Springett and Donnelly, 1966; Springett, 1967) lend great support to the idea that photocurrent is associated with bubbles trapped on vorticity (Grimes and Adams, 1990).

The vanishing of the signal related to the $1s$–$1p$ and $1s$–$2p$ transitions at low P has not yet received a satisfactory explanation. An interesting hypothesis (Grimes and Adams, 1990) is that the electron bubble in the $1p$ state is unstable against a non-radiative decay back into the ground state.

At first, the p-bubble elongates to follow the shape of the electron p-wave function, removing the degeneracy of the m_l manifold, with $m_l = 0$ being the lowest-lying state. The bubble keeps elongating until the "waist" of the hourglass-shaped bubble pinches off. The electron ultimately settles into the cavity on one side of the waist, while the empty one collapses with the emission of phonons.

The heat released in this process drives the photocurrent. For $P < 0.1\,\mathrm{MPa}$, the p-bubble is sufficiently stable to decay radiatively and no photocurrent is detected. We can immediately recognize the analogies with the scenario leading to the *electrino* hypothesis (Maris, 2000).

A very elegant experimental technique was later devised by Grimes and Adams (1992) to directly measure the infrared absorption and deserves a brief description. A sketch of the experimental apparatus is shown in Fig. 15.5.

The infrared radiation entering the cell is chopped at high frequency (1 kHz) and the field-emission current due to the high-voltage tips is modulated at low frequency (10 Hz). The infrared wavelength is swept while the output of the infrared detector is rectified by two lock-ins operating in series. The first lock-in takes its reference from the high-frequency light chopper, so its output is proportional to the photon flux incident on the detector. The second lock-in is phase-locked on the a.c. voltage that modulates the field-emission current, so its output is proportional to the infrared attenuation caused by the electrons and is recorded as a function of the wavelength.

In Fig. 15.6 the results for the energy of $1s$–$1p$ transition are reported together with previous results from the photoconduction technique (Grimes and Adams, 1990). The photoabsorption measurements extend up to the melting pressure, while the photoconduction ones quit earlier, validating the hypothesis that the photocurrent mechanism involves the trapping of bubbles on vorticity (Grimes and Adams, 1990).

The solid line is the result of the calculation within the simple square-well model with a constant surface tension $\sigma_{lv} = 0.341\,\mathrm{mN/m}$ (Grimes and Adams, 1990) and is a continuation of the solid line shown in Fig. 15.4. Consistent results

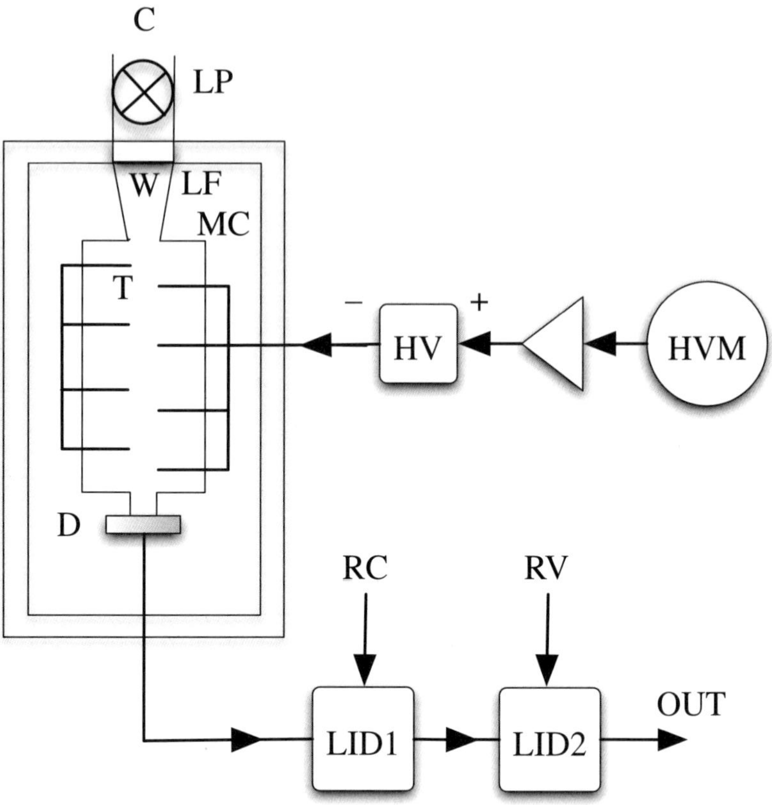

FIG. 15.5. Sketch of the experimental apparatus for photoabsorption measurements. (Grimes and Adams, 1992.) LP: light pipe, C: high-frequency (1 kHz) chopper, W: ZnSe window, LF: light funnel, MC: measuring cell, T: high-voltage tips for electron injection, HV: high-voltage supply, HVM: low-frequency (10 Hz) high-voltage modulator, D: Hg–Cd–Te infrared detector, LID1 and LID2: lock-in detectors, RC: reference from C, RV: reference from HVM, and OUT: signal out.

have also been obtained using direct absorption methods at fixed infrared wavelength by sweeping the pressure (Parshin and Pereverzev, 1990, 1992; Pereverzev and Parshin, 1994.)

In the (P, T) region in which both photoconduction and photoabsorption can be measured, the electron bubbles injected from the high-voltage tips spend a large fraction of their transit time through the cell trapped on vorticity. The hydrodynamic bubble-trapping potential (Donnelly and Roberts, 1969*b*) induces a longitudinal bubble distortion that lifts the m_l degeneracy of the $1p$ state. If one assumes that the shape of the distorted bubble is a prolate spheroid, its shape can be described by

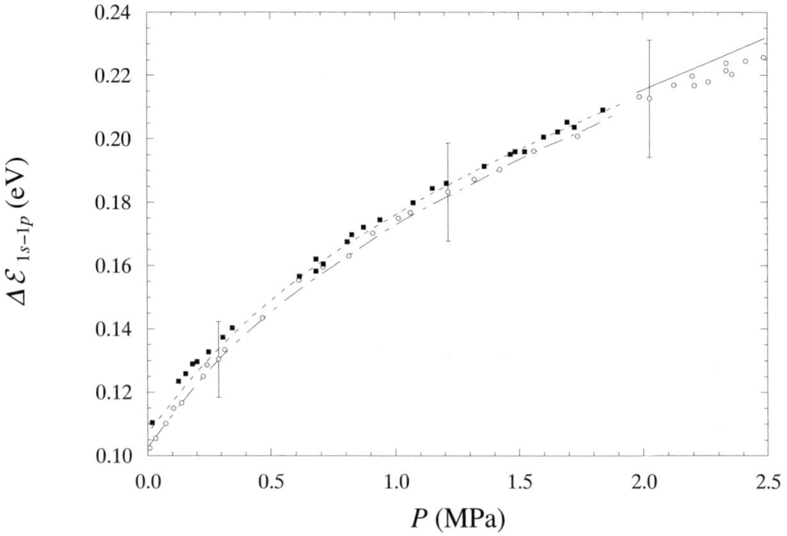

FIG. 15.6. $\Delta\mathcal{E}_{1s\text{–}1p}$ vs P at $T = 1.3\,\mathrm{K}$ (circles) (Grimes and Adams, 1992). Squares: photoconduction measurements (Grimes and Adams, 1990). Bars: measured linewidths. The meaning of the lines is explained in the text.

$$R_- = R_0\left[1+\left(\frac{\beta}{2}\right)\left(3\cos^2\theta-1\right)\right]. \tag{15.1}$$

The ground-state energy is assumed to be

$$\mathcal{E} = \mathcal{E}_0 + \mathcal{E}_v + \mathcal{E}_s + \mathcal{E}_{\mathrm{sub}}\,, \tag{15.2}$$

where $\mathcal{E}_{\mathrm{sub}}$ is the substitution energy (Donnelly and Roberts, 1969*b*).

The energies are corrected by the presence of the distortion factor β and are given by

$$\mathcal{E}_0 = \mathcal{E}_e\left(1+0.52\beta^2\right), \tag{15.3}$$

$$\mathcal{E}_v = \frac{4\pi}{3}R_0^3P\left(1+\frac{3}{5}\beta^2\right), \tag{15.4}$$

$$\mathcal{E}_s = 4\pi R_0^2\sigma_{lv}\left(1+\frac{4}{5}\beta^2\right), \tag{15.5}$$

$$\mathcal{E}_{\mathrm{sub}} = -2\pi\rho_s\left(\frac{\hbar}{m_4}\right)^2 R_0 \times\left[\left(1+\frac{a^2}{R_0^2}\right)^{1/2}\sinh^{-1}\left(\frac{R_0}{a}\right)-1\right](1+\beta)\,, \tag{15.6}$$

where $\mathcal{E}_e$ is the finite-well value with the elongation factor (Gross and Tung-Li, 1968) and $a = 1.46\,\text{Å}$ is the healing length.

The substitution energy favors elongations of the bubble along the vortex axis, i.e., the Bernoulli effect causes the pressure on the trapped bubble to be greatest at its equator and smallest near its poles, and this variation elongates the bubble into a prolate spheroid.

Using $V_0 = 1.02\,\mathrm{eV}$ and $\sigma_{lv} = 0.341 \times 10^{-3}\,\mathrm{N/m}$, one finds, for $\beta = 0$ and $P = 0$, that the equilibrium radius of a spherical bubble centered on the vortex line is expanded by 1.4%, to 17.4 Å, thereby lowering the calculated transition energy from 0.1061 eV for the free bubble to 0.1036 eV for a trapped bubble. This is close to the experimental value of 0.1028 eV. If, now, the total energy at $P = 0$ is minimized, one finds $\beta = 0.023$. The decrease of the ground-state energy is less than 1 meV.

Because bubble size and shape do not relax during an optical transition, the elongation removes the degeneracy of the m_l manifold of the $1p$ excited state. The $m_l = 0$ state is now lowered and the $m_l = \pm 1$ states are raised with respect to the spherical bubble. Using the results for a quadrupole deformation of the bubble (DuVall and Celli, 1969), the dashed and dash–dotted lines in Fig. 15.6 are obtained.

It should be noted, however, that the drift motion of the bubble induces a distortion of it toward the shape of an oblate spheroid, which produces a splitting of the $1p$ state in the opposite direction to that produced by the vortex line. Unfortunately, the splitting of the m_l manifold cannot be observed because the linewidths are too broad. The broadening is due to several causes, including thermally-activated breathing and quadrupole oscillations of the bubble.

The observed temperature dependence of the $1s$-$1p$ transition energy at constant P is also consistent with the prediction of the simple bubble model for a free bubble, as shown in Fig. 15.7. The decrease of the transition energy mainly reflects the increase in the bubble radius (see Fig. 3.7) due to the decrease of σ_{lv} with increasing T (Grimes and Adams, 1992). It is interesting to note how the bubble model that explains the properties of the electron state so well in the liquid is still valid up to the much higher densities of the solid. Actually, there are no qualitative differences between the spectroscopic properties and their theoretical description for the electron bubble in compressed superfluid helium and in solid hcp ^{4}He and bcc ^{3}He (Golov, 1995).

Solid helium is compliant enough for an electron bubble to form, whose radius is smaller than in the liquid phase for the obvious reason that the pressure is much higher.

As in the liquid, infrared absorption spectra of excess electrons in solid hcp ^{4}He and bcc ^{3}He show the $1s$-$1p$ transition (Golov, 1995), whose energy is reported in Fig. 15.8 together with the results obtained in the liquid and already shown in Fig. 15.6. The solid line is the prediction of the simple bubble model, which gives a very nice fit to the experimental data in both the superfluid and solid phases.

Also, the values of the electron bubble radius determined from infrared absorption spectra in solid He (Golov, 1995) agree well with those obtained in the

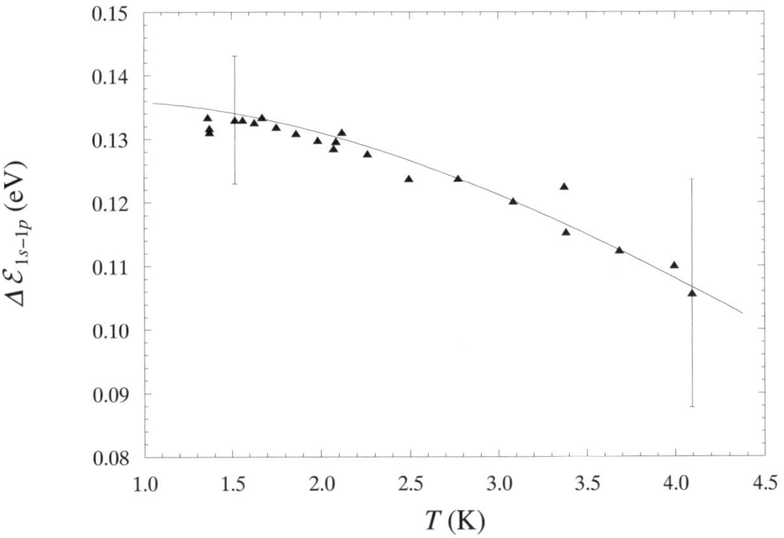

FIG. 15.7. $\Delta\mathcal{E}_{1s\text{–}1p}$ vs T at $P = 0.29$ MPa. (Grimes and Adams, 1992.) Bars: measured linewidths. Line: simple bubble model.

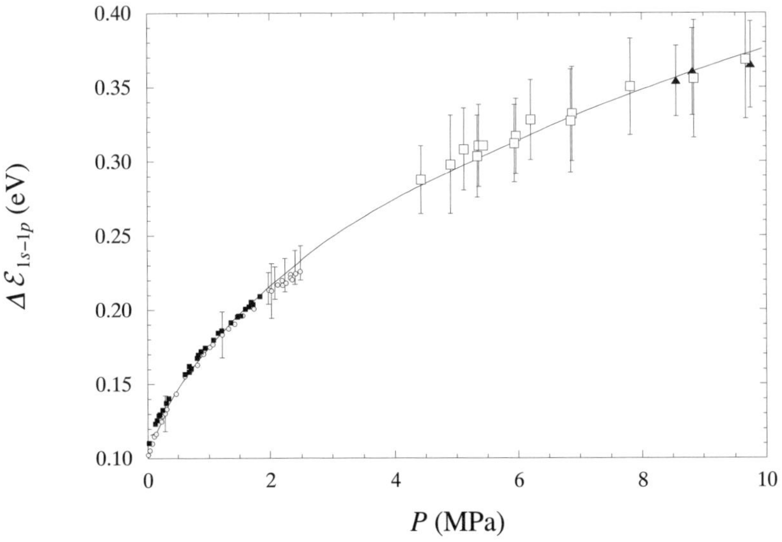

FIG. 15.8. $\Delta\mathcal{E}_{1s\text{–}1p}$ vs P in solid hcp ^{4}He (open squares) and in bcc ^{3}He (closed triangles) (Golov, 1995) together with the same liquid data of Fig. 15.6. Bars: measured linewidths. Line: simple bubble model.

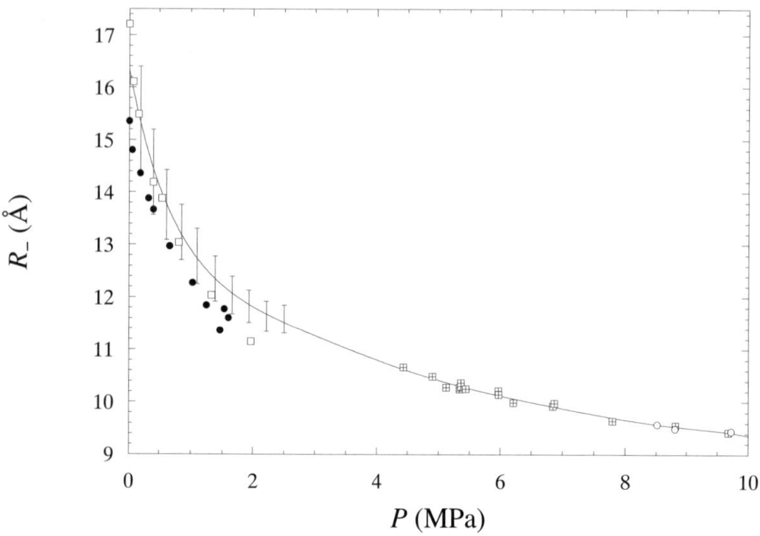

FIG. 15.9. R_- vs P in hcp ^{4}He (crossed squares) and in bcc ^{3}He (open circles). (Golov, 1995.) The data at lower P are the same as shown in Fig. 3.6. The line is eqn (3.24) extrapolated to the solid pressures.

liquid, in the sense that the data for the solid phase at high pressures lie on the extrapolation of the liquid data for lower P. This agreement is shown in Fig. 15.9, in which the high pressure determination of the negative ion radius is plotted together with the values in the liquid already reported in Fig. 3.6. The solid line in this figure is eqn (3.24).

15.2 Transmission of electrons through the liquid–vapor interface

The striking differences observed in the transport behavior of positive and negative ions in liquid He are basically related to their different structures. Electrons localize into a bubble because He appear to them as a barrier of ≈ 1 eV. Localization into an empty cavity leads to a lower-energy state. The radius of the cavity depends on the height of the potential barrier V_0 as well as on the thermodynamic parameters of the liquid, and determines to a large extent the transport properties of electron bubbles. The determination of V_0 and of the radius R_- is therefore very important.

In addition to the methods described in the previous sections, it is interesting to consider other ingenious experiments that investigate the transmission of electrons through the liquid–vapor interface in order to get information on the physical properties of the electron bubbles.

15.2.1 *Transmission into the liquid from the vapor*

The first experiment is the injection of electrons from the vapor into the liquid (Sommer, 1964). Electrons are produced in the vapor above the liquid and are attracted toward the interface by means of suitable electric fields. In order to prevent charging of the surface, a lateral electrode drains away the charge that accumulates on the surface.

By increasing the field that accelerates the electrons toward the surface, a threshold is reached, beyond which a fraction of the electrons impinging on the surface enter the liquid and are collected at an electrode at the bottom of the cell. By measuring the fraction of the current transmitted through the interface as a function of the accelerating field and comparing it to the electron-energy distribution function in the vapor, the barrier height V_0 can be determined.

The vapor is very dilute and the density-normalized field E/N, where N is the number density of the vapor, is quite large, and so the Druyvenstein approximation (Loeb, 1939; Wannier, 1966): for the distribution function can be used

$$n(\epsilon) = A\epsilon^{1/2} \exp\left\{-\left(\frac{3m_e}{m_4}\right)\left(\frac{N\sigma_{mt}\epsilon}{E}\right)^2\right\}, \tag{15.7}$$

where ϵ is the electron energy, E is the accelerating field, and $\sigma_{mt} \approx 6.2\,\text{Å}^2$ (Phelps *et al.*, 1960) is the electron–He momentum-transfer scattering cross-section.

Only those electrons with energy in excess of V_0 are transmitted over the barrier and their fraction f is easily calculated by integrating eqn (15.7) from V_0 upwards. In Fig. 15.10 the results of the calculation with a barrier $V_0 = 1.3\,\text{V}$ are compared to the fraction of the current detected at the submerged electrode. The data are plotted as a function of $(E/N\sigma_{mt}\sqrt{3m_e/m_4})/V_0$ that represents the ratio of the energy gained by the electrons from the electric field in a mean free path in the vapor to the barrier height.

Owing to the experimental uncertainties, especially in the knowledge of the field at the surface, and in view of the limited range of T and N, this result, $V_0 = 1.3\,\text{eV}$, can be considered to be in good agreement with other experimental determinations (Woolf and Rayfield, 1965; Broomall *et al.*, 1976).

In comparison with the measurements of photoejection or photoconduction (Woolf and Rayfield, 1965; Northby and Sanders, 1967), in which electrons are extracted from the inside of the bubble, this determination of V_0 might not be spoiled by the zero-point energy of the electron in the bubble.

15.2.2 *Transmission into the vapor from the liquid*

Differences between positive and negative ions also appear when ions are transmitted through the interface from the liquid into the vapor, as first observed by Careri *et al.* (1960). In particular, electron bubbles pulled toward the interface by an electric field E perpendicular to it, encounter an increasing difficulty to pass into the vapor as T is lowered below 1.7 K (Bruschi *et al.*, 1966*b*; Schoepe and Probst, 1970).

In Fig. 15.11 the current through the interface detected at the collector in the vapor is shown as a function of T^{-1} for several values of the electric field E across the surface.

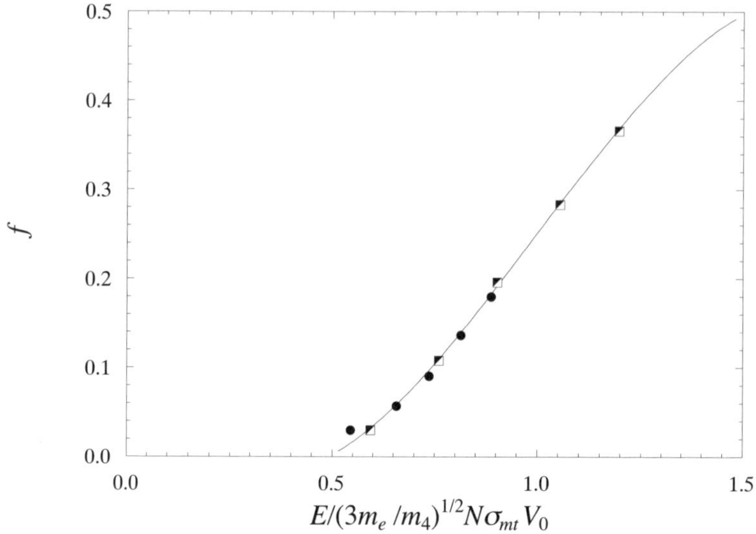

FIG. 15.10. Fraction f of current transmitted though the liquid–vapor interface vs $E/N\sigma_{mt}$. (Sommer, 1964.) Circles: $T = 1.48\,\mathrm{K}$ and $N \approx 1.1 \times 10^{25}\,\mathrm{m}^{-3}$. Squares: $T = 1.34\,\mathrm{K}$ and $N \approx 2.1 \times 10^{25}\,\mathrm{m}^{-3}$. Line: fraction of electrons with energy in excess of V_0 calculated from eqn (15.7).

In the temperature-dependent region, the current decreases exponentially with decreasing T as $\exp(-\mathcal{E}_b/k_BT)$. It is found that $\mathcal{E}_b$ is a few tens of K. It represents an activation energy that is found to depend slightly on the strength of the extracting field, as shown in Fig. 15.12.

The electron bubbles arriving at the surface from the liquid side are trapped into the potential well due to the combined effect of the pulling field and the image force, eqn (14.15) (Cole and Cohen, 1969; Cole, 1970; Cheng *et al.*, 1994*a*, 1994*b*; Jackson, 1998), i.e., by the same effect exploited for plasma resonance studies of ions below the liquid surface (Poitrenaud and Williams, 1972; Barenghi *et al.*, 1986).

If the massive electron bubbles are assumed to thermalize at the bottom of the well, they can escape into the vapor only by thermal diffusion over the barrier (Chandrasekhar, 1943). Physically, when the electron bubble "touches" the surface layer, assuming that the interface is very sharp, the bubble becomes mechanically unstable and bursts, allowing the electron to escape into the vapor (Ancilotto and Toigo, 1994, 1995).

If E is increased, the barrier height is lowered by an amount given by

$$\mathcal{E}_b = \mathcal{E}_b^0 - 2\sqrt{AeE}\,, \tag{15.8}$$

with A given by eqn (14.16). Equation (15.8) fits the experimental data quite well (see Fig. 15.12) with $\mathcal{E}_b^0 = 38.8\,\mathrm{K}$ and $2\sqrt{Ae} = 8.9 \times 10^{-2}\,\mathrm{K}\,(\mathrm{m/V})^{1/2}$,

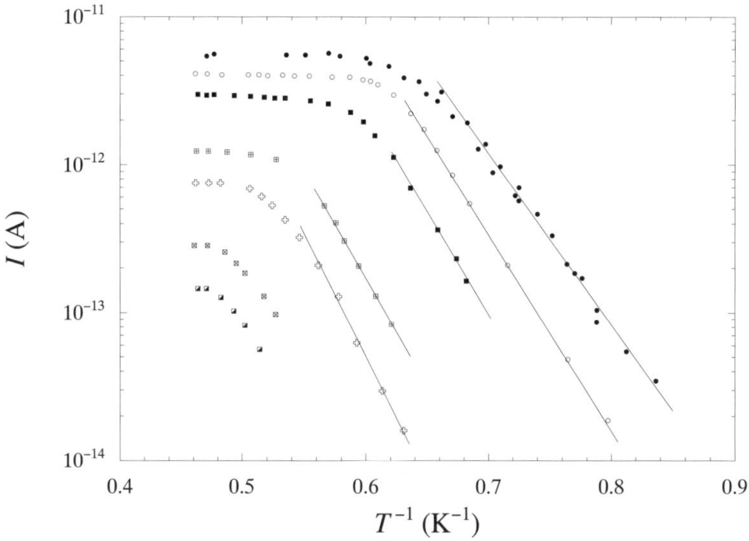

FIG. 15.11. Current I transmitted into the vapor through the liquid–vapor interface vs T^{-1} for different electric fields at the interface (Bruschi *et al.*, 1966*b*; Schoepe and Probst, 1970). $E\,(\mathrm{kV/m}) = 30$ (Bruschi *et al.*, 1966*b*), 15.4, 11.5, 3.8, 1.9, 0.54, and 0.26 (Schoepe and Probst, 1970) (from top).

in relatively good agreement with the value $2\sqrt{Ae} = 7.1 \times 10^{-2}\,\mathrm{K\,(m/V)}^{1/2}$ predicted by eqn (14.16) with $\epsilon_r = 1.057$.

Similar results are also obtained at the interface between the two liquids in a phase-separated ^{3}He–^{4}He solution (Kuchnir *et al.*, 1970). In this case, electron bubbles cross the interface coming from the ^{4}He-rich phase. The height of the barrier is $3.7\,\mathrm{K} < \mathcal{E}_b < 4.9\,\mathrm{K}$ for fields in the range $8.5\,\mathrm{kV/m} < E < 3.8\,\mathrm{kV/m}$.

A similar physical situation occurs in Ar when electrons are pulled from the liquid into the vapor. In this case, they are trapped below the liquid surface by the combined effect of the potential given by eqn (14.15) and of the negative value of V_0 in the liquid. The difference with electron bubbles below the liquid surface of He is that in Ar electrons thermalize at the bottom of the potential well and escape from it by thermionic emission, similar to the electrons entering liquid He in the experiment of Sommer (1964). From the analysis of the trapping time, a very good direct determination of $V_0 \approx -126\,\mathrm{meV}$ for electrons in liquid Ar is obtained (Borghesani *et al.*, 1990, 1991).

The analysis of the thermal diffusion of electron bubbles in an external field based on the Smoluchowski equation gives the trapping time of the electron bubbles below the surface (Chandrasekhar, 1943). In the thermal diffusion approximation, the trapping time τ of the bubbles is given by

$$\tau^{-1} = \alpha\,(E)\,\mu\,(T)\exp\left\{-\left(\mathcal{E}_b - 2\sqrt{AeE}\right)/T\right\}, \qquad (15.9)$$

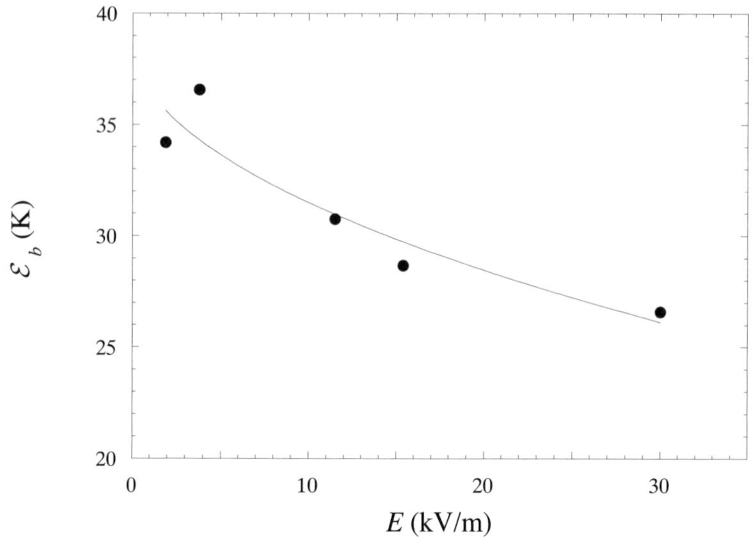

FIG. 15.12. $\mathcal{E}_b$ vs E. (Bruschi *et al.*,1966*b*; Schoepe and Probst, 1970.) Line: fit to eqn (15.8).

where $\alpha = CE^{3/4}\sqrt{V_c''}$, C is a constant, and V_c'' is essentially the curvature of the potential at the top of the well. $\mathcal{E}_b$ and V_c'' depend on the physical processes by which the negative carriers escape over the barrier (Rayfield and Schoepe, 1971*a*, 1971*b*). τ is measured directly, whereas $\mathcal{E}_b$ and V_c'' are left as adjustable parameters.

Trapping-time data for pure ^{4}He are reported in Fig. 15.13. The lines in the figure are calculated by means of eqn (15.9) with $\mathcal{E}_b = 43.8\,\mathrm{K}$ and with $V_c'' = 2.23 \times 10^3\,\mathrm{K}/\text{Å}^2$(Rayfield and Schoepe, 1971*b*).

Whereas the T-dependence of the observed lifetimes is reproduced correctly, the electric field dependence is only in fair agreement with the data. Moreover, the required cut-off of the image potential close to the surface is too sharp, thereby violating the underlying assumptions of the Smoluchowski equation, which is valid only for slowly-varying potentials (Schoepe and Rayfield, 1973).

A different mechanism must be responsible for the escape of electrons from bubbles through the interface into the vapor before the bubbles approach the interface close enough to burst. This mechanism is believed to be direct quantum tunneling of electrons into the vapor (Rayfield and Schoepe, 1971*c*; Schoepe and Rayfield, 1973).

When the bubbles approach quite close to the interface, assumed to be sharp, a thin liquid layer acts as the barrier for electron tunneling into the vapor. Such a situation is depicted schematically in Fig. 15.14. The barrier height for tunneling is the difference between the conduction band energy, V_0, of the delocalized electron in the liquid and the ground-state energy of the electron localized in the

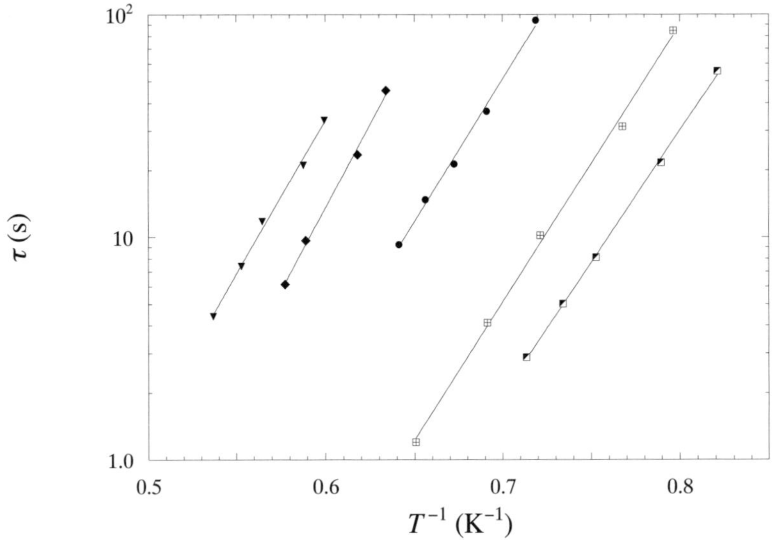

FIG. 15.13. τ vs T^{-1}. (Rayfield and Schoepe ,1971*b*.) $E\,(\mathrm{kV/m}) = 0.94$, 1.89, 4.86, 11.5, and 15.2 (from left). Lines: eqn (15.9).

bubble (Rayfield and Schoepe, 1971*c*).

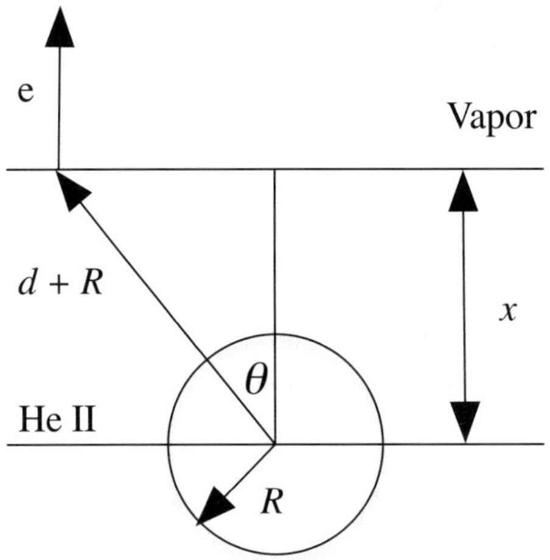

FIG. 15.14. Sketch of an electron bubble at a distance x from the interface. (Rayfield and Schoepe, 1971*c*.) The tunneling distance for the electron is $d = x/\cos\theta - R$, where R is the bubble radius.

The calculation of the tunneling frequency (or tunneling probability per unit time) $\mathcal{P}$ allows a determination of the bubble radius. $\mathcal{P}$ is directly measured in the experiment because the current from the surface is just $I \propto \mathcal{P} \exp(-\mathcal{P}t)$ and $\mathcal{P} = \tau^{-1}$, where τ is the trapping time (Schoepe and Rayfield, 1973).

Let $\mathcal{E}_1$ be the ground-state energy of the electron in the bubble, and $\alpha^2 = (2m_e/\hbar^2)(V_0 - \mathcal{E}_1)$, where m_e is the electron mass. The probability that an electron tunnels a distance d through the liquid is $\exp(-2\alpha d)$. The tunneling transition rate from the bubble at the solid angle $\mathrm{d}\Omega$ is

$$\nu \exp\left[-2\alpha\left(\frac{x}{\cos\theta} - R\right)\right]\frac{\mathrm{d}\Omega}{4\pi}, \tag{15.10}$$

where R is the bubble radius. $\nu \approx (2\mathcal{E}_1/m_e)^{1/2}/2R$ in a semi-classical sense is the frequency at which the electron "hits" the bubble wall. The total transition rate is obtained by integrating eqn (15.10) over the solid angle, yielding

$$\frac{1}{2}\nu e^{2\alpha R} W(x), \tag{15.11}$$

where

$$W(x) = e^{-2\alpha x} + 2\alpha x \,\mathrm{Ei}(-2\alpha x) \approx \left(\frac{e^{-2\alpha x}}{2\alpha x}\right) e^{-1/\alpha x}, \tag{15.12}$$

with x being the shortest distance between the bubble center and the interface. Ei is the exponential integral. The last approximation is justified because, in the range of interest, $2\alpha x \gg 1$ (Schoepe and Rayfield, 1973).

If the distribution of bubbles between x and $x+\mathrm{d}x$ is $n(x)\mathrm{d}x$, the current through the surface is $(1/2)\nu \exp(2\alpha R) W(x) n(x)\mathrm{d}x$. Although the Smoluchowski diffusion equation should be solved for $n(x)$, it is assumed that thermal equilibrium is established very rapidly in comparison with the escape time, so that an equilibrium distribution function can be used throughout the potential well. Therefore, one assumes

$$n(x) = \frac{N}{I_1} \exp\left[V(x)/k_{\mathrm{B}}T\right], \tag{15.13}$$

where N is the total number of bubbles and

$$I_1 = 2\left(\frac{A}{eE}\right)^{1/2} K_1\left(2\,(AeE)^{1/2}/T\right), \tag{15.14}$$

with K_1 a modified Bessel function. The total current across the surface is

$$j_s = PN = \frac{1}{2}\nu e^{2\alpha R} \int_0^\infty n(x) W(x)\mathrm{d}x = \frac{1}{2}\nu e^{2\alpha R} N \frac{I_2}{I_1}, \tag{15.15}$$

where

$$I_2 = \alpha^{-1} K_0(Z), \qquad Z = \left[4 \left(\frac{A}{T} + \alpha^{-1} \right) \left(2\alpha + \frac{eE}{T} \right) \right]^{1/2}, \tag{15.16}$$

and K_0 is a modified Bessel function.

Approximate expressions for K_0 and K_1 can be used, and one finally obtains

$$\mathcal{P}(\alpha, R, E, T) \approx \frac{1}{2} \nu e^{2\alpha R} e^{2(AeE)^{1/2}/T} e^{-(8A\alpha/T)^{1/2}} \mathcal{G}(E,T), \tag{15.17}$$

with

$$\mathcal{G} = (eE)^{3/4} \left\{ (2\alpha)^{5/4} T^{1/4} A^{1/2} \left(1 + \frac{3T}{16 (AeE)^{1/2}} \right) \right. \\ \left. \times \exp \left[\frac{eE}{\alpha T} \left(\frac{\alpha A}{2T} \right)^{1/2} + \left(\frac{2T}{\alpha A} \right)^{1/2} \right] \right\}. \tag{15.18}$$

$\mathcal{G}$ depends weakly on α and the dominant dependence of $\mathcal{P}$ is contained in the exponentials in eqn (15.17). A universal temperature dependence can be obtained for $\mathcal{P}$ at all E by normalizing:

$$\frac{\mathcal{P}}{\gamma} = \frac{1}{2} \nu e^{2\alpha R} e^{-(8A\alpha/T)^{1/2}}, \tag{15.19}$$

where

$$\gamma(E,T) = \mathcal{G}(E,T) e^{2(AeE)^{1/2}/T}. \tag{15.20}$$

From eqn (15.19) α and R can be determined from a best fit to the experimental data. It should be noted that nowhere in this derivation have the superfluid properties of He been used, so that this analysis is expected to be valid for all liquids in which electrons form bubbles.

In Fig. 15.15 P/γ is plotted against $T^{-1/2}$ for pure ^{4}He and ^{3}He, and for two different mixtures. For pure ^{4}He, a fit of eqn (15.19) to the experimental data yields $\alpha = 0.431\,\text{Å}^{-1}$, giving $V_0 - E_1 = 0.71\,\text{eV}$ and $R = 24.8\,\text{Å}$ (Schoepe and Rayfield, 1973). This value for the radius of the electron bubble is much larger than the accepted value $\approx 17\,\text{Å}$ at SVP. If this smaller value was used instead, trapping times a few hundred times larger than the experimental ones would be obtained.

This discrepancy has been partly solved by using a density-functional approach (Ancilotto and Toigo, 1994, 1995). The authors obtain an equilibrium bubble radius of 17.9 Å, in line with the accepted value. The trapping times are directly calculated from a solution of the bubble diffusion equation, including a loss term to account for electron tunneling. The main result is that the diffuseness of the interface hardly affects the tunneling matrix element, which depends exponentially on the distance from the surface.

However, it is found that the equilibrium distribution function of the bubbles close to the surface layer where tunneling is most effective in destroying them is

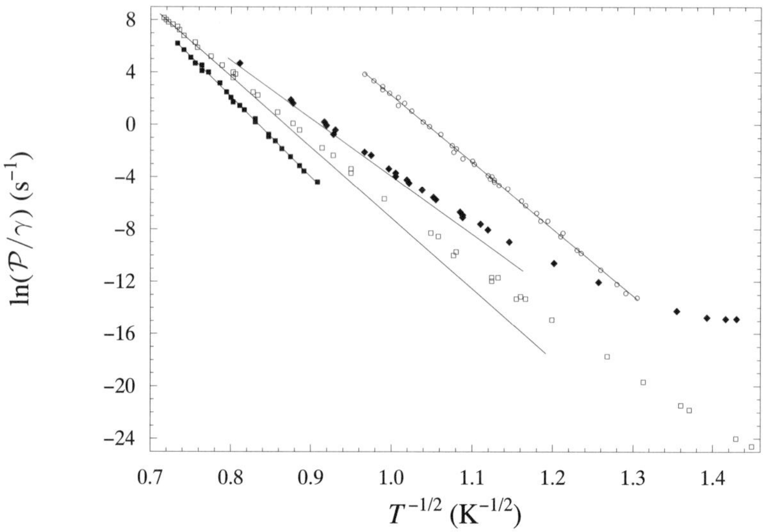

FIG. 15.15. Tunnel probability $\mathcal{P}/\gamma$ vs $T^{-1/2}$. (Schoepe and Rayfield, 1973.) Lines: eqn (15.19). Closed squares: pure ^{4}He, open circles: pure ^{3}He, open squares: $x_3 = 12\%$ ^{3}He–^{4}He mixture, and diamonds: $x_3 = 19.5\%$ ^{3}He–^{4}He mixture.

significantly different from the Boltzmann-type distribution in eqn (15.13). As a consequence, the computed trapping times are much smaller than those obtained by using eqn (15.13) and are now in semi-quantitative agreement with experiment. Unfortunately, there is still a great discrepancy between the estimated trapping times and the measurements of autodetachment times of electron bubbles in superfluid helium droplets (Fárník *et al.*, 1999).

The ground-state energy of the electron in the bubble has been directly probed in a similar experiment (Mitchell and Rayfield, 1972). It is known that negatively-charged vortex rings can propagate in a field-free region for quite long distances, so that they are exploited to carry the negative ions trapped on them toward the liquid surface at zero field.

It is also known that negatively-charged vortex rings release the electrons in the vapor upon hitting the liquid surface (Cunsolo *et al.*, 1968*a*; Surko and Reif, 1968). The energy spectrum of the released electrons is obtained by applying a retarding potential. The cut-off voltage, $\approx 150\,\mathrm{mV}$, is independent of the vortex ring energy and is a measure of the ground-state energy of the electron in the bubble.

If this amount of $0.15\,\mathrm{eV}$ is added to the $0.71\,\mathrm{eV}$ of the barrier height determined in the previously-described experiment, a total energy of $V_0 = 0.86\,\mathrm{eV}$ is obtained. This value is less than the value of $\approx 1\,\mathrm{eV}$ measured in other experiments (Woolf and Rayfield, 1965; Sommer, 1964; Northby and Sanders, 1967) but is in quite good agreement with the calculations of Fetter (1976).

In any case, all these experiments, in spite of the different numerical estimates they give, have confirmed without doubt that the negative He ion is an electron bubble.

16

ION TRANSPORT IN NORMAL LIQUID ^{4}HE

Charge transport experiments in the superfluid shed light on some physical characteristics of the solvated ion and of the self-trapped electron, such as their large, though microscopic, size and their spherical symmetry, which turn these complex charged structures into ideal microscopic probes to also carry out hydrodynamic experiments in the normal fluid.

The goals of this type of measurement are several. On one hand, both helium isotopes ^{3}He and ^{4}He have quite low density in the liquid phase owing to their zero-point motion and the transport properties of charges should be intermediate between the extremes of liquid- and gas-like behavior. It might also be interesting to observe if the boson and fermion nature of the two isotopes influences the transport phenomena, even in the non-degenerate case (Meyer *et al.*, 1962).

On the other hand, the study of excess charges in the superfluid phase of ^{4}He has clearly focused on the problem of the nature of the charge carriers, on their interaction with the elementary excitations making up the normal fraction of the superfluid, and on the quantum hydrodynamics of the superfluid flow past the ions (Ostermeier and Schwarz, 1972).

In any case, charge transport experiments in the normal liquid yield further pieces of information on the structure of the charge carriers. The connection between charge mobility and structure in the hydrodynamic case is obtained by using the Einstein equation relating the diffusion coefficient D to the ion mobility μ, $D/\mu = k_B T/e$ (eqn (3.1)), and the Stokes formula that gives the diffusion coefficient of a sphere of radius R moving in a medium of viscosity η:

$$D = \frac{k_B T}{6\eta\pi R}. \qquad (16.1)$$

Here k_B is the Boltzmann constant. It is obviously understood that the fluid is treated as a continuum and the calculation of the drag force on an ion is based on classical hydrodynamics, i.e., the drag force is calculated by solving the Navier–Stokes equations (5.35), implemented with the equation of continuity for an incompressible fluid, $\nabla \cdot \mathbf{v} = 0$.

Early measurements of the mobility of positive and negative ions (Meyer *et al.*, 1962) are shown for the sake of completeness in Figs 16.1 and 16.2. At the time of these measurements, the model of the two species of ion was not yet very clear and no microscopic information was gathered from these data.

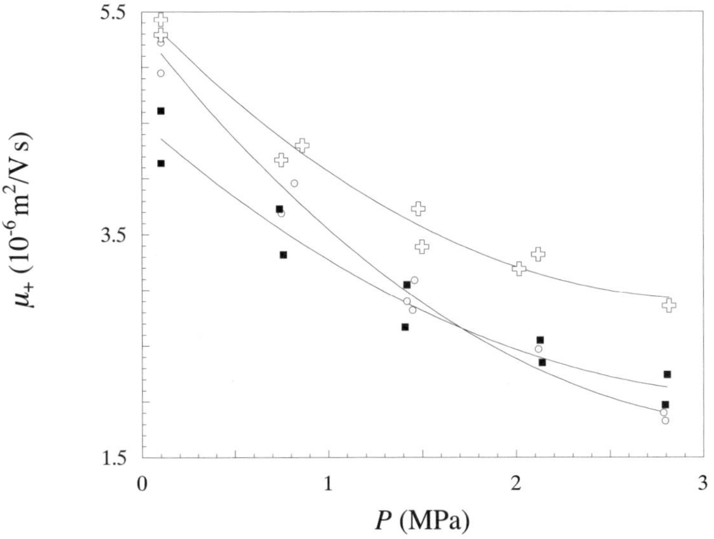

FIG. 16.1. μ_+ vs P in normal liquid ^{4}He. T (K) = 2.2 (crosses), 3.0 (squares), and 4.2 (circles). (Meyer *et al.*, 1962.)

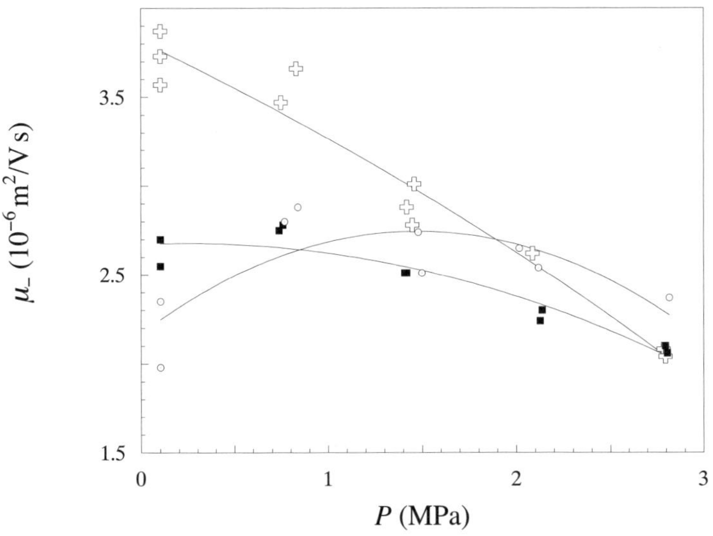

FIG. 16.2. μ_- vs P in normal liquid ^{4}He. T (K) = 2.2 (crosses), 3.0 (squares), and 4.2 (circles). (Meyer *et al.*, 1962.)

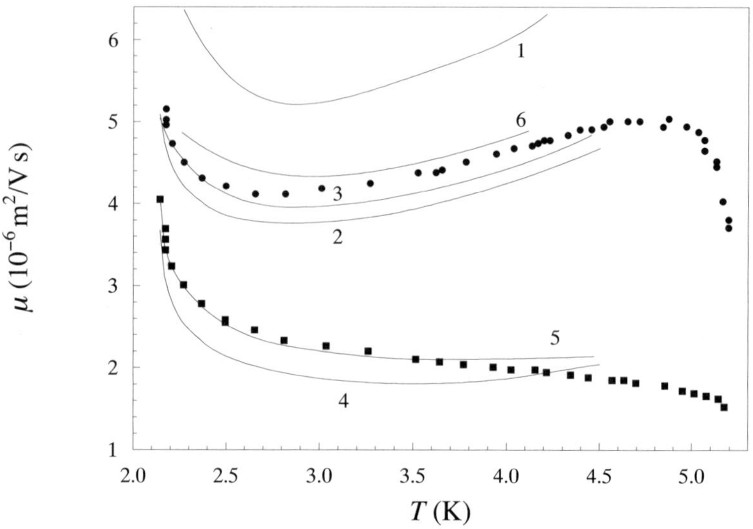

FIG. 16.3. μ_+ (circles, top) and μ_- (squares, bottom) vs T in normal liquid ^{4}He at SVP (Ostermeier and Schwarz, 1972). Curves 1 through 6 are explained in the text. Curve 6: Goodstein (1978).

16.1 Measurements at saturated vapor pressure

Accurate ion mobility measurements under saturated vapor pressure conditions above T_λ up to the critical temperature $T_c \approx 5.18\,\mathrm{K}$ have been performed by Ostermeier and Schwarz (1972) using the charge–pulse time-of-flight technique developed for ion mobility measurements in the superfluid (Schwarz, 1970, 1972*a*). The absolute accuracy of the data is better than 2%. The previous few determinations of lesser accuracy below the normal boiling point temperature $T_b = 4.2\,\mathrm{K}$ (Reif and Meyer, 1960; Meyer *et al.*, 1962) agree to within 20% with these much more accurate data. The experimental results for both positive and negative ions are shown in Fig. 16.3.

16.2 Hydrodynamics in the presence of electrostriction

The moving objects are charged species, either the positive ion or the electron bubble, and induce electrostriction in the surrounding fluid. Because of electrostrictive forces, ions are thus drifting in a fluid whose local density is enhanced with respect to the unperturbed fluid.

In order to correctly treat the viscous drag force it should also be realized that the shear viscosity $\eta(\rho, T)$ is locally enhanced as a consequence of the increase in the local density. If the viscosity η is known as a function of the density $\rho(T)$, the viscosity profile can be readily computed from that of the density determined by electrostriction according to eqn (3.9) (Atkins, 1959). Typical examples of density and viscosity profiles are shown in Fig. 16.4. Viscosity data from the literature have been used here (Welber, 1961; Goodwin, 1968).

The hydrodynamic problem is thus the determination of the drag force on a spherical object of radius R that moves at velocity v_D in such a way as to comply with the local density and viscosity profiles.

In the presence of non-uniform density and viscosity, which are denoted by $\rho(r)$ and $\eta(r)$ for the sake of simplicity, the Navier–Stokes equation takes on the following form:

$$\rho\left(\frac{\partial}{\partial t}+\mathbf{v}\cdot\boldsymbol{\nabla}\right)\mathbf{v}=-\boldsymbol{\nabla}P+\eta\left[\boldsymbol{\nabla}^2\mathbf{v}+\frac{1}{3}\boldsymbol{\nabla}\left(\boldsymbol{\nabla}\cdot\mathbf{v}\right)\right]$$
$$-\frac{2}{3}\boldsymbol{\nabla}\eta\left(\boldsymbol{\nabla}\cdot\mathbf{v}\right)+\boldsymbol{\nabla}\eta\times\left(\boldsymbol{\nabla}\times\mathbf{v}\right)+2\left(\boldsymbol{\nabla}\eta\cdot\boldsymbol{\nabla}\right)\mathbf{v}\,,\quad(16.2)$$

and must be solved together with the continuity equation

$$-\frac{\partial\rho}{\partial t}=\boldsymbol{\nabla}\cdot(\rho\mathbf{v})\,. \quad (16.3)$$

Pressure variations arising from the motion of the fluid have negligible effect on $\rho(r)$ and $\eta(r)$.

As usual, when tackling problems of hydrodynamics (Byron Bird *et al.*, 1960; Landau and Lifsits, 2000), the spherical obstacle of radius R is centered at the

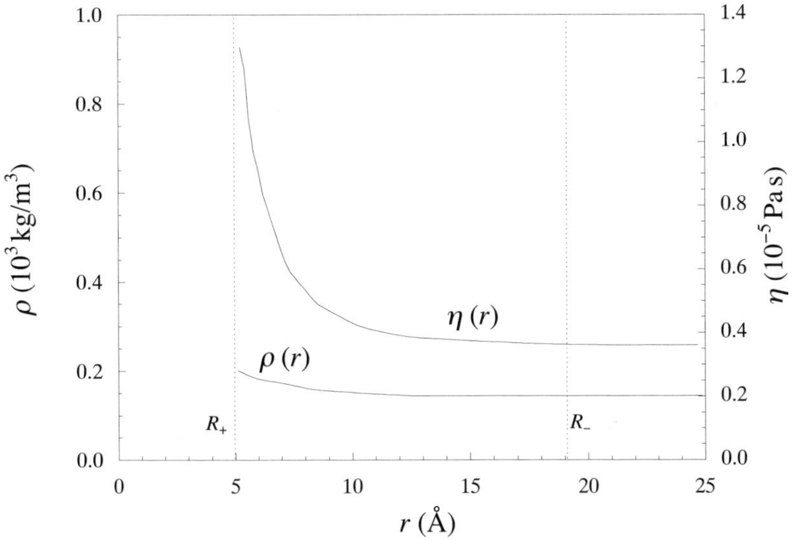

FIG. 16.4. Density and viscosity profiles due to electrostriction in bulk helium for $T=3.0\,\mathrm{K}$. (Ostermeier and Schwarz, 1972.) The vertical lines indicate the approximate radii of positive and negative ions.

origin and the fluid is assumed to flow past the sphere in such a way that at infinity its velocity is along the z-axis: $\mathbf{v} = v_D\hat{\mathbf{z}}$, where $\hat{\mathbf{z}}$ is the unit vector along the z-axis.

A steady-state solution is sought for very small Reynolds numbers owing to the small ion drift velocity in typical experiments. Thus, the left-hand sides of eqns (16.2) and (16.3) are both set equal to zero. The fluid flow is thus solenoidal and a generalized stream function $\mathbf{A}$ can be introduced, by analogy with the vector potential in electromagnetism, in such a way that the fluid velocity is obtained by taking its curl:

$$\rho\mathbf{v} = \boldsymbol{\nabla} \times \mathbf{A} \,. \tag{16.4}$$

The simplest stream function well suited for the spherically-symmetric problem at hand is

$$\mathbf{A} = -\frac{1}{2} r f\left(r\right) \hat{\mathbf{r}} \times \hat{\mathbf{z}} \,, \tag{16.5}$$

where $\hat{\mathbf{r}}$ and $\hat{\mathbf{z}}$ are unit vectors. The function $f(r)$ depends on the distance from the ion and must be determined by solving the Navier–Stokes equation (16.2) with proper boundary conditions. It is easy to show that the velocity field $\mathbf{v}$ takes the form

$$\rho\mathbf{v} = f\cos\theta\hat{\mathbf{r}} - \left(f + \frac{1}{2} r f'\right)\sin\theta\hat{\mathbf{z}} \,, \tag{16.6}$$

where the prime indicates differentiation with respect to r.

In order to directly calculate the drag force on the obstacle, eqn (16.6) is substituted directly into eqn (16.3), yielding (Ostermeier and Schwarz, 1972)

$$\boldsymbol{\nabla} P = g_1\left(r\right)\cos\theta\hat{\mathbf{r}} - g_2\left(r\right)\sin\theta\hat{\mathbf{z}} \,, \tag{16.7}$$

where g_1 and g_2 are given by

$$\begin{aligned} g_1\left(r\right) = {} & \left(\frac{\eta}{\rho}\right) f'' + \left(\frac{4\eta}{r\rho} - \frac{7\eta\rho'}{3\rho^2} + \frac{2\eta'}{\rho}\right) f' \\ & - \left(\frac{2\eta\rho'}{r\rho^2} + \frac{4\eta\rho''}{3\rho^2} - \frac{8\eta\rho'^2}{3\rho^3} + \frac{4\eta'\rho'}{3\rho^2}\right) f\left(r\right) \end{aligned} \tag{16.8}$$

and

$$\begin{aligned} g_2\left(r\right) = {} & \left(\frac{r\eta}{2\rho}\right) f''' + \left(\frac{3\eta}{\rho} - \frac{r\eta\rho'}{\rho^2} + \frac{r\eta'}{2\rho}\right) f'' + \left(\frac{2\eta}{r\rho} - \frac{4\eta\rho'}{\rho^2} - \frac{r\eta\rho''}{2\rho^2} + \frac{r\eta\rho'^2}{\rho^3}\right. \\ & \left. + \frac{\eta'}{\rho} - \frac{r\eta'\rho'}{2\rho^2}\right) f' - \left(\frac{7\eta\rho'}{3r\rho^2} + \frac{\eta\rho''}{\rho^2} - \frac{2\eta\rho'^2}{\rho^3} + \frac{\eta'\rho'}{\rho^2}\right) f \,. \end{aligned} \tag{16.9}$$

The pressure P is readily obtained upon integrating eqn (16.7), yielding

$$P = r g_2(r)\cos\theta + C \,, \tag{16.10}$$

where C is a constant. The consistency condition

$$g_1(r) = rg_2'(r) + g_2(r) \tag{16.11}$$

gives rise to the same fourth-order, homogeneous, linear differential equation for $f(r)$ as that obtained if the more traditional procedure of taking the curl of eqn (16.2) and substituting the expression for the velocity given in eqn (16.6) had been followed.

Conservation of linear momentum ensures that the total momentum flux into a spherical surface of radius $r > R$ is independent of r and equals the total drag force $\mathbf{F}$ exerted on the sphere. $\mathbf{F}$ is obtained by integrating the stress tensor over the spherical surface S of radius R: (Ostermeier and Schwarz, 1972)

$$\mathbf{F} = \hat{\mathbf{z}} \int_S (\sigma_{rr} \cos\theta - \sigma_{\theta r} \sin\theta)\, r^2 \sin\theta \mathrm{d}\theta \mathrm{d}\phi\,. \tag{16.12}$$

The components of the stress tensor are

$$\sigma_{rr} = -P + \frac{2}{3}\eta \left(2\frac{\partial v_r}{\partial r} - 2\frac{v_r}{r} - \frac{1}{r}\frac{\partial v_\theta}{\partial \theta} - \frac{v_\theta}{r \tan\theta} \right) \tag{16.13}$$

and

$$\sigma_{\theta r} = \eta \left(\frac{1}{r}\frac{\partial v_r}{\partial \theta} + \frac{\partial v_\theta}{\partial r} - \frac{v_\theta}{r} \right). \tag{16.14}$$

Equations (16.6) through (16.14) together yield the following linear differential equation for $f(r)$:

$$\begin{aligned} f''' &+ \left(\frac{4}{r} - \frac{2\rho'}{\rho} + \frac{\eta'}{\eta} \right) f'' - \left(\frac{4}{r^2} + \frac{6\rho'}{r\rho} + \frac{\rho''}{\rho} - \frac{2\rho'^2}{\rho^2} - \frac{2\eta'}{r\eta} + \frac{\eta'\rho'}{\eta\rho} \right) f' \\ &+ \left(\frac{2\rho'}{r^2\rho} - \frac{2\rho''}{r\rho} + \frac{4\rho'^2}{r\rho^2} - \frac{2\eta'\rho'}{r\eta\rho} \right) f - \frac{9\rho}{r^4\eta} F = 0\,. \end{aligned} \tag{16.15}$$

Equation (16.15) is made non-dimensional by scaling $\mathbf{v}$, η, and ρ by their asymptotic values v_D, η_∞, and ρ_∞, respectively, i.e., the ion drift velocity, and the viscosity and density of the unperturbed fluid at a large distance from the ion. r is measured in units of either ion radius, $R_\pm$. Finally, the drag force F is measured in units of $6\pi\eta_\infty v_D R_\pm$. Thus, the usual Stokes formula is obtained for a uniform fluid.

Equation (16.15) is the first integral of eqn (16.11) and F in eqn (16.15) is the integration constant that has to be determined by enforcing the proper boundary conditions. For the positive ion the no-slip boundary conditions are assumed, i.e., the fluid velocity is zero at the boundary. This condition yields, in non-dimensional form,

$$f(1) = 0\,, \tag{16.16}$$
$$f'(1) = 0\,, \tag{16.17}$$
$$\lim_{r\to\infty} f(r) = \,,1 \tag{16.18}$$
$$\lim_{r\to\infty} r f'(r) = 0\,. \tag{16.19}$$

For the electron bubble, in contrast, a perfect-slip condition is assumed, i.e., the tangential stress is zero at the boundary. In this case, the second equation of the previous set of boundary conditions has to be replaced by

$$f''(r) + \left(2 - \frac{\rho'}{\rho}\right) f'(r) = 0 \qquad \text{for } r = 1\,. \tag{16.20}$$

In order to solve eqn (16.15) with the given boundary conditions it is more convenient to introduce the non-dimensional variable $y = r^{-1}$. Equation (16.15) then becomes

$$y^2 f''' + y\left(2 - 2y\frac{\rho'}{\rho} + y\frac{\eta'}{\eta}\right) f'' + \left[-6 + y^2\left(2\frac{\rho'^2}{\rho^2} - \frac{\rho''}{\rho} - \frac{\eta'\rho'}{\eta\rho}\right)\right] f'$$
$$+ \left(6\frac{\rho'}{\rho} - 4y\frac{\rho'^2}{\rho^2} + 2y\frac{\rho''}{\rho} + 2y\frac{\eta'\rho'}{\eta\rho}\right) f + \frac{9\rho}{\eta} F = 0\,. \tag{16.21}$$

In terms of the new variable y, the boundary conditions for no slip become

$$f(1) = 0\,, \tag{16.22}$$
$$f'(1) = 0\,, \tag{16.23}$$
$$f(0) = 1\,, \tag{16.24}$$
$$\lim_{y\to 0} y f'(y) = 0\,. \tag{16.25}$$

For perfect slip, condition (16.25) must be replaced by

$$f''(y) - \frac{\rho'}{\rho} f'(y) = 0 \qquad \text{for } y = 1. \tag{16.26}$$

The easiest, though accurate, method to solve eqn (16.21) is to approximate the solution by a polynomial of the form

$$f(y) = \sum_{j=1}^{N} a_j\,(y-1)^j\,. \tag{16.27}$$

This polynomial is evaluated at the nodes $y_n = n/(N-1)$ for $n = 1, 2, \ldots, N-1$, and also the boundary conditions (16.24) and either (16.23) or (16.26) must be satisfied. This procedure yields $N+1$ algebraic equations for the N coefficients a_j and for the non-dimensional drag force F that can be solved by usual algebraic techniques. It is understood that the spatial dependence of η and ρ must be known in order for their derivatives in eqn (16.21) to be numerically evaluated at the nodes.

Once F has been determined in this way, the actual drag force is simply given by

$$\text{drag force} = (6\pi\eta_\infty R_\pm v_D)\, F\,, \tag{16.28}$$

and the ion mobility turns out to be

$$\mu_\pm = \frac{e}{6\pi\eta_\infty}\left(\frac{1}{R_\pm F}\right), \tag{16.29}$$

where the index $\pm$ indicates the sign of the ion. The product $R_\pm F$ can thus be considered as the ion hydrodynamic radius.

A detailed inspection of the solutions of this problem leads to the general conclusions that there are no qualitative differences between the no-slip and the perfect-slip cases, and that the drag force increases or decreases in the same way as the viscosity does. In contrast, a density enhancement around the sphere surprisingly leads to a reduction of the drag force, and vice versa. Moreover, it is observed that the largest increase of the drag is obtained for disturbances that extend farther out into the fluid. The more localized the density and viscosity profiles, the smaller the drag-force enhancement (Ostermeier and Schwarz, 1972).

16.2.1 *Results for $T \leqslant 4.2\,K$*

In order to compare the results of the hydrodynamic calculations with the experimental ion mobility data, a specific model of the charge carriers is required. First of all, it is to be observed upon inspection of Fig. 16.4 that the electrostriction-induced drag-force correction for the electron bubble is expected to be nearly negligible because of its very large radius. Moreover, it is assumed that the proper boundary condition for the electron bubble is the perfect-slip case because the bubble approximates a free surface.

The density profile around the positive ion is calculated below the normal boiling point temperature $T_b \approx 4.2\,\mathrm{K}$ according to the electrostrictive equation of Atkins (1959). The radius of the positive ion core is shown as a function of temperature in Fig. 3.3 for two different assumptions of the solid–liquid surface tension.

The radius of the negative ions is calculated by minimizing the free energy of electron localization with the inclusion of a small electrostrictive contribution (eqn (3.21)) (Miyakawa and Dexter, 1970; Ostermeier and Schwarz, 1972) and is shown as a function of temperature in Fig. 3.7. Although the experimental surface tension of liquid ^{4}He is available (Allen and Misener, 1938; Atkins and Narahara, 1965; Magerlein and Sanders, 1976; Iino *et al.*, 1985; Nakanishi and Suzuki, 1998; Vicente *et al.*, 2002), it is argued that the surface tension of the electron bubble might be greater. Therefore, the calculations of the drag force have also been carried out by using a surface tension value enhanced by a factor of 1.7 (Ostermeier and Schwarz, 1972) in order to get a bubble radius of about 15 Å at low temperature, in agreement with other experimental determinations of it (Parks and Donnelly, 1966; Springett, 1967).

In Fig. 16.3 the results of the hydrodynamic calculations of the drag force on ions in normal liquid ^{4}He at saturated vapor pressure are shown (Ostermeier and Schwarz, 1972).

The experimental mobility data are represented as circles for the positive ions and as squares for the negative ones. Curve 1 is the prediction of the pure Stokes formula for positive ions without the drag enhancement factor F. The temperature dependence of the experimental data is only reproduced qualitatively.

Curve 2 shows the prediction of the electrostriction model when hydrodynamics is treated correctly, if the solid–liquid surface tension is neglected. The prediction of the pure Stokes formula is drastically reduced by the inclusion of the effect of the density and viscosity profiles. The change is almost entirely due to the local increase of the viscosity because the local enhancement of the density increases the mobility slightly (Ostermeier and Schwarz, 1972). Owing to the fair agreement with the data, the conclusion is drawn that the physically important process is electrostriction. This, in turn, actually leads to a great local viscosity enhancement through the density increase that exerts an extra drag on the ions.

Curve 3 is obtained if a value $\sigma_{ls} = 10^{-4}\,\mathrm{J/m^2}$ is assumed for the liquid–solid surface tension. The slight improvement of the prediction lends some credibility to the assumption that the liquid–solid surface tension is not zero. In any case, the agreement between curve 2 and the experimental data constitutes a detailed verification of the electrostriction model (Ostermeier and Schwarz, 1972).

The results for the negative ions are presented in the same figure. Curve 4 represents the results of the hydrodynamic calculations if the experimental value for the liquid–vapor surface tension is used.

Curve 5 is obtained if σ_{lv} is enhanced by a factor of 1.7. Also, in this case the agreement between experiment and data is quite satisfactory owing to the uncertainty in viscosity and surface tension. The comparison between data and theory favors the idea that the surface tension of the bubble is larger than that in the bulk and that the boundary conditions for perfect slip are appropriate for the electron bubble (Ostermeier and Schwarz, 1972).

16.2.2 *Results for $T \leqslant 4.2\,K$ at very high pressure*

Measurements of the positive ion mobility in normal liquid ^{4}He under pressure up to solidification (Keshishev *et al.*, 1969) have been analyzed in terms of the approach described previously (Ostermeier and Schwarz, 1972). These experimental data and the hydrodynamic calculations at two temperatures are shown in Figs 16.5 and 16.6. Although the data at $T = 2.07\,\mathrm{K}$ are below the superfluid transition, hydrodynamics is still applicable because the normal fluid component acts as a viscous fluid even at a microscopic scale (Ostermeier and Schwarz,

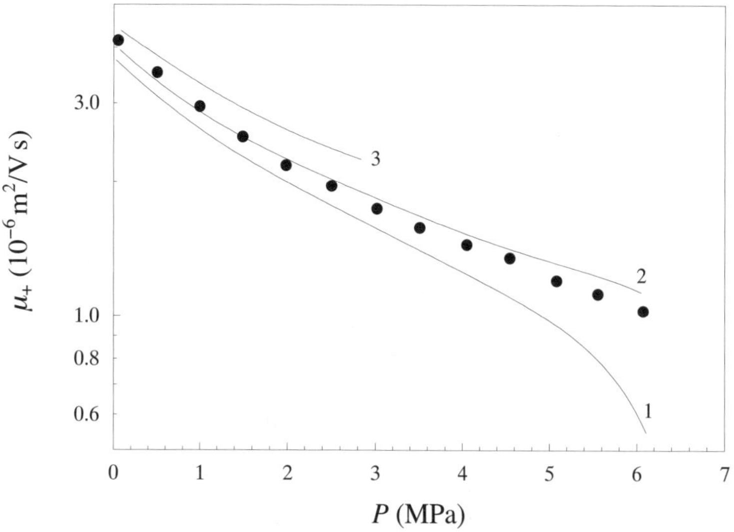

FIG. 16.5. μ_+ vs P up to solidification in normal liquid ^{4}He for $T = 2.63$ K (Keshishev *et al.*, 1969). Curve 1: hydrodynamic prediction with electrostriction and surface tension $\sigma_{ls} = 0$, and curve 2: $\sigma_{ls} = 10^{-4}$ J/m^2 (Ostermeier and Schwarz, 1972). Curve 3: Mathews model (Goodstein, 1978).

1972).

The mobility calculated with vanishing surface tension $\sigma_{ls} = 0$ shows a large drop upon approaching solidification. This drop can be explained by noting that near the melting pressure only a small additional electrostrictive pressure is required to produce solidification around the ion, thereby enormously increasing its radius and consequently decreasing the mobility. In contrast, the experimental data do not show such a drop. If a finite value of $\sigma_{ls} = (1 \pm 0.5) \times 10^{-4}$ J/m^2 is used in the computation, a much better agreement between data and theory is obtained. This fact would confirm, according to the authors, the existence of a solid core around the ion (Ostermeier and Schwarz, 1972).

16.2.3 *Does the liquid freeze around the ion?*

A somewhat different conclusion has been reached by Goodstein (1978). According to him, the measurements of the mobility cannot settle the question of whether the positive ion core actually solidifies or whether the local viscosity rises smoothly until the liquid is effectively immobilized around the ion.

The argument is based on analytic calculations of the drag on a slowly-moving sphere in fluid of variable viscosity (Mathews, 1978) and on more extended measurements of the shear viscosity and ion mobility close to the melting line (Scaramuzzi *et al.*, 1977*a*).

The Navier–Stokes equations are solved analytically for the case in which the viscosity varies with the distance r from the center of the sphere according to the following law:

$$\eta(r) = \eta_\infty \left[1 + \left(\frac{R_0}{r}\right)^s\right], \tag{16.30}$$

where R_0 is a cut-off radius that depends on the fluid and on the coupling between the ion and the fluid, s is an integer power-law index, and η_∞ is the viscosity of the unperturbed fluid at a large distance from the ion. This dependence of the viscosity is similar to that treated by Ostermeier and Schwarz (1972).

This approach can be pursued in the case of ions in He because in He I, not too close to either the λ-transition, the liquid–vapor critical point, or the melting line, the viscosity is approximately related to the pressure P by a linear relationship (Scaramuzzi *et al.*, 1977*a*):

$$\eta = aP + b, \tag{16.31}$$

where the coefficients a and b are functions of the temperature.

According to the model of electrostriction (Atkins, 1959), the local pressure around the ion, $P(r)$, rises above the value at a large distance from it, P_∞:

$$P(r) - P_\infty = \frac{\beta}{vr^4}, \tag{16.32}$$

where β depends on the polarizability of the liquid and v is the molar volume, which is approximately taken as a constant.

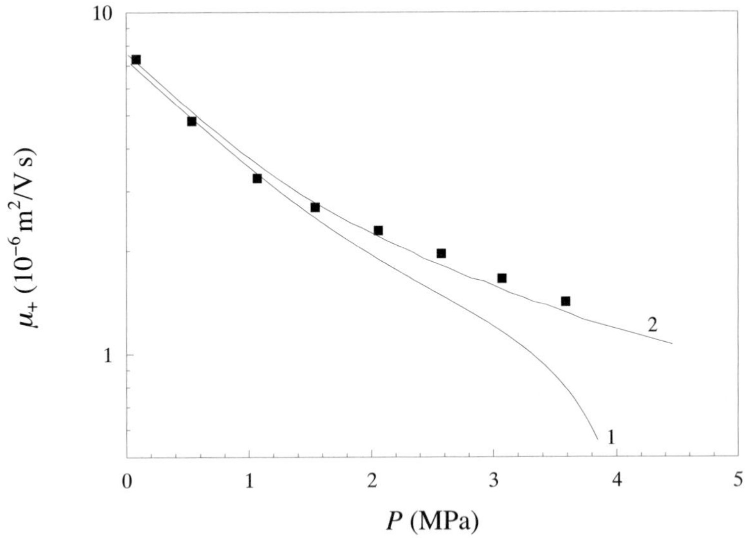

FIG. 16.6. μ_+ vs P up to solidification in liquid ^{4}He for $T = 2.07$ K (Keshishev *et al.*, 1969). Curves 1 and 2 are calculated in the same way as curves 1 and 2 in Fig. 16.5.

The cut-off radius is readily calculated by inverting eqn (16.30) and using eqns (16.31) and (16.32). For $s = 4$ one obtains

$$R_0^4 = \frac{a\beta}{v\eta_\infty} \,. \tag{16.33}$$

In case of no-slip conditions, suitable for the positive ions, the analytic solution for the drag on the sphere can be written as

$$\mu_+ = \frac{e}{6\pi\eta_\infty R_+ \phi} \,, \tag{16.34}$$

where R_+ is the true radius of the sphere (the positive ion, in this case). The factor ϕ is formally identical to the factor F_+ in eqn (16.29), but now it turns out to be a function of $y_+ = R_0/R_+$.

In the case of the negative ions, perfect-slip boundary conditions at the surface of the sphere are enforced and the analytical solution now gives the mobility as (Mathews, 1978)

$$\mu_- = \frac{e}{4\pi\eta_\infty R_- \Psi} \,. \tag{16.35}$$

The factor Ψ is formally equivalent to the factor $(2/3)F_-$ in eqn (16.29) and is now a function of $y_- = R_0/R_-$.

It must be stressed that this model, for which analytical solutions are provided, takes into account only the non-uniformity of the viscosity and assumes that the density is constant throughout. This assumption is consistent with the results of Ostermeier and Schwarz (1972), according to which the density variations have a much lesser influence on $F_\pm$ than those of the viscosity.

New insight into the question of the existence of a solidified ion core is gained by inspecting the limiting behavior of ϕ and Ψ when R_0 is greater than either R_+ or R_-, i.e., for $y_+ \gg 1$ or $y_- \gg 1$. It is shown (Mathews, 1978) that

$$\begin{aligned}\phi &= \frac{16\sqrt{2}}{9\pi} y_+^{1/4} \left(1 + \frac{4\sqrt{2}}{3\pi} y_+^{-3/4} + \dots\right) \\ &\approx 0.8 y_+^{1/4} \left(1 + 0.6 y_+^{-3/4}\right)\end{aligned} \tag{16.36}$$

and

$$\Psi = \frac{8\sqrt{2}}{3\pi} y_-^{1/4} \left(1 - \frac{8\sqrt{2}}{3\pi} y_-^{-3/4} + \dots\right) . \tag{16.37}$$

In the extreme case in which $y_- \to \infty$ or $y_+ \to \infty$, one obtains

$$6\pi R_+ \phi(\infty) = 6\pi \frac{16\sqrt{2}}{9\pi} \approx 6\pi\,(0.8)\,R_0 \tag{16.38}$$

for the positive ion and

$$4\pi R_- \Psi(\infty) = 4\pi \frac{8\sqrt{2}}{3\pi} R_0 \approx 6\pi\,(0.8)\,R_0 \tag{16.39}$$

for the negative ion.

Thus, if the ion core is small enough, the boundary conditions become irrelevant and the ion moves as though it were a solid sphere of hydrodynamic radius $0.8R_0$. It is evident that the quantity R_0 is a characteristic length that depends only on the properties of the bulk fluid and is independent of the ion structure.

R_0 varies between 6 and 7.3 Å in the range of interest (Goodstein, 1978). Thus, the estimate $0.8R_0$ for the effective hydrodynamic radius of the positive ion core is compatible with other estimates (Schwarz, 1972*c*). At the same time, it appears that the asymptotic results do not apply to the negative ion.

Quantitative results for the positive ions at SVP are reported in Fig. 16.3 as curve 6. In this case, the distance of closest approach of two ^{4}He atoms in the liquid has been chosen as the ion radius $R_+ \approx 2.6$ Å. The quality of the agreement with the experimental data is approximately similar to that of the hydrodynamic model of Ostermeier and Schwarz (1972).

In Fig. 16.6 the present calculations are compared with the data of Keshishev *et al.* (1969) and with the results of the calculations of Ostermeier and Schwarz (1972). The agreement of the present model with the experimental data is poorer than the results of Ostermeier and Schwarz, though the latter have been obtained by using the surface tension as an adjustable parameter.

The physical reason for the nice behavior of the Mathews' solutions is that the maximum energy dissipation is reached at a distance close to R_0. For $r < R_0$, the local viscosity is so large as to damp the ion motion. Thus, it makes no difference if He is a solid or an extremely viscous fluid at such small distances from the ion.

16.2.4 *Results for $T > 4.2\,K$*

The mobility behavior for temperatures above the normal boiling point deviates from the prediction of the electrostriction-modified hydrodynamic model (Ostermeier and Schwarz, 1972), as shown in Fig. 16.3. First of all, around $T \approx 4$ K, μ shows a downward deviation from the values predicted by using extrapolated viscosity data. The deviation is very evident for the negative ions, although it also occurs in the positive ion data.

A second, even more important feature, is the large drop in the mobility of the positive ions for $T > 4.7$ K upon approaching the liquid–vapor critical point, by analogy with data of positive ion mobility in pure ^{3}He near the critical point (Modena and Ricci, 1967; Cantelli *et al.*, 1968). A similar behavior has been observed in a different physical situation for the mobility of O_2^- ions in dense Ar (Borghesani *et al.*, 1997) and Ne (Borghesani *et al.*, 1993) gases near their critical points.

In order to exemplify how the electrostrictive hydrodynamic calculations improve the agreement of the experimental data but fail to reproduce the features near the critical point, the density-normalized zero-field mobility $\mu_0 N$ of O_2^- ions in dense gaseous Ne at $T = 45$ K are reported in Fig. 16.7 as a function of the density in a range, which encompasses the critical density (Borghesani *et al.*, 1993). The critical parameters of Ne are $T_c = 44.38$ K and $N_c \approx 14.43\,\mathrm{nm}^{-3}$ (McCarty and Stewart, 1965).

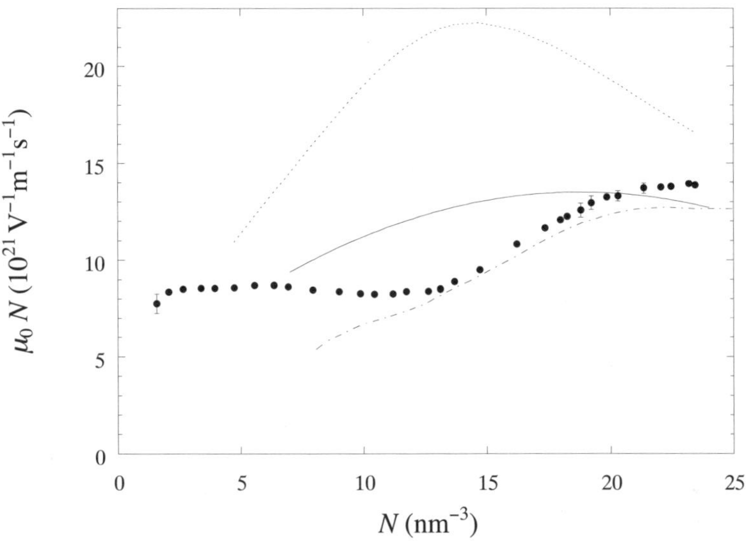

FIG. 16.7. Zero-field density-normalized mobility $\mu_0 N$ vs N of O_2^- ions in Ne gas across the critical density at $T = 45\,\mathrm{K}$ (Borghesani *et al.*, 1993). Dashed curve: pure Stokes formula with hydrodynamic radius $R_h = 4\,\text{Å}$. Solid line: electrostriction-modified hydrodynamic results. Dash–dotted line: Khrapak model (Volykhin *et al.*, 1995).

The structure of the negative oxygen molecular ion in Ne and He is quite complicated (Khrapak *et al.*, 1995, 1996; Volykhin *et al.*, 1995; Schmidt *et al.*, 1999) and resembles that of the alkali positive ions in liquid helium, as described previously in this book. The ion is locally surrounded by a small empty cavity that is formed because of the short-range repulsive exchange forces between the excess electron in the ion and the electronic shells of the surrounding atoms. Then, electrostriction induces a strong local density increase at larger distances.

At the high densities of this experiment hydrodynamics is expected to be applicable even in a gas (Mason and McDaniel, 1988). In spite of this expectation, it can be seen from Fig. 16.7 that the prediction of the pure Stokes formula, with an ion hydrodynamic radius $R_h = 4\,\text{Å}$ (upper dashed curve), is in strong disagreement with the experimental data.

In contrast, the electrostriction-modified hydrodynamic model of Ostermeier and Schwarz (1972) (solid curve in Fig. 16.7) improves the agreement of the Stokes formula with the experiment, especially at the highest densities, where it is plausible that the gas might be treated as a continuum. However, the large kink in the experimental data around the critical density is not reproduced at all. A reasonable agreement with the data at a density above the critical one has only been obtained by Volykhin *et al.* (1995) by taking as the hydrodynamic radius of the ion the distance, from the ion center, at which the electrostriction-induced density enhancement is a maximum, and evaluating the viscosity for this maximum density.

Except for some special gases, such as Xe and CO_2, whose critical temperature is very easy to reach, the viscosity in liquid He as well as in Ne gas and other gases has not been measured very accurately close to the critical point. Thus, its behavior is not very well known (Ostermeier and Schwarz, 1972), but it is known that the viscosity diverges at the critical point (Hohenberg and Halperin, 1977; Berg and Moldover, 1990). However, its strongest divergence, along the critical isochore, has such a small critical exponent that the viscosity may be considered finite for every practical purpose, unless the critical point is approached very closely.

Thus, one is allowed to make several reasonable assumptions about the behavior of the viscosity around the critical point in order to reproduce the experimental mobility data. This has been done for both liquid He (Ostermeier and Schwarz, 1972) and Ne gas (Borghesani *et al.*, 1993) without any significant improvements. The main reason is that, near T_c, we have $\rho(r) \approx \rho_c$ only at a very large distance from the ion, where electrostrictive effects are small anyway.

The failure of this electrostriction-modified hydrodynamic model in the neighborhood of the critical point may be attributed to several causes. In particular, it is expected that, somewhere close to the critical point, the hydrodynamic description ceases to be valid because the fluid correlation length ξ becomes too large (Ostermeier and Schwarz, 1972).

Significant deviations from the hydrodynamic results are expected when the correlation length becomes of the order of the size of the charge carrier. As $\xi \approx \xi_0[(T/T_c) - 1]^{-\nu}$, with $\xi_0 \approx 1$ Å and $\nu \approx 2/3$ (Stanley, 1971; Sengers and Levelt-Sengers, 1986), one expects deviations caused by critical-point fluctuations to appear 0.3 K below T_c for the positive ions and 0.06 K below T_c for the electron bubble.

By inspection of Fig. 16.3 one realizes that this prediction is qualitatively correct. The mobility drop of the positive ion is much larger than that of the electron bubble. This effect is very probably a consequence of the larger electrostriction influence exerted by a much smaller charge structure on the surrounding fluid.

Another reason for failure may be that the calculations of the ion radius do not consider the possibility that an additional layer of highly-correlated fluid, with thickness of order ξ, is dragged along by the ion during its motion, yielding an effective hydrodynamic radius that depends in some way on the compressibility of the fluid (Borghesani *et al.*, 1997). For these reasons, ions can be exploited to investigate critical-point fluctuations and dynamics on a scale some tens of Å long (Borghesani and Tamburini, 1999) .

17

ION TRANSPORT AT PHASE TRANSITIONS

Only a very few papers are devoted to the study of ion transport in fluid He in the neighborhood of one of its phase transitions. This fact is quite surprising because many pieces of information could be obtained from these measurements. The size of ions makes them the most suitable probes for investigating how the onset and growth of critical fluctuations close to the λ-transition and to the traditional liquid–vapor transition may induce the failure of the description of transport in terms of continuum hydrodynamics.

Moreover, the large increase of pressure around the ions because of electrostriction can be exploited to investigate the details of the ion structure close to the melting transition.

Several reasons might have hindered a thorough investigation of the interesting thermodynamical region around phase transitions. Among others, there may be the experimental difficulty in approaching the temperature and pressure values of the transition with high enough accuracy and stability. Conversely, a reason might be the lack of accurate enough knowledge of other related physical quantities, such as the viscosity that are necessary for a correct interpretation of the experimental results.

17.1 Ion mobility at the λ-transition

The λ-transition is known to belong to the class of critical phenomena that are characterized by the absence of latent heat of transition (Stanley, 1971; Sengers and Levelt-Sengers, 1978). This kind of transition is described by a suitable order parameter, whose average value vanishes as soon as the transition is approached from either side. In the case of He, the order parameter is taken to be the condensate fraction, which is related to the superfluid fraction as ρ_s/ρ, where ρ_s and ρ are the superfluid density and the total density of the liquid, respectively.

On either side of continuous transitions, and very close to them, the order parameter strongly fluctuates over all scales. The fluctuations may eventually become of macroscopic size (the phenomenon of *critical opalescence* is very well known, indeed) and they are correlated over long distances (Stanley, 1971). The thermodynamic properties of a fluid at the critical transition are thus dominated by these long-range fluctuations, whose correlation length ξ diverges to infinity if the critical point is asymptotically approached on either side of the transition. As a result, the measurable properties of the substance investigated may happen to be singular and the predictions of the strengths of these singularities are called *scaling laws* (Stanley, 1971; Goodstein, 1975).

From this point of view, it appears that ions should not be influenced by critical fluctuations because of the well-separated range of scales involved. Ions are microscopic objects, while critical fluctuations even extend to macroscopic distances.

It is argued (Watanabe, 1979) that the drift mobility of ions should be influenced more by small-wavelength fluctuations than by the long-wavelength ones and should be quite insensitive to the long-time tail of the decay of critical fluctuations. These predictions, however, are based on the assumption that the ions and the surrounding fluid interact via an extremely short-range contact potential.

This assumption is not quite reasonable as ions couple with the fluid by means of the (fairly) long-range induced dipole–charge interaction that leads to the well-known effect of electrostriction (Atkins, 1959). For this reason the ions should be sensitive to the critical fluctuations at least on a mid-range scale (Borghesani *et al.*, 1997) and can be used to probe the properties of He at the phase transition (Goodstein *et al.*, 1974).

Scaling laws predict the limiting behavior of the physical quantities of interest in terms of the distance from the transition temperature, T_λ in the present case. The figure of merit is the reduced temperature $\epsilon = |T - T_\lambda|/T_\lambda$. The range of critical behavior of the transport properties in He appears to be roughly $|\epsilon| \leqslant 10^{-3}$ (Goodstein *et al.*, 1974).

Early measurements of the ion mobility in the neighborhood of the λ-transition (Grimsrud and Scaramuzzi, 1966; Ahlers, 1971; Sitton and Moss, 1971; Ahlers and Gamota, 1972) did not show any anomaly in the mobility because the transition was only approached down to $|\epsilon| \simeq 5 \times 10^{-3}$, which is not close enough to test the scaling laws, as can be seen in Fig. 17.1.

Only the effective hydrodynamic radius of negative ions, defined by means of the Stokes formula for perfect slip $R_- = e/4\pi\eta\mu_-$, shows a very weak singularity, if any, for $T = T_\lambda$, while the hydrodynamic radius of the positive ions appears to be regular at T_λ, though it has a sharp maximum at $\approx 40\,\mathrm{mK}$ below T_λ, as shown in Fig. 17.2 (Ahlers and Gamota, 1972).

The Stokes formula is applicable in the hydrodynamic regime, in which the correlation length ξ of fluctuations in the order parameter is smaller than the ion radius. In a range of approximately $\pm 0.2\,\mathrm{K}$ around T_λ, $\xi \geqslant R$ and the Stokes formula should not work. Thus, any deviations from the Stokes formula should be a function of ξ/R and the (questionable) singularity of the ion radius should be caused by the divergence of ξ for $T = T_\lambda$. However, the accuracy of these measurements does not allow us to draw any definitive conclusions (Ahlers and Gamota, 1972).

Owing to the weakness of the mobility singularity at the λ-transition, if any, and to the rather low accuracy of conventional techniques for measuring ion mobilities, a differential technique capable of detecting mobility changes of the order of some parts in 10^4 has been devised (Goodstein *et al.*, 1974).

The apparatus consists of two identical diode-type cells of the type exploited

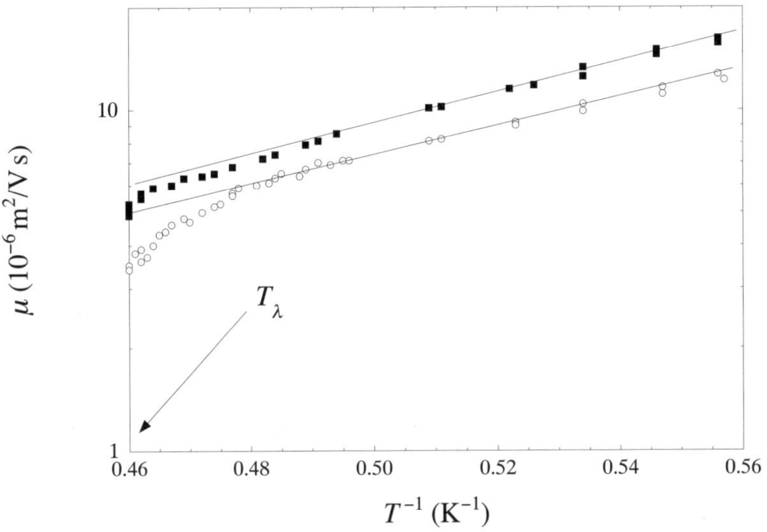

FIG. 17.1. μ vs T in He II below T_λ. Squares: positive ions. Circles: negative ions. (Sitton and Moss, 1971.)

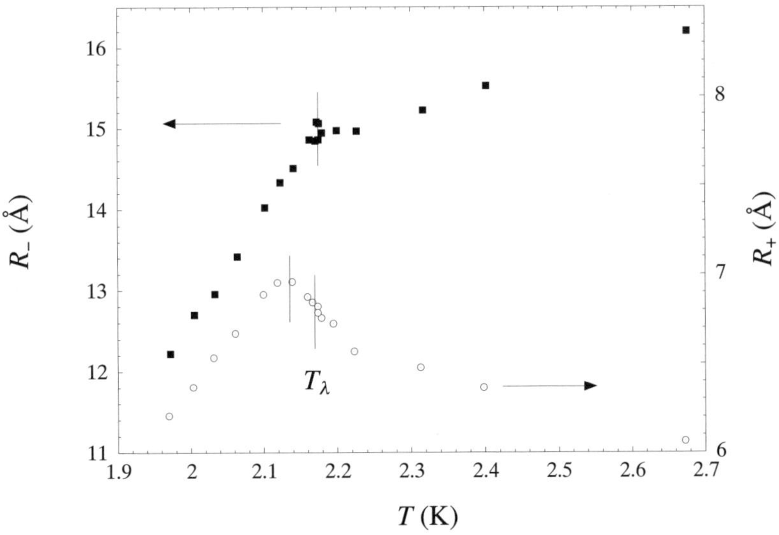

FIG. 17.2. Effective hydrodynamic radii R_+ (open symbols) and R_- (closed symbols) vs T. (Ahlers and Gamota, 1972.) Lines: eyeguides.

in traditional time-of-flight apparatuses for measuring the ion drift mobility (Meyer and Reif, 1958; Careri *et al.*, 1961). The two cells are immersed in the same helium bath and are in thermal contact with each other, but can be pressurized independently. The temperature of the helium bath is kept constant to within 10^{-5} K and the pressure to within $\delta P/P \approx 2 \times 10^{-5}$.

The technique exploited to measure the ion mobility is based on the *full-spacecharge* (FSC) method (Scaramuzzi, 1963). In a diode-type cell with planar electrodes, one of them is coated with an intense alpha-particle source that creates a thin ionized layer within the liquid. The outer surface of this plasma acts as an ion source. Ions of the desired polarity are extracted by applying a suitable potential difference between the electrodes.

If FSC conditions apply, the ion current from the ion source to the collector completely shields the source and the electric field there is zero. The Poisson equation is integrated easily, yielding the following relation between the ion current I and the applied voltage V:

$$I = \alpha \mu V^2 \,. \tag{17.1}$$

The proportionality constant $\alpha = (9/4)e\epsilon_0\epsilon_r(S/d^3)$ depends on the geometry of the diode cell through the electrode distance d and through its cross-sectional area S. ϵ_r is the dielectric constant of the liquid, e is the electron charge, and ϵ_0 is the vacuum permittivity. A plot of I versus V^2 is a straight line whose slope gives the mobility μ, provided that α is known. An example of such a plot is provided in Fig. 17.3.

This method can be used in this simple way to measure ion mobilities with an overall accuracy of approximately 10% and with a sensitivity to changes of 1%.

In order to increase the sensitivity by a factor of 100, the cell is operated at a constant current I_0. The electrometer connected to the collector is used as a zero meter by feeding into it a signal that nulls I_0. If the mobility changes in response to changes in the thermodynamic conditions of the He sample, the voltage applied to the electrodes is adjusted so as to maintain I_0 constant.

High-sensitivity relative-mobility measurements can be obtained if one removes the proportionality constant α by choosing a suitable reference point in the (P,T) plane where the mobility is μ_0, giving a voltage V_0. Thus, at any other point nearby,

$$\frac{\mu}{\mu_0} - 1 = \left(\frac{V_0}{V}\right)^2 - 1 \,. \tag{17.2}$$

If a point on the λ-transition is chosen as the reference, the quantity $(\mu - \mu_\lambda)/\mu_\lambda$ is measured.

Several measuring methods can be adopted with this experimental set-up. In particular, the two identical cells can be used as a *mobility bridge*. A reference point in the (P,T) plane is chosen and one of the two cells (reference cell) is kept fixed at that point at constant T and P. The pressure in the other cell (measuring cell) is changed isothermally. The collectors of the two cells are connected to a

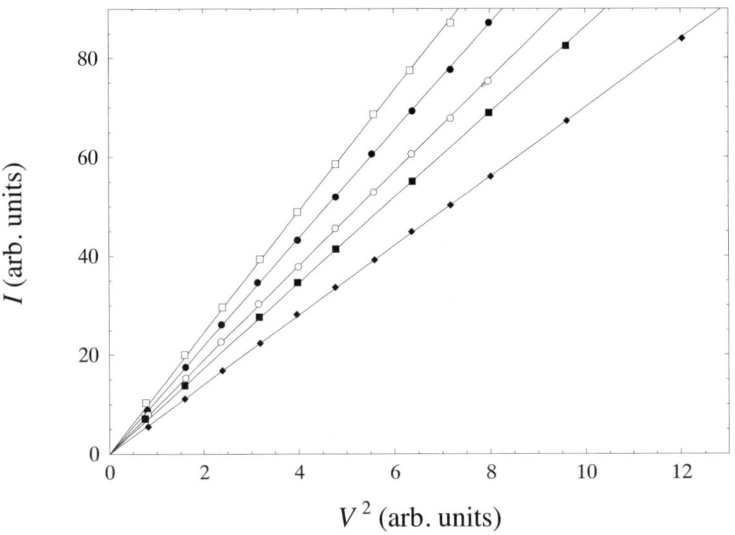

FIG. 17.3. I–V^2 characteristics of a diode in FSC operating mode in He at SVP below and close to T_λ. $T\,(\mathrm{K})$ = 2.003, 2.048, 2.095, 2.124, and 2.172 (from left). (Goodstein *et al.*, 1974.) $\mu \propto \mathrm{d}I/\mathrm{d}V^2$.

differential electrometer, which is zeroed at the reference point. Any changes in the mobility due to pressure changes in the measuring cell imbalance the electrometer output, which can be zeroed again by adjusting the voltage applied to the measuring cell. The other relevant methods are fully described in the original paper (Goodstein *et al.*, 1974). With these methods, relative-mobility changes as small as $|\Delta\mu/\mu| \approx 10^{-4}$ can be detected.

Generally speaking, it is important to know along which thermodynamic path crossing the λ-line the measurements are performed. A very schematic P–T diagram of liquid He is presented in Fig. 17.4.

Typical results for the mobility of positive ions measured along isobars and isotherms are shown in Figs 17.5 and 17.6, respectively. The mobility, either as a function of pressure along an isotherm or as a function of temperature along an isobar, decreases rapidly upon approaching the λ-transition from the superfluid side and crosses the transition without any apparent discontinuity. In the normal fluid phase, the mobility changes much less dramatically.

Taking into account the slope of the λ-line in the thermodynamic region investigated (Kierstead, 1967*b*), the behavior of the mobility as a function of the distance from the λ-transition is nearly the same, i.e., independent of the thermodynamic path chosen, either an isotherm or an isobar (Goodstein *et al.*, 1974).

In Fig. 17.7 the mobility of positive ions along the λ-line is presented. Most measurements are performed along isotherms. In Fig. 17.8 an expanded view of

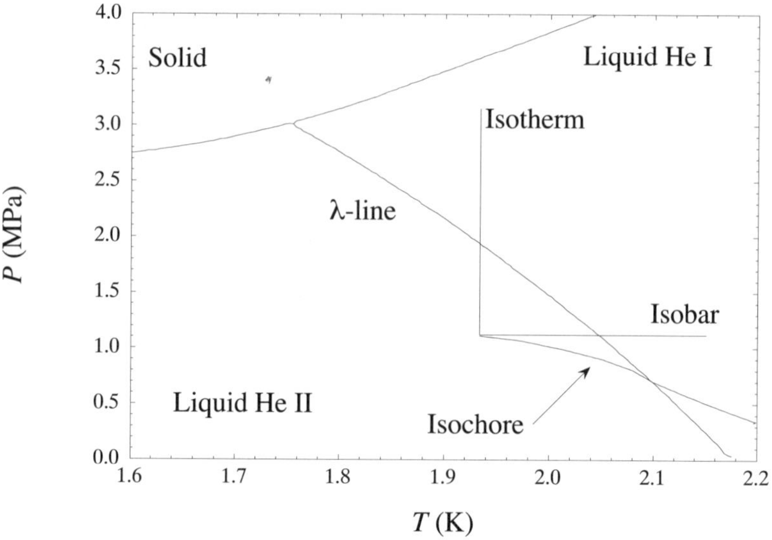

FIG. 17.4. Schematic phase diagram of He showing possible thermodynamic paths crossing the λ-line. (Goodstein *et al.*, 1974.)

the mobility along an isotherm approaching the transition from the superfluid side is reported in order to give indications on the accuracy with which the transition itself is located.

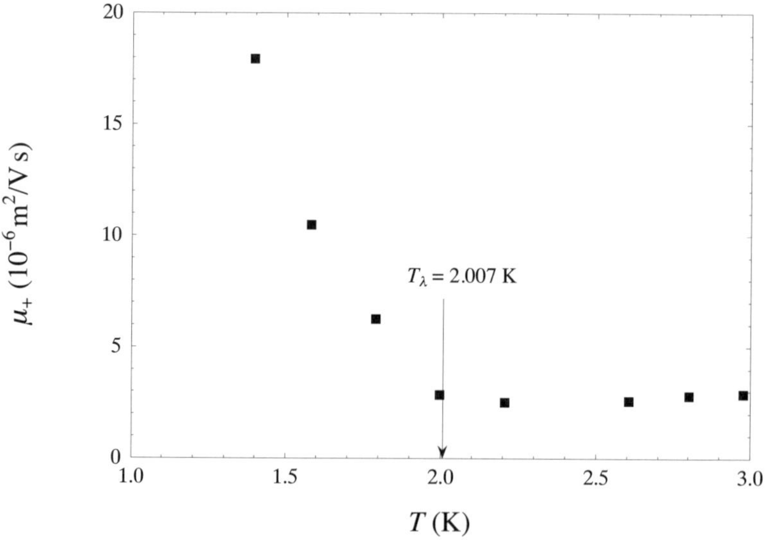

FIG. 17.5. μ_+ vs T along the isobar for $P = 1.438$ MPa. (Goodstein *et al.*, 1974.)

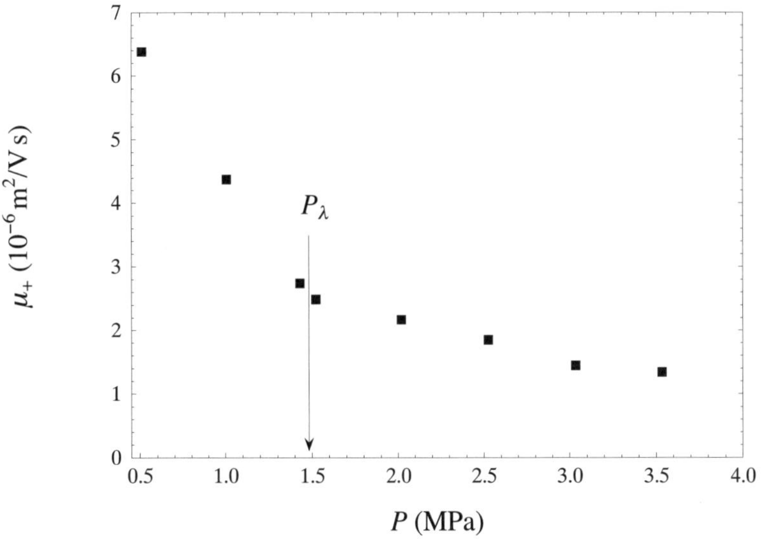

FIG. 17.6. μ_+ vs P along the isotherm for $T = 2.001\,$K, crossing the λ-line at $P_\lambda = 1.480\,$MPa. (Goodstein *et al.*, 1974.)

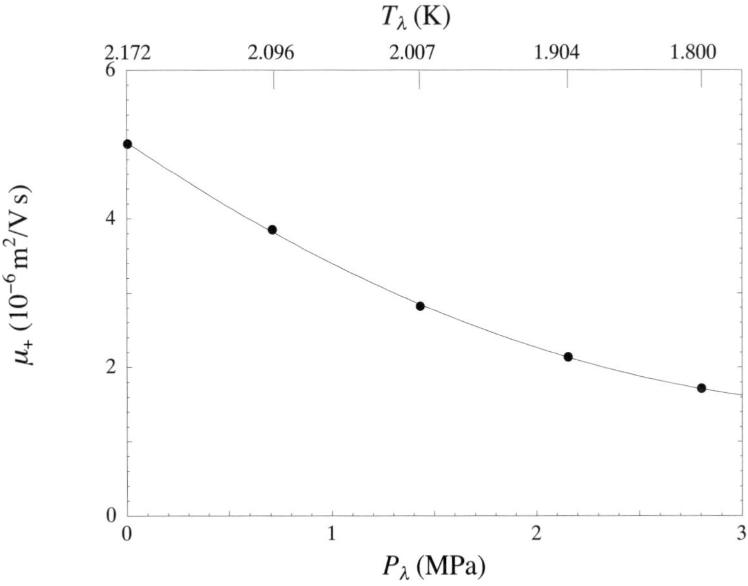

FIG. 17.7. μ_+ at the λ-line vs P_λ (lower scale) and T_λ (upper scale). (Goodstein *et al.*, 1974.) Line: eyeguide.

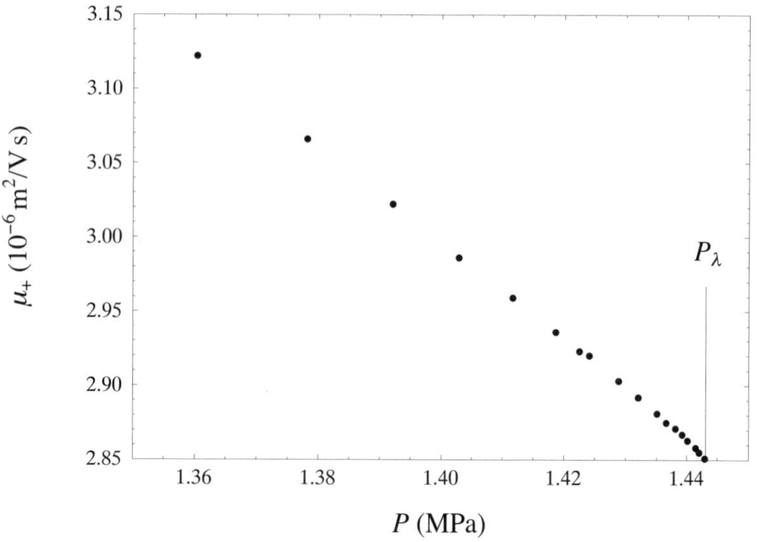

FIG. 17.8. μ_+ vs P along the $T - 2.006$ K isotherm, crossing the λ-line at $P_\lambda = 1.443$ MPa. (Goodstein *et al.*, 1974.)

The mobility is continuous at the transition with a slope whose magnitude increases upon approaching it. The slopes of the mobility along different paths approaching the λ-transition are related by the simple thermodynamic relationship

$$\left(\frac{\partial \mu}{\partial P}\right)_T = \left[\left(\frac{\partial \mu}{\partial T}\right)_V - \left(\frac{\partial \mu}{\partial T}\right)_P\right] \Big/ \left(\frac{\partial P}{\partial T}\right)_V . \tag{17.3}$$

As $(\partial P/\partial T)_V$ is finite at the λ-line (Kierstead, 1967*a*), any infinity detected in the slope $(\partial \mu/\partial P)_T$ along an isotherm would indicate a singularity in the mobility at the λ-line.

In Fig. 17.9 four sets of $(\Delta\mu_+/\mu_\lambda)_T = [\mu_+ - \mu_\lambda(T)]/\mu_\lambda(T)$ data recorded for positive ions on isotherms are shown as a function of $(\Delta P/P)_T = [P - P_\lambda(T)]/P_\lambda(T)$. Here $\mu_\lambda(T)$ and $P_\lambda(T)$ are the mobility and pressure on the λ-line at the temperature of the isotherm at hand.

The data plotted in log–log form are linear over nearly three orders of magnitude in $(\Delta P/P)_T$ and their dependence can be cast in the form

$$\left(\frac{\Delta\mu_+}{\mu_\lambda}\right)_T = A\left(-\frac{\Delta P}{P}\right)_T^{x'} , \tag{17.4}$$

where x' is the critical exponent. It is experimentally observed that $x' < 1$.

The slope is given by

$$\left(\frac{\partial \mu_+}{\partial P}\right)_T \propto \left(-\frac{\Delta P}{P}\right)_T^{x'-1} . \tag{17.5}$$

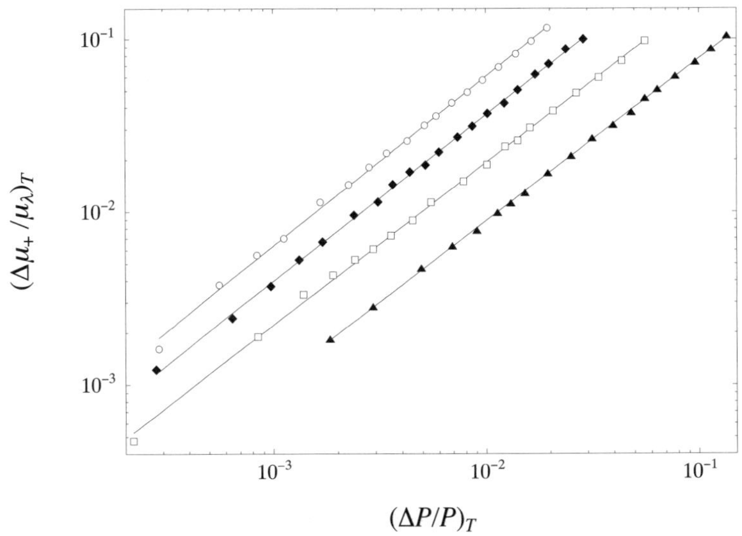

FIG. 17.9. $\Delta\mu_+/\mu_\lambda$ vs $\Delta P/P$ approaching the λ-transition along isotherms. (Goodstein *et al.*, 1974.) $T\,(\mathrm{K}) = 1.804$ (circles), 1.904 (diamonds), 2.006 (squares), and 2.094 (triangles).

As $x' < 1$, $(\partial\mu_+/\partial P)_T \to \infty$ as $\Delta P \to 0^-$, showing that the mobility has infinite slope at the λ-transition.

Negative ions behave in the same way and this behavior is followed even if the λ-line is approached by varying the temperature T along the saturated vapor pressure curve (Goodstein *et al.*, 1974).

It is customary (Stanley, 1971; Ahlers, 1976) to plot the data as a function of the distance, ϵ, in temperature from the transition. In the present case, ϵ is defined as

$$\epsilon\,(T, P) = [T - T_\lambda(P)]\,/T_\lambda \qquad (17.6)$$

where $T_\lambda(P)$ is the transition temperature at pressure P. Close to the λ-line, ϵ is given by

$$\epsilon = -\left(T_\lambda \frac{\mathrm{d}P_\lambda}{\mathrm{d}T}\right)(P - P_\lambda)\;. \qquad (17.7)$$

On approaching the transition from the superfluid side, ϵ, $P - P_\lambda$, and $\mathrm{d}P_\lambda/\mathrm{d}T$ are negative.

A comprehensive plot of all the mobilities of positive ions measured along isotherms is presented in Fig. 17.10 as a function of the reduced temperature ϵ. Similar plots are also obtained for the mobility of negative ions, as shown in Fig. 17.11, and for the mobility of positive ions along the vapor–pressure curve. All of the data appear to follow a universal scaling law as a function of the reduced temperature with critical exponent x'.

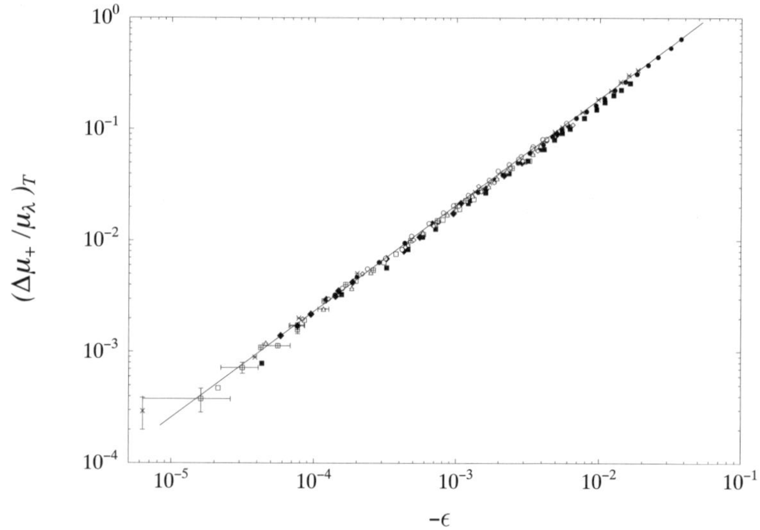

FIG. 17.10. $\Delta\mu_+/\mu_\lambda$ vs ϵ along isotherms. (Goodstein *et al.*, 1974.) T (K) = 1.804 (open circles), 1.904 (triangles), 2.094 (open diamonds), 2.006 (open and crossed squares), 1.800 (closed circles), 2.098 (closed diamonds), 1.999 (closed squares), and 1.804 (crosses).

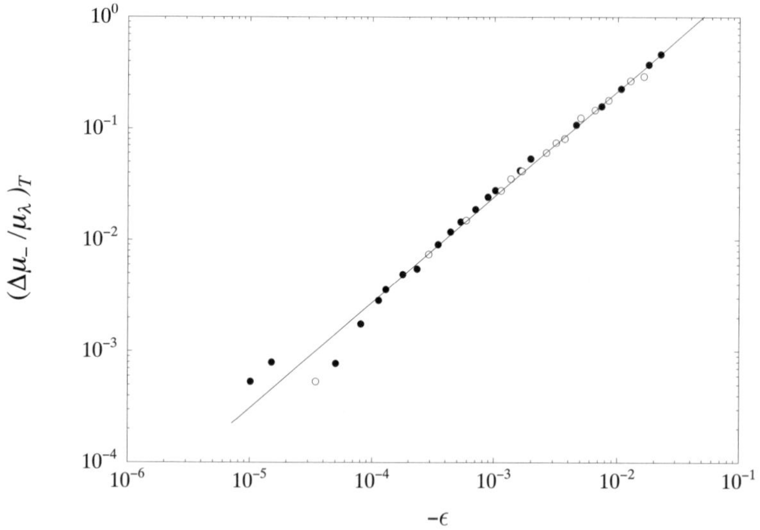

FIG. 17.11. $\Delta\mu_-/\mu_\lambda$ vs ϵ along isotherms. (Goodstein *et al.*, 1974.) T (K) = 2.000 (open circles), and 2.009 (closed circles).

If the mobility behavior along isotherms is known, the behavior along isobars in the same region close to the transition can be written, to first order in ϵ, as

$$\left(\frac{\Delta\mu}{\mu}\right)_P = \left(\frac{\Delta\mu}{\mu}\right)_T + \frac{T_\lambda}{\mu_\lambda}\frac{\mathrm{d}\mu_\lambda}{\mathrm{d}T}(-\epsilon)\,, \tag{17.8}$$

where T_λ, μ_λ, and the local derivative $\mathrm{d}\mu_\lambda/\mathrm{d}T$ may be considered as constants. In the same way, along isochores,

$$\left(\frac{\Delta\mu}{\mu}\right)_V = \left(\frac{\Delta\mu}{\mu}\right)_T + \frac{T_\lambda}{\mu_\lambda}\frac{\mathrm{d}\mu_\lambda}{\mathrm{d}T}(1-R)^{-1}(-\epsilon)\,, \tag{17.9}$$

where

$$R = T_\lambda\left(\frac{\mathrm{d}P_\lambda}{\mathrm{d}T}\right)^{-1}\epsilon\int_T^{T_\lambda}\left(\frac{\mathrm{d}P_\lambda}{\mathrm{d}T}\right)_V \mathrm{d}T\,, \tag{17.10}$$

with the integral evaluated along the isochore.

Along any paths different from an isotherm, a non-singular term proportional to ϵ is added. This term vanishes in the limit as $\epsilon \to 0$, leaving the leading behavior

$$\frac{\Delta\mu}{\mu} \propto (-\epsilon)^{x'} \tag{17.11}$$

along all kinds of path. The non-singular term should not affect the critical exponent x' measured along isotherms, although it makes the task of determining the critical exponent x' from the data difficult because the latter is very close to unity (Greywall and Ahlers, 1972).

A very careful analysis of the experimental data for the mobility of positive ions (Goodstein *et al.*, 1974) leads to the conclusion that the critical exponent x' has the value

$$x' = 0.940 \pm 0.021\,. \tag{17.12}$$

It can thus be concluded that the ion mobilities have infinite slope at the λ-transition if this is approached from the superfluid side.

Typically, the measurements of the mobility near the λ-transition have been interpreted in terms of the Stokes law (Ahlers and Gamota, 1972). On the basis of this assumption, the relative-mobility changes should scale as the relative viscosity changes, provided that the ion radius remains constant at the superfluid transition:

$$\frac{\Delta\mu}{\mu_\lambda} \propto \frac{\Delta\eta}{\eta_\lambda}\,, \tag{17.13}$$

where $\Delta\eta = \eta - \eta_\lambda$.

It is found (Dash and Taylor, 1956; Hammer and Webeler, 1965; Greywall and Ahlers, 1972; Bruschi *et al.*, 1975; Bruschi and Santini, 1978; Wang *et al.*, 1990) that the viscosity is finite at the superfluid transition, but has an infinite slope, with critical exponent $z' \approx 0.82$–0.85.[3] Within the experimental accuracy it is thus difficult to assess if the exponent 0.94 of the mobility is significantly different from the value 0.85 obtained for the viscosity (Goodstein *et al.*, 1974)

17.1.1 *The granular-fluid model*

The Stokes formula for the drag on a sphere moving in a viscous fluid relates the ion mobility to the viscosity of the liquid. The difference between the critical exponents of the ion mobility, x', and that of the shear viscosity measured in traditional hydrodynamic experiments, z', thus suggests the possibility that in the asymptotic region near the superfluid transition the measured viscosity depends on the size of the measuring probe.

Thus, in order to explain the difference between x' and z', a heuristic model has been developed that takes into account the presence of the critical fluctuations (Goodstein, 1977).

This model is based on the well-known result due to Einstein (1906), according to which a fluid containing a dilute dispersion of hard spheres with a volume concentration x has its viscosity increased in proportion to their volume fraction:

$$\eta = \eta_0 \left(1 + 2.5x\right) , \tag{17.14}$$

where η_0 is the viscosity of the pure liquid. This result is valid for not too high concentrations (Borghesani, 1985) and can be generalized to show that the viscosity of a fluid composed of a (fairly) dilute dispersion of one (immiscible) fluid in another is approximately given by the volume average of the viscosities of the two components.

For the He case in the asymptotic region close to the superfluid transition, the two components are naturally assumed to be the normal fluid and the superfluid. Critical fluctuations, occurring at any length scale, produce the appearance of *droplets* of normal fluid within the superfluid phase for $T < T_\lambda$ and close to it. The reverse happens upon approaching the transition from the normal fluid side.

The fluctuations are correlated over increasingly large distances as the transition is approached. The correlation length ξ is a measure of the longest distance over which fluctuations are correlated, and diverges as

$$\xi = \xi_0 \epsilon^{-\nu} , \tag{17.15}$$

where $\epsilon = |T_\lambda - T|/T_\lambda$. The critical exponent is $\nu \approx 2/3$ (Sengers and Levelt-Sengers, 1986).

The largest droplet has size $\sim \xi$. Within each droplet of normal fluid there is no quantum phase coherence. The existence of smaller and smaller fluctuations means that there are smaller patches of normal fluid and that within the normal regions there are smaller and smaller inclusions of superfluid, and so on, down to the dimension of an individual roton (Goodstein, 1977).

[3] Another experiment (Biskeborn and Guernsey, 1975), however, has given $z' \simeq 0.65$, which is very close to the critical exponent $\nu \simeq 2/3$ of the correlation length of critical fluctuations.

From this point of view, T_λ can be considered as the temperature, coming from below, at which the superfluid phase loses its spatial connectivity. For $T > T_\lambda$ the situation is reversed and there are inclusions of superfluid in the normal fluid.

For $T < T_\lambda$ an experiment with a characteristic length $L \gg \xi$ will not be influenced by the presence of *islands* of normal fluid with characteristic length ξ and will give results according to the homogeneous two-fluid model.

17.1.1.1 *Features of the model* The heuristic model thus developed (Goodstein, 1977) treats the fluid close to the λ-transition as a *granular* fluid, composed of islands of superfluid with inclusions of normal fluid. A complete solution of the hydrodynamic Navier–Stokes equations for such an inhomogeneous fluid is probably impossible, but an approximation is sought that gives the leading-order contributions to the singular parts of the mobility and viscosity in terms of ξ, i.e., of the size of the largest critical fluctuation.

It is assumed that, below T_λ, regions of normal fluid of size ξ are embedded in a background of connected superfluid, the reverse being observed for $T > T_\lambda$. Within each region of fluid, either normal fluid or superfluid, the fluid is homogeneous and endowed with finite viscosity even at the transition. The viscosity of the superfluid regions is assumed to be finite, too, because of the average effect of the enclosed sub-regions of normal fluid.

17.1.1.2 *Hydrodynamic determination of the effective viscosity* Let η_i and x_i be the viscosity and volume fraction of the two kinds of region, respectively. The effective large-scale viscosity of an inhomogeneous fluid, of background viscosity η_0, including regions of another fluid for a total volume fraction x_1 with viscosity η_1, is obtained by considering the Navier–Stokes equations for an incompressible fluid of uniform density in steady-state, laminar motion (Einstein, 1906; Landau and Lifsits, 2000):

$$\boldsymbol{\nabla} \cdot \mathbf{v} = 0\,, \tag{17.16}$$

$$-\boldsymbol{\nabla} P + \boldsymbol{\nabla} \cdot \left(\eta \overleftrightarrow{E}\right) = 0\,, \tag{17.17}$$

where η is the viscosity, P is the pressure, and $\overleftrightarrow{E}$ is a second-rank tensor with components

$$E_{ik} = \frac{\partial v_i}{\partial x_k} + \frac{\partial v_k}{\partial x_i}\,. \tag{17.18}$$

x_i are the Cartesian coordinates and v_i the corresponding velocity components.

In the case under investigation, in which η is non-uniform on a small scale, eqns (17.16) and (17.17) should be recovered on a larger scale by substituting P, η, and $\overleftrightarrow{E}$ with their volume averages, as, for instance,

$$\langle E_{ik} \rangle = \frac{1}{V} \int\limits_V E_{ik} \mathrm{d}V\,. \tag{17.19}$$

The integration volume V must be much larger than the scale of inhomogeneity.

The stress tensor, after averaging over a sufficiently large volume, becomes

$$\langle \sigma_{ik} \rangle = -\langle P \rangle \delta_{ik} + \eta_{\text{eff}} \langle E_{ik} \rangle \,, \tag{17.20}$$

where δ_{ik} is the Krőnecker symbol.

The effective viscosity η_{eff} is defined by eqn (17.20), and eqn (17.17) can be written, upon adopting the summation convention for repeated indices, as

$$\frac{\partial \langle \sigma_{ik} \rangle}{\partial x_k} = 0 \,. \tag{17.21}$$

The average stress tensor can be written as

$$\langle \sigma_{ik} \rangle = -\langle P \rangle \delta_{ik} + \eta_0 \langle E_{ik} \rangle + \frac{1}{V} \int_V (\sigma_{ik} - \eta_0 E_{ik} + P \delta_{ik}) \, \mathrm{d}V \,. \tag{17.22}$$

The integrand in eqn (17.22) vanishes in the homogeneous background fluid and thus gives a nonzero contribution stemming only from the small fraction of fluid with viscosity η_1.

Suppose that the velocity field can be written as

$$v_l = \alpha_{lk} x_k \,, \tag{17.23}$$

where $\overleftrightarrow{\alpha}$ is a constant symmetric tensor. The incompressibility of the fluid means that $\alpha_{ii} = 0$, as can be verified by direct substitution of eqn (17.23) into eqn (17.16). If eqn (17.23) is substituted into eqn (17.17), the uniform pressure P_0 is obtained in any region of uniform viscosity. The flow described by eqn (17.23) is thus considered as the unperturbed flow field.

If the regions of different viscosity do not appreciably perturb the flow, eqn (17.22) yields

$$\langle \sigma_{ik} \rangle = -\langle P \rangle \delta_{ik} + \eta_{\text{eff}} \langle E_{ik} \rangle = -P_0 \delta_{ik} + 2\eta_{\text{eff}} \alpha_{ik} \,, \tag{17.24}$$

with

$$\eta_{\text{eff}} = \eta_0 + (\eta_1 - \eta_0)\, x_1 = \eta_0 x_0 + \eta_1 x_1 \,, \tag{17.25}$$

where $x_0 = 1 - x_1$ and x_1 are the volume fractions of the regions with viscosities η_0 and η_1, respectively. Thus, to first order, the excess viscosity $\eta_{\text{eff}} - \eta_0$ is proportional to the volume fraction of the inhomogeneities.

In order to prove that the perturbation of the fluid flow induced by the inhomogeneities does not alter the previous result, consider that, if $x_1 < x_0$, the regions of viscosity η_1, embedded in regions of viscosity η_0, can be assumed to act independently of each other. The overall effect will thus be obtained by summing together the effects of each single region.

Consider a region of viscosity η_1 centered about the origin and surrounded to infinity by the region of viscosity η_0. The inner region can be assumed to be spherical without loss of generality. In the inner region the velocity field u_l is given by

$$u_l = v_l + u_{1l}\,, \tag{17.26}$$

while in the outer region it is

$$u_l = v_l + u_{0l}\,. \tag{17.27}$$

Moreover, both u_{0l} and u_{1l} must satisfy eqns (17.16)–(17.18) in their own regions and depend parametrically on the tensor $\overleftrightarrow{\alpha}$. The boundary conditions are that u_{0l} must vanish at infinity and that u_{1l} must remain finite at the origin.

The solutions for u_{0l} and u_{1l} are given by

$$u_{0l} = \left(\frac{3a}{r^2} - \frac{15b}{r^4}\right)\alpha_{ik}n_i n_k n_l + \frac{6b}{r^4}\alpha_{lk}n_k \tag{17.28}$$

$$P = P_0 + \eta_0\frac{6a}{r^3}\alpha_{ik}n_i n_k\,, \tag{17.29}$$

in the outer region and

$$u_{1l} = 48dr^3\alpha_{ik}n_i n_k n_l - \left(24cr + 120dr^3\right)\alpha_{lk}n_k\,, \tag{17.30}$$

$$P = P_0 - 504\eta_1 dr^2\alpha_{ik}n_i n_k \tag{17.31}$$

in the inner one (Goodstein, 1977). r is the distance from the origin, n_j, $j = 1, 2, 3$, are the components of the unit vector along $\mathbf{r}$, and a, b, c, and d are constants to be determined by imposing the boundary conditions at the interface between the two regions.

If the regions were of irregular shape, the interface might not be stationary. The only constraint would then be that the inner region preserves its volume. Nonetheless, the shape of the regions is irrelevant. In fact, by using the equation of motion $\partial\sigma_{ik}/\partial x_k = 0$ and the identity $\sigma_{ik} = \partial(\sigma_{il}x_k/\partial x_l)$, the integral in eqn (17.20) can be changed into a surface integral to be evaluated at a distance that is large in comparison to the size of the inner region.

By carrying out the integration, the only surviving term is the one containing ar^{-2} in eqn (17.28), yielding a correction to $\langle\sigma_{ik}\rangle$ proportional to a/V. Here a is a constant with the dimensions of volume and the only volume in the problem is that of the inner region. Thus, the effect of each region of viscosity η_1 is proportional to its relative volume and the total effect is simply obtained by weighting the effect of one region by its volume fraction x_1, yielding $\eta_{\text{eff}} - \eta_0 \propto x_1$.

It is possible to explicitly calculate the constants a, b, c, and d for a spherical inner region of radius R. The boundary conditions at $r = R$ are that u_l and $n_l\sigma_{lk}$ must be continuous in order to allow flow across the boundary. The constants are then

$$a = -\left[\frac{5\,(q-1)}{3\,(3+2q)}\right] R^3 \,, \tag{17.32}$$

$$b = -\left[\frac{q-1}{3\,(3+2q)}\right] R^5 \,, \tag{17.33}$$

$$c = \frac{q-1}{12\,(3+2q)} \,, \tag{17.34}$$

$$d = 0 \,, \tag{17.35}$$

where $q = \eta_1/\eta_0$. If the integral in eqn (17.20) is carried out, one obtains, as expected, eqn (17.24) with

$$\eta_{\text{eff}} = \eta_0 + \left[\frac{5}{3+2q}\right] (\eta_1 - \eta_0)\, x_1 \,. \tag{17.36}$$

This result is the same as eqn (17.25), except for an unimportant prefactor of order unity. As a check of internal consistency, it is to be noted that, if $q = \eta_1/\eta_0 \to \infty$, then Einstein's result for a dispersion of hard spheres, eqn (17.14), is recovered.

17.1.1.3 *Specific features of the model near the λ-transition* Upon approaching the transition from below, the growth of normal regions increasingly destroys the connectivity of the superfluid background. Regions of superfluid are included within regions of normal fluid of size $\simeq \xi$. These regions of included superfluid do not contribute to the volume of connected superfluid and are counted as part of the volume fraction of normal fluid. The remaining connected background participates in the large-scale superflow and has volume fraction $x_s \propto \rho_s$, the superfluid fraction.

For $T > T_\lambda$, the included superfluid regions have become the largest fluctuations that now include regions of normal fluid but they do not belong to the normal fluid background. The volume fraction of these isolated superfluid regions above T_λ is f. For this reason the viscosity of the normal fluid regions below T_λ and the viscosity of the normal fluid background above T_λ are in principle different.

The effective viscosity η can thus be written in such a way as to be continuous at the transition:

$$\eta = \eta_\lambda - x_s\,(\eta_\lambda - \eta_s) \qquad (T < T_\lambda) \,, \tag{17.37}$$

$$\eta = \eta_\lambda + (f_\lambda - f)\,(\eta_n - \eta_s) \qquad (T > T_\lambda) \,, \tag{17.38}$$

where η_s is the viscosity of the superfluid regions, η_n is the viscosity of the normal background fluid for $T > T_\lambda$, η_λ is the viscosity value at the transition and is also its value in the normal regions below T_λ, and f_λ is the value of f at the transition.

If one assumes that $x_s = \beta\rho_s/\rho$ and that β is finite but nonzero at the transition, though it may depend on T, the viscosity below T_λ can be written as

$$\frac{\eta_\lambda - \eta}{\eta_\lambda} = \beta\left(\frac{\eta_\lambda - \eta_s}{\eta_\lambda}\right)\frac{\rho_s}{\rho} . \tag{17.39}$$

17.1.1.4 *Prediction of the model* The mobility of positive ions is calculated from the drag on a sphere in a viscous medium. However, the ion charge produces electrostriction in the nearby fluid, setting its own characteristic scales.

The electrostriction model (Atkins, 1959) predicts that the local pressure P_r around the ion exceeds the applied pressure P according to

$$P_r - P = \frac{C_0}{r^4} , \tag{17.40}$$

where r is the distance from the positive ion and C_0 depends on the dielectric constant of the liquid.

Consider a situation in which the experimental conditions correspond to the point "1" in the P–T plane (Fig. 17.12). At some distance R_s the local pressure

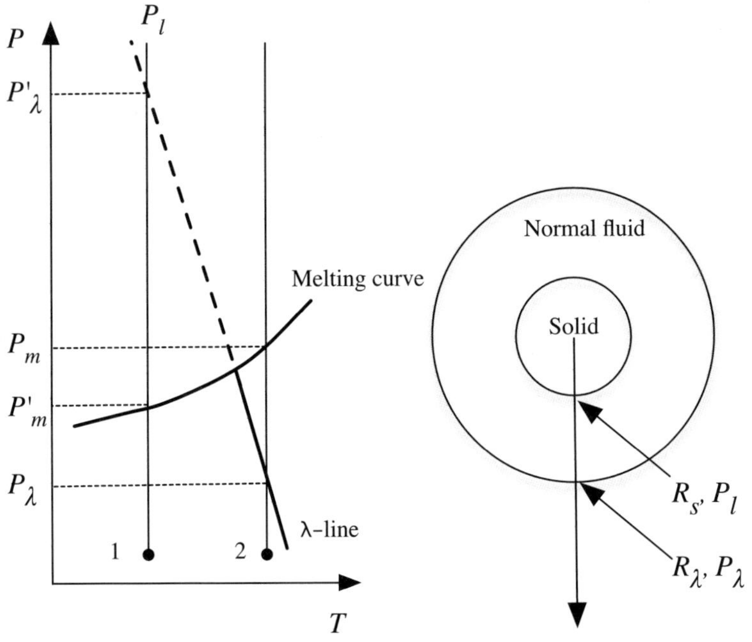

FIG. 17.12. Schematic picture showing the effect of electrostriction due to a positive ion in ^{4}He below the λ-transition. Left panel: (P, T) thermodynamic plane with the extrapolation of the λ-line into the supercooled region. Labels "1" and "2" correspond to different experimental conditions. Right panel: sketch of the ion structure [after Scaramuzzi *et al.* (1977*b*)].

is so high as to induce freezing of the liquid around the ion, that is thus a solid sphere of radius $R_s \approx 6 - 7\,\text{Å}$.

Below T_λ there is another characteristic length:

$$R_\lambda = \left(\frac{C_0}{P_\lambda(T) - P} \right)^{1/4} = C\epsilon^{-1/4}\,, \tag{17.41}$$

where $P_\lambda(T)$ is the pressure of the λ-transition at the temperature of the experiment and C relates $P_\lambda - P$ to $T_\lambda - T$ by means of the slope of the λ-line.

For $r > R_\lambda$ the fluid around the ion is the usual superfluid. For $R_s < r < R_\lambda$ the local pressure, for the given bath temperature, corresponds to that of bulk helium above the λ-transition. Eventually, for $r < R_s$ a crust of solid helium encompasses the ion. A sketch of the thermodynamic situation of the fluid around the ion is shown in Fig. 17.12.

Though R_λ obviously grows indefinitely as the transition is approached, it is always R_λ less than ξ. Equations (17.15) and (17.41) can be combined together so as to give

$$\frac{R_\lambda}{\xi} \propto \epsilon^{0.42}\,. \tag{17.42}$$

Thus, for $\epsilon = 10^{-5}$, $R_\lambda/\xi \approx 8 \times 10^{-3}$.

The central point of the present model is that it concentrates on the longest-range fluctuations of the order parameter present at the critical points without neglecting the shorter-range variations induced by electrostriction. Other models (Sobyanin, 1973) fail to reproduce the singular behavior of the mobility just because they explicitly rule out the possibility that features at a length scale shorter than ξ might influence the hydrodynamic ion transport.

If the ion is already in the normal fluid region, or just below T_λ, normal fluid completely surrounds the snowball. The ion is thus subjected to the effects of the normal viscosity η_n and its mobility is given by the Stokes formula:

$$\mu_n = \frac{e}{6\pi\eta_n R_s}\,. \tag{17.43}$$

In contrast, if the ion is within a region of connected superfluid, the hydrodynamic problem is much more complicated because of the presence of the shell of thickness $R_\lambda - R_s$ of normal fluid with viscosity η_n surrounding the sphere of solid He. Moreover, the fluid beyond R_λ has the macroscopic average η because it is the superfluid with the inclusions of normal fluid regions.

Again, the Navier–Stokes equations have to be solved in order to calculate the drag on a sphere of radius R_s, moving with velocity $\mathbf{v}$ in a fluid with viscosity $\eta = \eta_1$ for $R_s < r < R_\lambda$ and $\eta = \eta_0$ for $r > R_\lambda$. In this problem, flow is allowed across the interface.

For steady, laminar, incompressible flow, eqns (17.16)–(17.18) are still valid. Usually, a frame of reference centered on the center of the sphere is chosen in order that the fluid velocity at infinity is $\mathbf{U}$ along the z-direction.

Solutions are sought of the type

$$\mathbf{v} = \hat{\mathbf{r}} \cos\theta v_a(r) + \hat{\boldsymbol{\theta}} v_b(r)\,, \tag{17.44}$$

$$P = p(r)\cos\theta \tag{17.45}$$

in terms of the polar coordinates r and θ. $\hat{\mathbf{r}}$ and $\hat{\boldsymbol{\theta}}$ are the respective unit vectors. There is symmetry about the azimuthal angle ϕ.

According to Landau and Lifsits (2000), the components of the tensor E_{ik} are

$$E_{rr} = S_1(r)\cos\theta\,, \tag{17.46}$$

$$E_{\theta\theta} = E_{\phi\phi} = -\frac{1}{2}E_{rr}\,, \tag{17.47}$$

$$E_{r\theta} = S_2(r)\sin\theta\,, \tag{17.48}$$

with

$$S_1(r) = 2\frac{\partial v_a}{\partial r}\,, \tag{17.49}$$

$$S_2(r) = \frac{\partial v_b}{\partial r} - \frac{v_a + v_b}{r}\,. \tag{17.50}$$

The continuity equation (17.16) becomes

$$\frac{\partial v_a}{\partial r} + 2\frac{v_a + v_b}{r} = 0\,. \tag{17.51}$$

The radial and polar components of eqn (17.16) reduce, respectively, to

$$-\frac{\partial p}{\partial r} + \frac{\partial(\eta S_1)}{\partial r} + \frac{\eta(3S_1 + 2S_2)}{r} = 0\,, \tag{17.52}$$

$$\frac{p}{r} + \frac{\partial(\eta S_2)}{\partial r} + \frac{\eta}{r}\left(3S_2 + \frac{1}{2}S_1\right) = 0\,. \tag{17.53}$$

For $r > R_\lambda$, the solutions for v_a and v_b that satisfy the boundary conditions at infinity, $v_a = U$ and $v_b = -U$, are

$$v_a = U\left(1 - 2\frac{a}{r} + \frac{2b}{r^3}\right), \tag{17.54}$$

$$v_b = U\left(-1 + \frac{a}{r} + \frac{b}{r^3}\right), \tag{17.55}$$

yielding

$$p = -2\eta_0\left(\frac{a}{r^2}\right)U\,, \tag{17.56}$$

$$S_1 = \left(\frac{4a}{r^2} - \frac{12b}{r^4}\right)U\,, \tag{17.57}$$

$$S_2 = -\left(\frac{6b}{r^4}\right)U\,. \tag{17.58}$$

The solutions for the inner region $r < R_\lambda$ are given by

$$v_a = \left(\Gamma - 2\frac{A}{r} + 2\frac{B}{r^3} + Cr^2\right)U\,, \tag{17.59}$$

$$v_b = \left(-\Gamma + \frac{A}{r} + \frac{B}{r^3} - 2Cr^2\right)U\,, \tag{17.60}$$

and yield

$$p = -2\eta_1\left(\frac{A}{r^2} + 10Cr\right)U\,, \tag{17.61}$$

$$S_1 = \left(\frac{4A}{r^2} - \frac{12B}{r^4} + 4Cr\right)U\,, \tag{17.62}$$

$$S_2 = -\left(-\frac{6B}{r^4} - 3Cr\right)U\,. \tag{17.63}$$

The six unknown constants Γ, A, B, C, a, and b in eqns (17.54)–(17.63) are determined by enforcing the boundary conditions at $r = R_s$ and $r = R_\lambda$. The absence of unbalanced forces in the fluid allows us to write the drag force on the sphere as (Landau and Lifsits, 2000)

$$F = 8\pi\eta_0 aU\,, \tag{17.64}$$

and the boundary conditions can be exploited in order to eliminate the other constants in favor of a. Mass conservation implies $\Delta v_a = \Delta v_b$ for $r = R_\lambda$, where Δ means the difference between the values of a given quantity on either side of the interface.

The continuity of the radial (rr) and tangential ($r\theta$) components of the stress tensor for $r = R_\lambda$ yields the conditions $\Delta(-p + \eta S_1) = 0$ and $\Delta(\eta S_2) = 0$.

For $r = R_s$, assuming the positive ion to be a solid sphere, the no-slip conditions are applied, yielding $v_a = v_b = 0$. Setting $R = R_s/R_\lambda$ and $q = \eta_1/\eta_0$, the previous boundary conditions give

$$a = qA_+R_s\,, \tag{17.65}$$

where the index "+" indicates the positive ion. A_+ turns out to be

$$A_+^{-1} = \frac{4}{3}\frac{R+q-1}{R} - \frac{10}{3}\frac{\left(R^2-1\right)^2(1-q)}{R^5(3q+2)+2(q-1)}\,, \tag{17.66}$$

with $q = \eta_n/\eta$ and $R = R_s/R_\lambda$.

For the electron bubble, in contrast, it is customary to apply the perfect-slip conditions at the interface for $r = R_s$. This brings about the conditions $v_a = S_2 = 0$ and the constant A_+ is replaced by A_- given by

$$A_-^{-1} = 2\left\{1 + \frac{(q-1)\left[(2q+3)R^4 - 2(q-1)/R\right]}{(3q+2)R^5 - 3(q-1)}\right\}. \tag{17.67}$$

The drag coefficient ζ turns out to be

$$\zeta_\pm = 8\pi\eta_n A_\pm \tag{17.68}$$

and, as R_λ diverges at the λ-transition, it can be written, to leading order in R_s/R_λ, as

$$\zeta_\pm = M_\pm \pi\eta_n \left\{ 1 - \varphi \frac{R_s}{R_\lambda} + \mathcal{O}\left[\left(\frac{R_s}{R_\lambda} \right)^3 \right] \right\}, \tag{17.69}$$

where

$$\varphi = N_\pm \frac{(q-1)(2q+3)}{(3q+2)}. \tag{17.70}$$

The constants $M_\pm$ and $N_\pm$ are $M_+ = 6$ and $N_+ = 3/2$ for the positive ion, and $M_- = 4$ and $N_- = 1$ for the negative one.

Recalling that for a negative ion R_s is the bubble radius and that it is the radius of the snowball for the positive ion, the mobility is eventually found to be

$$\mu_s = \frac{e}{\zeta_\pm R_s}. \tag{17.71}$$

Here, μ_s is the mobility of an ion in the connected superfluid background.

In order to calculate the measured mobility, it should be noted that the ion crosses several patches of different composition during its flight across the experimental cell. Let d be the drift length and τ the time-of-flight of the ion. If $d \gg \xi$, the ion crosses a great number of normal regions, placed randomly along its path. Owing to the equivalence, on a statistical basis, of all possible paths through the liquid, the fraction of path covered by the ion in the normal fluid regions is $(1 - x_s)d$. If v_n is the mean velocity of an ion in these normal regions, the ion spends there a fraction $(1 - x_s)d/v_n$ of its drift time. An analogous argument is used for the paths in the superfluid regions.

The average drift velocity v_D can thus be immediately obtained as

$$\frac{1}{v_D} = \frac{\tau}{d} = \frac{1 - x_s}{v_n} + \frac{x_s}{v_s}, \tag{17.72}$$

where v_s is the mean velocity of an ion in a region of superfluid.

Recalling that the mobility is defined as the ratio of the drift velocity to the electric field, and defining μ_n and μ_s as the mobilities in the normal and superfluid regions, respectively, one obtains the following expression for the experimentally observed mobility μ:

$$\frac{1}{\mu} = \frac{1 - x_s}{\mu_n} + \frac{x_s}{\mu_s}. \tag{17.73}$$

μ is obtained by inverting eqn (17.73) and, to first order in x_s, is given by

$$\mu = \mu_n \left(1 + x_s \frac{\mu_s - \mu_n}{\mu_s} \right). \tag{17.74}$$

At the transition μ remains finite and its value is μ_λ. With the aid of eqns (17.43) and (17.71) and recalling that $x_s = \beta\rho_s/\rho$, eqn (17.74) can be rewritten to leading order in the singular terms as

$$\frac{\mu - \mu_\lambda}{\mu_\lambda} = \beta\left(\frac{\rho_s}{\rho}\right)\varphi\frac{R_s}{R_\lambda}\,. \tag{17.75}$$

Recalling that $\rho_s/\rho \sim \epsilon^{2/3}$ and $R_\lambda \sim \epsilon^{-1/4}$, the critical exponent in eqn (17.11) turns out to be

$$x' = \frac{2}{3} + \frac{1}{4} \approx 0.92\,,$$

in very good agreement with the experimental value 0.94 ± 0.02 (Goodstein *et al.*, 1974).

It is clear that the present model for the ion mobility must fail if the transition is approached closely enough. For T very close to T_λ, the correlation length ξ becomes so large that it is no longer possible to distinguish the behavior of ions in isolated superfluid islands and in connected superfluid patches.

It is assumed that electrostriction makes isolated superfluid islands disappear below T_λ, where their size is already smaller than ξ, while in the connected superfluid background it limits the growth of normal regions to a size R_λ. When ξ grows very large, it is possible that inclusions of superfluid are smaller than ξ, though still larger than R_λ.

For this reason, the mobility is larger than μ_n because there are regions of unsuppressed superfluid far away from the ion, at a distance larger than R_λ, but smaller than the size of the inclusion, which is small in comparison to ξ.

A detailed prediction of the temperature range in which the model is no longer applicable is not easy to do because there is no characteristic length for fluctuations smaller than ξ. A rough estimate of the temperature at which μ may depart from the behavior described by eqn (17.75) can be given by considering that the model fails when there is a significant probability that there are superfluid inclusions whose dimension D satisfies the condition $R_\lambda \ll D \ll \xi$. One can arbitrarily define that deviations from the predicted behavior appear as soon as $R_\lambda \sim 10^{-2}\xi$. This happens for $\epsilon < 10^{-5}$, at the lower extreme of the experimentally-investigated range (Goodstein *et al.*, 1974).

The main result of this model is that it gives a consistent picture of the flow dissipation near the superfluid transition in terms of the long-range fluctuations of the order parameter. The model clearly shows that the asymptotic behavior depends on the size of the measuring probe and that the λ-transition may be considered to be a transition in the connectivity of the superfluid regions.

17.2 Ion mobility near the melting transition

Attempts at settling the issue of whether the ion core is solid must necessarily rely on measurements carried out near the melting transition. In this thermodynamic situation it is easier for electrostriction to locally increase the pressure above the melting value.

If detailed pieces of information about the ion radius are to be deduced from the mobility by means of the Stokes formula, accurate viscosity data must be available for the same thermodynamic conditions of the mobility measurements.

With this aim, some researchers have carried out simultaneous measurements of mobility and viscosity in liquid He along and near the melting curve in the range $1.69\,\mathrm{K} < T < 1.95\,\mathrm{K}$ (Scaramuzzi *et al.*, 1977*a*; Cole *et al.*, 1978). Except for one preliminary work (Brody, 1975), this region was not previously investigated in detail. Mobilities are measured by the same differential spacecharge-limited technique used at the λ-transition (Goodstein *et al.*, 1974). Viscosities are measured by observing the resonance linewidth of a vibrating wire using the technique devised by Bruschi and Santini (1975) for investigating the He viscosity near the λ-transition at SVP (Bruschi *et al.*, 1975, 1977) and under pressure (Bruschi and Santini, 1978).

The vibrating wire viscometer was first introduced by Tough *et al.* (1963, 1964). A taut metallic wire in a transverse magnetic field is caused to vibrate by the passage of an electric current through it. The vibrations of the wire are damped by energy-dissipation mechanisms. Viscous friction overwhelms internal dissipation and the product $(\eta\rho)$ is measured.

In the original technique (Tough *et al.*, 1963, 1964) the wire is excited by a short current pulse and the decay time of vibrations is measured. Bruschi *et al.* (1975) greatly improved this technique by using continuous wave excitation of the wire. A small sinusoidal current of constant amplitude sets the wire into vibration. The frequency response of the wire is observed as a function of the excitation current. Lock-in techniques are exploited to measure the side frequencies. The difference between the quality factor Q *in vacuo* and in the liquid yields the contribution of the viscous damping in terms of the product $\eta\rho$. This technique has an absolute precision of 10^{-3} in $\eta\rho$ and a sensitivity of better than 10^{-4} to relative changes in $\eta\rho$. Owing to its high accuracy, this technique is very sensitive to impurities and a careful procedure to condense liquid He into the measuring cell has to be followed (Scaramuzzi *et al.*, 1977*a*).

In Fig. 17.13 the mobility of positive ions and in Fig. 17.14 the viscosity data along the melting curve are reported as a function of pressure (and temperature). The vertical line labeled "λ" in these figures indicates the pressure of the intersection between the melting curve and the λ-line. For $P \geqslant 3.3\,\mathrm{MPa}$, the viscosity is independent of T and can be fitted to a straight line as a function of P (Scaramuzzi *et al.*, 1977*a*).

The mobility of positive ions shown in Fig. 17.13 does not vanish when the pressure approaches the melting pressure, as also previously observed by other researchers (Ahlers and Gamota, 1972; Ostermeier and Schwarz, 1972). This means that the effective hydrodynamic radius R_{eff} of the self-induced solid sphere around the ion does not grow indefinitely, as deduced from the Stokes formula:

$$R_{\mathrm{eff}} = \frac{e}{6\pi\eta\mu_+}. \tag{17.76}$$

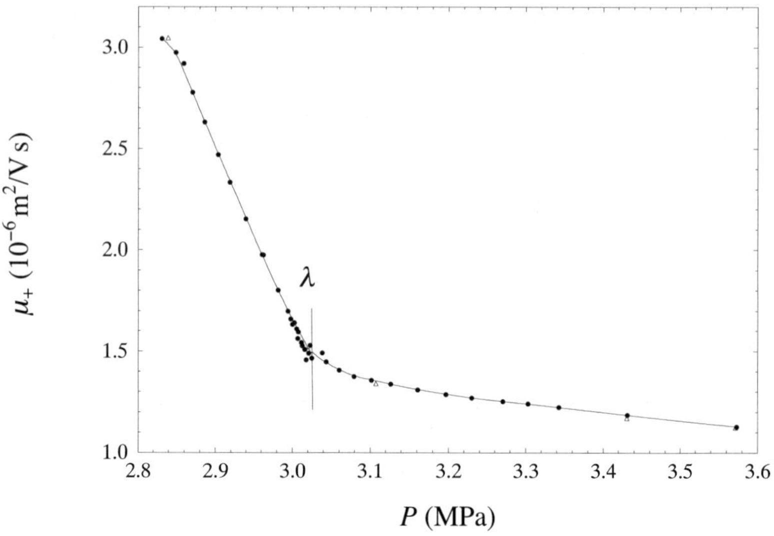

FIG. 17.13. μ_+ vs P along the melting curve (closed symbols). Triangles: end-points of isotherms. (Scaramuzzi *et al.*, 1977*a*.)

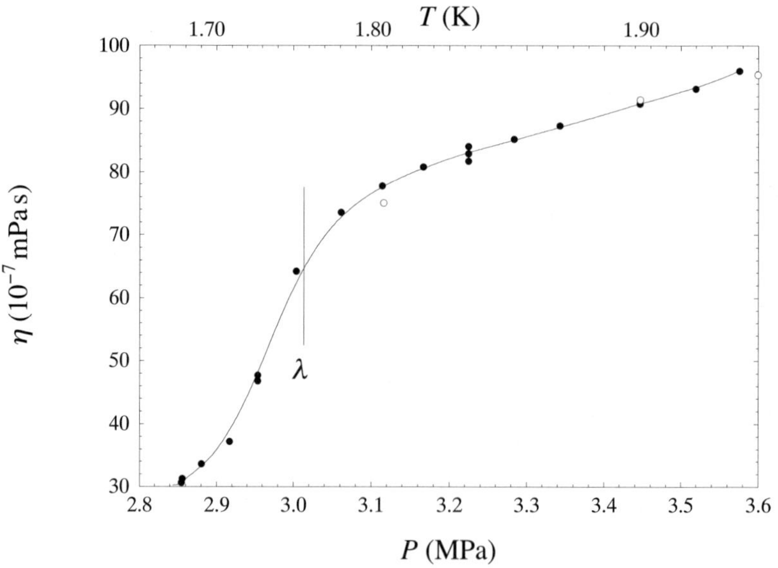

FIG. 17.14. η vs P (lower scale) and T (upper scale) along the melting curve (closed symbols). Open symbols: end-points of isotherms.(Scaramuzzi *et al.*, 1977*a*.)

The reason for such behavior is related to the existence of a solid–liquid surface tension at the interface between the solid sphere induced by electrostriction about the ion and the surrounding liquid. As a result, the local pressure at which He solidifies is higher than the bulk melting pressure P_m. The ion is therefore surrounded by a shell of supercooled, though stable, liquid between the solid surface and the distance R_m at which the local pressure equals P_m.

According to the electrostriction model of Atkins (1959), if the solid and liquid are assumed to be incompressible, the radius of the solid sphere R_s can be related to the externally-applied pressure by

$$P_m - P = \frac{\beta}{v_l R_s^4} - \frac{2\sigma_{ls}}{R_s}\frac{v_s}{v_l - v_s}, \qquad (17.77)$$

where $\beta = \alpha e^2/2(4\pi\epsilon_0\epsilon_r)^2$ depends on the He polarizability α and on the relative dielectric constant of the liquid ϵ_r. Here ϵ_0 is the vacuum permittivity, σ_{ls} is the solid–liquid surface tension, and v_s and v_l are the molar volumes of solid and liquid, respectively. R_s may not necessarily be the effective hydrodynamic radius R_{eff}.

If one assumes that $R_{\text{eff}} = R_s$, eqns (17.76) and (17.77) can be combined so as to yield

$$\Delta P \equiv P_m - P = \frac{\beta}{v_v}\left(\frac{6\pi\eta}{e}\right)^4 \mu - 2\sigma_{ls}\frac{v_s}{v_l - v_s}\left(\frac{6\pi\eta}{e}\right)\mu = a\mu^4 - b\mu, \qquad (17.78)$$

where a and b would be constants if the variation of η with P could be neglected. A plot of $\Delta P/\mu$ as a function of μ^3 would yield a straight line whose slope and intercept would provide the viscosity near melting (not previously measured) and the solid–liquid surface tension, which had not yet been measured.

However, the behavior of σ_{ls} deduced in this way appears to change too rapidly with temperature, in the very narrow range investigated, to be acceptable (Scaramuzzi *et al.*, 1977*a*). The main reason for this apparent inconsistency can be related to the validity of the assumption that the effective hydrodynamic radius of the ion and the radius of the self-induced solid sphere are the same. Actually, the work of Ostermeier and Schwarz (1972) had shown that R_{eff} and R_s are related by a multiplying factor. Moreover, the viscosity appears to remain finite at melting, whereas it is reasonable to assume that σ_{ls} remains nearly constant in the very narrow temperature range investigated (Scaramuzzi *et al.*, 1977*a*).

Thus, if viscosity and mobility are simultaneously measured along the same thermodynamic path (e.g., along isotherms), eqn (17.76) suggests that we construct the quantity

$$\frac{\mu - \mu_m}{\mu_m} - \frac{\eta_m - \eta}{\eta} \propto \mu\eta - \mu_m\eta_m \propto R_{\text{eff}}^{-1} - R_{\text{eff},m}^{-1}, \qquad (17.79)$$

where $R_{\text{eff},m}$ is the value of the effective hydrodynamic radius at melting. The resulting effective radius $R_{\text{eff},m}$ at melting is plotted in Fig. 17.15.

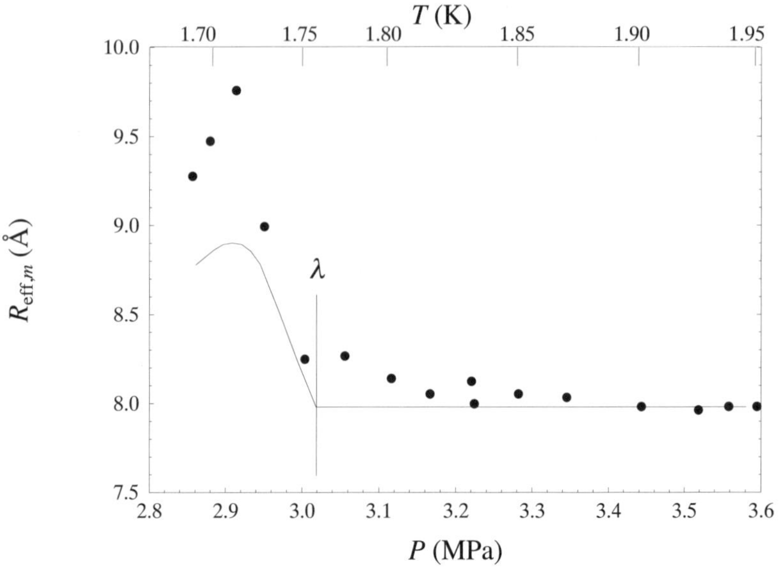

FIG. 17.15. $R_{\mathrm{eff},m}$ vs P (lower scale) and T (upper scale) for the positive ions along the melting curve. Curve: supercooled λ-transition model. (Scaramuzzi *et al.*, 1977*a*, 1977*b*).

Far above the λ-transition, R_{eff} rises smoothly as T is lowered toward the upper critical point. There it rises sharply, passes through a maximum for $T \approx 1.72$ K, then falls again. A similar behavior (see Fig. 17.2) has been observed along the vapor–pressure curve (Ahlers and Gamota, 1972), though it was unexplained then.

A model of electrostriction-induced λ-transition in the supercooled liquid has thus been developed that nicely explains the observed dependence of the effective radius as a function of pressure along the melting curve (Goodstein, 1977; Scaramuzzi *et al.*,1977*a*, 1977*b*.) This model applies anywhere along the λ-line and has been previously described in the context of the mobility at the λ-transition in Section 17.1.1.4.

Consider the liquid near to the melting line in the superfluid phase. This situation corresponds to the point "2" in the P–T plane shown in Fig. 17.12. At a certain distance R_λ from the ion, electrostriction makes the local pressure rise through the extrapolation of the λ-line into the supercooled liquid region. This extrapolation is shown as a dashed line in the figure. The radius of the solid sphere is quite irrelevant in this discussion and is assumed to be ≈ 8 Å.

Thus, two phase transitions occur moving away from the ion: the solid–liquid one, occurring for $r = R_s$ at the value P_l of the local pressure, and the normal liquid–superfluid transition at $r = R_\lambda$ for P_λ.

The existence of an He I/He II interface has been experimentally observed at the λ-transition in the gravitational field (Ahlers, 1968; Bennemann and Ketterson, 1976). *A fortiori*, it must exist at R_λ, where the field gradient due to the electric field of the ion is $4\beta/m_4 g R_\lambda^5 \approx 10^{12}$ times more intense than that due to gravity alone. Here g is the acceleration of gravity, m_4 is the He mass, and β depends on the He polarizability, as defined in eqn (17.77). R_λ is then obtained by inverting eqn (16.32) with $P = P_\lambda$.

In spite of the fact that the viscosity in the fluid surrounding the ion also varies continuously as a consequence of its pressure dependence, it is sufficient to use a simplified picture in order to qualitatively grasp the physical features of the model.

The most rapid variation of the viscosity occurs near R_λ. Elsewhere, the viscosity changes more slowly with the distance from the ion. Thus, the ion is assumed to be surrounded from R_s to R_λ by supercooled normal fluid of viscosity η_n. From R_λ to infinity the ion is surrounded by a liquid of the measured viscosity η.

At and above the λ-transition, $R_\lambda \to \infty$. The ion core is thus surrounded completely by the liquid of viscosity η_n, which is what is measured by the vibrating wire viscometer. In this situation, $R_{\text{eff}} = R_s$.

In contrast, below the superfluid transition, as $P_\lambda(T)$ increases, R_λ gradually shrinks until it becomes equal to R_s. Now, the ion is surrounded to infinity by fluid of viscosity η, once again measured by the viscometer. At both extremes, the effective radius equals the radius of the solid sphere. In between, there is an additional shell of fluid of thickness $R_\lambda - R_s$ and viscosity η_n contributing to the effective ion radius.

The hydrodynamic equations are solved in this situation (Goodstein, 1977) (see Section 17.1.1.4), yielding

$$R_{\text{eff}} = \frac{4}{3} q R_s A_+ \,, \tag{17.80}$$

where $q = \eta_n/\eta$. Here A_+ is given by eqn (17.66) with $R = R_\lambda/R_s$.

In order to simplify calculations for R_{eff} along the melting curve, it is assumed that η_n equals the measured η above the λ-transition, and that $\eta_n = \eta_\lambda$ (the measured value at the transition) below the transition. R_s is assumed constant and R_λ is calculated by means of the equation for the electrostriction 16.32. For η the measured values are exploited (Scaramuzzi *et al.*, 1977*a*).

The results of the calculations, shown by the solid line in Fig. 17.15, agree quite well with the experimental determination of the effective hydrodynamic radius of the positive ion. Above the upper critical point, i.e., the point at which the λ-line intersects the melting curve, $R_\lambda \to \infty$, $q = 1$, $A_+ = 4/3$, and $R_{\text{eff}} = R_s$. Below the λ-transition, the expected maximum in R_{eff} is semi-quantitatively reproduced.

Thus, the model strongly supports the physical picture that the maximum of R_{eff} observed along the melting curve as well as along the vapor–pressure curve (Ahlers and Gamota, 1972) is caused by a λ-transition induced by electrostriction in the fluid surrounding the ion. Thus, there is clear experimental evidence that the λ-line extends through the melting curve in the region of supercooled liquid. Only the presence of the strong field around the ion makes the supercooled liquid thermodynamically stable in its proximity.

Part II

Liquid helium-3

18

ION TRANSPORT IN NORMAL LIQUID ^{3}HE

It has previously been observed how the transport behavior of ions in superfluid ^{4}He strongly depends on their interaction with the elementary excitations of the liquid. When the thermal de Broglie wavelength λ_T of neighboring atoms is larger than their average distance, quantum statistics shows its importance. ^{4}He obeys Bose–Einstein statistics, whereas ^{3}He consists of fermions and is described by Fermi–Dirac statistics. In this case, the Pauli exclusion principle affirms that no more than one atom can occupy a given quantum state.

At low temperatures, according to Landau (1941), the normal ^{4}He fluid consists of quantized thermal excitations. However, in liquid ^{3}He, the atoms of the liquid interact strongly with each other, so one cannot apply the theory of an ideal Fermi gas in a straightforward way to predict the properties of liquid ^{3}He in the normal phase (Pines and Nozieres, 1966; Keller, 1969; Lee, 1997).

In any case, for ^{3}He a quasiparticle picture has also been introduced by Landau (1956, 1957), who developed a theory of strongly interacting Fermi liquids. Each quasiparticle is considered as an independent elementary excitation associated with a real particle that moves in a self-consistent field created by the interactions with all the surrounding particles. The quasiparticle is a state of the whole system and this state is uniquely specified by the spin and the momentum quantum numbers. Quasiparticles are envisioned as the bare fermions *dressed* by their interactions with all others. There is practically a one-to-one correspondence between atoms and quasiparticles (Keller, 1969; Lee, 1997; Guénault, 2003).

Owing to these reasons, one expects that the ion mobility should reflect the differences between the quasiparticles of the two He isotopes.

18.1 Ion mobility in liquid ^{3}He at high temperature

The first ion mobility measurements in liquid ^{3}He in the fairly-high temperature range $1.2\,\mathrm{K} < T < 3.2\,\mathrm{K}$ were performed by Meyer *et al.* (1962). Their results are presented in Fig. 18.1. The mobility of ions in liquid ^{3}He does not depend strongly on either T or N in this restricted range.

More interesting is the comparison of the ^{3}He ion mobility data obtained for $T \approx 3\,\mathrm{K}$ with the liquid ^{4}He data at the same temperature, as reported in Fig. 18.2 as a function of the liquid density. The values of the positive ion mobility μ_+ in ^{3}He are apparently located on a line drawn through the μ_+ data in ^{4}He and extrapolated to the much smaller ^{3}He densities. The same is not true for the negative ions, whose mobility appears to increase with increasing N. In any

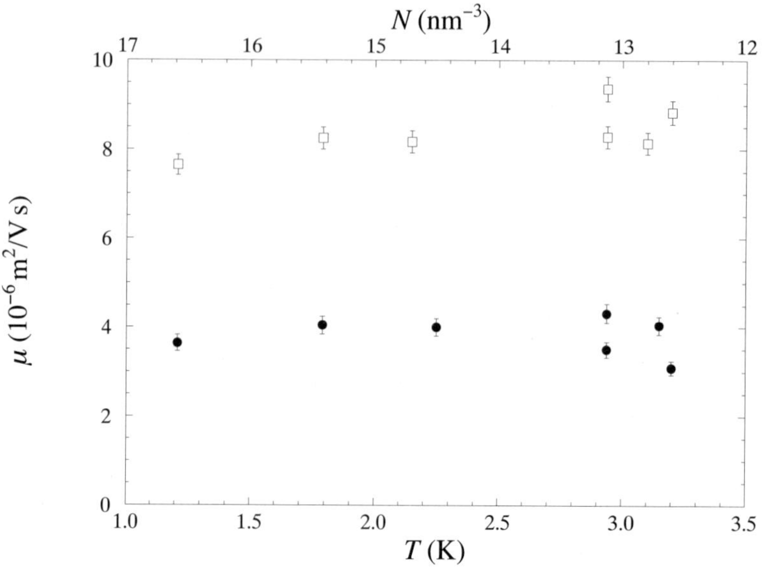

FIG. 18.1. μ vs T (lower scale) and N (upper scale) for negative (closed symbols) and positive (open symbols) ions in liquid ^{3}He at SVP. (Meyer *et al.*, 1962.)

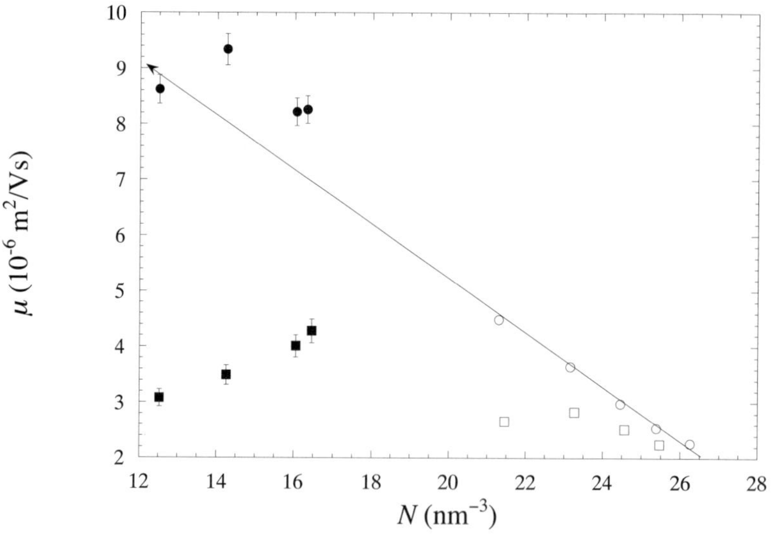

FIG. 18.2. μ vs N in ^{3}He (closed symbols) and ^{4}He (open symbols) for $T \approx 3\,$K. Circles: positive ions, and squares: negative ions. (Meyer *et al.*, 1962.)

case, it appears that the density is the most important factor influencing the ion mobility in the normal liquid.

The Nernst equation (3.1) can be used to infer the ion diffusion coefficient from the mobility. The diffusion coefficient for positive ions in liquid ^{3}He, $^3D_+$, is nearly five times smaller than the ^{3}He self-diffusion coefficient D_3 (Garwin and Reich, 1959), in accordance with the observation that positive ions are surrounded by a cluster of atoms induced by electrostriction and also in agreement with the relationship between positive ion-diffusion and self-diffusion in liquid Ar (Davis *et al.*, 1962*a*). Also, the weak temperature dependence of $^3D_+$ agrees with that of D_3, though the latter shows a strong increase for $T < 0.55\,\mathrm{K}$ (Hart and Wheatley, 1960).

Another interesting feature is that the mobility–viscosity product $\mu_+\eta$ for the positive ions for $T \approx 3\,\mathrm{K}$ is equal in both isotopic fluids within the experimental accuracy. The product $\mu_-\eta$ for the negative ions is nearly a factor three smaller than $\mu_+\eta$ and, again, is similar to the value in He I. The Stokes radius for positive ions in ^{3}He at $T = 3\,\mathrm{K}$ is $R_+ \approx 5\,\mathrm{Å}$, whereas for negative ions it turns out to be $R_- \approx 14\,\mathrm{Å}$ for $T \approx 3\,\mathrm{K}$ and $R_- = 9\,\mathrm{Å}$ for $T \approx 1.2\,\mathrm{K}$ (Meyer *et al.*, 1962).

It thus appears that the different statistics obeyed by the two isotopic fluids do not influence the ion mobility in the investigated temperature range. This is not surprising because quantum effects are expected to be relevant at much lower temperatures.

19

ION MOBILITY AT THE LIQUID–VAPOR TRANSITION IN ^{3}HE

A long-debated issue is the behavior of the transport coefficient of a fluid near the critical point of the liquid–vapor transition. The friction coefficient is a very important quantity in the realm of non-equilibrium statistical mechanics. It is therefore interesting to know whether its behavior is anomalous or not.

Mode-coupling theory (Kawasaki, 1966, 1970, 1976; Green and Sengers, 1966; Sengers, 1966, 1971; Kadanoff and Swift, 1968; Green, 1971; Stanley, 1971; Kawasaki and Gunton, 1972, 1978; Domb and Green, 1976) predicts that the diffusion coefficient should, at the critical point, behave according to the Stokes law:

$$D = \frac{k_{\mathrm{B}}T}{6\pi\eta\xi}, \tag{19.1}$$

where the correlation length ξ strongly diverges at the critical point, while the viscosity η is only weakly divergent (Berg and Moldover, 1990).

The Nernst–Einstein relationship between the diffusion coefficient and the ion mobility, $D/\mu = k_{\mathrm{B}}T/e$, and the well-defined structure of the ions in helium makes them very useful probes for studying hydrodynamics near the critical point. In particular, ^{3}He is a convenient substance because of the high purity available, a feature of paramount importance in proximity to the critical point, and also because there are experimental PVT data of very good quality (Sherman, 1965; Wallace and Meyer, 1970).

Moreover, it is worth recalling that quantum effects do not modify the critical behavior (Moldover and Little, 1965; Green and Sengers, 1966; Mistura and Sette, 1966; Roach and Douglass, 1966; Heller, 1967; Moldover, 1969; Greer and Moldover, 1981).

19.1 Experimental results

Measurements of ion mobility in ^{3}He near the critical point of the liquid–vapor transition (Modena and Ricci, 1967; Cantelli *et al.*, 1968) have been carried out with the aim of resolving the conflict between results for the diffusion coefficient obtained in different experiments (Noble and Bloom, 1965; Trappeniers and Oosting, 1966).

The mobility measurements are performed using the usual time-of-flight method (Cunsolo, 1961) with an absolute accuracy of $\pm 3\%$ (Modena and Ricci, 1967; Cantelli *et al.*, 1968). Gravity-induced density gradients are minimized by choosing a short drift distance. At the critical point, the maximum vertical density gradient is $\Delta\rho/\rho \approx 2\%$.

The critical parameters of ^{3}He are $T_c \approx 3.31\,\mathrm{K}$ and $\rho_c \approx 41.4\,\mathrm{kg\,m^{-3}}$, corresponding to a critical number density $N_c \approx 8.3\,\mathrm{atoms\,nm^{-3}}$. The critical pressure is $P_c \approx 0.115\,\mathrm{MPa}$ (Sherman, 1965; Zimmerman and Chase, 1967; Cantelli *et al.*, 1968; Wallace and Meyer, 1970).

The experimental results for the pressure dependence of the mobility of negative ions, which are electron bubbles as in the case of ^{4}He, are reported in Fig. 19.1 and the results for positive ions are shown in Fig. 19.2.

The $T = 3.271\,\mathrm{K}$ isotherm for negative ions and the $T = 3.258\,\mathrm{K}$ isotherm for positive ions show a discontinuous jump corresponding to the first-order liquid–vapor transition below T_c. Along the coexistence curve, the mobility of negative ions is always larger in the vapor than in the liquid, whereas the opposite is true for the positive ions. This is a consequence of the different ion structure and of the much greater influence of electrostriction in the case of positive carriers.

It is interesting to note that the points where the mobility is a minimum either lie on the coexistence curve for isotherms below the critical point or on its extrapolation for isotherms above the critical point, as shown in Fig. 19.3. In some way, the gas above the critical point is reminiscent of the nearby liquid–vapor transition. A similar behavior has also been observed in other critical phenomena experiments (Chase *et al.*, 1964) and also in the measurements of the mobility of O_2^- ions in near-critical Ar gas (Borghesani *et al.*, 1997).

The density dependence of the mobility for the two ionic species is shown in Fig. 19.4 and in Fig. 19.5. The data are presented here in the form of density-normalized mobility μN as functions of the reduced number density N/N_c, in-

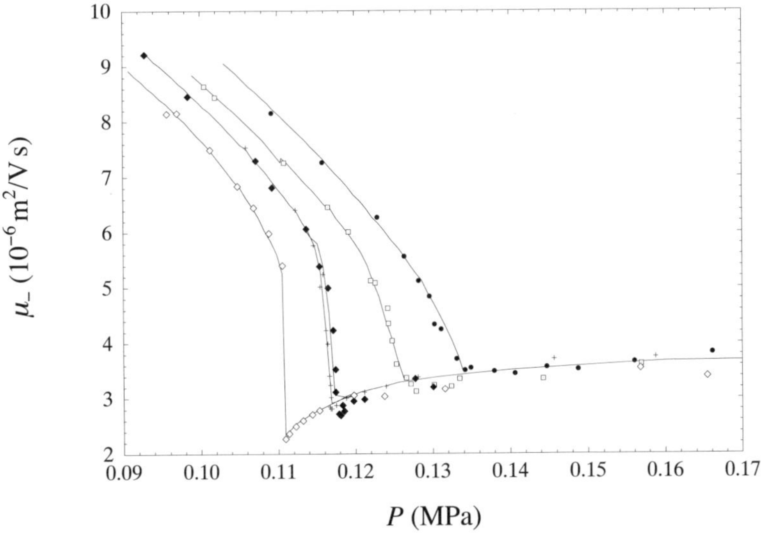

FIG. 19.1. μ_- vs P in liquid ^{3}He near the liquid–vapor critical point. (Cantelli *et al.*, 1968.) $T\,(\mathrm{K}) = 3.271$, 3.324, 3.333, 3.394, and 3.448 (from left). Lines: eyeguides.

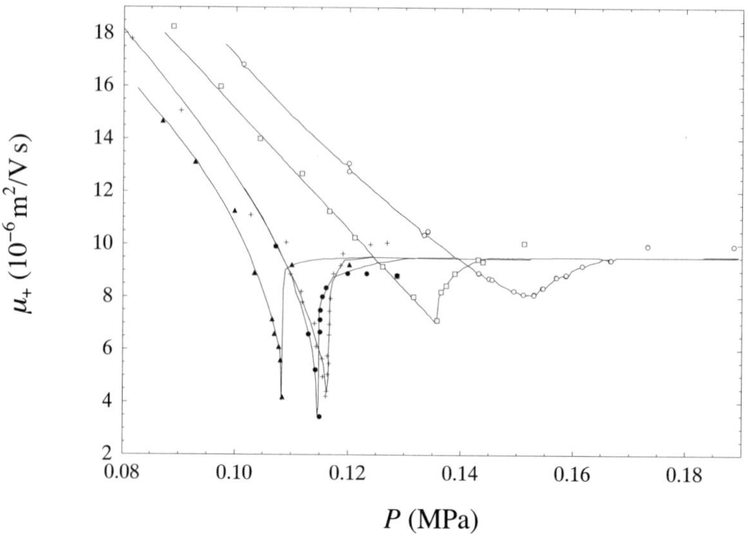

FIG. 19.2. μ_+ vs P in liquid ^{3}He near the liquid–vapor critical point. (Cantelli *et al.*, 1968.) T (K) = 3.258, 3.314, 3.320, 3.464, and 3.593 (from left). Lines: eyeguides.

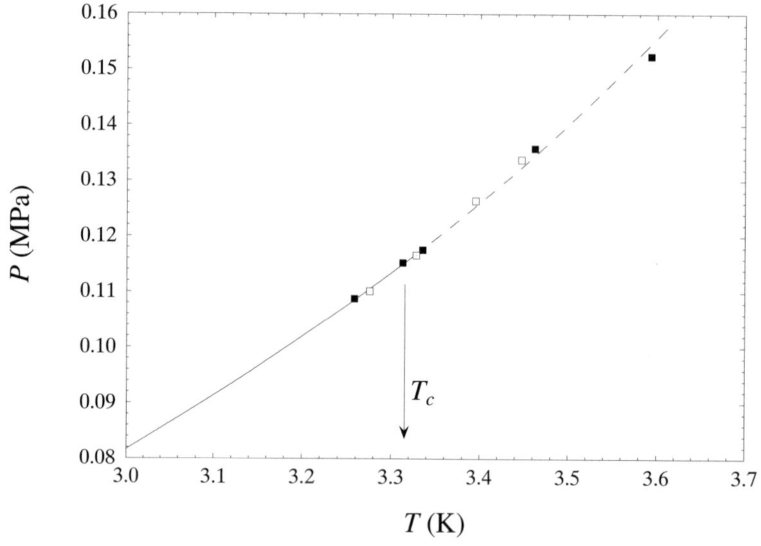

FIG. 19.3. Locus of mobility minima along isotherms. (Cantelli *et al.*, 1968.) Closed squares: positive ions, and open squares: negative ions. Gas–liquid coexistence curve of ^{3}He (solid line) and its extrapolation (dashed line).

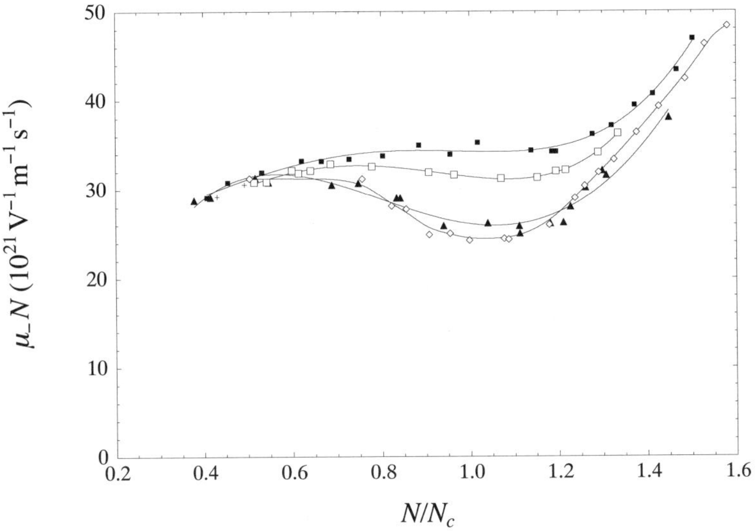

FIG. 19.4. $\mu_{-}N$ vs N/N_c. (Cantelli *et al.*, 1968.) $T\,(\mathrm{K}) = 3.448$, 3.394, 3.333, 3.324, and 3.271 (from top). Lines: eyeguides.

stead of the way used in the original paper, in which the mobility is plotted as a function of the mass density ρ of the gas (Modena and Ricci, 1967; Cantelli *et al.*, 1968).

The present way of showing the data is customary in the field of charge transport in gases (Huxley and Crompton, 1974) and has the advantage of allowing an immediate comparison with the data of O_2^- ion mobility in dense noble gases (Borghesani *et al.*, 1993, 1997), as shown in Fig. 16.7 for the Ne case.

It is apparent from Figs 19.4 and 19.5 that the mobility of both charge carriers does not vanish at the critical point. The temperature dependence of the mobility along (nearly) critical isopycnals is shown in Fig. 19.6.

According to an adaptation to ^{3}He of the bubble model (Kuper, 1961), the radius of the electron bubble should be nearly constant in the limited temperature range under investigation, whereas the radius of the cluster of ^{3}He atoms around the positive ion is proved to depend linearly on T at constant bulk ^{3}He density in the same limited temperature range (Cantelli *et al.*, 1968). The linear extrapolation into the critical region of this dependence is shown as a dashed line in Fig. 19.6. This can give a qualitative indication of the extent of the critical region.

Another way to show the width of the critical region, as far as mobilities are concerned, is to calculate the mobility defect $(\mu_n - \mu)/\mu_n$ as a function of the reduced density $(\rho - \rho_c)/\rho_c$, as shown in Fig. 19.7. μ_n is the mobility measured on an isotherm far away from the critical temperature and μ is the mobility on a nearly critical isotherm.

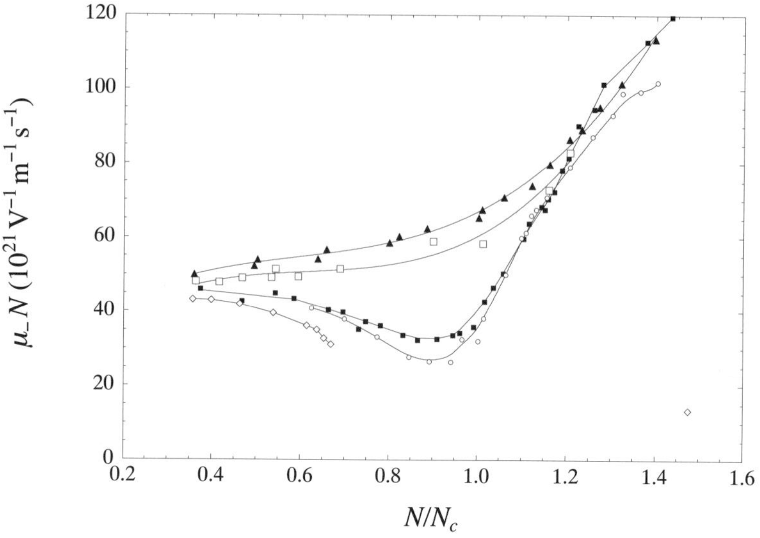

FIG. 19.5. μ_+N vs N/N_c. (Cantelli *et al.*, 1968.) $T\,(\mathrm{K}) = 3.593$, 3.464, 3.320, 3.314, and 3.258 (from top). Lines: eyeguides.

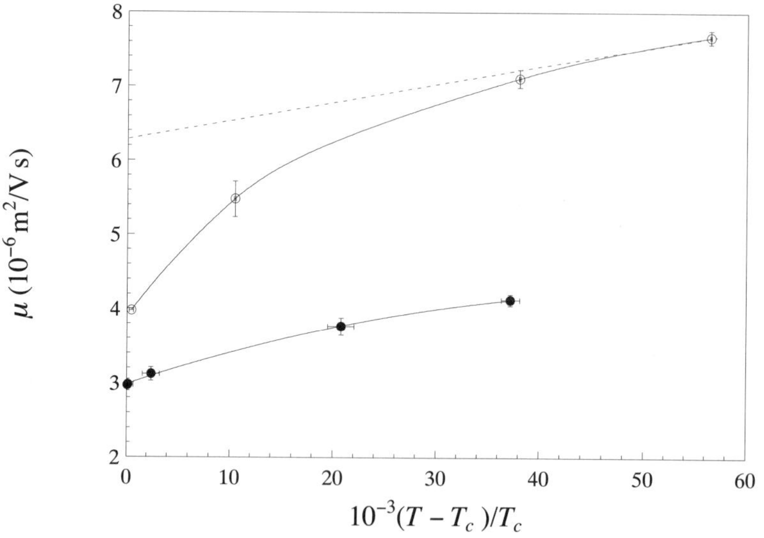

FIG. 19.6. μ vs $(T - T_c)/T_c$ for positive (open circles) and negative (closed circles) ions along the critical isopycnal. (Modena and Ricci, (1967); Cantelli *et al.*, 1968.) Solid lines: eyeguides. The dashed line is explained in the text.

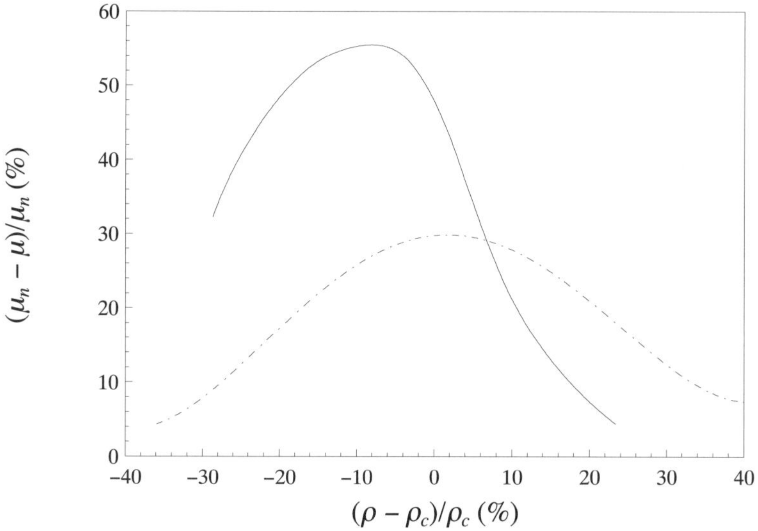

FIG. 19.7. Smoothed values of $(\mu_n - \mu)/\mu_n$ for negative (dash–dotted line) and positive (solid line) ions vs $(\rho - \rho_c)/\rho_c$. (Modena and Ricci, 1967; Cantelli *et al.*, 1968.) For negative ions, T (K) = 3.448 and 3.324 are chosen as the normal (μ_n) and nearly-critical (μ) isotherms, respectively. For positive ions, T (K) = 3.593 and 3.314 are used, respectively.

It should be noted that the mobility defect is approximately centered symmetrically about the critical density for the negative ions, whereas it is centered about a density smaller than the critical one for the positive ions and is strongly asymmetric. This difference is a consequence of the much stronger influence of electrostriction in the case of positive ions.

19.2 Hydrodynamic radius of ^{3}He at the critical point

The nearly 60% reduction of the mobility of positive charges in ^{3}He near the critical point shown in Fig. 19.7 can be qualitatively explained in terms of the usual electrostriction model (Achter, 1968).

The excess density profiles are calculated by using eqn (3.8). The perturbation of the density induced by electrostriction extends farther away into the liquid if the density of the unperturbed fluid at infinite distance from the ion is close to the critical value, as shown in Fig. 19.8.

If part of this excess mass is dragged along by the ion, its mobility must necessarily be lower near ρ_c than anywhere else and the mobility decrease is a consequence of the compressibility peak close to the critical point, which increases the range of the excess density.

The same conclusions are obtained (Springett, 1969) using the simplified

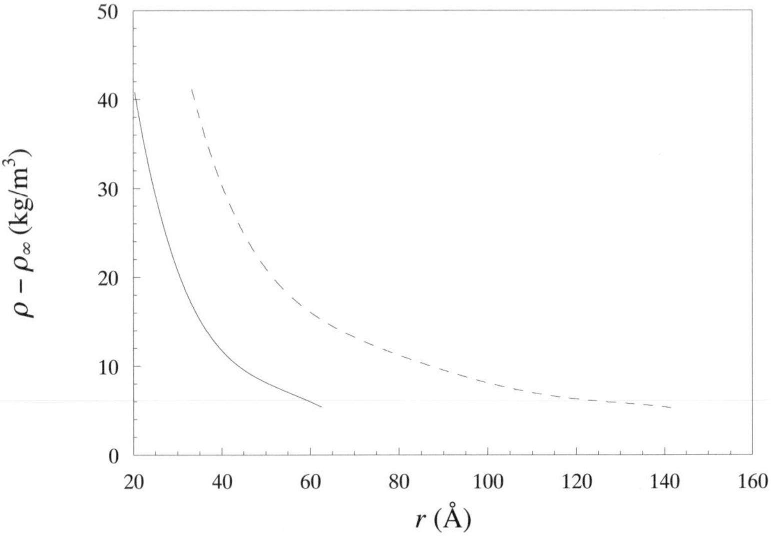

FIG. 19.8. Density profiles induced by electrostriction around positive ions. (Achter, 1968.) ρ_∞ is the density of the unperturbed fluid. Solid line: $\rho_\infty = 1.5\rho_c$, and dashed line: $\rho_\infty = \rho_c$.

equation of state valid near the critical point (Stanley, 1971; Sengers and Levelt-Sengers, 1978):

$$|1 - (P/P_c)| = B\left(|1 - (\rho/\rho_c)|\right)^\delta , \tag{19.2}$$

where P_c and ρ_c are the critical pressure and density, respectively. B and δ depend on the fluid. In particular, $\delta \approx 3.6$ for ^{3}He.

Figure 19.9 shows the radius at which a given excess density $\Delta\rho$ occurs for positive ions as a function of the reduced density. The results are not symmetric about the critical density and the maxima occur for $\rho < \rho_c$ as a consequence of the maximum compressibility at ρ_c.

In contrast, the electrostriction effect is smaller for negative ions because of their bubble structure. Owing to the large radius of the bubble, the electric field in the liquid is weaker and produces a smaller density increase (Achter, 1968). However, even the negative ion radius has a maximum, though not very pronounced, for $\rho > \rho_c$ as a consequence of the balance between the competing volume and surface contributions to the energy of the electron bubble (Springett, 1969). The electron bubble radius is shown in Fig. 19.10.

At the very high densities of the experiments (Modena and Ricci, 1967; Cantelli *et al.*, 1968) the interparticle mean free path is larger than the ion radius and the mobility is viscous limited. Thus, the inverse of the radius calculated numerically can be compared with the ion mobility, as shown in Fig. 19.11 for negative ions and in Fig. 19.12 for positive ions. The agreement with the negative ion data is quite poor, whereas that with the positive ion data is a bit better.

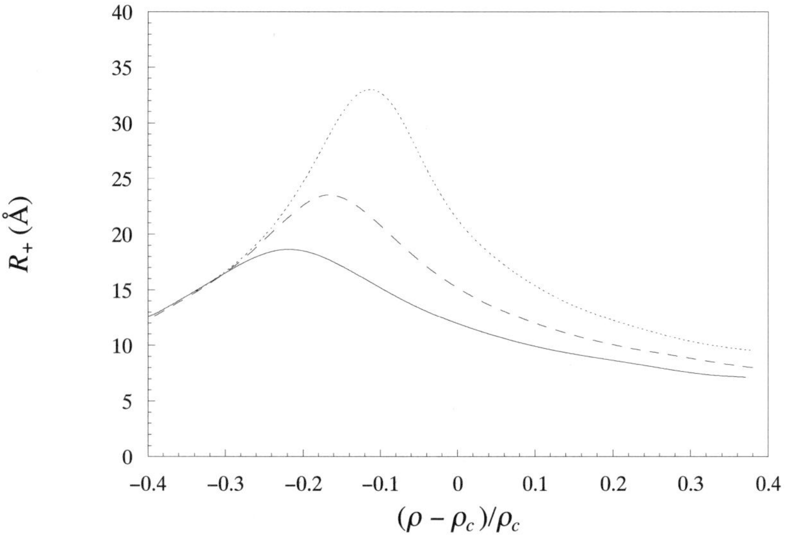

FIG. 19.9. Radius at which $\Delta\rho/\rho_c$ = constant for positive ions. (Springett, 1969.) $\Delta\rho/\rho_c = 0.4$, 0.3, and 0.2 (from bottom).

The fact that electrostriction produces the minimum excess density at nearly the same distance at which the mobility minimum occurs suggests that the effective hydrodynamic ion radius should be related to this distance (Springett, 1969).

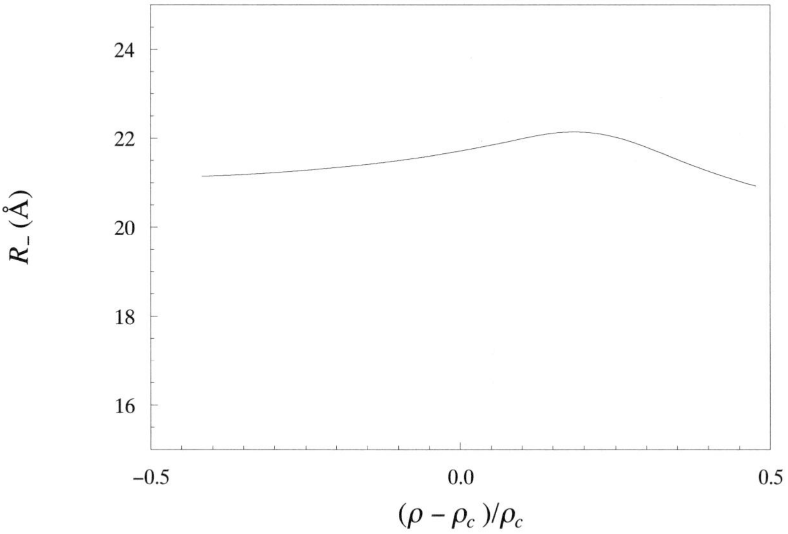

FIG. 19.10. R_- vs $(\rho - \rho_c)/\rho_c$ in liquid ^{3}He. (Springett, 1969.)

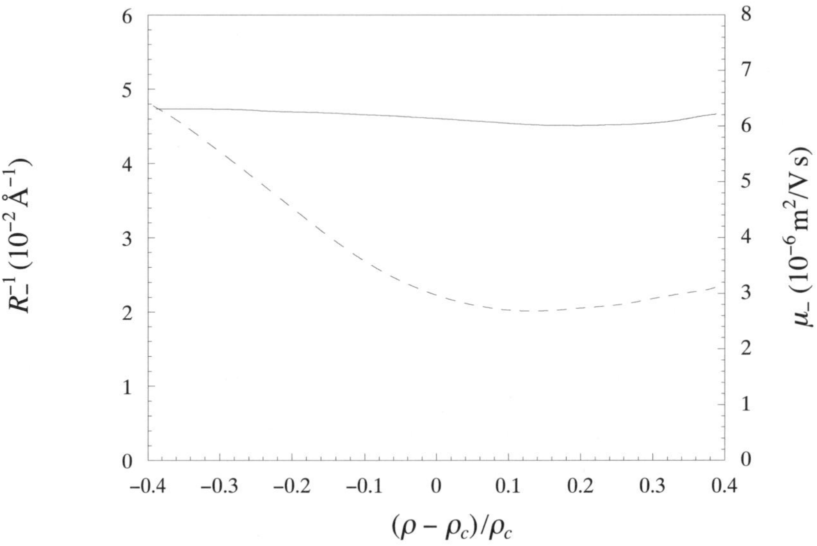

FIG. 19.11. Smoothed values of μ_- (dashed curve, right scale) (Modena and Ricci, 1967) and R_-^{-1} (solid curve, left scale) vs $(\rho - \rho_c)/\rho_c$ (Springett, 1969).

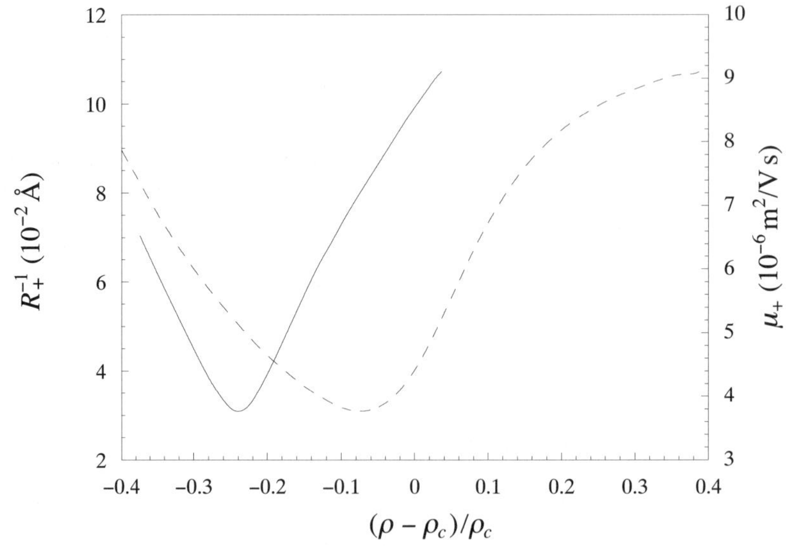

FIG. 19.12. R_+^{-1} vs $(\rho - \rho_c)/\rho_c$ (solid curve, left scale) (Springett, 1969) and smoothed values of μ_+ (dashed curve, right scale) (Cantelli *et al.*, 1968).

19.3 O_2^- ion mobility at the critical point of Ar

Strong analogies with the transport behavior of ions in ^{3}He at the liquid–vapor critical point are found in the behavior of O_2^- ions in dense Ar gas near the critical point (Borghesani *et al.*, 1997, 1999).

Electrons in a gas may get attached to electronegative impurities according to the well-known Bloch–Bradbury mechanism (Bradbury, 1933; Bloch and Bradbury, 1935). The electron affinity of O_2 is approximately 0.46 eV. Upon collision, electrons may produce a negative O_2^- ion in a vibrationally-excited state (Matejcik *et al.*, 1996). Collisions with the atoms of the host gas stabilize the negative ions, which are detected in a traditional time-of-flight experiment (Bruschi *et al.*, 1984; Neri *et al.*, 1997; Borghesani *et al.*, 1999).

The structure of the negative ion in several non-polar gases at high density has been extensively studied (Khrapak *et al.*, 1995, 1996, 1999; Volykhin *et al.*, 1995; Schmidt *et al.*, 1999; Berezhnov *et al.*, 2003). The negative ion structure in He and Ne resembles that of the electron bubble in liquid He: at short distance, a cavity is formed because of the short-range repulsion due to exchange forces, while electrostriction at larger distances increases the density locally around the ion. A much smaller cavity is formed in Ar gas. The complex structure formed can be termed an *ionic bubble* (Neri *et al.*, 1997).

A very small part of the energy absorbed by the ionic bubble during its drift motion in the dense gas under the action of the externally-applied electric field is radiated as sound waves as a consequence of the velocity-induced asymmetry in the collisions (Borghesani and Tamburini, 1999), but the main energy dissipation is through the usual friction mechanisms. The net result is that ions move at constant speed through the gas. At the usual electric field of the experiment (up to $\approx 0.3\,\mathrm{MV/m}$) (Borghesani *et al.*, 1997, 1999), the massive ionic bubble is in near thermal equilibrium with the host gas, and its mobility is independent of the field and equals its value at zero field, μ_0.

O_2^- mobility measurements have been performed along isotherms in the gas phase approaching T_c from above (Borghesani *et al.*, 1997, 1999). In Fig. 19.13 the zero-field density-normalized mobility $\mu_0 N$ is presented as a function of the gas density for the isotherm at $T = 157\,\mathrm{K}$, and in Fig. 19.14 for $T = 154.0\,\mathrm{K}$ and $T = 151.5\,\mathrm{K}$. The critical temperature of Argon is $T_c = 150.86\,\mathrm{K}$ and the critical number density is $N_c = 8.08$ atoms/nm^3 (Rabinovich *et al.*, 1988; Tegeler *et al.*, 1997).

Along the isotherm far away from the critical one, the mobility of O_2^- ions shows a very weak density dependence, as illustrated in Fig. 19.13, with only a shallow minimum around the reduced density $N_m/N_c \approx 0.9$.

Upon approaching the critical temperature, the mobility minimum becomes deeper and shifts to smaller values of the density. Eventually, for $T = 151.5\,\mathrm{K}$, the mobility minimum along the isotherm closest to the critical one T_c is very deep and occurs at a reduced density $N_m/N_c \approx 0.77$.

It is easily realized that the present behavior is very similar to that of the mobility of positive ions in near-critical ^{3}He described in the previous section

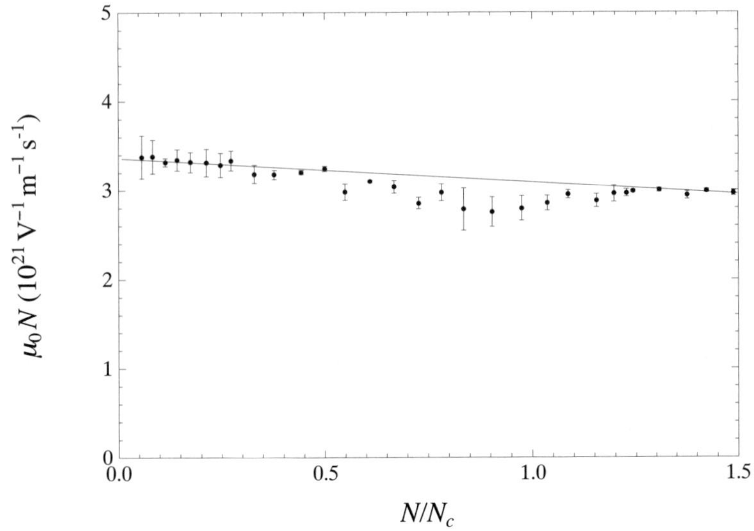

FIG. 19.13. $\mu_0 N$ vs N/N_c for O_2^- ions in Ar gas along the $T = 157\,K$ isotherm. (Borghesani *et al.*, (1997).) Line: linear fit of the data far from the critical density.

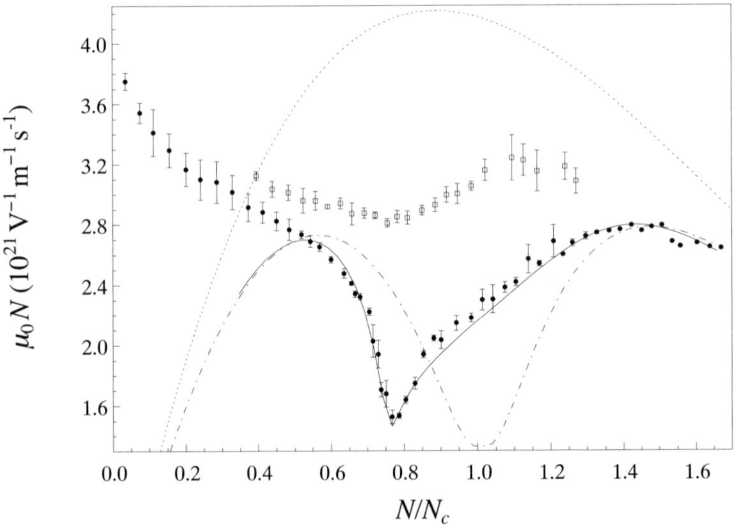

FIG. 19.14. $\mu_0 N$ vs N/N_c for O_2^- in Ar gas along two isotherms closer to T_c. $T\,(K) =$ 154.0 (squares) and 151.5 (circles). (Borghesani *et al.*, 1997.) The curves are explained in the text.

(Cantelli *et al.*, 1968). The only quantitative difference is that the density shift of the mobility minimum of O_2^- on the isotherm nearest to T_c is much larger than for ^{3}He, even though the relative distance in temperature from T_c is greater. In fact, the smallest reduced temperature attained in the ^{3}He experiment is $\epsilon = (T - T_c)/T_c \approx 9 \times 10^{-4}$ (Cantelli *et al.*, 1968), to be compared with the value in Ar of $\epsilon \approx 4 \times 10^{-3}$ (Borghesani *et al.*, 1997).

The greater distance of the mobility minimum density from ρ_c in Ar is obviously due to the much stronger influence that electrostriction exerts on Ar, whose polarizability is nearly eight times larger than that of He (Maitland *et al.*, 1981).

For $N/N_c \geqslant 0.5$, i.e., for $N \geqslant 4\,\mathrm{atoms/nm^3}$, the mean free path of Ar atoms is shorter than the ion radius and the mobility is viscous limited (Mason and McDaniel, 1988). The Stokes formula is thus inverted to yield the effective hydrodynamic radius R_h of the O_2^- ion.

Assuming that the Ar viscosity behaves smoothly in the density and temperature range investigated because its critical anomaly is very weak (Berg and Moldover, 1990), and using literature data for it (Trappeniers *et al.*, 1980; Rabinovich *et al.*, 1988), R_h is easily calculated and is shown in Fig. 19.15 for the $T = 151.5\,\mathrm{K}$ isotherm.

As expected, R_h is strongly peaked at the density N_m of the mobility minimum and decreases rapidly at both high and low density. At low density, for $N \approx 3\,\mathrm{atoms/nm^3}$ ($N/N_c \approx 0.37$), $R_h \approx 5\,\mathrm{Å}$. This value approximately corresponds to the radius $R_{1s} = 4.6\,\mathrm{Å}$ of the first complete solvation shell of Ar atoms surrounding the ion (Maitland *et al.*, 1981).

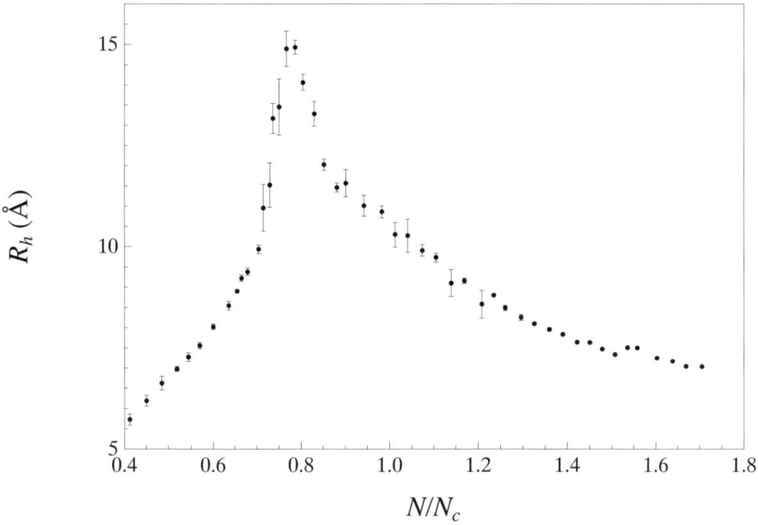

FIG. 19.15. R_h of O_2^- vs N/N_c for $T = 151.5\,\mathrm{K}$. (Borghesani *et al.*, 1997.)

On the high-density side, for $N \approx 14\,\text{atoms/nm}^3$ ($N/N_c \approx 1.73$), $R_h \approx 7.6\,\text{Å}$, which is very close to the value $R_{2s} = 8.2\,\text{Å}$ that corresponds to the radius of two complete solvation shells (Maitland *et al.*, 1981), in agreement with results of molecular dynamics simulations (Pollock and Alder, 1978).

If the effective hydrodynamic radius of the ion is assumed to vary linearly between the values at the extremes of the density range, the Stokes formula gives the dashed curve in Fig. 19.14 that is in strong disagreement with the experimental data.

Instead of pursuing the approach of the granular-fluid model developed for the λ-transition in ^{4}He (Goodstein, 1977), it is assumed that a correlated fluid layer of thickness proportional to the correlation length is dragged along by the ion, increasing its effective radius (Borghesani *et al.*, 1997; Borghesani and Tamburini, 1999). The surplus thickness has then to be determined by adjusting its value so as to give agreement with the experimental data.

In the critical region of Ar, the correlation length can be written as (Thomas and Schmidt, 1963)

$$\xi = \xi_0 \left[S\left(0\right)\right]^{1/2} . \tag{19.3}$$

$\xi_0 \approx 3.16\,\text{Å}$ is the *short-range correlation length.* $S(0)$ is the long-wavelength limit of the static structure factor $S(k)$ (van Hove, 1954; Cohen and Lekner, 1967) and is related to the isothermal compressibility of the gas $\chi_T = (1/V)(\partial V/\partial P)_T$ by (Stanley, 1971; Goodstein, 1975)

$$S\left(0\right) = N k_{\mathrm{B}} T \chi_T . \tag{19.4}$$

Thus, the hydrodynamic radius is written as

$$R_h = b_0 + b_1 N + b_2 \left[S\left(0\right)\right]^{1/2} , \tag{19.5}$$

where b_0, b_1, and b_2 are adjustable parameters. The linear term $b_0 + b_1 N$ interpolates between the radii of the first and second solvation shells, while the term proportional to $[S(0)]^{1/2}$ is only relevant close to the critical point.

If bulk values of the correlation length are used to calculate the hydrodynamic radius, as though electrostriction were absent, the dash–dotted curve in Fig. 19.14 is obtained. As expected, the contribution due to $S(0)$ is a maximum at the critical density, yielding an incorrect position of the mobility minimum.

On the other hand, it is known that electrostriction enhances density (and viscosity) in the fluid surrounding the ion. As an example, the electrostriction-induced density profile calculated for Ar at $T = 151.5\,\text{K}$ for the unperturbed density $N = 6\,\text{atoms/nm}^3$ ($N/N_c = 0.74$), close to the value $N_m = 6.25\,\text{atoms/nm}^3$ of the mobility minimum, is shown in Fig. 19.16. In order to better treat the shorter-range part of the Ar–O_2^- interaction, a 12–6–4 Lennard–Jones-type potential with hard-sphere radius $\sigma_{AO} = 2.9\,\text{Å}$ and depth $V_0 = 98\,\text{meV}$ (Mason and McDaniel, 1988) is used in the electrostriction equation (3.6), though the same conclusions could practically be drawn using the usual r^{-4} polarization potential.

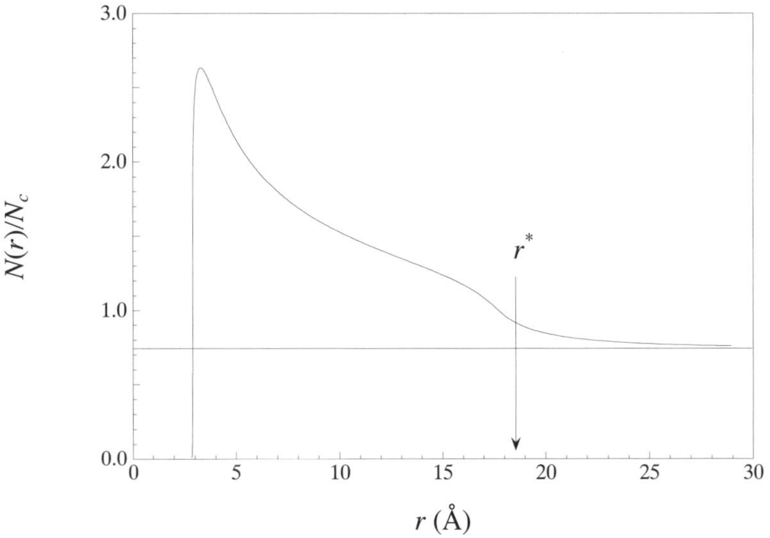

FIG. 19.16. Electrostriction-induced density profile in Ar for $N = 6\,\text{atoms/nm}^3$ at $T = 151.5\,\text{K}$. (Borghesani *et al.*, 1997.) $r^\star$ is the distance from the ion at which the thermodynamic properties of the gas are evaluated.

If the average density N of the unperturbed fluid is below N_c, there is a distance from the ion at which the local density $N(r)$ takes on the critical value N_c. At the same distance, the local value $S_r(r)$ of the static structure factor $S(0)$ is a maximum because the local compressibility is a maximum. For distances smaller or larger than this, $S_r(r)$ is smaller. In other words, the fluid goes through criticality on approaching the ion.

It is interesting to note that, by analogy with the case of ^{3}He (Cantelli *et al.*, 1968), the locus of the mobility minima of O_2^- in Ar is the extrapolation of the liquid–vapor coexistence curve, as shown in Fig. 19.17. In some sense, the gas is reminiscent of the nearby coexistence curve, even for $T > T_c$. A physical picture can thus be envisioned in which droplets of superheated liquid appear in the gas as a consequence of critical fluctuations. Electrostriction near the ion makes the local pressure rise through the extrapolation of the coexistence line into the region of superheated liquid that the strong ion field makes thermodynamically stable.

The analogy with the situation of the positive ions in superfluid ^{4}He near the melting transition (Goodstein, 1977; Scaramuzzi *et al.*, 1977*a*, 1977*b*;), in which it is assumed that the λ-line is extrapolated into a region of supercooled, electrostrictively-stabilized liquid surrounding the ion, is also evident.

The behavior of S_r as a function of the distance r from the ion and of the unperturbed density N suggests that the transport behavior of the ion is determined by the local fluid properties at a given distance from it, say $r^\star$, rather than

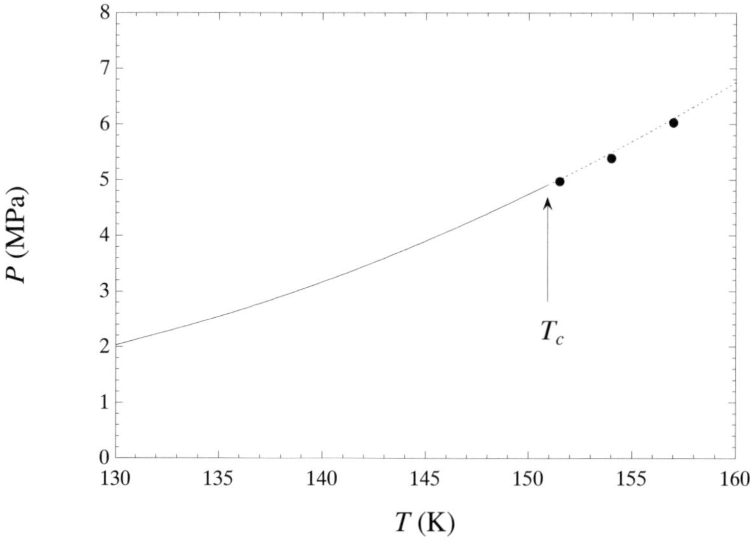

FIG. 19.17. Locus of the minima of $\mu_0 N$ along isotherms (closed symbols) (Borghesani *et al.*, 1997). Solid line: gas–liquid coexistence line (Rabinovich *et al.*, 1988), and dashed line: its extrapolation.

by the properties of the unperturbed fluid at a very large distance (Borghesani *et al.*, 1997). $r^\star$ is chosen by enforcing the condition that $S_r(r^\star)$ is a maximum at the same N at which $\mu_0 N$ is a minimum. This yields the value $r^\star = 18.5\,\text{Å}$, then used for all other densities and temperatures.

The viscosity η in the Stokes formula also depends on r through its dependence on N. Thus, its local value is used, $\eta_r^\star = \eta[N(r^\star)]$, i.e., the local value corresponding to the local density at $r = r^\star$.

The constants b_0, b_1, and b_2 in eqn (19.5) are adjusted so as to fit the mobility data, eventually yielding

$$R_h = 4.75 + 1.143 \times 10^{-1} N + 0.66 \left[S_r\left(r^\star\right)\right]^{1/2}, \qquad (19.6)$$

where R_h is expressed in Å and N in units of atoms/nm^3.

The solid curve in Fig. 19.14 shows the results of this calculation. The agreement of the data in the critical region is excellent. The reduction of the mobility due to local criticality in the fluid surrounding the ion is correctly located at $N_m < N_c$.

The density range of the influence of criticality-related mechanisms is very broad. Only for $N \geqslant 11$ atoms/nm^3 ($N/N_c \geqslant 1.36$) does the electrostrictively-modified Stokes formula (the dashed curve in Fig. 19.14) tend to merge with the pure Stokes formula (the solid curve in Fig. 19.14). In contrast, for $N \leqslant 4$ atoms/nm^3, the mobility is no longer viscous limited and the hydrodynamic description ceases to be valid, as expected.

As a final remark, the O_2^- ions in Ar behave most similar to positive than to negative ions in ^{3}He or in ^{4}He because they are not surrounded by the large empty cavity that surrounds electrons in He, thereby reducing the influence of electrostriction. It should be noted further that $R_h < r^\star$. Thus, it is unimportant if the cluster of Ar atoms surrounding the ion is liquid or not (Goodstein, 1978).

20

ION MOBILITY IN ^{3}HE AT INTERMEDIATE TEMPERATURES

In the same spirit as ions are used to probe the collective excitations of the liquid in superfluid He II, ions are exploited in liquid ^{3}He to investigate the nature of quasiparticles in a Fermi liquid. It is reasonably assumed that the same types of ionic structures are found in both liquid ^{3}He and liquid ^{4}He (Fetter, 1976; Ahonen *et al.*, 1978; Senbetu and Woo, 1979). The mobility of ions depends on their structure as in the He II case and on the nature of the scattering by the liquid. Differences in the statistics obeyed by the two isotopic fluids will appear as differences in the ion-scattering properties.

Quantumdegeneracy effects are expected to appear when the thermal energy of the sample is small with respect to the Fermi energy, which is related to the liquid density N by

$$\epsilon_F = \frac{\hbar^2}{2m^\star}\left(3\pi^2\right)^{2/3} N^{2/3}, \tag{20.1}$$

where $m^\star \approx 3m_3$ is the ^{3}He effective mass (Wheatley, 1968) and m_3 is its atomic mass. In the case of ^{3}He, a typical liquid density is of the order $N \simeq 17\,\mathrm{atoms/nm^3}$ (Meyer *et al.*, 1962), thus yielding a Fermi temperature $T_F = \epsilon_F/k_\mathrm{B} \approx 1.6\,\mathrm{K}$.

For this reason, at the fairly high temperatures of the early mobility experiments, $1\,\mathrm{K} < T < 3\,\mathrm{K}$, (Meyer *et al.*, 1962; Modena, 1963; de Magistris *et al.*, 1965), quantum-degeneracy effects are absent. The fluid behaves approximately in a classical way and ion transport occurs in the hydrodynamic regime (Clark, 1963). The validity of the Stokes formula for the mobility is experimentally confirmed by the observation that the Walden rule is reasonably well followed (Meyer *et al.*, 1962).

In an intermediate temperature range, around and below 1 K, the statistics are important in determining the fluid properties, but the heavy ion still suffers only small deflections upon collisions with the fluid quasiparticles. The motion of the ion can be described as a *quantum-mechanical Brownian motion* because it can be described by a Fokker–Planck equation in which the quantum-mechanical effects are included in a friction coefficient (Davis *et al.*, 1962*b*, 1965; Davis and Dagonnier, 1966).

The crossover from a hydrodynamic, statistics-independent regime to one in which quantum-mechanical effects start influencing the mobility is clearly observed in Fig. 20.1, in which the ion mobility measured at low pressure by using

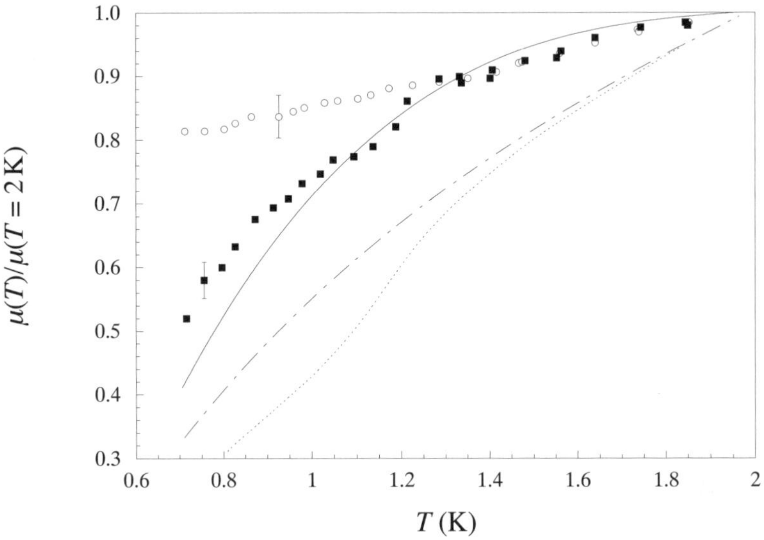

FIG. 20.1. μ vs T for positive (circles) and negative (squares) ions in liquid ^{3}He at low pressure. μ is normalized to unity at $T = 2.0\,\mathrm{K}$. (McClintock, 1973*b*.) Lines: behavior expected if the Walden rule is followed, using the viscosity data of Betts *et al.* (1963) (solid line), of Black *et al.* (1971) (dash–dotted curve), or of Webeler and Hammer (1966) (dotted curve).

a spacecharge field–emission diode technique is shown (McClintock, 1973*b*). Details about the technique exploited are described elsewhere (McClintock, 1973*a*).

Although it is not possible to conclude that the mobility μ strictly behaves according to hydrodynamics for $T = 2\,\mathrm{K}$ because of a significant disagreement between different authors as to the temperature dependence of the dynamic viscosity η in this region (Betts *et al.*, 1963; Webeler and Hammer, 1966; Black *et al.*, 1971; Beál-Monod, 1973), nonetheless the conclusion can be safely drawn that μ behaves non-classically below $T \approx 1.2\,\mathrm{K}$ (McClintock, 1973*b*).

Finally, in the limit of vanishingly-small temperature, $T \to 0$, the fermion momentum by far exceeds that of ions. In this case, ions may experience large deflections and a Fokker–Planck description of their motion is no longer valid. The Pauli exclusion principle, moreover, allows ions to collide only with fermions in a thin energy shell close to the Fermi energy. Thus, the ion mobility is expected to have a well-defined T^{-2} temperature dependence (Abe and Aizu, 1961; Clark, 1963; Davis and Dagonnier, 1966; Schappert, 1968), though this is not what is experimentally observed, as will be shown soon.

A number of experiments below 1 K have thus been carried out. In the following, the results for negative and positive ions will be treated separately. The reason is very simple. Whereas a unique ionic species is present in the case of negative ions and a direct comparison with the theoretical predictions can be made, in the case of positive ions there is the simultaneous existence of a number of distinct ionic species as a consequence of the contamination by ^{4}He impurities in colloidal suspension, whose solubility becomes so high as to modify the ^{4}He/^{3}He ratio in the solid ion snowball and in the surrounding liquid. This fact has bedevilled theory with contradictory and irreproducible experimental results for a long time.

21

NEGATIVE IONS IN NORMAL LIQUID ^{3}HE FOR $T < 1$ K

Measurements of the negative ion mobility μ_- in pure normal ^{3}He at low temperature have been carried out in several experiments over the years (Meyer *et al.*, 1962; Kuchnir *et al.*, 1970; McClintock, 1973*b*; Ahonen *et al.*, 1978; Long and Pickett, 1979). A summary of the results is plotted in Fig. 21.1, in which data at low as well as at high pressure are shown. The data extend down into the millikelvin region. The lowest temperature attained at low pressure is 17 mK (Kuchnir *et al.*, 1970), whereas the critical temperature for the superfluid transition into the A-phase, $T_c \approx 2.73$ mK, has been reached in pressurized liquid ^{3}He for $P = 2.84$ MPa (Ahonen *et al.*, 1978).

Some general qualitative features can be noticed in Fig. 21.1. At the lowest temperatures reached, no comparison with normal liquid ^{4}He can be evidently

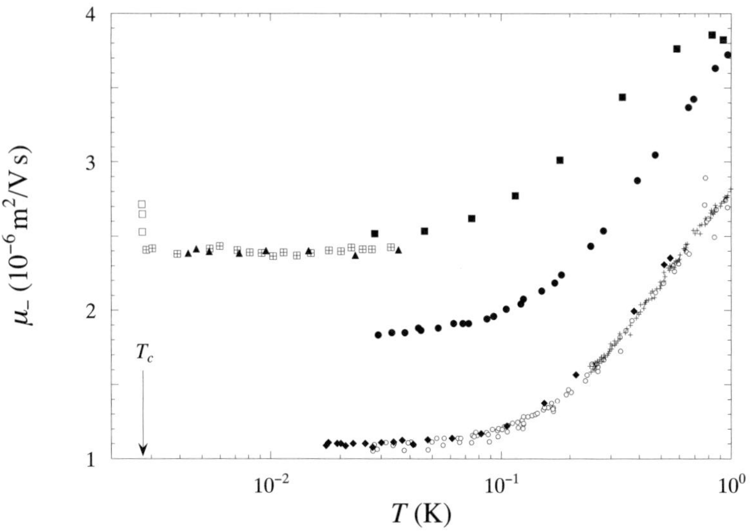

FIG. 21.1. μ_- vs T in normal liquid ^{3}He. Bottom curve: low-pressure data, 18 kPa $< P <$ 104 kPa. Open circles: Anderson *et al.* (1968), closed diamonds: Kuchnir *et al.* (1970), and crosses: McClintock (1973*b*). Middle curve: $P = 0.76$ MPa (Anderson *et al.*, 1968). Top curve: closed squares: $P = 2.83$ MPa (Anderson *et al.*, 1968), crossed squares: $P = 2.84$ MPa (Ahonen *et al.*, 1978), closed triangles: $P = 2.5$ MPa Long and Pickett (1979), and open squares: Superfluid ^{3}He, A phase (Ahonen *et al.*, 1978). The arrow shows the critical temperature $T_c \approx 2.73$ mK.

made. The comparison must be done with superfluid He II in the same temperature range (Schwarz, 1972*a*; Barenghi *et al.*, 1986, 1991). The data reported in Fig. 14.4 show that the negative ion mobility in He II at $T = 0.6\,\mathrm{K}$ and low pressure exceeds that in liquid ^{3}He by a factor of nearly 10^4. At still lower T, this factor increases up to 10^8 for $T \approx 50\,\mathrm{mK}$.

Further, the temperature dependence is quite different. In particular, it is much weaker in the case of ^{3}He. At all pressures, the mobility of negative ions levels off at around 50 mK and then remains constant. This fact is even more evident in the high-pressure case.

For a given T, μ_- increases with P. This is understood easily by recalling that the negative ion consists of an electron in an empty void and that the electron bubble is easily "squeezed" by increasing pressure in the same way as in He II (Springett, 1967; Zipfel, 1969; Ostermeier, 1973).

21.1 Analysis of the temperature dependence of the mobility

It is well known that the negative ion is a self-trapped electron in an empty cavity with radius $\approx 15\,\text{Å}$. When it moves in the liquid, it experiences a drag force due to the interaction with the excitations of the liquid. At relatively high T, the excitation mean free path is shorter than the ion radius and the hydrodynamic description of the drag force is approximately adequate (Davis *et al.*, 1962*b*; McClintock, 1973*b*) (see Fig. 20.1).

However, when the temperature is lowered, for $T \to 0$, the quasiparticle mean free path increases indefinitely, eventually becoming larger than the ion size. This situation is known as the *Knudsen limit*, in which the drag must be calculated by considering the collisions with individual quasiparticles. Thus, as in the case of He II, the mobility provides direct information on the excitation spectrum of the fluid.

The ion mobility is calculated as usual by solving the quantum-mechanical version of the Boltzmann transport equation derived by Uehling and Uhlenbeck (1933). Several authors performed these calculations for an ion in a Fermi liquid (Abe and Aizu, 1961; Clark, 1963; Davis and Dagonnier, 1966; Schappert, 1968; Kramer, 1970*a*, 1970*b*). The basic assumptions are as follows:

1. the applied field is very small, so that the motion of the ion caused by the field is small in comparison with its thermal motion;
2. the ion concentration is so low that ion–ion interactions are negligible;
3 the collisions between the impurity ion and the quasiparticles of the liquid are elastic;
4. the effective mass of the ion $M^\star$ is much larger than the effective mass of the quasiparticles $m^\star$, so that $(m^\star/M^\star)^{1/2} = \gamma \ll 1$;
5. the system is never far from equilibrium.

Condition 5 can be somewhat relaxed to $v_D \ll \bar{v}$, where v_D is the heavy-ion drift velocity and $\bar{v}$ is the average velocity of the quasiparticles taking part in the collision (Kramer, 1970*b*).

Conditions 4 and 5 can be used to characterize the temperature ranges in which the ion motion can be expected to be Brownian or not (Davis and Dagonnier, 1966; Ahonen *et al.*, 1978). According to condition 5, the average momentum and velocity of the heavy ions are thermal:

$$|\mathbf{V}| \sim \mathcal{O}\left(k_{\mathrm{B}}T/M^{\star}\right)^{1/2} ,$$
$$|\mathbf{P}| \sim \mathcal{O}\left(M^{\star}k_{\mathrm{B}}T\right)^{1/2} .$$

If T is so high that the liquid behaves classically, the average velocity and momentum of the fluid particles are also thermal:

$$|\mathbf{v}| \sim \mathcal{O}\left(k_{\mathrm{B}}T/m^{\star}\right)^{1/2} ,$$
$$|\mathbf{p}| \sim \mathcal{O}\left(m^{\star}k_{\mathrm{B}}T\right)^{1/2} .$$

In this case $|\mathbf{p}|/|\mathbf{P}| = \gamma \ll 1$. In other words, the average momentum of the heavy impurity is much larger than that of the fluid particles.

In contrast, at low temperature the quantum degeneracy of the fluid comes into play. Owing to the exclusion principle, the ion undergoes collisions only with quasiparticles near the Fermi surface, whose energy and momentum are ϵ_F and $p_F = (2m^{\star}\epsilon_F)^{1/2}$, respectively.

If one defines the Fermi temperature $T_F = \epsilon_F/k_{\mathrm{B}}$ and the parameter $\xi = (T_F/T)^{1/2}$, the ratio between the momenta of the the fluid particles and of the heavy ions becomes $|\mathbf{p}|/|\mathbf{P}| = \gamma\xi$. The ion still undergoes Brownian motion with a friction coefficient that depends on quantum-mechanical effects if $\gamma\xi \ll 1$ (Davis and Dagonnier, 1966).

As mentioned before, several theoretical approaches give equivalent formulas for the ion mobility. The approach of Kramer (1970*a*, 1970*b*) will be described here in detail in view of its close relationship with the calculation of the phonon- and roton-limited ion mobility in superfluid He II (Baym *et al.*, 1969; Barrera and Baym, 1972).

Let n and f be the distribution functions of the quasiparticles and of the impurity ions, respectively, that can be determined by a system of two coupled Boltzmann equations. Upon the action of the external force due to a vanishingly-small uniform electric field, $\mathbf{F} = -e\mathbf{E}$, the impurity ion acquires a limiting drift velocity proportional and parallel to the electric field strength: $\mathbf{v}_D = \mu\mathbf{E}$. This relationship defines the mobility μ. The drift velocity is the thermal average velocity of the ion: $\mathbf{v}_D = \langle\mathbf{v}\rangle \equiv \mathbf{V}$.

In steady state, the force acting on the ion equals the average rate of momentum transferred by the ion to the quasiparticles:

$$\frac{\partial\mathbf{P}}{\partial t} = \mathbf{F} = g\sum_{\mathbf{p}}\mathbf{p}\left(\frac{\partial n}{\partial t}\right)_c , \tag{21.1}$$

where $\mathbf{p}$ is the momentum of the quasiparticles and $(\partial n/\partial t)_c$ is the rate of change of the quasiparticle distribution function due to collisions with the ions. $g = 2$ is the spin degeneracy factor.

Assuming that the impurity ions are described by Boltzmann statistics (as a consequence of conditions 1, 2 and 5), the collisional term can be written as (Baym *et al.*, 1969; Kramer, 1970*b*)

$$\left(\frac{\partial n}{\partial t}\right)_c = \int \mathrm{d}^3 v\, \mathrm{d}^3 v' \sum_{\mathbf{p}} [f'n'(1-n) - fn(1-n')]\Gamma(\mathbf{pv} \to \mathbf{p'v'}) \,. \tag{21.2}$$

The shorthand notation $f' \equiv f(\mathbf{v}')$ and $n' \equiv n(\mathbf{p}')$ has been introduced here. $\Gamma(\mathbf{pv} \to \mathbf{p'v'})$ is the scattering probability that the initial state characterized by the ion velocity $\mathbf{v}$ and quasiparticle momentum $\mathbf{p}$ ends up in the final state identified by $\mathbf{v}'$ and $\mathbf{p}'$.

Detailed balance enforces the condition that the scattering probability is invariant with respect to time reversal:

$$\Gamma(\mathbf{pv} \to \mathbf{p'v'}) = \Gamma(\mathbf{p'v'} \to \mathbf{pv}) \,. \tag{21.3}$$

A great simplification can be achieved if the recoil of the impurity in a collision can be neglected. In this case, collisions turn out to be elastic and one can set

$$|\mathbf{v}'| = |\mathbf{v}| \,, \tag{21.4}$$

$$f = \delta(\mathbf{v} - \mathbf{V}) \,. \tag{21.5}$$

If the quasiparticle spectrum is of the form $\epsilon = p^2/2m^\star$, the condition of absence of recoil can be written as $\gamma \ll 1$.

By inserting eqns (21.4) and (21.5) into eqns (21.1) and (21.2), one obtains

$$\mathbf{F} = g \sum_{\mathbf{pp}'} \mathbf{p}\,(n' - n)\,\Gamma(\mathbf{V}; \mathbf{p} \to \mathbf{p}') \,, \tag{21.6}$$

where the notation $\Gamma(\mathbf{pv} \to \mathbf{p'v'}) = \delta(\mathbf{v} - \mathbf{V})\Gamma(\mathbf{V}; \mathbf{p} \to \mathbf{p}')$ has been used.

Equation (21.6) is valid for a Fermi gas in the limit of vanishingly-small temperature and is valid for any arbitrary $\mathbf{V}$. It is also trivial to show that, upon interchanging the summation order, eqn (21.6) can be cast in the following form:

$$\mathbf{F} = g \sum_{\mathbf{pp}'} n\,(\mathbf{p}' - \mathbf{p})\,\Gamma(\mathbf{V}; \mathbf{p} \to \mathbf{p}') \,. \tag{21.7}$$

The equation in this form is prone to a very simple physical interpretation: the force on the ion is given by the average momentum transfer during a single collision summed over all scattering events. The same equation has been used by several authors in order to compute the phonon-limited mobility of negative ions in superfluid He II (Arkhipov and Shalnikov, 1960; Schwarz and Stark, 1968; Baym *et al.*, 1969).

The transition rate Γ may depend in an intrinsic way on the ion drift velocity $\mathbf{V}$. One can remove this dependence by noting that, in the frame of reference in which the ion is at rest, $\partial \mathbf{P}/\partial t$ vanishes (Baym *et al.*, 1969). In this frame of reference the kinematic constraint for elastic collision is simply given by the energy conservation law of the quasiparticles: $\tilde{\epsilon}' = \tilde{\epsilon}$, where the tilde means that a quantity is evaluated in the rest frame of the ion.

If the quasiparticle spectrum is

$$\epsilon = \frac{p^2}{2m^\star}, \tag{21.8}$$

Galilean relativity is valid for the quasiparticle gas. One can thus write $\tilde{\epsilon} = \epsilon(\tilde{p})$ and the elastic collision constraint becomes $\tilde{p}' = \tilde{p}$. In the case of collective excitations, for which the energy spectrum is no longer given by eqn (21.8), $p' \neq p$ in general, and the kinematics built in Γ must be treated with great care (Kramer, 1970*b*).

In eqn (21.6) the difference $n' - n$ is explicitly of first order in $\mathbf{V}$. Thus, any intrinsic dependence of Γ on $\mathbf{V}$ can be neglected for small ion drift velocities as long as a linear expansion of $\mathbf{F}$ on $\mathbf{V}$ is sought.

Energy conservation in the frame of reference of the liquid yields

$$\epsilon' - \mathbf{p}' \cdot \mathbf{V} = \epsilon - \mathbf{p} \cdot \mathbf{V}\,. \tag{21.9}$$

Inserting eqn (21.9) into eqn (21.6), one obtains to leading order in $\mathbf{V}$

$$\mathbf{F} = g \sum_{\mathbf{pp}'} \mathbf{p} \left\{ n\left[\hat{p}', \epsilon + (\mathbf{p}' - \mathbf{p}) \cdot \mathbf{V}\right] - n\left(\hat{p}, \epsilon\right) \right\} \Gamma\left(0; \mathbf{p} - \mathbf{p}'\right). \tag{21.10}$$

Here $\hat{p}$ and $\hat{p}'$ are unit vectors and the possible anisotropy of the quasiparticle distribution function arising as a consequence of the distortion of the isotropic equilibrium distribution function n_0 induced by the ion motion has been explicitly taken into account.

To leading order in $\mathbf{V}$, the distortion of the distribution function can be described in terms of a correction $\delta n(p)$ to n as

$$n\left(\hat{p}', \epsilon\right) - n\left(\hat{p}, \epsilon\right) = \mathbf{V} \cdot \left(\mathbf{p}' - \mathbf{p}\right) \delta n\left(p\right). \tag{21.11}$$

In the Knudsen limit, i.e., when the quasiparticle mean free path largely exceeds the ion size, δn can be neglected.

Γ is expressed in terms of the incoming quasiparticle current per unit volume, of the density of the final states, and of the differential scattering cross-section $\mathrm{d}\sigma/\mathrm{d}\Omega$ as

$$\Gamma\left(0; \mathbf{p} \to \mathbf{p}'\right) = \left|\frac{\partial \epsilon}{\partial p}\right| \frac{\left(2\pi\hbar\right)^3}{p^2} \delta\left(p - p'\right) \frac{\mathrm{d}\sigma}{\mathrm{d}\Omega}\left(p, \theta\right). \tag{21.12}$$

Inserting eqn (21.12) into eqn (21.10) and expanding the distribution function difference to leading order in $\mathbf{V}$, the following expression for $\mathbf{F}$ is obtained if the parallelism between $\mathbf{V}$ and $\mathbf{F}$ is taken into account:

$$\mathbf{F} = \frac{g}{h^3} \int \left(\hat{V} \cdot \mathbf{p}\right) \left[V \cdot \left(\mathbf{p}' - \mathbf{p}\right)\right] \left(\frac{\mathrm{d}n_0}{\mathrm{d}\epsilon} + \delta n\right) \left|\frac{\partial \epsilon}{\partial p}\right| \frac{\mathrm{d}\sigma}{\mathrm{d}\Omega} \mathrm{d}\Omega\, \mathrm{d}^3 p, \tag{21.13}$$

where $\hat{V}$ is a unit vector and n_0 is the isotropic Fermi distribution function:

$$n_0(\epsilon) = \{\exp[(\epsilon - \epsilon_F)/k_B T] + 1\}^{-1}\,. \tag{21.14}$$

Equation (21.13) is valid provided that $V \ll \bar{v}$, where $\bar{v}$ is the average velocity of the quasiparticles participating in the scattering. This is a less stringent condition than the previous condition 5.

After integration over the angular variables, eqn (21.13) becomes

$$\mu_-^{-1} \equiv \frac{F}{V} = -\frac{g}{6\pi^2\hbar^3}\int p^4\left(\frac{\mathrm{d}n_0}{\mathrm{d}\epsilon} + \delta n\right)\left|\frac{\mathrm{d}\epsilon}{\mathrm{d}p}\right| \sigma_{mt}(p)\,\mathrm{d}p\,, \tag{21.15}$$

where the momentum-transfer scattering cross-section $\sigma_{mt}(p)$ is defined as

$$\sigma_{mt}(p) = \int \frac{\mathrm{d}\sigma}{\mathrm{d}\Omega}(1 - \cos\theta)\,\mathrm{d}\Omega\,. \tag{21.16}$$

In the Knudsen limit, $\delta n = 0$ and eqn (21.15) becomes

$$\mu_-^{-1} = -\frac{g}{6\pi^2\hbar^3}\int p^4 \frac{\mathrm{d}n_0}{\mathrm{d}\epsilon}\sigma_{mt}(p)\,\mathrm{d}p\,. \tag{21.17}$$

This equation is identical to the equation that gives the phonon-limited mobility of the negative ion in superfluid He II (see eqn (5.32)) (Baym *et al.*, 1969) and has been derived by Davis and Dagonnier (1966) by a different method based on the Boltzmann transport equation under more stringent conditions than the present ones.

It should be noted that, in the Knudsen limit and in the Boltzmann region, the drag force $\mathbf{F}$ is proportional to the number density of the quasiparticles N, as can be easily verified by inserting the Maxwell–Boltzmann distribution function $n_0 = (1/2)N(2\pi\hbar^2\beta/m^\star)^{3/2}\exp-(\beta p^2/2m^\star)$ into eqn (21.17). For this reason, the product $\mu_- N$ is independent of N, or, equivalently, the mobility is inversely proportional to the number of available scatterers (Kramer, 1970*b*).

If it is assumed that the differential cross-section neither depends on the velocity of the quasiparticles nor on the velocity of the impurity ions, the momentum transport cross-section is a constant σ_0, and the integral in eqn (21.17) can be evaluated in the limit of low temperature by means of the Sommerfeld expansion (Ashcroft and Mermin, 1976), yielding to leading order in T

$$\mu_- = \frac{3\pi^2 e\hbar^3}{\sigma_0 p_F^4}\left[1 + \frac{1}{3}\left(\frac{\pi k_B T}{\epsilon_F}\right)^2\right]^{-1}\,. \tag{21.18}$$

By recalling that the density of quasiparticles in the $T \to 0$ limit is $N = p_F^3/3\pi^2\hbar^3$, eqn (21.18) can be rewritten as

$$\mu_- = \frac{e}{N\sigma_0 p_F}\left[1 + \frac{1}{3}\left(\frac{\pi k_B T}{\epsilon_F}\right)^2\right]^{-1}\,. \tag{21.19}$$

It is interesting to note that the ion mobility in the $T \to 0$ limit is finite:

$$\mu_0 = \frac{e}{N\sigma_0 p_F} \,. \tag{21.20}$$

The physics is the same as for the case of impurity scattering in metals, in which electrons are scattered by defects bound to a solid lattice that provides the infinite mass necessary to neglect recoil (Ashcroft and Mermin, 1976).

If the condition $T \ll T_F$ is no longer satisfied, there are corrections to the equations derived in the Knudsen limit because of the finite mean free path of the fermions. The quasiparticles can now be considered to form a weakly-interacting Fermi gas. The corrections are calculated in terms of the distortion of the fermionic distribution function induced by the ion motion (Kramer, 1970*a*, 1970*b*) or in terms of the Friedel density oscillations around the impurity ion (Gould and Ma, 1968, 1969).

In the former case, the corrections are given in terms of the ratio of the ion radius R_- to the quasiparticles mean free path, leading to a correction of the coefficient of the T^2 term in eqn (21.19):

$$\mu_- = \frac{e}{N\sigma_0 p_F}\left\{1 - \left[\frac{\pi^2 a m m^{\star 3} J_1}{2^6\pi^5 3\hbar^6 p_F} - \frac{1}{3}\left(\frac{\pi}{\epsilon_F}\right)^2 (k_B T)^2\right]\right\}^{-1} , \tag{21.21}$$

where J_1 is an angular integral over the transition probabilities for collisions between quasiparticles of like and unlike spins (Kramer, 1970*a*) given in terms of the Landau parameters (Landau, 1956, 1957; Wheatley, 1968; Dy and Pethick, 1969). m and $m^\star$ are the bare and effective mass of the atoms of the liquid, respectively.

In the latter case, Friedel density oscillations give rise to a non-analytical correction of the type $T^2 \ln T$:

$$\mu_- = \frac{3\pi}{k_F^2 (k_F R_-)^2}\left[1 + \left(\frac{8}{315}\right)(k_F R_-)\, m^\star V\,(2k_F) \times \left(\frac{k_B T}{\epsilon_F}\right)^2 \ln\left(\frac{\epsilon_F}{k_B T}\right) + \mathcal{O}\left(T^2\right)\right] , \tag{21.22}$$

where the geometrical cross-section of the ion $\sigma_0 = \pi R_-^2$ has been used. $V(2k_F)$ represents the backward scattering amplitude for two quasiparticles at the Fermi surface with opposite momenta.

This correction yields a positive temperature dependence for the mobility, in contrast with eqn (21.19). Unfortunately, the temperature range of the experiments is so restricted that no significant differences between a $T^2 \ln T$ behavior and a pure T^2 behavior can be detected. Moreover, it has been pointed out that this last result has been obtained by making inconsistent approximations, and that it is physically unlikely that a quantum correction should be more important for a large impurity than for a small one (Kramer, 1970*a*; Kramer, 1970*b*).

The previous analysis is actually based on the neglect of recoil. After a collision with a ^{3}He quasiparticle the ion would acquire an additional momentum $\hbar\mathbf{K}$ if it were behaving as a free particle. The ion average recoil energy is then

$$\langle E_{\text{rec}} \rangle = \frac{\hbar^2 K^2}{2M^\star} \,. \tag{21.23}$$

In a Fermi liquid $K \simeq k_F$ and a recoil temperature T_{rec} can be defined as the temperature at which the ion thermal energy equals its recoil energy:

$$T_{\text{rec}} = \frac{\hbar^2 k_F^2}{2k_{\text{B}} M^\star} = \left(\frac{m^\star}{M^\star}\right) T_F \equiv \gamma^2 T_F \ll T_F \,. \tag{21.24}$$

For $T \gg T_{\text{rec}}$, the thermal energy is much larger than the recoil energy, recoil can be neglected, and the elastic scattering model previously described may be used.

A different regime, in contrast, occurs in the opposite limit in which $T \ll T_{\text{rec}}$, or, stated according to Davis *et al.* (1962*b*), when $\xi^{-1} \ll \gamma \ll 1$. A very rough estimate of the behavior of the mobility in this temperature range can be obtained by applying the Drude model for the mobility (Kittel, 1986):

$$\mu \propto \tau \propto \frac{1}{N\sigma V_r} \,, \tag{21.25}$$

where τ is the mean time between collisions, or the *relaxation time*, N is the number density of the scatterers, σ is the momentum-transfer cross-section, and V_r is the mean relative velocity of the ion–quasiparticle system. At the low temperatures under investigation, the velocity of fermions at the Fermi surface is much larger than the thermal velocity of ions. Hence, $V_r \simeq p_F/m^\star$.

In a Fermi liquid, the exclusion principle allows ions to scatter off quasiparticles close to the Fermi surface, i.e., those with momenta in a shell δp_F at p_F, thereby leading to a reduction of the effective cross-section by both the probability of the fermions being in the shell δp_F before the collision and the probability of them being scattered into an empty final state in the same shell:

$$\sigma_{\text{eff}} \propto \sigma \left(\frac{\delta p_F}{p_F}\right)^2 \,. \tag{21.26}$$

Eventually, by recalling that the scatterer density is $N \propto p_F^3$, one obtains

$$\mu \propto \frac{m^\star}{\sigma \left(p_F \delta p_F\right)^2} \propto T^{-2} \,. \tag{21.27}$$

Detailed calculations have been performed by Clark (1963), who solved the Boltzmann transport equation, and by Schappert (1968) within the formalism of linear response theory (Kubo, 1957; Nakano, 1957). The two calculations agree to within a numerical factor of order unity, though the linear response theory approach appears to include polarizability effects in the scattering cross-section. The mobility in the low-temperature regime is thus given as (Schappert, 1968)

$$\mu_-(T \to 0) \propto \frac{e\hbar^3}{\sigma\,(k_{\mathrm{B}}T)^2\,(m^\star + M^\star)^2}\,, \tag{21.28}$$

where the scattering cross-section is calculated in the Born approximation as

$$\sigma = 4\left[a - \frac{1}{2}\alpha_0\left(\frac{e}{4\pi R_-^2}\right)^2 \frac{M_{\mathrm{red}} R_-^3}{\hbar^2}\,,\right]^2 \tag{21.29}$$

where α_0 is the static polarizability of ^{3}He, a is the hard-core radius of the quasiparticle–ion interaction and is taken equal to the ion radius R_-, and $M_{\mathrm{red}} = m^\star M^\star/(m^\star + M^\star)$ is the reduced mass of the ion–quasiparticle system.

If one recalls the definition of the recoil temperature T_{rec}, eqn (21.28) can be cast in the form (Ahonen *et al.*, 1978)

$$\mu_- \propto \frac{8}{\pi^2}\frac{e}{N\sigma p_F}\left(\frac{T_{\mathrm{rec}}}{T}\,.\right)^2 \tag{21.30}$$

A further effect that has been theoretically conjectured to appear at low temperature is the so-called *drag* effect. If the He–He and the ion–He interaction is not very strong, the momentum increase in the ions caused by the applied electric field is partially trapped into the He–ion system. He is not able to remove the excess momentum delivered by the accelerating impurities. As a consequence, the mobility increases when such conditions are met. A variational calculation performed by Bailyn and Lobo (1968) yields for the ion mobility an additional term proportional to T^{-2}, which does not alter the overall temperature dependence of the mobility given by eqn (21.30).

The recoil temperature depends on the effective mass of the ions. The size of the negative ions is calculated by means of the bubble model previously explained in Section 3.2, with the obvious use of the thermodynamic parameters of ^{3}He (Anderson *et al.*, 1968). The electron bubble radius varies between 20.3 Å at vapor pressure and 10.8 Å at 2.88 MPa. The respective values of the effective ion mass are 290 and 91 ^{3}He atomic masses (Ahonen *et al.*, 1978). Thus, T_{rec} is expected to be $T_{\mathrm{rec}} \approx 17\,\mathrm{mK}$ at vapor pressure and $T_{\mathrm{rec}} \approx 100\,\mathrm{mK}$ for $P = 2.88\,\mathrm{MPa}$ (Ahonen *et al.*, 1978).

By inspecting Fig. 21.1, it is easy to realize that the theoretical predictions, at least for the negative ions, fail to reproduce the experimental data. First of all, in the limit $T \to 0$, the mobility remains constant for all pressures without showing the T^{-2} divergence predicted by eqn (21.30). Moreover, at higher T, μ_- increases with increasing T, whereas eqn (21.19) predicts that it should decrease instead.

The actual temperature independence of the negative ion mobility at very low temperatures cannot be understood unless recoil is treated in a more fundamental way in terms of the van Hove scattering function that takes into proper account the dynamics of the ion motion (Josephson and Lekner, 1969).

The impurity ion, having finite mass, is not fixed and the scattering process turns out to be inelastic. The inelastic scattering cross-section per unit solid angle is related to the differential cross-section $\sigma(\theta)$ by

$$\sigma_{\text{in}} = \sigma(\theta) S_{\mathbf{V}} \left(\mathbf{K}, \omega\right) , \tag{21.31}$$

where $S_{\mathbf{V}}(\mathbf{K}, \omega)$ is the dynamic structure factor that enters into the similar problems of neutron or light scattering (van Hove, 1954; Cohen and Lekner, 1967; Stanley, 1971; Lovesey, 1984), $\hbar\mathbf{K}$ and $\hbar\omega$ are the momentum and energy, respectively, transferred in the collision, and $\mathbf{V}$ is the mean ion velocity, i.e., the drift velocity. It is also assumed implicitly that the differential cross-section depends only on the angle θ because the Fermi velocity is much larger than any other velocities in the problem. At the low temperatures involved, the ion thermal momentum is negligible with respect to p_F. Thus, $K = 2k_F \sin(\theta/2)$.

The dynamic structure factor gives the spectrum of the energy absorbed when the ion is instantaneously given momentum hK. Its dependence on the ion velocity is deduced by assuming that the only effect of the external force acting on the ion because of the applied electric field is to superimpose a uniform velocity $\mathbf{V}$ upon the random ion motion. A simple application of Galilean relativity yields

$$S_{\mathbf{V}} \left(\mathbf{K}, \omega\right) = S_0 \left(K, \omega + \mathbf{V} \cdot \mathbf{K}\right) , \tag{21.32}$$

where S_0 is the dynamic structure factor in the absence of the external field.

The rate at which momentum is transferred to the quasiparticle can then be expressed as an integral over the solid angles $\Omega_{i(f)}$ and the energies $\epsilon_{i(f)}$ of the initial and final states of the quasiparticles (Josephson and Lekner, 1969):

$$\frac{\mathrm{d}\mathbf{P}}{\mathrm{d}t} = \frac{3N}{4k_F} \int \mathrm{d}\Omega_i \mathrm{d}\epsilon_i \mathrm{d}\Omega_f \mathrm{d}\epsilon_f \mathbf{K} \sigma(\theta) S_{\mathbf{V}} \left(\mathbf{K}, \omega\right) f\left(\epsilon_i\right) \left[1 - f\left(\epsilon_f\right)\right] , \tag{21.33}$$

where the suffixes i and f indicate the initial and final states of the scattered quasiparticle, respectively, and N is the scatterer density. $\hbar\omega = \epsilon_i - \epsilon_f$ and $\mathbf{K} = \mathbf{k}_i - \mathbf{k}_f$.

Microscopic reversibility and thermal equilibrium enforce detailed balance conditions requiring that $S_0(K, \omega)$ satisfies the relationship

$$S_0\left(K, \omega\right) + S_0\left(K, -\omega\right) = S_0\left(K, \omega\right) \left(1 + e^{-\hbar\beta\omega}\right) , \tag{21.34}$$

where S_0 is actually independent of the direction $\mathbf{K}/|\mathbf{K}|$ and, as usual, $\beta = 1/k_{\mathrm{B}}T$ (Cohen and Lekner, 1967).

Finally, the mobility, or, strictly speaking, the *static* mobility μ_0, i.e., the mobility in the limit of vanishingly-small ion drift velocity $\mathbf{V}$, can be expressed in terms of the intermediate scattering function $F_0(K,t)$ that is the Fourier transform of $S_0(K,\omega)$ (van Hove, 1954; Josephson and Lekner, 1969; Egelstaff, 1994):

$$\begin{aligned}\frac{1}{\mu_0} &= \frac{\pi^2 N p_F}{2e}\int_0^{\pi}\sin\theta\,(1-\cos\theta)\,\sigma(\theta)\,\mathrm{d}\theta\int_{-\infty}^{\infty}\frac{F_0(K,\hbar\beta\tau)}{\cosh^2(\pi\tau/2)}\mathrm{d}\tau\\ &= \frac{\pi^2 N p_F}{4ek_F^4}\int_0^{2k_F}K^3\sigma(K)\,\mathrm{d}K\int_{-\infty}^{\infty}\frac{F_0(K,\hbar\beta\tau)}{\cosh^2(\pi\tau/2)}\mathrm{d}\tau\,. \end{aligned}\tag{21.35}$$

The information on the ion motion during the time interval t is contained in $F_0(K,t)$, which is given as the ensemble average (van Hove, 1954; Lovesey, 1984; Egelstaff, 1994):

$$F_0(K,t) = \langle\exp\{i\mathbf{K}\cdot[\mathbf{r}(t)-\mathbf{r}(0)]\}\rangle\,, \tag{21.36}$$

where $\mathbf{r}(t)$ is the Heisenberg operator for the ion position at time t. Equation (21.35) takes into account in a more general way the effect of the surrounding fluid on the recoil of the ion.

Before proceeding with the calculation of the mobility of the ion in a Fermi liquid, it is instructive to calculate $F_0(K,t)$ from the Langevin equation for a classical Brownian particle in order to see how different regimes of motion are reflected into it and, consequently, in the mobility.

Consider a classical Brownian particle with effective mass $M^\star$ subjected to a stochastic force $\mathbf{F}(t)$ that is uncorrelated with both the position and velocity of the particle. The Langevin equation for the velocity of this particle can be written as (March and Tosi, 1976; Egelstaff, 1994)

$$\frac{\mathrm{d}\mathbf{v}(t)}{\mathrm{d}t} + \zeta\mathbf{v}(t) = \frac{1}{M^\star}\mathbf{F}(t)\,, \tag{21.37}$$

where $M^\star\zeta$ is the friction coefficient.

If both sides are multiplied by $\exp(\zeta t)$ and integrated twice, one easily obtains

$$\begin{aligned}\mathbf{r}(\tau)-\mathbf{r}(0) &= \frac{\mathbf{v}(0)}{\zeta}\left[\exp(-\zeta\tau)-1\right] - \frac{1}{\zeta M^\star}\int_0^{\tau}\mathbf{F}(t)\left\{\exp\left[-\zeta(\tau-t)\right]-1\right\}\mathrm{d}t\\ &= \mathbf{A}+\mathbf{B}\,,\end{aligned}\tag{21.38}$$

where $\mathbf{A}$ and $\mathbf{B}$ are a shorthand notation for the two terms on the right-hand side of this equation.

The intermediate scattering function is easily related to $\mathbf{A}$ and $\mathbf{B}$ by means of eqn (21.36), yielding

$$F_0(\mathbf{K},\tau) = \langle\exp\left[i\mathbf{K}\cdot(\mathbf{A}+\mathbf{B})\right]\rangle\,. \tag{21.39}$$

In the Gaussian approximation, according to the Bloch identity (Lovesey, 1984; Egelstaff, 1994),

$$\langle \exp{(i\mathbf{K}\cdot\mathbf{u})}\rangle = \exp\left[-\frac{1}{2}\langle(\mathbf{K}\cdot\mathbf{u})^2\rangle\right], \tag{21.40}$$

where $\mathbf{u}$ is a normally-distributed variable of zero mean.

By taking the suitable ensemble averages of $\mathbf{A}$ and $\mathbf{B}$, one immediately gets

$$\frac{1}{2}\langle(\mathbf{K}\cdot\mathbf{A})\rangle = \frac{k_{\mathrm{B}}T}{2M^\star}\frac{\left[\exp{(-\zeta\tau)}-1\right]^2}{\zeta^2}K^2\,, \tag{21.41}$$

$$\frac{1}{2}\langle(\mathbf{K}\cdot\mathbf{B})\rangle = DK^2\left\{-\frac{\exp{(-2\zeta\tau)}-1}{2\zeta}+2\frac{\exp{(-\zeta\tau)}-1}{\zeta}+\tau\right\}, \tag{21.42}$$

where the diffusion coefficient $D = k_{\mathrm{B}}T/M^\star\zeta$ has been introduced. A discussion about the unsatisfactory time-reversal asymmetry of eqns (21.41) and (21.42) can be found in Egelstaff (1994).

Nonetheless, the limiting forms of the intermediate scattering function are

$$F_0(K,t) \to \exp{\left(-k_{\mathrm{B}}TK^2\tau^2/2M^\star\right)} \qquad \text{for } \tau\zeta \ll 1\,, \tag{21.43}$$

$$F_0(K,t) \to \exp{\left(-DK^2\tau\right)} \qquad \text{for } \tau\zeta \gg 1\,. \tag{21.44}$$

For $t = 0$, $F_0(K,0) = 1$. In the case that the particle motion has to be treated according to quantum mechanics, the intermediate scattering function for small times is simply multiplied by the phase factor $\exp{(-i\hbar K^2 t/2M^\star)}$ arising from the Heisenberg representation of the position operator in eqn (21.36) (Ahonen *et al.*, 1978).

The friction coefficient can be related to the mobility by means of the Drude relationship $\zeta^{-1} = M^\star\mu_0/e$. For time intervals smaller than ζ^{-1}, starting at some time $t = 0$, the ion moves ballistically as if it were a free particle. However, for later times, the motion is diffusive.

If F_0 decays rapidly for times $|t| < \hbar\beta$, its integral over τ in eqn (21.35) is small and the mobility increases. In contrast, if F_0 decays more slowly with time, its integral contributes more to the inverse mobility and the mobility itself may not increase.

It is convenient to summarize here, in Table 21.1, the forms of the intermediate scattering functions, of the dynamic structure factor, and of the width of the absorption peak in the Gaussian approximation for the three interesting cases that are of relevance here. The regime indicated as *elastic scattering* is appropriate for an indefinitely-massive ion ($M^\star \to \infty$). The width of the absorption peak ΔE_{rec} gives an indication of the elasticity of the collision. It is zero in the case of pure elastic scattering. In the case of diffusive motion, the width is proportional to the temperature:

$$\Delta E_{\mathrm{rec}} \sim \frac{\hbar K^2\mu}{e}k_{\mathrm{B}}T \approx \frac{\hbar k_F^2\mu_0}{e}k_{\mathrm{B}}T\,, \tag{21.45}$$

Table 21.1 *Intermediate scattering function $F_0(K,t)$, dynamic structure factor $S_0(K,\omega)$, and width of the absorption peak $\Delta E_{\rm rec}$ for different regimes of ion motion. (Ahonen et al., 1978.)*

Regime	$F_0(K,t)$	$S_0(K,\omega)$	$\Delta E_{\rm rec}$
Ballistic	$\exp\left(-\frac{k_{\rm B}TK^2}{2M^\star}t^2 - i\frac{\hbar K^2}{2M^\star}t\right)$	$\left(\frac{M^\star}{2\pi k_{\rm B}TK^2}\right)^{1/2} \times \exp\left[-\frac{M^\star\hbar^2}{2k_{\rm B}TK^2}\left(\omega - \frac{K^2}{2M^\star}\right)^2\right]$	$\sim\left(\frac{k_{\rm B}T\hbar^2K^2}{M^\star}\right)^{1/2}$
Elastic scattering	1	$\delta(\omega)$	0
Diffusive	$\exp\left(-K^2D\,\lvert t\rvert\right)$	$\frac{K^2D}{\pi\left[\omega^2+(K^2D)^2\right]}$	$\sim \hbar K^2D$

where the Nernst–Einstein relation $D/\mu_0 = k_{\rm B}T/e$ has been used. The fact that at very low temperatures the recoil ion momentum is practically k_F has also been taken into account.

It can be shown that the results of the theories mentioned above (Abe and Aizu, 1961; Clark, 1963; Davis and Dagonnier, 1966; Schappert, 1968) can be considered as special cases of eqn (21.35) in which the actual F_0 has been replaced by that of an ensemble of free particles of mass $M^\star$ described by a velocity distribution corresponding to thermal equilibrium (Josephson and Lekner, 1969).

The behavior of the mobility strictly depends on the time dependence of the integrand in eqn (21.35). The most relevant time interval corresponds to $|t| \leqslant \hbar\beta$. If the average motion of the ion through the surrounding fluid over an interval $\hbar\beta$ is far from resembling that of a free particle, the effective mass approximation turns out to be a poor one (Bowley and Lekner, 1970).

Using an estimate of the relaxation time derived from the experimental values of the mobility $t^\star \sim \zeta^{-1} = M^\star\mu_0/e$, with the value of $M^\star \approx 390m_3$ estimated by Anderson *et al.* (1968), one obtains that $\hbar\beta > t^\star$ for $T \leqslant T^\star = 0.3\,{\rm K}$. Thus, for $T \leqslant T^\star$, a better approximation than the free-ion model is required for F_0.

The ion can recoil as a free particle only for a very short time before another scattering event occurs. The absorbed energy is not well defined because of the uncertainty principle leading to a shortening of the time interval between two successive scattering events. The fluid is acting so as to reduce the average distance traveled by an ion in a time interval of order $t \sim \hbar\beta$, which is the time required by the uncertainty principle to assign an energy $k_{\rm B}T$ to the ion (Kuchnir *et al.*, 1975). This fact leads to an increase of the real part of F_0, thus reducing the mobility with respect to the prevision of the effective mass approximation (Josephson and Lekner, 1969; Bowley and Lekner, 1970; Chen and Prokof'ev, 1990).

If the coupling of the ion with the fluid is very strong, so that the distance traveled by the ion in a time $\hbar\beta$ is small compared with k_F^{-1}, then $F_0 \approx 1$ for $|t| < \hbar\beta$ and μ reduces to the temperature-independent value given by eqn (21.20). The physical picture is that the recoil energy, in the case of a diffusive motion, may become much less than the thermal energy of the ion (see Table 21.1), so that the restriction on the recoil energies imposed by the Fermi statistics is irrelevant and the mobility remains independent of temperature (Wőlfle *et al.*, 1980).

If the geometrical ion cross-section πR_-^2 is taken for the momentum-transfer cross-section σ_0, the estimated mobility is within a factor of order unity with respect to the experimental low-temperature value (Josephson and Lekner, 1969).

In the Gaussian approximation (Josephson and Lekner, 1969; Bowley, 1971*b*; Lovesey, 1984; Egelstaff, 1994), the intermediate scattering function can be written as

$$F_0\left(K,t\right) = \exp\left[-\frac{1}{2}K^2\gamma\left(t\right)\right], \tag{21.46}$$

where the fluctuation–dissipation theorem (Kubo, 1966) has been used to express the fluctuation velocity spectrum of the ion motion in terms of the frequency-dependent mobility

$$\gamma\left(t\right) = \frac{2\hbar}{\pi e}\int\limits_0^\infty \mathsf{Re}\left[\mu\left(\omega\right)\right]\frac{\cosh\hbar\beta\omega/2 - \cos\omega\left(t - i\hbar\beta/2\right)}{\hbar\beta\omega\sinh\hbar\beta\omega/2}\,\mathrm{d}\omega\,. \tag{21.47}$$

For both t and $\hbar\beta$ larger than the inverse of the lowest frequency at which $\mu(\omega)$ departs significantly from μ_0, the following limiting form of $\gamma(t)$ can be used:

$$\gamma\left(t\right) = \frac{2\hbar\mu_0}{\pi e}\left[\ln\sinh\frac{\pi t}{\hbar\beta} + \ln\frac{\hbar\beta\omega_c}{\pi} + \Gamma\right] - i\mu_0\,\mathrm{sgn}\,t\,, \tag{21.48}$$

where $\Gamma \approx 0.577$ is Euler's constant and the cut-off frequency ω_c is defined in such a way as to satisfy the following relationship:

$$\lim_{\Omega\to 0}\int\limits_\Omega^\infty \mathsf{Re}\left[\mu\left(\omega\right)\right]\mathrm{d}\ln\omega + \mu_0\ln\frac{\Omega}{\omega_c} = 0\,. \tag{21.49}$$

If the lowest frequency is $(\mu_0 M^\star/e)^{-1} \approx 10^{11}\,\mathrm{s}^{-1}$ for both kinds of ion, eqn (21.48) can be used quite safely, though with some precautions, for $T \leqslant 0.1\,$K (Josephson and Lekner, 1969).

Equations (21.35) and (21.47) must be solved self-consistently for μ_0, provided that a specific model for $\mu(\omega)$ is adopted (Bowley, 1971*b*, 1973, 1977*a*). In fact, the rate at which quasiparticles scatter off the ion is proportional to F_0 and $\mu(\omega)$ depends itself on the rate of quasiparticle scattering.

Though none of the models is flawless (Wőlfle *et al.*, 1980), it is interesting to note that, as the mobility is very weakly dependent on the temperature, the dominant contribution to the temperature dependence of $\gamma(\hbar\beta\tau)$ for fixed relaxation time $\tau = \mu_0 M^\star/e$ is originated by the explicit temperature dependence of the term $(2\mu_0/\pi)\ln(\hbar\beta\omega_c/\pi)$ in eqn (21.48). As a result one obtains (Josephson and Lekner, 1969)

$$\frac{\partial F_0\left(K, \hbar\beta\tau\right)}{\partial \ln\beta} \approx \mu_0 K^2 \pi F_0\left(K, \hbar\beta\tau\right) \tag{21.50}$$

and, by means of eqn (21.35),

$$\frac{\partial \mu_0^{-1}}{\partial \ln\beta} \approx -\frac{\langle K^2\rangle}{\pi}, \tag{21.51}$$

where $\langle K^2\rangle$ is the mean square momentum transfer per collision, weighted by a factor $1 - \cos\theta$. It is expected that $0 < \langle K^2\rangle < (8/3)k_F^2$, which is the weighted mean expected for isotropic scattering. Thus, the temperature coefficient of the

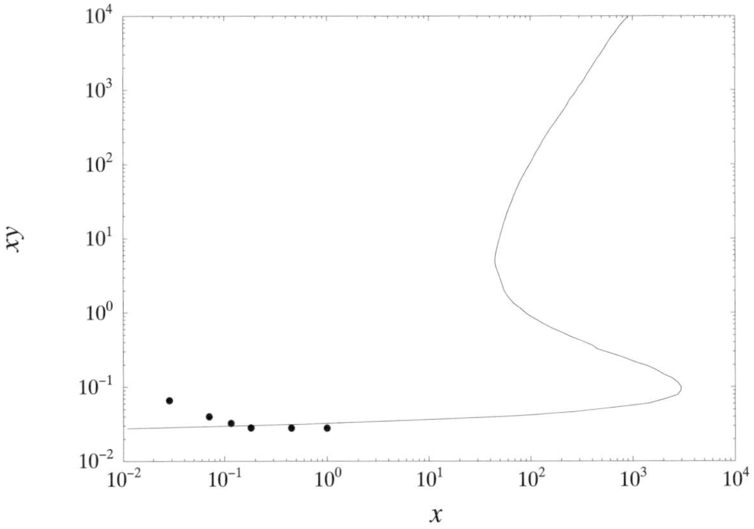

FIG. 21.2. Scaled negative ion mobility $xy\,(= 2\hbar k_F^2\mu_0/\pi e)$ vs inverse temperature $x\,(= 2\hbar^2 k_F^2/M^\star k_B T)$ in liquid ^{3}He (Bowley, 1971*b*). The solid curve is calculated by setting $(k_F R_-)^2 = 210$ and $M^\star = 390 m_3$ in order to match the low-T experimental data (Kuchnir *et al.*, 1970).

mobility $\alpha = \partial \ln \mu_0 / \partial \ln T$ is expected to be negative. An estimate for negative ions is $|\alpha| \leqslant 0.1$, showing a weak, possibly logarithmic, increase of the mobility with decreasing temperature, in much better agreement with the experimental results than the T^{-2} prediction of the elastic scattering models.

In Fig. 21.2 the results of the model with one single relaxation time (Bowley, 1971*b*) are compared with the low-pressure experimental data of Kuchnir *et al.* (1970). The theory of Bowley (1971*b*) agrees quite well with the experiment down to $T \approx 17\,$mK, although the calculated mobility shows a weak increase with decreasing temperature, as also obtained with a more refined approach by Wőlfle *et al.* (1980). For even lower temperatures, the instability of the self-consistent solution for μ_0 is believed to reflect the crude approximation of a single relaxation time.

21.2 Pressure dependence of the mobility

As is easily recognized by inspecting Fig. 21.1, the low-T limiting value of the negative ion mobility in liquid ^{3}He depends strongly on pressure. The reason is to be attributed to the shrinking of the electron bubble as the liquid is pressurized. The situation is identical to the case of He II, as discussed previously (see Section 3.2).

Accurate and extensive measurements of the pressure dependence of the negative ion mobility have been carried out by Ahonen *et al.* (1978) and supplement earlier experimental results (Anderson *et al.*, 1968; Kuchnir *et al.*, 1970).

In Fig. 21.3 the mobility μ_- of negative ions as a function of pressure for $T \approx 20\,$mK in the normal phase of ^{3}He is shown and compared with earlier results. The different data sets agree well to within the experimental accuracy.

Equation (21.20) shows that the main pressure dependence of μ_- stems from the scattering cross-section σ. It is reasonable to assume, owing to the symmetry of the problem, that the differential cross-section is isotropic and that the momentum-transfer cross-section equals the geometrical cross-sectional area of the ion: $\sigma = \pi R_-^2$. The mobility can thus be used to infer the negative ion radius that is plotted in Fig. 21.4.

The ion radius is calculated by adopting the bubble model described earlier in Section 3.2. The equilibrium bubble radius is obtained by minimizing the bubble free energy, eqn (3.21), by assuming a pressure-independent liquid–vapor surface tension $\sigma_{ls} = 1.5 \times 10^{-4}\,$J/m^2 (Lovejoy, 1954, 1955; Zinov'eva, 1956; Atkins and Narahara, 1965; Wilks, 1967).

The ground-state energy of the delocalized electron is calculated as a function of pressure within the modified optical model (Broomall *et al.*, 1976; Fetter, 1976) as

$$V_0\left(P_{\text{loc}}\right) = \left(\frac{2\pi\hbar^2 N_{\text{loc}} a}{m_e}\right)\left[1 + \left(\frac{9\pi N_{\text{loc}}}{2}\right)^{1/3}\right], \qquad (21.52)$$

where N_{loc} and P_{loc} are the local density and pressure, respectively, of the liquid near the wall of the electron bubble. $a = 0.63\,$Å is the electron–atom scattering

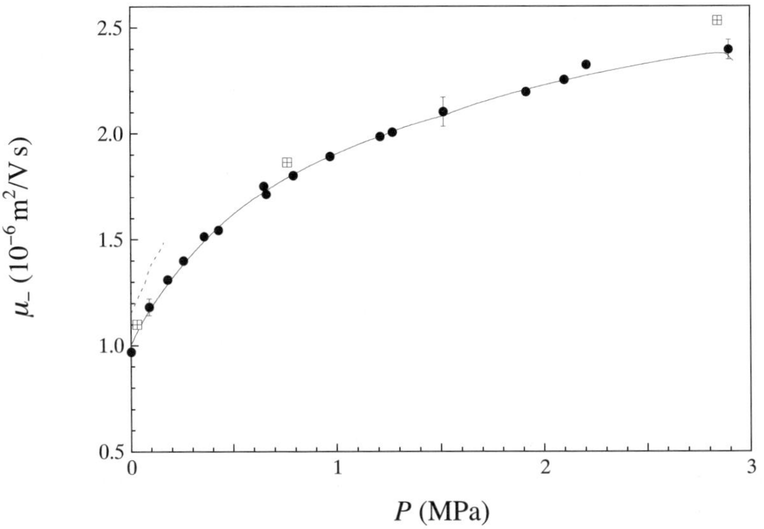

FIG. 21.3. μ_- vs P in normal liquid ^{3}He for $T = 20$ mK (closed circles) (Ahonen *et al.*, 1978). Solid line: eyeguide. Dotted line: Anderson *et al.* (1968). Crossed squares: Kuchnir *et al.* (1970).

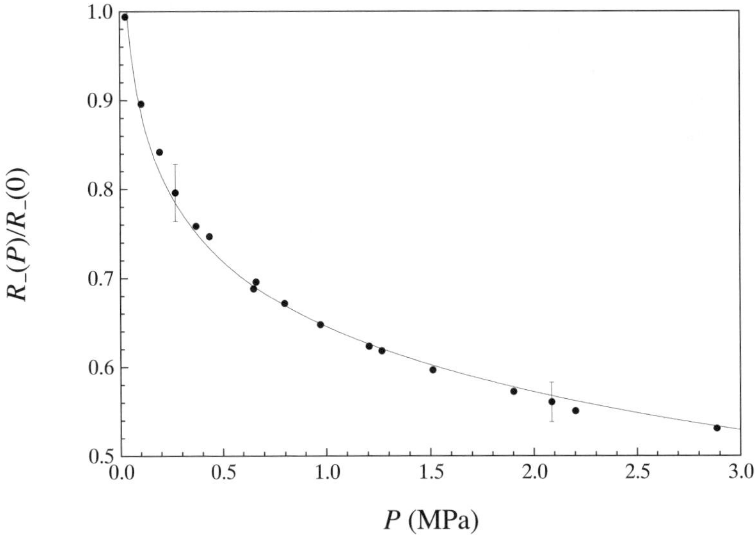

FIG. 21.4. R_- vs P in normal liquid ^{3}He for $T = 20$ mK (Ahonen *et al.*, 1978.) The data are normalized to unity for $P = 0$. Solid line: bubble model.

length (O'Malley, 1963; Tankersley, 1973).

The local values of the pressure and density of the liquid near the ion are used in order to account for the distortion of the fluid near the ion induced by electrostriction. The relationship between the local pressure and the hydrodynamic pressure P in the unperturbed fluid is given by the electrostriction equations (3.6) and (3.7) (Atkins, 1959). In the case of ^{3}He, whose polarizability is quite small, $\alpha \approx 2.8 \times 10^{-40}$ F m, the local pressure $P_{\rm loc}$ can be written in terms of P as

$$P = P_{\rm loc} - \frac{1}{2} \frac{e^2 \alpha N (P_{\rm loc})}{(4\pi\epsilon_0)^2 \left[1 + (\alpha/\epsilon_0) N (P_{\rm loc})\right] R_-^4} . \tag{21.53}$$

The solid line in Fig. 21.4 represents the results of the bubble model and agrees very nicely with the experimental data. Moreover, the theoretical absolute value at $P = 0$, $R_-^t(0) = 21.3$ Å, agrees quite well with the experimental value $R_-^e(0) = (20.3 \pm 2)$ Å (Ahonen *et al.*, 1978).

The results for the electron bubble radius as a function of pressure in ^{3}He are very similar to those obtained for He II (Springett, 1967; Zipfel, 1969; Ostermeier, 1973) (see Fig. 3.6). The slight differences, especially at relatively small pressure, are actually due to the different thermodynamic properties of the two different isotopic liquids and are not related to the different statistics obeyed by them.

21.3 Electric field dependence of the drift velocity

The theories described in the previous sections apply in the zero-velocity limit. A generalization to finite velocities is due to Fetter and Kurkijärvi (1977) within a more general approach developed to treat the case of superfluid ^{3}He.

The starting point is eqn (21.33) for the force imparted to the quasiparticles in the theory of Josephson and Lekner (1969). It can be cast in the form (Fetter and Kurkijärvi, 1977)

$$\frac{\mathrm{d}\mathbf{P}}{\mathrm{d}t} = \frac{2k_F^2}{\pi\hbar} \int \frac{\mathrm{d}\Omega}{4\pi} \int \frac{\mathrm{d}\Omega'}{4\pi} \mathbf{K} \frac{\mathrm{d}\sigma}{\mathrm{d}\Omega} \mathcal{I} , \tag{21.54}$$

where $\mathbf{K} = \mathbf{k} - \mathbf{k}'$, as usual. $\mathcal{I}$ is given by

$$\mathcal{I} = \left(e^{\beta\hbar\mathbf{K}\cdot\mathbf{V}} - 1\right) \int_0^\infty S_0 \left[K, \frac{(\epsilon - \epsilon' - \hbar\mathbf{K}\cdot\mathbf{V})}{\hbar}\right] f(\epsilon) \left[1 - f(\epsilon')\right] \mathrm{d}\epsilon \, \mathrm{d}\epsilon' , \tag{21.55}$$

where $\mathrm{d}\sigma/\mathrm{d}\Omega$ is the differential scattering cross-section and f is the equilibrium Fermi distribution function.

Recalling that the dynamic structure factor S is the Fourier transform of the intermediate scattering function F:

$$S_{\mathbf{V}}(\mathbf{K}, \omega) = (2\pi)^{-1} \int_{-\infty}^{\infty} e^{i\omega t} F_{\mathbf{V}}(\mathbf{K}, t) \, \mathrm{d}t , \tag{21.56}$$

the integrand $\mathcal{I}$ can be evaluated, thus yielding

$$\mathcal{I} = \frac{\pi}{2\beta^2} \tanh\left(\frac{1}{2}\beta\hbar\mathbf{K}\cdot\mathbf{V}\right) \int_{-\infty}^{\infty} \frac{F_0\,(K,t)}{\cosh^2(\pi t/2\hbar\beta)} \left[1 + \frac{\sin^2\left(\frac{1}{2}\mathbf{K}\cdot\mathbf{V}t\right)}{\sinh^2(\pi t/2\beta\hbar)}\right] \mathrm{d}t\,, \tag{21.57}$$

where only the real, even part of F_0 is involved. An expansion to first order in $\mathbf{V}$ yields the results of Josephson and Lekner (1969).

As the mobility of negative ions is quite small, of order $10^{-6}\,\mathrm{m^2/V\,s}$ (Anderson *et al.*, 1968; Kuchnir *et al.*, 1970; Ahonen *et al.*, 1978), the non-dimensional relaxation time $\hbar\beta = \hbar k_F^2\mu_-/e \approx 0.05 \ll 1$ is also small, thus indicating that the ion motion should be treated in the diffusion approximation. The appropriate intermediate scattering function $F_0(K,t)$ is then given by eqn (21.44) and its related $S_0(K,\omega)$ has a Lorentzian form with a finite width K^2D in frequency.

It is convenient to rewrite $F_0(K,t)$ in eqn (21.57) as $1+[F_0(K,t)-F_0(K,0)]$, where 1 is the elastic scattering contribution to F_0 (see Table 21.1), and the terms in brackets represent the diffusive correction. The integral in eqn (21.57) now becomes

$$\mathcal{I} = \hbar^2\mathbf{K}\cdot\mathbf{V} + \frac{\pi}{2\beta^2}\tanh\left(\frac{1}{2}\hbar\beta\mathbf{K}\cdot\mathbf{V}\right) \times \int_{-\infty}^{\infty} \frac{F_0\,(K,t) - F_0\,(K,0)}{\cosh^2(\pi t/2\hbar\beta)} \left[1 + \frac{\sin^2\frac{1}{2}\mathbf{K}\cdot\mathbf{V}t}{\sinh^2(\pi t/2\hbar\beta)}\right] \mathrm{d}t\,. \tag{21.58}$$

The drag force acting on the ion, calculated by inserting this expression for $\mathcal{I}$ into eqn (21.54), contains a term linear in the drift velocity $\mathbf{V}$ resulting from the elastic scattering model, whereas any nonlinear corrections stem from the finite width of S_0.

An estimate of the nonlinearities can be obtained by evaluating the integral in eqn (21.58) to leading order in $D = k_\mathrm{B}T\mu_-/e$ and to second order in $\mathbf{K}\cdot\mathbf{V}$:

$$\mathcal{I} \approx \hbar^2\mathbf{K}\cdot\mathbf{V}\left[1 - \frac{2\ln 2}{\pi}\hbar\beta K^2 D + (\hbar\beta\mathbf{K}\cdot\mathbf{V})\frac{\hbar\beta K^2 D}{6\pi}\left(2 - \frac{9\zeta(3)}{2\pi^2}\right)\right], \tag{21.59}$$

where $\zeta(3) \approx 1.2$ is the Riemann zeta function and $\beta = 1/k_\mathrm{B}T$.

With this approximation for $\mathcal{I}$ the drag force on the ions becomes

$$eE = N\sigma_1 p_F V\left\{1 - \frac{4\ln 2}{\pi e\hbar}p_F^2\mu_-\frac{\sigma_2}{\sigma_1} + \frac{5}{\pi e\hbar}\left[\ln 2 - \frac{9\zeta(3)}{2\pi^2}\right]\beta^2 p_F^4\mu_-\frac{\sigma_3}{\sigma_1}\right\}, \tag{21.60}$$

where $N = p_F^3/3\pi^2\hbar^3$ is the ^{3}He number density and the σ_js are defined as a particular weighted mean of the differential cross-section:

$$\sigma_j = \int \frac{\mathrm{d}\sigma}{\mathrm{d}\Omega}(1-\cos\theta)^j\,\mathrm{d}\Omega\,. \tag{21.61}$$

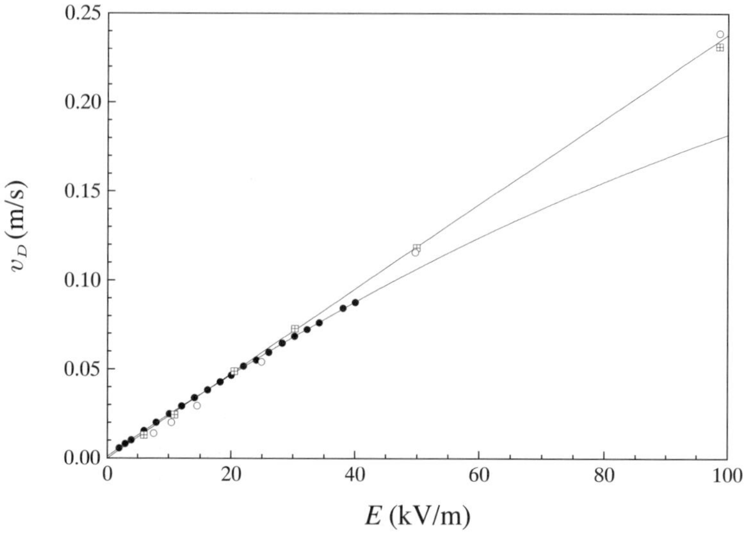

FIG. 21.5. v_D vs E in liquid ^{3}He for $T = 3\,$mK and $P = 2.88\,$MPa (Ahonen *et al.*, 1978) (closed circles), and for $T\,$(mK) $= 4.36$ (open circles) and 35.2 (crossed squares) (Long and Pickett, 1979). Lines: linear and parabolic fits of the data.

The issue of the existence of nonlinearities in the V–E relationship is still to be settled experimentally. In Fig. 21.5 the drift velocity–electric field data for negative ions taken by two different groups are shown (Ahonen *et al.*, 1978; Long and Pickett, 1979). Whereas Ahonen *et al.* (1978) have observed a weak nonlinearity in a limited electric field range for $T = 3\,$mK and $P = 2.88\,$MPa, Long and Pickett (1979) did not observe any nonlinearities in an extended electric field range for all temperatures from $T = 35.2\,$mK down to $T = 4.36\,$ mK for $P = 2.5\,$MPa.

Though the nonlinearity observed in the experiment of Ahonen *et al.* (1978) is of the right order of magnitude as predicted by the theory, nonetheless the electric field and drift velocity ranges are too restricted to make a quantitative comparison with the theory. Even worse, as nonlinearities are not observed by Long and Pickett (1979) in a wider field range and in an extended temperature range, the doubt remains that the nonlinearities observed by Ahonen *et al.* (1978) might be an experimental artifact.

22

POSITIVE IONS IN NORMAL LIQUID ^{3}HE FOR $T < 1$ K

In some respects, the behavior of the positive ions in normal liquid ^{3}He is more complicated than that of negative ions. The early measurements of Anderson *et al.* (1968), reported in Figs 22.1 and 22.2, gave puzzling results both at low as well as at high pressure.

The main features of the experimental results can be summarized as follows. The positive ion mobility μ_+ decreases slowly with decreasing temperature down to ≈ 300 mK and passes through a shallow minimum around $T = 200$ mK. In the temperature range $70\,\text{mK} < T < 100\,\text{mK}$, μ_+ shows a sharp transition to values higher by a factor of approximately 2. Eventually, for $T < 70$ mK, μ_+ increases logarithmically with decreasing T.

Although the early measurements were carried out with a double-gate velocity spectrometer, in which ions are produced by a radioactive source (Anderson *et al.*, 1968), subsequent measurements with a spacecharge-limited field-emission/ionization technique (McClintock, 1973*b*) confirmed those results.

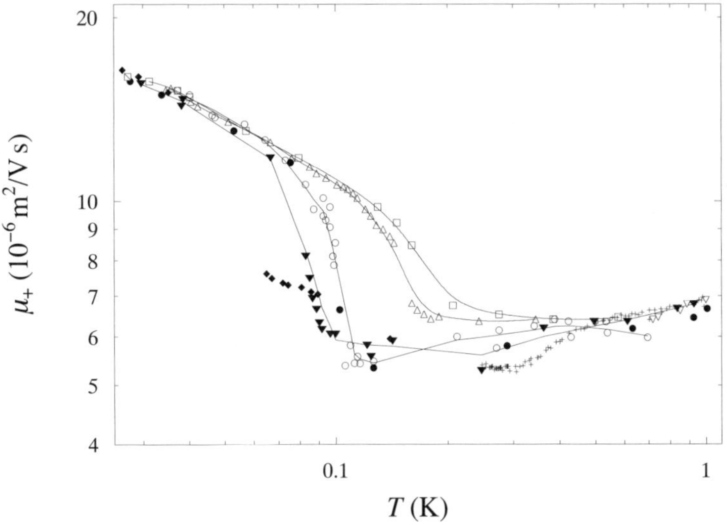

FIG. 22.1. μ_+ vs T in normal liquid ^{3}He. The different symbols refer to pressures in the range $18\,\text{kPa} < P < 104\,\text{kPa}$ for fields in the range $6.2\,\text{kV/m} < E < 20\,\text{kV/m}$ (Anderson *et al.*, 1968). The crosses refer to field-emission measurements (McClintock, 1973*b*). Lines: eyeguides.

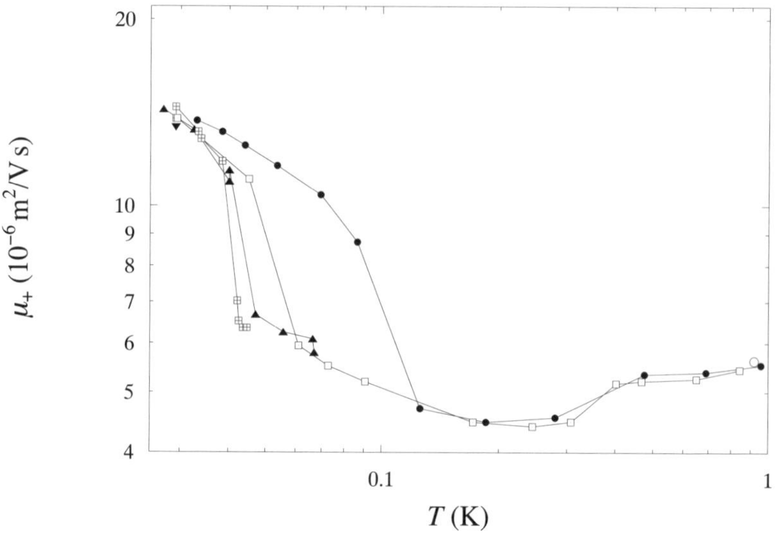

FIG. 22.2. μ_+ vs T in normal liquid ^{3}He for $P = 0.76$ MPa. The different symbols refer to various runs with $6\,\text{kV/m} < E < 20\,\text{kV/m}$. (Anderson *et al.*, 1968.) Lines: eyeguides.

To even greater surprise came more accurate measurements in the range $50\,\text{mK} < T < 500\,\text{mK}$ for $P = 8$ kPa (Kuchnir *et al.*, 1975) and for $P = 0.1$ MPa (Barber *et al.*, 1975). These show a mobility maximum near 200 mK, though, below 200 mK, they exhibit the curiously inconsistent behavior shown in Fig. 22.3. The several curves measured by Barber *et al.* (1975) correspond to different repeated traverses across the mobility maximum near 200 mK.

The puzzling characteristic of the experimental results is the hysteresis at the transition at which the mobility changes its temperature dependence. In the range $70\,\text{mK} < T < 200\,\text{mK}$, one of two distinct temperature-dependent branches could be observed depending on the experimental conditions. Below $T \simeq 70$ mK, a unique mobility is again observed. Thus, the mobility apparently depends on the past history of the sample, on the possibility of permanent polarization of the insulating parts of the experimental cell, or on thermal gradients in it, and so on.

In spite of the inconsistency of the data for $T < 200$ mK, there is no doubt as to the existence of the mobility peak, whose shape suggests a cooperative transition of some sort. As the thermodynamic properties of ^{3}He in this temperature range vary smoothly, it appears that the most probable phenomenon is a temperature-dependent modification of the structural properties of the positive ion. In particular, it has been suggested that ^{4}He impurities may condense in a layer on the surface of the ^{3}He snowball because the ^{4}He atoms occupy a smaller volume than the ^{3}He atoms (Bowley, 1977*a*, 1978).

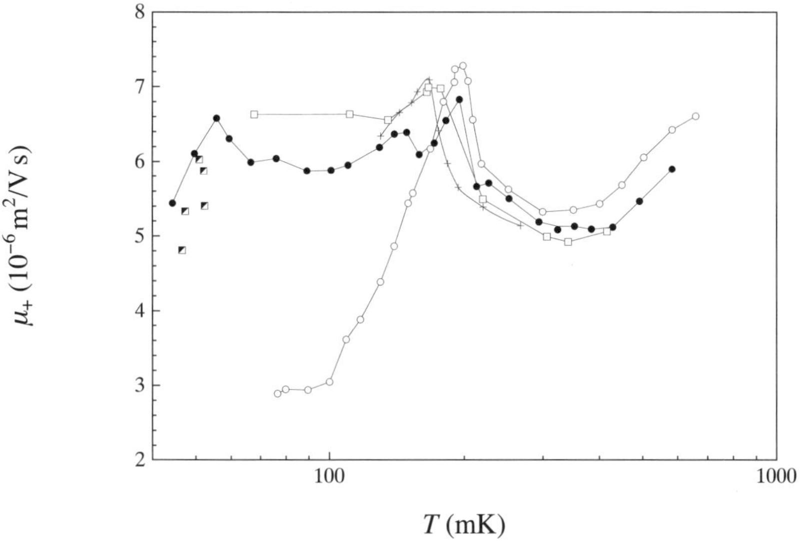

FIG. 22.3. μ_+ vs T in normal liquid ^{3}He for $P = 8$ kPa (open circles) (Kuchnir *et al.*, 1975) and for $P = 0.1$ MPa (Barber *et al.*, 1975). Lines: eyeguides.

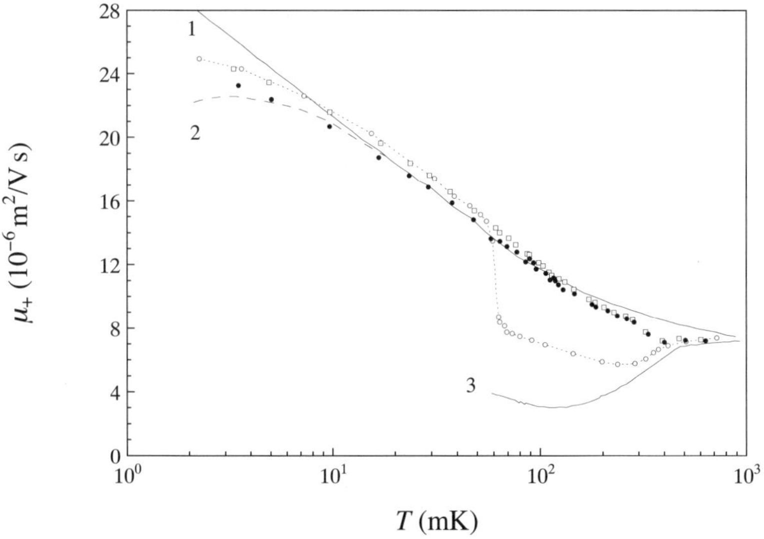

FIG. 22.4. μ_+ vs T in ^{3}He–^{4}He mixtures of different ^{4}He concentrations x_4 for $P = 8$ kPa. x_4 (ppm) = less than 20 (squares), 300 (open circles) (Roach *et al.*, 1977*a*), 30 (closed circles) (Roach *et al.*, 1979). Dotted line: eyeguide. The lines labeled 1, 2 (Bowley, 1977*a*, 1978), and 3 (Sluckin, 1977) are explained in the text.

This phenomenon is believed to also explain the rapid change in the mobility around 70 mK. The hysteresis in the mobility at this transition is a consequence of the presence of multiple stable configurations of the electrostricted snowball around the positive ions contaminated by different amounts of ^{4}He impurities.

Measurements in pure ^{3}He, with a ^{4}He impurity concentration $x_4 < 20\,$ppm, and in a $x_4 = 300\,$ppm mixture have conclusively demonstrated that the anomalous behavior of the ion mobility is due to ^{4}He contamination of the sample (Roach *et al.*, 1977*b*, 1979). In Fig. 22.4, μ_+ is presented as a function of T for three samples of different purity at low pressure. In particular, the pure liquid data correspond to the higher mobility branch, whereas the mixture data correspond to the lower mobility species (Roach *et al.*, 1977*b*, 1979).

Thus, the mobility does not depend only on the interaction with the medium but it depends also on the structure of the positive ion, which will be discussed next.

22.1 The structure of the positive ion in normal liquid ^{3}He

The accepted picture for the positive ion in liquid He is the snowball model (Atkins, 1959). The liquid is treated as a continuum in which a point charge carrier is embedded. The ion–liquid interaction is described by the polarization induced by the point charge on the atoms of the liquid. Polarization and, hence, electrostriction depend on the strength of the atomic polarizability.

If ^{4}He impurities are present in the liquid, they are subjected to a net effective attractive polarization potential in the neighborhood of the ion that can be cast in the form (Bowley and Lekner, 1970; Bowley, 1977*a*)

$$V(r) = \frac{\delta_{43}\alpha e^2}{2\left(4\pi\epsilon_0\right)^2 r^4}, \tag{22.1}$$

where α is the atomic polarizability and $\delta_{43} = \left(v_4 - v_3\right)/v_3 < 0$ is the fractional excess volume of the two isotopes. It is therefore energetically more convenient for a ^{4}He atom to replace a ^{3}He atom near the ion because it can experience more of the attractive polarization potential.

A naive application of equilibrium statistical mechanics would predict that the condensation of ^{4}He atoms should be more effective at low T, implying that the ion radius should increase with decreasing T, the opposite of what one would guess from the observed temperature dependence of the mobility.

More refined calculations based on thermodynamics explain semi-quantitatively the physical situation (Sluckin, 1977). It is predicted that a *halo* of ^{4}He-rich fluid forms around the snowball as the phase-separation curve is reached at low temperatures.

The changes of the local pressure $P(r)$ and ^{4}He concentration $c(r)$ induced by the presence of the ion are calculated by noting that, at equilibrium, the chemical potential must be constant throughout the fluid (Landau and Lifsits, 1958). The case of a pure fluid has already been considered in Section 3.1 and will be repeated here briefly.

The local chemical potential $\tilde{g}_3(r)$ is related to the chemical potential of the unperturbed fluid g_3 and to the ion–fluid interaction potential $\phi_3(r)$ by the following equation:

$$\tilde{g}_3(r) = g_3 - \phi_3(r) \; . \tag{22.2}$$

The interaction potential is the polarization potential $\phi_3(r) = -\gamma/r^4$, where $\gamma = \alpha_3 e^2/[2\epsilon_r(4\pi\epsilon_0)^2] \approx 1.91 \times 10^{-59}\,\mathrm{J/m^4}$. Here, α_3 is the atomic polarizability of ^{3}He and $\epsilon_r \approx 1$ is its dielectric constant.

At constant T and P, the thermodynamic equilibrium condition yields

$$\mathrm{d}P = -n_3 \mathrm{d}\phi_3 \, , \tag{22.3}$$

where n_3 is the number density of ^{3}He. This equation is formally identical to eqn (3.5).

As the ion is approached, the local pressure may exceed the melting pressure P_m and around the ion a solid snowball forms, whose radius R_m is obtained by integrating eqn (22.3) and solving for the snowball radius:

$$R_m = \left\{ \frac{\alpha_3 e^2}{2\left(4\pi\epsilon_0\right)^2} \left[\int\limits_{P_0}^{P_m} \frac{\mathrm{d}P}{n_3(P)} \right]^{-1} \right\}^{1/4} \; . \tag{22.4}$$

Here, P_0 is the pressure in the unperturbed fluid at a large distance from the ion, i.e., the external pressure. For pure ^{3}He, $R_m \approx 6.2\,\text{Å}$ for $P_0 = 0$ (Roach *et al.*, 1979).

In the case of a mixture, electrostriction also induces a concentration gradient in the neighborhood of the ion. In this case, a ^{3}He-rich liquid phase, a ^{4}He-rich liquid phase, a solid mixture phase, and, possibly, solid ^{3}He- and ^{4}He-rich phases may appear depending on the thermodynamic conditions of the sample (Sluckin, 1977). A very schematic phase diagram of liquid ^{3}He–^{4}He mixtures is given in Fig. 22.5. As the mobility measurements are carried out at constant ^{4}He concentration, the structure of the ion is sought as a function of temperature under this constraint.

The tricritical point occurs for $T_t = 0.867\,\mathrm{K}$ at zero pressure. For $T > T_t$, the concentration of ^{4}He around the ion can increase but no phase separation takes place. The local pressure keeps increasing on approaching the ion until P_m is reached for $R = 6.2\,\text{Å}$ and the solid snowball appears.

Below T_t, both the local pressure and concentration follow a path in the thermodynamic space as the ion is approached. They both increase as the distance from the ion decreases, and approach the freezing curve and the phase separation curve. If the critical concentration for phase separation is reached first before the freezing one, the liquid will phase separate and a ^{4}He-rich *halo* will form around the solid snowball.

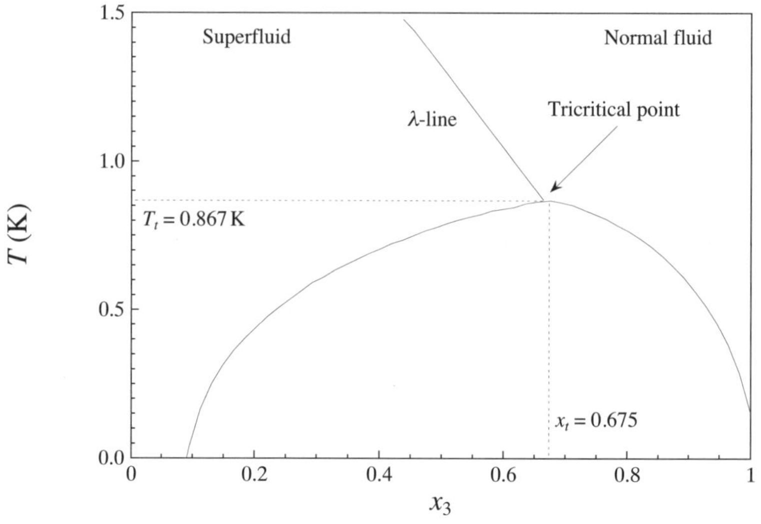

FIG. 22.5. Schematic phase diagram of ^{3}He–^{4}He mixtures.

The Gibbs free energy G of a dilute solution containing N_4 ^{4}He atoms, N_3 ^{3}He atoms, and one ion can be written as (Landau and Lifsits, 1958; Guggenheim, 1977)

$$G = N_3 g_{3,0}(P,T) + N_4 k_B T \left[\ln \frac{N_4}{N_3} - 1 \right] + N_4 \psi(P,T) + (N_3 + N_4)\,\phi\,, \quad (22.5)$$

where it has been assumed that $\phi_3 \approx \phi_4 \equiv \phi$ (Roach *et al.*, 1979). $g_{3,0}$ is the chemical potential in the absence of ^{4}He atoms, and ψ has yet to be specified.

The chemical potential for the two He isotopes are obtained, as usual, by differentiating the Gibbs free energy with respect to the number of each species:

$$g_3 = \left(\frac{\partial G}{\partial N_3} \right)_{P,T} = g_{3,0}(P,T) - k_B T x + \phi(r)\,, \quad (22.6)$$

$$g_4 = \left(\frac{\partial G}{\partial N_4} \right)_{P,T} = k_B T \ln x + \psi(P,T) + \phi(r)\,, \quad (22.7)$$

where $x = N_4/N_3$ is the concentration.

At equilibrium, T, g_3, and g_4 are all constants. Thus, the equilibrium condition yields

$$v_3 \mathrm{d}P - k_B T \mathrm{d}x + \mathrm{d}\phi = 0\,, \quad (22.8)$$

$$v_4 \mathrm{d}P + k_B T \mathrm{d}\ln x + \mathrm{d}\phi = 0\,, \quad (22.9)$$

where $v_3 = \partial g_{3,0}/\partial P$ and $v_4 = \partial\psi/\partial P$ are the specific volumes of the two isotopic atoms.

As a dilute solution is considered, $N_3 \gg N_4$. Thus, one can write $v_3 = 1/n_3$, where n_3 is the number density of the ^{3}He atoms. The specific ^{4}He volume is defined as

$$v_4 = \frac{1}{n_3\left(1+\alpha\right)}, \tag{22.10}$$

where $\alpha \approx -0.32$ (Kerr, 1954; Boghosian *et al.*, 1966) is the fractional excess volume occupied by a ^{4}He atom in liquid ^{3}He (Saam and Laheurte, 1971).

By alternately eliminating $\mathrm{d}x$ and $\mathrm{d}P$ in eqns (22.8) and (22.9), the following differential equations are obtained for P and x, respectively:

$$\mathrm{d}P = -n_3\frac{\left(1+x\right)\left(1+\alpha\right)}{1+\alpha+x}\mathrm{d}\phi\,, \tag{22.11}$$

$$\mathrm{d}x = -\frac{\alpha x}{k_{\mathrm{B}}T\left(1+\alpha+x\right)}\mathrm{d}\phi\,. \tag{22.12}$$

Equation (22.11) is a generalization of eqn (22.3) to the case of mixtures and describes an approximate r^{-4} dependence of the local pressure. Similarly, eqn (22.12) predicts that the concentration x increases nearly exponentially with r^{-4}. Actually, recalling that $x \ll 1$, the integration of eqn (22.12) is easily performed, yielding

$$x\left(r\right) = x\left(\infty\right)\exp\left[\alpha\gamma/\left(1+\alpha\right)k_{\mathrm{B}}Tr^4\right], \tag{22.13}$$

where $x(\infty)$ is the bulk ^{4}He concentration.

At lower temperatures, the phase-separation curve is reached before the freezing curve is met. Thus, a halo of ^{4}He-rich fluid forms around the solid snowball (Sluckin, 1977). Its radius is determined by the condition

$$x\left(R_H\right) = x_0\left(T\right), \tag{22.14}$$

where $x_0(T)$ is the concentration at which phase separation occurs for a given temperature. x_0 is given by

$$x_0\left(T\right) = 0.85T^{3/2}\exp\left(-T_x/T\right), \tag{22.15}$$

where $T_x = 0.56\,\mathrm{K}$ (Saam and Laheurte, 1971). The insertion of eqn (22.15) into eqn (22.13) immediately gives the halo radius as

$$R_H = \left\{\frac{\alpha\gamma}{\left(1+\alpha\right)k_{\mathrm{B}}\left[\frac{3}{2}T\ln T/\left(0.85x_0\right)^{2/3} - T_x\right]}\right\}^{1/4}, \tag{22.16}$$

where $x_0 = x(\infty)$.

R_H would diverge as $x \to x_0$ if it were not prevented from doing so by the surface tension σ_{34} between the two phases. An increase of the halo radius would increase the surface energy. So, ^{4}He is energetically favored to keep condensing onto the cell walls rather than accreting the halo, whose radius remains practically constant.

In the presence of a free surface the grand thermodynamic potential Ω has to be minimized:

$$\mathrm{d}\Omega = P\mathrm{d}V + S\mathrm{d}T + N\mathrm{d}g + \sigma_{34}\mathrm{d}A\,, \qquad (22.17)$$

where $\sigma_{34}\mathrm{d}A$ is the surface energy contribution. In equilibrium, T and the chemical potential g are constant in both phases.

Owing to the inhomogeneous pressure caused by the ion, eqn (22.17) is integrated at constant T and g, yielding

$$\Omega_H = \int_0^{R_s} P_s(r)\,\mathrm{d}V + \int_{R_s}^{R_H} P_4(r)\,\mathrm{d}V + \int_{R_H}^{\infty} P_3(r)\,\mathrm{d}V + 4\pi\left[R_s^2\sigma_{4s} + R_H^2\sigma_{34}\right] \qquad (22.18)$$

and

$$\Omega_s = \int_0^{R_s} P_s(r)\,\mathrm{d}V + \int_{R_s}^{\infty} P_4(r)\,\mathrm{d}V + 4\pi R_s^2\sigma_{3s}\,. \qquad (22.19)$$

Ω_H and Ω_s are the grand thermodynamic potential in the presence and absence of the halo, respectively. The suffixes s, 4, and 3 refer to the solid, ^{4}He-rich, and ^{3}He-rich phases, respectively. σ_{ij} are the appropriate surface tensions separating the different phases. Equation (22.18) describes a snowball surrounded by the halo, whereas eqn (22.19) describes a bare snowball.

The halo radius is obtained by minimizing Ω_H with respect to r, yielding

$$P_4(R_H) - P_3(R_H) = 2\frac{\sigma_{34}}{R_H}\,. \qquad (22.20)$$

Equation (22.20) has to be solved by investigating the behavior of the local pressure (Cole and Sluckin, 1977)

$$P(r) = P\left[\tilde{g}_3(r), \tilde{g}_4(r), T\right]\,, \qquad (22.21)$$

with the condition of constant chemical potential:

$$\tilde{g}_{3,4}(r) + \phi(r) = g_{3,4}\,, \qquad (22.22)$$

where $g_{3,4}$ is the chemical potential of the appropriate species for $r \to \infty$: $g_{3,4} = g_{3,4}[x(\infty), T, P(\infty)]$.

P is a multivalued function. It takes on different values in the various phases: P_s in the solid, P_4 in the ^{4}He-rich phase, and P_3 in the ^{3}He-rich phase. The halo forms as soon as eqn (22.21) is satisfied. However, the solution does not exist for all temperatures. For instance, no halo forms for temperatures above those for which phase separation occurs. In this case, there is only a solid snowball whose radius R_s is obtained by minimizing Ω_s in eqn (22.19) with respect to R_s:

$$P_s(R_s) - P_3(R_s) = 2\frac{\sigma_{3s}}{R_s}\,. \qquad (22.23)$$

Well above these temperatures, as the r^{-1} dependence of the surface pressure contribution is much weaker than the r^{-4} dependence of the electrostrictive pressure, surface tension can be neglected and the radius is given by eqn (22.16).

In the contrast, near or below the phase-separation temperature, the surface tension must be taken into account. The pressure due to electrostriction in the two phases is approximately given by

$$P_{3,4}(r) = P_{3,4}(\infty) + \frac{\gamma n_{3,4}}{r^4}, \tag{22.24}$$

where $P_3(\infty) = P_4(\infty)$. Equations (22.24) and (22.20) together yield

$$R_H = \left[\frac{\gamma(n_4 - n_3)}{2\sigma_{34}}\right]^{1/3}. \tag{22.25}$$

The temperature dependence of the radius R_H at vapor pressure for a ^{4}He concentration $x = 3 \times 10^{-4}$ is shown in Fig. 22.6. The phase separation for this concentration occurs for $T_0(x) \approx 108\,\text{mK}$. Below T_0, R_H is described by eqn (22.25) and is a constant, independent of temperature. For $T > T_0$, R_H is described by eqn (22.16) and decreases rapidly with increasing temperature (Sluckin, 1977).

A rough comparison with the experimental mobility data can be made by assuming that the ion–^{3}He scattering cross-section is purely geometrical: $\sigma = \pi R_H^2$. By assuming the validity of eqn (21.20), the mobility in pure ^{3}He and in the mixture should scale according to the ratio $(R_s/R_H)^2$. Thus, the mobility $\mu_{+,x}$ in the mixture is approximately given by

$$\mu_{+,x} = \mu_+ \left(\frac{R_s}{R_H}\right)^2, \tag{22.26}$$

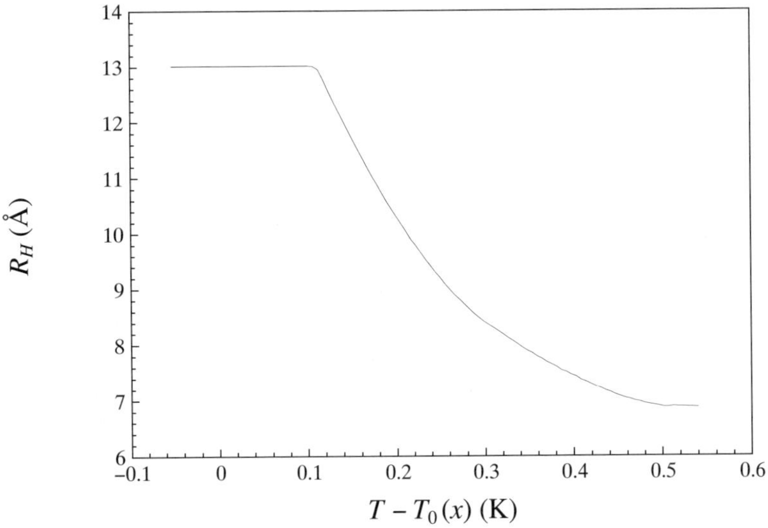

FIG. 22.6. R_H vs $T - T_0(x)$ at vapor pressure for a ^{4}He concentration $x = 3 \times 10^{-4}$. (Sluckin, 1977.) The phase-separation temperature is $T_0(x) = 0.108\,\text{K}$.

where μ_+ is the ion mobility in pure liquid ^{3}He.

The results of this calculation, using the data of Roach *et al.* (1977*b*) for μ_+ in the pure liquid, are shown as curve 3 in Fig. 22.4 and agree semi-quantitatively with the experimental data for the corresponding mixture (Sluckin, 1977).

The disappearance of the halo for $T < 60\,$mK is probably related to the amount of ^{4}He available for the formation of the halo and to the ion generation technique specifically adopted in that experiment (Alexander *et al.*, 1977, 1979). It may also happen that, at low temperatures, the ^{4}He content in the sample is so depleted by adsorption on the cell walls that the ion may not acquire a full halo during its too short transit time and the ion may not be in equilibrium with ^{4}He (Bowley, 1977*a*; Roach *et al.*, 1977*b*).

A direct, qualitative confirmation of this halo model for the structure of the positive ion in a ^{4}He$-^3$He mixture is given by Leiderer and Wanner (1978), who measured the mobility of both positive and negative ions in a tricritical mixture using a time-of-flight method (Ahonen *et al.*, 1976, 1978).

In a tricritical mixture (see Fig. 22.5) the ^{4}He concentration is quite high, $x = 0.325$, but the behavior of the halo radius is expected to be similar to that in low-concentration mixtures. At such high concentration, the bulk phase separation occurs at the tricritical temperature $T_t = 0.867\,$K. x remains constant for $T > T_t$ and decreases along the ^{3}He-rich side of the phase diagram down to $x = 0.09$ for $T = 0.5\,$K.

The results for the mobilities for both positive and negative ions in the tricritical mixtures are presented in Fig. 22.7. Whereas the mobility of negative ions passes smoothly through the tricritical point without any influence due to critical fluctuations (Leiderer *et al.*, 1974) and approaches the behavior of the pure liquid at lower temperatures because the ^{4}He concentration diminishes with decreasing T, the mobility of positive ions decreases rapidly on approaching T_t from above and remains nearly constant for $T < T_t$. The change of slope in the temperature dependence of the mobility indicates that the phase separation has occurred. It should be noted that this change of slope appears to be slightly shifted to lower T with respect to T_t at vapor pressure. This fact is attributed to the elevated pressure near the ion surface.

The positive ion radius R_+ can be estimated by again assuming that $\mu_+ \propto R_+^2$. The mobility drop yields a radius change of $\approx 20\%$. Assuming that the ion radius in pure liquid ^{3}He far above T_t is $R_+ \approx 7\,\text{Å}$, below T_t it turns out to be $R_+ \approx 8.5\,\text{Å}$ (Leiderer and Wanner, 1978). This value is nearly 35% smaller than the radius calculated for $T < 0.3\,$K according to eqn (22.25) (Sluckin, 1977). At higher T, the surface tension decreases and the computed halo radius is still higher. As an example, $R_H \approx 15\,\text{Å}$ at $T = 0.5\,$K and $R_H \approx 30\,\text{Å}$ at $T = 0.8\,$K (Leiderer and Wanner, 1978). In particular, on approaching T_t, the surface tension is expected to vanish as $\sigma_{34} \propto t^2$ and the density difference as $n_4 - n_3 \propto t$, where $t = 1 - T/T_t$ (Widom, 1972). According to eqn (22.25), the halo radius ought to diverge as $t^{-1/3}$.

The experimental data, however, do not show any such divergence below T_t.

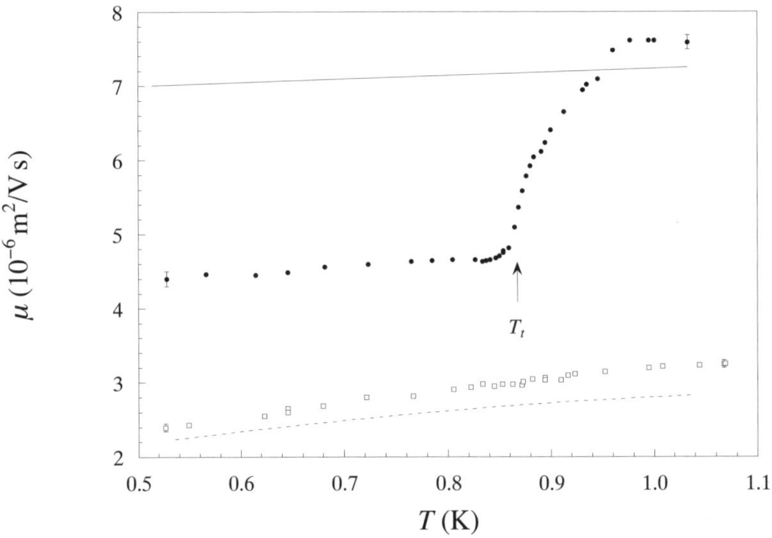

FIG. 22.7. μ_+ (circles) and μ_- (squares) vs T in a tricritical ^{3}He–^{4}He mixture at SVP (Leiderer and Wanner, 1978). Solid line: μ_+ in pure liquid ^{3}He (Roach *et al.*, 1977*b*). Dashed line: μ_- in pure liquid ^{3}He (Anderson *et al.*, 1968). T_t: tricritical temperature.

Instead they show a temperature-independent radius, which is ≈ 1.5 Å larger than the ion radius at 1 K (Leiderer and Wanner, 1978). This difference amounts to less than one atomic layer. Moreover, it must be recalled that the finite thickness of the interface between the halo and the surrounding liquid has not been taken into account, thus making the quantitative agreement with the halo model even worse.

Thus, the conclusion is that the formation of a halo around the positive ion according to the halo model is very probable, though the prediction of the model does not agree well quantitatively with the experimental findings (Leiderer and Wanner, 1978).

23

MULTIPLE SPECIES OF POSITIVE IONS IN NORMAL LIQUID ^{3}HE

The physical picture of the structure of the positive ion in liquid ^{3}He described by the halo model allows us to understand the hysteretic behavior of ion mobility at the transition for temperatures in the range $70\,\mathrm{mK} < T < 200\,\mathrm{mK}$ (see, for instance, Fig. 22.1). The obvious conclusion is that measurements were not always performed on pure samples, but on mixtures of different and uncontrolled concentration, which is dependent on the past history of the sample.

The apparent erratic behavior of the mobility disappears if measurements are performed in samples of known concentration. The discontinuity in the mobility occurs at different temperatures for different concentrations, as shown in Fig. 23.1 for measurements at low pressure $P = 8\,\mathrm{kPa}$ (Roach *et al.*, 1979). A similar shift of the mobility discontinuity to lower temperatures for samples of increasing ^{4}He concentration is also observed at higher pressure up to nearly 3 MPa (Alexander *et al.*, (1977, 1979); Roach *et al.*, 1979).

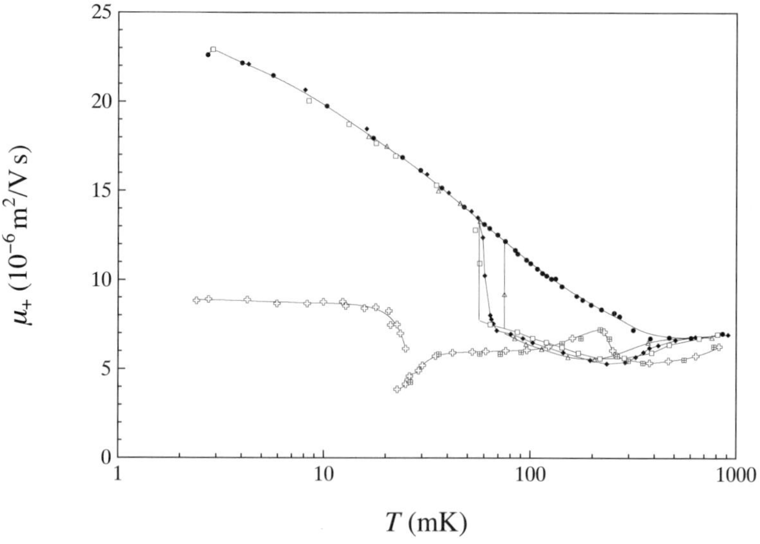

FIG. 23.1. μ_+ vs T for $P = 8\,\mathrm{kPa}$ in ^{3}He–^{4}He mixtures of ^{4}He concentration x_4(ppm) $= 30$ (circles), 100 (triangles), 300 (closed diamonds), 1000 (open squares), 3000 (crosses), and 10 000 (crossed squares). (Roach *et al.*, 1979.) Lines: eyeguides.

Below $T \approx 200$–300 mK, the mobility in the more dilute samples lies below the values of the pure liquid. This situation remains the same down to temperatures in the range $30\,\mathrm{mK} < T < 70\,\mathrm{mK}$, depending on the concentration and pressure. In this range, the mobility in the mixtures rejoins abruptly the pure liquid data.

The interpretation of the behavior of the mobility is quite straightforward (Roach *et al.*, 1979). Near $T \approx 300$ mK a phase separation occurs and the halo forms around the solid ion core. By lowering T the halo radius increases, thereby also increasing the scattering cross-section of the ion and reducing the mobility in the mixtures with respect to the pure liquid.

When T is further lowered down to 30–70 mK, a diminishing bulk ^{4}He content (Saam and Laheurte, 1971) and a decrease in the ion absorption cross-section (Bowley, 1977*a*) eventually induce a starvation of the ion to a point where it cannot gather enough ^{4}He atoms to form an equilibrium halo during the drift time, which is relatively short in comparison to the relaxation time for adsorption (Alexander *et al.*, 1977; Roach *et al.*, 1979). Once the temperature at which this condition is met is reached, the temperature dependence of the mobility in the mixture rapidly reverts to the pure liquid behavior.

However, an even more intriguing and unexpected phenomenon, first observed by Alexander *et al.* (1977), occurs for $T \leqslant 30$ mK at higher pressures ($P \geqslant 2$ MPa), namely, the appearance of additional ion species whose mobilities are lower than the pure liquid value but have similar temperature dependence (Alexander *et al.*, 1977, 1979; Alexander and Pickett, 1978; Roach *et al.*, 1979). The number of different species increases with increasing P. Some of these observations are presented in Figs 23.2 and 23.3.

This peculiar behavior implies the existence of a family of metastable halos, possibly arising from stable shell structures about the ion core (Namaizawa, 1978). In the presence of ^{4}He impurities in the liquid, there may be competition between the two isotopic species for incorporation in the solid snowball. Alternatively, the condensation of a solid ^{4}He layer around the ^{3}He solid core or even an attraction of ^{4}He atoms in the high-density, but still liquid, halo region outside the snowball may occur (Alexander *et al.*, 1979).

The existence of so many additional ionic species with apparently different radii, whose appearance depends in a complicated way on the sample history, is related to the particular kind of ion injection technique used (Alexander *et al.*, (1977, 1979); Roach *et al.*, 1979.)

These measurements have been carried out with a field emission/ionization spacecharge-limited technique. It is conjectured (Alexander *et al.*, 1979) that the roughness of the emission tip causes large inhomogeneities in the emission current. As a consequence, there may be spots around the tip where the local current density is very high. In these regions there is a local increase of the liquid temperature and of the ^{4}He solubility. The solid helium electrostricted around the emission tip by the strong local field may be richer in ^{4}He than the bulk liquid. The snowballs around the ions are formed here and may contain a non-equilibrium amount of ^{4}He. Upon entering the bulk liquid, the ^{4}He concen-

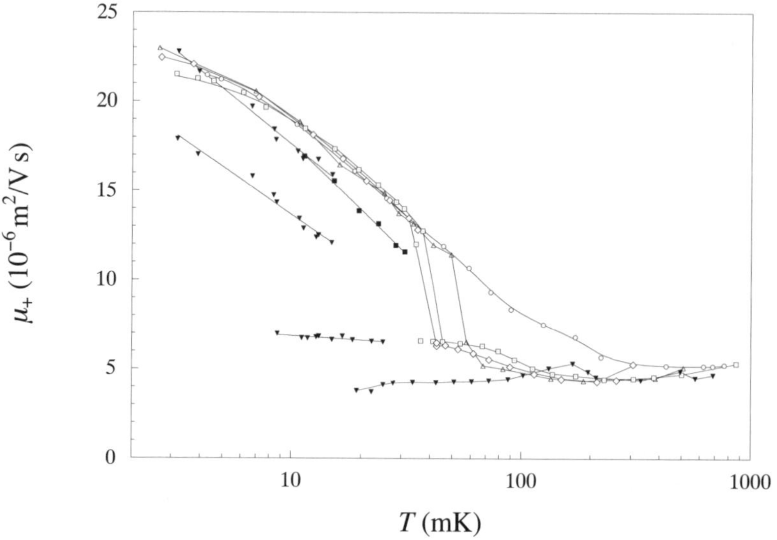

FIG. 23.2. μ_+ vs T in ^{3}He–^{4}He mixtures for P = 1 MPa. (Roach *et al.*, 1979.) x_4(ppm) = 30 (open circles), 100 (open triangles), 300 (open diamonds), 1000 (squares), and 3000 (inverted triangles). Lines: eyeguides.

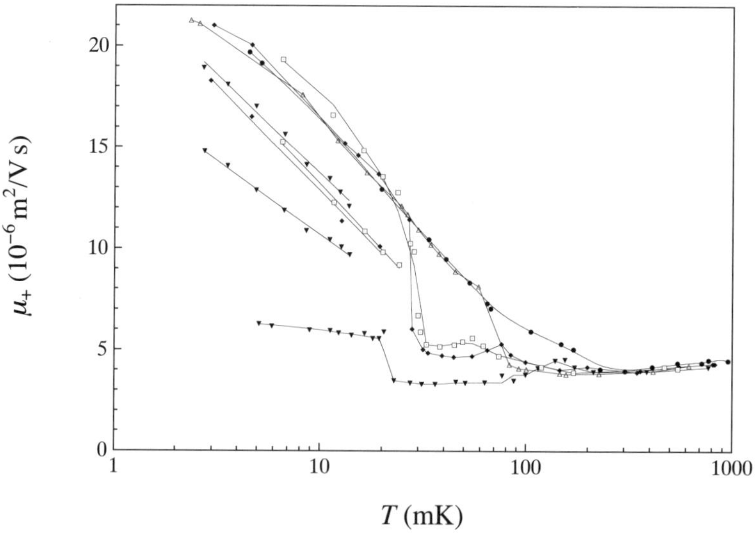

FIG. 23.3. μ_+ vs T in ^{3}He–^{4}He mixtures for P = 2MPa. (Roach *et al.*, 1979.) x_4(ppm) = 30 (circles), 100 (open triangles), 300 (open squares), 1000 (closed diamonds), and 3000 (inverted triangles). Lines: eyeguides.

tration in the snowball cannot change because of its vanishing solubility at low temperature.

The spectrum of different ionic species is thus related to the ^{3}He/^{4}He ratio in the snowball and in the halo. Moreover, the presence of multiple ionic species is observed at high pressure, at which a ^{4}He film coats all the surfaces. It is thus conjectured that its behavior influences the production of the different ions. If it is mobile, the ^{4}He around the tip is in equilibrium with the ^{4}He in the rest of the cell and no large local increase of its concentration in the tip region is possible. In contrast, if the film solidifies at higher pressures, there may be such a local increase.

At higher temperatures, for which the ^{4}He solubility is much larger and cannot be neglected, some ^{4}He atoms in the snowball may dissolve in the bulk liquid, and some of them in the liquid may condense on the ion, realizing a dynamical equilibrium of the ^{4}He/^{3}He ratio in the snowball. Depending on the dominant process, the mixture of different ions is expected to relax to an equilibrium configuration with a stable ^{4}He/^{3}He ratio at a rate that depends on the ^{4}He solubility.

However, as this solubility is such a rapidly-varying function of T, the equilibrium ions should appear abruptly at a critical temperature at which the relaxation time is comparable to a typical experimental drift time, and a discontinuity will appear if the mobilities before and after relaxation are quite different (Alexander *et al.*, 1979).

This picture is supported by the observation that, if condensation of ^{4}He around the snowball is responsible for the relaxation to an equilibrium configuration, the critical temperature of the discontinuity would depend on the sample concentration, as appears to be the case (for instance, see Figs 23.2 and 23.3).

At higher temperatures, above the discontinuity, the ^{4}He solubility is large enough for the ^{3}He ions to be completely glazed with ^{4}He, yielding the observed temperature independence of the mobility (Alexander *et al.*, 1977; Bowley, 1977*a*).

23.1 Temperature dependence of the mobility of multiple positive ions

As multiple ions appear only at low temperature and high pressure, and because the fastest of them approach the pure liquid value, it is assumed, albeit not conclusively, that the single ionic species present at lower pressure below the discontinuity consists of a solid snowball of pure ^{3}He (Alexander *et al.*, 1979). This is a very reasonable hypothesis for accounting for the experimental observations.

Once the existence and nature of the multiple positive ions is recognized and is related to the amount of ^{4}He impurities in the liquid sample, it is very interesting to investigate how their mobility depends on the temperature.

Bowley (1977*a*) has developed a theory that leads to a generalization of the theory of Josephson and Lekner (1969) than can be applied to finite frequencies. It yields a prediction for the real part of the frequency-dependent inverse mobility:

$$\mathrm{Re}\frac{1}{\mu(\omega)} = \frac{\hbar\left[\tanh(\hbar\beta\omega/2)\right]}{6e(\hbar\beta\omega/2)} \times \int_0^{2k_F} k^3\sigma(k)\,\mathrm{d}k \int_{-\infty}^{\infty} F_0(k,t)\,\frac{\cosh(\pi t/\hbar\beta) - \cos(\omega t)}{\sinh^2(\pi t/\hbar\beta)}\frac{\mathrm{d}t}{\hbar\beta}. \quad (23.1)$$

Equations (21.35), (21.47), and (23.1) form a set of coupled equations. There is, however, the need for a closure relation relating $\mathrm{Re}\mu(\omega)$ in eqn (21.47) to $\mathrm{Re}[1/\mu(\omega)]$ in eqn (23.1). A possible accomplishment of this goal has been suggested by Bowley (1977*a*), who used the following *Ansatz*:

$$\frac{1}{\mu(\omega)} = \mathrm{Re}\frac{1}{\mu(\omega)} + i\omega\frac{M^\star}{e} \quad (23.2)$$

or, equivalently,

$$\mathrm{Re}\mu(\omega) = \frac{\mathrm{Re}\left[\mu(\omega)^{-1}\right]}{\mathrm{Re}\left[\mu(\omega)^{-1}\right]^2 + \left(\frac{\omega M^\star}{e}\right)^2}, \quad (23.3)$$

where $M^\star$ is the ionic effective mass, and contributions to $\mathrm{Re}\mu(\omega)$ may come from radiation and scattering of phonons (Bowley, 1973).

$\mathrm{Im}[\mu(\omega)]^{-1}$ provides a cut-off for $\mathrm{Re}\mu(\omega)$ and eqn (23.2) assumes that it is identical to the single-relaxation-time model introduced earlier (Bowley, 1971*b*):

$$\frac{1}{\mu(\omega)} = \frac{1}{\mu_+} + i\omega\frac{M^\star}{e}, \quad (23.4)$$

where μ_+ is the static mobility, i.e., the mobility measured in time-of-flight experiments.

The theory expresses the non-dimensional quantity $(\hbar/e)k_F^2\mu_+ = \Delta E_{\mathrm{rec}}/k_\mathrm{B}T$, which is the width of the absorption peak and is a measure of the elasticity of the collisions, as a function of the reduced temperature T_0/T, with $T_0 = \hbar e/k_\mathrm{B}M^\star\mu_+$. T_0 is a measure of the temperature at which the relaxation time $\mu_+M^\star/e$ equals the time $\hbar/k_\mathrm{B}T_0$ required by the uncertainty principle to assign the ion an energy $k_\mathrm{B}T_0$.

In Fig. 23.4 some model results of the calculation of the mobility are shown for several values of the parameter $c = 12/(k_F R_+)^2$ (Bowley, 1977*a*). For $T_0/T \geqslant 10$, the theory predicts that the ion mobility is linear in $\ln T$ (Bowley, 1977*a*). A similar logarithmic dependence is predicted by more refined calculations (Wőlfle *et al.*, 1980). A good agreement with the data is obtained by suitably choosing the values of the effective ion mass and its radius.

The mobility calculated for $c = 0.344$, corresponding to $R_- \approx 7.5$ Å, and with $T_0 \approx 0.95$ K, corresponding to $M^\star \approx (44 \pm 4)m_3$, is plotted as curve 1 in Fig.

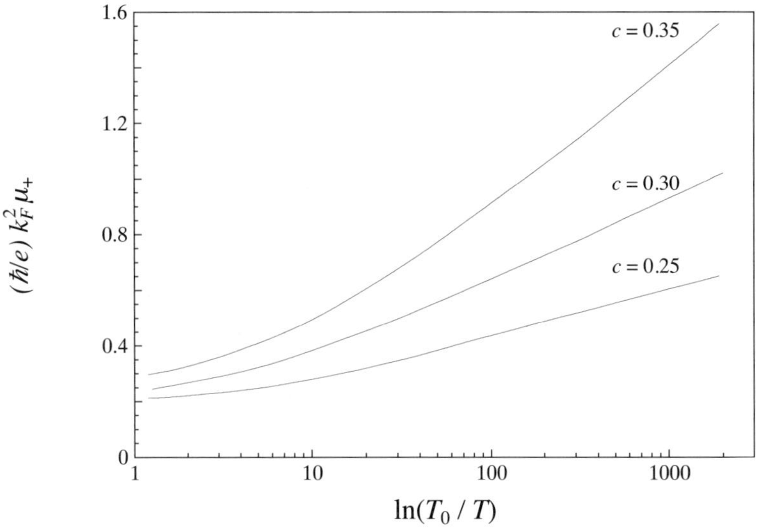

FIG. 23.4. $(\hbar/e)k_F^2\mu_+$ vs T_0/T for several values of the parameter c. (Bowley, 1977*a*.)

22.4. The agreement with the experiment is remarkably good. Only at the lowest temperatures does the theoretical prediction deviate from the experimental data. The reason for this behavior is that the previous theory is valid in the limit of vanishingly-small drift velocities, i.e., equivalently, in the limit of electric fields $E \to 0$.

There is, actually, an important velocity scale in the problem in addition to the drift velocity v_D and the Fermi velocity v_F (Alexander and Pickett, 1979). Suppose that the ion can be modeled as a smooth sphere endowed with infinite mass, traveling at speed v_D, and colliding with the quasiparticles moving at speed v_F. Upon head-on collisions, the quasiparticles are backscattered specularly in the forward direction and gain an energy $\Delta\epsilon = 2v_D p_F$. If $\Delta\epsilon \approx k_B T$, the restriction on the final scattering states because of the Pauli principle is lifted and more states are available for the ion to scatter. Thus, it experiences a large increase in the drag. This is equivalent to a decrease of the ratio v_D/E that falls below its zero-field value, which is usually taken as the ion mobility (Alexander *et al.*, 1977; Roach *et al.*, 1977*a*).

The onset of nonlinear behavior is expected to occur as soon as the ion drift velocity exceeds a threshold velocity $\tilde{v} \sim 2k_B T/\hbar k_F$, which is large enough to allow energy transfers of order $k_B T$ during ion–quasiparticle scattering events (Alexander *et al.*, 1978). In the extreme high-field limit, for which $v_D \gg \tilde{v}$, only a small fraction of the scattered quasiparticles either starts or ends up within the region of partially-populated states at the Fermi surface, the temperature of the liquid becomes irrelevant, and a temperature-independent behavior of the mobility is expected (Alexander *et al.*, 1978; Bowley, 1978; Alexander and

Pickett, 1979).

The mobility data of Roach *et al.* (1979) presented in Fig. 22.4 are actually v_D/E values obtained in an electric field range around 10^2 kV/m in which non-linear effects are present. As the reduction of the mobility below its zer–field value is largest at the lowest temperatures, a downward bending of the apparent mobility with decreasing temperature is expected and actually observed. The behavior of the electric field dependence of the mobility will, however, be described later.

23.1.1 *Below the discontinuity*

Accurate measurements of the positive ion mobility below the temperature T_D of the discontinuity for several pressures in pure ^{3}He samples or in mixtures of controlled impurity content have been carried out by different groups, yielding a consistent picture of the temperature dependence of μ_+ for temperatures down to less than 2 mK.

Below T_D and for pressures below 2 MPa, only one ionic species is observed both in the pure liquid ($x_4 < 30$ ppm) (Roach *et al.*, 1977*b*; Kokko *et al.*, 1978) and in a controlled-concentration mixture ($x_4 = 500$ ppm) (Alexander and Pickett, 1978; Alexander *et al.*, 1979). Particular care must be taken to determine the drift mobility as the v_D/E ratio in the linear velocity regime down to $E \approx 1$ kV/m. In fact, nonlinear effects are increasingly strong particularly, at low temperatures for a field as low as 3 kV/m (Kokko *et al.*, 1978), as presented in Fig. 23.5.

The mobility of positive ions behaves according to the $\ln T_0/T$ prediction (Bowley, 1977*a*; Wőlfle *et al.*, 1980) if measurements are actually carried out at low enough electric field strength. In Fig. 23.6 the zero-field mobility is shown as a function of temperature for some pressures in a pure sample with $x_4 < 30$ ppm (Kokko *et al.*, 1978). The mobility data of Kokko *et al.* (1978) show a nice linear relationship with $\ln T$, whereas the data of Roach *et al.* (1977*b*) show a downward bending due to the fact that their measurements are performed for $E \approx 20$ kV/m, well inside the nonlinear region in which the mobility is below its zero-field limit.

The way in which pressure influences the logarithmic temperature dependence of the mobility has been investigated accurately in mixtures of constant concentration, $x_4 \approx 500$ ppm (Alexander *et al.*, 1978, 1979). Below $P = 2$ MPa, only one ionic species is present. Above this pressure, several ionic species may appear in a way that depends on the sample history. Depending on the voltage of the emission tip, first a single ion is observed that is termed a *fast ion.* If the emission tip voltage is increased, a second, slower ion appears that is called a *slow ion.* By further increasing the tip voltage, several ionic species of different mobility are created. In any case, the dominant species are the slow and fast ions. The mobility of these multiple ions as a function of temperature is shown in Fig. 23.7.

For $P > 2$ MPa, the history of the sample can be manipulated in such a way that this species, termed a *fast ion*, is also present together with the other ions

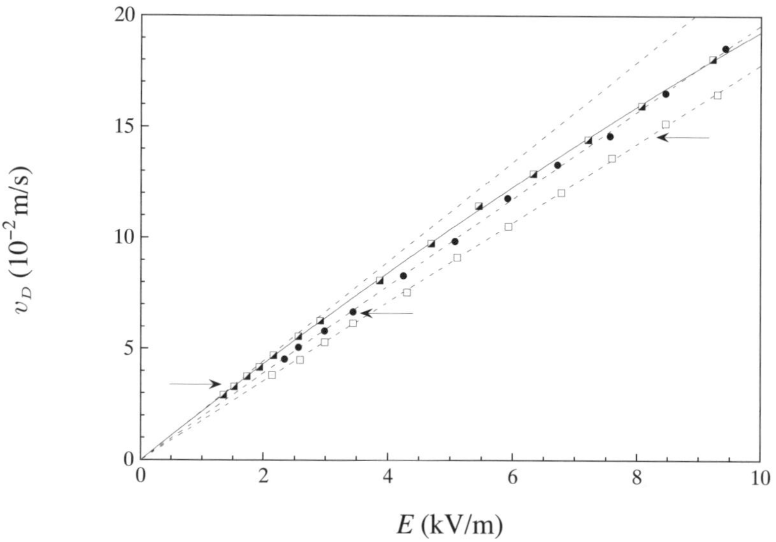

FIG. 23.5. v_D vs E in pure normal liquid ^{3}He for $P = 0.6\,$MPa showing nonlinear behavior. (Kokko *et al.*, 1978.) $T\,$(mK) $= 2.66$ (half-filled squares), 5.14 (circles), and 8.42 (dotted squares). Dashed lines: extrapolations from the linear velocity region. Arrows indicate the inelastic threshold velocity $k_\mathrm{B}T/p_F$.

of different mobility. As the temperature and pressure dependencies of the fast ion join smoothly with those of the single ions for $P < 2\,$MPa, it is conjectured that both species are the same ion and that they are made up of snowballs of pure ^{3}He (Alexander *et al.*, 1979).

The mobility of the fast or single ions is plotted as a function of T for several P in Fig. 23.8. Owing to the nonlinearity of v_D as a function of E, the drift velocity is measured in a field range $9\,\mathrm{kV/m} < E < 20\,\mathrm{kV/m}$. As v_D can be empirically expanded as a power series in E, $v_D = \mu_+ \left(1 - aE + bE^2 + \ldots\right)$, in which the quadratic term is negligible in this field range, one can write $v_D^{-1} = \mu_+^{-1} E^{-1} (1 + aE)$ and obtain the mobility as $\mu_+ \approx \mathrm{d}E^{-1}/\mathrm{d}v_D^{-1}$.

For all pressures, the mobility obeys a logarithmic law as a function of T that can be described by

$$\mu_+ (P, T) = \mu_0 + A\,(P) \ln{(T_0/T)}\,. \tag{23.5}$$

By fitting eqn (23.5) to the experimental data, the values $T_0 = 0.7\,$mK and $\mu_0 = 32.3 \times 10^{-6}\,\mathrm{m^2/V\,s}$ are obtained (Alexander and Pickett, 1978; Alexander *et al.*, 1979). The slope coefficient $A(P)$ is shown in Fig. 23.9 as a function of P.

Interestingly, μ_+ decreases as P is increased, as directly seen in Fig. 23.10. This behavior is the exact opposite of that of negative ions (see, for instance, Fig. 21.1).

The radius of the ion can be guessed by using the theory of Bowley (1977*a*)

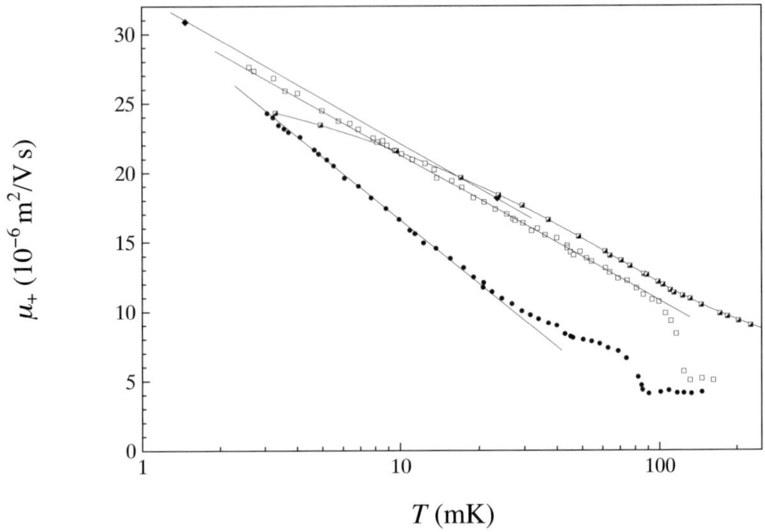

FIG. 23.6. μ_+ vs T in pure normal liquid ^{3}He (Kokko *et al.*, 1978). P (MPa) = 0.05 (diamonds), 0.6 (dotted squares), and 2.8 (circles). Straight lines: logarithmic fits. Half-filled squares: low-P data (Roach *et al.*, 1977*b*). Solid curve: eyeguide.

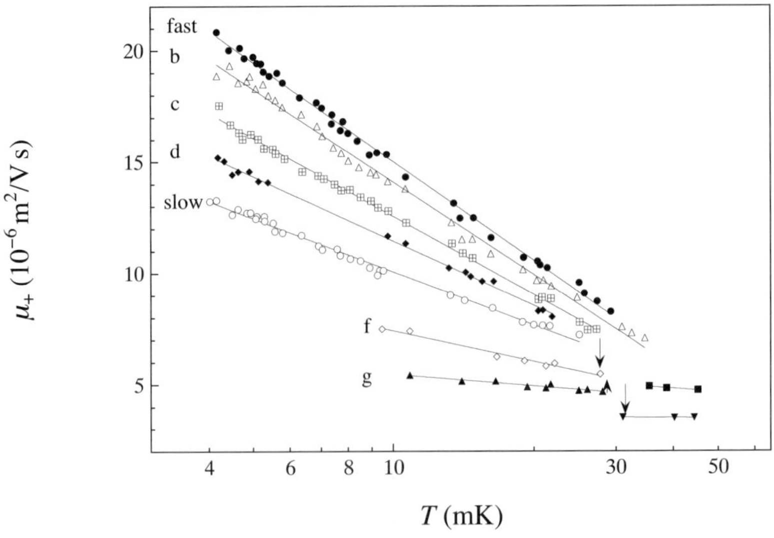

FIG. 23.7. μ_+ vs T for multiple positive ions in normal liquid ^{3}He at P = 2.9 MPa. (Alexander *et al.*, 1979.) Arrows indicate the discontinuity. Lines below the discontinuity: logarithmic fits.

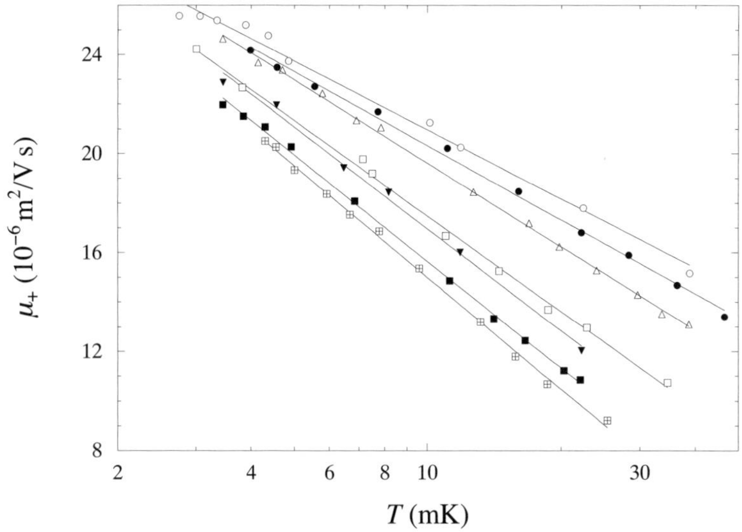

FIG. 23.8. Mobility of the fast or single ions vs T. ((Alexander and Pickett, 1978), 1978; Alexander *et al.*, 1979.) From top: P = 0.16, 0.5, 1.2, 2.0, 2.3, 2.7, and 2.9 MPa.

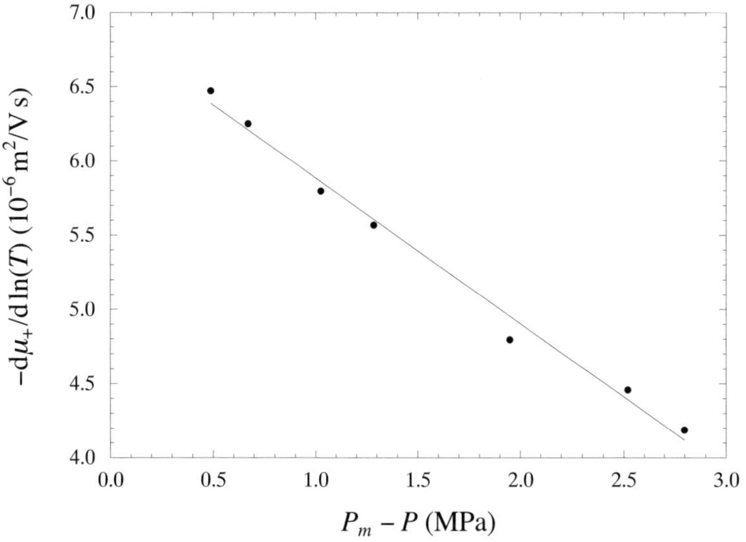

FIG. 23.9. $A(P)$ vs $P_m - P$. The melting pressure of ^{3}He is $P_m \approx 3.44$ MPa. (Alexander and Pickett, 1978; Alexander *et al.*, 1979.) Line: eyeguide.

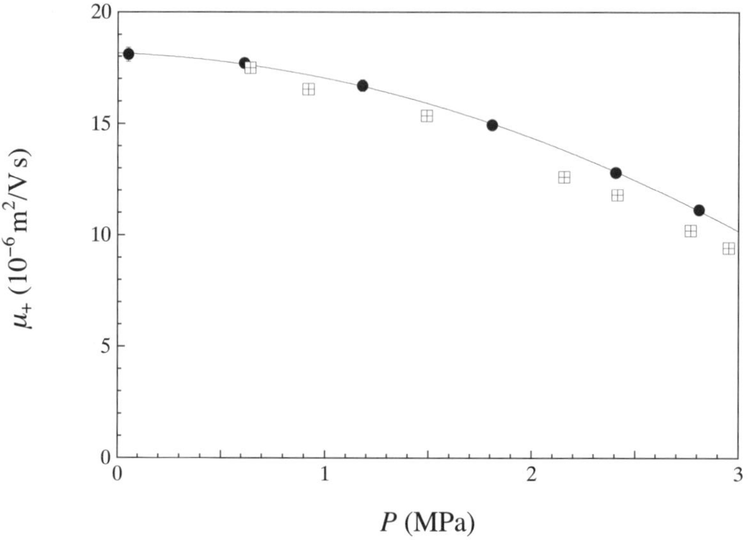

FIG. 23.10. μ_+ vs P for $T = 24$ mK. Circles: experiment (Kokko *et al.*, 1978). Squares: data calculated with eqn (23.5) using the $A(P)$ values of Fig. 23.9. Line: eyeguide.

that relates it to the slope of the mobility as a function of $\ln T$.

Surprisingly, the ion radius decreases from ≈ 7.5 Å at low pressure down to ≈ 6 Å for $P = 2.9$ MPa (Alexander and Pickett, 1978; Kokko *et al.*, 1978; Alexander *et al.*, 1979). This is contrary to the prediction of the snowball model (Atkins, 1959) that predicts an increase of the ion radius upon approaching melting.

This contradiction has not yet been settled. It has been suggested that the ion may have a layered structure and that an increase of pressure might not be sufficient to establish an additional layer, but it simply compresses the core further (Alexander *et al.*, 1979). On the other hand, the Fermi momentum and the density of quasiparticles increase toward higher pressures leading, to an increased scattering rate and to a further reduction of the ion mobility.

Thus, inferring the ion radius from the mobility data is not as straightforward as it might be supposed. Evidently, the observed effect might be an artifact of the theory, whose assumption of a constant differential scattering cross-section might not be reasonable. However, it must be recalled (Alexander *et al.*, 1978) that the work of Scaramuzzi *et al.* (1977*a*) may be interpreted as evidence for the decrease of the positive ion radius in liquid ^{4}He as the melting pressure is approached.

Of the other ionic species, only for the so-called *slow* ion is there sufficient experimental sensitivity to investigate the temperature dependence as a function of pressure (Alexander *et al.*, 1979). The results are shown in Fig. 23.11. There are some differences with the *fast* ions. In particular, the slope $\mathrm{d}\mu_+/\mathrm{d}\ln T$ decreases

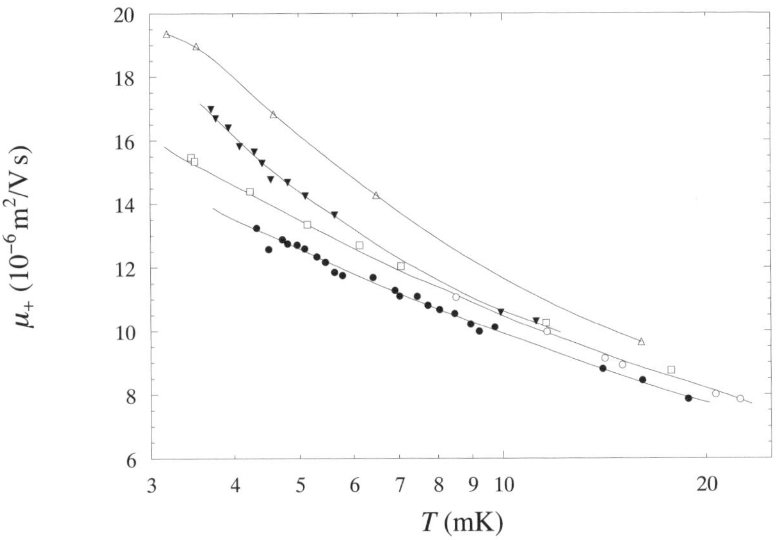

FIG. 23.11. μ_+ vs T for the *slow* ions. (Alexander *et al.*, 1979.) P (MPa) $= 2.3$ (triangles), 2.5 (inverted triangles), 2.7 (squares), and 2.9 (circles). Lines: eyeguides.

with increasing P. According to the theory of Bowley (1977*a*), this means that the ion radius increases with P, as the snowball model would predict (Atkins, 1959).

Once more, nothing can be stated about these ions with certainty. Probably, their structure might be layered and the increase of pressure might progressively stabilize the outermost layer.

23.1.2 *At the discontinuity*

The behavior of the positive ion mobility at the discontinuity is quite complicated. The temperature T_D at which, upon decreasing temperature, ions make a transition to higher mobility values depends on pressure, sample purity, and, unfortunately, on the pressure history of the sample itself.

A careful analysis of the ionic signal reveals that a transition between ions of different mobilities occurs at constant drift field when the temperature crosses the discontinuity temperature (Alexander *et al.*, 1977; Alexander *et al.*, 1979.) The high-mobility ions convert to low-mobility ones during the time-of-flight. This transition can be observed by inspecting the drift time as a function of the inverse electric field if a non-gating technique is used in the velocity spectrometer apparatus.

In Fig. 23.12 the ionic transit time τ is plotted as a function of the inverse drift field E^{-1} for $T = 58.8\,\mathrm{mK}$ and $P = 0.5\,\mathrm{MPa}$ in the discontinuity region of the mobility (Alexander *et al.*, 1977; Alexander *et al.*, 1979).

In this temperature region the nonlinear behavior of the drift velocity is

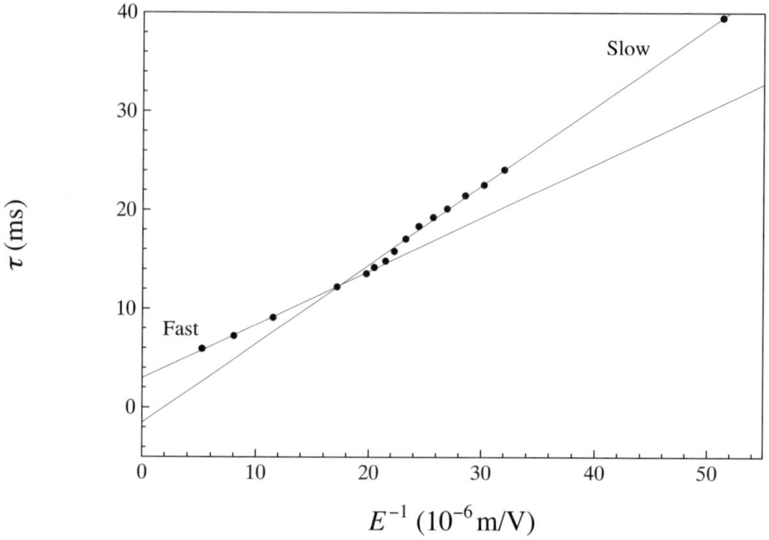

FIG. 23.12. τ vs E^{-1} for $T = 58.8$ mK and $P - 0.5$ MPa in the discontinuity region. (Alexander *et al.*, 1977; Alexander *et al.*, 1979.) Labels "Slow" and "Fast" refer to the type of ions.

absent and the mobility can be obtained as $\mu_+ = d[1/(\mathrm{d}\tau/\mathrm{d}E^{-1})]$, where d is the drift length and τ is the transit time. A large slope means a small mobility and vice versa.

Whereas τ is a linear function of E^{-1} with a unique slope either for $T < T_D$ or $T > T_D$, at $T = T_D$ there is a kink at intermediate transit times and τ shows two different limiting values of the slope.

If ions travel at the speed $v_f = \mu_f E$ for a time τ_f in the fast state and travel the remaining distance $d_s = d - d_f$ at the speed $v_s = \mu_s E$ for a time τ_s in the slow state, the total transit time is

$$\tau = \frac{d}{\mu_s E} + \tau_f \left(1 - \frac{\mu_f}{\mu_s}\right). \tag{23.6}$$

If the transit time is short, the conversion from fast to slow ions takes places only partially, if not at all. Thus, $\tau_f \approx \tau$, $d_f \approx d$, and eqn (23.6) reduces to $\tau = d/\mu_f E$, yielding a large apparent mobility.

Conversely, for drift times longer than the inverse conversion rate, the conversion to slow ions is complete, eqn (23.6) reduces to $\tau = d/\mu_s E$, and a slow mobility is measured. The kink appears when the transit time is comparable to the inverse rate of ion conversion.

Moreover, the extrapolation to $E^{-1} = 0$ of the slow ion portion of the curve intersects the transit time axis at the time $\tau = t_f(1 - \mu_f/\mu_s) < 0$, thus indicating that ions reaching the collector with that mobility have not had enough time to cross the whole cell and must have started with a larger mobility (Alexander *et al.*, 1977).

The positive intercept of the fast ion portion of the curve at $E^{-1} = 0$ only means that the transit time also includes the time spent by the ion crossing the distance between the gating grids before entering the true drift space.

From curves like the one shown in Fig. 23.12, the ion conversion time, defined as the transit time for which the slope variation is largest, can be extracted as a function of temperature. The result is presented in Fig. 23.13 (Alexander *et al.*, 1979).

The temperature dependence of the inverse of the ion conversion time, i.e., the ion conversion rate, is described by an exponential law of the inverse temperature:

$$\tau^{-1} = 4.4 \times 10^8 \, T^{3/2} \exp\left(-T_i/T\right), \qquad (23.7)$$

with τ^{-1} expressed in s^{-1} and with $T_i = 0.47\,\mathrm{K}$ (Alexander *et al.*, 1979).

Not surprisingly, this is nearly the same exponential dependence as for the ^{4}He solubility, eqn (22.15) (Saam and Laheurte, 1971). The small difference between $T_x = 0.56\,\mathrm{K}$ and $T_i = 0.47\,\mathrm{K}$ may be attributed to the not very accurate determination of T_i, owing to the limited temperature range in which the ion conversion time has been determined. The similarity between the exponential dependence of the conversion rate and of the solubility lends some credibility to the idea that the conversion process is triggered by the solubility of ^{4}He.

Unfortunately, the extreme dependence of the discontinuity on the sample history, and on the conditions of the emission tip and its related hot spot (up to seventeen different ionic species have been recorded in some conditions) does

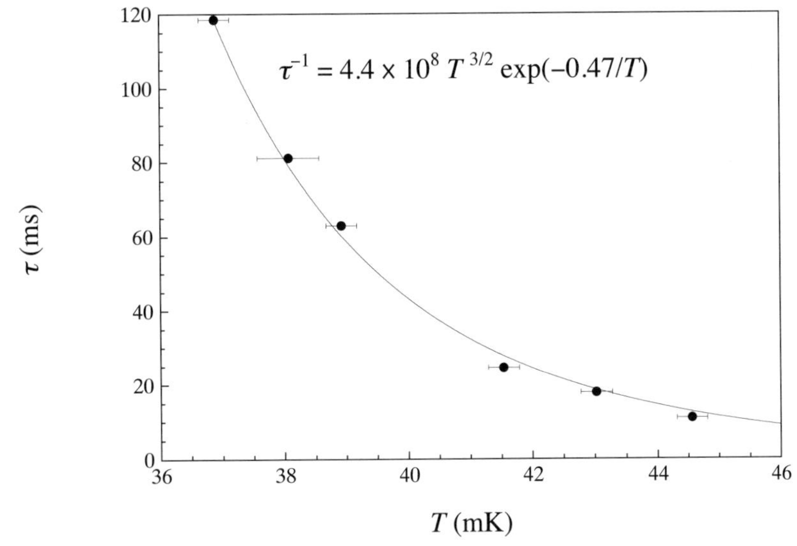

FIG. 23.13. Transit time for ion conversion τ vs T. (Alexander *et al.*, 1979.)

not allow us to draw any definite conclusions about the structure of the positive ion in a dilute ^{4}He–^{3}He mixture, whether it is a solid ^{3}He snowball clad with a ^{4}He halo, or a solid with a range of ^{3}He/^{4}He ratios trapped into it during the creation, which then relax to equilibrium at a temperature at which the solubility of ^{4}He in the bulk allows it, or whatever (Alexander *et al.*, 1979).

23.2 Field dependence of the positive ion mobility

Initially, the study of the motion of ions in the highly-degenerate phase of liquid ^{3}He was limited to the behavior of the so-called *zero-field* mobility, i.e., the ratio of the drift velocity and electric field in the limit of vanishingly-small fields: $\mu_+ = \lim_{E\to 0} (v_D/E)$.

From an experimental point of view, the apparent mobility $\mu_a = v_D/E$ is measured as a function of the field. As soon as it apparently becomes field independent, its value is taken as representative of the zero-field limit, owing to the great experimental difficulties in performing measurements in this limit. However, the discrepancies that emerged between the low-temperature behavior of μ_+ predicted by theory (Bowley, 1977*a*) and the outcome of some experimental measurements (Roach *et al.*, 1977*b*) have made it clear that the drift velocity has a superlinear dependence on E, whose strength increases as T is lowered. A detailed investigation of the electric field dependence is thus required in order to obtain reliable values of the zero-field mobility that are used to test theories.

The reason for the onset of a nonlinear behavior of the drift velocity v_D as a function of the applied electric field E is quite simple. At high v_D large energy transfers can be achieved during ion–quasiparticle scattering events. These transfers are of order $\Delta E \sim v_D p_F$. If ΔE becomes comparable to $k_B T$, the restriction on the final state due to the Pauli exclusion principle is lifted and the ion is subjected to an increased drag force that reduces its speed. The detection of the nonlinearity of $v_D(E)$ is also a confirmation of the necessity of including the finite width of the structure factor $S_0(q, \omega)$ at high enough speeds (Fetter and Kurkijărvi, 1977).

Whereas nonlinearities are not detected in the case of negative ions for T as low as $\approx 4\,$mK and for E as high as $0.11\,$MV/m (Long and Pickett, 1979), evident nonlinearities in the case of positive ions are observed at low temperatures for fields as low as $E \approx 3\,$kV/m or larger and at all pressures (Alexander and Pickett, 1978; Kokko *et al.*, 1978; Roach *et al.*, 1979).

Detailed experimental investigations of the electric field dependence of the drift velocity have been performed on both the fast and slow ions described previously (Alexander *et al.*, 1978; Alexander and Pickett, 1979). The results are shown in Figs 23.14 to 23.16, in which the apparent mobility v_D/E is plotted.

First of all, it should be noted that the field dependences of v_D/E for the fast and slow ions are similar. The different structure of the two types of ion is irrelevant. This is not surprising, insofar as ions are considered as semi-macroscopic, smooth spheres that move at high enough speed through the liquid.

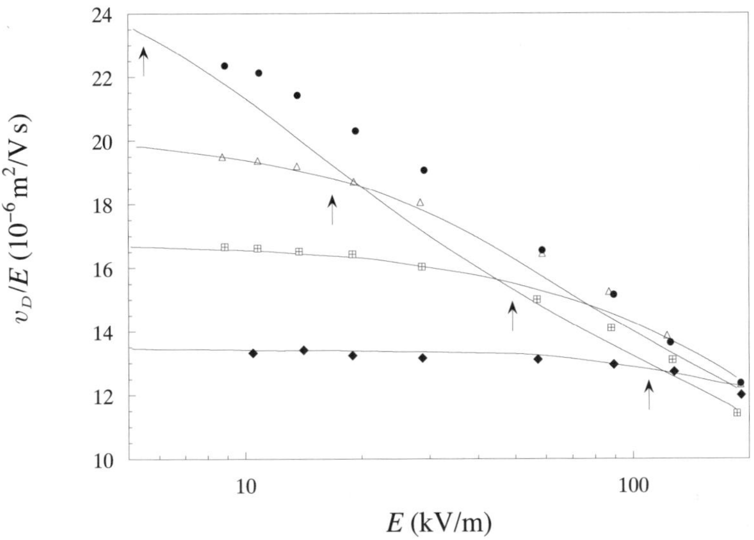

FIG. 23.14. v_D/E vs E for fast positive ions at $P = 0.5\,\text{MPa}$ for $T\,(\text{mK}) = 4.0$ (circles), 11.0 (triangles), 22.3 (squares), and 46.7 (diamonds) (Alexander *et al.*, 1978). Lines: theory with adjustable parameters: $c = 0.35$ and $T_0 = 0.80\,\text{K}$, corresponding to $R_+ = 7.5\,\text{Å}$ and $M^\star = 50m_3$ (Bowley, 1978). Arrows show for what E we have $\hbar k_F v_D = 2k_B T$.

The nonlinearities becomes progressively more pronounced as T is lowered. The arrows in the figures indicate the value of the field E for which the condition $v_D = 2k_BT/\hbar k_F$ is met. The lower T, the higher is the field value at which the nonlinear behavior sets in.

Finally, for any given pressure, the high-field data converge to a common curve, forming a temperature-independent envelope that is different for slow and fast ions (Alexander *et al.*, 1978).

The field dependence of the drift velocity can be described in terms of the theory developed by Bowley (1977*a*), that is a generalization of those of Josephson and Lekner (1969) and of Fetter and Kurkijärvi (1977).

The average force that acts on the ion moving at the average drift velocity $\mathbf{v}_D$ as a consequence of the scattering off the quasiparticles, and which must be balanced by the electric force $e\mathbf{E}$, is given by eqns (21.54)–(21.58). It can be cast in the following form (Bowley, 1977*a*):

$$e\mathbf{E} = \frac{k_F^2}{8\pi^2\beta^2\hbar}\int \mathrm{d}\Omega \int \mathrm{d}\Omega' \mathbf{q}\sigma(\mathbf{q}) \tanh\left[\left(\beta\hbar\mathbf{q}\cdot\mathbf{v}_D\right)/2\right]$$
$$\times \int_{-\infty}^{\infty} \frac{F(q,t)}{\sinh^2(\pi t/\hbar\beta)}\left[\cosh(\pi t/\hbar\beta) - \cos(\mathbf{q}\cdot\mathbf{v}_D)\right]\mathrm{d}t\,, \qquad (23.8)$$

where $\beta^{-1} = k_B T$, $\hbar\mathbf{q}$ is the transferred momentum, σ is the differential cross-section, and the solid angles Ω and Ω' refer to the directions of the quasiparticle momenta $\mathbf{k}$ and $\mathbf{k}'$ in the initial and final states, respectively.

By choosing the direction of $\mathbf{k}$ as the polar axis for the integration over Ω', the integral in eqn (23.8) can be transformed into an integral over d$\mathbf{q}$:

$$\frac{eE}{v_D} = \frac{\hbar}{4}\int_0^{2k_F} q^3\sigma(q)\,\mathrm{d}q \int_{-1}^{1} y^2 \frac{\tanh(\hbar\beta\omega/2)}{\hbar\beta\omega/2}\,\mathrm{d}y$$
$$\times \int_{-\infty}^{\infty} \frac{\cosh(\pi t/\hbar\beta) - \cos(\pi t/\hbar\beta)}{\sinh^2(\pi t/\hbar\beta)} F(q,t)\,\frac{\mathrm{d}t}{\hbar\beta}, \qquad (23.9)$$

where $\cos^{-1}(y)$ is the angle between $\mathbf{q}$ and $\mathbf{v}_D$, and $\omega = \hbar q v_D y$.

The similarity between eqns (23.9) and (23.1) for $\mathrm{Re}\mu(\omega)^{-1}$ should be noted. This is not surprising as $\mathbf{q}\cdot\mathbf{v}_D$ represents a Doppler shift in frequency: as an energy $\hbar\omega$ can be exchanged in a scattering event by an oscillating ion, so an energy $\hbar\mathbf{q}\cdot\mathbf{v}_D$ can be exchanged by a drifting ion (Bowley, 1977*a*).

Equation (23.9) can be evaluated by introducing the suitable dimensionless variables $\tau = t/\hbar\beta$, $w = \hbar\beta\omega$, $x = k^2/4k_F^2$, and $c = 12/(k_F R_+)^2$, and by

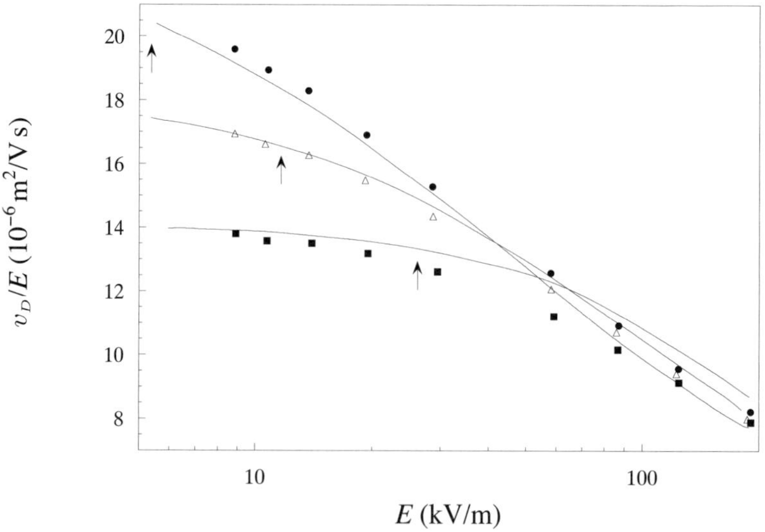

FIG. 23.15. v_D/E vs E for fast positive ions at P = 2.7 MPa for T (mK) = 3.4 (circles), 6.5 (triangles), and 12.4 (squares) (Alexander *et al.*, 1978). Lines and arrows have the same meaning as in Fig. 23.14. Here c = 0.41 and T_0 = 0.15 K, corresponding to R_+ = 6.17 Å and to $M^\star = 250 m_3$.

assuming a constant differential-cross section $\sigma(q) = R_+^2/4$. By so doing, one obtains (Bowley, 1977*a*)

$$\frac{E}{v_D} = \left(\frac{6\hbar k_F{}^2}{ec}\right) \int_0^1 x\mathrm{d}x \int_{-1}^{1} y^2 \left[\frac{\tanh\left(\hbar\beta k_F v_D x^{1/2} y\right)}{\hbar\beta k_F v_D x^{1/2} y}\right] \mathrm{d}y$$

$$\times \int_{-\infty}^{\infty} F(x,\tau)\, \frac{\cosh(\pi\tau) - \cos\left(2\hbar\beta k_F v_D x^{1/2} y\tau\right)}{\sinh^2(\pi\tau)} \mathrm{d}\tau\,. \qquad (23.10)$$

The intermediate scattering function is given by

$$F(x,\tau) = \exp\left\{-cx \int_0^{\infty} f(w,T) \left[\frac{1-\cos(w\tau)}{\tan(w\tau/2)} - i\sin(w\tau)\right] \mathrm{d}\ln w\right\}, \qquad (23.11)$$

where

$$f(w,T) = \frac{g(w)}{[g(w)]^2 + (w\mu_0 M^\star T)^2} \qquad (23.12)$$

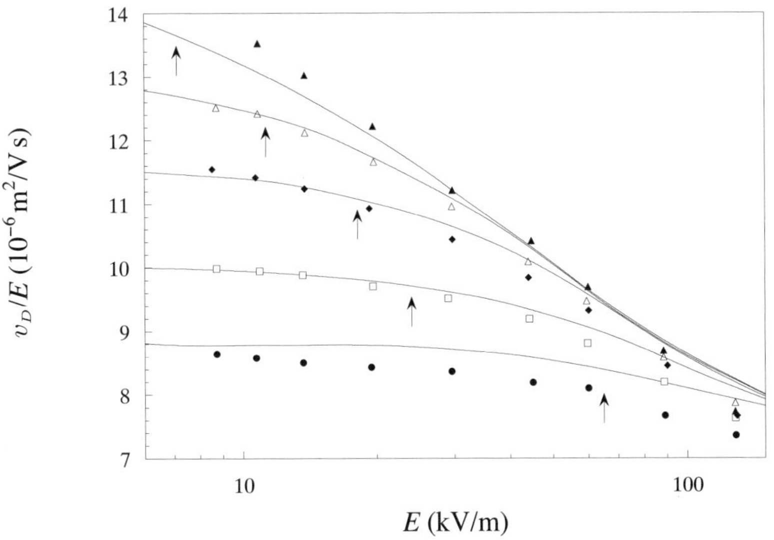

FIG. 23.16. $\mu = v_D/E$ vs E for slow positive ions at $P = 2.7\,\mathrm{MPa}$ for T (mK) = 3.4 (closed triangles), 5.1 (open triangles), 7.0 (diamonds), 11.5 (squares), and 17.6 (circles) (Alexander *et al.*, 1978). Lines and arrows have the same meaning as in Fig. 23.14. Here $c = 0.35$ and $T_0 = 0.12\,\mathrm{K}$, corresponding to $R_+ = 6.68\,\mathrm{Å}$ and to $M^\star = 370 m_3$.

with

$$g(w) = 2\pi \frac{\tanh(w/2)}{w} \int_0^1 x\,\mathrm{d}x \int_{-\infty}^{\infty} F(x,\tau)\,\frac{\cosh(\pi\tau) - \cos(w\tau)}{\sinh^2(\pi\tau)}\,\mathrm{d}\tau \qquad (23.13)$$

and $\mu_0 = 3\pi/\hbar\left({k_F}^2 R_+\right)^2$. $M^\star$ is the effective ion mass, first introduced by Josephson and Lekner (1969). Equation (23.13) is a different formulation of the closure relation (eqn (23.1)) introduced by Bowley (1977*a*).

Equations (23.11)–(23.13) are to be solved self-consistently for the intermediate scattering function $F(x,\tau)$ that is inserted into eqn (23.10) to evaluate v_D/E. This procedure is carried out numerically in terms of the adjustable parameters $c = 12/(k_F R_+)^2$ and $T_0 = 4(\hbar k_F)^2/\pi k_B c M^\star$.

The results of these calculations are the solid lines in Figs 23.14 to 23.16. The agreement of the theory of Bowley (1978) with the experimental data is particularly good and also accounts for the downward bending of the T-dependence of μ_+ at low P reported in Fig. 22.4 as curve 2 (Roach *et al.*, 1977*b*, 1979).

The values of the adjustable parameters used for the fit are reported in the captions to the figures. It turns out that the ion radius decreases upon increasing P, as already noted in the measurements of the temperature dependence of the mobility under pressure (Alexander and Pickett, 1978; Alexander *et al.*, 1979).

The rapid variation of $M^\star$ with P ($M^\star = 50 m_3$ at $P = 0.5\,\mathrm{MPa}$ and $M^\star = 250 m_3$ at $P = 2.7\,\mathrm{MPa}$ for the fast ions) casts some doubt on the interpretation of $M^\star$ as the effective mass of the ion. As $M^\star$ enters through the expression $\mathtt{Im}\mu(\omega)^{-1} = i\omega M^\star$, implying that the high-frequency cut-off in $\mathtt{Re}\mu(\omega)$ arises from the inertia of the ion only (Josephson and Lekner, 1969; Bowley, 1977*a*), it may be conjectured that the mechanisms producing the cut-off are more complicated than this, and probably include damping due to a coupling of the ion and the zero-sound modes of the liquid (Alexander *et al.*, 1978).

23.2.1 *Nonlinear behavior of positive ions: equivalence of T and v_D*

In the nonlinear region, the drift velocity v_D and temperature T are related in a simple, though unexpected, way (Alexander and Pickett, 1979). The results for the fast positive ions at $P = 2.7\,\mathrm{MPa}$ are shown in Figs 23.17 and 23.18. The arrows in the plots indicate either the temperature or the drift velocity at which the condition $0.31\hbar k_F v_D = k_B T$ is met. The solid lines are the predictions of the model of Bowley (1978). The curves labeled as $E = 0$ and $T = 0$ mean that the values of T and E are the lowest reached experimentally.

Figures 23.17 and 23.18 show that the ratio v_D/E becomes independent of T at low T and high E, and independent of v_D at high T and small v_D. The onset of nonlinearities is determined by the condition $0.31\hbar k_F v_D = k_B T$.

The curves plotted in these two figures look very similar to each other. In particular, the limiting lines for $E = 0\,(v_D = 0)$ and $T = 0$ have the same slope. This similarity bears on a fundamental, though yet unknown, symmetry between velocity and temperature. This symmetry shows up by making contour

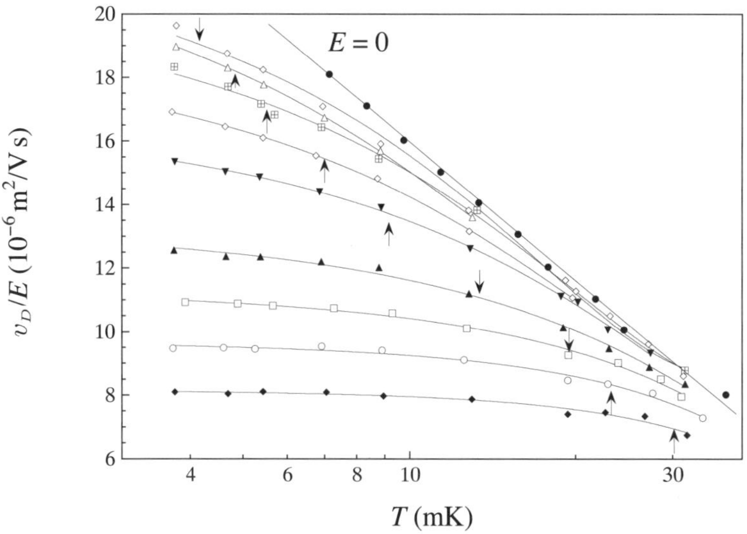

FIG. 23.17. v_D/E vs T for positive ions at $P = 2.7\,\mathrm{MPa}$ (Alexander and Pickett, 1979). $E\,(\mathrm{kV/m}) = 0$, 8.79, 10.74, 13.67, 19.53, 29.3, 58.59, 87.89, 127, and 195.3 (from top). Arrows indicate the value of T for which $v_D = k_\mathrm{B}T/0.31\hbar k_F$. Lines: theory (Bowley, 1978).

plots of constant v_D/E as a function of T and v_D. The contour plots become straight lines, as illustrated in Fig. 23.19, if they are plotted against $v_D^{3/2}$ and $T^{3/2}$, yielding the relationship

$$(0.31\hbar k_F v_D/k_\mathrm{B})^{3/2} + T^{3/2} = \text{constant}. \tag{23.14}$$

Unfortunately, there is no theoretical argument to explain this 3/2-law.

The intercepts with the T- and v_D-axes yield the $E = 0$ and $T = 0$ lines of Figs 23.17 and 23.18, respectively.

In the high $v_D/E > 10 \times 10^{-6}\,\mathrm{m^2/V\,s}$ region, the mobility is linear in $\ln(T)$. In this logarithmic region, eqn (23.14) allows v_D/E to be described over the whole range of T and v_D by the following relationship:

$$\frac{v_D}{E} = A + B \ln\left[T^{3/2} + (Cv_D)^{3/2}\right], \tag{23.15}$$

where A, B, and C are constants. For the data plotted in Fig. 23.19, the values $A = -13.11 \times 10^{-6}\,\mathrm{m^2/V\,s}$, $B = 4.167 \times 10^{-6}\,\mathrm{m^2/V\,s}$, and $C = 2.05 \times 10^{-2}\,\mathrm{K\,s/m}$ are obtained (Alexander and Pickett, 1979).

If the 3/2-relationship is valid, the data of v_D/E for all v_D and T at any given pressure should fall on a single curve if plotted as a function of T_R, where

$$T_R^{3/2} = T^{3/2} + (0.31\hbar k_F v_D/k_\mathrm{B})^{3/2}. \tag{23.16}$$

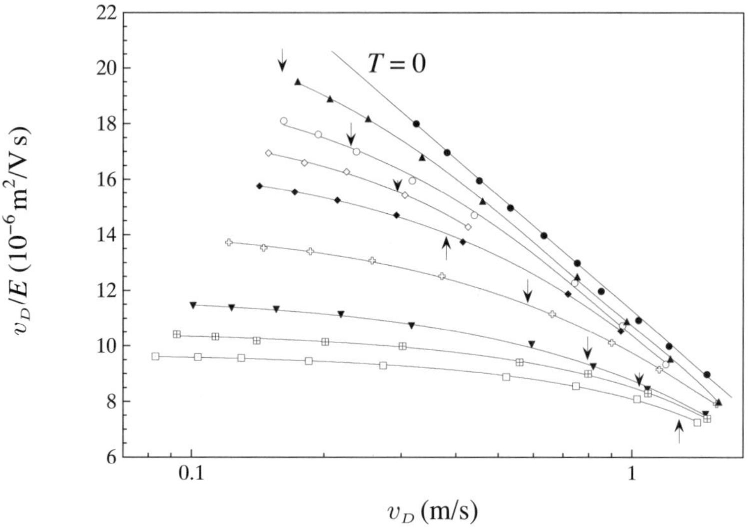

FIG. 23.18. v_D/E vs v_D for positive ions at $P = 2.7$ MPa (Alexander and Pickett, 1979). T (mK) = 0, 3.39, 4.99, 6.5, 8.35, 12.4, 18.6, 22.4, and 27 (from top). Arrows indicate the value of v_D for which $T = 0.31\hbar k_F v_D/k_B$. Lines: theory (Bowley, 1978).

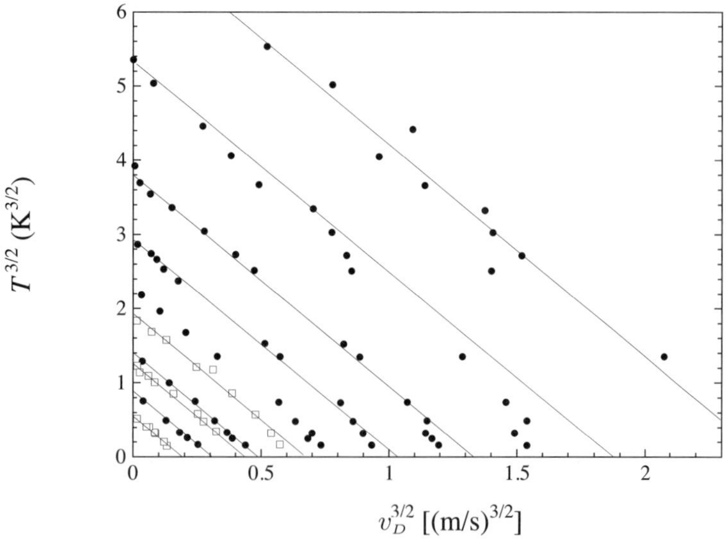

FIG. 23.19. Contour plots of constant v_D/E as a function of $v_D^{3/2}$ and $T^{3/2}$ for positive ions at $P = 2.7$ MPa. (Alexander and Pickett, 1979.) Lines: eqn (23.14).

This is, actually, the case, as shown in Fig. 23.20 for $P = 0.5$, 2.0, and 2.7 MPa for the fast ions (curves 1, 2, and 3) and for $P = 2.7$ MPa for the slow ions (curve 4). A nice logarithmic fit is obtained for all curves. An analysis of the data in Fig. 23.20 confirms the validity of eqn (23.15) for all pressures if the constants A and B are now functions of P.

It is also reassuring that the data for the slow positive ions (curve 4) are well fitted by eqn (23.15) with the same T_R as for the fast ions, i.e., with the same $0.31\hbar k_F/k_\mathrm{B}$ factor normalizing v_D to T. This fact indicates that the 3/2-functional form is independent of the ion species, and hence of the ion radius, and should be a general property of the transport of a smooth spherical object in a highly-degenerate Fermi liquid.

At low T, the characteristic energies transferred to quasiparticles largely exceed $k_\mathrm{B}T$. As the energy that an ion can exchange to a quasiparticle is a linear function of v_D, the quantity $0.31\hbar k_F v_D/k_\mathrm{B}$ can be interpreted as the effective ion temperature. The two observed transport regimes, i.e., the one in which v_D is linear in E and the one in which v_D is superlinear in E, correspond to situations in which the effective ion temperature is lower or higher than the liquid temperature. In other words, the ion is nearly in equilibrium with the quasiparticles of the liquid if $0.31\hbar k_F v_D/k_\mathrm{B}T \ll 1$, whereas it is very epithermal if $0.31\hbar k_F v_D/k_\mathrm{B}T > 1$. This situation closely resembles what happens to a charge carrier moving through a gas of hard spheres when the reduced electric field E/N, where N is the gas number density, is varied (Huxley and Crompton, 1974).

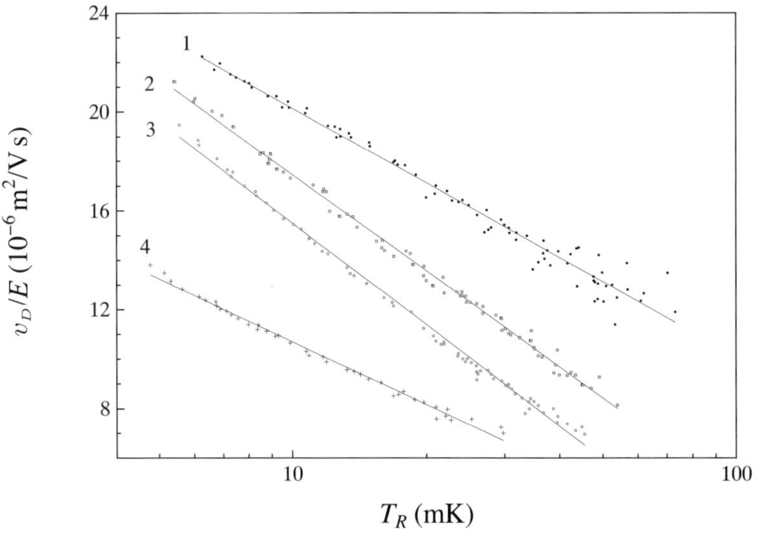

FIG. 23.20. v_D/E vs T_R for positive ions. (Alexander and Pickett, 1979.) Fast ions: $P = 0.5$ MPa (curve 1), $P = 2.0$ MPa (curve 2), and $P = 2.7$ MPa (curve 3). Slow ions: $P = 2.7$ MPa (curve 4). Lines: logarithmic fits.

24

ION TRANSPORT IN SUPERFLUID ^{3}HE

24.1 Fundamentals of superfluid ^{3}He

In a Bose system, single-particle states may be multiply occupied and at low temperatures the system may condense into the lowest of them, giving origin to a condensate in which particles are strongly correlated and that shows superfluidity. In contrast, the appearance of superfluidity in a Fermi system, such as ^{3}He, occurs as a result of the formation of correlated pairs of fermions of zero total momentum, which then undergo Bose–Einstein condensation.

According to the theory of supercondutivity (Bardeen *et al.*, 1957), any Fermi systems with attractive interactions should undergo a phase transition at low enough temperature into a state in which strongly-correlated pairs of particles can form. This general argument leads to the expectation that ^{3}He also makes a transition to such a state because, at a large distance, atoms attract each other by means of long-range van der Waals or dispersion forces (Maitland *et al.*, 1981).

However, at short range, ^{3}He atoms repel each other because of exchange forces due to the Pauli exclusion principle. The presence of the repulsive hard-core interaction makes pair correlations in ^{3}He different from the superconductor case. Owing to the repulsive short-range interaction, the wave function of a pair must vanish when the individual particles in the pair come within a short distance of each other, thus requiring that the Cooper pairs have nonzero relative angular momentum (Pitaevskii, 1960). Thus, atom pairing in liquid ^{3}He is expected to be anisotropic in the real space (Wőlfle, 1979).

The discovery of superfluid ^{3}He (Osheroff *et al.*, 1972*b*), and the NMR measurements on this substance (Osheroff *et al.*, 1972*a*) and their interpretation (Leggett, 1972, 1973) confirmed these predictions.

At very low temperatures, in the milliKelvin region, liquid ^{3}He undergoes a phase transition and becomes superfluid. A very schematic phase diagram in zero magnetic field is shown in Fig. 24.1. At zero magnetic field two superfluid phases appear, depending on the applied pressure: ^{3}He–A and ^{3}He–B. Their features are different from each other, and will be described very shortly. Complete reviews of different levels are found in the literature (Keller, 1969; Leggett, 1975; Wheatley, 1975; Salomaa and Volovik, 1987; Lee, 1997; Guénault, 2003). Here, the approach of Guénault (2003) is followed.

The phase transition between the normal liquid and the superfluid is characterized by a jump of the heat capacity, typical of the continuous, or second-order, transitions. In contrast, the transition between the two superfluid phases A and B is a first-order one with latent heat. Heat is absorbed in the B$\rightarrow$A passage

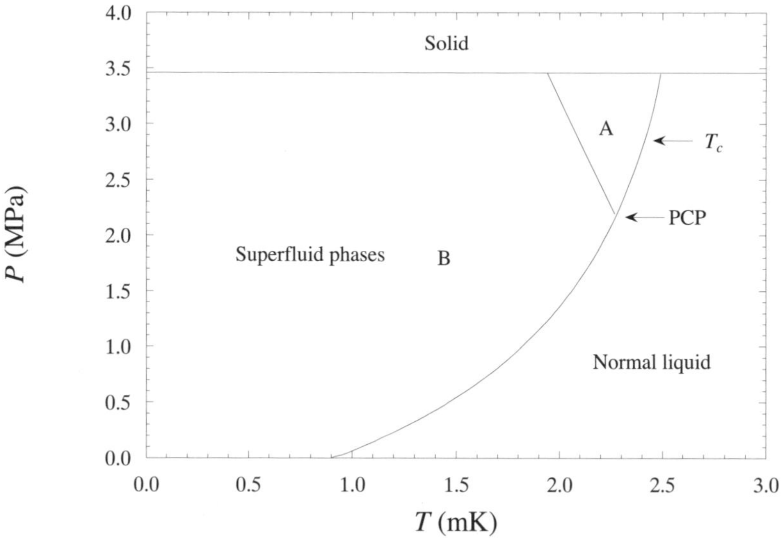

FIG. 24.1. Schematic phase diagram of ^{3}He at low T showing the superfluid phases in zero magnetic field. T_c is the continuous transition line. PCP is the poly-critical point at which the normal fluid–superfluid A boundary meets the first-order A–B boundary. (Guénault, 2003.)

and released in the opposite way.

The phase diagram changes significantly when a magnetic field B of a few mT is applied, as shown in Fig. 24.2. The poly-critical point (PCP), i.e., the intersection between the second-order normal-to-superfluid line and the first-order A-to-B line, disappears. As the magnetic field is increased, the A-phase grows at the expense of the B-phase (Guénault, 2003). A slice of a new superfluid phase, called A_1, appears (Gully *et al.*, 1973) and separates the normal liquid and the A-superfluid. The temperature width of the A_1-phase is proportional to the magnetic field strength at constant pressure (Lawson *et al.*, 1973, 1975.) Eventually, for $B > 500\,\mathrm{mT}$, the B-phase disappears, and nearly all the fluid is in the A-phase, with only the thin slice of A_1 left.

The observed magnetic properties of superfluid ^{3}He are explained in terms of a spin $S = 1$ state of the Cooper pairs. As the total wave function of the pairs must be antisymmetric with respect to the exchange of the coordinate because of the Pauli principle, the angular momentum L of the pairs must be odd. The simplest choice is $L = 1$. For this reason, the formation of the Cooper pairs is called *p-wave pairing* (Guénault, 2003).

Thus, there are triplets both in real and spin space. The spin states with $S_z = \pm 1, 0$ are conventionally written as $\uparrow\uparrow$, $\downarrow\downarrow$, and $\uparrow\downarrow$, respectively, the last symbol being a shorthand notation for the $(|\uparrow\downarrow\rangle + |\downarrow\uparrow\rangle)/\sqrt{2}$ spin wave function. The meaning of the symbols $\uparrow$ and $\downarrow$ is obvious.

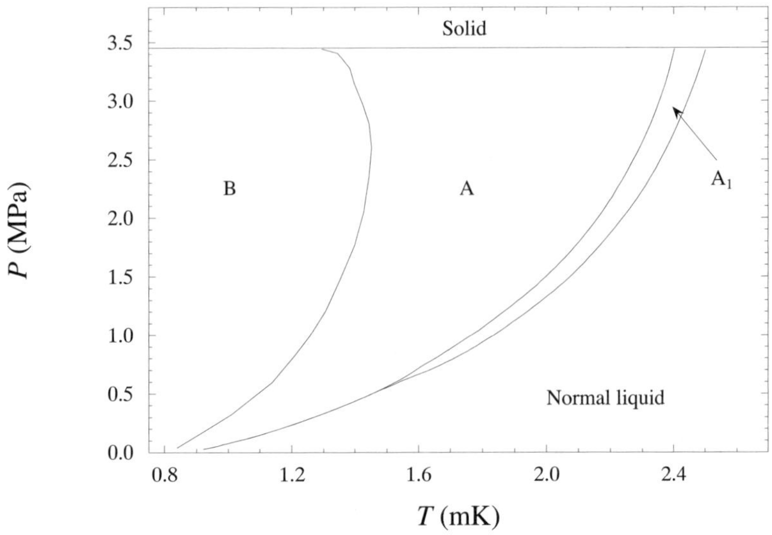

FIG. 24.2. Phase diagram of ^{3}He showing the superfluid phases in a magnetic field. (Guénault, 2003.)

The A-phase has the same magnetic susceptibility as the normal liquid, in which no pairing occurs. Therefore, only $\uparrow\uparrow$ and $\downarrow\downarrow$ states are involved and phase A is called a state of *equal-spin pairing*. The A_1 phase appears only near the critical line and should consist of spin states aligned along the axis of the magnetic field, i.e., those states with $\uparrow\uparrow$ spins. Finally, the B-phase has a lower susceptibility than the normal state and consists of $\uparrow\downarrow$ spin states.

The A-phase is highly anisotropic because only the spin states $\uparrow\uparrow$ and $\downarrow\downarrow$ are involved, and corresponds to the p-wave equal-spin pairing first proposed by Anderson and Morel (1961) and Anderson and Brinkman (1973). This state is an orbital $m_l = 1$ state along some direction $\hat{\mathbf{l}}$, which defines the common direction of the orbital angular momentum of the pairs, and a spin $m_s = 0$ state along some direction $\hat{\mathbf{d}}$, which is defined as the direction of zero-spin projection.

The order parameter is expressed as the product of an orbital part in the configuration or momentum space and of a part in the spin space. If only the angular dependence is considered, the order parameter is defined as

$$\psi_{AM} \propto e^{i\varphi} \sin\theta \left[\frac{1}{\sqrt{2}} (| \uparrow\downarrow\rangle + | \downarrow\uparrow\rangle) \right]. \tag{24.1}$$

The spin part of the order parameter is independent of any orbital variables. This means that every point on the Fermi surface in the k-space has the same $\hat{\mathbf{d}}$. In this state, the vectors $\hat{\mathbf{l}}$ and $\hat{\mathbf{d}}$ are always parallel to each other because this configuration leads to a minimization of the repulsive energy of the pair. This configuration is termed *spontaneously-broken spin–orbit symmetry* (Guénault, 2003).

The behavior of the BCS gap follows that of the order parameter with its strong orbital anisotropy, with nodes at $\theta = 0$ and $\theta = \pi$. The gap is described by

$$\Delta(\mathbf{k}) = \Delta_0 \sin\theta\,, \tag{24.2}$$

where θ is the angle between the symmetry axis of the order and the direction $\mathbf{k}$ of a quasiparticle, as shown in Fig. 24.3.

The anisotropy stems from the fact that a particle with wave vector $\mathbf{k}$ aligned along the z-axis cannot couple with a particle with wave vector $-\mathbf{k}$ to yield a Cooper pair with an angular momentum contribution along this axis. Such quasiparticles do not contribute to the A-phase ordering and so have $\Delta = 0$. In contrast, pairs of particles with $\mathbf{k}$ lying in the equatorial plane have their angular momentum already aligned along the z-axis and give a maximum contribution to the superfluid state, i.e., their order parameter $\Delta = \Delta_0$ is a maximum.

The patterns of $\hat{\mathbf{l}}$ in the fluid as a function of position have strong analogies with the patterns found in liquid crystals and are termed *textures*. As the orientation of $\hat{\mathbf{l}}$ is perpendicular to the wall of the container (Ambegaokar *et al.*, 1974, 1975), the boundary conditions are very important for determining the texture patterns in liquid ^{3}He–A, as well as the magnetic and/or the flow field.

A rotation in the spin space of the spin state $(1/\sqrt{2})(|\uparrow\downarrow\rangle + |\downarrow\uparrow\rangle)$ yields the equal-spin pairing of the Anderson–Morel order parameter:

$$\psi_{AM} \propto e^{i\varphi} \sin\theta \left(|\uparrow\uparrow\rangle + e^{i\Phi}|\downarrow\downarrow\rangle\right)\,, \tag{24.3}$$

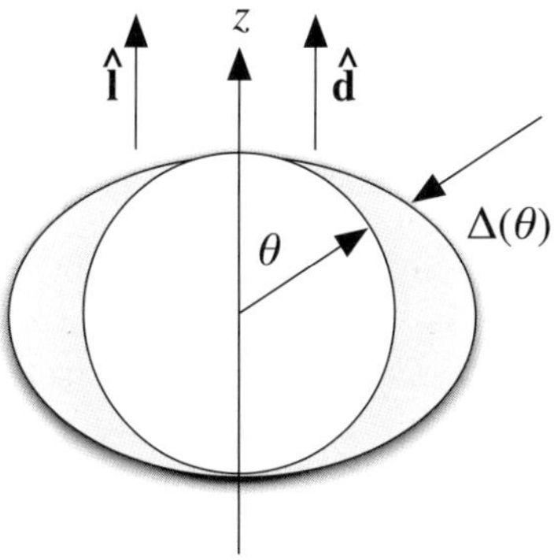

FIG. 24.3. The angular dependence of the anisotropic energy gap of the A-phase around the Fermi surface. There is revolution symmetry around the z-axis.

where Φ is a phase factor. This representation allows us to think of the A-phase as composed of only $|\uparrow\uparrow\rangle$ and $|\downarrow\downarrow\rangle$ states, whereas the A_1-phase consists only of $|\uparrow\uparrow\rangle$ states.

On the other hand, the B-phase has quite different properties. It is believed that the stable state of a weakly-coupled p-wave superfluid is one state of the highest possible symmetry, as described by the model of Balian and Werthamer (1963). Of all the possible combinations of $L_z = \pm 1$, 0 and of $S_z = \pm 1, 0$ of the Cooper pairs, the simplest state is the $^3P_0$ state described by the wave function

$$\psi_{BW} \propto Y_1^{-1} |\uparrow\uparrow\rangle + Y_1^0 |\uparrow\downarrow + \downarrow\uparrow\rangle + Y_1^1 |\downarrow\downarrow\rangle , \qquad (24.4)$$

where the Y_l^m are the usual spherical harmonics. All the spin states are included and no equal-spin pairing state occurs.

The state described by the wave function ψ_{BW} is spherically symmetric because the total angular momentum is $J = 0$. Again, a Cooper pair consisting of $(\mathbf{k}, -\mathbf{k})$ one-particle states is characterized by assigning $\hat{\mathbf{l}}$ and $\hat{\mathbf{d}}$, as in the case of the A-phase. In this case, however, $\hat{\mathbf{l}} \perp \mathbf{k}$, because the particles of the pair are circling around each other, and also $\hat{\mathbf{l}} \perp \hat{\mathbf{d}}$ in order to give $J = 0$. As a consequence, $\hat{\mathbf{d}} \| \hat{\mathbf{k}}$, where $\hat{\mathbf{k}}$ is a unit vector in the direction of $\mathbf{k}$.

The resulting phase is spherically symmetric with an isotropic energy gap Δ that is independent of $\mathbf{k}$, as shown in Fig. 24.4.

Any arbitrary rotation of $\hat{\mathbf{d}}$ about some axis $\hat{\mathbf{n}}$ leaves the state unaltered. This degeneracy is removed if the spin–orbit interaction is taken into account, resulting in a rotation of the spin coordinates with respect to the orbital ones by an angle of $\vartheta = 104°$. This rotated state is believed to correctly describe the B-phase, which is considered a pseudo-isotropic phase because the structure of the superfluid is governed by $\hat{\mathbf{n}}$, thus allowing textures with liquid crystal-like behavior to be observed, but the overall orbital symmetry of the order parameter is still spherical.

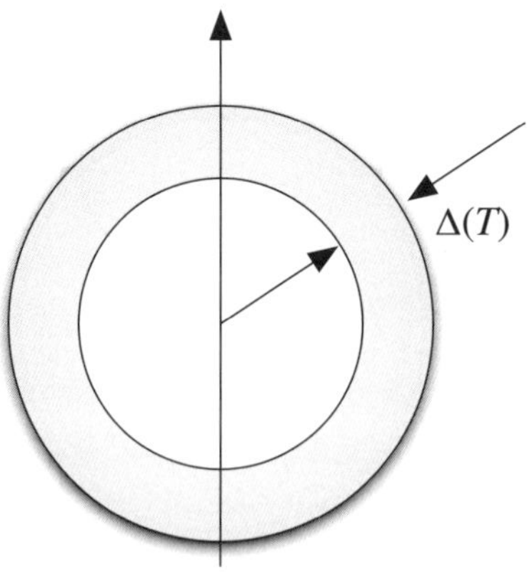

FIG. 24.4. The isotropic energy gap of the B-phase.

It is evident that the complexity of the order parameter of ^{3}He and its anisotropy should influence the transport properties of ions. Simply stated, one expects the ion mobility to be anisotropic in the A-phase and isotropic in the B-phase (Bowley, 1976*a*).

24.2 Phenomenology of negative ion transport in superfluid ^{3}He

As for the case of ^{4}He, ions are also very useful probes for investigating the features of the superfluid state in ^{3}He. The size of the negative ion in ^{3}He is of order 15 Å, depending on pressure, and is thus smaller than the mean free path of quasiparticles at milliKelvin temperatures and shorter than the coherence length of the superfluid state of ^{3}He (Guénault, 2003).

The ion mobility obviously depends on the density of the excitations available for scattering, but it may also yield information on how the transport cross-section is affected by the nature of the superfluid (Baym *et al.*, 1979).

At such low temperatures only the field emission/ionization technique is acceptable for producing ions in the liquid as far as the heat input into the sample is concerned (Ahonen *et al.*, 1976). Typical numbers are 1 μW/pA heat leak for a tritiated titanium ionization source of a few mCi (Anderson *et al.*, 1968; Kuchnir *et al.*, 1970) and 16 nW/pA for thin-film electron emitters (Onn *et al.*, 1974), to be compared with the value 0.5 nW/pA of the field-emission technique (Ahonen *et al.*, 1976). For this reason all measurements exploit this ion production technique in a single-pulse time-of-flight method.

A general feature of the experimental results Ahonen *et al.*, 1976, 1978; Roach *et al.*, 1977*a*) is that the nonlinear behavior of the drift velocity as a function of the applied electric field is more pronounced than in the normal liquid. There is a low-field range in which the drift velocity v_D is linear with the field strength E, and a high-field region in which v_D depends superlinearly on E.

24.2.1 *Zero-field measurements*

The constraint of measuring at small fields in order to obtain the mobility limits the measurement range to $\mu/\mu_N \leqslant 10$, where μ_N is the value of the mobility in the normal phase, and to $T/T_c > 0.8$. It is worth recalling that the drift mobility of negative ions in the normal liquid ^{3}He becomes independent of T below approximately 20-100 mK, depending on pressure (see Fig. 21.1).

The general behavior of the zero-field mobility is shown in Fig. 24.5 as a function of the reduced temperature T/T_c at a high enough pressure to first cross the normal–superfluid A boundary and then cross the A–B boundary by lowering T (Ahonen *et al.*, 1976, 1978).

Below T_c, the ion mobility increases rapidly with decreasing T in both A- and B-phases, though it is greater in the B-phase than in the A-phase. Well below T_c the increase is nearly exponential in T^{-1}. A mobility jump appears for $P = 2.84$ MPa as the A–B boundary is crossed from the supercooled A-phase into the B-phase.

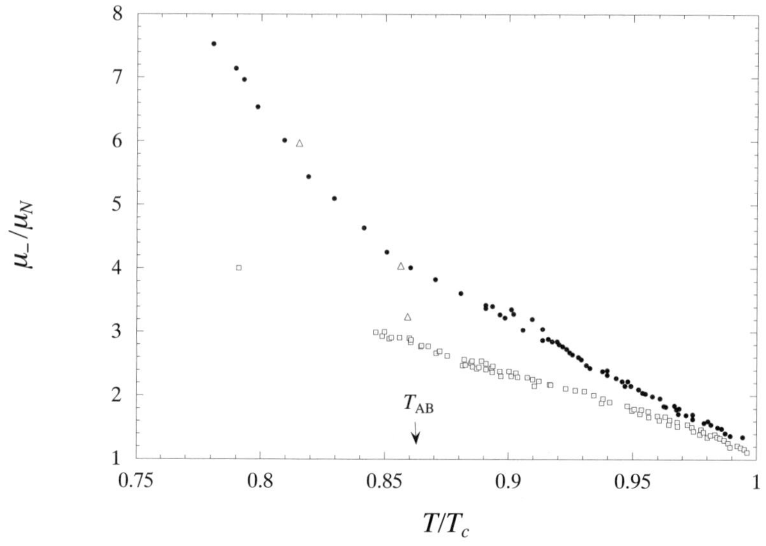

FIG. 24.5. μ_-/μ_N vs T/T_c in superfluid ^{3}He for P = 2.84 MPa in the A-phase (squares) and in the B-phase (triangles), and for P = 1.80 MPa in the B-phase (circles). (Ahonen *et al.*, 1976, 1978.) T_{AB} is the A–B transition temperature.

The increase in the mobility below T_c is associated qualitatively with the decrease in the number of still-unpaired particles in the liquid, and the lower mobility in the A-phase with respect to the B-phase is attributed to the different angular dependence of the energy gaps of the two phases.

It is worth recalling that in these measurements ions are moving perpendicularly to the external, static magnetic field $H = 28.4\,\mathrm{mT}$ [4] required by NMR thermometry (Ahonen *et al.*, 1976, 1978). In this experimental configuration, in which the angle between the drift field **E** and the magnetic field **H** cannot be varied, the vector $\hat{\mathbf{l}}$ may lie anywhere in a plane perpendicular to **H** and, consequently, at any angle with respect to **E**. Thus, no information can be obtained about the effects of anisotropy on the mobility in the A-phase.

These anisotropy effects show up if the tilting angle θ between **E** and **H** is varied, as in the experiment of Roach *et al.* (1977*a*). Here, an auxiliary field of strength $H = 1.8\,\mathrm{mT}$ can be rotated so as to make an angle θ with the direction of the drift field. The results for negative ions at $P = 2.6\,\mathrm{MPa}$ are shown in Fig. 24.6.

In the A-phase, the negative ion mobility μ_- decreases as the tilting angle θ is increased from 0° to 90°. If $\theta = 0$, μ_- appears to be continuous across the A–B transition, indicated by a dashed line in the figure. The angular dependence

[4]Here, the symbol H instead of B is used for the magnetic field in order to avoid confusion with the superfluid phase B.

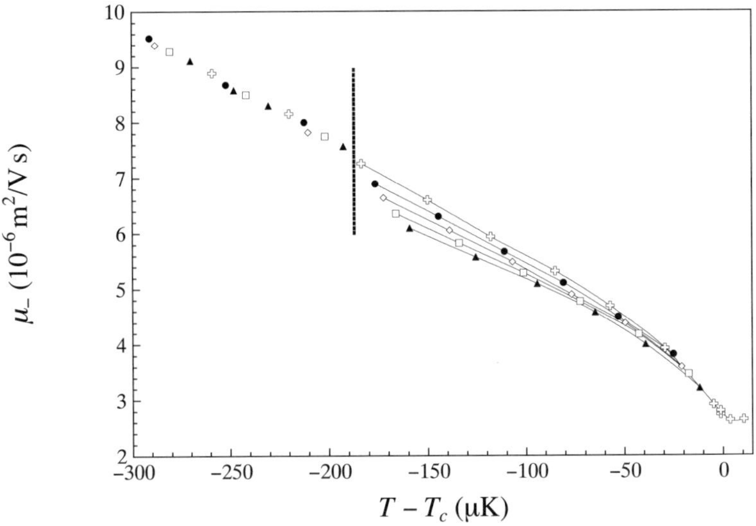

FIG. 24.6. μ_- vs $T - T_c$ in superfluid ^{3}He–A at temperatures below T_c for $P = 2.6$ MPa. (Roach *et al.*, 1977*a*.) θ (deg.) = 0 (crosses), 30 (circles), 45 (diamonds), 60 (squares), and 90 (triangles) is the tilting angle of the 1.8 mT magnetic field H with respect to the drift field E. Dashed line: A–B transition.

of the measured mobility explains the sudden drop in μ_- at the A–B transition (Ahonen *et al.*, 1976, 1978).

This behavior is interpreted in terms of the so-called *fan-averaged texture* model (Roach *et al.*, 1977*a*). It is assumed that the liquid is described by a texture in which the vector $\hat{\mathbf{l}}$ lies, on average, uniformly distributed in the plane perpendicular to $\mathbf{H}$. This leads to a mobility tensor with principal values $\mu_{11} = \mu_{22} = (1/2)(\mu_{\parallel} + \mu_{\perp})$ and $\mu_{33} = \mu_{\perp}$. The magnetic field is aligned parallel to the 3 axis and $\mu_{\parallel}$ and $\mu_{\perp}$ refer to the components of the mobility tensor parallel and perpendicular to the A-phase gap axis, respectively, for a perfectly aligned sample.

The measured mobility $\mu(\theta) = |\overleftrightarrow{\mu} \cdot \hat{\mathbf{E}}|$ is the component along the direction $\hat{\mathbf{E}}$ of the electric field, which is tilted by an angle θ with respect to the magnetic field. Thus, the mobility depends on the tilting angle in the following way:

$$\mu(\theta) = \mu_{\perp} \cos^2\theta + \frac{1}{2}\left(\mu_{\parallel} + \mu_{\perp}\right)\sin^2\theta . \tag{24.5}$$

The observed mobility for $H = 0$ would be $\langle\mu\rangle = (2/3)\mu_{\perp} + (1/3)\mu_{\parallel}$. An inspection of Fig. 24.6 reveals that $\mu_-(90°) < \mu_-(0°)$, yielding $\mu_{\perp} > \mu_{\parallel}$. This behavior is qualitatively understood by recalling that the energy gap vanishes along the gap axis. This fact leads to a relative increase in the number of normally-excited quasiparticles that move in the same direction as the ion motion where the maximum momentum transfer can occur.

In Fig. 24.7 the mobility μ_- in the phase A for $T - T_c = -160\,\mu$K, $P = 2.6$ MPa, and $H = 1.8$ mT is plotted as a function of the tilting angle θ (Roach *et al.*, 1977*a*). The solid line is eqn (24.5) with $\mu_\perp = 6.75 \times 10^{-6}\,\mathrm{m^2/V\,s}$ and $\mu_\parallel = 5.50 \times 10^{-6}\,\mathrm{m^2/V\,s}$.

Negative ion mobility data recorded in an extended temperature range with a more refined lock-in detection technique are presented in Fig. 24.8 for a field configuration that allows the measurement of $\mu_\perp$ in the A-phase (Nummila *et al.*, 1989). Just below T_c, the ion mobility can be used as a sensitive secondary thermometer, allowing a resolution of $0.1\,\mu$K above $0.95T_c$. This kind of thermometer is fast because the ion probes the temperature of the fluid directly.

An investigation closer to the critical line (Ahonen *et al.*, 1978) shows that μ_- does not depend on pressure in the B-phase, whereas in the A-phase it shows a small pressure dependence, as shown in Figs 24.9 and 24.10, respectively.

Owing to the much higher temperature resolution that can be achieved in the A-phase because of NMR thermometry (Osheroff *et al.*, 1972*a*), the region close to T_c can be studied very accurately (Ahonen *et al.*, 1978). It is found that the quantity $(\mu_N/\mu_- - 1)^2$ is a linear function of the reduced temperature $1 - T/T_c$, according to the theoretical prediction (Soda, 1975)

$$\mu_-^{-1} = \mu_N^{-1}\left[1 - c\left(1 - \frac{T}{T_c}\right)^{1/2}\right] + \mathcal{O}\left(\Delta^2\right), \qquad (24.6)$$

where c is a coefficient that depends on the mutual orientation of the vectors $\hat{\mathbf{d}}$

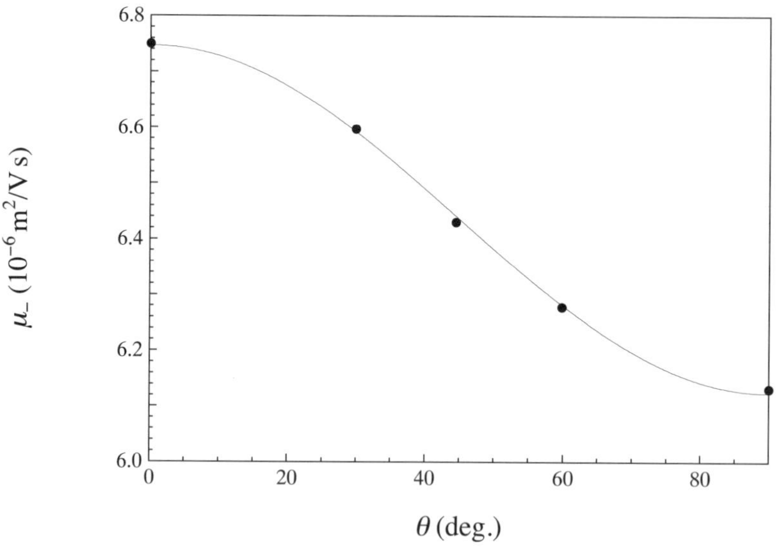

FIG. 24.7. μ_- vs θ in the A-phase for $T - T_c = -160\,\mu$K, $P = 2.6$ MPa, and $H = 1.8$ mT. (Roach *et al.*, 1977*a*.)

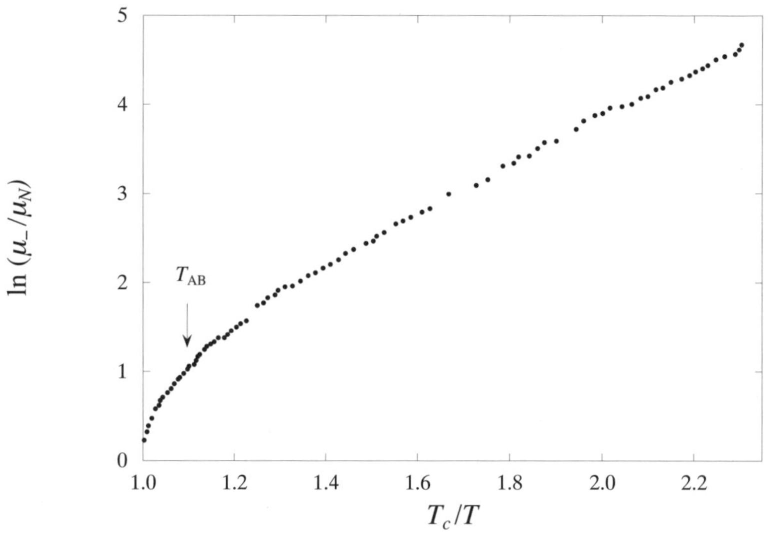

FIG. 24.8. μ_-/μ_N vs T_c/T in superfluid ^{3}He for $P = 2.93\,\mathrm{MPa}$. (Nummila *et al.*, 1989.) μ_N is the mobility in the normal phase.

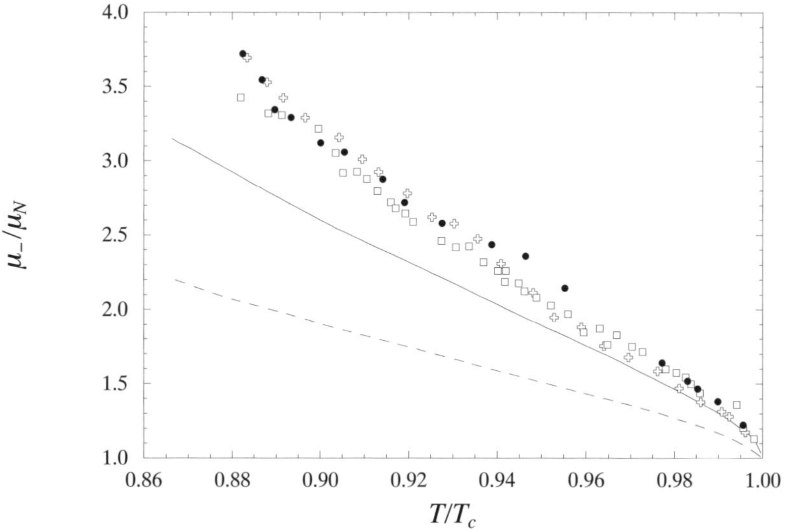

FIG. 24.9. μ_-/μ_N vs T/T_c in the B-phase for $P = 0.6\,\mathrm{MPa}$ with $E = 4\,\mathrm{kV/m}$ (crosses), and $P = 1.8\,\mathrm{MPa}$ with $E = 4\,\mathrm{kV/m}$ (squares) and $E = 2\,\mathrm{kV/m}$ (circles). (Ahonen *et al.*, 1978.) The lines are explained in the text.

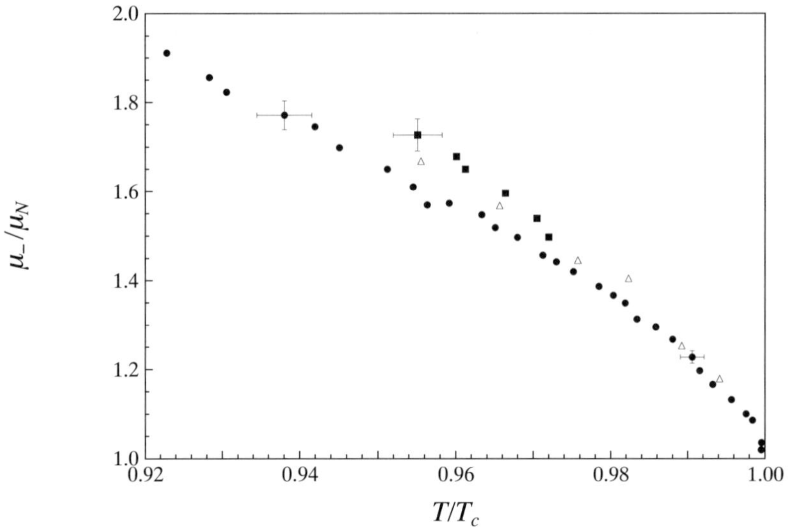

FIG. 24.10. μ_-/μ_N vs T/T_c in the A-phase for $P = 2.0\,\mathrm{MPa}$ with $E = 4\,\mathrm{kV/m}$ (triangles) and $E = 2\,\mathrm{kV/m}$ (squares), and for $P = 2.88\,\mathrm{MPa}$ with $E = 4\,\mathrm{kV/m}$ (circles). (Ahonen *et al.*, 1978.)

and $\hat{\mathbf{l}}$. Equation (24.6) can be considered a sort of a scaling law. The experimental results for $P = 2.84\,\mathrm{MPa}$ are presented in Fig. 24.11.

It is interesting to note that this law has the scaling form typical of critical phenomena (Stanley, 1971). Within the experimental accuracy, the slope turns out to be pressure independent in the range $2.0\,\mathrm{MPa} < P < 2.88\,\mathrm{MPa}$ (Ahonen *et al.*, 1978).

24.2.2 *Nonlinear velocity regime*

At higher electric fields, the drift velocity of ions is no longer proportional to the field strength and enters the nonlinear regime also observed in the normal liquid. As in the case of the normal liquid, the lower the temperature, the smaller is the field at which the departure from the linear behavior appears.

In Fig. 24.12 the nonlinear behavior of v_D as a function of E in both superfluid ^{3}He–A and ^{3}He–B as a function of temperature for $P = 2.6\,\mathrm{MPa}$ is shown (Roach *et al.*, 1977*a*). Similar results are also obtained at different pressures, the drift velocity being smaller in the A-phase than in the B-phase (Ahonen *et al.*, 1976, 1978).

The general shape of the curves is the same in both phases and does not depend on the presence of a static magnetic field. Quite interestingly, at higher fields all curves, both in the A- and B-phases, become nearly parallel to the normal fluid curve. The theoretical calculations for the nonlinear regime in superfluid ^{3}He–B based on the Born approximation (Fetter and Kurkijärvi, 1977)

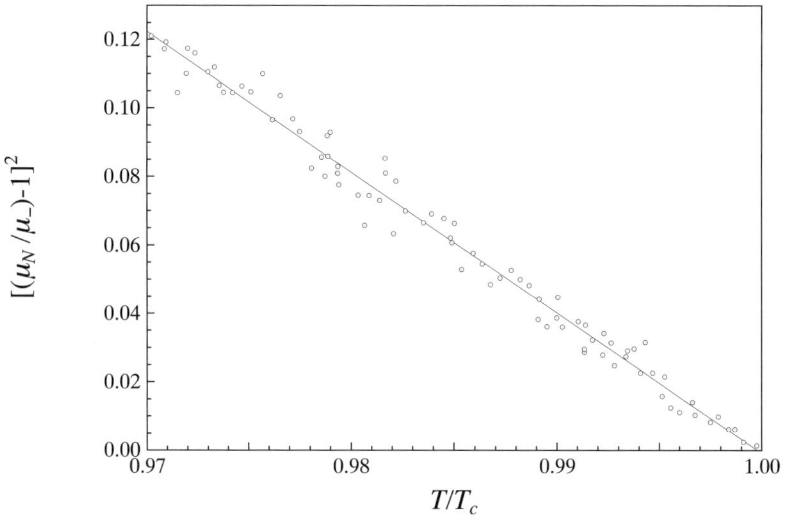

FIG. 24.11. $[(\mu_N/\mu_-)-1]^2$ vs T/T_c in superfluid ^{3}He–A near T/T_c for $P = 2.84$ MPa and $E = 2$ kV/m. (Ahonen *et al.*, 1978.) Line: the function $4.1(1 - T/T_c)$.

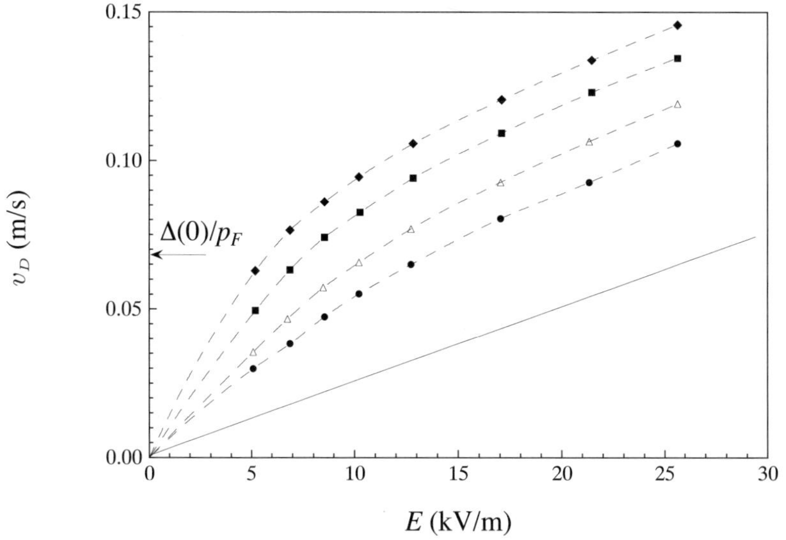

FIG. 24.12. v_D vs E for negative ions in superfluid ^{3}He for $P = 2.6$ MPa. (Roach *et al.*, 1977*a*.) $T_c - T$ (μK) = 438 (diamonds) and 284 (squares) for the B-phase; 159 (triangles), and 91 (circles) for the A-phase. The arrow shows $v_c = \Delta(0)/p_F$. Dashed lines: eyeguides. Solid line: normal liquid.

only give a qualitative agreement with the experimental data (Ahonen *et al.*, 1978).

The crossover from the linear to the nonlinear regime occurs for velocities close to the critical velocity $v_c \simeq \Delta(0)/p_F$, for which the energy $2p_F v_D$ gained by the quasiparticles in the collisions with ions exceeds the value $2\Delta(0)$, where $\Delta(0)$ is the BCS value of the energy gap.Thus, the high-velocity behavior is mainly determined by the pair-breaking effect of the drifting ion. It is interesting to note that this behavior closely resembles that of ions in superfluid ^{4}He when their velocity exceeds the Landau critical velocity for roton emission $v_L \approx \Delta/p_0$, where Δ and p_0 are the roton energy gap and momentum, respectively (Phillips and McClintock, 1974*a*; Allum *et al.*, 1976*a*; Bowley and Sheard, 1977).

For $P = 1.8\,\mathrm{MPa}$, $\Delta(0) \approx 1.764 k_B T_c$ and the corresponding Landau velocity is $v_c \approx 6.7 \times 10^{-2}\,\mathrm{m/s}$, i.e., the largest velocity observed in the linear regime (Ahonen *et al.*, 1976). At the lowest T, μ_- becomes so large that the weakest field is strong enough to drive the ion to this limit (Nummila *et al.*, 1989).

24.3 Phenomenology of positive ion transport in superfluid ^{3}He

Only a few papers are devoted to the mobility μ_+ of positive ions in superfluid ^{3}He (Roach *et al.*, 1977*a*; Kokko *et al.*, 1978). Some experimental data are presented in Figs 24.13 to 24.15. The remarkable differences between positive and negative ions observed in the normal liquid persist in the superfluid. μ_+ increases with decreasing T below T_c, though not as rapidly as for the negative ions, and shows anisotropy effects in the A-phase.

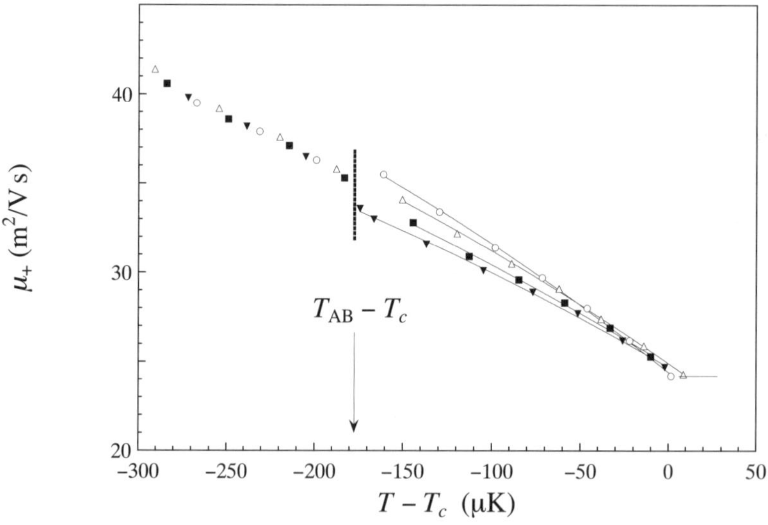

FIG. 24.13. μ_+ vs $T - T_c$ for $P = 2.6\,\mathrm{MPa}$. A field $H = 1.8\,\mathrm{mT}$ forms an angle θ with E. (Roach *et al.*, 1977*a*.) θ (deg.)= 0, 30, 60, and 90 (from top).

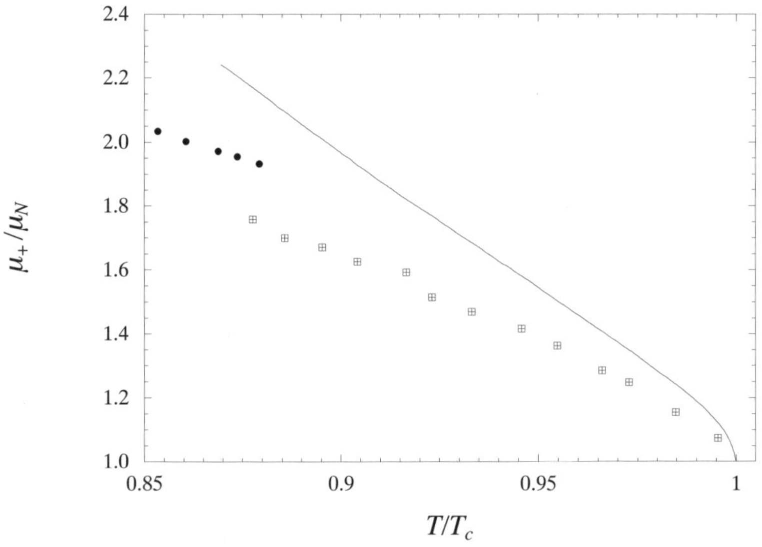

FIG. 24.14. μ_+/μ_N vs T/T_c in ^{3}He–B for P (MPa) = 2.8 (circles) and 1.8 (squares) (Kokko *et al.*, 1978). Line: theory (Bowley, 1976*a*).

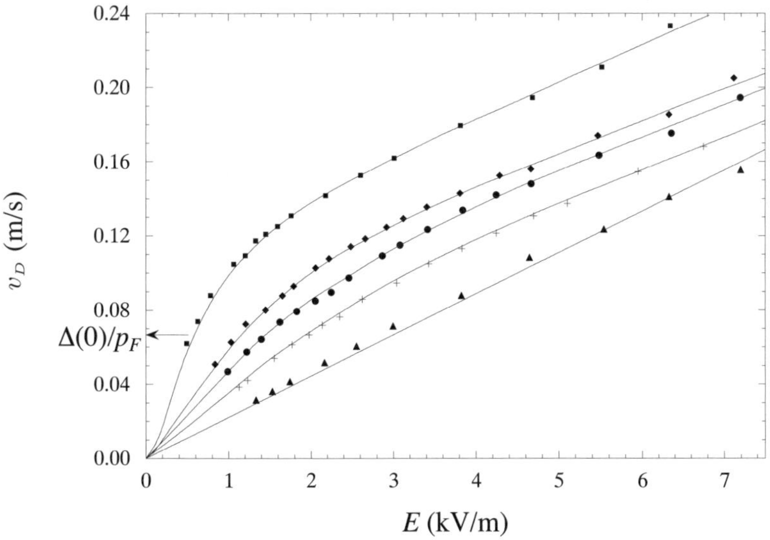

FIG. 24.15. v_D vs E for positive ions. T/T_c = 0.62 (squares), 0.81 (diamonds) for phase B; 0.89 (circles), 0.93 (crosses) for phase A; normal fluid at T = 3.11 mK (triangles) (Kokko *et al.*, 1978).

In Fig. 24.13 μ_+ is presented as a function of temperature for $P = 2.6\,\mathrm{MPa}$ (Roach *et al.*, 1977*a*). Data at different P (Kokko *et al.*, 1978) agree well with these. The anisotropy of μ_+ in the A-phase can be observed by adding a small magnetic field forming an angle θ with the electric drift field. The *fan-averaged texture* model proposed for the negative ions (eqn (24.5)) describes the data well at constant T, yielding $\mu_\perp = 35.2 \times 10^{-6}\,\mathrm{m^2/V\,s}$ and $\mu_\parallel = 30.4 \times 10^{-6}\,\mathrm{m^2/V\,s}$ (Roach *et al.*, 1977*a*). Small discrepancies with the data of Kokko *et al.* (1978) may be attributed to texture effects.

In the B–phase μ_+ lies below any theoretical model (see Fig. 24.14). This probably means that the major difference between the normal and the superfluid is related to the number of thermally-excited quasiparticles (Kokko *et al.*, 1978).

Finally, in Fig. 24.15, v_D is shown as a function of E (Kokko *et al.*, 1978). Similar results have also been obtained by other researchers (Roach *et al.*, 1977*a*). The nonlinearity resembles that of negative ions and its onset is consistent with the critical velocity for pair–breaking, $v_c \sim \Delta(0)/p_F$.

24.4 Theory of the negative ion mobility in superfluid ^{3}He

The theoretical analysis of the mobility of negative ions is very complicated for several reasons. In addition to the problems encountered in the case of the normal liquid, issues specific to the superfluid nature of the liquid, including its anisotropy in the A-phase, come into play. The temperature and field dependence of the mobility, as well as its anisotropy, must reflect the excitation spectrum of the quasiparticles. Here, only a brief sketch of the theory will be outlined because its details are beyond the scope of this book.

24.4.1 *Superfluid ^{3}He–B*

Baym *et al.* (1977) have derived a general equation for the mobility in superfluid ^{3}He–B that takes into account the excitation spectra of the liquid and of the ion within the theoretical framework established by Josephson and Lekner (1969). This general approach recovers the results obtained in previous works, in which different approximations were used (Bowley, 1976*a*, 1977*b*; Fetter and Kurkijärvi, 1977; Soda, 1975, 1977).

The ion–liquid interaction is assumed to depend on the local ^{3}He and ion densities. The drag acting on the ion is determined from the rate $\mathrm{d}\mathbf{P}/\mathrm{d}t$ at which an ion transfers momentum to the quasiparticles. If the ion concentration is low, the ^{3}He atoms may be assumed to be at rest and in equilibrium during the ion motion. In the superfluid phase, in addition to the usual scattering with the ^{3}He quasiparticles, the ion motion is altered by the creation and annihilation of pairs (Baym *et al.*, 1977; Fetter and Kurkijärvi, 1977).

The derivation of $\mathrm{d}\mathbf{P}/\mathrm{d}t$ follows the general lines used, for instance, for the calculation of the phonon-limited mobility of the electron bubble in superfluid He II (Baym *et al.*, 1969):

$$\frac{\mathrm{d}\mathbf{P}}{\mathrm{d}t} = n_3 \sum_{\mathbf{k}} \int_{-\infty}^{\infty} \hbar\mathbf{k} S_3(\mathbf{k},\omega) S_i(-\mathbf{k},-\omega) |t(\mathbf{k},\omega,\mathbf{v})|^2 \frac{\mathrm{d}\omega}{2\pi}. \tag{24.7}$$

$\hbar\mathbf{k}$ and $\hbar\omega$ are the momentum and energy, respectively, transferred during the collision. n_3 is the density of quasiparticles and S_3 is the equilibrium structure function for ^{3}He density fluctuations. S_i describes the spectrum of possible energy transfers $-\hbar\omega$ to the moving ion for a given momentum-transfer $\hbar\mathbf{k}$. $\mathbf{v}$ is the ion velocity. It is also assumed that the t-matrix element for this scattering process depends only on the variables quoted in its parentheses (Baym *et al.*, 1979). The t-matrix formalism is adopted to avoid the difficulty encountered in using the Born approximation (Soda, 1977).

By exploiting the fact that there would be no drag on the ion if the ^{3}He quasiparticles were drifting at the same speed as the ion and in equilibrium with it, it is possible to calculate the momentum-transfer to leading order in $\mathbf{v}$ for small electric fields:

$$\frac{\mathrm{d}\mathbf{P}}{\mathrm{d}t} = n_3 \sum_{\mathbf{k}} \int_{-\infty}^{\infty} \hbar\mathbf{k} \left[S_3(\mathbf{k},\omega) - S_3(\mathbf{k},\omega,\mathbf{v})\right] S_i(-\mathbf{k},-\omega,\mathbf{v}) |t(\mathbf{k},\omega,\mathbf{v})|^2 \frac{\mathrm{d}\omega}{2\pi}. \tag{24.8}$$

Here $S_3(\mathbf{k},\omega,\mathbf{v})$ is the equilibrium dynamic structure factor for ^{3}He moving with speed $\mathbf{v}$.

To leading order in $\mathbf{v}$, one can set $\mathbf{v} = 0$ in S_i and in $|t|^2$. Equation (24.8) can be simplified by using the detailed balance conditions for the Ss:

$$S_3(-\mathbf{k},-\omega,\mathbf{v}) = S_3(\mathbf{k},\omega,\mathbf{v}) \exp\left[-\hbar\beta(\omega - \mathbf{k}\cdot\mathbf{v})\right], \tag{24.9}$$

$$S_i(\mathbf{k},\omega) = S_i(-\mathbf{k},-\omega) \exp(\hbar\beta\omega), \tag{24.10}$$

where $S_i(\mathbf{k},\omega) = S_i(\mathbf{k},\omega,\mathbf{v}=0)$ and $t(\mathbf{k},\omega) = t(\mathbf{k},\omega,\mathbf{v}=0)$, and the time-reversal symmetry of the collision process:

$$|t(-\mathbf{k},-\omega)|^2 = |t(\mathbf{k},\omega)|^2, \tag{24.11}$$

thus yielding

$$\frac{\mathrm{d}\mathbf{P}}{\mathrm{d}t} = \frac{n_3}{2} \sum_{\mathbf{k}} \hbar\mathbf{k} \left[\exp(\hbar\beta\mathbf{k}\cdot\mathbf{v}) - 1\right] \times \int_{-\infty}^{\infty} S_3(\mathbf{k},\omega,\mathbf{v}) S_i(-\mathbf{k},-\omega) |t(\mathbf{k},\omega)|^2 \frac{\mathrm{d}\omega}{2\pi}. \tag{24.12}$$

At steady state, the momentum loss equals the rate $e\mathbf{E}$ at which momentum is supplied by the external electric field. By expanding eqn (24.12) to leading order in $\mathbf{v}$, one obtains the mobility tensor as

$$\frac{e}{\mu_{ij}} = \frac{\hbar^2 \beta n_3}{2} \sum_{\mathbf{k}} k_i k_j \int_{-\infty}^{\infty} S_3(\mathbf{k}, \omega) S_i(-\mathbf{k}, -\omega) |t(\mathbf{k}, \omega)|^2 \frac{d\omega}{2\pi}, \tag{24.13}$$

where μ_{ij} is defined by $v_i = \sum_j \mu_{ij} E_j$.

In normal ^{3}He and in its superfluid B-phase, the mobility is isotropic and eqn (24.13) reduces to

$$\frac{e}{\mu_-} = \frac{\hbar^2 \beta n_3}{2} \sum_{\mathbf{k}} k^2 \int_{-\infty}^{\infty} S_3(\mathbf{k}, \omega) S_i(-\mathbf{k}, -\omega) |t(\mathbf{k}, \omega)|^2 \frac{d\omega}{2\pi}. \tag{24.14}$$

In normal ^{3}He, collective modes have too small a density to be relevant, and the dominant momentum-transfer mechanism is scattering off ^{3}He quasiparticles. Thus, in normal ^{3}He one has

$$S_3(\mathbf{k}, \omega) = \frac{1}{n_3} \sum_{\mathbf{p}, \mathbf{p}'} \delta_{\mathbf{p}'-\mathbf{p}, \hbar\mathbf{k}} n_p (1 - n_{p'}) 2\pi\delta(\hbar\omega + \xi_p - \xi_{p'}), \tag{24.15}$$

where $\mathbf{p}$ and $\mathbf{p}'$ are the momenta of the initial and final states, respectively, of the ^{3}He quasiparticles, and ξ_p and ξ'_p are their energies as measured with respect to the ^{3}He chemical potential. $n_p = [\exp(\beta\xi_p) + 1]^{-1}$ is the Fermi equilibrium distribution function.

Near the Fermi surface the quasiparticle energy in the normal state is $\xi_p \approx (p - p_F)v_F$, where p_F and v_F are the Fermi momentum and velocity, respectively. Inserting eqn (24.15) into eqn (24.14) the drag, and hence the mobility, becomes

$$\frac{e}{\mu_-} = \frac{1}{6} \sum_{\mathbf{p}, \mathbf{p}'} (\mathbf{p}' - \mathbf{p})^2 \beta n_p (1 - n_{p'}) S_i(\mathbf{p} - \mathbf{p}', \xi_p - \xi_{p'}) |t_{\mathbf{p}, \mathbf{p}'}|^2. \tag{24.16}$$

Here, the dependence of the scattering matrix element on the initial and final momentum states of quasiparticles is explicitly shown.

If the ion were recoiling as a free particle in collisions, and for temperatures much higher than the recoil temperature defined in Chapter 21 (eqn (21.24)), $S_i(\mathbf{k}, \omega) \approx 2\pi\delta(\omega)$. In this case the mobility becomes

$$\frac{e}{\mu_-} = \frac{\pi}{3} \sum_{\mathbf{p}, \mathbf{p}'} (\mathbf{p}' - \mathbf{p})^2 \left(-\frac{\partial n_p}{\partial \xi_p} \right) \delta(\xi_p - \xi_{p'}) |t_{\mathbf{p}, \mathbf{p}'}|^2. \tag{24.17}$$

For $T \ll T_F$, eqn (24.17) gives

$$\frac{e}{\mu_-} = \frac{m^{\star 2}}{12\pi^3} \int_0^{2p_F} k^3 |t(k)|^2 dk = n_3 p_F \sigma_{mt}, \tag{24.18}$$

where $|t(k)|^2 = |t_{\mathbf{p}+\hbar\mathbf{k}, \mathbf{p}}|^2$ for $\mathbf{p}$ on the Fermi surface and the momentum-transfer cross-section is given by

$$\sigma_{mt} = \int_{-1}^{1} (1 - \cos\theta) \left(\frac{m^\star}{2\pi\hbar^2} \right)^2 |t\left[k = 2k_F \sin(\theta/2)\right]|^2 \, \mathrm{d}\cos\theta \,. \tag{24.19}$$

Equation (24.17) has to be compared to eqn (21.17), and eqn (24.18) to eqn (21.20).

In spite of the fact that the ion motion is diffusive, as discussed previously, the ion recoil spectrum is modified by the scattering process in such a way that the scattering can be considered elastic because, for negative ions, the characteristic energy transfer (eqn (21.45)) is typically much smaller than the thermal energy $\hbar\bar{\omega} \approx \hbar^2 k_F^2 \mu_- k_\mathrm{B} T/e \approx 0.1 k_\mathrm{B} T$ (Anderson *et al.*, 1968; Kuchnir *et al.*, 1970; Ahonen *et al.*, 1976). This is a consequence of the large radius of the negative ion and of its large effective mass.

The approximation of elastic scattering may also be retained in superfluid ^{3}He–B in a region not too far below T_c. However, as μ_- increases rapidly with decreasing T below T_c, deviations between theory and experiment should appear below T_c owing to the gradual breakdown of the elastic scattering approximation (Baym *et al.*, 1979).

In addition to the scattering off ^{3}He quasiparticles, ions can transfer momentum by creating or destroying Cooper pairs. To do so, an energy transfer $\geqslant 2\Delta$ is required for a slowly-drifting ion. If $2\Delta > \hbar k_F^2 \mu_- k_\mathrm{B} T/e$, the scattering is elastic enough to neglect pair creation or destruction processes. Just below T_c the situation is reversed as $2\Delta < \hbar k_F^2 \mu_- k_\mathrm{B} T/e$, but such processes, though allowed kinematically, have negligible amplitude.

For these reasons, as long as the elastic approximation $S_i = 2\pi\delta(\omega)$ can be assumed to be valid, the mobility is given by the generalization of eqn (24.16) to superfluid quasiparticles:

$$\frac{e}{\mu_-} = \frac{1}{6} \sum_{\mathbf{p}\sigma, \mathbf{p}'\sigma'} \left(\mathbf{p}' - \mathbf{p}\right)^2 \beta n_p \left(1 - n_{p'}\right) 2\pi\delta\left(\epsilon_p - \epsilon_{p'}\right) |t_{\mathbf{p}'\sigma', \mathbf{p}\sigma}|^2 \,. \tag{24.20}$$

The quasiparticle energy in the B-phase is

$$\epsilon_p = \left({\xi_p}^2 + \Delta^2\right)^{1/2} \tag{24.21}$$

and the distribution function is

$$n_p = \frac{1}{\exp\left(\beta\epsilon_p\right) + 1} \,. \tag{24.22}$$

In eqn 24.20 the sum is over all quasiparticle spin (σ) and momenta. $t_{\mathbf{p}'\sigma', \mathbf{p}\sigma}$ is the amplitude for scattering of a quasiparticle from the state $(\mathbf{p}, \sigma)$ to the state $(\mathbf{p}', \sigma')$.

By defining a spin-averaged differential scattering cross-section as

$$\frac{\mathrm{d}\sigma}{\mathrm{d}\Omega_{p'}} = \frac{1}{2\left(2\pi\right)^3 v_p} \sum_{\sigma, \sigma'} \int_0^\infty 2\pi {p'}^2 \delta\left(\epsilon_p - \epsilon_{p'}\right) |t_{\mathbf{p}'\sigma', \mathbf{p}\sigma}|^2 \mathrm{d}p' \,, \tag{24.23}$$

v_p is the superfluid quasiparticle velocity:

$$v_p = \frac{\partial \epsilon_p}{\partial p} = v_F \left(\frac{\xi_p}{\epsilon_p} \right) . \qquad (24.24)$$

A quasiparticle of momentum p can scatter elastically to a state with the same p for which $\xi_{p'} = \xi_p$, or to one for which $p' = 2p_f - p$ with $\xi_{p'} = -\xi_p$ because the quasiparticle energy is an even function of $p - p_F$. So, one can write (Merzbacher, 1970)

$$\frac{\mathrm{d}\sigma\,(\xi)}{\mathrm{d}\Omega} = \left(\frac{m^\star}{2\pi\hbar^2} \right)^2 \left(\frac{\epsilon_p}{\xi_p} \right)^2 \langle |t|^2 \rangle \,, \qquad (24.25)$$

with

$$\langle |t|^2 \rangle = \frac{1}{2} \sum_{\sigma,\sigma'}{}' \left| t_{\mathbf{p}'\sigma',\mathbf{p}\sigma} \right|^2 . \qquad (24.26)$$

The summation is restricted to those final states with $\epsilon_{p'} = \epsilon_p$ and momenta $\mathbf{p}'$ lying in the solid angle $\mathrm{d}\Omega$.

Now, the drag becomes (Baym *et al.*, 1977, 1979)

$$\frac{e}{\mu_-} = -n_3 p_F \int_{-\infty}^{\infty} \left| \frac{\partial \epsilon}{\partial \xi} \right| \left(\frac{\partial n}{\partial \epsilon} \right) \sigma_{mt}\,, (\xi)\, \mathrm{d}\xi \qquad (24.27)$$

where the "p" index has been dropped. The momentum-transfer cross-section is given by

$$\sigma_{mt}(\xi) = \int (1 - \cos\theta) \frac{\mathrm{d}\sigma(\xi)}{\mathrm{d}\Omega} \mathrm{d}\Omega \,, \qquad (24.28)$$

where θ is the angle between the directions of the initial and final momenta. Equation (24.27) in the normal fluid gives

$$\frac{e}{\mu_N} = n_3 p_F \sigma_{mt}^N (\xi = 0) \,. \qquad (24.29)$$

Two sources give origin to the differences between the mobility in the normal fluid and in the superfluid. The first one is due to the changes of the quasiparticle velocity and thermal distribution function, whereas the second one is due to modifications of the scattering cross-section. Bowley (1976*a*) assumed that the cross-section is unaffected by the superfluid transition. Equation (24.29) then simply gives

$$\frac{\mu_B}{\mu_N} = \frac{1}{2} \left[1 + \exp\left(\beta \Delta \right) \right] . \qquad (24.30)$$

In Fig. 24.16 the experimental data of Ahonen *et al.* (1978) at $P = 1.8$ MPa in superfluid ^{3}He–B are compared to the theory. Equation (24.30) that describes only the increase of the mobility with decreasing temperature as a result of the reduction in the number of thermal quasiparticles and in their velocities (Bowley,

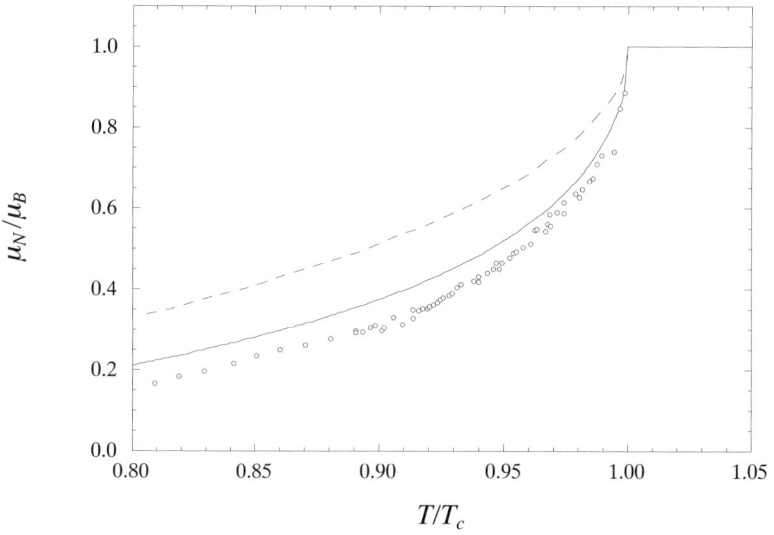

FIG. 24.16. μ_N/μ_B vs T/T_c in superfluid ^{3}He–B for $P = 1.8$ MPa (Ahonen *et al.*, 1978). Dashed line: eqn (24.30) (Bowley, 1976*a*). Solid line: theory (Baym *et al.*, 1977).

1976*a*, 1977*b*) is shown as a dashed line and only accounts for half of the total variation of the mobility (Ahonen *et al.*, 1976, 1978; Roach *et al.*, 1977*a*). This is also true at lower pressures, as shown in Fig. 24.9, in which eqn (24.30) is also shown as a dashed line. It is clear that a correct description of the mobility must include the modification of the intermediate scattering states induced by the superfluidity (Baym *et al.*, 1977, 1979).

A first attempt at modifying the momentum-transfer cross-section because of superfluidity is made by assuming that the scattering amplitude in the superfluid phase t^S can be obtained by performing a Bogoliubov transformation on the scattering amplitude in the normal phase t^N (Soda, 1975, 1977; Fetter and Kurkijärvi, 1977):

$$t^S_{\mathbf{p}',\mathbf{p}} = uu' t^N_{\mathbf{p}',\mathbf{p}} - vv' t^{N\star}_{\mathbf{p}',\mathbf{p}}, \tag{24.31}$$

where the primes refer to the final states and the spin indices have been dropped. u and v are the coherence factors.

Squaring and averaging over spins, one obtains

$$|t^S_{\mathbf{p}',\mathbf{p}}|^2 = \frac{1}{2}\left(1 + \frac{\xi\xi'}{\epsilon\epsilon'}\right)|t^N_{\mathbf{p}',\mathbf{p}}|^2 - \frac{1}{2}\frac{\mathbf{\Delta}\cdot\mathbf{\Delta}'}{\epsilon\epsilon'}\mathrm{Re}\left[\left(t^N_{\mathbf{p}',\mathbf{p}}\right)^2\right], \tag{24.32}$$

where $\mathbf{\Delta}(\hat{p})$ is the vector describing the p-wave pairing gap: $\mathbf{\Delta}\cdot\mathbf{\Delta}' = \Delta^2\cos\theta$. Inserting eqn (24.32) into eqn (24.25), the resulting cross-section diverges for $\xi \to 0$, thus violating unitarity and leading to a logarithmic divergence of the mobility.

The origin of this behavior is related to the divergent density of quasiparticles $\propto |\epsilon/\xi|$ at the gap edge (Soda, 1977; Bromley, 1981).

These problems are solved by taking into account modifications of the intermediate scattering states by solving a full Lippman–Schwinger equation for the t-matrix for quasiparticle–ion scattering (Baym *et al.*, 1977, 1979). These calculations are very difficult and long, and are far beyond the scope of this book. The interested reader will find them in the literature. It suffices to quote here the main results.

For any partial waves up to a given value $l_0 = k_F R_-$, where R_- is the ion radius, for $|\xi| \ll \Delta$, the scattering amplitudes are all equal. As a consequence, the quasiparticle–ion-scattering cross-section is strongly peaked in the forward direction, as in diffraction, with an angular width $\sim l_0^{-1/2}$. At the same time, it is greatly depressed at large angles because of destructive interference. As forward-scattering processes do not transfer a great deal of momentum, the transport cross-section is greatly reduced for $|\xi| \ll \Delta$ with respect to its normal-state value. This is a result of a large density of quasiparticle states near the gap edge.

Figure 24.17 shows how this forward-scattering effect due to interference influences the spin-averaged momentum-transfer and total cross-sections for ^{3}He quasiparticle–ion scattering. The cross-sections are calculated by assuming $l_0 = 10.3$, a value appropriate for the experimental value of the electron bubble radius for $P = 1.8\,\mathrm{MPa}$ (Ahonen *et al.*, 1978). Distortions of the order parameter near the bubble surface (Rainer and Vuorio, 1977) are neglected owing to the large coherence length of the superfluid.

The momentum-transfer cross-section $\sigma_{mt}(\xi)$ has a minimum at low energy of order σ_{mt}^N/l_0, where σ_{mt}^N is the normal fluid value. Once the cross-section is known, the mobility ratio between the normal and superfluid state can be calculated by using eqn (24.27) as

$$\frac{\mu_N}{\mu_B} = \int_{-\infty}^{\infty} \left|\frac{\partial \epsilon}{\partial \xi}\right| \left(\frac{\partial n}{\partial \epsilon}\right) \frac{\sigma_{mt}(\xi)}{\sigma_{mt}^N} \mathrm{d}\xi \,. \tag{24.33}$$

The mobility data of Ahonen *et al.* (1978) are used to determine the ion radius and the normal fluid value of the cross-section. The BCS weak-coupling gap $\Delta_{BCS} \approx 3.06 k_\mathrm{B} T(1 - T/T_c)^{1/2}$ near T_c, enhanced by a strong-coupling factor of 1.05 (Paulson *et al.*, 1976), is used as Δ. The results are plotted as the solid lines in Figs 24.16 and 24.9, showing how the present theory improves on that of Bowley (1976*a*). The deviations at lower temperatures are expected because of the breakdown of the elastic scattering approximation based on the no-recoil assumption $\hbar k_F^2 \mu_-/e \ll 1$. For instance, at $T/T_c \approx 0.9$ and $P = 1.8\,\mathrm{MPa}$, $\hbar k_F^2 \mu_-/e \approx 1/3$.

In contrast, the pressure dependence of the ratio μ_N/μ_B at constant T/T_c is rather small. It arises from a pressure dependence of the ion radius, which leads only to small changes because the relative cross-section is practically independent of it. There may also be a pressure dependence of the strong-coupling

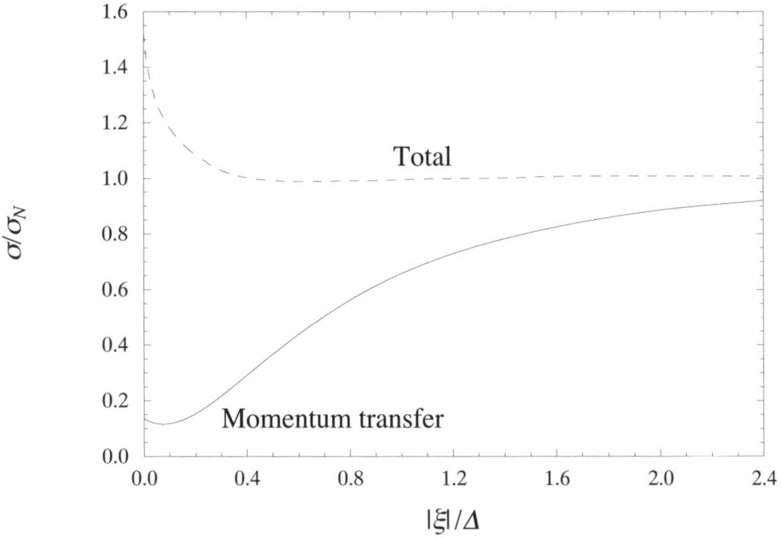

FIG. 24.17. Spin-averaged momentum-transfer (solid line) and total (dashed line) cross-sections for the ion–^{3}He quasiparticle scattering as a function of $|\xi|/\Delta$ for $l_0 = 10.3$, and normalized to their values in the normal phase. (Baym *et al.*, 1977.)

enhancement factor of the energy gap, but it is also small (Baym *et al.*, 1979). As a result, both the experiment and theory do not show any appreciable pressure dependence of the mobility, as demonstrated by Fig. 24.9.

Finally, we briefly mention the conjecture that ballistic coherent motion of ions in superfluid ^{3}He–B for $T < 0.2T_c$–$0.3T_c$ may occur in concurrence with a giant mass renormalization (Prokof'ev, 1995; Rosch and Kopp, 1998).

24.4.2 *Superfluid ^{3}He–A*

In ^{3}He–A the orbital anisotropy of the elementary excitations about the $\hat{\mathbf{l}}$-axis leads to a tensor structure of the ion mobility, as first observed by Roach *et al.* (1977*a*), who applied an auxiliary magnetic field to change the direction of $\hat{\mathbf{l}}$ in the liquid with respect to the drift field.

The theory for the mobility in the A-phase (Salomaa *et al.*, 1980*a*, 1980*b*) is a generalization of the theory developed for the B-phase (Baym *et al.*, 1977, 1979) that now takes into account the anisotropy of the order parameter.

As for the B-phase case, the scattering of the ion is assumed to be elastic. Hence, at small drift velocities, collisions that alter the number of superfluid excitations cannot occur and the mobility is determined by the scattering of the ion off ^{3}He quasiparticles. Their spectrum in the A-phase is given by

$$\epsilon_p^2 = \xi_p^2 + |\Delta(\hat{p})|^2 , \tag{24.34}$$

where $\xi_p \approx v_F(p - p_F)$ is the energy of the quasiparticles in the normal state. The energy gap is now given by

$$\Delta(\hat{p}) = \Delta(T)\sin\theta e^{i\varphi}\hat{\mathbf{l}} \tag{24.35}$$

for quasiparticles in the direction $\hat{p} = \mathbf{p}/p$ relative to the direction of $\hat{\mathbf{l}}$.

The components of the mobility tensor are given by an obvious generalization of eqn (24.17):

$$\frac{e}{\mu_{\|}} = \sum_{\mathbf{p}',\mathbf{p}} (p_z - p'_z)^2 \langle |t_{\mathbf{p}'\sigma',\mathbf{p},\sigma}(\epsilon_p)|^2\rangle \left(-\frac{\partial n_p}{\partial \epsilon_p}\right) 2\pi\delta(\epsilon_p - \epsilon_{p'}) \,, \tag{24.36}$$

$$\frac{e}{\mu_{\perp}} = \frac{1}{2}\sum_{\mathbf{p}',\mathbf{p}} \left[(p_x - p'_x)^2 + (p_y - p'_y)^2\right] \langle |t_{\mathbf{p}'\sigma',\mathbf{p},\sigma}(\epsilon_p)|^2\rangle \times \left(-\frac{\partial n_p}{\partial \epsilon_p}\right) 2\pi\delta(\epsilon_p - \epsilon_{p'}) \,. \tag{24.37}$$

$\hat{\mathbf{l}}$ is directed along the z-axis and the subscripts $\|$ and $\perp$ refer to the directions parallel and perpendicular to it. n_p is the Fermi distribution function and $t_{\mathbf{p}',\mathbf{p}}(\epsilon_p)$ is the amplitude for scattering a superfluid quasiparticle of energy ϵ_p from a state with momentum $\mathbf{p}$ into one with momentum $\mathbf{p}'$.

By analogy with the case of the B-phase (eqn (24.23)), spin-averaged differential scattering cross-sections can be defined that take into account the anisotropy of the A-phase:

$$v_p \frac{\mathrm{d}\sigma}{\mathrm{d}\Omega'}(\hat{p}', \hat{p}, \epsilon_p) = \frac{1}{2}\sum_{\sigma,\mathbf{p}',\sigma'} |t_{\mathbf{p}'\sigma',\mathbf{p},\sigma}(\epsilon_p)|^2 2\pi\delta(\epsilon_p - \epsilon_{p'}) \,, \tag{24.38}$$

where the use of the same symbol σ for the cross-section and the spin indices should not confuse the reader. v_p is the group velocity of the quasiparticles in the superfluid:

$$v_p = \left|\frac{\partial \epsilon_p}{\partial \mathbf{p}}\right| = v_F \frac{|\xi_p|}{\epsilon_p} = v_F \frac{\left[\epsilon_p^2 - |\Delta(\hat{p})|^2\right]^{1/2}}{\epsilon_p} \,, \tag{24.39}$$

neglecting the velocity component perpendicular to $\hat{p}$:

$$(v_p)_{\perp} = \frac{v_F}{2}\frac{\Delta^2(T)}{\epsilon_F \epsilon_p}\sin\theta\cos\theta \,, \tag{24.40}$$

which is almost always small with respect to the parallel component.

$\epsilon_{p'}$ is an even function of $\xi_{p'}$ and the sum in eqn (24.38) extends over states on the branch of the energy spectrum with $\xi_{p'} > 0$ and $\xi_{p'} < 0$. Thus, the spin-averaged differential cross-section becomes

$$\frac{d\sigma}{d\Omega'}\left(\hat{p}',\hat{p},\epsilon_p\right)=\left(\frac{m^\star}{2\pi\hbar^2}\right)^2\frac{\epsilon_{p'}}{\left[\epsilon_{p'}^2-|\Delta\left(\hat{p}'\right)|^2\right]^{1/2}}\langle|t_{\hat{p}',\hat{p}}|^2\rangle\frac{\epsilon_p}{\left[\epsilon_p^2-|\Delta\left(\hat{p}\right)|^2\right]^{1/2}} \tag{24.41}$$

where $m^\star$ is the effective mass of the quasiparticles at the Fermi surface and $\langle|t_{\hat{p}',\hat{p}}|^2\rangle$ is the branch-averaged squared transition matrix element defined by

$$\langle|t_{\hat{p}',\hat{p}}|^2\rangle=\frac{1}{2}{\sum_{\sigma',\sigma}}'\left|t_{\mathbf{p}'\sigma',\mathbf{p},\sigma}\left(\epsilon_p\right)\right|^2, \tag{24.42}$$

where the summation is restricted to the states in the final branches.

The total cross-section is obtained by integrating eqn (24.41) over the final momenta of the quasiparticles:

$$\sigma_t\left(\hat{p},\epsilon\right)=\int\limits_{\Omega'(\epsilon)}\frac{d\sigma}{d\Omega'}\left(\hat{p}',\hat{p},E\right)d\Omega', \tag{24.43}$$

where, for the sake of simplicity, the index p has been dropped from $\epsilon_p=\epsilon$. The total cross-section σ_t depends on the angle between the initial momentum of the quasiparticle and the orientation of the anisotropy axis.

The parallel and perpendicular mobilities can thus be written in terms of the spin-averaged momentum-transfer cross-section as

$$\frac{e}{\mu_{\|,\perp}}=n_3p_F\int\frac{d\Omega}{4\pi}\int\limits_{|\Delta(\hat{p})|}\left(-\frac{\partial n\left(\epsilon\right)}{\partial\epsilon}\right)\sigma^{mt}_{\|,\perp}\left(\hat{p},\epsilon\right)d\epsilon, \tag{24.44}$$

where $n_3=p_F^3/3\pi^2\hbar^3$ is the number density of the quasiparticles and $n\left(\equiv n_p\right)$ is the Fermi distribution. The momentum-transfer cross-sections for a stationary ion are defined as

$$\sigma^{mt}_{\|}\left(\hat{p},\epsilon\right)=\frac{3}{2}\int\limits_{\Omega'(\epsilon)}\left(\Delta\hat{p}_z\right)^2\frac{d\sigma}{d\Omega'}\left(\hat{p}',\hat{p},\epsilon\right)d\Omega', \tag{24.45}$$

$$\sigma^{mt}_{\perp}\left(\hat{p},\epsilon\right)=\frac{3}{2}\int\limits_{\Omega'(\epsilon)}\left(\Delta\hat{p}^+\Delta\hat{p}^-\right)^2\frac{d\sigma}{d\Omega'}\left(\hat{p}',\hat{p},\epsilon\right)d\Omega', \tag{24.46}$$

with $\Delta\mathbf{p}=\mathbf{p}-\mathbf{p}'$ and $p^\pm=\left(p_x\pm ip_y\right)/\sqrt{2}$.

In eqn (24.44) there is a lower limit of integration that is the lowest energy a quasiparticle with momentum in the direction $\hat{p}$ can have. Moreover, it should be noted that for $\epsilon>\Delta$ all initial and final directions for the quasiparticle momentum are allowed in the scattering process, whereas for $\epsilon<\Delta$ the allowed momenta lie within cones of semi-vertical angle $\theta_0=\sin^{-1}\left(\epsilon/\Delta\right)$ with axes along $\hat{\mathbf{l}}$ and $-\hat{\mathbf{l}}$. For $\theta_0<\theta<\pi-\theta_0$ there are no physical quasiparticle states with energy ϵ, as can be immediately recognized by intersecting the anisotropic energy gap in Fig. 24.3 with a circle of radius $\epsilon < \Delta$.

Finally, the mobilities can be conveniently written in terms of angular-averaged momentum-transfer cross-sections $\langle\sigma^{mt}_{\|,\perp}\rangle$ as

$$\frac{e}{\mu_{\|,\perp}} = n_3 p_F \int_{-\infty}^{\infty} \langle\sigma^{mt}_{\|,\perp}\rangle \left(\frac{\partial n(\epsilon)}{\partial \epsilon}\right) \mathrm{d}\epsilon \,, \tag{24.47}$$

where the cross-sections $\langle\sigma^{mt}_{\|,\perp}\rangle$ are defined by

$$\langle\sigma^{mt}_{\|,\perp}(\epsilon)\rangle = \frac{1}{4\pi} \int_{\Omega(\epsilon)} \sigma^{mt}_{\|,\perp}(\hat{p},\epsilon)\,\mathrm{d}\Omega \,. \tag{24.48}$$

If the coherent nature of the superfluid excitations were disregarded, a constant differential cross-section could be used (Bowley, 1976*a*, 1977*b*). Different assumptions about how to calculate the cross-sections can be made, either by assuming an energy-independent differential cross-section (Bromley, 1981) or by performing a Bogoliubov transformation on the normal-state interaction (Soda, 1975).

The results of Bowley (1976*a*) can again be obtained by replacing $\mathrm{d}\sigma/\mathrm{d}\Omega'$ by a constant, yielding

$$\frac{\mu_N}{\mu_\|} = 2\int_0^1 \left(1+3s^2-4s^3\right) n\left[\Delta(\theta)\right] \mathrm{d}s \,, \tag{24.49}$$

$$\frac{\mu_N}{\mu_\perp} = 2\int_0^1 \left(\frac{5}{2}-3s-\frac{3}{2}s^2+2s^3\right) n\left[\Delta(\theta)\right] \mathrm{d}s \,, \tag{24.50}$$

where $s = \cos\theta$ and $\mu_N = e/n_3 p_F \sigma^N_{mt}$. Here σ^N_{mt} is the momentum-transfer cross-section in the normal state.

In the neighborhood of T_c, Δ is small and an expansion of eqns (24.49) and (24.50) to leading order in $\Delta/k_\mathrm{B}T$ yields

$$\frac{\mu_N}{\mu_\|} \approx 1 - 0.4206\frac{\Delta(T)}{k_\mathrm{B}T} \,, \tag{24.51}$$

$$\frac{\mu_N}{\mu_\perp} \approx 1 - 0.4678\frac{\Delta(T)}{k_\mathrm{B}T} \,, \tag{24.52}$$

with the parallel mobility lower than the perpendicular one because of the larger number of quasiparticles moving in a direction parallel to $\hat{\mathbf{l}}$, where the energy gap has a node, than in the perpendicular direction. Moreover, as the A-phase energy gap is, on average, smaller than in the B-phase for any given maximum Δ, the mobility in the A-phase is smaller than in the B-phase, as experimentally observed (Ahonen *et al.*, 1976, 1978; Roach *et al.*, 1977*a*, 1977*b*).

A comparison of the predictions of the constant cross-section model can be made by evaluating the mobility increase near T_c (Bowley, 1976*a*; Salomaa *et al.*, 1980*a*). Taking into account strong-coupling effects for Δ, the averaged mobility at the melting pressure is given by

$$\left(\frac{\mu_{\parallel}+\mu_{\perp}}{2\mu_N}-1\right)^2 = 3.6\left(1-\frac{T}{T_c}\right). \tag{24.53}$$

At $P = 2\,\text{MPa}$, the coefficient is estimated to be 2.7. The functional form given by eqn (24.53) is correct, as shown in Fig. 24.11 for $P = 2.84\,\text{MPa}$ (Ahonen *et al.*, 1978). However, the experimental coefficient is larger, indicating that the effects of superfluidity on the cross-sections must be taken into account.

The calculation of the influence of superfluidity on the scattering amplitudes and cross-sections in the A-phase shares many common features with the B-phase calculation (Baym *et al.*, 1977; Salomaa *et al.*, 1980*a*). However, the presence of anisotropy makes it much more complicated and requires extensive numerical work (Salomaa *et al.*, 1980*a*). It suffices here to say that the cross-section displays strong energy and angular dependencies, leading to large differences in the calculated $\mu_{\parallel}$ and $\mu_{\perp}$, as shown in Fig. 24.18. The anisotropy is more pronounced than in the simpler approximations.

A comparison with the experimental data of Ahonen *et al.* (1978) must take into account the presence of a magnetic field perpendicular to the electric field

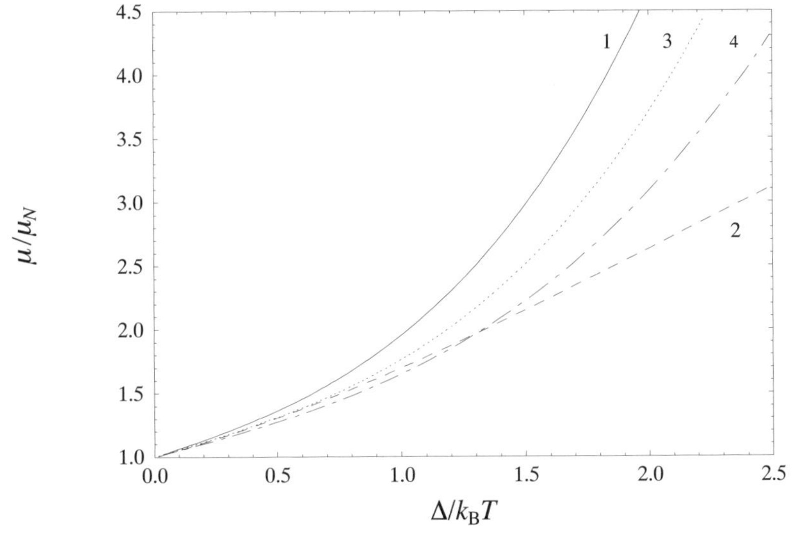

FIG. 24.18. $\mu_{\perp}/\mu_N$ (curve 1) and $\mu_{\parallel}/\mu_N$ (curve 2) vs $\Delta(T)/k_BT$ (Salomaa *et al.*, 1980*a*, 1980*b*). Curves 3 and 4 show the same components calculated with a constant differential cross-section (Bowley, 1976*a*, 1977*b*).

used for drift. Thus, $\hat{\mathbf{l}}$ is expected to be randomly oriented in a plane perpendicular to the magnetic field and the observed mobility should be an equal-weight average of $\mu_{\|}$ and $\mu_{\perp}$:

$$\mu_{-} = \frac{1}{2}\left(\mu_{\|} + \mu_{\perp}\right) . \tag{24.54}$$

In Fig. 24.19 the calculated mobility is compared with the experimental data for $P = 2.88$ and 2.84 MPa (Ahonen *et al.*, 1976, 1978), and for $P = 2.6$ MPa (Roach *et al.*, 1977*a*). The agreement with the experimental data is very good. The use of a constant differential cross-section (Bowley, 1976*a*, 1977*b*) also gives fairly good results, but underestimates the average mobility. Only the data for $P = 2.6$,MPa appear to be systematically higher than predicted and also higher than those of the other group. The discrepancy may be attributed to texture effects. At the same time, however, the anisotropy measured by Roach *et al.* (1977*a*) is semi-quantitatively reproduced by this theory.

The pressure dependence in the calculated mobility enters via the pressure dependence of the product $p_F R_-$ and of the energy gap Δ. The value of $p_F R_-$ does not change very much with P because the ion radius R_- decreases, whereas the Fermi momentum p_F decreases so as to keep their product nearly constant. On the other hand, Δ decreases with increasing P, thus leading to a smaller mobility. Unfortunately, the experimental data show the opposite tendency (Ahonen *et al.*, 1978). This is an effect that has not yet received any explanation.

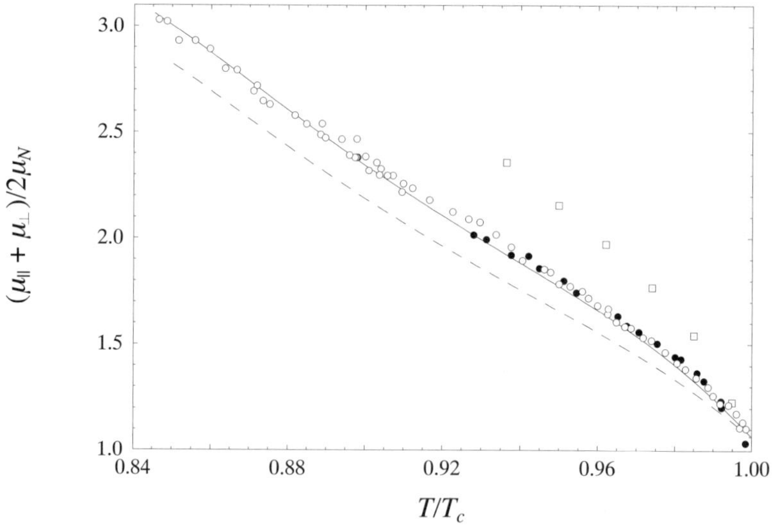

FIG. 24.19. Computed mobility (eqn (24.54), solid line) vs T/T_c in superfluid ^{3}He–A (Salomaa *et al.*, 1980*a*, 1980*b*). P (MPa) $= 2.88$ (closed circles), 2.84 (open circles) (Ahonen *et al.*, 1976, 1978), and $P = 2.6$ MPa (squares) (Roach *et al.*, 1977*a*). The dashed line is the calculation of Bowley (1976*a*).

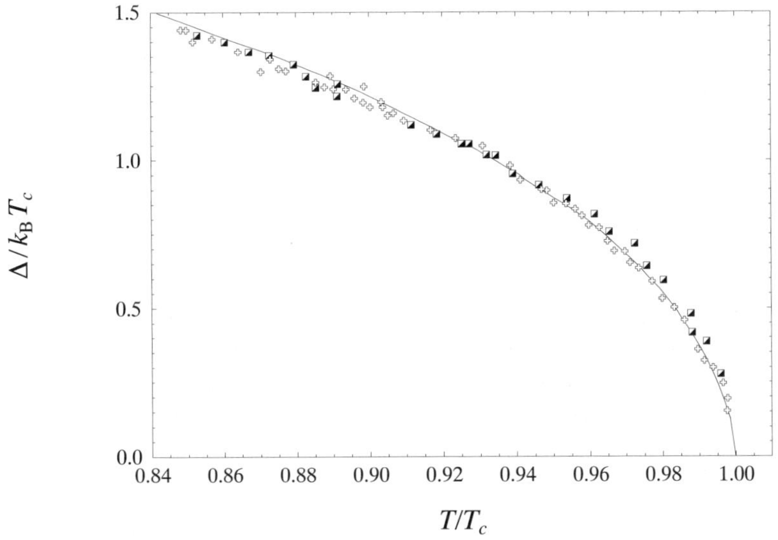

FIG. 24.20. Δ/k_BT vs T/T_c in the anisotropic A-phase determined from mobility data (Salmelin and Salomaa, 1987*a*). Crosses: $P = 2.84\,\text{MPa}$ (Ahonen *et al.*, 1978), and squares: $P = 2.93\,\text{MPa}$ (Simola *et al.*, 1986, 1987*b*). Solid line: $1.32\Delta_{BCS}$.

An improved numerical scheme (Salmelin and Salomaa, 1987*a*, 1990) for the calculation of the anisotropy of the ion mobility within the formalism of Salomaa *et al.* (1980*a*) allows us to deduce the anisotropic energy gap by inverting the experimental data (Ahonen *et al.*, 1976, 1978; Simola *et al.*, 1986, 1987*a*). It turns out that all the data fit on a single BCS-like gap $\Delta(T) = 1.32\Delta_{BCS}(T)$, as shown in Fig. 24.20. The multiplication factor accounts for strong-coupling effects. This estimate is consistent with thermodynamic measurements (Archie, 1978). A similar conclusion is obtained, though with less accuracy, if the modification of the scattering cross-section due to superfluidity is neglected (Bowley, 1976*b*).

24.5 Ion mobility in extremely strong magnetic fields

24.5.1 *Normal liquid* ^{3}*He*

The logarithmic increase in the mobility of positive ions observed in normal ^{3}He has been explained in terms of self-consistent theories that include inelastic scattering processes between the ion and the fluid quasiparticles (Anderson *et al.*, 1968; Bowley, 1977*a*, 1978; Wőlfle *et al.*, 1980; Prokof'ev, 1993, 1994). Prokof'ev (1993) has given the following analytical formula:

$$\mu_+ = \frac{e}{gp_F^2}\left(1 + \frac{20C_4}{g\pi}\ln\frac{\gamma\Gamma}{2\pi k_B T}\right), \tag{24.55}$$

with $g = p_F^2\sigma/3\pi^2$, $\gamma \approx 1.57$, and $\Gamma = n_3 p_F \sigma/M^\star$. Here n_3 is the density of ^{3}He quasiparticles, σ is the cross-section, C_4 is a constant related to σ, and $M^\star$ is the ion mass.

It has been conjectured that the positive ion can exchange quasiparticles with the helium atoms from the ice-like shell surrounding it and that *Kondo*-like exchange spin scattering of ^{3}He quasiparticles with the localized nuclear spins on the snowball surface must be taken into account (Edel'shtein, 1983).

If this were the case, exchange spin scattering should be suppressed in a magnetic field H strong enough to overcome thermal energies, i.e., for $\mu_n H \gg k_B T$, where μ_n is the nuclear ^{3}He magneton (Obara *et al.*, 2001).

A number of experiments have thus been performed with strong magnetic fields up to 15 T in order to test this prediction (Golov *et al.*, 1997; Obara *et al.*, 2000, 2001; Yamaguchi *et al.*, 2002, 2003*a*, 2003*b*, 2004; Ishimoto *et al.*, 2003).

In these experiments ions are produced by field emission and their drift velocity is measured with a standard gated time-of-flight technique. A uniform magnetic field H of strength up to 15 T can be set in a direction parallel to the ion velocity. This has been measured for electric field strengths as low as $E \approx 1\,\mathrm{kV/m}$, well inside the linear velocity region. The sample is purified well down to a ^{4}He concentration of less than 1 ppm.

Some mobility results for $P = 3.23\,\mathrm{MPa}$ for various magnetic field strengths H are shown in Fig. 24.21. Similar results are also obtained for $P = 2.86\,\mathrm{MPa}$. Below $T = 20\,\mathrm{mK}$, the logarithmic dependence on $1/T$ is obeyed for all H. μ_+ can be fitted by the two-parameter formula

$$\mu_+(H,T) = \alpha(H) + A(H)\ln\left(\frac{1}{T}\right). \tag{24.56}$$

The slope coefficient $A(H)$ exhibits a broad peak, whose maximum depends on P. It occurs for $H \approx 7\,\mathrm{T}$ for $P = 2.86\,\mathrm{MPa}$ and for $H \approx 3\,\mathrm{T}$ for $P = 3.23\,\mathrm{MPa}$, as shown in Fig. 24.22. According to Obara *et al.* (2001), the maximum would indicate the existence of an exchange magnetic scattering, though it is not the main effect producing the logarithmic temperature dependence of the mobility.

The dependence of μ_+ on H in the normal fluid varies with T at constant P. For $T \approx 20\,\mathrm{mK}$, μ_+ decreases slightly with increasing H. This behavior is practically the same at all pressures. As T is lowered at constant P, μ_+ becomes progressively more dependent on H. Below $T \approx 3.2\,\mathrm{mK}$, a broad peak appears. This behavior is depicted in Fig. 24.23.

This broad peak also persists when the temperature is lowered below 3 mK and the normal fluid to superfluid A_1 boundary is crossed, as clearly shown in Fig. 24.24.

The broad peak gradually appears, strengthens, and moves to lower magnetic field strengths as P is increased at constant $T = 3.2\,\mathrm{mK}$. The peak position H_P turns out to be a linear function of the applied pressure P:

$$H_P(P) = A_P - B_P P, \tag{24.57}$$

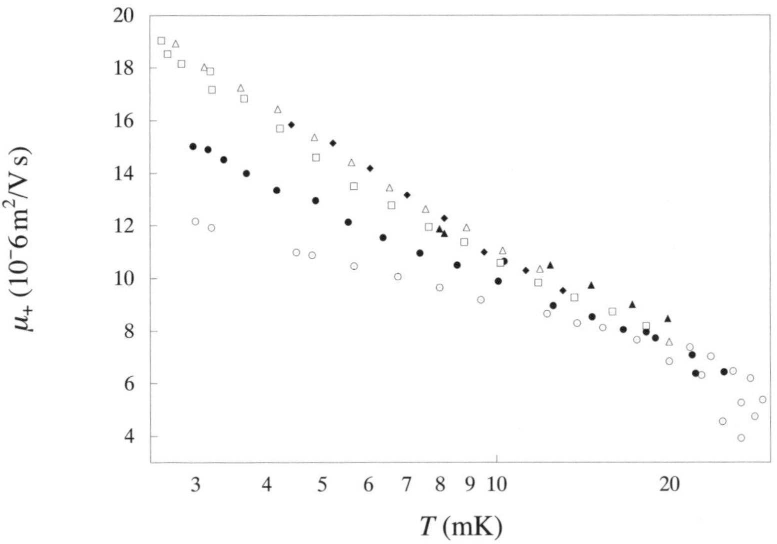

FIG. 24.21. μ_+ vs T in normal liquid ^{3}He at P = 3.23 MPa for several magnetic field strengths H (T) = 0 (closed triangles), 0.6 (diamonds), 3 (open triangles), 5.9 (squares), 9.9 (closed circles), and 13.8 (open circles). (Obara *et al.*, 2001.)

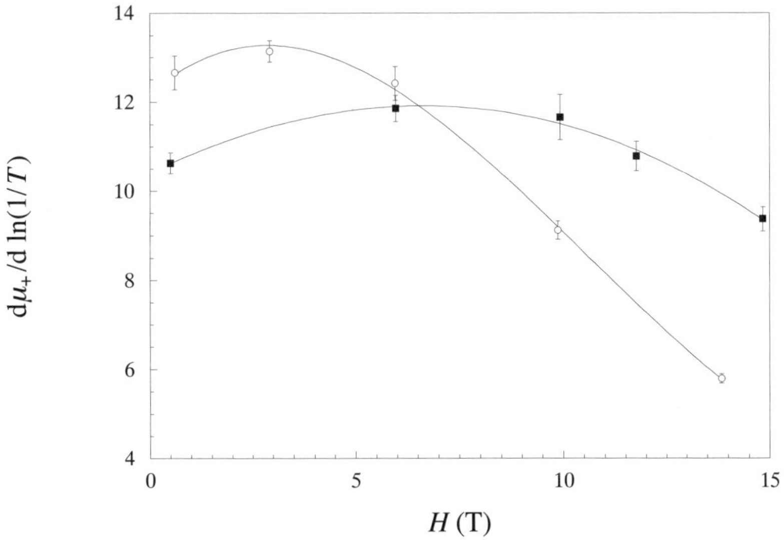

FIG. 24.22. $A(H) = \mathrm{d}\mu_+/\mathrm{d}\ln(1/T)$ vs H in normal liquid ^{3}He for P (MPa) = 2.86 (squares) and 3.23 (circles) (Obara *et al.*, citeyearNPobara2001.)

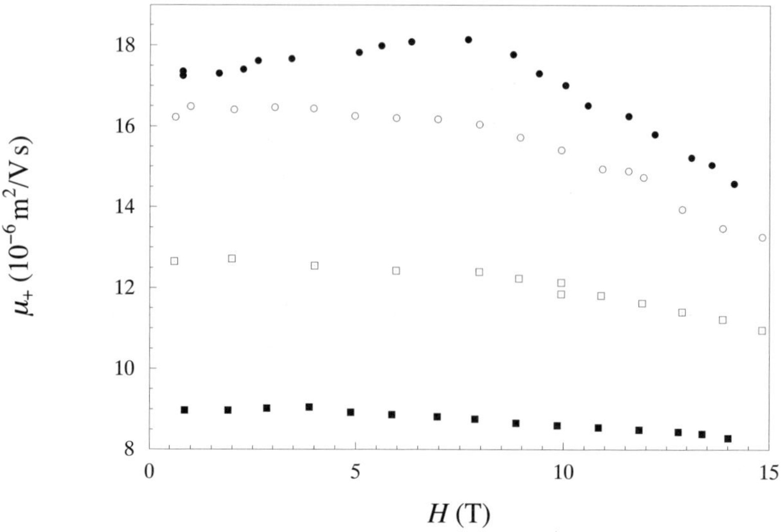

FIG. 24.23. μ_+ vs H at $P = 2.86\,\text{MPa}$ for $T\,(\text{mK}) = 20$, 9.1, 5.1, and 3.2 (from bottom). (Obara *et al.*, 2001.)

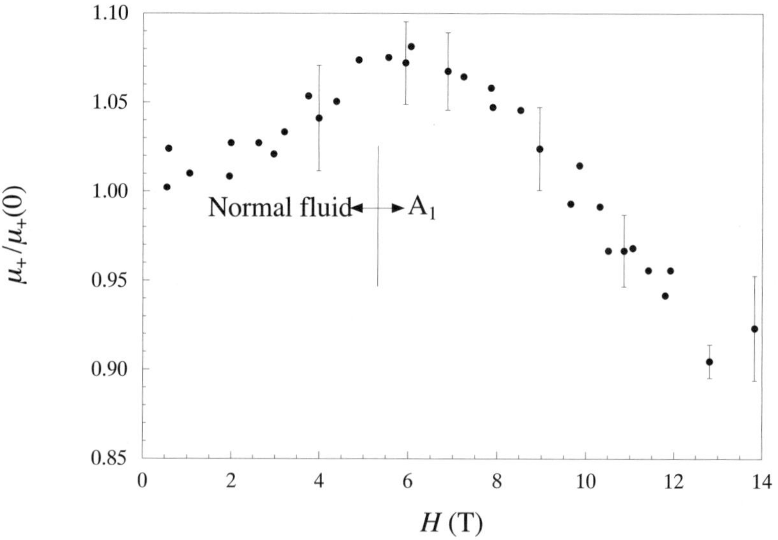

FIG. 24.24. $\mu_+/\mu_+(0)$ (normalized at $H = 0.6\,\text{T}$) vs H for $T = 2.6\,\text{mK}$ and $P = 2.86\,\text{MPa}$. (Obara *et al.*, 2001.) The normal fluid–superfluid A_1 boundary is shown.

with $A_P = 28.26\,\mathrm{T}$ and $B_P = 7.597\,\mathrm{T/MPa}$. This behavior is inherent to the positive ions, as the mobility of the negative ones remains independent of the field at all temperatures (Obara *et al.*, 2001). For this reason, one might argue that this behavior might be due to the magnetic interaction between the spins localized on the snowball surface and the ^{3}He quasiparticles. Unfortunately, the mechanism leading to the observed anomalous behavior of the mobility of the positive ions is still unknown in spite of the many hypotheses suggested (Obara *et al.*, 2001).

24.5.2 *Superfluid* ^{3}*He*

The anomalous behavior of the ion mobility in strong magnetic fields associated with Kondo-like exchange spin scattering with ^{3}He quasiparticles due to the existence of ^{3}He spins localized on the snowball surface might be affected by the transition to a superfluid phase characterized by an anisotropic order parameter (Yamaguchi *et al.*, 2002).

Measurements aimed at a better understanding of the magnetic scattering mechanism were performed in the same apparatus used for the normal fluid (Obara *et al.*, 2001). The experimental set-up is such that the magnetic field is parallel to the ion velocity and the $\hat{\mathbf{l}}$-vector in the A-phase lies in the plane perpendicular to the drift electric field, and $\mu_\perp$ is measured. In this configuration, the mobility is a maximum in the A-phase.

In the superfluid phase the mobility rises rapidly with decreasing T below T_c (Roach *et al.*, 1977*a*; Kokko *et al.*, 1978) and the measurements must be performed at the lowest possible electric field in order to remain in the linear region (Alexander *et al.*, 1978). Below 1 mK, the amount of charge injected into the drift space by means of an emission tip is so large as to create a spacecharge electric field comparable with the drift field itself. Other experimental issues make this kind of experiment at high pressure difficult (Yamaguchi *et al.*, 2002).

In Fig. 24.25 the behavior of the positive ion mobility is shown as a function of temperature for two values of the applied magnetic field H. At variance with the slow logarithmic behavior in the normal phase, μ_+ increases steeply with decreasing T in the superfluid phase. Two kinks in the mobility data indicate the crossing of the two boundaries at T_{A_1} and at T_{A_2}, i.e., the normal-A_1– and the A_1–A_2–phase transitions, respectively (A_2 is the name for the A-phase when the presence of the magnetic field makes the A_1-phase appear).

In order to compare the experimental results with the theoretical prediction of Baym *et al.* (1977) (eqn (24.47)), the inverse mobility normalized at its value at T_{A_1} is plotted in Fig. 24.26 as a function of T for $P = 2.88\,\mathrm{MPa}$ and for several magnetic field strengths H. The same behavior is also observed for $P = 2.0$ and $3.23\,\mathrm{MPa}$, indicating that pressure has little influence on the temperature dependence of the mobility.

$1/\mu_+$ starts decreasing with respect to its normal value as soon as the superfluid A_1-phase is entered at the field-dependent temperature T_{A_1}. The slope in the A_1-phase depends on H. However, as soon as the A_2-phase is reached for

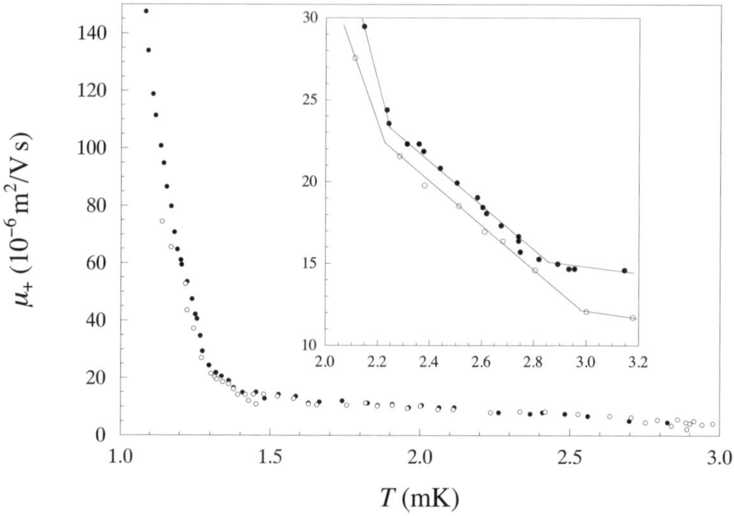

FIG. 24.25. μ_+ vs T at $P = 3.23\,\text{MPa}$. (Yamaguchi *et al.*, 2002.) $H\,(\text{T}) = 9.86$ (closed circles) and 13.8 (open circles). The inset is an expanded view showing the kinks at the normal–A_1 and A_1–A_2 transitions.

$T = T_{\text{A}_2}$, μ_+ no longer depends on H and is described by a universal function. As the normal fraction in the A_2-phase does not depend on H, this universality suggests that the momentum-transfer cross-section has the same magnetic field dependence across the superfluid transition.

In the A_1-phase, however, the behavior of μ_+ is no longer universal and depends strongly on H because the transition temperature T_{A_1} is a function of the field strength. Only if the normalized inverse mobility is plotted as a function of the reduced temperature T/T_{A_1} do the data for different fields belong to a universal curve approaching the value 1/2, as shown in Fig. 24.27.

The explanation of this behavior can be given in terms of the *two-fluid model* developed to describe the viscosity in the A_1-phase (Roobol *et al.*, 1997). The model assumes that only up-spins in the A_1-phase (the quantization axis is defined antiparallel to the magnetic field) condense into the superfluid, while down-spins keep behaving as a normal Fermi liquid. Thus, the normalized inverse mobility is approximately given by

$$\frac{\mu_+(T_{\text{A}_1})}{\mu_+(T)} = \frac{\mu_+(T_{\text{A}_1})}{2}\left[\frac{1}{\mu_\uparrow} + \frac{1}{\mu_\downarrow}\right] \approx \frac{1}{2}\frac{\mu_+(T_{\text{A}_1})}{\mu_\uparrow} + 0.5\,, \tag{24.58}$$

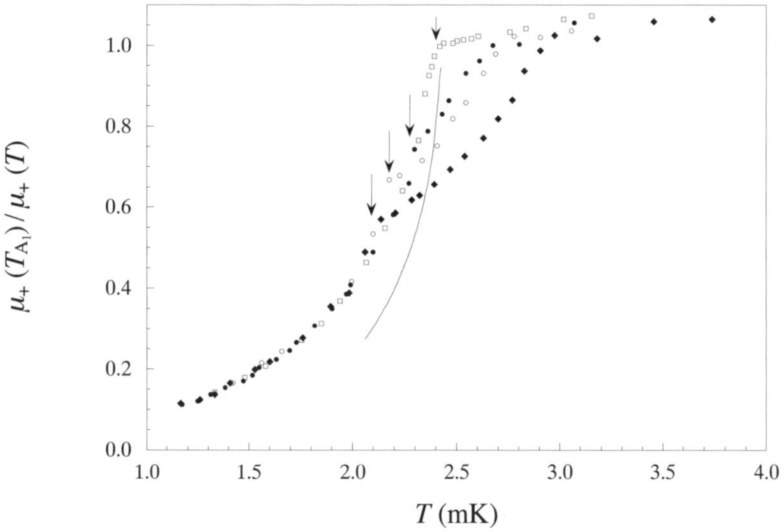

FIG. 24.26. $\mu_+(T_{A_1})/\mu_+(T)$ vs T for $P = 2.88\,\text{MPa}$ for $H\,(\text{T}) = 2.96$ (squares), 5.92 (closed circles), 9.86 (open circles), and 13.8 (diamonds) (Yamaguchi *et al.*, 2002). Arrows indicate the A_1–A_2 transition temperatures T_{A_2} for each H. Line: theory (Salomaa *et al.*, 1980*a*).

where $\mu_{\uparrow,\downarrow}$ are the contributions to the mobility from the up- and down-spins, respectively.

There are no theoretical calculations for the mobility of positive ions in the superfluid phase except near the transition (Soda, 1984), whereas the calculations based on the elastic scattering approximation apparently explain the experimental results for negative ions, even at such low T for which inelastic scattering is no longer negligible (Nummila *et al.*, 1989). Thus, the solid lines in Figs 24.26 and 24.27 represent the elastic scattering model calculations for negative ions at $H = 0$ and $P = 2.8\,\text{MPa}$ (Salomaa *et al.*, 1980*a*, 1980*b*; Salmelin and Salomaa, 1990). The calculated prediction has a stronger temperature dependence than the experimental data, clearly indicating that inelastic scattering cannot be neglected in the case of positive ions.

The magnetic field dependence of the normalized inverse mobility in the A_2-phase is shown for several pressures at $T = 1.89\,\text{mK}$ in Fig. 24.28. The curves for the different pressures are each given a 0.1 offset with respect to the immediately lower one for visual clarity. Similar results are also obtained in the normal liquid for $T = 3.20\,\text{mK}$ and in the A_1-phase, thus indicating that the exchange spin scattering between the positive ions and the quasiparticles is not affected by the superfluid transition (Yamaguchi *et al.*, 2002).

Unfortunately, there is as yet no explanation of the overall behavior of μ_+^{-1} as a function of H, in spite of the several hypotheses suggested that include, among

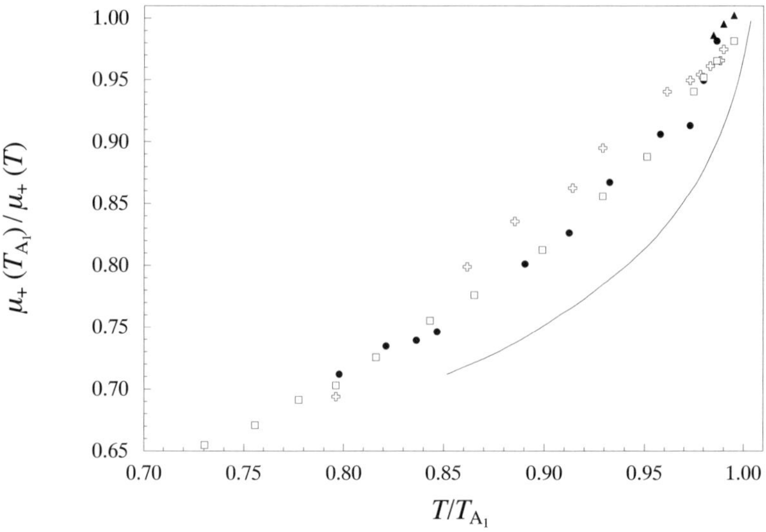

FIG. 24.27. $\mu_+(T_{A_1})/\mu_+(T)$ vs T/T_{A_1} at $P = 2.88\,\text{MPa}$ in the A_1-phase for $H\,(\text{T}) = 2.96$ (triangles), 5.92 (crosses), 9.86 (circles), and 13.8 (squares) (Yamaguchi *et al.*, 2002). The line is the prediction of the elastic scattering model (Salomaa *et al.*, 1980*a*).

others, mechanisms such as Ruderman–Kittel–Kasuya–Yoshida (RKKY)-type local spin density oscillations near the snowball or spin–flip exchange scattering suppression due to magnetic-field enhancement of the snowball magnetization (Yamaguchi *et al.*, 2002).

At variance with positive ions, negative ions consist of an empty cavity in which an electron is localized. For this reason, there are no localized spins on the bubble surface and magnetic scattering (Edel'shtein, 1983) should be absent, though the electron spin is strongly coupled to its environment via an electron–nuclear hyperfine interaction (Reichert *et al.*, 1979).

The mobility of negative ions in superfluid ^{3}He at strong magnetic fields has been measured by Yamaguchi *et al.* (2004). The configuration of the magnetic and electric fields is such that $\mu_\perp$ is measured. Results obtained at $P = 2.86\,\text{MPa}$ for several magnetic field strengths are presented in Fig. 24.29.

The behavior of μ_- is very similar to that of μ_+ shown in Fig. 24.25. The two transition temperatures T_{A_1} and T_{A_2} can clearly be seen only for the data taken with the highest $H = 9.9$ and $13.9\,\text{T}$, whereas only one is observed for the lowest fields owing to the narrow temperature region occupied by the A_1-phase. Most of the $0.1\,\text{T}$ data lie higher than the others because they correspond to the B-phase; However, no sign of the transition can be seen, though it is expected at about $2.1\,\text{mK}$.

The rapid rise in the mobility below the normal to superfluid transition is

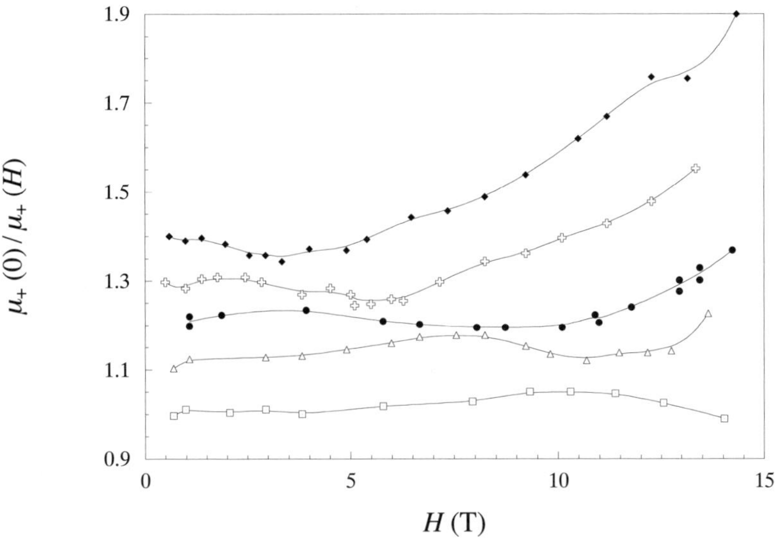

FIG. 24.28. $\mu_+(H = 0)/\mu_+(H)$ vs H at $T = 1.89\,\text{mK}$ in the A_2-phase for $P\,(\text{MPa}) = 2.0$ (squares), 2.21 (triangles), 2.56 (circles), 2.86 (crosses), and 3.29 (diamonds). (Yamaguchi *et al.*, 2002.)

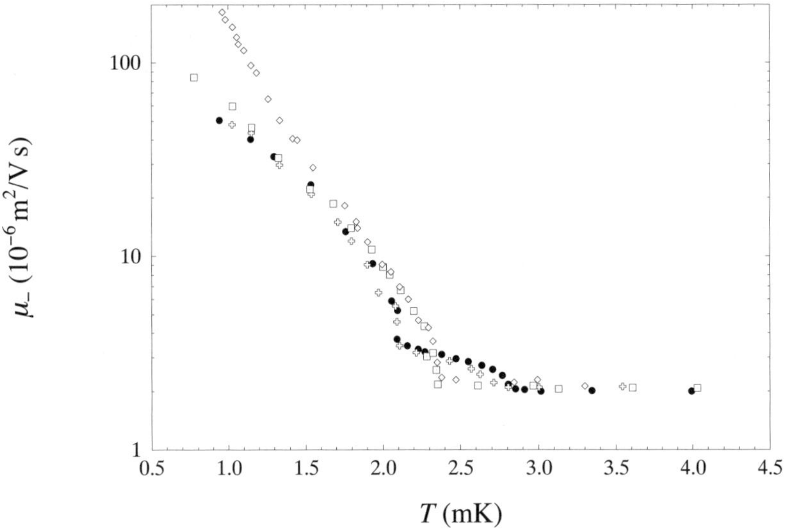

FIG. 24.29. μ_- vs T for $P = 2.86\,\text{MPa}$ in superfluid ^{3}He under high magnetic fields $H\,(\text{T}) = 0.1$ (diamonds), 0.6 (squares), 9.9 (crosses), and 13.9 (circles). (Yamaguchi *et al.*, 2004.)

due to the formation of the superfluid energy gap that reduces the number of quasiparticles available for scattering and to a change in the cross-section (Baym *et al.*, 1977; Salomaa *et al.*, 1980*a*).

In Fig. 24.30 the data for $P = 2.86\,\text{MPa}$ are plotted as μ_-^{-1} as a function of T to make the comparison with the prediction of the elastic scattering model given by eqn (24.47) easier (Baym *et al.*, 1977; Salomaa *et al.*, 1980*a*). All the data in the A$_2$-phase lie on the same curve, independently of the magnetic field. In contrast, the data in the A$_1$-region for $H = 9.9$ and $13.9\,\text{T}$, just below T_{A_2}, are different from the $0.6\,\text{T}$ data. The difference arises because the energy gaps for the spin-up and spin-down states, $\Delta_\uparrow$ and $\Delta_\downarrow$, are different in the A$_2$-phase close to T_{A_2} in the presence of a magnetic field (Leggett, 1975).

Moreover, at low temperatures such that $T_c - T \gg T_{\text{A}_1} - T_{\text{A}_2}$, the gaps become nearly equal and independent of the magnetic field. So, the mobility data also become field independent and fall on a common curve.

In the A$_1$-phase the inverse mobility apparently extrapolates to $\mu_N/\mu(T \to 0) = 0.5$, as positive ions do (see Fig. 24.27), and the explanation is once again given in terms of the two-fluid model that leads to eqn (24.58) (Roobol *et al.*, 1997).

It is interesting to note, as shown in Fig. 24.31, that, if the normalized inverse mobility both in the A$_2$- and A$_1$-phases is plotted as a function of the reduced temperature, either T/T_c or T/T_{A_1}, a universal curve is obtained. This curve agrees well with the theoretical prediction of Salomaa *et al.* (1980*a*) obtained

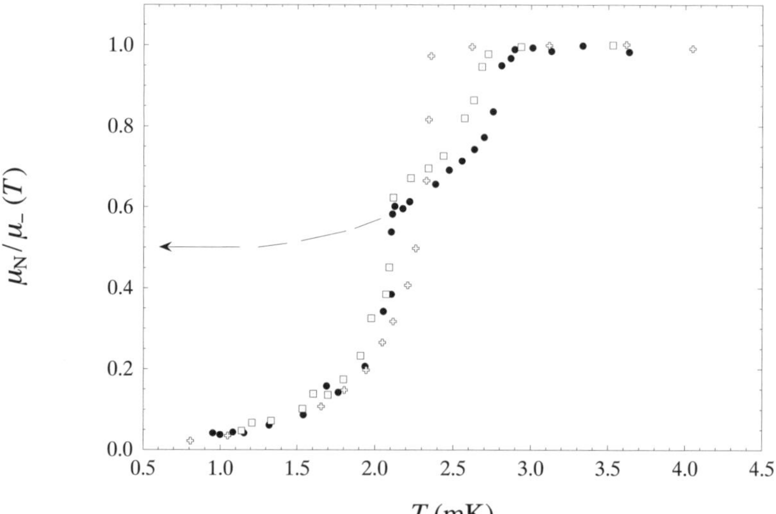

FIG. 24.30. μ_N/μ_- vs T in superfluid ^{3}He at $P = 2.86\,\text{MPa}$ under strong magnetic field H. (Yamaguchi *et al.*, 2004.) H (T) = 0.6 (crosses), 9.9 (squares), and 13.9 (circles). μ_N is the mobility in the normal phase at the transition.

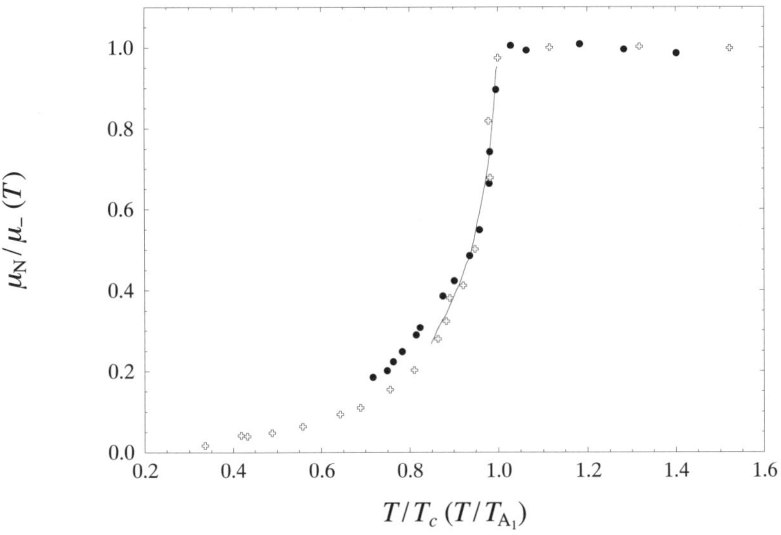

FIG. 24.31. μ_N/μ_- vs $T/T_c\,(T/T_{A_1})$ at $P = 2.86$ MPa in a magnetic field (Yamaguchi *et al.*, 2004). H (T) $= 0.6$ (crosses) and 13.9 (circles). Line: elastic scattering model (Salomaa *et al.*, 1980*a*).

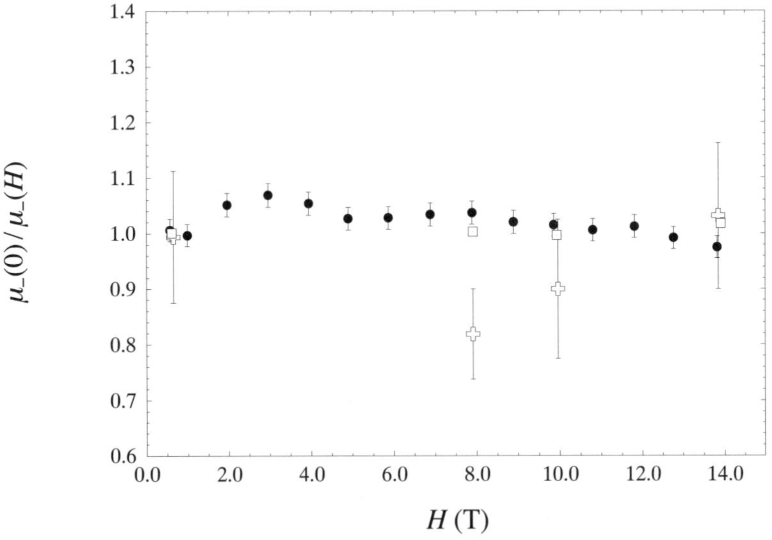

FIG. 24.32. $\mu_-(0)/\mu_-(H)$ vs H for $P = 2.86$ MPa in the normal phase at $T = 3$ mK (squares), in the A_1-phase at $T = 2.41$ mK (circles), and in the A_2-phase at $T = 1.53$ mK (crosses). (Yamaguchi *et al.*, 2004.)

for the A-phase at zero field and 2.8 MPa with $\Delta = 1.32\Delta_{BCS}$ (Salmelin and Salomaa, 1987*a*).

Experimental data in the A_2-region for $P = 2.0$ and 3.08 MPa with $H = 0.6$ T agree with the data shown in Fig. 24.31 and therefore do not confirm the small pressure dependence predicted by the theory (Salomaa *et al.*, 1980*a*).

Finally, it appears that in both the normal phase and in the superfluid A_1- and A_2-phases the negative ion mobility is independent of the magnetic field, as can be seen in Fig. 24.32. Thus, the conclusion is drawn that the momentum-transfer cross-section for scattering between negative ions and quasiparticles is independent of the magnetic field both in the normal and in the superfluid phase, in contrast to its anomalous dependence for the positive ions. This difference is a strong indication that magnetic scattering arising because of spins localized on the surface of the snowball surrounding the positive ions is an important factor that influences their motion in superfluid ^{3}He.

24.6 Ions and vortices in superfluid ^{3}He

The *p*-wave pairing of ^{3}He atoms leading to superfluidity can take place in all three spin-triplet states. The projection of the macroscopic wave function of the superfluid onto these states is represented by an order parameter A_{ij} consisting of a complex 3×3 matrix. Whereas in superfluid ^{4}He only one wave function has to be dealt with, a set of nine wave functions is to be considered for superfluid ^{3}He. This fact is the fundamental cause of the existence of the two superfluid phases A and B, and is also the reason for the existence of a very rich "zoolog" of quantized vortex types.

The magnetic moment of the ^{3}He nuclei allows us to easily investigate vortices in superfluid ^{3}He by exploiting NMR techniques, which are not useful in the case of ^{4}He. In contrast with the case of ^{4}He, ions have rarely been used to investigate vortices in superfluid ^{3}He. For this reason, a thorough description of the vortices that can occur in superfluid ^{3}He is far beyond the scope of this book. A very useful introductive review is that of Lounasmaa and Thuneberg (1999), whereas a very complete one is that of Salomaa and Volovik (1987).

Here, only a brief sketch of the main types of vortices in ^{3}He is given in order to gain a better understanding the results of the scant number of experimental data obtained with ions. With this aim, an excellent review is given by Guénault (2003).

24.6.1 *Superfluid ^{3}He–B*

The isotropy of the B-phase of ^{3}He makes this liquid simpler and more similar to superfluid He II. The wave function has a well-defined phase factor and the phase $\phi(\mathbf{r})$ must be a well-behaved, single-valued function. As in the case of the heavier isotope, this fact leads to a quantized circulation $\oint \mathbf{v}_s \cdot \mathbf{dl} = \kappa_0 = h/2m_3 = 6.65\times10^{-8}\,\mathrm{m^2/s}$, as verified experimentally (Zieve *et al.*, 1993). Note the factor 2 multiplying the atomic mass of ^{3}He because the superfluid quasiparticles consist of Cooper pairs of ^{3}He atoms.

Vortices are nucleated by setting into rotation the cryostat containing the superfluid ^{3}He sample. They are generated when a critical counterflow velocity between the normal fluid and the superfluid fractions is reached. The Magnus force pulls them toward the center of the cell. The creation of each individual vortex is detected as a jump in the NMR signal.

As the order parameter has both spin and space properties, so there are more ways to relax the topological constraints of a simple uniform texture than with a simple ^{4}He-type vortex structure. For this reason, transitions to a different type of vortex with a different core structure are observed.

In the low-temperature side of the ^{3}He–B phase diagram, the stable vortex is a singly-quantized line but with a highly-structured double core consisting of two half-quantum vortices bound together, although they might be separated by a distance as large as 1 μm (Lounasmaa and Thuneberg, 1999).

At higher temperatures, but still in the B-phase, there is a region in which the order parameter in the core of the vortex is similar to that of the bulk A-phase. This is an example of a first-order transition in a quantized topological object (Lounasmaa and Thuneberg, 1999). There is a spontaneously-broken symmetry in the order parameter A_{ij}. The high-temperature vortex type is rotationally symmetric around its axis, but broken parity allows its core to have an A-phase-like order parameter. Similarly, circular symmetry is broken for the low-T vortex types, which split into two half-quantum vortices (Kondo *et al.*, 1991). Both types of vortices are magnetic and are observed in NMR experiments (Hakonen *et al.*, 1983).

This behavior is a consequence of the fact that the superfluid wave function in ^{3}He–B is a multi-component one. For the order parameter to be a scalar as in ^{4}He, the superfluid wave function Ψ must vanish at the vortex line. In ^{3}He–B not all nine components of the wave function vanish simultaneously, leading to a nonzero order parameter everywhere.

In addition to these types of "traditional" hydrodynamic vortices, there is a third, more exotic type of vortex, the spin–mass vortex (Kondo *et al.*, 1992; Korhonen *et al.*, 1993). Cooper pairs can sustain not only mass flow supercurrents but also spin supercurrents. The spin–mass vortex is a hybrid entity consisting of a linear mass vortex, a linear spin vortex with the same core, and a planar defect (a soliton) that travels to the rotating wall.

24.6.2 *Superfluid ^{3}He–A*

In addition to the phase, the order parameter in the A-phase depends on the two unit vectors $\hat{\mathbf{l}}$ and $\hat{\mathbf{d}}$. As a consequence, although ^{3}He–A is superfluid, $\boldsymbol{\nabla} \times \mathbf{v}_s$ can be nonzero. Any inhomogeneities of the texture of $\hat{\mathbf{l}}$ lead to $\boldsymbol{\nabla} \times \mathbf{v}_s \neq 0$. It is not necessary that the vortex core is normal for the wave function to vanish because it is automatically zero in the superfluid along the direction of $\hat{\mathbf{l}}$.

Different types of vortices are then found in ^{3}He–A. There are *singular* vortices with a highly-distorted core, similar to the normal core vortices found in superfluid ^{3}He–B and ^{4}He, and that are singly quantized.

There are *continuous* vortices with a *soft* core, characterized by the fact that superfluidity is retained everywhere, but the $\hat{\mathbf{l}}$-vector changes direction across the core (Salomaa and Volovik, 1987; Lounasmaa and Thuneberg, 1999). A theorem relates the texture of $\hat{\mathbf{l}}$ to the circulation in the superfluid $\kappa = n\kappa_0$: if $\hat{\mathbf{l}}$ goes through all directions on the unit sphere, then $n = 2$ (Mermin and Ho, 1976). Three basic forms of vortices with $n = 2$ have been identified (Lounasmaa and Thuneberg, 1999). Other more complicated structures with $n = 4$ are also allowed.

Finally, there are *vortex sheets* associated with planar soliton defects (or singularities).

24.6.3 *Experiments with ions*

The influence of the rotation on the textures of a ^{3}He–A sample has been studied with negative ions. In the experimental set-up of Simola *et al.* (1986), the experimental cell has cylindrical symmetry and can be set into rotation about its axis with angular speed $\boldsymbol{\Omega}$. Negative ions produced by field emission move along the cell axis, either parallel or antiparallel to $\boldsymbol{\Omega}$.

It is found that the motion of the ions is strongly influenced if the sample is rotating. In particular, the ions are much slower than in the non rotating case and the shape of the ion pulse is strongly modified, being stretched in time.

The measurements are carried out at $P = 2.93\,\text{MPa}$ in a temperature range $T/T_c \geqslant 0.7$ (Simola *et al.*, 1986). A small, constant, and uniform magnetic field $H = 28.4\,\text{mT}$ is aligned in the same direction as the drift field.

The ion drift velocity is given by

$$\mathbf{v} = \mu_\perp \mathbf{E} - \Delta\mu \left(\mathbf{E} \cdot \hat{\mathbf{l}}\right) \hat{\mathbf{l}}, \tag{24.59}$$

where $\mathbf{E}$ is the drift field and $\Delta\mu = \mu_\perp - \mu_\parallel$ is the difference between the components of the mobility tensor perpendicular and parallel to the anisotropy axis $\hat{\mathbf{l}}$. In superfluid ^{3}He–A, $\mu_\perp > \mu_\parallel$ and the anisotropy $\Delta\mu/\mu_\perp$ increases as T is lowered below T_c (Roach *et al.*, 1977*b*; Ahonen *et al.*, 1978).

In a stationary sample, $\hat{\mathbf{l}}$ is oriented by the cooperative effect of the magnetic anisotropy and dipolar energies:

$$\mathcal{E}_M + \mathcal{E}_D \propto \left[(\mathbf{H}/H_c) \cdot \hat{\mathbf{d}}\right]^2 - \left(\hat{\mathbf{l}} \cdot \hat{\mathbf{d}}\right)^2 . \tag{24.60}$$

This expression, in which the critical field is $H_c \approx 2.5\,\text{mT}$, is minimized for $\hat{\mathbf{l}} \| \hat{\mathbf{d}} \perp \mathbf{H}$ (Simola *et al.*, 1987*b*). Thus, $\hat{\mathbf{l}}$ is locked to the plane parallel to $\mathbf{E}$. Because of the configuration of the electric and magnetic fields relative to the anisotropy axis, $\mathbf{E} \| \mathbf{H} \perp \hat{\mathbf{l}}$, $\mu_\perp$ is measured for $\Omega = 0$.

The ion time-of-flight $\tau^\star$ measured during the rotation is compared with the time-of-flight τ obtained with the cell at rest. It is found that the ion drift time is longer when the cell is rotating. In Fig. 24.33 the ratio $\tau^\star/\tau$ is presented as a function of the angular velocity Ω. It should be noted that the drift time

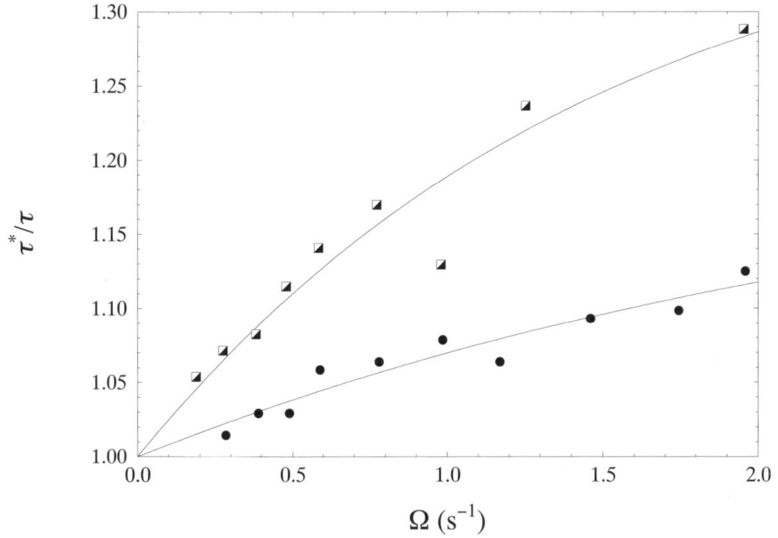

FIG. 24.33. Relative drift time of negative ions $\tau^\star/\tau$ vs Ω in superfluid ^{3}He–A. (Simola *et al.*, 1986.) $T/T_c = 0.78$ (squares) and 0.89 (circles). The number of ions per pulse is $N = 7 \times 10^4$. Lines: fits to eqn (24.71).

increase is not linear with Ω, which is proportional to the areal density of vortices. Moreover, it turns out that it depends on the size of the ion pulses (Simola *et al.*, 1986).

The hypothesis is proposed (Simola *et al.*, 1986) that this behavior is caused by a deflection of the ion trajectories by the texture of $\hat{\mathbf{l}}$ in the soft cores of the vortices that are present in the rotating liquid. The anisotropy of the mobility in the A-phase produces a deflection of the ion paths that are focused along the vortex cores (Williams and Fetter, 1979).

In the geometry of the experiment, with $\hat{\mathbf{l}}$ uniformly locked orthogonal to $\mathbf{E}\|\mathbf{H}$, $\mu_\perp$ is measured for $\Omega = 0$. Rotation of the sample pushes the $\hat{\mathbf{l}}$-vector out of the horizontal plane in the region of the soft vortex cores thereby changing the effective mobility and the measured drift time of the ion pulses (Nummila *et al.*, 1989).

When the ^{3}He–A sample is set into rotation in a strong magnetic field $H > 10\,\mathrm{mT}$, doubly-quantized vortices ($p = 2$) are generated (Hakonen *et al.*, 1982*a*, 1982*b*, 1983, 1985). These vortices are non-singular because they contain the A-phase throughout the core and the superfluid velocity is distributed continuously in a macroscopic region of size $r_c \simeq 5\xi_D = 30\,\mu\mathrm{m}$ around the vortex axis. ξ_D is the temperature-dependent *dipolar healing length* (Vulovic *et al.*, 1984). Within the soft core of the vortex, $\hat{\mathbf{l}}$ is unlocked from the uniform $\hat{\mathbf{d}}$-vector field and is distributed over the whole solid angle of 4π. Vortices with one quantum of circulation ($p = 1$) have a soft core in which $\hat{\mathbf{l}}$ is distributed over only half of the total solid angle (Nummila *et al.*, 1989).

Because the vorticity in ^{3}He–A is continuously distributed over the large soft core around the vortex axis, the hydrodynamic suction is not sufficiently strong to trap ions on vortices as in the case of He II (Donnelly, 1965; Parks and Donnelly, 1966; Pratt and Zimmermann, 1969).

In ^{3}He–A the vorticity $\nabla \times \mathbf{v}_s$ can be written in terms of the gradients of the $\hat{\mathbf{l}}$-vector (Mermin and Ho, 1976) as

$$(\nabla \times \mathbf{v}_s)_z = \frac{\hbar}{2m_3}\hat{\mathbf{l}} \cdot \left(\frac{\partial \hat{\mathbf{l}}}{\partial \phi} \times \frac{\partial \hat{\mathbf{l}}}{\partial r} \right) . \tag{24.61}$$

According to this equation, non-singular continuous vortices consist only of a bending of the $\hat{\mathbf{l}}$-field. $\hat{\mathbf{l}}$ bends out of the xy-plane within the soft core so that $E\hat{\mathbf{z}} \cdot \hat{\mathbf{l}} \neq 0$, where the electric field $\mathbf{E} = E\hat{\mathbf{z}}$ is directed along the z-axis.

As the mobility is anisotropic, $\mu_\parallel < \mu_\perp$, the drift velocity along the lines of the electric field is reduced and the ion trajectories are modified by the presence of a transverse velocity $\mathbf{v}_\perp = -\Delta\mu(\mathbf{E} \cdot \hat{\mathbf{l}})\hat{\mathbf{l}}_\perp$, where $\Delta\mu = \mu_\perp - \mu_\parallel$ (Nummila *et al.*, 1989).

A calculation of the delay measured when the sample is rotated must take into account several facts. The ion pulse is continuously deformed by vortices and this deformed charge distribution interacts with the texture of $\hat{\mathbf{l}}$. The distortion of the charge distribution is counteracted by the Coulomb repulsion between the ions.

The balance between the focusing force due to the $\hat{\mathbf{l}}$-texture and the Coulomb repulsion is easily found (Nummila *et al.*, 1989). From eqn (24.59) one has

$$\nabla \cdot \mathbf{v} = \mu_\perp \nabla \cdot \mathbf{E} - \Delta\mu \nabla \cdot \left[\left(\mathbf{E} \cdot \hat{\mathbf{l}} \right) \hat{\mathbf{l}} \right] . \tag{24.62}$$

For $\mathbf{E} = E\hat{\mathbf{z}}$, the continuity equation in the Eulerian view becomes

$$\frac{\mathrm{d}\rho}{\mathrm{d}t} + \rho \left[\mu_\perp \frac{\rho}{\epsilon} - \Delta\mu E \nabla \cdot \left(\hat{\mathbf{l}}\hat{l}_z \right) \right] = 0 \, , \tag{24.63}$$

where ρ is the charge density, ϵ is the dielectric constant, and the Maxwell equation $\nabla \cdot \mathbf{E} = \rho/\epsilon$ is used.

In the focusing region of the soft core, the balance between the focusing and the repulsive Coulomb force is obtained for

$$\rho_{\mathrm{core}} \approx \epsilon E \frac{\Delta\mu}{\mu_\perp} \nabla \cdot \left(\hat{\mathbf{l}}\hat{l}_z \right) , \tag{24.64}$$

showing that ions are focused in the regions in which $\nabla \cdot (\hat{\mathbf{l}}\hat{l}_z) > 0$.

For typical experiments the ion density is $n \approx 10^{12} < \rho_{\mathrm{core}}/e \approx 10^{15}$ ions/m^3, leading to the trapping of nearly all the ions in the soft-core region (Nummila *et al.*, 1989).

The quantitative analysis of the increased drift time can be performed by considering the geometrical aspects of the focusing process in order to correctly set-up the necessary continuity equations. The geometry of the problem is shown in Fig. 24.34.

The vortex lattice is periodic. So, only one unit cell of cross-sectional area A_u has to be considered. The electric field is directed along the vertical direction. The pulse of thickness d consists of ions in the bulk where $\hat{\mathbf{l}} \perp \mathbf{E}$ and $(\mathbf{v})_z \simeq \mu_\perp E$, and in the soft core where $\{(\mathbf{v})_z\}_c \simeq \{\mu_\perp - \Delta\mu \hat{l}_z^2\}_c E = (\mu_\perp - \Delta\mu\{\hat{l}_z^2\}_c)E$. Here $\{\ldots\}_c$ denotes a proper average over the focal area in the core. ρ_b and ρ_c are the ion densities in the bulk and in the core, respectively. The mobility along the vortex core is defined as $\mu_c \equiv \mu_\perp - \alpha\Delta\mu$, with $\alpha \equiv \{\hat{l}_z^2\}_c$.

The averaged drift time $\tau^\star$ of the ion pulse during rotation is given by the following relationship:

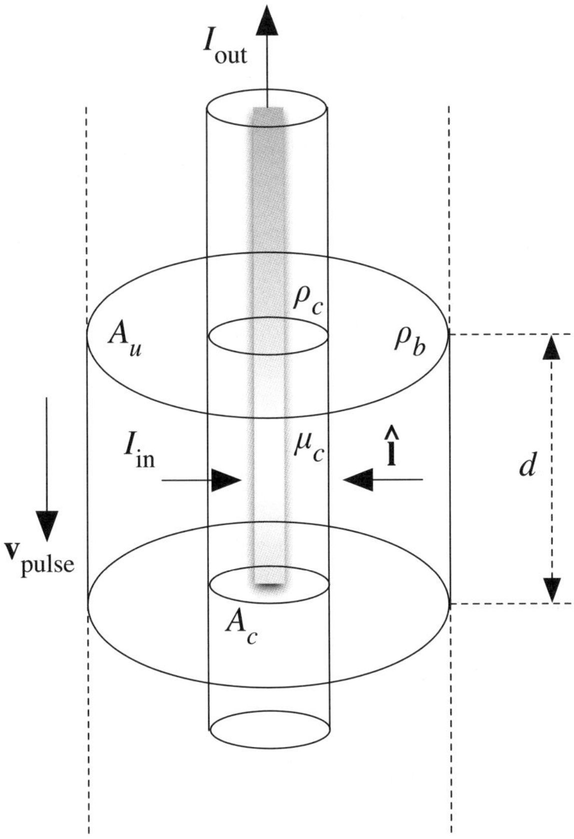

FIG. 24.34. Geometry for the focusing model. (Nummila *et al.*, 1989.)

$$L = \int_0^{\tau^\star} \{(\mathbf{v}_D)_z\}_u \, \mathrm{d}t = \frac{E}{Q_0} \int_0^{\tau^\star} [\mu_\perp Q_b(t) + \mu_c Q_c(t)] \,, \tag{24.65}$$

where L is the drift distance, $\{\ldots\}_u$ denotes averaging over the unit cell, and Q_b/Q_0 and Q_c/Q_0 are the fractions of the total pulse charge in the bulk liquid and in the soft-core regions, respectively. The total charge is obviously $Q_0 = (A_u d)\rho_0$, where ρ_0 is the initial charge density of the pulse. This relationship holds true all the time if Coulomb repulsion is neglected.

For focusing to occur, a current $I_{\mathrm{in}}(t)$ must flow into the core, corresponding to an ion flow $\rho \mathbf{v}_\perp$ for $r \to r_c$. The amount of charge flowing into the core is

$$Q_c(t) \approx \int_0^t I_{\mathrm{in}}(t')\mathrm{d}t' \,, \tag{24.66}$$

and the charge in the bulk is just $Q_b(t) \approx Q_0 - Q_c(t)$.

In a typical experiment, the angular velocity is such that the areal density of vortices is not very large. Thus, the ratio of the areas of the core A_c and of the unit cell $A_u = p\pi\hbar/2m_3\Omega$ is $A_c/A_u \lessapprox 0.05$ and the initial charge in the core can be neglected (Nummila *et al.*, 1989).

Because $|\mu_c| < |\mu_\perp|$, a current

$$I_{\mathrm{out}} = E\left\{\rho\Delta\mu\hat{l}_z^2\right\}_c A_c \tag{24.67}$$

is flowing along each vortex core through the trailing edge of the drifting ion pulse.

If Coulomb repulsion is neglected and if it is assumed that the ions focused in the vortex core still propagate along the core even though they are left behind the ion pulse moving in the bulk liquid, continuity gives $I_{\mathrm{in}} = I_{\mathrm{out}}$ in the frame of reference moving with the ions in the bulk. If it is further assumed that the ion bulk density is constant and uniform, and that the average density of ions in the core at the site of the propagating pulse is $\propto \{\rho\}_c = c\rho_b$, with $c \approx 1$, then eqns (24.66) and (24.67) and the conservation of the total charge give

$$I_{\mathrm{in}}(t) = c\left[\rho_0 - \frac{1}{A_u d}\int_0^t I_{\mathrm{in}}(t')\,\mathrm{d}t'\right] A_c E\alpha\Delta\mu \,. \tag{24.68}$$

Solving eqn (24.68) for I_{in} yields

$$I_{\mathrm{in}}(t) = I_0 \exp(-t/t_0) \,, \tag{24.69}$$

in which $I_0 = c\rho_0 A_c E\alpha\Delta\mu$ and $t_0 = (\Omega/\Omega_0)(\alpha\Delta\mu/\mu_\perp)^{-1}\tau$. Here τ is the ion drift time in the absence of rotation and the characteristic angular frequency Ω_0 is defined as

$$\Omega_0 \equiv \frac{A_u d}{c A_c L}\Omega\,. \tag{24.70}$$

Eventually, the implicit equation for the rotation-induced increase in the drift time is obtained by inserting eqn (24.69) into eqn (24.65):

$$\frac{\tau^\star}{\tau} = 1 + \alpha\frac{\Delta\mu}{\mu_\perp}\frac{\tau^\star}{\tau} - \frac{\Omega_0}{\Omega}\left[1 - \exp\left(-\alpha\frac{\Delta\mu}{\mu_\perp}\frac{\Omega_0}{\Omega}\frac{\tau^\star}{\tau}\right)\right]. \tag{24.71}$$

This equation fits the experimental data nicely, as shown by the solid lines in Fig. 24.33, and also the temperature dependence of the retardation effect, as shown in Fig. 24.35. Here, the $\tau^\star/\tau$ data are presented as a function of the ratio $\mu_\perp/\mu_N$ because this quantity provides the most convenient temperature scale, obtained by inverting the data shown in Fig. 24.8.

An analysis of the fitting parameters (Nummila *et al.*, 1989) leads to some interesting conclusions, such as, for instance, the decrease in the core size as the applied magnetic field H is reduced.

The focusing process is thus a geometrical phenomenon related to the delicate balance between Coulomb repulsion and $\hat{\mathbf{l}}$-texture-induced attraction. The good agreement between model and experimental data supports the conclusion that the vortex cores are actually continuous.

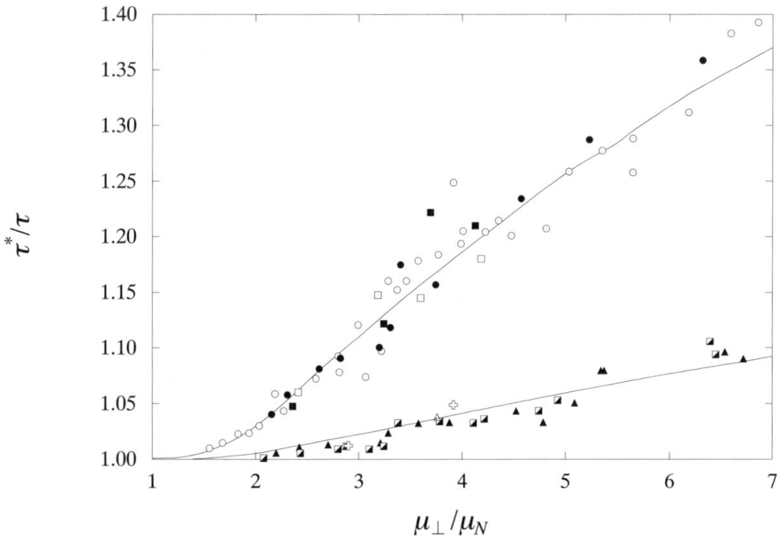

FIG. 24.35. $\tau^\star/\tau$ vs $\mu_\perp/\mu_N$ for $\Omega\,(\mathrm{s}^{-1}) = \pm 1.95$ (upper curve) and $\pm\,0.27$ (lower curve). (Simola *et al.*, 1986.) A large fraction of the data is collected in the supercooled A-phase. Open and closed symbols refer to two slightly different values of the drift field. The lines are a fit of eqn (24.71) to data with the same number $N = 7\times10^4$ of ions/pulse.

Moreover, in the same experiment but for high magnetic fields, vortex states have also been observed in which there is no change in the drift time and almost no change in the pulse shape is observed (Simola *et al.*, 1987*b*). These states are detected if the rotating state is created adiabatically, i.e., by a continuous rotation at constant Ω through T_c or by a gradual stepwise acceleration at constant $T < T_c$. The observations are consistent with the hypothesis that for this new vortex state the ion mobility in the core is $\mu_{\text{core}} \approx \mu_\perp$.

The absence of delay suggests that these vortices are singular with some other superfluid phase in the core. Salmelin and Salomaa (1987*b*) have indeed proposed the explanation that in this new rotating state the negative ions are focused and move along singular quantized vortex lines whose cores consist of the superfluid polar phase ^{3}He–P, with the Cooper pairs in the $L_z = 0$, $S_z = 0$ state (Barton and Moore, 1974; Muzikar, 1976; Fetter *et al.*, 1983; Salomaa and Volovik, 1985*a*). Detailed calculations show that singular polar-core vortices in rotating ^{3}He–A are viewed as *holes* in the superfluid and the ions move preferentially along these tubes of bulk polar phase.

In any case, both types of vortices focus the ion flow along the vortex axis. For $H < 0.4\,\text{mT}$ only continuous vortices are observed, whereas both are observed for higher H (Simola *et al.*, 1987*b*).

To conclude with the ion–vortex interaction in superfluid ^{3}He, it should be mentioned that the detection of the trapping of ions on vortices in the B-phase is outside the reach of experiment (Nummila *et al.*, 1989). In fact, the mechanisms of ion trapping in He II and in ^{3}He–B are very different. In superfluid He II the coherence length ξ of the superfluid state is quite short, much shorter than the ion radius. The superfluid velocity gradients are thus enormous and lead to the strong Bernoulli hydrodynamic suction $F_s \propto \nabla(\rho_s v_s)$ that efficiently traps ions (Donnelly, 1965; Parks and Donnelly, 1966).

In contrast, in ^{3}He–B, $\xi \propto \xi_0(1 - T/T_c)^{1/2}$, with $\xi_0 \simeq 20\,\text{nm}$, is orders of magnitude larger than in He II. As a consequence, the superfluid velocity gradients stay small (Mineev and Salomaa, 1984; Salomaa and Volovik, 1985*b*) and no strong enough hydrodynamic forces are available for ion trapping.

The dominant ion–vortex interaction in ^{3}He–B thus occurs in the vortex core for $r < \xi$. An ion can be regarded as a pair-breaking center (Rainer and Vuorio, 1977) that destroys the superfluid state in a volume of size $\approx \xi_0^3$. So, there is a loss of condensation energy of order $\sigma_{mt}^N k_F^2 \Delta^2(T)/T_c$ near T_c, where σ_{mt}^N is the momentum-transfer cross-section in the normal phase, $\hbar k_F$ is the Fermi momentum, and $\Delta(T)$ is the amplitude of the energy gap. Hence, ions are attracted toward regions of vanishingly-small order parameter to minimize the loss of condensation energy. Most of the ion attraction is thus supplied by the dependence of Δ on the radial distance from the vortex axis inside the core.

The big difference with the He II case is that the maximum depth of the trapping potential in ^{3}He–B is $U_0 \approx 50 k_B T_c \approx 5\,\mu\text{eV}$ (Thuneberg *et al.*, 1981), whereas it is $\approx 7\,\text{meV}$ in He II (Pratt and Zimmermann, 1969). More detailed calculations by Salomaa and Volovik (1983) have shown that vortices in the

B-phase have a superfluid core that reduces U_0 further.

Moreover, at low pressures, where the increase of the radius R_- of the ion might favor trapping because the condensation energy loss is proportional to $\sigma^N_{mt} \propto R^2_-$, the proposed non-axisymmetric structure of the vortices may offer the ions an easy channel to detrap (Salomaa and Volovik, 1986; Thuneberg, 1986). These simple considerations put too stringent requirements onto experiments and no trapping has actually been observed (Nummila *et al.*, 1989).

Part III

Dense helium gas

25

ELECTRONS IN DENSE HE GAS

Electrons injected into liquid He get rapidly self-trapped in empty cavities whose transport properties have been discussed in the previous chapters. The size of these electron bubbles is determined by the delicate balance between the kinetic energy increase due to localization, the deformation energy spent in expanding the cavity to its equilibrium radius, and the electron energy in the delocalized state.

In liquid He the energy of the bottom of the conduction band $V_0 \approx 1\,\mathrm{eV}$ is so high, the deformation energy so small, and the mean thermal energy is so low that the bubbles are fairly large, empty, and stable with respect to thermal fluctuations.

The phenomenon of self-trapping, however, is not limited to liquid He (Khrapak and Yakubov, 1975; Castellani *et al.*, 1981; Hernandez, 1991), liquid Ne (Bruschi *et al.*, 1972), liquid H_2 (Grimm and Rayfield, 1975; Sakai *et al.*, 1981; Levchenko and Mezhov-Deglin, 1992; Berezhnov *et al.*, 2003), or liquid normal hydrocarbons (Ichikawa and Yoshida, 1981; Holroyd and Schmidt, 1989), but can occur whenever the conditions mentioned above are fulfilled. In particular, self-trapping may occur in the gas phase in He (Levine and Sanders, 1962) and Ne (Borghesani *et al.*, 1988, 1990*a*, 1994*b*), but also in polar gases (Krebs, 1984) such as ammonia (Krebs *et al.*, 1980; Krebs and Heintze, 1982), water vapor (Giraud and Krebs, 1982), methanol (Asaad *et al.*, 1996; Krebs and Lang, 1996), and hydrogen cyanide (Klahn and Krebs, (1998*a*, 1998*b*)).

If the gas density N is sufficiently low or the temperature T high enough, it may occur that V_0 is no longer able to support bound states and the free energy $\Delta\mathcal{A}$ in eqn (3.21) cannot be minimized for any bubble radius. Or it may happen that the minimum free energy is not negative enough to ensure the stability of the bubbles with respect to thermal excitation.

The gas phase offers the possibility of varying both N and T in a broad range so that the details of localization can be investigated thoroughly. By increasing N at constant T, or by decreasing T at constant N, the free energy becomes progressively more favorable to localized than to extended states. As the mobility of electron bubbles is much lower than that of quasi-free electrons and the observed mobility is a weighted average of the mobility of both states, a continuous transition is expected to occur from high mobility values in the dilute gas phase to a low mobility value in the dense gas. Thus, the investigation of the density and temperature dependence of the electron mobility in the dense gas yields important pieces of information about self-trapping.

In the following chapters, the question of the dynamics of the bubble formation will be first addressed and then the results of the experiments on electron mobility in dense He gas will be discussed.

26

DYNAMICS OF THE FORMATION OF THE ELECTRON BUBBLE

The energetics of the electron bubble in liquid helium, i.e., the conditions for which the self-localized state of an electron in a non-polar liquid may exist, is a well-studied subject, both theoretically (Jortner *et al.*, 1965; Springett *et al.*, 1967, 1968; Fowler and Dexter, 1968; Miyakawa and Dexter, 1970; Nieminen *et al.*, 1980; Hernandez, 1991) and experimentally (Northby and Sanders, 1967; Grimes and Adams, 1990, 1992; Parshin and Pereverzev, 1990, 1992; Pereverzev and Parshin, 1994), to quote just a few authors, as it has also been described in the previous chapters.

Localized electron states can exist if a large configurational modification of the liquid structure surrounding the electron leads to a decrease of the free energy of the system with respect to the case of the delocalized electron in the conduction band of the liquid (Hernandez, 1991). For this reason, electron bubble states also exist in other fluid systems, such as Ne (Bruschi *et al.*, 1972), D_2 (Springett *et al.*, 1968), and even in clusters (Jiang *et al.*, 1993; Northby *et al.*, 1994). Also, the structural properties of the localized state are very well known.

Localized states can also occur for exotic atoms such as positronium (Iakubov and Khrapak, 1982) and for alkali atoms (Cole and Toigo, 1978) because of the strong repulsive exchange interaction between electrons in the atomic shells.

However, the dynamics of the formation of electron bubbles is not equally well investigated. The experimental information is scarce for the obvious reason that the dynamical process of bubble formation is related to the rearrangement and relaxation of the fluid around the electron, a process that occurs in the picosecond time scale.

Experiments on positronium annihilation in liquid helium and their theoretical interpretation (Ferrell, 1957; Iakubov and Khrapak, 1978, 1982; Mikhin *et al.*, 2003) suggest that the localized positronium bubble is formed on a time scale that is short compared to the lifetime of Ps, $\tau > 100\,\mathrm{ps}$. Muon spin relaxation experiments in Ne (Storchak *et al.*, 1996) have shown that part of the excess electrons localize rapidly within less than 10^3 ps, whereas a fraction of them do not localize even on the 10^3 ps scale. Electrons escaping the fate of localization have also been observed in drift mobility experiments in liquid Ne at high electric fields (Sakai *et al.*, 1992).

Information on the time scale of the relaxation of a quasi-free electron to form a bubble in liquid helium is obtained from experiments of electron injection (Hernandez and Silver, 1970, 1971). A time $\tau \sim 2\,\mathrm{ps}$ is apparently needed for an electron of 1 eV injected into liquid helium to relax and form the bubble state. Another rough piece of information about the characteristic time of bubble formation is obtained by calculating the time taken for sound to propagate a distance of the order of the electron bubble radius. By taking $R_- \approx 17\,\text{Å}$ and the sound velocity $c_s \approx 240\,\text{m/s}$, an estimate of 7 ps is obtained (Jiang *et al.*, 1993).

In any case, the process leading to the formation of the electron bubble can be described in the way schematically depicted in Fig. 26.1 (Hernandez, 1991; Sakai *et al.*, 1992). An electron with an energy in excess of a few tenths of an eV is injected into the conduction band of the liquid, in which it is endowed with a high mobility because its wave function is delocalized. In the liquid, the electron

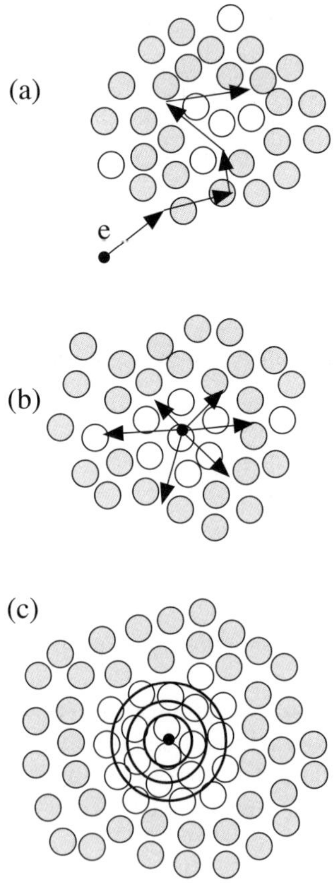

FIG. 26.1. Schematic view of the process of formation of the electron bubble (after a graphical intuition of Sakai *et al.* (1992)). (a) Scattering of an electron in the delocalized state. (b) Temporary trapping over a local density fluctuation. (c) Stable electron bubble. Shaded circles represent He atoms and unshaded circles represent voids.

undergoes elastic collisions with atoms of the host (see Fig. 26.1 (a)). It loses energy until it eventually reaches the bottom of the conduction band.

During its motion in the liquid, the electron can find, rather than cause, density fluctuations as these have a very broad spectrum (Hernandez, 1991). The density fluctuations also drive fluctuations of the potential acting on the electrons because the electron energy at the bottom of the conduction band $V_0(N)$ depends on the density (Broomall *et al.*, 1976; Plenkiewicz *et al.*, 1989, 1991). If the density in the fluctuation is lower than average, the local electron energy is lower than in the bulk fluid and the density fluctuation acts as a trapping well.

The electron may couple to the fluctuation because the potential well can have resonant or virtual states in the continuum (Gasiorowicz, 1974; Hernandez, 1991; Landau and Lifsits, 2003). The electron spends more time above the well being reflected back and forth between the walls of the incipient bubble, losing its excess energy (see Fig. 26.1 (b)). An incipient bubble is formed via the non-radiative electron localization process originating from non-adiabatic crossing of the potential energy surfaces of the quasi-free and localized excess electron states (Jortner, (1971); Rosenblit and Jortner, 1995, 1997).

The longer the time the electron spends above the well, the greater are the chances it has of pushing away more and more atoms, so that the well becomes deeper and more attractive, eventually leading to a fully-fledged electron bubble (see Fig. 26.1 (c)). Once the electron is localized in the bubble, its mobility is low and comparable to that of negative ions (Jahnke *et al.*, 1971; Borghesani and Santini, 1990*a*, 2002).

The capture of the electron is possible if it loses its excess kinetic energy because of elastic scattering events during its residence time τ_f above the fluctuation. The cross-section for capture σ_c can thus be written as

$$\sigma_c = \sigma_{\rm el} P_c \approx \sigma_f u_{-1} (\tau_f - \tau_k) \ , \tag{26.1}$$

where $\sigma_{\rm el}$ is the cross-section for elastic scattering, σ_f is the cross-section for elastic scattering above the well, P_c is the probability for electron capture, and τ_k is the time required to transfer the kinetic energy to the walls of the fluctuation (Artem'ev and Khrapak, 1986; Sakai *et al.*, 1992). u_{-1} is the unit step function.

The scattering cross-section for an electron on a fluctuation with resonance character is given by (Merzbacher, 1970; Gasiorowicz, 1974; Landau and Lifsits, 2003)

$$\sigma_f = \frac{\pi}{k^2 + \kappa_1^2} \ , \tag{26.2}$$

where $\hbar k = mv$ is the electron momentum, and m and v are its rest mass and velocity, respectively. κ_1 is related to the energy of the resonant state ($\epsilon_1 < 0$) or of the virtual state ($\epsilon_1 > 0$) by the usual relationship

$$\kappa_1^2 = \frac{2m}{\hbar^2} |\epsilon_1| \ . \tag{26.3}$$

On symmetry grounds and for the sake of simplicity, the density fluctuations can be considered as voids whose spherical shape is retained throughout the whole process. In Fig. 26.2 a schematic potential diagram of the problem is shown. For almost-resonant voids of radius R, $|\epsilon_1| \ll V_0$ with ϵ_1 given by

$$\epsilon_1 \simeq \frac{m}{2\hbar^2} R^2 \left(V_0 - \frac{\hbar^2 \pi^2}{8mR^2} \right)^2 . \tag{26.4}$$

From eqns (26.1) and (26.3), it is clear that the capture cross-section σ_c exhibits a sharp maximum for $\epsilon_1 \approx 0$, i.e., for

$$R \approx R_0 = \frac{\pi \hbar}{2\sqrt{2mV_0}} . \tag{26.5}$$

The residence time τ_f, i. e., the time the electron spends above the well, is given by (Merzbacher, 1970)

$$\tau_f = \frac{2R}{v} \frac{2k^2}{k^2 + \kappa_1^2} = \frac{2m}{\hbar} kR\sigma_f . \tag{26.6}$$

Also, this residence time is a maximum for $R = R_0$, i.e., when $\epsilon_1 \approx 0$. This is equivalent to the statement that the incipient bubble forms as soon as the energy

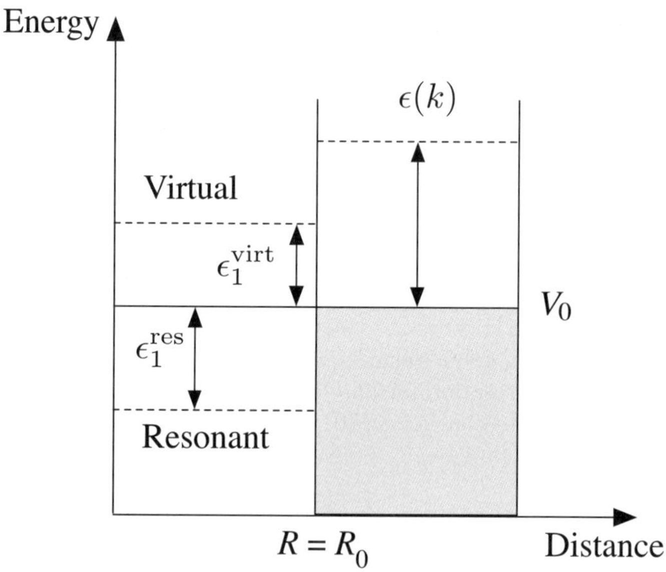

FIG. 26.2. Schematic potential diagram of the spherical voids. (Sakai *et al.*, 1992.) $\epsilon(k)$ is the energy of an electron in the conduction band of the liquid. ϵ_1^{virt} and ϵ_1^{res} represent the energy levels of a virtual or resonant state, respectively.

of the localized state equals that of the delocalized one. τ_f is much greater than the classical transit time $\tau_{\rm cl} = 2R_0(m/2V_0)^{1/2} = (8m/h)R_0^2$, i.e., the time a classical particle of energy V_0 would take to cross the region of the potential well.

In order to estimate the time required by the electron to lose its excess kinetic energy at the void and become trapped, two qualitatively different processes must be clearly discriminated.

The first one is a very rapid vertical transition process consisting of a fast transition of the electron to the virtual or resonant state, during which there is no time for the surrounding fluid to relax to cope with the changes of the electron wave function. A guess of the time associated with this process leads to an estimate of the fraction of electrons injected into the liquid in a traditional time-of-flight experiment that can be localized as a function of the drift field (Sakai *et al.*, 1992). It also explains the electric field dependence of the drift velocity of electrons at extremely high fields in liquid rare gases (Sakai *et al.*, 1981, 1982; Schmidt *et al.*, 2001) and in gaseous He (Schwarz and Prasad, 1976), and, presumably, also the existence of delocalized electrons, which do not relax on a time scale of 10^3 ps, as observed in muon spin relaxation experiments in condensed rare gases (Storchak *et al.*, 1996; Eshchenko *et al.*, 2002).

The second process corresponds to a "slow" adiabatic motion of the newly-occupied resonant level due to the pressure P_e the electron is exerting on the walls of the void because of the repulsion with the electronic shells of the surrounding atoms. The electron pressure is related to the kinetic energy of localization and amounts to

$$P_e = -\left[\int\limits_{r\leqslant R} |\psi(r)|^2 \, \mathrm{d}r\right] \frac{1}{4\pi R^2}\frac{\partial \mathcal{E}_e^{1s}(R)}{\partial R}\,, \tag{26.7}$$

where $\mathcal{E}_e^{1s}$ is the energy eigenvalue of the ground state of the electron in the bubble given by eqn (3.22). The prefactor in brackets takes into account the fact that the well is not indefinitely high so that part of the electron wave function ψ spills over the walls and does not contribute to the pressure on them (Iakubov and Khrapak, 1982).

At the instant of the bubble rise the electron radius is $R_0 \approx 3$ Å, as obtained from eqn (26.5), and the pressure amounts to

$$P_e \approx \frac{\pi\hbar^2}{16mR_0^5} \approx 10^3\,\mathrm{MPa}\,. \tag{26.8}$$

The bubble thus expands against the external pressure. The expansion is quite slow because the force on the bubble walls is only $\propto R^{-3}$ although the pressure is $\propto R^{-5}$, and because the effective mass for the radial movement of the walls amounts to $M_R = M/3$, where $M = (4/3)\pi R^3\rho$ is the mass of the liquid displaced by the void and ρ is the mass density of the liquid.

Further expansion of the bubble increases the binding energy of the electron and eventually the equilibrium state of the electron bubble is reached that minimizes the free energy $\Delta\mathcal{A}$ given by eqn (3.21), as explained in Section 3.2. Thus, the stabilization time can be roughly estimated to be given by the final equilibrium radius R_- divided by the sound velocity, yielding the 7 ps evaluation mentioned above.

However, during the expansion of the bubble, only a fraction of the energy is spent irreversibly for work done against viscous forces and for the generation of sound and shock waves. A relevant fraction of the energy is converted to kinetic energy imparted to the displaced liquid. As a consequence, the bubble keeps expanding by inertia after the equilibrium radius is reached until the electron pressure inside the bubble reduces so much below the value of the external pressure that the reverse process, i.e., contraction, begins, and so on.

The relaxation process can thus be envisaged as an attenuated vibration in which viscosity and radiation of sound are the main damping processes (Iakubov and Khrapak, 1982; Rosenblit and Jortner, 1995, 1997). Similar physical processes are believed to occur in the phenomenon of cavitation nucleating on electrons in liquid helium (Hall *et al.*, 1995; Maris, 1995; Classen *et al.*, 1998; Grinfeld and Kojima, 2003).

The hydrodynamic problem of the bubble dynamics can be envisaged as the reverse Rayleigh model for a cavity collapse in a liquid induced by external pressure (Strutt, 1917) developed for the description of the solvation dynamics of the solvated electron in water (Rips, 1997). The model can be adapted to the present case of the electron bubble dynamics by assuming the electron pressure to be the driving force for the expansion, which is counteracted by the contractible force on the bubble resulting from surface tension and by the pressure–volume work involved in the creation of the bubble itself (Iakubov and Khrapak, 1982; Rosenblit and Jortner, 1995). The same approach has been followed in order to describe the collision-induced parametric oscillations of the O_2^- ionic bubble in near-critical argon gas (Borghesani and Tamburini, 1999).

It is assumed (Rosenblit and Jortner, 1997) that the initial localization time in the incipient bubble is short on the time scale of the equilibrium bubble formation. Surface-hopping calculations give an estimate of 50–100 fs for this process to occur (Space and Coker, 1991, 1992). It is also assumed that the bubble retains its spherical shape throughout and that the liquid is incompressible.

The dynamics of the cavity boundary is determined by the flow of the surrounding liquid, which follows the contraction or expansion of the electronic density. In the absence of energy dissipation, the previous assumptions result in a potential flow of the surrounding liquid that is described by the Euler equation for the radial component of the liquid velocity $v(r)$ (Landau and Lifsits, 2000):

$$\frac{\partial v}{\partial t} + v\frac{\partial v}{\partial r} = -\frac{1}{\rho}\frac{\partial P}{\partial r}, \tag{26.9}$$

where ρ is the liquid density and P is the net pressure acting on the bubble. r is the coordinate from the center of the bubble.

The continuity condition for the incompressible fluid allows one to express the radial component of the liquid velocity in terms of the radial velocity of the bubble boundary $U \equiv U(R)$:

$$v(r) = U(R)\frac{R^2}{r^2}\,, \tag{26.10}$$

where R becomes the reaction coordinate.

By inserting eqn (26.10) into eqn (26.9), the Rayleigh–Plesset equation (Plesset, 1949) is obtained for the cavity boundary velocity (Rips, 1997):

$$\frac{\mathrm{d}U}{\mathrm{d}t} = -\frac{3U^2}{2R} + \frac{1}{\rho R}\left[P(R) - P_0\right]\,, \tag{26.11}$$

where P_0 is the applied external pressure and $P(R)$ is the pressure on the cavity boundary. The net pressure acting on the bubble, $P(R) - P_0$, can be expressed in terms of the free-energy change in the electron solvation process as (Rips, 1997)

$$P(R) - P_0 = -\frac{1}{4\pi R^2}\frac{\partial \Delta\mathcal{A}}{\partial R}\,, \tag{26.12}$$

where $\Delta\mathcal{A}$ is given by eqn (3.21).

The solution is defined by imposing the condition of a vanishing initial velocity of the boundary wall, i.e., $U(R_0) = 0$.

The integration of the Rayleigh–Plesset equation can be carried out by defining the auxiliary variables (Rips, 1997)

$$R \equiv \exp(x)\,, \tag{26.13}$$

$$U^2(R) \equiv \phi(x)\,, \tag{26.14}$$

$$\frac{2}{\rho}\left[P(R) - P_0\right] \equiv \zeta(x)\,. \tag{26.15}$$

Equation (26.11) then becomes

$$\frac{\mathrm{d}\phi}{\mathrm{d}x} + 3\phi(x) = \zeta(x)\,. \tag{26.16}$$

The solution of eqn (26.16) when the initial boundary velocity is zero for $R = R_0$ is given by

$$\phi(x) = \exp(-3x)\int_{x_0}^{x} \exp(3y)\zeta(y)\mathrm{d}y\,. \tag{26.17}$$

Transforming back to the original variables, the velocity of the void boundary is obtained as

$$U(R) = \left(\frac{2}{\rho}\right)^{1/2} R^{-3/2}\left\{\int_{R}^{R_0} r^2\left[P_0 - P(r)\right]\right\}^{1/2} \mathrm{d}r\,, \tag{26.18}$$

where r is a dummy integration variable.

Inserting eqn (26.12) into eqn (26.18), one finally obtains

$$U(R) = (2\pi\rho)^{-1/2} R^{-3/2} \left[\Delta\mathcal{A}(R_0) - \Delta\mathcal{A}(R)\right]^{1/2} . \tag{26.19}$$

In the absence of dissipation mechanisms the cavity pulsates around the equilibrium radius R_-. Within the approximations introduced, the time scale for the solvation of the electron can be identified with the *first passage time*, i.e., with the time necessary for the expansion of the bubble from the initial to the equilibrium radius:

$$\tau(R) = \int_{R_0}^{R_-} U^{-1}(r)\mathrm{d}r . \tag{26.20}$$

Substitution of eqn (26.19) into eqn (26.20) yields

$$\tau(R) = \left(\frac{2}{\rho}\right)^{1/2} \int_{R_0}^{R_-} r^{3/2} \left[\Delta\mathcal{A}(R_0) - \Delta\mathcal{A}(r)\right]^{-1/2} \mathrm{d}r . \tag{26.21}$$

τ is an estimate of the lower bound to the actual cavity relaxation process.

In this dissipationless approximation, the condition of the fluid incompressibility is violated because the boundary velocity exceeds the sound velocity (Rosenblit and Jortner, 1997). Thus, realistically, damping mechanisms must be considered. The most important ones are sound emission and viscous damping (Iakubov and Khrapak, 1982; Artem'ev and Khrapak, 1986; Rips, 1997; Rosenblit and Jortner, 1997).

The dissipation caused by the emission of sound waves is treated by assuming that their wavelength is much larger than the equilibrium radius of the cavity, i.e., $\lambda \gg R_-$. Standard fluid mechanics expresses the rate of energy emission $\mathcal{I}$ as (Landau and Lifsits, 2000)

$$\mathcal{I} = \frac{\rho}{4\pi c_s} \left\langle \left[\ddot{\mathcal{V}}\left(t - \frac{r}{c_s}\right)\right]^2 \right\rangle . \tag{26.22}$$

$\mathcal{V} = (4\pi/3)R^3$ is the bubble volume, c_s is the sound velocity in the liquid, and the angular brackets denote averaging over one oscillation period of the cavity. The second time derivative of the volume is easily expressed in terms of the velocity of the boundary as

$$\ddot{\mathcal{V}} = 4\pi\left(2RU^2 + R^2\dot{U}\right) . \tag{26.23}$$

For the incompressible fluid the boundary velocity is given by eqn (26.19) and its derivatives are easily obtained by differentiating this expression. By using eqns (26.23) and (26.19), eqn (26.22) becomes (Rips, 1997)

$$\mathcal{I} = \frac{1}{4\pi\rho c_s}\left\langle \frac{1}{R^4}\left[\Delta\mathcal{A}(R_0) - \Delta\mathcal{A}(R) - R\frac{\partial\Delta\mathcal{A}(R)}{\partial R}\right]^2\right\rangle. \tag{26.24}$$

The energy dissipated during the expansion cycle is perturbatively given by

$$\Delta E_D^{SE} \approx \mathcal{I}\tau\,, \tag{26.25}$$

where τ is the first-passage time calculated in the absence of dissipation.

In Fig. 26.3 the boundary velocity is shown for liquid He for $P_0 = 0\,\mathrm{MPa}$ and $T = 0.4\,\mathrm{K}$ with and without sound damping. The damping mechanism due to the emission of sound waves greatly reduces the boundary velocity, that now always remains below the sound velocity, and also the maximum value of the radius reached in the expansion cycle is limited (Rosenblit and Jortner, 1997).

It should be mentioned that the appropriate theoretical description of the bubble boundary that takes into account the finite sound velocity c_s is given by the Herring equation (Trilling, 1952):

$$\begin{aligned}\left(1 - \frac{2U}{c_s}\right)\frac{\mathrm{d}U}{\mathrm{d}t} = &-\left(1 - \frac{4U}{3c_s}\right)\frac{3U^2}{2R} \\ &+\frac{1}{\rho R}\left[P(R) - P_0 + \frac{U}{c_s}\left(1 - \frac{U}{c_s}\right)R\frac{\partial P(R)}{\partial R}\right].\end{aligned} \tag{26.26}$$

Another damping mechanism is viscous energy dissipation. If bulk viscosity is neglected (Byron Bird *et al.*, 1960), only the shear viscosity η is responsible for energy dissipation. The standard approach (Neppiras, 1980) consists of adding to the free energy a viscous term of the form (Rips, 1997; Benderskii *et al.*, 2002)

$$\mathcal{A}_v = 16\pi\eta\left(\frac{\rho_n}{\rho}\right)\int^{R} rU(r)\mathrm{d}r\,, \tag{26.27}$$

where ρ_n is the normal fraction in the two-fluid model of the superfluid. The effect of this additional contribution is to change the pressure $P(R)$ acting on the bubble walls by the additional term $-4\eta(\rho_n/\rho)U(R)/R$ (Benderskii *et al.*, 2002).[5] The addition of this term into eqn (26.26) clearly damps the bubble oscillations. However, the choice of η is not as straightforward as desired because the use of a macroscopic shear viscosity for a microscopic object is questionable (Rips, 1997).

A similar hydrodynamic approach has been followed to calculate the time evolution of the radius of the positronium bubble in liquid He (Iakubov and Khrapak, 1982). In this case, the approximation known as the Kirkwood–Bethe hypothesis (Kirkwood and Bethe, 1967; Wood, 1967; Knapp *et al.*, 1970) has been used. The qualitative features of the dynamics of the bubble in the absence

[5] A slightly different expression for the viscous drag has to be used in the ballistic regime (Benderskii *et al.*, 2002).

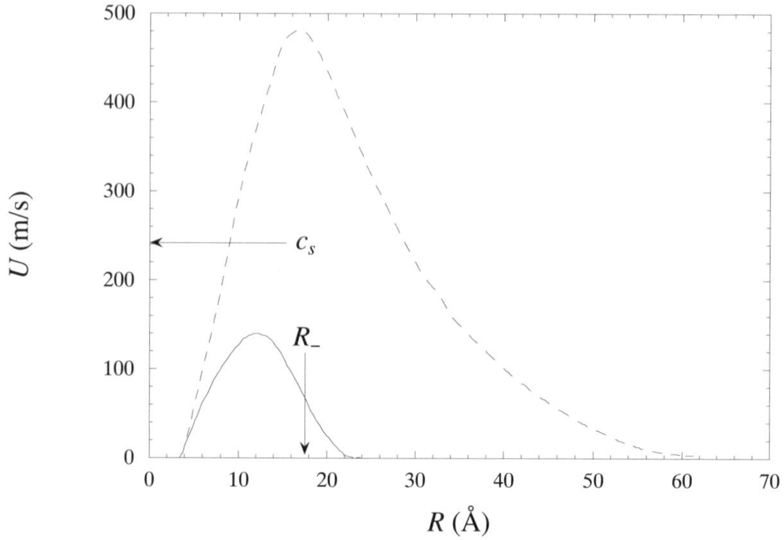

FIG. 26.3. Velocity of the boundary U vs bubble radius R in He at zero pressure for $T = 0.4\,\text{K}$. (Rosenblit and Jortner, 1997.) Dashed line: no dissipation. Solid line: dissipation by sound wave emission.

or in the presence of damping are very similar to those calculated for an electron in water (Rips, 1997) and can be observed by inspecting Fig. 26.4 (Iakubov and Khrapak, 1982).

In the absence of dissipation (curve 1), the boundary oscillation has a very large amplitude and the bubble pulsates forever with a very long period. If the damping due to sound wave emission is taken into account, the oscillation is strongly attenuated and only a couple of rebounds are observed (curve 2). Finally, if viscous damping is also accounted for, the bubble radius rapidly attains its equilibrium value and rebounds are hardly observed at all.

It should be noted that viscous damping in superfluid He II should be much less than in the normal liquid, leading to longer relaxation times. In any case, the time scale of the measurements of ion transport described throughout this book is several orders of magnitude larger than the typical bubble relaxation times. So, the bubble dynamics is not relevant in this kind of experiment.

Once the electron bubble is in its equilibrium state, its electron can be ejected from it by infrared absorption, as in the experiments of photoabsorption (Grimes and Adams, 1990, 1992), leaving behind an empty cavity that will collapse under the action of the external pressure and surface tension. The calculated dynamics of a collapsing empty bubble in liquid ^{4}He and ^{3}He for $P = 0$ MPa and $T = 0.4\,\text{K}$ is shown in Fig. 26.5.

It should be noted that the rapid dynamics of the electron bubble expansion leads to a time evolution of the electron transition energies on the picosecond

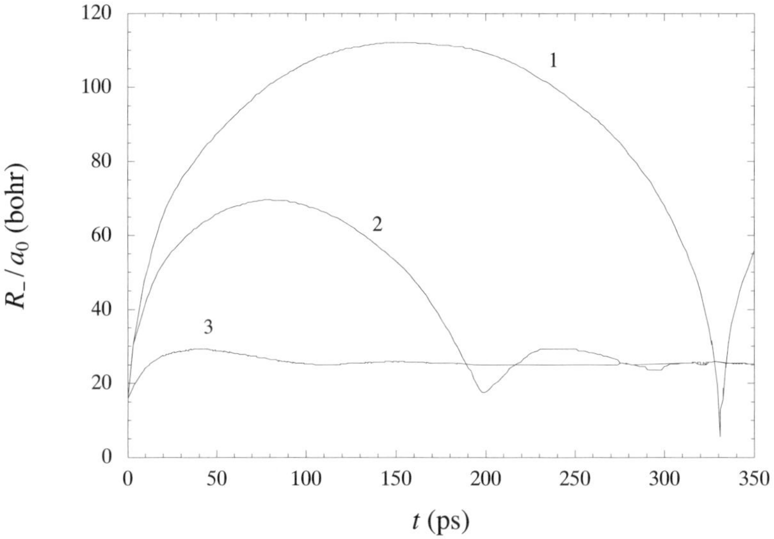

FIG. 26.4. Time evolution of the positronium bubble in liquid He for $T = 4.2\,\text{K}$. (Iakubov and Khrapak, 1982.) Curve 1: no dissipation. Curve 2: dissipation by sound wave emission. Curve 3: sound wave emission and viscous damping.

time scale that could possibly be observed in ultrafast time-resolved electron spectroscopy experiments (Martin *et al.*, 1993).

The oscillations of the bubble that forms around the He_2 excimer rather than those of the "true" electron bubble are exploited to investigate the dissipative superfluid dynamics in bulk He II on a molecular scale in time-resolved, pump–probe experiments (Benderskii *et al.*, 2002). The Rydberg transitions of the triplet $He_2^\star$ excimers, which solvate in bubble states in liquid He, are used as nanoscale transducers to initiate and monitor the motion of the superfluid.

The starting point is the production of metastable $He_2^\star\ a\,^3\Sigma_u^+(2s)$ excimers, which are solvated in a spherical bubble of $\approx 7\,\text{Å}$ in diameter. After complete thermalization, a short laser pulse promotes the $2s$ electron from the ^3a state. Vertical excitation from the equilibrium geometry of the bubble necessarily leads to a repulsive interaction with the liquid. The bubble expands to accommodate the excited electron and undergoes damped oscillations until the new equilibrium on the excited state is reached. A time-delayed copy of the pulse probes the effect of the excitation, relying on the ^3d–^3b fluorescence as a signal. As the energy gap between the several Rydberg states depends on the bubble radius, the electronic resonances are modulated by the bubble oscillations.

It is found that damping is so strong that the bubble oscillations are damped out after one period, whose temperature dependence follows that of the normal fluid fraction, as shown in Fig. 26.6, in accordance with both the solution of the Herring equation, eqn (26.26), and the time-dependent density functional treat-

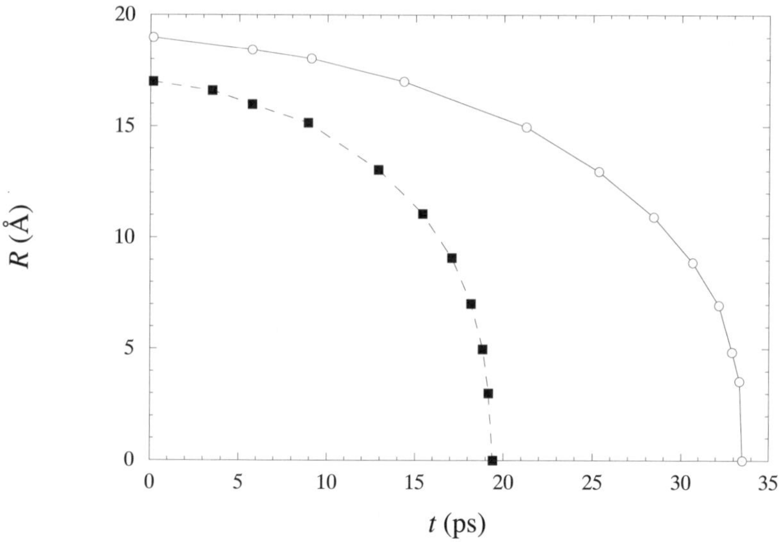

FIG. 26.5. Calculated collapse dynamics of an empty bubble in liquid ^{4}He (squares) and ^{3}He (circles) for $T = 0.4\,\mathrm{K}$ at $P = 0$ MPa. (Rosenblit and Jortner, 1997.)

ment of the superfluid dynamics (Eloranta and Apkarian, 2002). The temperature dependence of the breathing period tracks that of the normal fraction in the superfluid as measured with the macroscopic disk viscometer (Andronikashvili, 1946).

The pressure dependence of the breathing period is shown in Fig. 26.7. The linear dependence is due to the pressure-dependent quenching of the ^3d state fluorescence (Keto *et al.*, 1974) arising from the penetration of the liquid into the nodal region of the Rydberg $3s$ electron in this state (Eloranta *et al.*, 2002).

It can therefore be concluded that radiation of sound in the far field, driven by the acceleration on the cavity boundary in a compressible fluid, and a temperature-dependent drag in the the near field are the main dissipation mechanisms that drive the relaxation process of the bubble. The experiment, however, cannot discriminate if the drag is strictly of viscous origin or if it is due to ballistic scattering of rotons from the bubble edge (Benderskii *et al.*, 2002).

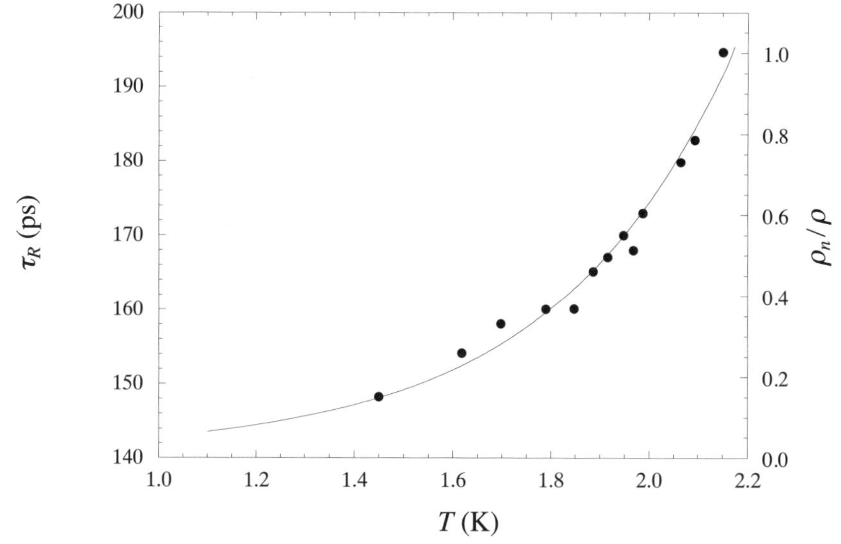

FIG. 26.6. Breathing period of the bubble vs T in superfluid He II (left scale) (Benderskii *et al.*, 2002). Line: normal fluid fraction (right scale) (Andronikashvili, 1946).

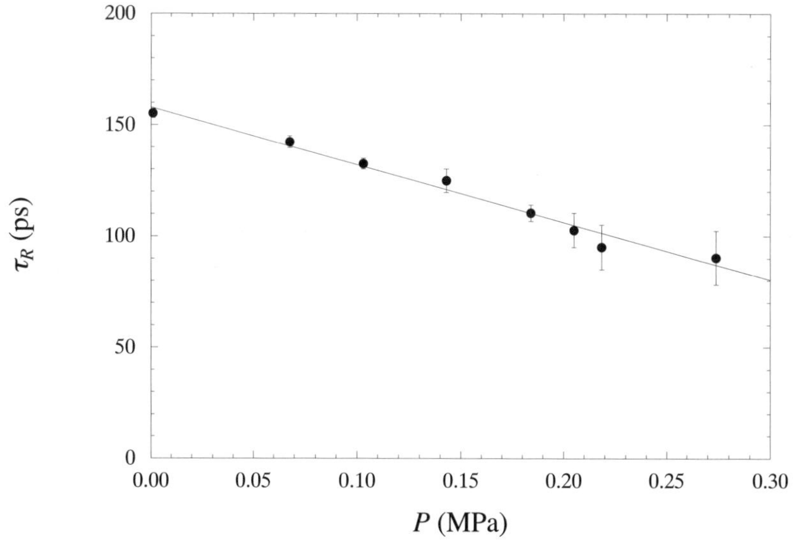

FIG. 26.7. τ_R vs P in superfluid He II at $T = 1.46$ K. (Benderskii *et al.*, 2002.) Line: eyeguide.

27

ELECTRON MOBILITY IN DENSE HE GAS

Classical kinetic theory predicts that electrons drifting in a dilute gas under the action of a vanishingly small electric field move with a constant velocity v_D proportional to the electric field E as a consequence of the occasional events of elastic scattering off atoms of the gas (Huxley and Crompton, 1974).

If the momentum-transfer electron–atom scattering cross-section $\sigma_{mt}(\epsilon)$ is known as a function of the electron energy ϵ, the zero-field electron mobility $\mu_0 = \lim_{E\to 0}(v_D/E)$ can be easily calculated as

$$(\mu_0)_{\rm cl} = \frac{4e}{3N(2\pi m)^{1/2}(k_{\rm B}T)^{5/2}} \int_0^\infty \frac{\epsilon}{\sigma_{mt}(\epsilon)} \exp(-\epsilon/k_{\rm B}T)\,{\rm d}\epsilon\,, \qquad (27.1)$$

where N is the gas number density. Until the conditions for the applicability of kinetic theory are fulfilled, the density-normalized zero-field mobility $(\mu_0 N)_{\rm cl}$ depends only on the gas temperature T, but is independent of N.

On the other hand, if it is assumed that, in addition to high-mobility extended states, electrons can be found in low-mobility localized states whose existence depends on the thermodynamic conditions of the gas, then deviations from the prediction of kinetic theory are expected to appear as the density is increased so as to favor the appearance of a significant fraction of localized states.

27.1 Experimental results

The first experimental evidence of the existence in He of localized states even in the gas phase has been obtained by measuring the drift velocity of electrons with the pulsed-Townsend photoinjection technique (Levine and Sanders, 1962).

A typical experimental set-up is shown in Fig. 27.1 (Borghesani *et al.*, 1986*a*, 1988). A short UV-pulse impinges on a semi-transparent photocathode that acts as an electron emitter. Electrons are injected into the gas, in which they thermalize rapidly in a few picoseconds, i.e., on a time scale several orders of magnitude shorter than the typical drift times, which range from microseconds upward. An adjustable potential difference applied between the emitter and the anode (collector) produces the electric field that pulls the electrons through the gas. The induction current in the collector is converted to voltage either by feeding it to a resistor or to a capacitor, depending on the signal-to-noise ratio. The signal is then suitably amplified and registered to an oscilloscope. The analysis of the waveform easily yields the drift time (Borghesani and Santini, 1990*b*).

The earliest zero-field mobility data in dense He gas were obtained by Levine and Sanders (1962) in a temperature range just below the temperature of the

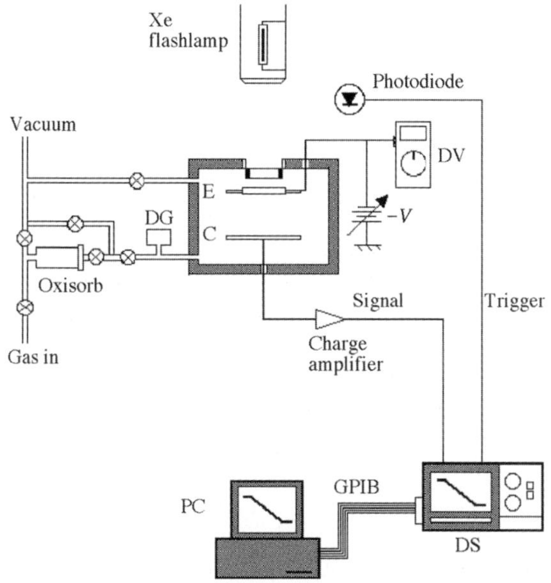

FIG. 27.1. Schematic experimental set-up for the pulsed-Townsend photoinjection technique. (Borghesani *et al.*, (1988).) E: emitter, C: collector, −V: potential difference, DG: digital pressure gauge, DS: digital oscilloscope, DV: digital voltmeter, and GPIB: general-purpose interface bus.

normal boiling point. Results for $T = 3.65$, 3.902, and 4.19 K are reported in Fig. 27.2 as a function of the reduced pressure P/P_s, where P_s is the saturated vapor pressure (Levine and Sanders, 1967).

In Fig. 27.3 the mobility measured for $P = P_s$ is shown as a function of temperature. μ_0 decreases enormously in a very narrow pressure or temperature range. To make the comparison with the prediction of classical kinetic theory, eqn (27.1), easier, the data are shown as a function of the gas density N in Fig. 27.4.

Owing to the small polarizability of the He atom, the electron–atom scattering cross-sections are very weakly energy dependent, as shown in Fig. 27.5. The He atom scatters electrons very much in the same way as a hard sphere of radius equal to its scattering length $a \approx 0.62\,\text{Å}$, corresponding to a constant cross-section $\sigma_{mt} = \sigma_0 = 4\pi a^2 \approx 4.9 \times 10^{-20}\,\text{m}^2$ (O'Malley, 1963; Crompton *et al.*, 1970; Zecca *et al.*, 1996). In this case, eqn (27.1) is easily integrated so as to yield

$$\mu_0 = \frac{4e}{3N\sigma_0\,(2\pi m k_{\text{B}} T)^{1/2}}\,. \tag{27.2}$$

Equation (27.2) predicts that the quantity $(T/T_0)^{1/2}\mu_0$ is a universal function of

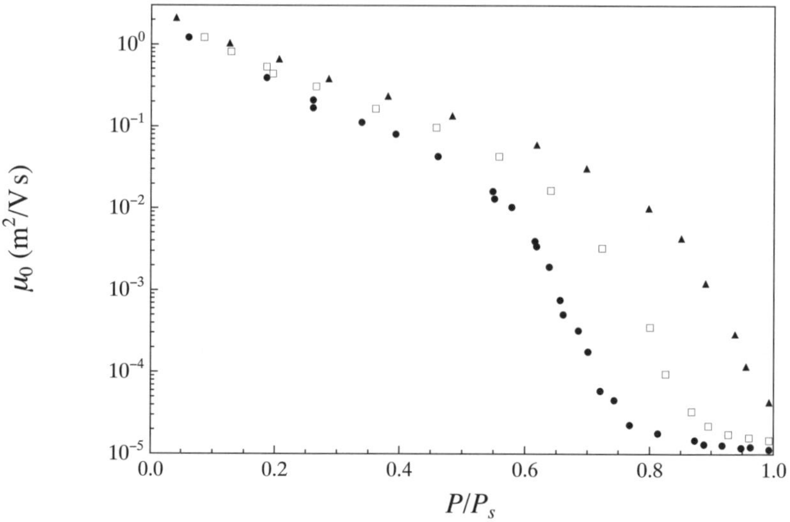

FIG. 27.2. μ_0 vs P/P_s for T (K) = 4.19 (circles), 3.902 (squares), and 3.650 (triangles). (Levine and Sanders, 1967.) P_s is the saturated vapor pressure.

the gas density N. The data in Fig. 27.4 are plotted in this fashion. In the same figure, the classical prediction, eqn (27.2), is plotted as a solid line.

At low density, the experimental data approach the expected value predicted

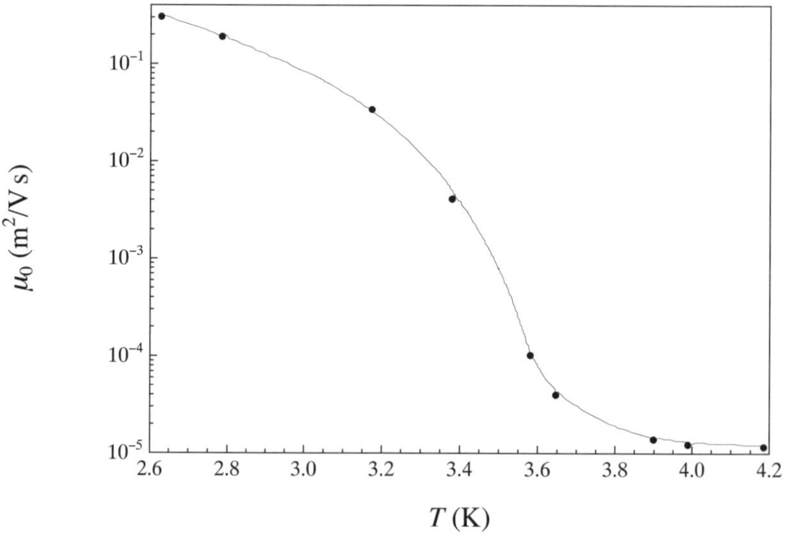

FIG. 27.3. μ_0 vs T at SVP. (Levine and Sanders, 1967.) Line: eyeguide.

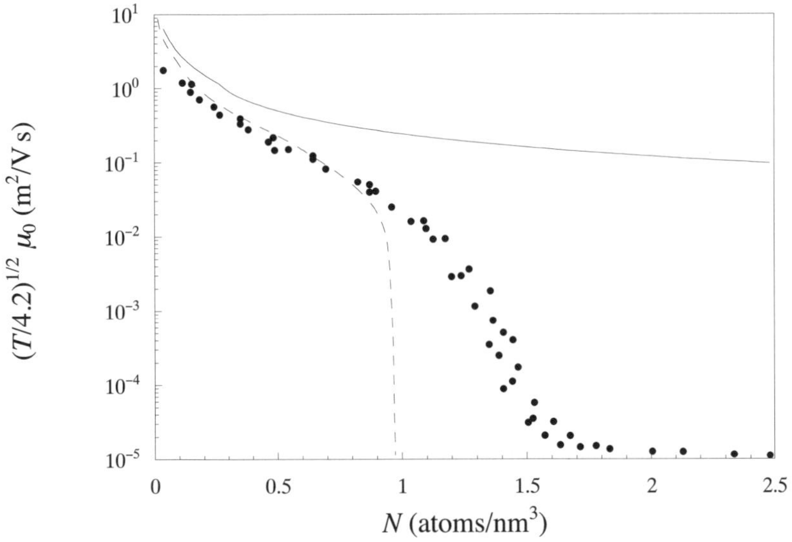

FIG. 27.4. $(T/4.2)^{1/2}\mu_0$ vs N for the same temperatures as in Fig. 27.2. (Levine and Sanders, 1967.) Solid line: eqn (27.1) with constant momentum-transfer scattering cross-section. Dashed line: optimum fluctuation model.

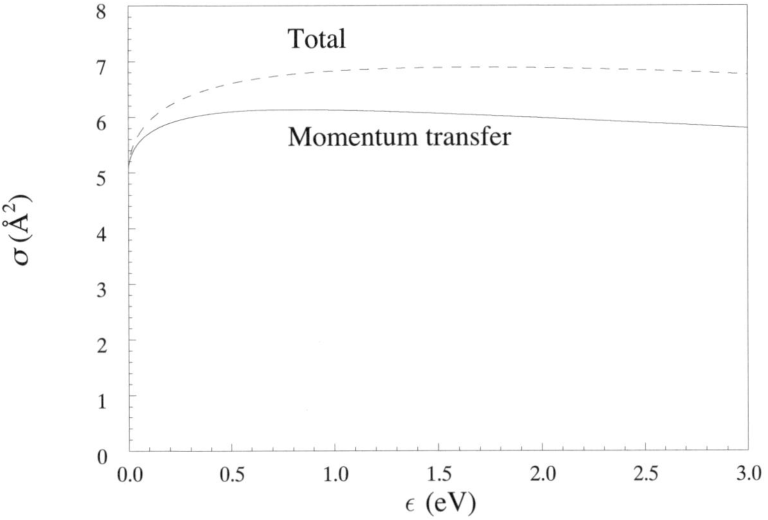

FIG. 27.5. Electron–He atom scattering cross-sections σ vs ϵ. (O'Malley, 1963; Crompton *et al.*, 1970; Zecca *et al.*, 1996.)

by classical kinetic theory for quasi-free (delocalized) electrons. However, at higher densities the disagreement between theory and experiment becomes huge. It is also clear that the observed behavior of the mobility is determined by the gas density, rather than by temperature or pressure individually.

The anomalous behavior of the electron drift mobility as a function of the gas density has been confirmed in several more experiments based on different experimental techniques that either extend the temperature range investigated (Harrison *et al.*, 1973; Jahnke *et al.*, 1975), or improve the experimental accuracy (Schwarz and Prasad, 1976; Schwarz, 1978, 1980), or both (Borghesani and Santini, 2002). The experimental techniques used include Cunsolo-type (Cunsolo, 1961), single-gate drift velocity spectrometer (Harrison *et al.*, 1973), double-gate drift velocity spectrometer (Jahnke *et al.*, 1971, 1975), photoinjection (Levine and Sanders, 1962, 1967; Borghesani and Santini, 2002), and charge–pulse time-of-flight (Schwarz and Prasad, 1976; Schwarz, 1978, 1980).

The zero-field mobility μ_0 obtained in these experiments for $T < 26\,\mathrm{K}$ is shown as a function of N in Fig. 27.6. In this figure the predictions of classical kinetic theory are also presented and appear to be in strong disagreement with the experimental data.

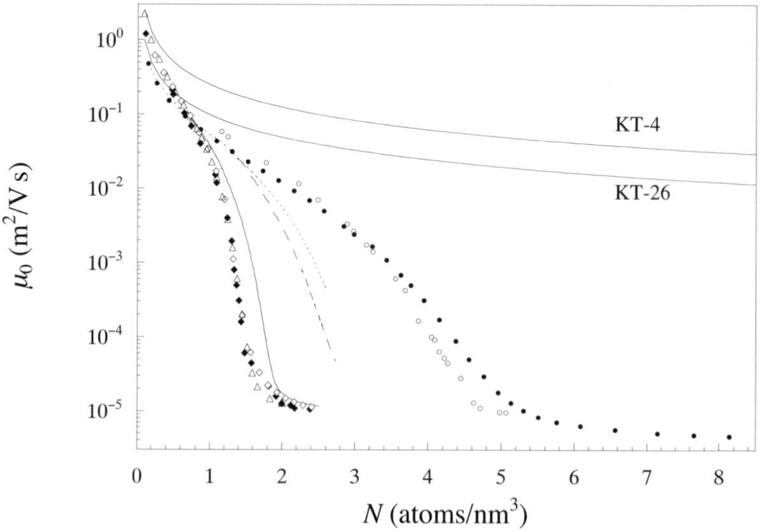

FIG. 27.6. μ_0 vs N in dense He gas for $T < 26\,\mathrm{K}$. $T = 4.2\,\mathrm{K}$: triangles (Schwarz, 1980), closed diamonds (Levine and Sanders, 1967), and open diamonds (Harrison *et al.*, 1973). Smoothed data by Harrison *et al.* (1973): $T\,(\mathrm{K}) = 7.3$ (solid line), 11.6 (dashed line), 18.1 (dotted line). $T(K) = 20.3$ (open circles) (Jahnke *et al.*, 1975), and 26.1 (closed circles) (Borghesani and Santini, 2002). Lines KT-4 and KT-26: classical kinetic theory for $T = 4$ and $26\,\mathrm{K}$, respectively.

Mobility measurements in the low-T region, in which small temperature differences are accurately resolved, are presented in Fig. 27.7, and accurate mobility measurements showing the drop-off also at a very high temperature are shown in Fig. 27.8. The data are presented here in the form of the zero-field density-normalized mobility $\mu_0 N$ to remove the explicit density dependence in eqn (27.1). In these last two figures the prediction of classical kinetic theory would be a density-independent constant corresponding to the zero-density extrapolation of the experimental data.

It is evident that the mobility decreases gradually from the low-density value predicted by classical kinetic theory to the high-density value comparable to that of the electron bubble in liquid He II. The measured mobility is a weighted average over all charge carriers present in the sample. If high-mobility quasi-free electrons and low-mobility localized electrons are present at the same time, the observed value of the mobility will be intermediate between the two extremes. The gradual crossover between the two limiting behaviors is a manifestation of the fact that the relative proportion of the species changes in favor of the low-mobility one as the density is increased.

As anticipated, this kind of behavior is also observed in polar gases such as NH_3 (Krebs *et al.*, 1980), as shown in Fig. 27.9, and in the other noble gas Ne, shown in Fig. 27.10 (Borghesani and Santini, 1990*a*), whose electron–atom interaction is dominated by short-range repulsive exchange forces (O'Malley *et al.*, 1979; O'Malley and Crompton, 1980).

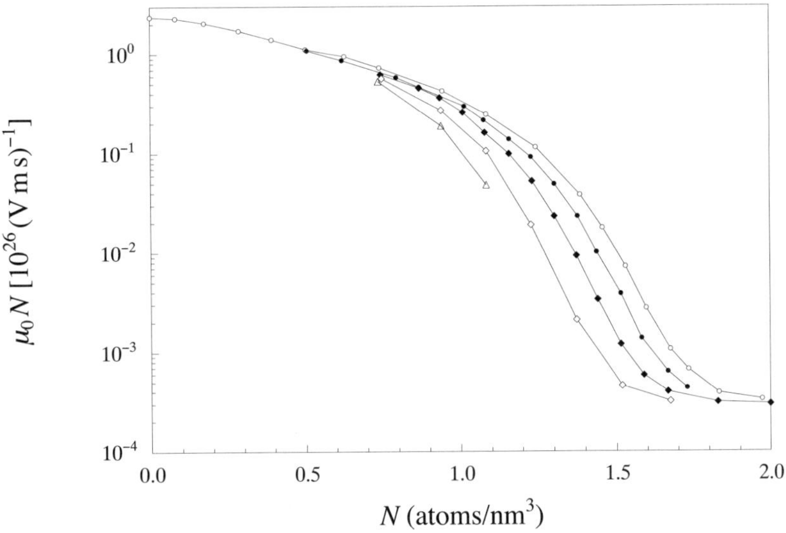

FIG. 27.7. $\mu_0 N$ vs N in the low-T region. (Schwarz, 1980.) T (K) = 3.40, 3.80, 4.20, 4.60, and 4.98 (from left to right). Lines: eyeguides.

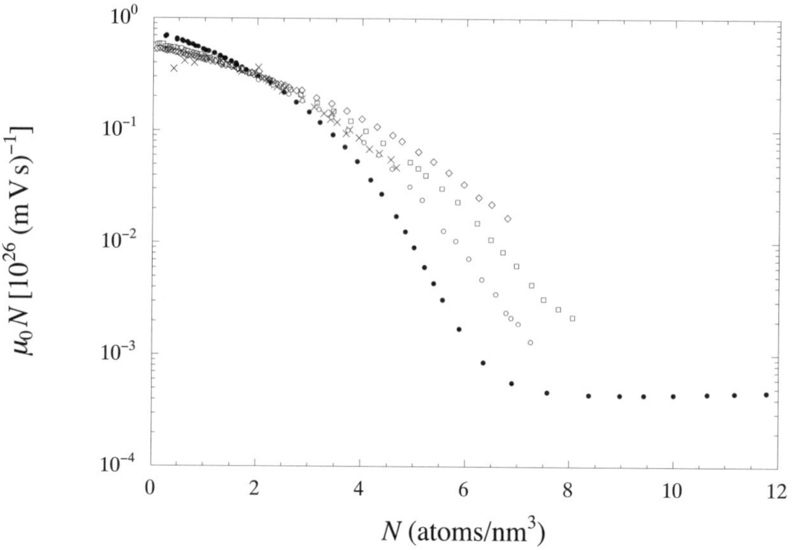

FIG. 27.8. $\mu_0 N$ vs N for $T\,(\mathrm{K}) = 34.5$ (closed circles), 45.0 (open circles), 54.5 (squares), 64.4 (diamonds) (Borghesani and Santini, 2002), and 52.8 (crosses) (Jahnke *et al.*, 1975).

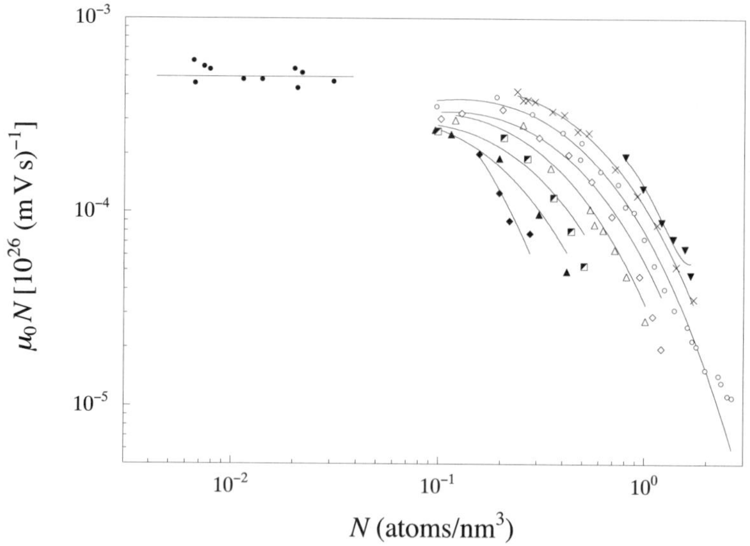

FIG. 27.9. $\mu_0 N$ vs N in NH_3. (Krebs *et al.*, 1980.) Closed circles: data at coexistence, $210\,\mathrm{K} < T < 238\,\mathrm{K}$. The data at high N are obtained for $T\,(\mathrm{K}) = 296$, 320, 340, 367, 380, 400, 420, and 440 (from left to right).

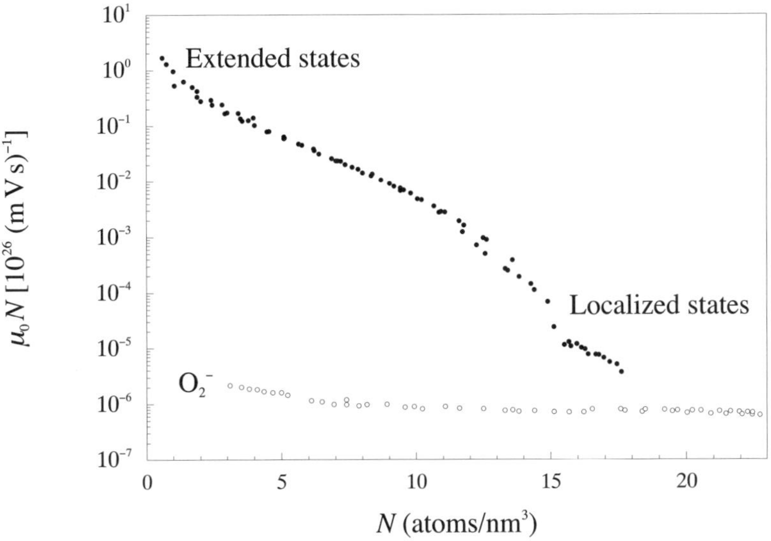

FIG. 27.10. $\mu_0 N$ vs N in Ne for $45.0\,\mathrm{K} < T < 47.9\,\mathrm{K}$. (Borghesani and Santini, 1990*a*.) ($T_c \approx 44.4\,\mathrm{K}$ and $N_c \approx 14.44\,\mathrm{atoms/nm^3}$.) Open symbols refer to O_2^- ions. The sample purity for the electron mobility measurements must be below 1 ppb, whereas it is of some ppm for the O_2^- mobility measurements.

From a phenomenological point of view, three density regions can be characterized by observing the behavior of the drift mobility as a function of the electric field. The first one is the low-density region, in which the $\mu_0(N)$ curve has an upward concavity. The second region is the transition region at intermediate density, in which the concavity changes. Finally, there is is the high-density region, in which only low-mobility, localized states are observed.

In the first region, at low density, the drift mobility is field independent at low electric field strength E and then decreases with increasing E because electrons are no longer in thermal equilibrium with the atoms of the host gas. This behavior is plotted in Fig. 27.11. It is customary to plot the density-normalized mobility μN as a function of the reduced electric field strength E/N because, for a constant scattering cross-section σ_0, the energy gained by electrons in one mean free path $\ell = 1/N\sigma_0$ is $eE\ell = eE/N\sigma_0 \propto E/N$ (Huxley and Crompton, 1974).

According to classical kinetic theory (Huxley and Crompton, 1974), the density-normalized mobility μN as a function of the reduced electric field E/N is given by

$$\mu N = -\left(\frac{e}{3}\right)\left(\frac{2}{m}\right)^{1/2} \int_0^\infty \frac{\epsilon}{\sigma_{mt}(\epsilon)} \left[\frac{\mathrm{d}g(\epsilon)}{\mathrm{d}\epsilon}\right] \mathrm{d}\epsilon\,. \qquad (27.3)$$

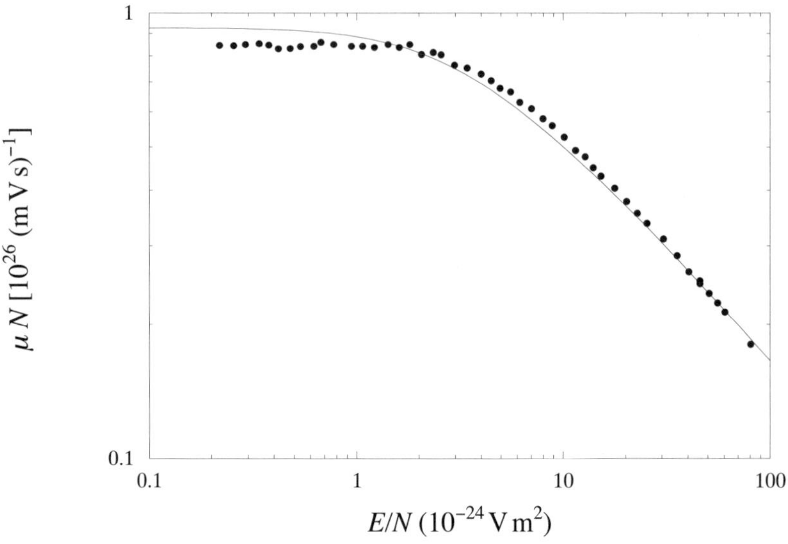

FIG. 27.11. μN vs E/N in He for $N = 0.188\,\text{atoms/nm}^3$ and $T = 26.1\,\text{K}$. (Borghesani and Santini, 2002.) Line: classical kinetic theory.

where $g(\epsilon)$ is the Davydov–Pidduck distribution function, the two-term solution of the Boltzmann equation, given by

$$g(\epsilon) = A \exp\left\{ -\int_0^{\epsilon} \left[k_B T + \left(\frac{M}{6m}\right)\left(\frac{eE}{N}\right)^2 \frac{1}{z\sigma_{mt}^2(z)} \right]^{-1} \mathrm{d}z \right\}, \tag{27.4}$$

where M is the He atom mass and A is a normalization constant defined by the condition

$$\int_0^{\infty} \epsilon^{1/2} g(\epsilon)\, \mathrm{d}\epsilon = 1\,. \tag{27.5}$$

The solid line in Fig. 27.11 is calculated by means of eqns (27.3) and (27.4) with the proper scattering cross-section.

In the intermediate or transition region, where the mobility is already strongly depressed by localization effects, the drift velocity curves exhibit rather spectacular nonlinearities as a function of the electric field (Levine and Sanders, 1967; Schwarz, 1980; Borghesani and Santini, 2002), as shown in Fig. 27.12.

In terms of the mobility, μ is constant at small field strengths, though its value is much lower than that predicted by classical kinetic theory. Then, above a certain threshold field, μ increases until it approaches the classical field dependence for hot electrons predicted by eqns (27.3) and (27.4). Exactly the same behavior, shown in Fig. 27.13, has been observed in Ne (Borghesani and Santini, 1990*a*).

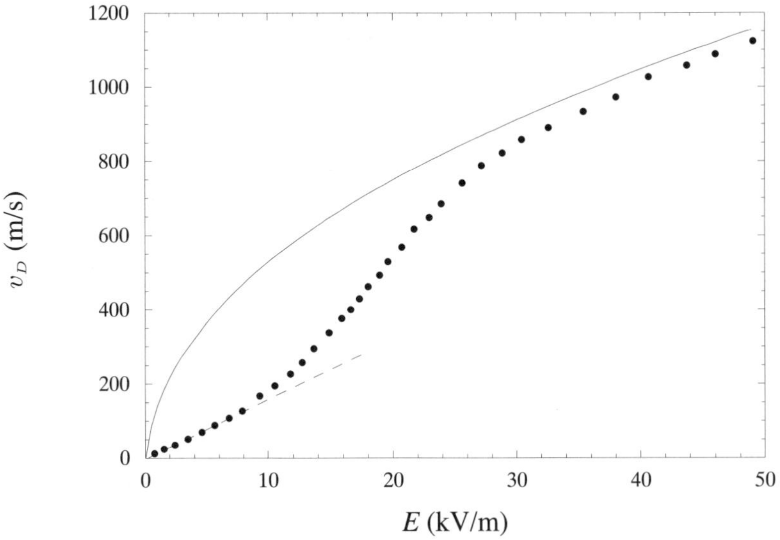

FIG. 27.12. v_D vs E in He for $N = 1.089\,\text{atoms/nm}^3$ and $T = 4.2\,\text{K}$. (Schwarz, 1980.) Solid line: classical kinetic theory. Dashed line: linear zero-field extrapolation yielding μ_0.

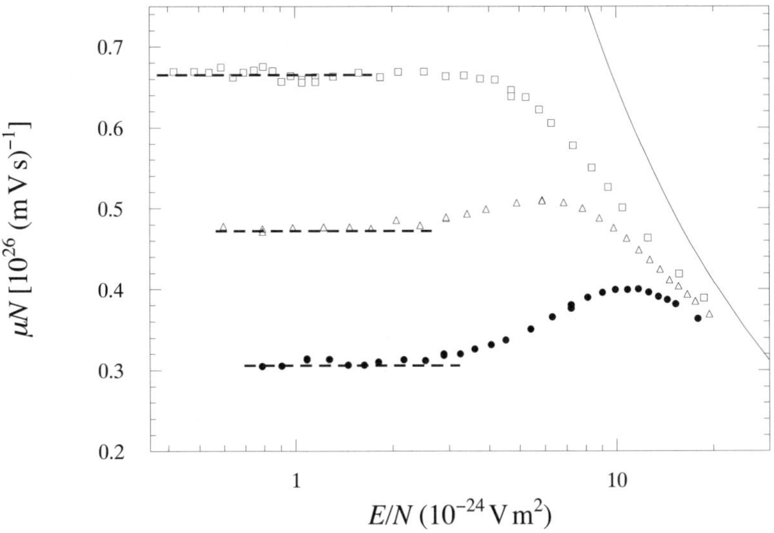

FIG. 27.13. μN vs E/N in Ne for $N = 9.56$, 10.21, and $11.08\,\text{atoms/nm}^3$ (from top) and $T \approx 46.5\,\text{K}$ (Borghesani and Santini, 1990a.) Solid line: classical kinetic theory. Dashed lines: zero-field value $\mu_0 N$.

The supralinear deviations of the drift velocity, i.e., the rise of the mobility above the low-field limiting value to join the classical hot-electron field dependence, are interpreted as the result of electron-heating effects due to the strong electric field (Schwarz, 1980; Borghesani and Santini, 1990*a*). It may either happen that free electrons are too energetic to become self-trapped and, thence, remain delocalized (Schwarz and Prasad, 1976; Atrazhev, 1984; Sakai *et al.*, 1992; Schmidt *et al.*, 2001), or that a field-assisted escape process might lead to a depletion of the trapped-electron population and to a consequent increase of the highly-mobile electron fraction. Interestingly enough, the superlinearity in the $\mu N(E/N)$ curves in Ne starts appearing in correspondence to the threshold density $N^{\star} \approx 9.5\,\text{atoms/nm}^3$ for which the $\mu_0 N(N)$ curve changes its concavity (Borghesani and Santini, 1990*a*).

Finally, in the high-N region, in which only localized electrons are present, the drift velocity again becomes linear with the electric field, yielding a constant mobility even at the highest reduced fields investigated, namely $E \approx 150\,\text{kV/m}$ corresponding to $E/N \approx 80 \times 10^{-24}\,\text{V m}^2$ at $T = 4.2\,\text{K}$ for $N = 1.824\,\text{atoms/nm}^3$ (Schwarz, 1980), or $E \approx 1\,\text{MV/m}$ corresponding to $E/N \approx 83 \times 10^{-24}\,\text{V m}^2$ at $T = 26\,\text{K}$ for $N = 12\,\text{atoms/nm}^3$ (Borghesani and Santini, 1990*a*).

The reason why μ is field independent at such high densities is that the electron bubbles are too massive to be heated by the electric field above the gas temperature and thus remain in thermal equilibrium with the atoms of the environment at all times. An alternative explanation is that the large electron bubbles interact by means of viscous forces with the extremely dense gas and that the drag is so large that they move at the constant limiting speed allowed by the combination of electric field pull and hydrodynamic resistance. In any case, the linearity of v_D versus E at high N does not give any additional pieces of information about the structure of the electron bubble and is therefore not worth being displayed here.

As can be noted by inspecting the previous figures, the deviations from the classical behavior already occur at low density for all temperatures and are not limited to the highest densities. As a consequence, two distinct areas of investigation, both experimental and theoretical, have appeared with relatively little overlap in spite of the fact that the same phenomenon is dealt with (Schwarz, 1980).

On one hand, traditional swarm experiments in He and other noble gases have been pushed to high densities in order to test the classical description (Grűnberg, 1969; Bartels, 1972, 1973, 1975; Robertson, 1977; Huang and Freeman, 1978; Borghesani *et al.*, 1985, 1988, 1992; Floriano *et al.*, 1987; Jacobsen *et al.*, 1989; Borghesani and Santini, 1994*a*, 1994*b*; Lamp and Buschhorn, 1994; Borghesani, 2001).

On the other hand, other experiments are connected to the existence of the bubble in liquid He and are aimed at investigating the conditions under which localization also takes place in the dense gas (Levine and Sanders, 1962, 1967; Harrison *et al.*, 1973; Jahnke *et al.*, 1975; Schwarz, 1980; Borghesani and Santini,

1990*a*).

The same division appearing in the experimental field also occurs on the theoretical side, where some authors have focused on the extension of kinetic theory to higher densities (Legler, 1970; Atrazhev and Yakubov, 1977; Braglia and Dallacasa, 1978, 1982; Dallacasa, 1979; O'Malley, 1980, 1992; Iakubov and Polishuk, 1982; Polishuk, 1983*a*, 1983*b*, 1984; Atrazhev, 1984; Kalia *et al.*, 1989; Bagheri *et al.*, 1994), while other authors have devoted their efforts to investigating the transition of the electrons to the bubble states (Khrapak and Iakubov, 1973, 1975, 1979; Ebner and Punyanitya, 1979; Nieminen *et al.*, 1980; Ichikawa and Yoshida, 1981; Yuan and Ebner, 1981; Bartholomew *et al.*, 1985; Coker *et al.*, 1987; Miller and Reese, 1989; Miller and Fang, 1990; Hernandez and Martin, 1991; Chandler and Leung, 1994; Leung and Chandler, 1994; Miller, 1994; Cao and Berne, 1995).

Finally, several authors have taken the approach of treating the gas as the source of a random potential acting on electrons and have applied concepts of the theory of disordered systems (Anderson, 1958), such as mobility edge (Mott, 1967, 1974; Cohen, 1973) and percolation (Stauffer, 1985), to explain the overall mobility behavior (Eggarter and Cohen, 1970, 1971, 1974; Eggarter, 1972; Hernandez and Ziman, 1973; Simon *et al.*, 1990; Adams *et al.*, 1992; Borghesani and Santini, 1992; Borghesani and O'Malley, 2003).

It is therefore convenient to also adopt this schematic division in the following.

27.2 Mobility at low and intermediate density

Anomalous density effects on the electron mobilities are observed in a number of compressed gases (Bowe, 1960; Levine and Sanders, 1962, 1967; Lowke, 1963; Lehning, 1968; Grűnberg, 1969; Allen and Prew, 1970; Bartels, 1972, 1973, 1975; Schwarz, 1978, 1980; Huang and Freeman, 1981; Borghesani *et al.*, 1985, 1988).The density effects can be divided into two types and show up in different gases according to the main character of the electron–atom interaction.

The *negative density effect*, i.e., the decrease of $\mu_0 N$ with increasing N, occurs in gases such as He and Ne (but also in CO_2), whose interaction with electrons is dominated by the short-range repulsive exchange forces so that their scattering length is $a > 0$.

In contrast, the *positive density effect*, i.e., the increase of $\mu_0 N$ with N, appears in gases such as the heavier noble gases Ar, Kr, and Xe, in which the electron–atom interaction is primarily due to long-range polarization forces that endow the atoms with a negative scattering length $a < 0$ (Zecca *et al.*, 1996).

The classical single-scattering picture predicts that $\mu_0 N$ is independent of N according to eqn (27.1). This picture breaks down at higher densities because the mean free path for electrons becomes comparable to their thermal wavelength and multiple-scattering corrections must be included in the collision frequency (Legler, 1970; Atrazhev and Yakubov, 1977; Dallacasa, 1979).

Several multiple-scattering theories have been developed in the limit of vanishingly-small reduced electric field E/N (Legler, 1970; Atrazhev and Yakubov,

1977; Braglia and Dallacasa, 1978, 1982; Dallacasa, 1979; O'Malley, 1980, 1992; Iakubov and Polishuk, 1982; Polishuk, 1983*a*, 1983*b*). All of them treat the gas as a continuum and apply the multiple-scattering theory for the refraction of the electron wave in a homogeneous medium (Foldy, 1945; Lax, 1951).

There is general consensus that multiple-scattering effects in a dense disordered medium induce a shift of the electron kinetic energy (Fermi, 1934). However, different authors treat the multiple-scattering effects in different ways (Polishuk, 1984). For instance, there is no agreement as to whether the energy shift has to be considered before or after electron thermalization (O'Malley, 1980; Borghesani and O'Malley, 2003) or if different mechanisms have to be invoked to explain the two kinds of density effects (Atrazhev and Yakubov, 1977) in spite of the obvious consideration that we are faced with the common physical phenomenon of electron scattering off atoms and that this process deserves a unified description.

For this reason, a heuristic model has been devised that incorporates all multiple-scattering effects within a single framework, still retaining the single-scattering picture of classical kinetic theory (Borghesani *et al.*, 1988, 1992). Owing to the heuristic approach, the goodness of the model has to be judged on the basis of the agreement with the experimental data. Until now, this model has accurately described the anomalous density effects in all noble gases, irrespective of the sign of the scattering length, as well as the electric field dependence of the mobility.

In the following, the heuristic model will be described with emphasis on its agreement with the data and on the relationship with the multiple-scattering theories available.

27.2.1 *The heuristic model*

The deviations of the mobility from the expected classical behavior can be explained in terms of three main multiple-scattering effects that arise as a consequence of the fact that the wavelength of electrons becomes comparable with their mean free path and with the interatomic spacing. This situation occurs more easily at low temperature and at high density, but it is not necessarily limited to this case.

27.2.1.1 *Density-dependent energy shift* The first multiple-scattering effect is a shift of the ground-state energy of the electron above its thermal value. For small densities, the energy shift V_0 has been calculated by Fermi (1934) and is proportional to the density itself:

$$V_0(N) = \frac{2\pi\hbar^2}{m} Na\,, \tag{27.6}$$

where m is the electron mass and a is the scattering length for the electron–atom scattering. The sign of V_0 depends on the sign of the scattering length, which is positive for repulsive interactions and negative for attractive ones. For this reason, V_0 has to be considered as a shift of the total energy.

Springett *et al.* (1968) have shown that the total energy shift can be separated into two contributions of kinetic and potential energy:

$$V_0(N) = E_K(N) + U_P(N)\,. \tag{27.7}$$

$U_P(N)$ is a negative potential energy contribution due to the polarization of the surrounding gas. Being a sum of individual polarization potentials weighted by the statistical distribution of atoms, the leading dependence of U_P on N is $U_P \propto N^{4/3}$ (Fermi, 1934; Springett *et al.*, 1968; Hernandez and Martin, 1991), but it is not relevant in this context.

On the other hand, $E_K(N)$ is a zero-point kinetic energy contribution arising from excluding the electron from the hard-core volume of the atoms, and is intrinsically positive. In order to calculate it for s-wave states, density fluctuations are neglected, a Wigner–Seitz unit cell of radius r_s and volume

$$(4/3)\pi r_s^3 = N^{-1} \tag{27.8}$$

is assigned to each atom (Wigner and Seitz, 1933), and the gas is replaced by an average ordered fluid, in which electrons are described by Bloch-type wave functions (Hernandez and Martin, 1991).

The average translational symmetry requires that the electron wave function is invariant under a translation across the unit Wigner–Seitz cell:

$$\psi\,(\mathbf{r} + 2\mathbf{r}_s) = \psi(\mathbf{r})\,, \tag{27.9}$$

and leads to the following eigenvalue equation for the wave vector k_0 of the s-wave state:

$$\tan\left\{k_0\left[r_s - \tilde{a}\,(k_0)\right]\right\} = k_0 r_s\,, \tag{27.10}$$

where $\tilde{a}$ is the hard-core radius of the Hartree–Fock potential for rare gas atoms and is defined in terms of the total scattering cross-section σ_t by the relationship $\sigma_t = 4\pi\tilde{a}^2$ (Springett *et al.*, 1968; Borghesani *et al.*, 1988, 1992). For repulsive gases, such as He, $\tilde{a}$ is the usual scattering length a. Here k_0 depends on r_s by means of eqn (27.10) and, hence, on N through eqn (27.8).

The kinetic energy spectrum of this state of the propagating electron is given by (Wigner and Seitz, 1933)

$$\epsilon' = E_K + \epsilon = \frac{\hbar^2 k_0^2}{2m} + \frac{\hbar^2 k^2}{2m}\,. \tag{27.11}$$

This equation uniquely defines the kinetic energy shift $E_K(N)$ as

$$E_K(N) = \frac{\hbar^2 k_0^2}{2m}\,. \tag{27.12}$$

It should be recalled that it is the group velocity

$$v = \frac{\hbar k}{m} = \left\{\frac{2}{m}\left[\epsilon' - E_K(N)\right]\right\}^{1/2} \tag{27.13}$$

that contributes to the energy equipartition value arising from the gas temperature (Wannier, 1966) and to the electron propagation.

27.2.1.2 *Quantum self-interference* The second multiple-scattering effect considered by the model is a quantum self-interference effect occurring when the electron mean free path and its wavelength become comparable. This effect causes an enhancement of the electron backscattering because the electron wave packet scattered off atoms along paths connected by time-reversal symmetry may interfere with itself (Bergmann, 1983; Ascarelli, 1992).

The quantum self-interference effect correction to electron conductivity has been calculated within different theoretical frameworks, either by people of the gas plasma community (Yakubov, 1970, 1973; Atrazhev and Yakubov, 1977; Braglia and Dallacasa, 1978; Dallacasa, 1979; O'Malley, 1980, 1992; Iakubov and Polishuk, 1982; Yakubov and Polishuk, 1982), or by people of the disordered systems community (Abrahams *et al.*, 1979; Adams and Paalanen, 1987, 1988; Adams *et al.*, 1992; Herman *et al.*, 2001), or both (Polishuk, 1983*a*, 1983*b*, 1984).

In any case, when the electron wavelength at thermal energies is smaller than the mean free path ℓ, i.e., when $k\ell < 1$, where k is the electron wave vector, a perturbative treatment of the problem leads to an enhancement of the scattering rate ν that can be cast in the form (Atrazhev and Yakubov, 1977)

$$\nu(\epsilon) = \nu_0 \left(1 + 2\frac{\lambdabar}{\ell_0}\right), \tag{27.14}$$

where $\lambdabar = \lambda/2\pi$, $\lambda = h/(2m\epsilon)^{1/2}$ is the electron wavelength, and $\ell_0 = 1/N\sigma_{mt}$ is the electron classical mean free path. The scattering rate in the dilute gas limit ν_0 is related to the electron velocity v and to ℓ_0 by

$$\nu_0 = \frac{v}{\ell_0} = \left(\frac{2\epsilon}{m}\right)^{1/2} N\sigma_{mt}(\epsilon) . \tag{27.15}$$

At thermal energies, this enhancement factor of the scattering rate corresponds to a reduction of the mobility with respect to its classical value (Atrazhev and Yakubov, 1977):

$$\frac{\mu}{\mu_0} = 1 - cNT^{-1/2}, \tag{27.16}$$

where $c = \hbar/(2\pi m k_B)^{1/2}$.

In the jargon of the theory of disordered systems, this regime, for which $k\ell < 1$, is called *weak localization* (Adams and Paalanen, 1987). If the disorder is made increasingly stronger by increasing the gas density in such a way that the condition $k\ell < 1$ is no longer fulfilled, electrons become completely localized with exponentially decaying wave functions (Anderson, 1958). Strong disorder leads to the appearance of a mobility edge at finite energies (Mott, 1974). The position of the mobility edge corresponds to the Ioffe–Regel criterion for localization: $\lambda \approx \ell$ (Ioffe and Regel, 1960; Gee and Freeman, 1986).

In correspondence to the mobility edge, the scattering rate diverges (Polishuk, 1984). A very simple and intuitive argument can be given in order to obtain. in a self-consistent way, this result without summing to all orders the energy diagrams leading to eqn (27.14) (Neri, 1996). Equation (27.14) can be easily rewritten in the following way:

$$\nu = \nu_0 \left(1 + \frac{\hbar \nu_0}{\epsilon}\right) . \tag{27.17}$$

The correction factor in parentheses, however, should contain the actual scattering rate experienced by the electron, i.e., ν_0 should be replaced by ν in the parentheses:

$$\nu = \nu_0 \left(1 + \frac{\hbar \nu}{\epsilon}\right) . \tag{27.18}$$

Equation (27.18) can be easily solved for ν, yielding

$$\nu = \frac{\nu_0}{1 - \hbar\nu_0/\epsilon} = \frac{\nu_0}{1 - 2\lambda/\ell_0} ; \tag{27.19}$$

this is exactly the result obtained by Polishuk (1984) with a factor of $2\pi/3 \approx 2.094$ instead of 2. Equation (27.19) can be expanded for $\lambda/\ell_0 < 1$ to obtain eqn (27.14) again, but predicts a diverging scattering rate in correspondence to the energy ϵ_c of the mobility edge, defined as

$$\epsilon_c = \frac{2}{m} \left(\hbar \sigma_{mt} (\epsilon_c) N\right)^2 . \tag{27.20}$$

27.2.1.3 *Correlation among scatterers* The last multiple-scattering effect considered in the heuristic model is the correlation among scatterers (Lekner, 1968). At low temperature and high density, the electron wave function spans a region containing several atoms at once. The amplitude of the scattered wave is thus a coherent sum of partial contributions scattered off each individual atom.

The correlation among atoms, which is particularly strong near the critical point, leads to an enhancement of the scattering cross-section by a wave vector-dependent factor $\mathcal{F}$ given by (Lekner, 1968)

$$\mathcal{F}(k) = \frac{1}{4k^4} \int_0^{2k} q^3 S(q) \, \mathrm{d}q , \tag{27.21}$$

where $S(q)$ is the static structure factor (Goodstein, 1975).

Theories for small-angle X-ray scattering give the following expression for $S(q)$ (Fixman, 1960; Thomas and Schmidt, 1963; Stanley, 1971):

$$S(q) = \frac{S(0) + (qL)^2}{1 + (qL)^2} , \tag{27.22}$$

where $S(0)$ is the long-wavelength limit of the structure factor that is related to the gas isothermal compressibility χ_T by the relationship $S(0) = N\chi_T k_B T$.

The correlation length L is expressed by the equation

$$L^2 = 0.1 l^2 \left[S(0) - 1 \right] , \tag{27.23}$$

where l is the so-called *short-range correlation length*, a few Å long.

By substituting eqn (27.22) into eqn (27.21), the enhancement factor due to correlations can be written as (Borghesani and Santini, 1992)

$$\mathcal{F}(k) = 1 + \left[S(0) - 1 \right] \left\{ \frac{(2kL)^2 - \ln\left[1 + (2kL)^2\right]}{8(kL)^4} \right\} . \tag{27.24}$$

By recalling that the energy ϵ is related to the wave vector k by the relationship $\epsilon = [\hbar^2/(2mL^2)](kL)^2$, the enhancement factor can be easily expressed in terms of the electron energy as

$$\mathcal{F}(\epsilon) = 1 + \left[S(0) - 1 \right] \left\{ \frac{\left(8mL^2/\hbar^2\right)\epsilon - \ln\left[1 + \left(8mL^2/\hbar^2\right)\epsilon\right]}{8\left[\left(2mL^2/\hbar^2\right)\epsilon\right]^2} \right\} . \tag{27.25}$$

At low energies, $\lim_{\epsilon \to 0} \mathcal{F}(\epsilon) = S(0)$ means that scattering is strongly affected by the gas fluctuations, whereas at larger energies $\mathcal{F} \to 1$, so that correlations can be neglected.

27.2.1.4 *The equations of the heuristic model* Once the three multiple-scattering effects are specified, it is a straightforward job to implement the modified kinetic equations for the electron mobility. The mobility is simply given by the kinetic formulas (27.3) and (27.4), provided that the momentum-transfer scattering cross-section $\sigma_{mt}(\epsilon)$ is replaced everywhere by the effective cross-section dressed by the multiple-scattering effects previously described (Borghesani and Santini, 1992; Borghesani, 2006):

$$\sigma^{\star}_{mt}\left(\epsilon'\right) = \sigma_{mt}\left(\epsilon'\right)\mathcal{F}\left(\epsilon'\right)\left[1 + \frac{2\hbar N \mathcal{F}\left(\epsilon'\right)\sigma_{mt}\left(\epsilon'\right)}{\left(2m\epsilon'\right)^{1/2}}\right] , \tag{27.26}$$

where $\epsilon' = \epsilon + E_K(N)$ is the shifted kinetic energy.

This heuristic approach describes the deviations of the mobility from the classical prediction in the noble gases, irrespective of the sign of their scattering length, without any adjustable parameters and within a unique framework.

The three multiple-scattering effects have different weights in different gases because the cross-sections have different values and energy dependencies. For instance, in Ne σ_{mt} is small but increases very rapidly with the electron energy. Thus, most of the observed effect on the mobility comes from the density-dependent shift of the energy. In contrast, the cross-section of He is large and practically independent of energy, and the most relevant contribution to the deviation from the classical behavior stems from the quantum self-interference effect. In Ar, the cross-section is large and very strongly dependent on the energy, so that all the effects are equally important (Borghesani and Santini, 1992). In particular, in He the ratio λ/ℓ may become large enough so as to lead to the appearance of a mobility edge, as will be discussed later in this chapter.

This heuristic model allows a unique description of the mobility of excess electrons in a dense rare gas that is independent of the nature of the gas itself. The peculiar features of each type of gas are embedded in the specific energy dependence of the scattering cross-section.

In order to show the generality of this heuristic approach, the predictions of the model are compared with the mobility data measured in different dense, rare gases. In Fig. 27.14 we compare the model with the experimental results obtained in dense He gas at $T = 26.1\,\mathrm{K}$ (Borghesani and Santini, 2002). The case of Ne is illustrated in Fig. 27.15 for $T \approx 46.5\,\mathrm{K}$ (Borghesani *et al.*, 1988). Finally, the experimental zero-field, density-normalized mobility measured in dense Ar gas at $T = 162.7\,\mathrm{K}$ (Borghesani *et al.*, 1992) is presented in Fig. 27.16. The prediction of the heuristic model is always represented by the solid line.

The agreement of this model with experiment is very good. The behavior of the zero-field mobility in the low and intermediate density range is reproduced with great accuracy. It has also been proved that it also reproduces the electric field dependence of the mobility very well (Borghesani *et al.*, 1988; Borghesani and Santini, 1990*a*, 1992, 1994*a*, 1994*b*, 2002; Borghesani, 2001), including the superlinearities observed in He (Levine and Sanders, 1967; Schwarz, 1980).

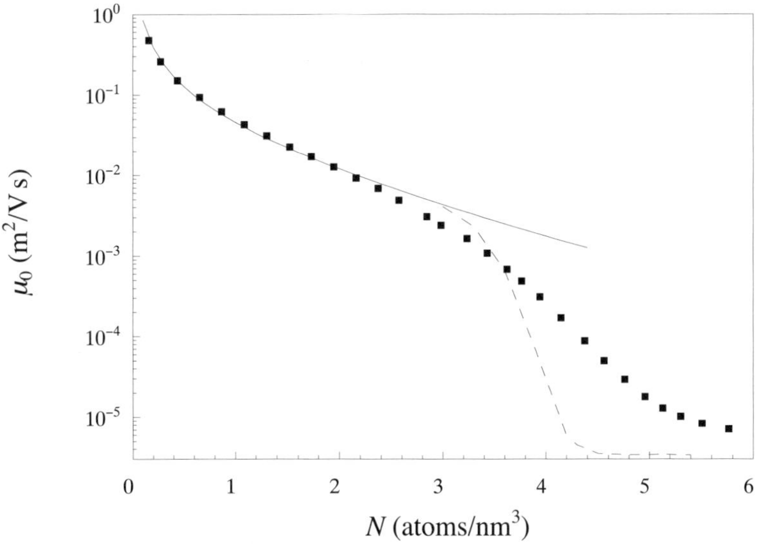

FIG. 27.14. μ_0 vs N in He for $T = 26.1\,\mathrm{K}$. (Borghesani and Santini, 2002.) Solid line: heuristic model with the self-interference factor calculated according to O'Malley (1980). Dashed line: optimum fluctuation model.

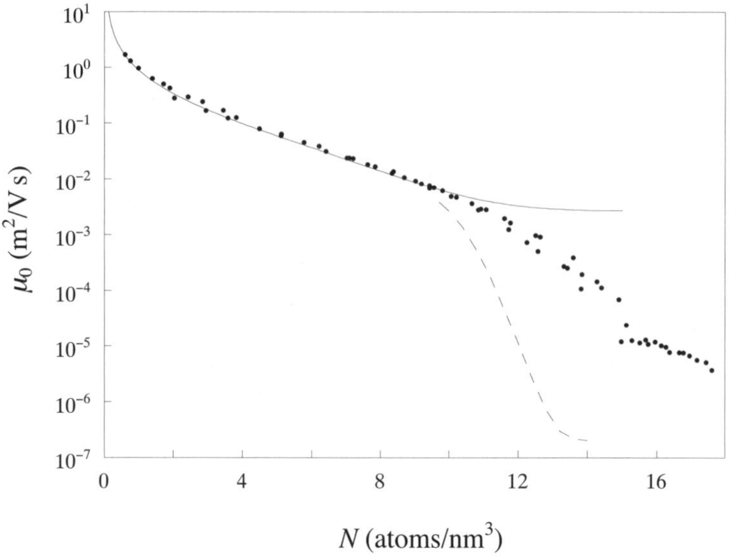

FIG. 27.15. μ_0 vs N in Ne for $T \approx 46.5\,\mathrm{K}$. (Borghesani and Santini, 1990*a*.) Solid line: heuristic model. Dashed line: optimum fluctuation model.

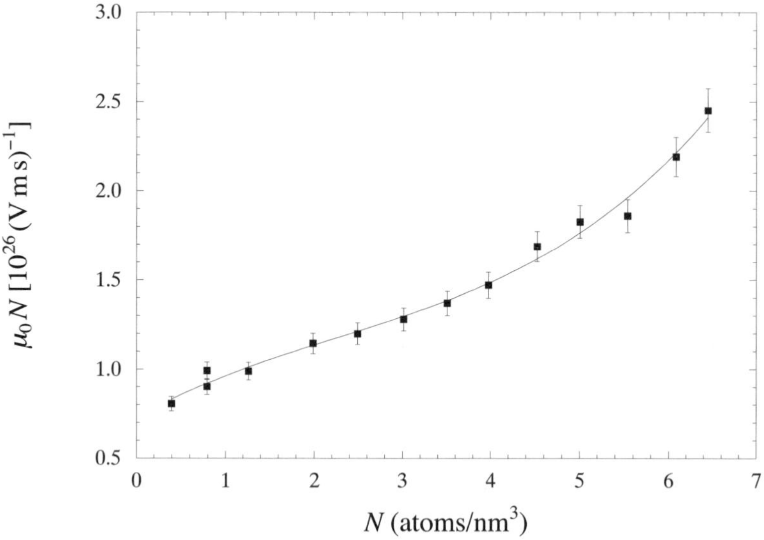

FIG. 27.16. $\mu_0 N$ vs N in Ar for $T = 162.7\,\mathrm{K}$. (Borghesani *et al.*, 1992.) Line: heuristic model.

In He and Ne the data depart from the prediction of the heuristic model at a density that approximately corresponds to the concavity change of the mobility curve, i.e., at the density of the transition between the two transport regimes of extended and localized electron states. The high-density region has thus to be treated in a different way.

27.3 Mobility at high density: optimum fluctuation model

If one calculates the free-energy change $\Delta\mathcal{A}$ between the extended and the localized states within the simple square-well model according to eqn (3.21), or with more sophisticated self-consistent models (Hernandez, 1991), one finds that, for each given temperature, there is a threshold density below which no localized states exist. Above this density, the measured mobility is a weighted sum of the mobility of the extended and localized states (Levine and Sanders, 1967; Young, 1970; Schwarz and Prasad, 1976):

$$\mu = \frac{n_f \mu_f + n_b \mu_b}{n_f + n_b}, \tag{27.27}$$

where μ_f is the mobility of quasi-free electrons, i.e., the mobility calculated in the previous section of this chapter, whereas μ_b is the mobility of the electron bubbles, and n_f and n_b are the relative populations of quasi-free and self-trapped electrons.

μ_b is not known, but is surely very much smaller than μ_f. For this reason, it is sufficient to give even a rough estimate of it. At the density at which self-trapping occurs, the mean free path for atom–atom collisions is $\ell_g \approx 10\,\text{Å}$, of the same order of magnitude as the radius of the electron bubble determined in the experiments in liquid He. Thus, neither kinetic theory nor classical hydrodynamics can describe μ_b.

A useful interpolation formula is given by Tyndall (1938):

$$\mu_b = \frac{e}{6\pi\eta R}\left[1 + \frac{9\pi\eta}{4NR\left(2\pi M k_{\mathrm{B}} T\right)^2}\right], \tag{27.28}$$

where η is the viscosity, R is the radius of the electron bubble, and M is the mass of one atom of the gas. This formula interpolates between the kinetic and the hydrodynamic regime.

The relative populations of the two species of electrons, quasi-free and self-trapped, are calculated by means of the *optimum fluctuation method* (Lifsits, 1968), i.e., the equilibrium ratio is given by

$$n_b/n_f = \exp\left[\left(-\Delta\mathcal{A}\right)_{\min}/k_{\mathrm{B}}T\right], \tag{27.29}$$

where $(\Delta\mathcal{A})_{\min}$ is the value of the free energy minimized with respect to the bubble radius at constant N and T.

The results of the optimum fluctuation model at low temperature are plotted as a dashed line in Fig. 27.4. It is clear that the qualitative features of the mobility are reproduced, but the transition to the low-mobility regime is far too rapid because the transition region is not described well by this model (Levine and Sanders, 1967; Harrison *et al.*, 1973; Schwarz, 1980).

An attempt at improving the description of the mobility in the transition region is made with the introduction of the model of a *partially-filled bubble*, or *pseudobubble* (Borghesani and Santini, 1990*a*, 2002), to account for the fact that in the gas phase there is a lack of a well-defined interface separating the bubble interior from the outside environment and that the higher thermal energies in the gas experiments with respect to the experiment in the liquid may allow atoms to penetrate the cavity.

It is assumed that the simple square well approximates the physical problem (Hernandez, 1991), but it is also assumed that the interior of the bubble is partially filled with atoms with local density N_i. The filling factor is $F = N_i/N$, where N is the bulk gas density.

The Schrődinger equation is to be solved in order to find the energy eigenvalue of the electron in the ground-state in the bubble. If the bubble was empty, the well would be $V_0(N)$ deep, where $V_0(N)$ is the energy of the delocalized electrons in the bulk gas. As the bubble is now partially filled, the potential energy to be used must account for the fact that the presence of He atoms in the bubble offsets the ground-state energy by an amount $V_i = V_0(FN)$, so that the net potential well is now $V_0(N) - V_i$. Here $V_0(N)$ can be safely approximated by $E_K(N)$.

The polarization contribution to the potential energy can be written, to leading order in F, as

$$\mathcal{E}_P = -\frac{\alpha e^2}{2(4\pi\epsilon_0)^2 R}(1-F)N\,, \tag{27.30}$$

where R is the bubble radius.

Now, the Schrődinger equation is solved as usual, but now the ground-state eigenvalue $\mathcal{E}_1$ also depends on the filling fraction F in addition to N, R, and T.

The free-energy change is calculated as (Borghesani and Santini, 1990*a*, 2002)

$$\Delta\mathcal{A} = \mathcal{E}_1 + V_0(FN) + \mathcal{E}_P + \mathcal{W} - V_0(N)\,, \tag{27.31}$$

where

$$\mathcal{W} = \frac{4\pi}{3}R^3 P\left[1 - F - \frac{FN}{P}\int\limits_{NF}^{N}\frac{P(n)}{n^2}\,\mathrm{d}n\right] \tag{27.32}$$

is the mechanical work spent in expanding the bubble. $P(N)$ is the gas pressure given by the equation of state as a function of the density. If the bubble was empty, the work would have simply been $(4\pi/3)PR^3$.

The free-energy change now has to be minimized with respect to R and F at constant N and T. Its minimum value is used for calculating the equilibrium populations of quasi-free and self-trapped electrons and, hence, the mobility. The results are plotted as dashed lines in Fig. 27.14 for He at $T = 26.1\,\mathrm{K}$ (Borghesani

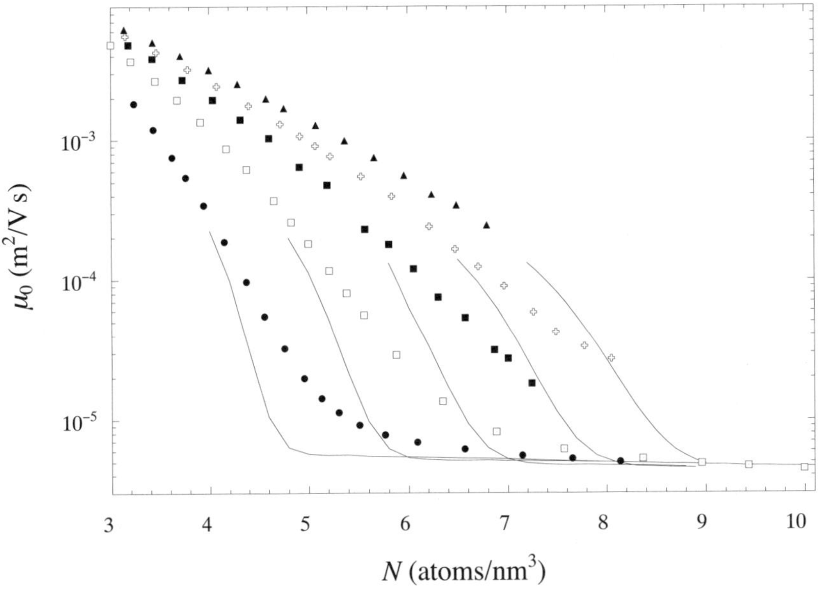

FIG. 27.17. μ_0 vs N in He for T = 26.1, 34.5, 45.0, 54.5, and 64.4 K (from left to right). (Borghesani and Santini, 2002.) Lines: optimum fluctuation model.

and Santini, 2002) and in Fig. 27.15 for Ne at $T \approx 46.5$ K (Borghesani and Santini, 1990*a*).

The transition to the low-mobility regime is a bit smoother than in the empty bubble case shown in Fig. 27.4, probably also because the temperatures of the experiments reported in Figs 27.14 and 27.15 are higher than the temperature of the experiment of Levine and Sanders (1967). In any case, the transition is so rapid because the optimum fluctuation method gives the most probable bubble state, whereas there may be a continuous distribution of bubbles of different radii and filling fractions (Borghesani and Santini, 2002).

In spite of its crudeness, the partially-filled bubble model performs an excellent job at predicting the shift of the localization transition to higher densities as the temperature is increased. If attention is focused only on the region in which self-trapped states dominate, it can be noted by inspecting Fig. 27.17 that the prediction of the partially-filled bubbles tracks quite accurately the behavior of the mobility as a function of T (Borghesani and Santini, 2002).

27.4 Mobility edge and percolation

27.4.1 *Mobility edge*

As previously observed, the quantum self-interference effect due to multiple scattering leads to a correction to the scattering rate that is related linearly to the density as expressed by eqn (27.14) (Atrazhev and Yakubov, 1977). Similar results, with a factor of $2\pi/3$ instead of 2 in eqn (27.14), have also been obtained by different methods (Polishuk, 1984; Kirkpatrick and Belitz, 1986).

The actual scattering rate after all diagrams have been summed up to all order of perturbations is given by eqn (27.19), which is valid both in the low-density limit (weak localization) as well as in the limit of strong localization at high density (Polishuk, 1984).

At the so-called *mobility edge*, the scattering rate diverges and the electron no longer propagates. The mobility edge ϵ_c is defined as the energy threshold below which no conduction takes place. In the present case, it is obtained by equating to zero the denominator of eqn (27.19), yielding eqn (27.20). For this value of the electron energy, the Ioffe–Regel criterion for the localization, $\lambda \approx \ell$, is fulfilled, i.e., the mean free path and the wavelength of electrons are comparable.

Electron states below the mobility edge are not mobile and do not contribute to the measured mobility. In order to take into account this restriction on the electron population that contributes to the mobility, the integral in eqn (27.1) must be restricted to energies above ϵ_c.

If correlations among scatterers are neglected (a good assumption unless the critical region is explored), and if the density-dependent energy shift of the electron ground-state energy is disregarded (this is a fairly reasonable assumption for He, whose cross-sections are weakly dependent on the electron energy, but it is a very bad approximation for Ne because its cross-sections depend strongly on the energy), then the ratio of the actual to the classical mobility can be written as (Polishuk, 1984; Adams *et al.*, 1992)

$$\frac{\mu_0}{(\mu_0)_{\mathrm{cl}}} = \exp\left(-y_c^2\right) - \frac{\sqrt{\pi}}{2} y_c \,\mathrm{erfc}\,(y_c) \,, \tag{27.33}$$

where erfc is the complementary error function (Bevington, 1969). y_c is given by

$$y_c = \left(\frac{2\pi}{3}\right) \frac{\hbar}{\sqrt{2mk_{\mathrm{B}}T}} N\sigma_0 = \left(\frac{1}{3}\right) \frac{\lambda_T}{\ell_0} \,, \tag{27.34}$$

where σ_0 is the scattering cross-section, which is assumed to be constant for He, and $\lambda_T = \hbar/\sqrt{2mk_{\mathrm{B}}T}$ is the electron thermal wavelength. If the quantum self-interference factor is written as $(1 - f\lambda/\ell_0)^{-1}$ with $f = 2$ (Atrazhev and Yakubov, 1977) or $2\pi/3$ (Polishuk, 1984), y_c can be interpreted as the square root of the mobility-edge energy divided by the thermal energy:

$$y_c = \frac{f}{2} \left(\frac{\epsilon_c}{k_{\mathrm{B}}T}\right)^{1/2} . \tag{27.35}$$

If the correlations among scatterers and the energy shift have to be accounted for, it is easy to show that the mobility edge has to be calculated by self-consistently solving the following equation (Neri, 1996):

$$\epsilon_c = \frac{1}{2m} \left\{ f\hbar S(0)\sigma_{mt} \left[\epsilon_c + E_K(N)\right] N \right\}^2 . \tag{27.36}$$

Finally, the density-normalized zero.field mobility is given by (Neri, 1996)

$$\mu_0 N = \frac{4e}{3\,(2\pi m)^{1/2}\,(k_{\mathrm{B}}T)^{5/2}} \int_{\epsilon_c}^{\infty} \epsilon \frac{1-(\epsilon_c/\epsilon)^{1/2}}{S(0)\sigma_{mt}\left[\epsilon + E_K(N)\right]} e^{-\epsilon/k_{\mathrm{B}}T}\,\mathrm{d}\epsilon\,. \qquad (27.37)$$

At the temperatures of the experiments in He gas, it is sufficient to consider only the long-wavelength limit $S(0)$ of the static structure factor that happens to be slightly less than 1 (Rabinovich *et al.*, 1988).

In Fig. 27.18 the results of the mobility-edge predictions (eqns (27.37) and (27.36)) are compared with the experimental data obtained in He for $T = 34.5\,\mathrm{K}$ (De Riva, 1993; Neri, 1996; Borghesani and Santini, 2002). Curve 1 is obtained by setting $f = 2\pi/3$, $S(0) = 1$, and $E_K = 0$. The predicted mobility is much lower than the experimental data because the mobility edge is $\propto S(0)^2$, yielding too large a lower integration limit in eqn (27.36), so that too many extended states are not counted properly.

Curve 2 is obtained by using the correct value of $S(0)$, by introducing the energy shift E_K, and by using the Polishuk value of the factor $f = 2\pi/3$. This curve is indistinguishable from the prediction of Polishuk (1984), eqn (27.33). The practically identical results obtained with two different formulas are due to two competing effects. On one hand, the use of the actual thermodynamical value of $S(0) < 1$ (Rabinovich *et al.*, 1988) reduces the mobility edge, thereby

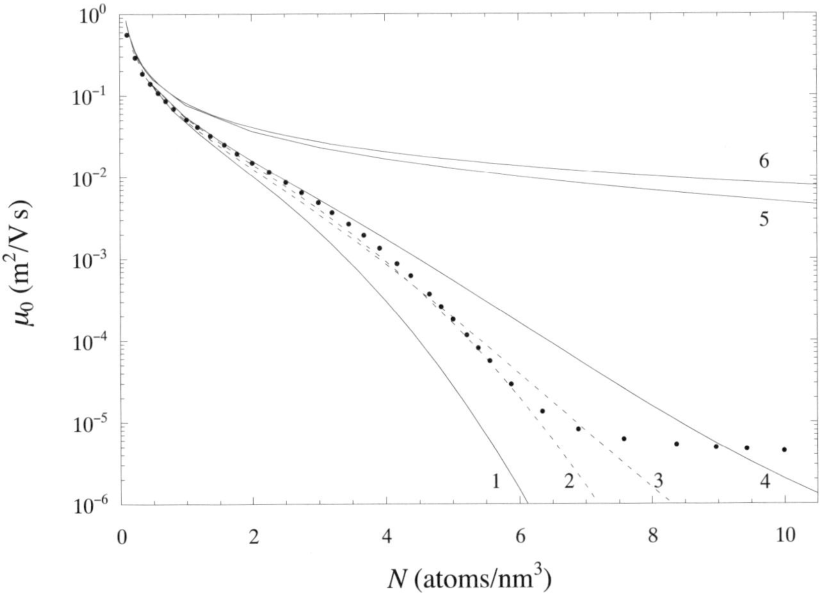

FIG. 27.18. μ_0 vs N in He for $T = 34.5\,\mathrm{K}$. (Neri, 1996, 2002.) The meaning of the curves is explained in the text.

increasing the contribution of the highly-mobile extended states. On the other hand, the scattering cross-section of He has to be evaluated at the shifted energy $\epsilon + E_K(N)$. As σ_{mt} increases slightly with energy, its value at the shifted energy is larger than that at the unshifted energy, thus leading to an increase of the mobility edge and to a reduction of the contribution of the mobile states. As a result, the two effects practically cancel each other.

The sensitivity of the results on the value chosen for the factor f can be perceived by inspecting curve 3, which is calculated with the same parameters as curve 2, except for $f = 2$. The mobility edge turns out to be a little bit smaller than for $f = 2\pi/3$ and the contribution of the mobile states is larger.

The importance of the energy shift can be grasped by observing curve 4, which is calculated by setting $E_K(N) = 0$. In spite of its weakness, the energy dependence of the scattering cross-section of He is large enough to produce observable effects. If the energy shift is not accounted for, the mobility edge computed according to eqn (27.36) is too small and the contribution of the extended states is overestimated.

Finally, curve 5 is calculated by setting $\epsilon_c = 0$ while retaining the weak-localization correction factor for the scattering rate (eqn 27.19), and curve 6 is the classical prediction without multiple-scattering effects. The analysis of the data at different temperatures (Polishuk, 1984; Adams *et al.*, 1992) yields similar agreement.

It should be noted that the mobility-edge theory predicts that the electron mobility vanishes at high enough density, whereas even the most deeply-localized electrons, whether they are trapped in density fluctuations, in bubbles, or whether they are localized in the Anderson sense, will have a finite mobility owing to the compliance of the environment and to the gas atoms' diffusion (Adams *et al.*, 1992), thus accounting for the observed saturation of the drift mobility at high density.

It should also be pointed out that mobility-edge theory does not work in the case of Ne (Borghesani and Santini, 1990*a*, 1992), although the physics is the same as in the case of He. In Ne the cross-section is far too small to give a large enough mobility edge. In spite of this, the transition to a low-mobility regime in Ne is a proven experimental fact and is phenomenologically very similar to what occurs in He. Thus, the Ne data cast some doubt on the validity of the mobility-edge approach.

27.4.2 *Percolation*

The problem of electron transport in dense He gas can be considered as a practical realization of the more general problem of the study of the electronic states in a disordered medium. In fact, He gas represents a simple example of a three-dimensional random medium.

Models of disordered systems (Mott, 1967; Cohen *et al.*, 1969; Economou and Cohen, 1970; Cohen, 1973) have shown that the energy eigenvalues are grouped into bands whose edges are smeared out, in contrast with what happens in crystalline materials. Inside each band there exist critical energies ϵ_c and ϵ'_c (see Fig. 27.19) that separate electronic states of different nature (Eggarter, 1972). Those states whose energy lies between ϵ_c and ϵ'_c are extended, whereas those in the tails of the density of states are localized. Electrons with energies outside the central portion of the density of state of Fig. 27.19 have vanishing mobility.

If the disorder is increased, the mobility edges approach the band center, leading eventually to the disappearance of the region of extended states. The electron wave function is localized and its amplitude decreases exponentially. This is the celebrated *Anderson transition* (Anderson, 1958). It is assumed that localization can also occur in the case of structural disorder (Hernandez, 1982).

The actual potential experienced by an electron in He gas closely resembles that of a gas of randomly-placed hard spheres. The average distance between the sphere is changed by changing the gas density, and the gas is considered either dilute or dense, depending on the extension of the electron wave function with respect to the average interatomic distance. The gas is considered dense if the electron wave functions overlap many scattering centers at once.

Eggarter has addressed the problem of calculating the electronic density of state in this type of disordered medium (Eggarter and Cohen, 1970, 1971; Eggarter, 1972). The transport of electrons is treated as a semi-classical percolation problem. This view has been recently supported by path-integral Monte Carlo studies (Coker *et al.*, 1987; Laria and Chandler, 1987).

The semi-classical model is built by replacing the set of randomly-located

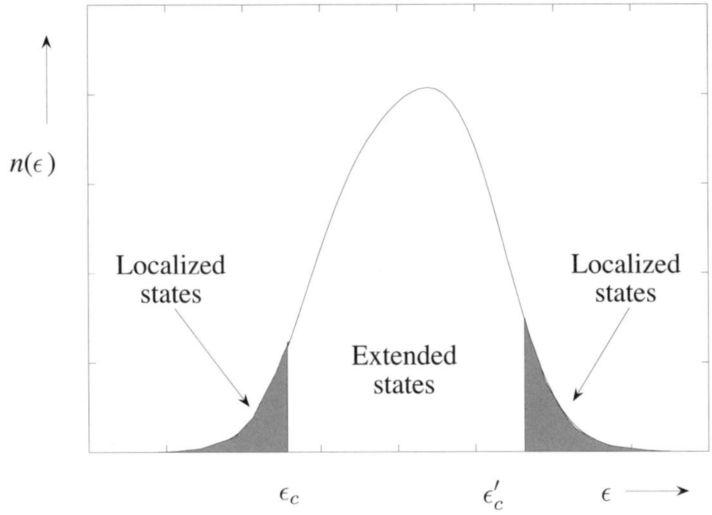

FIG. 27.19. Sketch of the electronic density of states in a disordered medium. ϵ_c and ϵ'_c are the mobility edges that separate the region of extended states from those of localized states.

hard spheres by an equivalent, random, smooth effective potential. Then, statistical fluctuations in the average properties, arising from statistical fluctuations in the density of the scatterers, lead to the appearance of a mobility edge that is intimately related to the percolation threshold. The appearance of localized states is the result of unusually deep fluctuations in the random effective potential.

In order to transform the hard-sphere potential into an equivalent one in which electrons move in a smooth effective potential, the gas is divided into boxes, or cells, of side L. This sampling length is the crucial point of the model. The electron samples the scatterer distribution within such a box.

An estimate of the spatial extent of the wave function of an electron of energy ϵ can be obtained by invoking the uncertainty principle:

$$\Delta x \geqslant \frac{\hbar}{2\Delta p_x} = \frac{\sqrt{3}\hbar}{2\,(2m\epsilon)^{1/2}} \propto \lambda\,, \tag{27.38}$$

where $\lambda = h/\sqrt{2m\epsilon}$ is the electron wavelength and where the relationships $\langle p^2 \rangle = 2m\epsilon$ and $\Delta p_x = \sqrt{\langle p_x^2 \rangle}$ are used.

For electrons of low enough energy, $\Delta x \gg r_s$, where r_s is the radius of the Wigner–Seitz cell. In this case, electrons interact with many scatterers simultaneously. The average potential experienced by the electron is thus given by the local, density-dependent Wigner–Seitz energy, eqn (27.12), $V(N) \approx E_K(N)$.[6] In this way, the effects of higher-order multiple-scattering corrections to the electron energy are taken into approximate account (Schwarz, 1980). More refined calculations of the average potential use a better description of the electron wave functions (Simon *et al.*, 1990) without modifying the general features of the problem at hand, but with much poorer final results for the mobility.

As the gas density fluctuates from box to box, so does the potential. The sampling length L sets the scale of the autocorrelation of the potential.

The calculation of the density of states proceeds via quasi-classical counting. It is assumed that within each box of side L the local electron density of states $n_i(\epsilon)$ is that pertaining to a free particle with energy ϵ above the average Wigner–Seitz energy $V_i \equiv V(N_i)$, where N_i is the local gas density in the box:

$$n_i(\epsilon) = \frac{L^3}{2\pi^2}\left(\frac{2m}{\hbar^2}\right)^{3/2} (\epsilon - V_i)^{1/2} \qquad \text{for } \epsilon \geqslant V_i\,. \tag{27.39}$$

It should, however, be recalled that this way of computing the local density of states has given origin to strong criticism (Hernandez and Ziman, 1973; Eggarter and Cohen, 1974).

The total density of states is obtained by carrying out a weighted sum over all cells. If the cell side L is large enough, the number M_i of atoms in the cell is a stochastic variable normally distributed about the average value

[6] For He, the polarization contribution U_P to the ground-state energy of the electron can be safely neglected (Springett *et al.*, 1968).

$$M = L^3 N\,, \tag{27.40}$$

where N is the average gas density, with variance

$$\sigma_M^2 = MS(0)\,, \tag{27.41}$$

where $S(0)$ is the long-wavelength limit of the static structure factor that accounts for the non-ideal behavior of the gas. As a consequence, the potentials inside the box are also described by a Gaussian distribution with variance

$$\sigma_V^2 = \frac{NS(0)}{L^3}\left(\frac{\partial V}{\partial N}\right)^2 . \tag{27.42}$$

Thus, the ensemble-averaged density of states is given by

$$n(\epsilon) = \frac{1}{2\pi^2}\left(\frac{2m}{\hbar^2}\right)^{3/2} \frac{1}{(2\pi)^{1/2}\sigma_V} \int (\epsilon - \tilde{V})^{1/2} e^{-(\tilde{V}-V)^2/2\sigma_V^2}\,\mathrm{d}\tilde{V}\,. \tag{27.43}$$

After changing variables by setting $z = (\epsilon - \tilde{V})/\sigma_V$ and $x = (\epsilon - V)/\sigma_V$, eqn (27.43) can be suitably rewritten as

$$n(\epsilon) = \frac{1}{2\pi^2}\left(\frac{2m}{\hbar^2}\right)^{3/2} \sigma_V^{1/2}\mathcal{F}(x,z)\,, \tag{27.44}$$

with

$$\mathcal{F}(x,z) = \frac{1}{(2\pi)^{1/2}} \int_0^\infty z^{1/2} e^{-(x-z)^2/2}\,\mathrm{d}z\,. \tag{27.45}$$

It is worth recalling that $\tilde{V}$ is a stochastic variable with average value V.

The procedure of averaging over the cells gives origin to a density of states with a low-energy tail. The states located deeply in the tail, $x \ll 0$ or, equivalently, $\epsilon \ll V$, come from those cells in which the local density and potential are much lower than average. The surrounding cells act as a nearly-impenetrable barrier for the electron wave function that is thus effectively trapped. In this way, space is divided into *allowed* and *forbidden* regions.

According to the model (Eggarter and Cohen, 1970), at some threshold energy, the allowed regions start to percolate, i.e., they form a connected conductive path for electrons, in what is known as a *lake-to-ocean percolation transition* (Simon *et al.*, 1990). Electrons may, or may not, percolate through this fluctuating landscape, depending on how strong the fluctuations are with respect to the electron energy. A measure of the strength of the fluctuating potential is given by its variance σ_V^2.

If the electron energy is above the *percolation threshold*, electrons percolate from cell to cell and thus propagate with high mobility. In contrast, if their energy is below the percolation threshold, they remain localized and have a low mobility allowing for atomic motion. The percolation threshold ϵ_c is identified with the mobility edge of the previous section and is also naturally associated with the energy of the bottom of the conduction band (Zallen and Scher, 1971; Simon *et al.*, 1990).

In order to calculate the percolation threshold, space is divided into accessible and forbidden regions depending on the condition that the electron energy is larger or smaller than the local Wigner–Seitz energy. The fraction of space available to electrons is simply given by

$$\phi(\epsilon) = \int_{-\infty}^{x} e^{-w^2/2}\, \mathrm{d}w = \frac{1}{2}\left[1 + \mathrm{erf}\left(\frac{x}{\sqrt{2}}\right)\right] \qquad \text{for } x = (\epsilon - V)/\sigma_V\,, \tag{27.46}$$

which is the probability that a randomly-chosen point belongs to an accessible region. For a simple cubic lattice of side L, $\phi(\epsilon_c) = 0.30$, but the choice of the type of lattice is irrelevant (Eggarter and Cohen, 1970).

Equation (27.46) can be inverted so as to give

$$\epsilon_c \simeq V - 0.52\sigma_V\,(\epsilon_c)\,. \tag{27.47}$$

It is clear that the value of the mobility edge strongly depends on the sampling length L through σ_V, as can be seen from eqn (27.42).

The mobility of the electrons is given by a weighted average over percolating and non-percolating states:

$$\mu = \frac{1}{Q}\int \mu(\epsilon) n(\epsilon) e^{-\epsilon/k_{\mathrm{B}}T}\, \mathrm{d}\epsilon\,, \tag{27.48}$$

where

$$Q = \int n(\epsilon) e^{-\epsilon/k_{\mathrm{B}}T}\, \mathrm{d}\epsilon \tag{27.49}$$

is the partition function.

The energy-dependent mobility of the free electrons is assumed to be given by the Drude result (Ashcroft and Mermin, 1976):

$$\mu_f(\epsilon) = \frac{2e}{3m}\tau(\epsilon)\,, \tag{27.50}$$

where $\tau(\epsilon)$ is the mean free time between collisions and is expressed in terms of the electron mean free path $\ell(\epsilon)$ and velocity v as (Wannier, 1966)

$$\tau(\epsilon) = \frac{\ell(\epsilon)}{v}\,. \tag{27.51}$$

This last result is strictly valid for hard spheres, but is also a very good approximation for He (Eggarter and Cohen, 1970) and Ne (Borghesani and Santini, 1992).

Two scattering mechanisms determine the electron mean free path. The first one is the usual electron–atom scattering described by the momentum-transfer scattering cross-section σ_{mt} that yields the mean free path ℓ_c given by

$$\ell_c^{-1}(\epsilon) = N\sigma_{mt}(\epsilon)\,, \tag{27.52}$$

where the energy ϵ incorporates the energy shift V and the multiple-scattering effects described in Section 27.2 (Borghesani *et al.*, 1988, 1992; Borghesani and Santini, 1992.)

The second scattering mechanism is scattering off prohibited "islands". The resulting mean free path is

$$\ell_p = L\phi(\epsilon)\left[1 - \phi(\epsilon)\right]^{-1}\,. \tag{27.53}$$

The total mean free path ℓ is given by

$$\ell^{-1}(\epsilon) = \ell_c^{-1} + \ell_p^{-1} = \frac{1 - \phi + LN\sigma_{mt}\phi}{L\phi}\,. \tag{27.54}$$

The inverse velocity $1/v$ to be used in eqn (27.51) must be replaced by an average of the inverse group velocity $1/v_g = [m/2(\epsilon - V)]^{1/2}$ over the allowed regions:

$$\langle v^{-1}\rangle = \frac{1}{\phi(\epsilon)}\int\limits_{\epsilon\geqslant V}\left(\frac{1}{v}\right)\frac{1}{\sqrt{2\pi}\sigma_V}e^{-(\tilde{V}-V)^2/2\sigma_V^2}\,\mathrm{d}\tilde{V}\,. \tag{27.55}$$

The final expression for the mean time between collisions is

$$\tau(\epsilon) = \left(\frac{2m}{\sigma_V}\right)^{1/2} p(\epsilon)\frac{\ell(\epsilon)}{\phi(\epsilon)}\left(\frac{\mathrm{d}\mathcal{F}(x)}{\mathrm{d}x}\right), \tag{27.56}$$

where the factor $p(\epsilon)$ is the conditional probability that a point belongs to an allowed and unbounded region, and describes the fact that an electron, even though in an allowed region, may not propagate unless this region spans the whole volume.

The choice of p is arbitrary. Eggarter and Cohen (1970) have chosen explicitly

$$p(\epsilon) = \begin{cases} 0\,, & \epsilon \leqslant \epsilon_c\,,\\ 1 - \exp\left\{-25\left[\phi\left(\epsilon\right) - \phi\left(\epsilon_c\right)\right]\right\}, & \epsilon > \epsilon_c\,. \end{cases} \tag{27.57}$$

In order to account for the gas compliance, the localized states are given a finite mobility μ_b using the Stokes–Cunningham interpolation formula (27.28). Finally, the energy-dependent mobility to insert into eqn (27.48) is obtained as

$$\mu(\epsilon) = \left[1 - p\left(\epsilon\right)\right]\mu_b + p(\epsilon)\mu_f(\epsilon)\,. \tag{27.58}$$

In the original papers (Eggarter and Cohen, 1970, 1971) the electron wavelength $\lambda = h/\sqrt{2m\epsilon}$ is chosen as the scale for the sampling length, $L = c\lambda$. Here $c \sim 1$ is an adjustable parameter of the model. It is clear that the larger the fluctuations are, the smaller is the sampling length.

In Fig. 27.20 the Eggarter model is compared to the most accurate mobility data for $T = 4.2\,\mathrm{K}$ (Schwarz, 1980). Curves 1 to 4 are the results of the Eggarter model with $c = 0.8$, 1.0, 1.2, and 1.4, respectively (Schwarz, 1980). A value of $c \sim 1$ gives a good agreement with the data and the smooth transition between the high- and low-mobility regimes is described quite accurately. A slight change with T of the value of c also allows a good description of the data of Harrison *et al.* (1973) to be obtained up to approximately 18 K. The results of the percolation model of Simon *et al.* (1990) are not reported because their agreement with the data is much worse.

In the original percolation model (Eggarter and Cohen, 1970) the mobility of the quasi-free electrons did not yet include the multiple-scattering corrections for obvious historical reasons. As the explored density range in the low-temperature region is not very wide, these corrections are not too important and the final results are not spoiled, albeit they have not been included.

At higher temperatures, however, the transition to the low-mobility regime occurs at much larger densities than at low T, and the multiple-scattering corrections to the mobility of the quasi-free electrons must be treated more carefully (De Riva, 1993; Borghesani and Santini, 2002). In He, the main correction is due to the quantum self-interference effect, which is treated by multiplying μ_f by the factor $\exp(-2N\sigma_{mt}\lambda_T)$ (O'Malley, 1980).

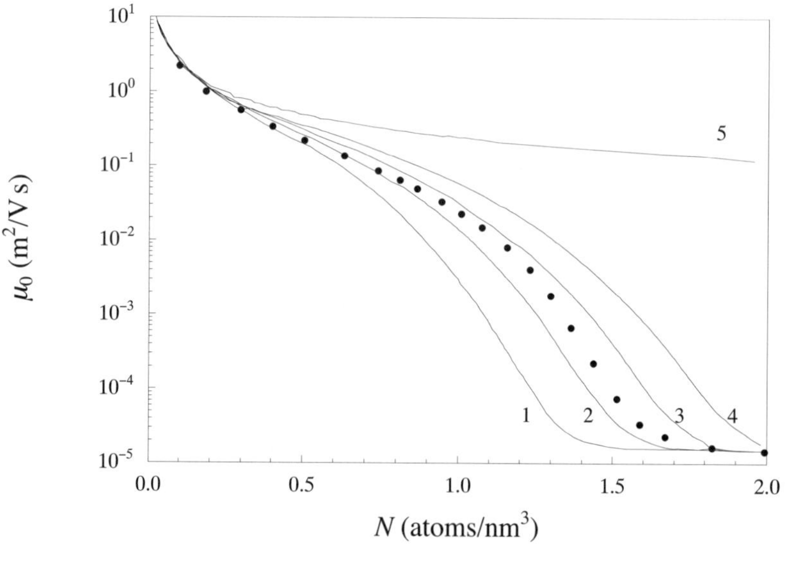

FIG. 27.20. μ_0 vs N in He for $T = 4.2\,\mathrm{K}$. (Schwarz, 1980.) Curves 1 to 4: Eggarter model with adjustable parameter $c = 0.8$, 1.0, 1.2, and 1.4, respectively. Curve 5: classical prediction.

The results of the percolation model with $c = 3.0$ for $T = 26.1\,\mathrm{K}$ are reported in Fig. 27.21 (De Riva, 1993; Borghesani and Santini, 2002). The agreement between experiment and theory is very good. It should be noted, however, that the electron wavelength must be increased by a factor of $c = 3$ if its use as the sampling length is to be successful.

The percolation model of Eggarter and Cohen (1970) is simple and intuitively appealing. It is quite successful, but it is also phenomenological and controversial. A delicate issue is the quasi-classical counting of the states. Actually, the main criticism raised against this model is concerned with the choice of the electron wavelength as the sampling length (Hernandez, 1973; Hernandez and Ziman, 1973; Schwarz, 1980).

In the fluctuating density problem, the effective potential depends on the averaging volume. If the amplitude of the potential fluctuations in the sampling cell is too small and if the cell side is too short, bound states may not even exist. Thus, the use of the density of states of a free particle in a box may lead to an overestimation of the number of localized states in the tail of the distribution (Borghesani and Santini, 1992).

It has been suggested that a better sampling length would be the classical mean free path ℓ_c (Hernandez, 1973). After all, the electron is affected by the gas density upon collisions and, thus, the correct choice for L seems to be ℓ_c. The reason why this distinction is not important in He at low temperature is that the thermal wavelength $\lambda_T = h/\sqrt{2\pi m k_{\mathrm{B}} T}$ is comparable to the mean free path $\ell_c = 1/N\sigma_{mt}$.

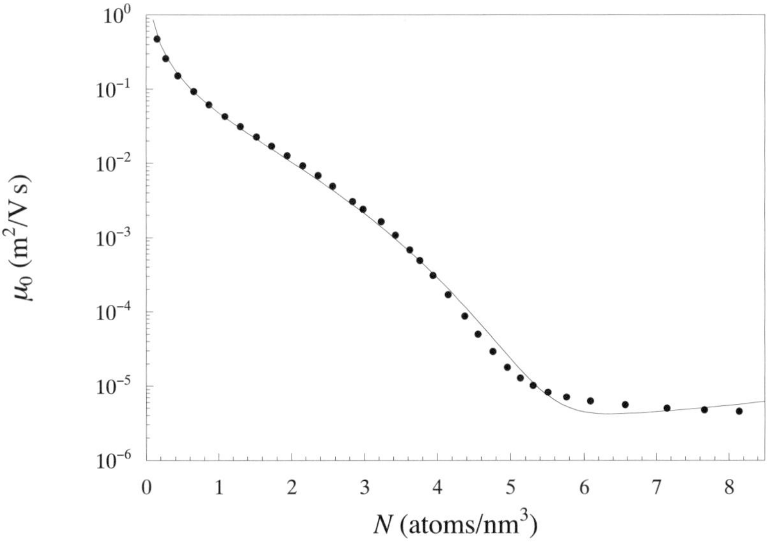

FIG. 27.21. μ_0 vs N in He for $T = 26.1\,\mathrm{K}$ (De Riva, 1993; Borghesani and Santini, 2002). Solid line: percolation model with $c = 3.0$ (Eggarter and Cohen, 1970).

At thermal energies, for $T = 4.2\,\mathrm{K}$, $\lambda_T \approx 360\,\text{Å}$ and $\ell_c \approx 200\,\text{Å}$ for $N = 1\,\mathrm{atoms/nm}^3$. However, at higher temperatures, e.g., at $T = 26\,\mathrm{K}$, a sampling length approximately three times as large as the thermal wavelength must be chosen, albeit the mean free path does not change very much because of the weak energy dependence of the scattering cross-section. The use of this multiplication factor quite interestingly makes the sampling length approximately independent of temperature and, thus, again comparable with the mean free path.

Experiments in Ne have apparently settled the issue in favor of the use of the mean free path as the sampling length of the problem (Borghesani and Santini, 1992). In the neon case, the temperature $T \approx 46.5\,\mathrm{K}$ is much higher than in the experiments with He (Levine and Sanders, 1962, 1967; Schwarz, 1980; Borghesani and Santini, 2002), so that the thermal wavelength is shorter, $\lambda_T \approx 100\,\text{Å}$, but the mean free path is much larger owing to the smallness of the scattering cross-section (O'Malley, 1963; O'Malley and Crompton, 1980). For instance, at thermal energy, and for $N = 1\,\mathrm{atoms/nm}^3$, $\ell_c = 1/N\sigma_{mt} \approx 3.8 \times 10^{-7}\,\mathrm{m}$!

Thus, the mean free path ℓ_c is chosen as the scale for the sampling length: $L = \alpha\ell_c$, where $\alpha \approx 1$ is an adjustable parameter of the model. Within a cell the intrinsic density fluctuations must be neglected because the electron averages adiabatically over them (Hernandez, 1973). This decoupling of the electron from the intrinsic fluctuations occurs only when the density of states has to be calculated. However, correlations among cells do affect the electron motion when it is propagating from cell to cell.

The dominant multiple-scattering effect that for Ne is the energy shift is accounted for by evaluating the cross-section at the shifted energy. Moreover, the correlation among scatterers is very important because the experiment is performed near the critical temperature. This correlation is accounted for by dividing the classical mean free path by the long-wavelength limit of the structure factor $S(0)$:

$$\ell_c' = \frac{\ell_c}{S(0)} = \frac{1}{N\sigma_{mt}S(0)}\,. \tag{27.59}$$

Then, it is easy to show that eqn (27.54) is simply replaced by

$$\ell(\epsilon) = \frac{\alpha\phi(\epsilon)\ell_c'(\epsilon)}{1 + (\alpha - 1)\phi(\epsilon)}\,. \tag{27.60}$$

In Fig. 27.22 the results of the percolation model are compared with the Ne data at $T \approx 46.5\,\mathrm{K}$ (Borghesani *et al.*, 1988; Borghesani and Santini, 1992). The solid line represents the modified percolation model with a sampling length $L = \alpha\ell_c$, where $\alpha = 1.05$. The good agreement with the data at low N is due to the correct treatment of the multiple-scattering effects (Borghesani *et al.*, 1988, 1992). The large sampling length allows a correct counting of the localized states and the model closely tracks the transition to the low-mobility region.

In contrast, the dotted line is calculated by using the electron wavelength as the sampling length, $L = c\lambda$, with $c = 1.5$. For $c < 1.5$, the number of localized

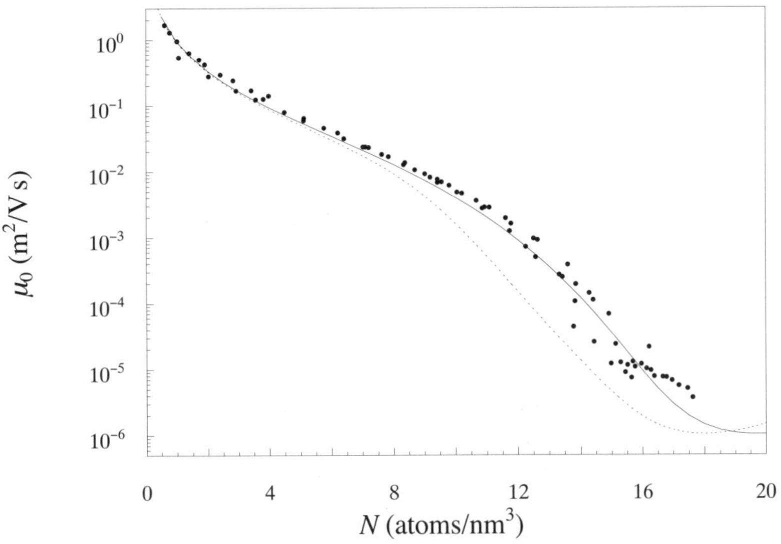

FIG. 27.22. μ_0 vs N in Ne for $T \approx 46.5$ K. (Borghesani *et al.*, 1988, 1992.) Solid line: modified percolation model with $L = 1.05\ell_c$. Dotted line: original percolation model with $L = 1.5\lambda$.

states is grossly overestimated and the calculated mobility drops off even more precipitously at small N. For larger $c > 1.5$, in the low-mobility region, the calculated mobility remains orders of magnitude higher than the experimental data because the number of localized states is strongly underestimated. There is no way, by using the wavelength as the sampling length, of obtaining such good an agreement as with the mean free path.

Thus, when the difference between the values of the mean free path and the thermal wavelength is large, as in the case of Ne, a clear distinction between the roles of these two quantities can be made. In any case, it is proved that a correct treatment of the mobility of the extended states including the corrections due to multiple-scattering effects is absolutely necessary.

As a final remark, it should be noted that it is not yet understood why a quantum-percolation problem can be treated in this semi-classical way (Soukoulis *et al.*, 1987; Root *et al.*, 1988; Root and Skinner, 1988). Moreover, the percolation approach does not offer any possibility of distinguishing between classically-localized (classically-nonpercolating) and Anderson-localized (wave-interference localized) states at the bottom of the conduction band (Simon *et al.*, 1990). Whatever the mechanism of localization may be, the compliance of the gas may always lead to the formation of the electron bubble or pseudobubble.

Fortunately, however, useful suggestions about the nature of the localized states can be obtained by molecular dynamics simulations that yield good agreement with the experimental mobility data and, at the same time, provide snapshots of the environment of the electron in the dense gas (Ancilotto and Toigo, 1992).

REFERENCES

Abe, R. and Aizu, K. (1961). Mobility of ions in a system of interacting Bose particles. *Phys. Rev.*, **123**, 10–18.

Abraham, B. M., Eckstein, Y., Ketterson, J. B., Kuchnir, M., and Roach, P. R. (1970). Velocity of sound, density, and Grűneisen constant in liquid ^{4}He. *Phys. Rev. A*, **1**, 250–7.

Abrahams, E., Anderson, P. W., Licciardello, D. C., and Ramakrishnan, T. V. (1979). Scaling theory of localization: Absence of quantum diffusion in two dimensions. *Phys. Rev. Lett.*, **42**, 673–6.

Achter, E. (1968). Ion mobility in ^{3}He in the critical region. *Phys. Lett.*, **27 A**, 687–8.

Adams, P. W., Browne, D. A., and Paalanen, M. A. (1992). Evidence for a first-order correction to the Boltzmann conductivity of a disordered three-dimensional electron gas. *Phys. Rev. B*, **45**, 8837–40.

Adams, P. W. and Paalanen, M. A. (1987). Localization in a Nondegenerate Two–Dimensional Electron Gas. *Phys. Rev. Lett.*, **58**, 2106–2109.

Adams, P. W. and Paalanen, M. A. (1988). Anderson localization of electrons in dense ^{4}He gas. *Phys. Rev. Lett.*, **61**, 451–4.

Ahlers, G. (1968). Effect of the gravitational field on the superfluid transition in He^4. *Phys. Rev.*, **171**, 275–82.

Ahlers, G. (1971). On the viscosity of ^{4}He near the superfluid transition. *Phys. Lett.*, **37 A**, 151–2.

Ahlers, G. (1976). Experiments near the superfluid transition in ^{4}he and ^{3}he–^{4}he mixtures. In *The physics of liquid and solid helium* (ed. K. H. Bennemann and J. B. Ketterson), Volume 1, Chapter 2, pp. 85–206. Wiley, New York.

Ahlers, G. and Gamota, G. (1972). Ion mobilities in ^{4}He near the superfluid transition and the melting line. *Phys. Lett.*, **38 A**, 65–6.

Ahonen, A. I., Kokko, J., Lounasmaa, O. V., Paalanen, M. A., Richardson, R. C., and Schoepe, W. (1976). Mobility of negative ions in superfluid ^{3}He. *Phys. Rev. Lett.*, **37**, 511–15.

Ahonen, A. I., Kokko, J., Paalanen, M. A., Richardson, R. C., Schoepe, W., and Takano, Y. (1978). Negative ion motion in normal and superfluid ^{3}He. *J. Low Temp. Phys.*, **30**, 205–28.

Alexander, P. W., Barber, C. N., McClintock, P. V. E., and Pickett, G. R. (1977). Positive-ionic mobility and a hierarchy of ions in normal Liquid ^{3}He. *Phys. Rev. Lett.*, **39**, 1544–7.

Alexander, P. W., McClintock, P. V. E., Pickett, G. R., and Bowley, R. M. (1978). Nonlinear behaviour of positive ions in normal liquid ^{3}He: A comparison between experiment and a parameterless theoretical prediction. *J. Phys. C: Solid State Phys.*, **11**, L881–5.

Alexander, P. W., McClintock, P. V. E., Pickett, G. R., and Bowley, R. M. (1979). Multiple positive ionic mobilities in normal liquid ^{3}He contaminated with ^{4}He. *J. Low Temp. Phys.*, **34**, 695–721.

Alexander, P. W. and Pickett, G. R. (1978). Pressure and temperature dependence of the positive ionic mobility in normal liquid ^{3}He. *Phys. Lett.*, **67 A**, 391–3.

Alexander, P. W. and Pickett, G. R. (1979). The equivalence of temperature and velocity in determining the non-linear positive ionic motion in normal liquid ^{3}He. *J. Phys. C: Solid State Phys.*, **12**, L495–L9.

Allen, J. F. and Misener, A. D. (1938). The surface tension of liquid helium. *Proc. Camb. Philos. Soc.*, **34**, 299–300.

Allen, N. L. and Prew, B. A. (1970). Some measurements of electron drift velocities in compressed gases. *J. Phys. B: At. Mol. Phys.*, **3**, 1113–26.

Allum, D. R., Bowley, R. M., and McClintock, P. V. E. (1976*a*). Evidence for roton pair creation in superfluid ^{4}He. *Phys. Rev. Lett.*, **36**, 1313–16.

Allum, D. R. and McClintock, P. V. E. (1976*a*). Condensation of isotopic impurities on the negative ion in He II. *J. Phys. C: Solid State Phys.*, **9**, L273–6.

Allum, D. R. and McClintock, P. V. E. (1976*b*). Field emission in a liquid ^{3}He–^{4}He solution. *Phys. Lett.*, **56 A**, 199–200.

Allum, D. R. and McClintock, P. V. E. (1976*c*). Observation of a second critical field for negative ions in He II. *J. Phys. C: Solid State Phys.*, **9**, L371–5.

Allum, D. R. and McClintock, P. V. E. (1977). Atoms in contact. IX Breaking the superfluid speed limit. *New Scientist*, **73**, 651–3.

Allum, D. R. and McClintock, P. V. E. (1978*a*). Nucleation of negatively charged vortex rings in superfluid Helium-4 near its solidification pressure. *J. Low Temp. Phys.*, **31**, 321–38.

Allum, D. R. and McClintock, P. V. E. (1978*b*). Rings among rotons: Measurements of the vortex nucleation rate for negative ions whose drift velocities are limited by roton emission. *J. Phisique Colloque (Paris)*, **39**(C–6), 174–5.

Allum, D. R., McClintock, P. V. E., and Phillips, A. (1975). Anomalous roton-induced drag on negative ions in He II. *Phys. Lett.*, **54 A**, 385–6.

Allum, D. R., McClintock, P. V. E., Phillips, A., and Bowley, R. M. (1976*b*). The breakdown of superfluidity in liquid ^{4}He: An experimental test of Landau's theory. *Phil. Trans. Roy. Soc. (London)*, **284 A**, 179–224.

Ambegaokar, V., de Gennes, P. G., and Rainer, D. (1974). Landau–Ginsburg equations for an anisotropic superfluid. *Phys. Rev. A*, **9**, 2676–85.

Ambegaokar, V., de Gennes, P. G., and Rainer, D. (1975). Erratum: Landau–Ginsburg equations for an anisotropic superfluid. *Phys. Rev. A*, **12**, 345.

Amit, D. and Gross, E. P. (1966). Vortex rings in a Bose fluid. *Phys. Rev.*, **145**, 130–45.

Ancilotto, F. and Toigo, F. (1992). Computer simulations of excess electron transport in neon. *Phys. Rev. A*, **45**, 4015–22.

Ancilotto, F. and Toigo, F. (1994). Properties of an electron bubble approaching the surface of liquid helium. *Phys. Rev. B*, **50**, 12820–30.

Ancilotto, F. and Toigo, F. (1995). Theory of electron escape from the surface of liquid helium. *Z. Phys.*, **B 98**, 309–11.

Anderson, A. C., Kuchnir, M., and Wheatley, J. C. (1968). Experimental ion mobilities in liquid He^3 below 1 °K. *Phys. Rev.*, **168**, 261–70.

Anderson, P. W. (1958). Absence of diffusion in certain random lattices. *Phys. Rev.*, **109**, 1492–505.

Anderson, P. W. and Brinkman, W. F. (1973). Anisotropic superfluidity in He^3: A possible interpretation of its stability as a spin–fluctuation effect. *Phys. Rev. Lett.*, **31**, 1108–11.

Anderson, P. W. and Morel, P. (1961). Generalized Bardeen–Cooper–Schrieffer states and the proposed low-temperature phase of liquid He^3. *Phys. Rev.*, **132**, 1911–34.

Andersson, N. and Comer, G. L. (2001). Slowly rotating general relativistic superfluid neutron stars. *Class. Quantum Grav.*, **18**, 969–1002.

Andreev, A. F. (1966). Surface tension of weak helium solutions. *Sov. Phys.-JETP*, **23**, 939–41.

Andronikashvili, E. L. (1946). Temperature dependence of the normal density of helium II. *Zhur. Eksp. Theor. Fiz.*, **18**, 424–8.

Andronikashvili, L. and Mamaladze, Yu. G. (1966). Quantization of macroscopic motions and hydrodynamics of rotating helium II. *Rev. Mod. Phys.*, **38**, 567–625.

Archie, C. N. (1978). *The viscosity and superfluid density of He–3 to one millikelvin.* Ph.D. thesis, Cornell University, Ithaca, NY.

Arfken, G. (1985). *Mathematical methods for physicists.* Academic Press, Orlando, FL.

Arkhipov, R. G. and Shalnikov, A. I. (1960). On the problem of motion of charges in liquid helium. *Sov. Phys.-JETP*, **37**, 888–90.

Artem'ev, A. A. and Khrapak, A. G. (1986). Kinetics of electron self-trapping in nonpolar liquids. *Sov.-Tech. Phys. Lett*, **12**, 424–5.

Asaad, A. N., Klahn, Th., and Krebs, P. (1996). The influence of multiple scattering processes on the electron mobility in low density methanol gas. *J. Chem. Phys.*, **105**, 8633–8.

Ascarelli, G. (1992). Hall mobility of electrons in liquid xenon. *J. Phys. C: Condensed Matter*, **4**, 6055–72.

Ashcroft, N. W. and Mermin, N. D. (1976). *Solid state physics.* HRW, Philadelphia, PA.

Ashton, R. A. and Glaberson, W. I. (1979). Vortex waves in superfluid ^4He. *Phys. Rev. Lett.*, **42**, 1062–4.

Atkins, K. R. (1959). Ions in liquid helium. *Phys. Rev.*, **116**, 1339–43.

Atkins, K. R. and Narahara, Y. (1965). Surface tension of liquid He^4. *Phys. Rev.*, **138**, A437–41.

Atkins, K. R. and Stasior, R. A. (1953). First sound in liquid helium at high pressures. *Can. J. Phys.*, **31**, 1156–64.

Atrazhev, V. M. (1984). Delocalisation of electrons in dense helium gas by external electric field. *J. Phys. D: Appl. Phys.*, **17**, 889–902.

Atrazhev, V. M., Berezhnov, A. V., and Timoshkin, I. V. (2002). Electron scattering in atomic liquids: Application to the maximum of electron mobility. *Phys. Rev. B*, **66**, 205106.

Atrazhev, V. M. and Yakubov, I. T. (1977). The electron drift velocity in dense gases. *J. Phys. D: Appl. Phys.*, **10**, 2155–63.

Awschalom, D. D. and Schwarz, K. W. (1984). Observation of a remanent vortex-line density in superfluid Helium. *Phys. Rev. Lett.*, **52**, 49–52.

Bagheri, A., Baluji, K. L., and Datta, S. M. (1994). Density dependence of electron mobility in dense gases. *Z. Phys. D*, **32**, 211–17.

Bailyn, M. and Lobo, R. (1968). Helium drag in the theory of impurity mobility in liquid helium. *Phys. Rev.*, **176**, 222–5.

Balian, R. and Werthamer, N. R. (1963). Superconductivity with pairs in a relative p wave. *Phys. Rev.*, **131**, 1553–64.

Barber, C. N., McClintock, P. V. E., Miller, I. E., and Pickett, G. R. (1975). Anomalous positive ions mobility in liquid ^{3}He. *Phys. Lett.*, **54 A**, 241–2.

Bardeen, J., Cooper, L. N., and Schrieffer, J. R. (1957). Theory of superconductivity. *Phys. Rev.*, **108**, 1175–1204.

Barenghi, C., Mellor, C. J., Muirhead, C. M., and Vinen, W. F. (1986). Experiments on ions trapped below the surface of superfluid ^{4}He. *J. Phys. C: Solid State Phys.*, **19**, 1135–44.

Barenghi, C. F., Mellor, C. J., Meredith, J., Muirhead, C. M., Sommerfeld, P. K. H., and Vinen, W. F. (1991). Ions trapped below the surface of superfluid helium. I. The observation of plasma resonances, and the measurement of effective masses and ionic mobilities. *Phil. Trans. Roy. Soc. (London): Physical Sci. Eng.*, **334**, 139–72.

Barmatz, M. and Rudnick, I. (1968). Velocity and attenuation of first sound near the lambda point of helium. *Phys. Rev.*, **170**, 224–38.

Barrera, R. and Baym, G. (1972). Roton-limited mobility of ions in superfluid He^4. *Phys. Rev. A*, **6**, 1558–66.

Bartels, A. K. (1972). Pressure dependence of electron drift velocity in hydrogen at 77.8 K. *Phys. Rev. Lett.*, **28**, 213–15.

Bartels, A. K. (1973). Density dependence of the electron drift velocity in argon. *Phys. Lett.*, **44 A**, 403–4.

Bartels, A. K. (1975). Density dependence of electron drift velocities in helium and hydrogen at 77.6 K. *Appl. Phys.*, **8**, 59–64.

Bartholomew, J., Hall, R., and Berne, B. J. (1985). Behavior of an electron in helium gas. *Phys. Rev. B*, **32**, 548–50.

Barton, G. and Moore, M. A. (1974). Some p-wave phases of superfluid helium-3 in strong-coupling theory. *J. Phys. C: Solid State Phys.*, **7**, 4220–35.

Bauer, H., Beau, M., Bernhardt, A., Friedl, B., and Reyher, H. J. (1989). Implantation of atoms into liquid helium for the purpose of impurity spectroscopy. *Phys. Lett.*, **137 A**, 217–24.

Bauer, H., Beau, M., Fischer, J., Reyher, H. J., Rosenkranz, J., and Venter, K. (1990). Optical spectroscopy of alkali and alkaline–earth ions in superfluid ^{4}He. *Physica B*, **165–6**, 137–8.

Bauer, H., Hausmann, M., Mayer, R., Weber, E., and Winnacker, A. (1985). Implantation of impurity ions into He II for optical spectroscopy purposes. *Phys. Lett.*, **110 A**, 279–82.

Baym, G., Barrera, R. G., and Pethick, C. J. (1969). Mobility of the electron bubble in superfluid helium. *Phys. Rev. Lett.*, **22**, 20–3.

Baym, G., Pethick, C. J., and Salomaa, M. (1977). Mobility of the electron bubble in superfluid ^{3}He. *Phys. Rev. Lett.*, **38**, 845–8.

Baym, G., Pethick, C. J., and Salomaa, M. (1979). Mobility of negative ions in superfluid ^{3}He–B. *J. Low Temp. Phys.*, **36**, 431–66.

Beál-Monod, M. T. (1973). High-temperature transport properties of liquid He^3. *Phys. Rev. Lett.*, **31**, 513–17.

Beijersbergen, J. H. M., Hui, Q., and Takami, M. (1993). Spectroscopy and dynamics of trapped alkaline earth atoms in superfluid helium. *Phys. Lett.*, **181 A**, 393–403.

Benderskii, A. V., Eloranta, J., Zadoyan, R., and Apkarian, V. A. (2002). A direct interrogation of superfluidity on molecular scales. *J. Chem. Phys.*, **117**, 1201–13.

Bendt, P. J., Cowan, R. D., and Yarnell, J. L. (1959). Excitations in liquid helium: Thermodynamic calculations. *Phys. Rev.*, **113**, 1386–95.

Bennemann, K. H. and Ketterson, J. B. (ed.) (1976). *The physics of liquid and solid helium.* Volume 1. Wiley, New York.

Berezhnov, A. V., Khrapak, A. G., Illenberger, E., and Schmidt, W. F. (2003). Anomalous mobility of negative charges in liquid hydrogen: A model of ionic bubble. *High Temp.*, **41**, 425–7.

Berg, R. F. and Moldover, M. R. (1990). Critical exponent for viscosity. *Phys. Rev. A*, **42**, 7183–6.

Bergmann, G. (1983). Physical interpretation of weak localization: A time-of-flight experiment with conduction electrons. *Phys. Rev. B*, **28**, 2914–20.

Berloff, N. G. (2000). Vortex nucleation by a moving ion in a Bose condensate. *Phys. Lett.*, **277 A**, 240–4.

Berloff, N. G. and Roberts, P. H. (2000*a*). Capture of an impurity by a vortex line in a Bose condensate. *Phys. Rev. B*, **63**, 024510.

Berloff, N. G. and Roberts, P. H. (2000*b*). Motions in a Bose condensate. VII. Boundary-layer separation. *J. Phys. A: Math. Gen.*, **33**, 4025–38.

Berloff, N. G. and Roberts, P. H. (2001). Motion in a Bose condensate: VIII. The electron bubble. *J. Phys. A: Math. Gen.*, **34**, 81–91.

Betts, D. S., Osborne, D. W., Welber, B., and Wilks, J. (1963). The viscosity of liquid helium 3. *Phil. Mag.*, **8**, 977–87.

Bevington, P. R. (1969). *Data reduction and error analysis for the physical sciences.* McGraw–Hill, New York.

Bhatti, M. H. and McClintock, P. V. E. (1979). Optimal positioning of field emitters for ion injection in liquid helium. *Cryogenics*, **19**, 535–6.

Biskeborn, R. and Guernsey, R. W. (1975). Critical exponents for the shear viscosity of ^{4}He at T_λ. *Phys. Rev. Lett.*, **34**, 455–8.

Black, M. A., Hall, H. E., and Thomson, K. (1971). The viscosity of liquid helium 3. *J. Phys. C: Solid State Phys.*, **4**, 129–42.

Bloch, F. and Bradbury, N. E. (1935). On the mechanism of unimolecular electron capture. *Phys. Rev.*, **48**, 689–95.

Blount, E. I. and Varma, C. M. (1976). Hydrodynamics of vortex generation in flowing superfluid helium. *Phys. Rev. B*, **14**, 2888–99.

Boghosian, C., Meyer, H., and Rives, J. E. (1966). Density, coefficient of thermal expansion, and entropy of compression of liquid helium–3 under pressure below 1.2 °K. *Phys. Rev.*, **146**, 110–19.

Boltnev, R. E., Frossati, G., Gordon, E. B., Krushinskaya, I. N., Popov, E. A., and Usenko, A. (2002). Embedding impurities into liquid helium. *J. Low Temp. Phys.*, **127**, 245–58.

Bondarev, V. N. (1973). Localized states of rotons near ions in helium II. *Sov. Phys.–JETP Lett.*, **18**, 405–7.

Bondarev, V. N. and Fisher, I. Z. (1974). Mobility of positive ions in He^3-He^4 solutions. *Sov. Phys.–JETP*, **38**, 597–9.

Borghesani, A. F. (1985). Concentration-dependent behavior of the shear viscosity of coal–fuel oil suspensions. *Rheol. Acta*, **24**, 189–97.

Borghesani, A. F. (2001). Electron mobility maximum in dense argon gas at low temperature. *J. Electrostatics*, **53**, 89–106.

Borghesani, A. F. (2006). Electron and ion transport in dense rare gases. *IEEE–TDEI*, **13**, 492–502.

Borghesani, A. F., Bruschi, L., Santini, M., and Torzo, G. (1985). Density dependence of the electronic mobility in high density Neon gas. *Phys. Lett.*, **108 A**, 255–8.

Borghesani, A. F., Bruschi, L., Santini, M., and Torzo, G. (1986*a*). Simple photoelectronic source for swarm experiments in high-density gases. *Rev. Sci. Instrum.*, **57**, 2234–7.

Borghesani, A. F., Bruschi, L., Santini, M., and Torzo, G. (1986*b*). The influence of the cell geometry on the signal shape in electron drift velocity measurements. *Z. Naturforsch.*, **41 A**, 912–20.

Borghesani, A. F., Bruschi, L., Santini, M., and Torzo, G. (1988). Electron mobility in neon at high densities. *Phys. Rev. A*, **37**, 4828–35.

Borghesani, A. F., Carugno, G., Cavenago, M., and Conti, E. (1990). Electron transmission through the Ar liquid–vapor interface. *Phys. Lett.*, **149 A**, 481–4.

Borghesani, A. F., Carugno, G., and Santini, M. (1991). Experimental determination of the conduction band of excess electrons in liquid Ar. *IEEE Trans. EI*, **26**, 615–22.

Borghesani, A. F., Neri, D., and Barbarotto, A. (1997). Mobility of O_2^- ions in near critical Ar gas. *Chem. Phys. Lett.*, **267**, 116–22.

Borghesani, A. F., Neri, D., and Barbarotto, A. (1999). Critical behavior of O_2^-

ions in argon gas. *Int. J. Thermophys.*, **20**, 899–909.

Borghesani, A. F., Neri, D., and Santini, M. (1993). Low-temperature O_2^- mobility in high-density neon gas. *Phys. Rev. E*, **48**, 1379–89.

Borghesani, A. F. and O'Malley, T. F. (2003). Inhomogeneous gas model for electron mobility in high density neon gas. *J. Chem. Phys.*, **118**, 2760–6.

Borghesani, A. F. and Santini, M. (1990*a*). Electron mobility and localization effects in high-density Ne gas. *Phys. Rev. A*, **42**, 7377–88.

Borghesani, A. F. and Santini, M. (1990*b*). Electron swarms experiments in fluids–signal waveform analysis. *Meas. Sci. Technol.*, **1**, 939–47.

Borghesani, A. F. and Santini, M. (1992). Electron localization–delocalization transition in high-density neon gas. *Phys. Rev. A*, **45**, 8803–10.

Borghesani, A. F. and Santini, M. (1994*a*). Density and field dependence of excess electron mobility in high-density noble gases. In *Linking the gaseous and condensed phases of matter. The behavior of slow electrons* (ed. L. G. Christophorou, E. Illenberger, and W. F. Schmidt), Volume B 326, NATO ASI Series, pp. 259–79. Plenum Press, New York.

Borghesani, A. F. and Santini, M. (1994*b*). Excess electron localization in high-density neon gas. In *Linking the gaseous and condensed phases of matter. The behavior of slow electrons* (ed. L. G. Christophorou, E. Illenberger, and W. F. Schmidt), Volume B 326, NATO ASI Series, pp. 281–301. Plenum Press, New York.

Borghesani, A. F. and Santini, M. (2002). High-temperature electron localization in dense He gas. *Phys. Rev. E*, **65**, 056403.

Borghesani, A. F., Santini, M., and Lamp, P. (1992). Excess electron mobility in high-density argon gas. *Phys. Rev. A*, **46**, 7902–9.

Borghesani, A. F. and Tamburini, F. (1999). Parametric resonance in the drift motion of an ionic bubble in near critical Ar gas. *Phys. Rev. Lett.*, **83**, 4546–9.

Bowe, J. C. (1960). Drift velocity of electrons in nitrogen, helium, neon, krypton, and xenon. *Phys. Rev.*, **117**, 1411–15.

Bowley, R. M. (1971*a*). Mobility of a heavy impurity in ^{3}He and in ^{3}He–^{4}He mixtures: I. *J. Phys. C: Solid State Phys.*, **4**, 853–69.

Bowley, R. M. (1971*b*). Mobility of an impurity in a Fermi liquid: analogy with a phase transition. *J. Phys. C: Solid State Phys.*, **4**, L207–11.

Bowley, R. M. (1971*c*). Mobility of an impurity in liquid Helium II. *J. Phys. C: Solid State Phys.*, **4**, 1645–51.

Bowley, R. M. (1973). Microscopic theory of impurities in liquid helium. III. *J. Low Temp. Phys.*, **10**, 481–501.

Bowley, R. M. (1976*a*). Mobility of negative ions in superfluid ^{3}He. *J. Phys. C: Solid State Phys.*, **9**, L151–4.

Bowley, R. M. (1976*b*). Motion of negative ions in superfluid ^{3}He. *J. Phys. C: Solid State Phys.*, **9**, L653–6.

Bowley, R. M. (1976*c*). The motion of negative ions in liquid helium in strong electric fields. *J. Low Temp. Phys.*, **24**, 41–51.

Bowley, R. M. (1976*d*). Vortex nucleation by negative ions in liquid ^{4}He. *J.*

Phys. C: Solid State Phys., **9**, L367–70.

Bowley, R. M. (1977*a*). Mobility of positive ions in normal liquid ^{3}He. *J. Phys. C: Solid State Phys.*, **11**, 75–84.

Bowley, R. M. (1977*b*). Motion of negative ions in superfluid ^{3}He. *J. Phys. C: Solid State Phys.*, **10**, 4033–44.

Bowley, R. M. (1978). Motion of positive ions in normal ^{3}He. *J. Phys. C: Solid State Phys.*, **11**, 1361–6.

Bowley, R. M. (1984). Nucleation of vortex rings by negative ions in liquid helium at low temperatures. *J. Phys. C: Solid State Phys.*, **17**, 595–613.

Bowley, R. M. and Lekner, J. (1970). Positive ion mobility in ^{3}He–^{4}He mixtures. *J. Phys. C: Solid State Phys.*, **3**, L127–30.

Bowley, R. M., McClintock, P. V. E., Moss, F. E., Nancolas, G. G., and Stamp, P. C. E. (1982). The breakdown of superfluidity in liquid ^{4}He. III. Nucleation of quantized vortex rings. *Phil. Trans. Roy. Soc. (London)*, **307 A**, 201–60.

Bowley, R. M., McClintock, P. V. E., Moss, F. E., and Stamp, P. C. E. (1980). Vortex nucleation in isotopically pure superfluid ^{4}He. *Phys. Rev. Lett.*, **44**, 161–4.

Bowley, R. M., Nancolas, G. G., and McClintock, P. V. E. (1984). Vortex nucleation in ultradilute superfluid ^{3}He–^{4}He solutions. *Phys. Rev. Lett.*, **52**, 659–62.

Bowley, R. M. and Sheard, F. W. (1977). Motion of a negative ion at supercritical velocities in liquid ^{4}He at low temperatures. *Phys. Rev. B*, **16**, 244–54.

Bradbury, N. E. (1933). Electron attachment and negative ion formation in oxygen and oxygen mixtures. *Phys. Rev.*, **48**, 883–90.

Braglia, G. L. and Dallacasa, V. (1978). Theory of the density dependence of electron drift velocity in gases. *Phys. Rev. A*, **18**, 711–17.

Braglia, G. L. and Dallacasa, V. (1982). Theory of electron mobility in dense gases. *Phys. Rev. A*, **26**, 902–14.

Brody, B. (1975). Charge-carrier mobilities in superfluid helium under pressure. *Phys. Rev. B*, **11**, 170–7.

Bromley, D. J. (1981). Negative-ion drag in superfluid ^{3}He. *Phys. Rev. B*, **23**, 4503–22.

Broomall, J. R., Johnson, W. D., and Onn, D. G. (1976). Density dependence of the electron surface barrier for fluid ^{3}He and ^{4}He. *Phys. Rev. B*, **14**, 2819–25.

Brubaker, N. R., Edwards, D. O., Sarwinski, R. E., Seligmann, P., and Sherlock, R. A. (1970). Investigation by second sound of the inertial mass of ^{3}He in superfluid ^{4}He at low temperatures. *Phys. Rev. Lett.*, **25**, 715–18.

Bruschi, L., Maraviglia, B., and Mazzoldi, P. (1966*a*). Second-sound attenuation associated with hot ions in liquid He II. *Phys. Rev.*, **143**, 84–90.

Bruschi, L., Maraviglia, B., and Moss, F. E. (1966*b*). Measurement of a barrier for the extraction of excess electrons from liquid helium. *Phys. Rev. Lett.*, **17**, 682–4.

Bruschi, L., Mazzi, G., and Santini, M. (1970). Discontinuities of ionic mobilities in simple liquids. *Phys. Rev. Lett.*, **25**, 330–3.

Bruschi, L., Mazzi, G., and Santini, M. (1972). Localized electrons in liquid neon. *Phys. Rev. Lett.*, **28**, 1504–6.

Bruschi, L., Mazzi, G., Santini, M., and Torzo, G. (1975). Measurements of ^{4}He viscosity near the superfluid transition. *J. Low Temp. Phys.*, **18**, 487–500.

Bruschi, L., Mazzi, G., Santini, M., and Torzo, G. (1977). The behavior of the ^{4}He viscosity near the superfluid transition. *J. Low Temp. Phys.*, **29**, 63–88.

Bruschi, L., Mazzoldi, P., and Santini, M. (1966*c*). Temperature dependence of the critical velocity of positive ions in liquid helium. *Phys. Rev. Lett.*, **17**, 292–4.

Bruschi, L., Mazzoldi, P., and Santini, M. (1968*a*). Periodic critical velocities of ions in liquid helium II. Temperature dependence. *Phys. Rev.*, **167**, 203–14.

Bruschi, L., Mazzoldi, P., and Santini, M. (1968*b*). Positive ions in liquid helium II. The critical velocity for creation of vortex rings. *Phys. Rev. Lett.*, **21**, 1738–40.

Bruschi, L. and Santini, M. (1970). Methods for the measurement of the drift velocity of ions in liquids. *Rev. Sci. Instrum.*, **41**, 102–7.

Bruschi, L. and Santini, M. (1975). Vibrating wire viscometer. *Rev. Sci. Instrum.*, **46**, 1560–8.

Bruschi, L. and Santini, M. (1978). The ^{4}He viscosity near the superfluid transition under pressure. *J. Low Temp. Phys.*, **33**, 357–68.

Bruschi, L., Santini, M., and Torzo, G. (1984). Resonant electron attachment to oxygen molecules in dense helium gas. *J. Phys. B: At. Mol. Phys.*, **17**, 1137–54.

Burdick, B. (1965). Negative ions in liquid helium II. *Phys. Rev. Lett.*, **14**, 11–13.

Byron Bird, R., Stewart, W. E., and Lightfoot, E. N. (1960). *Transport phenomena.* Wiley, New York.

Cade, A. G. (1965). Binding of positive–ion complexes to vortex rings in liquid helium II. *Phys. Rev. Lett.*, **15**, 238–9.

Calogero, F. (1967). *Variable phase approach to potential scattering.* Academic Press, New York.

Cantelli, R., Modena, I., and Ricci, F. P. (1968). Mobility of positive and negative charges in ^{3}He at the critical point. *Phys. Rev.*, **171**, 236–42.

Cao, J. and Berne, D. J. (1995). A scaling and mapping theory for excess electrons in simple fluids. *J. Chem. Phys.*, **102**, 432–6.

Careri, G., Cunsolo, S., and Mazzoldi, P. (1961). Critical drift velocity of ions in liquid helium. *Phys. Rev. Lett.*, **7**, 151–3.

Careri, G., Cunsolo, S., and Mazzoldi, P. (1964*a*). Periodic discontinuities in the drift velocity of ions in liquid helium II. *Phys. Rev.*, **136**, A303–10.

Careri, G., Cunsolo, S., Mazzoldi, P., and Santini, M. (1965). Experiments on the creation of charged quantized vortex rings in liquid helium at 1 °K. *Phys. Rev. Lett.*, **15**, 392–6.

Careri, G., Cunsolo, S., and Vicentini–Missoni, M. (1964*b*). Critical velocities of ions in liquid helium II detected by heat flush. *Phys. Rev.*, **136**, A311–15.

Careri, G., Fasoli, U., and Gaeta, F. S. (1960). Experimental behavior of ionic structures in liquid He–II. *Nuovo Cimento*, **15**, 774–83.

Careri, G., McCormick, W. D., and Scaramuzzi, F. (1962). Ions in rotating liquid helium II. *Phys. Lett.*, **1**, 61–3.

Careri, G., Scaramuzzi, F., and Thomson, J. O. (1959). Heat flush and mobility of electric charges in liquid helium. I. Non-turbulent flow. *Nuovo Cimento*, **13**, 186–96.

Castellani, C., Di Castro, D., and Peliti, L. (ed.) (1981). *Disordered systems and localization.* Springer, Berlin.

Celli, V., Cohen, M. H., and Zuckerman, M. J. (1968). Vibrational modes of a negative ion in liquid helium. *Phys. Rev.*, **173**, 253–6.

Chandler, D. and Leung, K. (1994). Excess electrons in liquids: Geometrical perspectives. *Annu. Rev. Phys. Chem.*, **45**, 557–91.

Chandrasekhar, S. (1943). Stochastic problems in physics and astronomy. *Rev. Mod. Phys.*, **15**, 1–89.

Chang, C. C. and Cohen, M. H. (1973). Structure of the surface of liquid He^4 at zero temperature. *Phys. Rev. A*, **8**, 1930–6.

Chapman, S. and Cowling, T. G. (1970). *The mathematical theory of non-uniform gases.* Cambridge University Press.

Chase, C. E. (1959). The velocity of sound in liquid helium near the lambda point. *Phys. Rev. Lett.*, **2**, 197–9.

Chase, C. E., Williamson, R. C., and Tisza, L. (1964). Ultrasonic propagation near the critical point in helium. *Phys. Rev. Lett.*, **13**, 467–9.

Chen, Y. and Prokof'ev, N. (1990). Classical Mechanics at $t = 0$: Mobility of negative ion in ^{3}He. *Phys. Rev. Lett.*, **65**, 1761–4.

Cheng, D. K., Cromar, M. W., and Donnelly, R. J. (1973). Influence of an axial current on negative-ion trapping in rotating helium I. *Phys. Rev. Lett.*, **31**, 433–6.

Cheng, E., Cole, M. W., and Cohen, M. H. (1994*a*). Binding of electrons to the surface of liquid helium. *Phys. Rev. B*, **50**, 1136–42.

Cheng, E., Cole, M. W., and Cohen, M. H. (1994*b*). Erratum: Binding of electrons to the surface of liquid helium [*Phys. Rev. B*, **50**, 1136 (1994)]. *Phys. Rev. B*, **50**, 16134.

Chrètien, M., Gross, E. P., and Deser, S. (ed.) (1968). *Statistical physics. Phase transitions and superfluidity.* Brandeis University Summer School in Theoretical Physics 1966. Gordon and Breach, New York.

Christophorou, L. G. (1984). *Electron–molecule interactions and their applications.* Volume 1 & 2. Academic Press, Orlando, FL.

Clark, C. (1965). Self-trapped electrons in liquid helium II. *Phys. Lett.*, **16**, 42–3.

Clark, R. C. (1963). The mobility of an ion in a Fermi liquid. *Proc. Phys. Soc.*, **82**, 785–93.

Classen, J., Su, C. K., Mohazzab, M., and Maris, H. J. (1998). Electrons and cavitation in liquid helium. *Phys. Rev. B*, **57**, 3000–10.

Cohen, M. H. (1973). The electronic structure of disordered systems. In *Electrons in fluids. The nature of metal–ammonia solutions* (ed. J. Jortner and N. R. Kestner), pp. 257–85. Springer Verlag, Heidelberg.

Cohen, M. H., Fritzsche, H., and Ovshinsky, S. R. (1969). Simple band model for amorphous semiconducting alloys. *Phys. Rev. Lett.*, **22**, 1065–8.

Cohen, M. H. and Lekner, J. (1967). Theory of hot electrons in gases, liquids, and solids. *Phys. Rev. Lett.*, **158**, 305–9.

Coker, D. F., Berne, B. J., and Thirumalai, D. (1987). Path integral Monte Carlo studies of the behavior of excess electrons in simple fluids. *J. Chem. Phys.*, **86**, 5689–702.

Cole, M. W. (1970). Properties of image-potential-induced surface states of insulators. *Phys. Rev. B*, **2**, 4239–52.

Cole, M. W. and Bachman, R. A. (1977). Structure of positive impurity ions in liquid helium. *Phys. Rev. B*, **15**, 1388–94.

Cole, M. W. and Cohen, M. H. (1969). Image-potential-induced surface bands in insulators. *Phys. Rev. Lett.*, **23**, 1238–41.

Cole, M. W., Goodstein, D. L., Savoia, A., and Scaramuzzi, F. (1978). Report on the use of ion mobilities to study phase transitions in liquid ^{4}He. *J. Electrostatics*, **5**, 121–34.

Cole, M. W. and Sluckin, T. J. (1977). Nucleation of freezing by charged particles. I. Thermodynamics. *J. Chem. Phys.*, **67**, 746–50.

Cole, M. W. and Toigo, F. (1978). Structural transition for positive impurity in fluids. *Phys. Rev. B*, **17**, 2054–6.

Cope, J. A. and Gribbon, P. W. F. (1965). Vibrating ions in liquid He II. *Phys. Lett.*, **16**, 128–30.

Cope, J. A. and Gribbon, P. W. F. (1970*a*). Ions in He II at 1 °K. I. Velocity discontinuities. *J. Phys. C: Solid State Phys.*, **3**, 460–71.

Cope, J. A. and Gribbon, P. W. F. (1970*b*). Ions in He II at 1 °K. II. Heat flush effects. *J. Phys. C: Solid State Phys.*, **3**, 472–4.

Crompton, R. W., Elford, M. T., and Robertson, A. G. (1970). The momentum transfer cross-section for electrons in helium derived from drift velocities at 77 degrees K. *Aust. J. Phys.*, **23**, 667–81.

Cross, M. C. (1974). Impulse given to a plate by a quantized vortex ring. *Phys. Rev. A*, **10**, 1442–1445.

Cunsolo, S. (1961). A method for the measurements of the mobility of electric charges in liquids. *Nuovo Cimento*, **21**, 76–83.

Cunsolo, S., Dall'Oglio, G., Maraviglia, B., and Ricci, M. V. (1968*a*). Ejection of electrons from superfluid helium surface by annihilation of quantized vortex rings. *Phys. Rev. Lett.*, **21**, 74–5.

Cunsolo, S. and Maraviglia, B. (1969). Charged-vortex-ring creation by ions in superfluid helium. *Phys. Rev.*, **187**, 292–6.

Cunsolo, S., Maraviglia, B., and Ricci, M. V. (1968*b*). Formation of charged vortex rings in liquid helium below 0.8 °K. *Phys. Lett.*, **26 A**, 605–6.

Cunsolo, S. and Mazzoldi, P. (1961). The Influence of pressure on the mobility of electric charges in liquid helium II. *Nuovo Cimento*, **20**, 949–55.

Dahm, A. J. (1969). Evidence for condensation of He^3 atoms on the surface of bubbles in liquid He^4. *Phys. Rev.*, **180**, 259–62.

Dahm, A. J. and Sanders, T. M. (1966). Relaxation time, effective mass, and structure of ions in liquid helium. *Phys. Rev. Lett.*, **17**, 126–30.

Dallacasa, V. (1979). The drift of thermal electrons in dense gases. *J. Phys. B: At. Mol. Phys.*, **12**, 3125–34.

Dash, J. G. and Taylor, R. D. (1956). Hydrodynamics of oscillating disks in viscous fluids: Density and viscosity of normal fluid in pure He^4 from 1.2 °K to the lambda point. *Phys. Rev.*, **105**, 7–24.

Davis, H. T. and Dagonnier, R. (1966). On Brownian motion in a Fermi liquid. *J. Chem. Phys.*, **44**, 4030–5.

Davis, H. T., Hiroike, K., and Rice, S. A. (1965). On the quantum-mechanical Fokker–Planck equation. *J. Chem. Phys.*, **43**, 2633–42.

Davis, H. T., Rice, S. A., and Meyer, L. (1962*a*). On the kinetic theory of dense fluids. XIII. The mobility of negative ions in liquid Ar, Kr, Xe. *J. Chem. Phys.*, **37**, 2470–2.

Davis, H. T., Rice, S. A., and Meyer, L. (1962*b*). Theory of electronic and ionic mobility in liquid He^4 I and liquid He^3. *Phys. Rev. Lett.*, **9**, 81–3.

de Magistris, P., Modena, I., and Scaramuzzi, F. (1965). Dependence of ionic mobilites in liquid ^3He on temperature density. In *Proceedings of the ninth international conference on low temperature physics* (ed. J. A. Daunt), Volume 9, p. 349. Plenum Press, New York.

De Riva, A. M. (1993). *Misure di mobilità di extraelettroni in gas He denso tra 25 e 80 K. Il fenomeno del self-trapping.* Ph.D. thesis, University of Padua.

De Toffol, G., Ancilotto, F., and Toigo, F. (1996). Excited states of alkali atoms in liquid ^4He from density functional calculations. *J. Low Temp. Phys.*, **102**, 624–53.

DeConde, K., Williams, G. A., and Packard, R. E. (1974). New factors affecting the measured lifetime of electrons trapped on vortex lines in He II. *Phys. Rev. Lett.*, **33**, 683–5.

Dexter, D. L. and Fowler, W. B. (1969). Instability of two-electron bubbles and bubble formation in liquid helium. *Phys. Rev.*, **183**, 307–9.

Di Castro, C. (1966). A phenomenological model for creation of vortices by ions in liquid helium II. *Nuovo Cimento*, **42 B**, 251–6.

Dietrich, O. W., Graf, E. H., Huang, C. H., and Passell, L. (1972). Neutron scattering by rotons in liquid helium. *Phys. Rev. A*, **5**, 1377–91.

Doake, C. S. M. and Gribbon, P. W. F (1969). Fast Ions in liquid helium. *Phys. Lett.*, **30 A**, 251–2.

Doake, C. S. M. and Gribbon, P. W. F. (1971). A hydrodynamic approach to the roton-limited mobility of ions in liquid helium. *J. Phys. C: Solid State Phys.*, **4**, 2457–65.

Doake, C. S. M. and Gribbon, P. W. F. (1972). Vorticity at a grid in liquid He II. *J. Phys. C: Solid State Phys.*, **5**, 2998–3011.

Domb, C. and Green, M. S. (ed.) (1976). *Phase transitions and critical phenomena.* Volume 5A. Academic Press, London.

Donnelly, R. J. (1965). Theory of the interaction of ions and quantized vortices in rotating helium II. *Phys. Rev. Lett.*, **14**, 39–41.

Donnelly, R. J. (1967). *Experimental superfluidity.* University of Chicago Press.

Donnelly, R. J. (1972). The Landau parameters in helium II. *Phys. Lett.*, **39 A**, 221–2.

Donnelly, R. J. (1991). *Quantized vortices in helium II.* Cambridge Studies in Low Temperature Physics. Cambridge University Press.

Donnelly, R. J. and Roberts, P. H. (1969*a*). Nucleation of quantized vortex rings by ions in helium II. *Phys. Rev. Lett.*, **23**, 1491–5.

Donnelly, R. J. and Roberts, P. H. (1969*b*). Stochastic theory of the interaction of ions and quantized vortices in helium II. *Proc. Roy. Soc. (London)*, **A 312**, 519–51.

Donnelly, R. J. and Roberts, P. H. (1971). Stochastic theory of the nucleation of quantized vortices in superfluid helium. *Phil. Trans. Roy. Soc. (London)*, **271 A**, 41–100.

Donnelly, R. J. and Roberts, P. H. (1977). Need for more precise thermodynamic and neutron scattering data on liquid helium. *J. Phys. C: Solid State Phys.*, **10**, L683–5.

Douglass, R. L. (1964). Ion trapping in rotating helium II. *Phys. Rev. Lett.*, **13**, 791–4.

Douglass, R. L. (1966). Capture width of vortex lines for negative ions in helium II. *Phys. Rev.*, **141**, 192–7.

Douglass, R. L. (1968). Mobility of negative ions along vortices in rotating superfluid helium. *Phys. Rev.*, **174**, 255–63.

Douglass, R. L. (1969). Escape of ions trapped in rotating superfluid helium below 1 °K. *Phys. Lett.*, **28 A**, 560–1.

DuVall, B. and Celli, V. (1969). Electronic properties of negative ions in liquid helium. *Phys. Rev.*, **180**, 276–86.

Dy, K. S. and Pethick, C. J. (1969). Transport coefficients of a normal Fermi liquid: Application to liquid He^3. *Phys. Rev.*, **185**, 373–84.

Ebner, C. and Edwards, D. O. (1970). The low temperature thermodynamic properties of superfluid solutions of ^{3}He in ^{4}He. *Phys. Rep.–Phys. Lett.*, **2 C**, 77–154.

Ebner, C. and Punyanitya, C. (1979). Density-functional theory of simple classical fluids. II. Localized excess electron states. *Phys. Rev. A*, **19**, 856–65.

Economou, E. N. and Cohen, M. H. (1970). Localization in disordered materials: Existence of mobility edge. *Phys. Rev. Lett.*, **25**, 1445–8.

Edel'shtein, V. M. (1983). Exchange scattering of quasiparticles by a positive ion in ^{3}He. *Sov. Phys.–JETP*, **58**, 317–29.

Eden, V. L. and McClintock, P. V. E. (1984). The effect of strong electric fields on exotic negative ions in He II: Possible evidence for the nucleation of charged vortex rings. *Phys. Lett.*, **102 A**, 197–200.

Edwards, D. O., Brewer, D. F., Seligman, P., Skertic, M., and Yaqub, M. (1965). Solubility of He^3 in liquid He^4 at 0 °K. *Phys. Rev. Lett.*, **15**, 773–5.

Edwards, D. O. and Saam, W. F. (1978). The free surface of liquid helium. In *Progress in low temperature physics* (ed. D. F. Brewer), Volume VII A, pp. 283–369. North Holland, Amsterdam.

Egelstaff, P. A. (1994). *An introduction to the liquid state.* Clarendon Press, Oxford.

Eggarter, T. P. (1972). Semiclassical theory of electron transport properties in a disordered material. *Phys. Rev. A*, **5**, 2496–509.

Eggarter, T. P. and Cohen, M. H. (1970). Simple model for density of states and mobility of an electron in a gas of hard-core scatterers. *Phys. Rev. Lett.*, **25**, 807–10.

Eggarter, T. P. and Cohen, M. H. (1971). Mobility of excess electrons in gaseous He: A semiclassical approach. *Phys. Rev. Lett.*, **27**, 129–32.

Eggarter, T. P. and Cohen, M. H. (1974). Comment on ''Electron localization by density fluctuations". *J. Phys. C: Solid State Phys.*, **7**, L103–6.

Einstein, A. (1906). Eine neue Bestimmung der Molekűldimensionen. *Ann. Phys.*, **19**, 289–306.

Ellis, T., Jewell, C. L., and McClintock, P. V. E. (1980*a*). Measurements of the Landau velocity in He II. *Phys. Lett.*, **78 A**, 358–60.

Ellis, T. and McClintock, P. V. E. (1982). Effective mass of the normal negative-charge carrier in bulk He II. *Phys. Rev. Lett.*, **48**, 1834–7.

Ellis, T. and McClintock, P. V. E. (1985). The breakdown of superfluidity in liquid ^{4}He. V. Measurements of the Landau critical velocity for roton creation. *Phil. Trans. Roy. Soc. (London)*, **315 A**, 259–300.

Ellis, T., McClintock, P. V. E., and Bowley, R. M. (1983). Pressure dependence of the negative ion effective mass in He II. *J. Phys. C: Solid State Phys.*, **16**, L485–9.

Ellis, T., McClintock, P. V. E., Bowley, R. M., and Allum, D. R. (1980*b*). The breakdown of superfluidity in liquid ^{4}He. II. An investigation of excitation emission from negative ions travelling at extreme supercritical velocities. *Phil. Trans. Roy. Soc. (London)*, **296 A**, 581–95.

Eloranta, J. and Apkarian, V. A. (2002). A time dependent density functional treatment of superfluid dynamics: Equilibration of the electron bubble in superfluid ^{4}He. *J. Chem. Phys.*, **117**, 10139–50.

Eloranta, J., Schwentner, N., and Apkarian, V. A. (2002). Structure and energetics of $He_2^\star$ bubble-states in superfluid ^{4}He. *J. Chem. Phys.*, **116**, 4039–53.

Esel'son, B. N., Kovdrya, Yu. Z., and Shikin, V. B. (1971). Mobility of positive and negative ions in He^3–He^4 superfluid solutions. *Sov. Phys.–JETP*, **32**, 37–43.

Esel'son, B. N., Kovdrya, Yu. Z., and Shikin, V. B. (1972). Some features of charged particle motion in He^3–He^4 solutions located in strong electric fields. *Sov. Phys.–JETP*, **35**, 344–8.

Eshchenko, D. G., Storchak, V. G., Brewer, J. H., Morris, G. D., Cottrell, S. P., and Cox, S. F. J. (2002). Excess electron transport and delayed muonium formation in condensed rare gases. *Phys. Rev B*, **66**, 035105.

Fárník, M., Samelin, B., and Toennies, J. P. (1999). Measurements of the lifetimes of electron bubbles in large size selected $^4He_N^-$ droplets. *J. Chem. Phys.*, **110**, 9195–201.

Fermi, E. (1934). Sopra lo spostamento per pressione delle righe elevate delle serie spettrali. *Nuovo Cimento*, **11**, 157–66.

Ferrell, R. A. (1957). Long lifetime of positronium in liquid helium. *Phys. Rev.*, **108**, 167–8.

Fetter, A. L. (1963). Vortices in an mmperfect Bose gas. I. The condensate. *Phys. Rev.*, **138**, A429–37.

Fetter, A. L. (1967). Quantum theory of superfluid vortices. I. Liquid helium II. *Phys. Rev.*, **162**, 143–53.

Fetter, A. L. (1971). Core structure of a quantized vortex. *Phys. Rev. Lett.*, **27**, 986–8.

Fetter, A. L. (1972). Boundary pressures and forces generated by moving vortices. *Phys. Rev. A*, **6**, 402–13.

Fetter, A. L. (1976). Vortices and ions in helium. In *The physics of liquid and solid helium* (ed. K. H. Bennemann and J. B. Ketterson), Volume 1, Chapter 3, pp. 207–305. Wiley, New York.

Fetter, A. L. and Iguchi, I. (1970). Low-temperature mobility of trapped ions in rotating He II. *Phys. Rev. A*, **2**, 2067–74.

Fetter, A. L. and Kurkijärvi, J. (1977). Motion of ions in normal and superfluid ^{3}He. *Phys. Rev. B*, **15**, 4272–80.

Fetter, A. L., Sauls, J. A., and Stein, D. L. (1983). Vortices in rotating superfluid ^{3}He–A. *Phys. Rev. B*, **28**, 5061–74.

Feynman, R. P. (1953). Atomic theory of liquid helium near absolute zero. *Phys. Rev.*, **91**, 1301–8.

Feynman, R. P. (1954). Atomic theory of the two-fluid model of liquid helium. *Phys. Rev.*, **94**, 262–77.

Feynman, R. P. and Cohen, M. (1956). Energy spectrum of the excitations in liquid helium. *Phys. Rev.*, **102**, 1189–204.

Főrste, M., Gűnther, H., Riediger, O., Wiebe, J., and zu Putlitz, G. (1997). Atoms and ions in superfluid helium (^{4}He). IV. Temperature dependence of the mobility of alkali earth ions. *Z. Phys. B*, **104**, 317–22.

Főrste, M., Gűnther, H., Riediger, O., Wiebe, J., and zu Putlitz, G. (1998). Temperature dependent mobility of alkali–earth ions in superfluid helium (^{4}He). *J. Low. Temp. Phys.*, **110**, 231–6.

Findlay, J. C., Pitt, A., Smithand, H. G., and Wilhelm, J. O. (1938). The velocity of sound in liquid helium. *Phys. Rev.*, **54**, 506–9.

Findlay, J. C., Pitt, A., Smithand, H. G., and Wilhelm, J. O. (1939). The velocity of sound in liquid helium under pressure. *Phys. Rev.*, **56**, 122.

Fixman, M. (1960). Density correlations, critical opalescence, and the free energy of nonuniform fluids. *J. Chem. Phys.*, **33**, 1357–62.

Fletcher, J. R. and Bowley, R. M. (1973). The photomobility of electron bubbles in liquid helium. *Phys. Lett.*, **44 A**, 121–2.

Floriano, M. A., Gee, N., and Freeman, G. R. (1987). Electron mobility in dense gases of deuterated methanes as functions of electric field strength, temperature, and density. *J. Chem. Phys.*, **87**, 4829–34.

Foldy, L. L. (1945). The multiple scattering of waves. I. General theory of isotropic scattering by randomly distributed scatterers. *Phys. Rev.*, **67**, 107–19.

Fowler, W. B. and Dexter, D. L. (1968). Electronic bubble states in liquid helium. *Phys. Rev.*, **176**, 337–43.

Gaeta, F. S. (1962). Ionic currents in β-irradiated liquid helium II. *Nuovo Cimento*, **26**, 1173–81.

Gamota, G. (1972). Growing collective modes in quantized vortex ring beams in He II. *Phys. Rev. Lett.*, **28**, 1691–93.

Gamota, G. and Barmatz, M. (1969). Impulse of a beam of quantized vortex rings. *Phys. Rev. Lett.*, **22**, 874–6.

Gamota, G., Hasegawa, A., and Varma, M. C. (1971). Evolution of a pulse of charged vortex rings in superfluid helium. *Phys. Rev. Lett.*, **26**, 960–3.

Gamota, G. and Sanders, T. M. (1965). Direct measurement of the size of charged quantized vortex rings in He II. *Phys. Rev. Lett.*, **15**, 949–51.

Gamota, G. and Sanders, T. M. (1968*a*). Erratum: Vortex-ring interactions in superfluid liquid helium. *Phys. Rev. Lett.*, **21**, 589.

Gamota, G. and Sanders, T. M. (1968*b*). Vortex-ring interactions in superfluid liquid helium. *Phys. Rev. Lett.*, **21**, 200–2.

Gamota, G. and Sanders, T. M. (1971). Size of quantized vortex rings in liquid helium II. *Phys. Rev. A*, **4**, 1092–9.

Garwin, R. L. and Reich, H. A. (1959). Self-diffusion and nuclear relaxation in He^3. *Phys. Rev.*, **115**, 1478–92.

Gasiorowicz, S. (1974). *Quantum Physics.* Wiley, New York.

Gavin, P. J. and McClintock, P. V. E. (1973). Field emission and field ionisation in liquid helium under pressure. *Phys. Lett.*, **43 A**, 257–8.

Gee, N. and Freeman, G. R. (1986). Electron transport in dense gases: Limitations on the Ioffe–Regel and Mott criteria. *Can. J. Chem.*, **64**, 1810–16.

Gűnther, H., Foerste, M., zu Putlitz, G., and von Stein, U. (1996). Atoms and ions in superfluid helium (^4He). III. Mobility measurements of alkali earth ions. *Z. Phys. B*, **101**, 613–17.

Giraud, V. and Krebs, P. (1982). The onset of electron localization in subcritical water vapor. *Chem. Phys. Lett.*, **86**, 85–90.

Glaberson, W. I. (1969). Trapped-ion motion in helium II. *J. Low Temp. Phys.*, **4**, 289–311.

Glaberson, W. I. and Johnson, W. W. (1975). Impurity ions in liquid helium. *J. Low Temp. Phys.*, **20**, 313–38.

Glaberson, W. I. and Steingart, M. (1971). Temperature dependence of the vortex-core parameter in He II. *Phys. Rev. Lett.*, **26**, 1423–5.

Glaberson, W. I., Strayer, D. M., and Donnelly, R. J. (1968*a*). Model for the core of a quantized vortex line in helium II. *Phys. Rev. Lett.*, **21**, 1740–4.

Glaberson, W. I., Strayer, D. M., and Donnelly, R. J. (1968*b*). Motion of negative ions along quantized vortex lines. *Phys. Rev. Lett.*, **20**, 1428–31.

Golov, A. (1995). Spectroscopic study of injected electrons in solid helium. *Z. Phys.*, **B 98**, 363–6.

Golov, A., Obara, K., Ikegami, H., Akimoto, H., and Ishimoto, H. (1997). Positive ion mobility in normal and superfluid ^{3}He at high magnetic field. *J. Low Temp. Phys.*, **110**, 225–30.

Gomer, R. (1961). *Field emission and field ionization.* Volume 9, Harvard Monographs in Applied Science. Harvard University Press, Cambridge, MA.

Goodstein, D. L. (1975). *States of matter.* Prentice–Hall, Englewood Cliffs, NJ.

Goodstein, D. L. (1977). Viscosity, ion mobility, and the λ transition. *Phys. Rev. B*, **15**, 5210–16.

Goodstein, D. L. (1978). A note on the structure of positive ions in liquid ^{4}He. *J. Low Temp. Phys.*, **33**, 137–44.

Goodstein, D. L., Buontempo, U., and Cerdonio, M. (1968). Ion mobility discontinuities in superfluid helium: A test of the Huang–Olinto theory. *Phys. Rev.*, **168**, 181–6.

Goodstein, D. L., Savoia, A., and Scaramuzzi, F. (1974). Critical behavior of ion mobilities near the superfluid transition. *Phys. Rev. A*, **9**, 2151–66.

Goodwin, J. M. (1968). *The pressure dependence of viscosity in liquid helium.* Ph.D. thesis, University of Washington, Seattle, WA.

Gould, H. and Ma, S. K. (1968). Low-temperature mobility of heavy impurities in Fermi liquids. *Phys. Rev. Lett.*, **21**, 1379–82.

Gould, H. and Ma, S. K. (1969). Low-temperature ion mobility in interacting Fermi liquids. *Phys. Rev.*, **183**, 338–48.

Green, M. S. (ed.) (1971). *Critical phenomena.* Volume LI, Varenna Lectures. Academic Press, New York.

Green, M. S. and Sengers, J. V. (ed.) (1966). *Critical phenomena. Proceedings of a conference.*, Washington, DC. National Bureau of Standards: US, GPO. Misc. Note No. 273.

Greer, S. C. and Moldover, M. R. (1981). Thermodynamic anomalies at critical points of fluids. *Annu. Rev. Phys. Chem.*, **32**, 233–65.

Greywall, D. S. and Ahlers, G. (1972). Second-sound velocity, scaling, and universality in He II under pressure near the superfluid transition. *Phys. Rev. Lett.*, **28**, 1251–4.

Grűnberg, R. (1969). Measurements of the electron drift velocity in helium gas at high pressures. *Z. Naturforsch.*, **24 A**, 1838–9.

Grilly, E. R. (1966). Compressibility of liquid He^4 as a function of pressure. *Phys. Rev.*, **149**, 97–101.

Grimes, C. C. and Adams, G. (1976). Observation of two-dimensional plasmons and electron–ripplon scattering in a sheet of electrons on liquid helium. *Phys. Rev. Lett.*, **36**, 145–8.

Grimes, C. C. and Adams, G. (1990). Infrared spectrum of the electron bubble in liquid helium. *Phys. Rev. B*, **41**, 6366–71.

Grimes, C. C. and Adams, G. (1992). Infrared-absorption spectrum of the electron bubble in liquid helium. *Phys. Rev. B*, **45**, 2305–10.

Grimm, G. F. and Rayfield, G. W. (1975). Evidence for an electronic bubble in liquid hydrogen. *Phys. Lett.*, **54 A**, 473–5.

Grimsrud, D. T. and Scaramuzzi, F. (1966). Measurements of the mobilities of positive and negative ions near the lambda transition. In *Proceedings of the tenth international conference on low temperature physics* (ed. M. P. Malkov), Volume I, pp. 197–205. Moscow Press, Moscow.

Grinfeld, P. and Kojima, H. (2003). Instability of the 2S electron bubbles. *Phys. Rev. Lett.*, **91**, 105301.

Gross, E. P. and Tung-Li, H. (1968). Vibrations of an electron bubble in liquid helium. *Phys. Rev.*, **170**, 190–3.

Guénault, T. (2003). *Basic superfluids.* Masters Series in Physics and Astronomy. Taylor & Francis, London.

Guggenheim, E. A. (1977). *Thermodynamics.* North Holland, Amsterdam.

Gully, W. J., Osheroff, D. D., Lawson, D. J., Richardson, R. C., and Lee, D. M. (1973). Effects of magnetic field on the "A" transition in liquid ^{3}He. *Phys. Rev. A*, **8**, 1633–7.

Guo, H. M., Edwards, D. O., Sarwinski, R. E., and Tough, J. J. (1971). Fermi-liquid behavior of ^{3}He adsorbed on liquid helium. *Phys. Rev. Lett.*, **27**, 1259–63.

Hakonen, P. J., Ikkala, O. T., and Islander, S. T. (1982*a*). Experiments on vortices in rotating superfluid ^{3}He–A. *Phys. Rev. Lett.*, **49**, 1258—61.

Hakonen, P. J., Ikkala, O. T., Islander, S. T., Lounasmaa, O. V., Markkula, T. K., Roubeau, P., Saloheimo, K. M., Volovik, G. E., Andronikashvili, E. L., Garibashvili, D. I., and Tsakadze, J. S. (1982*b*). NMR experiments on rotating superfluid ^{3}He–A: Evidence for vorticity. *Phys. Rev. Lett.*, **48**, 1838—41.

Hakonen, P. J., Krusius, M., Salomaa, M. M., Simola, J. T., Bunkov, Yu. M., Mineev, V. P., and Volovik, G. E. (1983). Magnetic vortices in rotating superfluid ^{3}He–B. *Phys. Rev. Lett.*, **51**, 1362–5.

Hakonen, P. J., Krusius, M., and Seppälä, H. K. (1985). NMR studies on vortices in rotating ^{3}He–A. *J. Low. Temp. Phys.*, **60**, 187–210.

Hall, H. E. (1960). The rotation of liquid helium II. *Adv. Phys.*, **9**, 89–146.

Hall, H. E. (1974). *Solid state physics.* Wiley, New York.

Hall, H. E. and Vinen, W. F. (1956). The rotation of liquid helium II. experiments on the propagation of second sound in uniformly rotating helium II. *Proc. Roy. Soc. (London)*, **A 238**, 204–14.

Hall, S. C., Classen, J., Su, C. K., and Maris, H. J. (1995). Experimental evidence for bubble nucleation on electrons in liquid ^{4}He. *J. Low Temp. Phys.*, **101**, 793–7.

Halley, J. W. and Cheung, A. (1968). Theory of vortex resonance in liquid helium. *Phys. Rev.*, **168**, 209–21.

Halley, J. W. and Ostermeier, R. M. (1977). A new calculation of the ion velocity shift in a vortex resonance experiment in He II. *J. Low Temp. Phys.*, **26**, 877–98.

Halpern, B. and Gomer, R. (1969*a*). Field emission in liquids. *J. Chem. Phys.*, **51**, 1031–47.

Halpern, B. and Gomer, R. (1969*b*). Field ionization in liquids. *J. Chem. Phys.*, **51**, 1048–56.

Hammer, D. C. and Webeler, R. W. H. (1965). Viscosity × normal density of liquid helium in a temperature interval about the lambda point. *Phys. Lett.*, **15**, A437–41.

Harrison, H. R., Sander, L. M., and Springett, B. E. (1973). Electron mobility and localization in dense ^{4}He gas. *J. Phys. B: At. Mol. Phys.*, **6**, 908–17.

Harrison, H. R. and Springett, B. E. (1971). Electron mobility variation in dense hydrogen gas. *Chem. Phys. Lett.*, **10**, 418–21.

Hart, H. R. and Wheatley, J. C. (1960). Self-diffusion in liquid He^3. *Phys. Rev. Lett.*, **4**, 3–5.

Heller, P. (1967). Experimental investigations of critical phenomena. *Rep. Prog. Phys.*, **30**, 731–826.

Hendry, P. C., Lawson, N. S., McClintock, P. V. E., Williams, C. D. H., and Bowley, R. M. (1988*a*). Macroscopic quantum tunnelling of vortices in He II. *Phys. Rev. Lett.*, **60**, 604–7.

Hendry, P. C., Williams, C. D. H., and McClintock, P. V. E. (1988*b*). Comment on nature of exotic negative carriers in superfluid ^{4}He. *Phys. Rev. Lett.*, **60**, 865.

Henshaw, D. G. (1959). Energy–momentum relation in liquid helium by inelastic scattering of neutrons. *Phys. Rev. Lett.*, **1**, 127–9.

Henshaw, D. G. and Woods, A. D. B. (1961*a*). Modes of atomic motions in liquid helium by inelastic scattering of neutrons. *Phys. Rev.*, **121**, 1266–74.

Henshaw, D. G. and Woods, A. D. B. (1961*b*). Modes of atomic motions in liquid helium by inelastic scattering of neutrons. In *Proceedings of the seventh international conference on low temperature physics* (ed. G. M. Graham and A. C. Hollis-Hallett), pp. 539–42. North Holland, Amsterdam.

Henson, B. L. (1964). Mobility of positive ions in liquefied Argon and Nitrogen. *Phys. Rev.*, **135**, A1002–8.

Herman, D., Mathur, H., and Dahm, A. J. (2001). Dephasing of electrons on helium by collisions with gas atoms. *Phys. Rev. B*, **63**, 115418.

Hernandez, J. P. (1973). Self-trapped states of an electron in structurally disordered systems. *Phys. Rev. A*, **2**, 1755–65.

Hernandez, J. P. (1975). Self-trapped states of electrons in dense fluids. *Phys. Rev. B*, **11**, 1289–96.

Hernandez, J. P. (1982). One-electron states in materials with structural disorder. *J. Phys. C: Solid State Phys.*, **15**, 1923–37.

Hernandez, J. P. (1991). Electron self-trapping in liquids and dense gases. *Rev. Mod. Phys.*, **63**, 675–97.

Hernandez, J. P. and Martin, L. W. (1991). Analysis of excess electron states in neon gas. *Phys. Rev. A*, **43**, 4568–71.

Hernandez, J. P. and Silver, M. (1970). Dynamics of electron-bubble formation in helium. *Phys. Rev. A*, **2**, 1949–54.

Hernandez, J. P. and Silver, M. (1971). Dynamics of electron-bubble formation in helium (E). *Phys. Rev. A*, **3**, 2152.

Hernandez, J. P. and Ziman, J. M. (1973). Electron localization by density fluctuations. *J. Phys. C: Solid State Phys.*, **6**, L251–3.

Hess, G. B. (1967). Angular momentum of superfluid helium in a rotating cylinder. *Phys. Rev.*, **161**, 189–93.

Hess, G. B. (1972). On focusing beams of charged vortex rings. *Phys. Lett.*, **41 A**, 275–6.

Hess, G. B. and Fairbank, W. M. (1967). Measurements of angular momentum in superfluid helium. *Phys. Rev. Lett.*, **19**, 216–18.

Hickman, A. P. and Lane, N. F. (1971). Localized excited states of helium in liquid helium. *Phys. Rev. Lett.*, **26**, 1216–19.

Hicks, W. M. (1884). On the steady motion and small vibrations of a hollow vortex. *Phil. Trans. Roy. Soc. (London)*, **A175**, 161–95.

Hicks, W. M. (1885). Researches on the theory of vortex rings. Part II. *Phil. Trans. Roy. Soc. (London)*, **A176**, 725–80.

Hilt, O., Schmidt, W. F., and Khrapak, A. G. (1994). Ionic mobilities in liquid xenon. *IEEE Trans. Diel.-El. Ins.*, **1**, 648–56.

Hiroike, K., Kestner, N. R., Rice, S. A., and Jortner, J. (1965). Study of the properties of an excess electron in liquid helium. II. A refined description of configuration changes in the liquid. *J. Chem. Phys.*, **43**, 2625–32.

Hohenberg, P. C. and Halperin, B. I. (1977). Theory of dynamic critical phenomena. *Rev. Mod. Phys.*, **49**, 435–79.

Holroyd, R. A. and Schmidt, W. F. (1989). Transport of electrons in nonpolar fluids. *Annu. Rev. Phys. Chem.*, **40**, 439–68.

Huang, K. (1963). *Statistical mechanics.* Wiley, New York.

Huang, K. and Olinto, A. C. (1965). Quantized vortex rings in superfluid helium. A phenomenological theory. *Phys. Rev.*, **139**, A1441–51.

Huang, S. S. S. and Freeman, G. R. (1978). Electron mobilities in gaseous, critical, and liquid xenon. Density, electric field, and temperature effects: Quasilocalization. *J. Chem. Phys.*, **68**, 1355–62.

Huang, S. S. S. and Freeman, G. R. (1981). Electron transport in gaseous and liquid argon: Effects of density and temperature. *Phys. Rev. A*, **24**, 714–24.

Huang, W. and Dahm, A. J. (1976). Mobilities of ions trapped on vortex lines in dilute ^{3}He–^{4}He solutions. *Phys. Rev. Lett.*, **36**, 1466–9.

Huggins, E. R. (1970). Exact Magnus-Force formula for three-dimensional Fluid-core vortices. *Phys. Rev. A*, **1**, 327–31.

Huggins, E. R. (1972). Force exerted by charged vortex rings. *Phys. Rev. Lett.*, **29**, 1067–70.

Huxley, L. G. H. and Crompton, R. W. (1974). *The diffusion and drift of electrons in gases.* Wiley, New York.

Iakubov, I. T. and Khrapak, A. G. (1978). On the annihilation rate of positronium in dense rare gases. *Appl. Phys.*, **16**, 179–83.

Iakubov, I. T. and Khrapak, A. G. (1982). Self-trapped states of positron and positronium in dense gases and liquids. *Rep. Prog. Phys.*, **45**, 697–751.

Iakubov, I. T. and Polishuk, A. Ya. (1982). Influence of multiple scattering processes on the electron mobility in moderately dense gases. *Phys. Lett.*, **91 A**, 67–9.

Ichikawa, T. and Yoshida, H. (1981). Localization of excess electrons in a compressible nonpolar fluid. *J. Chem. Phys.*, **75**, 5432–7.

Iguchi, I. (1971). Model for quantum scattering of rotons. I. Hard-sphere interaction. *J. Low Temp. Phys.*, **4**, 637–56.

Iguchi, I. (1972). Roton trapping by a vortex line. *Phys. Rev. A*, **6**, 1087–91.

Ihas, G. G. and Sanders, T. M. (1970). Injection and mobility of potassium ions in liquid helium. *Phys. Lett.*, **31 A**, 502–3.

Ihas, G. G. and Sanders, T. M. (1971). Exotic negative charges carriers in liquid helium. *Phys. Rev. Lett.*, **27**, 383–6.

Iino, M., Suzuki, M., and Ikushima, A. J. (1985). Surface tension of liquid ^{4}He. Surface energy of the Bose–Einstein condensate. *J. Low Temp. Phys.*, **61**, 155–69.

Ioffe, A. F and Regel, A. R. (1960). Non-crystalline, amorphous, and liquid electronic semiconductors. *Prog. Semicond.*, **4**, 237–91.

Iordanskii, S. V. (1965). Vortex ring formation in a superfluid. *Sov. Phys.–JETP*, **21**, 467–71.

Ishimoto, H., Yamaguchi, A., Obara, K., Efimov, V., Ueno, D., Masutomi, R., and Golov, A. (2003). Positive ion mobility in liquid He-3 under very high magnetic fields. *Physica B*, **329**, 342–5.

Ivanenko, D. and Kolesnikov, N. (1952). The electrino hypothesis. *Doklady Akad. Nauk (USSR)*, **87**, 923–25.

Jackson, J. D. (1998). *Classical electrodynamics.* Wiley, New York.

Jacobsen, F. M., Gee, N., and Freeman, G. R. (1989). Electron mobility in krypton dense gas as functions of electric field strength, temperature, and density. *J. Chem. Phys.*, **91**, 6943–47.

Jahnke, J. A., Meyer, L., and Rice, S. A. (1971). Zero-field mobility of an excess electron in fluid argon. *Phys. Rev. A*, **3**, 734–52.

Jahnke, J. A., Silver, M., and Hernandez, J. P. (1975). Mobility of excess electrons and O_2 formation in dense fluid helium. *Phys. Rev. B*, **12**, 3420–7.

Jiang, T., Kim, C., and Northby, J. A. (1993). Electron attachment to helium microdroplets: Creation induced magic? *Phys. Rev. Lett.*, **71**, 700–03.

Johnson, W. W. and Glaberson, W. I. (1972). Positive impurity ions in He II. *Phys. Rev. Lett.*, **29**, 214–17.

Johnson, W. W. and Glaberson, W. I. (1974). Size measurement of positive impurity ion in liquid helium. *Phys. Rev. A*, **10**, 868–71.

Jones, B. K. (1969). Critical velocities of ions in liquid helium. *Phys. Rev.*, **186**,

168–74.

Jortner, J. (1971). Theoretical studies of excess electron states in liquids. *Ber. Bunseng. Phys. Chem.*, **75**, 696–703.

Jortner, J., Kestner, N. R., Rice, S. A., and Cohen, M. H. (1965). Study of the properties of an excess electron in liquid helium. I. The nature of the electron–helium interactions. *J. Chem. Phys.*, **43**, 2614–25.

Josephson, B. D. and Lekner, J. (1969). Mobility of an impurity in a Fermi liquid. *Phys. Rev. Lett.*, **23**, 111–3.

Kadanoff, L. P. and Swift, J. (1968). Transport coefficients near the liquid–gas critical point. *Phys. Rev.*, **166**, 89–101.

Kalia, R. K., Vashishta, P., and de Leeuw, S. W. (1989). Quantum molecular dynamics study of electron transport in an external field. *J. Chem. Phys.*, **90**, 6802–3.

Kanorsky, S. I., Arndt, M., Dziewior, R., Weis, A., and Hãnsch, T. W. (1994). Optical spectroscopy of atoms in solid helium. *Phys. Rev. B*, **49**, 3645–47.

Kawasaki, K. (1966). Diffusion constants near the critical point for time-dependent Ising models. III. Self-diffusion constant. *Phys. Rev.*, **150**, 285–90.

Kawasaki, K. (1970). Kinetic equations and time correlation functions of critical fluctuations. *Ann. Phys. (NY)*, **61**, 1–56.

Kawasaki, K. (1976). Mode coupling and critical dynamics. In *Phase transitions and critical phenomena* (ed. C. Domb and M. S. Green), Volume 5A, Chapter 4, pp. 166–403. Academic Press, London.

Kawasaki, K. and Gunton, J. D. (1972). Dynamical behavior of He^3–He^4 mixtures near the tricritical point. *Phys. Rev. Lett.*, **29**, 1661–4.

Kawasaki, K. and Gunton, J. D. (1978). Critical dynamics. In *Progress in liquids physics* (ed. C. A. Croxton), pp. 175–211. Wiley, Chichester.

Keller, W. E. (1969). *Helium-3 and Helium-4*. Plenum Press, New York.

Kerr, E. C. (1954). Orthobaric densities of He^3 1.3 °K to 3.2 °K. *Phys. Rev.*, **96**, 551–4.

Keshishev, K. O., Kovdrya, Y. Z., Mezhov-Deglin, L. P., and Shal'nikov, A. I. (1969). Mobility of charges in liquid helium up to its solidification. *Sov. Phys.-JETP*, **29**, 53–7.

Kestner, N. R. (1973). Theory of electrons in polar fluids. In *Electrons in fluids. The nature of metal–ammonia solutions* (ed. J. Jortner and N. R. Kestner), pp. 1–28. Springer Verlag, Heidelberg.

Keto, J. W., Soley, F. J., Stockton, M., and Fitzsimmons, W. A. (1974). Dynamic properties of neutral excitations produced in electron-bombarded superfluid helium. II. Afterglow fluorescence of excited helium molecules. *Phys. Rev. A*, **10**, 887–96.

Khalatnikov, I. M. and Zharkov, V. N. (1957). Theory of diffusion and thermal conductivity for dilute solutions of He^3 in helium II. *Sov. Phys.-JETP*, **5**, 905–19.

Khrapak, A. G. and Iakubov, I. T. (1973). Theory of a bubble state of an electron in dense gases. *High Temp.*, **11**, 993–6.

Khrapak, A. G. and Iakubov, I. T. (1979). Electrons and positrons in dense gases. *Sov. Phys. Usp.*, **22**, 703–26.

Khrapak, A. G., Schmidt, W. F., and Volykhin, K. F. (1995). Structure of O_2^- in dense helium gas. *Phys. Rev. E*, **51**, 4804–06.

Khrapak, A. G., Tegeder, P., Illenberger, E., and Schmidt, W. F. (1999). The energy of a negative ion in a non-polar liquid. *Chem. Phys. Lett.*, **310**, 557–60.

Khrapak, A. G., Volykhin, K. F., and Schmidt, W. F. (1996). Structure and mobility of negative ions in dense rare gases. In *Imaging detectors in high energy, astroparticle and medical physics* (ed. J. Park), pp. 46–51. World Scientific, Singapore.

Khrapak, A. G. and Yakubov, I. T. (1975). Contribution to the theory of the states of electrons injected into a dense gas. *Sov. Phys.–JETP*, **42**, 1036–40.

Kierstead, H. A. (1967*a*). Lambda curve of liquid He^4. *Phys. Rev.*, **162**, 153–61.

Kierstead, H. A. (1967*b*). Lambda transformation of liquid He^4 at high pressures. *Phys. Rev.*, **153**, 258–62.

Kirkpatrick, T. R. and Belitz, D. (1986). Nonanalytic behavior of ultrasonic attenuation in disordered electronic systems. *Phys. Rev. B*, **34**, 2168–75.

Kirkwood, J. G. and Bethe, H. A. (1967). The pressure wave produced by an underwater explosion. In *Shock and detonation waves* (ed. W. W. Wood), Chapter 1, pp. 1–34. Gordon Breach, New York.

Kittel, C. (1986). *Introduction to solid state physics.* Wiley, New York.

Klahn, Th. and Krebs, P. (1998*a*). Electron and anion mobility in low density hydrogen cyanide gas. I. Dipole-bound electron ground states. *J. Chem. Phys.*, **109**, 3959–70.

Klahn, Th. and Krebs, P. (1998*b*). Electron and anion mobility in low density hydrogen cyanide gas. II. Evidence for the formation of the anion clusters. *J. Chem. Phys.*, **109**, 543–50.

Knapp, R. T., Daily, J. W., and Hammitt, F. G. (1970). *Cavitation.* McGraw–Hill, New York.

Kokko, J., Paalanen, M. A., Schoepe, W., and Takano, Y. (1978). Positive ion mobility in normal and superfluid ^{3}He. *J. Low Temp. Phys.*, **33**, 69–78.

Kondo, Y., Korhonen, J. S., Krusius, M., Dmitriev, V. V., Mukharsky, Y. M., Sonin, E. B., and Volovik, G. E. (1991). Direct observation of the nonaxisymmetric vortex in superfluid ^{3}B. *Phys. Rev. Lett.*, **67**, 81–4.

Kondo, Y., Korhonen, J. S., Krusius, M., Dmitriev, V. V., Thuneberg, E. V., and Volovik, G. E. (1992). Combined spin–mass vortex with soliton tail in superfluid ^{3}B. *Phys. Rev. Lett.*, **68**, 3331–4.

Konstantinov, D. and Maris, H. J. (2003). Detection of excited-state electron bubbles in superfluid helium. *Phys. Rev. Lett.*, **90**, 173–204.

Korhonen, J. S., Kondo, Y., Krusius, M., Thuneberg, E. V., and Volovik, G. E. (1993). Observation of combined spin–mass vortices in rotating ^{3}B. *Phys. Rev. B*, **47**, 8868–85.

Kramer, L. (1970*a*). Low-temperature ion mobility in Fermi liquids. *Phys. Rev. A*, **2**, 2063–67.

Kramer, L. (1970*b*). Low-temperature mobility of the electron bubble in dilute He^3–He^4 mixtures and in pure He^3. *Phys. Rev. A*, **1**, 1517–27.

Krebs, P. (1984). Localization of excess electrons in dense polar vapors. *J. Phys. Chem.*, **88**, 3702–9.

Krebs, P., Giraud, V., and Wantschik, M. (1980). Electron localization in dense ammonia vapor. *Phys. Rev. Lett.*, **44**, 211–13.

Krebs, P. and Heintze, M. (1982). Migration of excess electrons in high density supercritical ammonia. *J. Chem. Phys.*, **76**, 5484–92.

Krebs, P. and Lang, U. (1996). Electron mobility and multiple scattering effects in dense methanol gas. *J. Chem. Phys.*, **100**, 10482–9.

Kubo, R. (1957). Statistical-mechanical theory of irreversible processes, I. General theory and simple applications to magnetic and conduction problems. *J. Phys. Soc. Jpn.*, **12**, 570–86.

Kubo, R. (1966). The fluctuation–dissipation theorem. *Rep. Prog. Phys.*, **29**, 255–84.

Kuchnir, M., Ketterson, J. B., and Roach, P. R. (1971). Discontinuous field-velocity relation for vortex rings in superfluid solutions. *Phys. Lett.*, **36 A**, 287–8.

Kuchnir, M., Ketterson, J. B., and Roach, P. R. (1972). Ion motion in dilute ^{3}He–^{4}He solutions at ultralow temperatures. *Phys. Rev. A*, **6**, 341–55.

Kuchnir, M., Ketterson, J. B., and Roach, P. R. (1975). Anomalous temperature dependence of the positive ion mobility in liquid ^{3}He. *J. Low Temp. Phys.*, **19**, 531–5.

Kuchnir, M., Roach, P. R., and Ketterson, J. B. (1970). Low-temperature negative-ion mobility in liquid ^{3}He. *Phys. Rev. A*, **2**, 262–3.

Kuper, C. G. (1961). Theory of negative ions in liquid helium. *Phys. Rev.*, **122**, 1007–11.

Lamb, H. (1945). *Hydrodynamics*. Dover, New York.

Lamp, P. and Buschhorn, G. (1994). Electron transport in fluid argon in combined electric and magnetic fields. *Phys. Rev. B*, **50**, 16824–34.

Landau, L. D. (1941). The theory of superfluidity of helium II. *J. Phys. USSR*, **5**, 71–90.

Landau, L. D. (1947). On the theory of superfluidity of helium II. *J. Phys. USSR*, **11**, 91–2.

Landau, L. D. (1956). The theory of a Fermi liquid. *Sov. Phys.–JETP*, **3**, 920–27.

Landau, L. D. (1957). Oscillations in a Fermi liquid. *Sov. Phys.–JETP*, **5**, 101–11.

Landau, L. D. and Khalatnikov, I. M. (1949). The theory of the viscosity of helium II. I. Collisions of elementary excitations in helium II. *Sov. Phys.–JETP*, **19**, 637–53.

Landau, L. D. and Lifsits, E. M. (1958). *Statistical physics*. Addison–Wesley, New York.

Landau, L. D. and Lifsits, E. M. (2000). *Fluid mechanics.* Butterworth–Heinemann, Oxford.

Landau, L. D. and Lifsits, E. M. (2003). *Quantum mechanics: Non–relativistic theory.* Butterworth–Heinemann, London.

Langer, J. S. and Fisher, M. E. (1967). Intrinsic critical velocity of a superfluid. *Phys. Rev. Lett.*, **19**, 560–3.

Langlois, D., Sedraikan, D. M., and Carter, B. (1998). Differential rotation of relativistic superfluid in neutron stars. *Mon. Not. R. Astron. Soc.*, **297**, 1189–201.

Laria, D. and Chandler, D. (1987). Comparative study of theory and simulation calculations for excess electrons in simple fluids. *J. Chem. Phys.*, **87**, 4088–92.

Lattimer, J. M. and Prakash, M. (2001). Neutron star structure and the equation of state. *ApJ*, **550**, 426–42.

Lawson, D. T., Bozler, H. M., and Lee, D. M. (1975). Anisotropy in superfluid ^{3}He and the attenuation of zero sound. *Phys. Rev. Lett.*, **34**, 121–4.

Lawson, D. T., Gully, W. J., Goldstein, S., Richardson, R. C., and Lee, D. M. (1973). Attenuation of zero sound and the low-temperature transitions in liquid ^{3}He. *Phys. Rev. Lett.*, **30**, 541–4.

Lax, M. (1951). Multiple scattering of waves. *Rev. Mod. Phys.*, **23**, 287–310.

Lee, D. M. (1997). The extraordinary phases of liquid ^{3}He. *Rev. Mod. Phys.*, **69**, 645–65.

Leggett, A. J. (1972). Interpretation of recent results on He^3 below 3 mK: A new liquid phase? *Phys. Rev. Lett.*, **29**, 1227–30.

Leggett, A. J. (1973). Microscopic theory of NMR in an anisotropic superfluid (^{3}He–A). *Phys. Rev. Lett.*, **31**, 352–55.

Leggett, A. J. (1975). A theoretical description of the new phases of liquid ^{3}He. *Rev. Mod. Phys.*, **47**, 331–414.

Legler, W. (1970). Multiple scattering and drift of electrons in gases of high density. *Phys. Lett.*, **31 A**, 129–31.

Lehning, H. (1968). Resonance capture of very slow electrons in CO_2. *Phys. Lett.*, **28 A**, 103–4.

Leiderer, P. and Wanner, W. (1978). Positive-ion structure in a tricritical ^{3}He–^{4}He mixture. *Phys. Rev. Lett.*, **41**, 480–2.

Leiderer, P., Watts, D. R., and Webb, W. W. (1974). Light scattering by ^{3}He–^{4}He mixtures near the tricritical point. *Phys. Rev. Lett.*, **33**, 483–5.

Lekner, J. (1968). Scattering of waves by an ensemble of fluctuating potentials. *Philos. Mag.*, **18**, 1281–6.

Leung, K. and Chandler, D. (1994). Phase diagram for excess electrons in simple fluids. *Phys. Rev. E*, **49**, 2851–65.

Levchenko, A. A. and Mezhov-Deglin, L. P. (1992). Negative charges in liquid hydrogen and deuterium. *J. Low Temp. Phys.*, **89**, 457–63.

Levine, J. L. and Sanders, T. M. (1962). Anomalous electron mobility and complex negative ion formation in low-temperature helium vapors. *Phys. Rev. Lett.*, **8**, 159–61.

Levine, J. L. and Sanders, T. M. (1967). Mobility of electrons in low-temperature helium gas. *Phys. Rev.*, **154**, 138–49.

Levy, B. R. and Keller, J. B. (1963). Low-energy expansion of scattering phase shifts for long-range potentials. *J. Math. Phys.*, **4**, 54–64.

Lifsits, E. M. (1968). Theory of fluctuation levels in disordered systems. *Sov. Phys.–JETP*, **26**, 462–9.

Lifsits, E. M. and Pitaevskii, L. P. (1957). Absorption of second sound in rotating helium II. *Sov. Phys.–JETP*, **6**, 418–19.

Loeb, L. B. (1939). *Fundamental processes of electric discharge in gases.* Wiley, New York.

London, F. (1964). *Superfluids.* Volume II. Dover, New York.

Long, G. W. and Pickett, G. R. (1979). No non-linearity in the negative ionic mobility of normal liquid ^{3}He at low temperatures. *J. Phys. C: Solid State Phys.*, **12**, L9–11.

Lounasmaa, O. V. and Thuneberg, E. (1999). Vortices in rotating superfluid ^{3}He. *Proc. Natl. Acad. Sci. USA*, **96**, 7760–7.

Lovejoy, D. R. (1954). *Methods of concentrating helium-3 in helium-3–helium-4 mixtures.* Ph.D. thesis, University of Toronto.

Lovejoy, D. R. (1955). Surface tension of He^3. *Can. J. Phys.*, **33**, 49–53.

Loveland, R. J., le Comber, P. G., and Spear, W. E. (1972). Experimental evidence for electronic bubble states in liquid Ne. *Phys. Lett.*, **39 A**, 225–6.

Lovesey, S. W. (1984). *Theory of neutron scattering from condensed matter.* Clarendon Press, Oxford.

Lowke, J. J. (1963). The drift velocity of electrons in hydrogen and nitrogen. *Aust. J. Phys.*, **16**, 115–35.

Magerlein, J. H. and Sanders, T. M. (1976). Surface tension of H^4 near T_λ. *Phys. Rev. Lett.*, **36**, 258–61.

Maitland, G. C., Rigby, M., Smith, E. B., and Wakeham, W. A. (1981). *Intermolecular forces. Their origin and determination.* Clarendon Press, Oxford.

March, N. H. and Tosi, M. P. (1976). *Atomic dynamics in liquids.* McMillan Press, London.

Maris, H. J. (1995). Theory of quantum nucleation of bubbles in liquid helium. *J. Low Temp. Phys.*, **98**, 403–24.

Maris, H. J. (2000). On the fission of elementary particles and the evidence for fractional electrons in liquid helium. *J. Low Temp. Phys.*, **120**, 173–204.

Martin, J. L., Migus, A., Mourou, G. A., and Zewail, A. H. (ed.) (1993). *Ultrafast Phenomena.* Volume VIII. Springer–Verlag, Berlin.

Mason, E. A. and McDaniel, E. W. (1988). *Transport properties of ions in gases.* Wiley, New York.

Matejcik, S., Kiendler, A., Stampfli, P., and Mắrk, T. D. (1996). Vibrationally resolved electron attachment to oxygen clusters. *Phys. Rev. Lett.*, **77**, 3771–4.

Mathews, J. (1978). Drag force on a slowly moving sphere in a medium with variable viscosity. *Phys. Fluids*, **21**, 876–82.

McCarty, R. D. and Stewart, R. B. (1965). Technical report, National Bureau of Standards, US GPO, Washington, DC. Misc. Note No. 8726.

McCauley, J. and Onsager, L. (1975*a*). Electrons and vortex lines in He II, I. Brownian motion theory of capture and escape. *J. Phys. A: Math. Gen.*, **8**, 203–13.

McCauley, J. and Onsager, L. (1975*b*). Electrons and vortex lines in He II, II. Theoretical analysis of capture and release experiments. *J. Phys. A: Math. Gen.*, **8**, 882–90.

McClintock, P. V. E. (1969). Field emission in liquid helium. *Phys. Lett.*, **29 A**, 453–54.

McClintock, P. V. E. (1971). Field-emission and field-ionisation in liquid ^{3}He. *Phys. Lett.*, **35 A**, 211–12.

McClintock, P. V. E. (1973*a*). Field emission and field ionization in liquid ^{3}He above 0.25 K. *J. Low Temp. Phys.*, **11**, 15–31.

McClintock, P. V. E. (1973*b*). Ionic mobility in liquid ^{3}He above 0.25 K. *J. Low Temp. Phys.*, **11**, 277–88.

McClintock, P. V. E. (1973*c*). Temperature dependence of the field emission current in He II. *J. Phys. C: Solid State Phys.*, **6**, L186–90.

McClintock, P. V. E. (1995). Ions and the Landau critical velocity in He II. *Z. Phys. B*, **98**, 429–34.

McClintock, P. V. E. (1999). Vortex nucleation in superfluid ^{4}He. *J. Phys.: Condens. Matter*, **11**, 7695–709.

McClintock, P. V. E., Moss, F. E., and Bowley, R. M. (1980). Inhibition of vortex nucleation in He II by strong electric fields. *J. Phys. C: Solid State Phys.*, **13**, L503–9.

McClintock, P. V. E., Moss, F. E., Nancolas, G. G., and Stamp, P. C. E. (1981). Profound influence of isotopic impurities on vortex nucleation of He-II. *Physica*, **107 B**, 573–4.

McClintock, P. V. E. and Read-Forrest, H. (1973). Angular variation of current from field-emission and field-ionization sources in liquid helium. *Cryogenics*, **13**, 371–3.

McClintock, P. V. E. and Webster, A. G. (1974). Size effects in superfluid field emission. *Cryogenics*, **14**, 650–3.

Mermin, N. D. and Ho, T. L. (1976). Circulation and angular momentum in the A phase of superfluid helium-3. *Phys. Rev. Lett.*, **36**, 594–7.

Merzbacher, E. (1970). *Quantum mechanics.* Wiley, New York.

Meyer, L. (1966). Behavior of ions and quantized vortex rings of low energy in helium II around 0.3 °K. *Phys. Rev.*, **148**, 145–50.

Meyer, L., Davis, H. T., Rice, S. A., and Donnelly, R. J. (1962). Mobility of ions in liquid He^4 I and He^3 as a function of pressure and temperature. *Phys. Rev.*, **126**, 1927–34.

Meyer, L. and Reif, F. (1958). Mobilities of He ions in liquid helium. *Phys. Rev.*, **110**, 279–80.

Meyer, L. and Reif, F. (1960). Scattering of thermal energy ions in superfluid liquid He by phonons and ^{3}He atoms. *Phys. Rev. Lett.*, **5**, 1–3.

Meyer, L. and Reif, F. (1961). Ion motion in superfluid liquid helium under pressure. *Phys. Rev.*, **123**, 727–31.

Meyer, L. and Soda, Y. (1965). Behavior of quantized vortex rings near a wall. *Phys. Rev.*, **137**, A428–31.

Mikhin, K. V., Stepanov, S. V., and Byakov, V. M. (2003). Formation of the Ps bubble in liquid media. *Rad. Phys. Chem.*, **68**, 415–17.

Miller, B. N. (1994). Self trapping of light particles in fluids: the path integral approach. In *Linking the gaseous and condensed phases of matter. The behavior of slow electrons* (ed. L. G. Christophorou, E. Illenberger, and W. F. Schmidt), Volume B 326, NATO ASI Series, pp. 141–148. Plenum Press, New York.

Miller, B. N. and Fang, Y. (1990). Localization in fluids: A comparison of competing theories and their application to positron annihilation. *Phys. Rev. A*, **42**, 2228–34.

Miller, B. N. and Reese, T. (1989). Self-trapping of a light particle in a dense fluid: A mesoscopic model. *Phys. Rev. A*, **8**, 4735–48.

Mills, R. L. (2001). The nature of free electrons in superfluid helium–a test of quantum mechanics and a basis to review its foundations and make a comparison to classical theory. *Int. J. Hydrogen Energy*, **26**, 1059–96.

Mills, R. L. (2002). The grand unified theory of classical quantum mechanics. *Int. J. Hydrogen Energy*, **27**, 565–95.

Mineev, V. P. and Salomaa, M. M. (1984). Interaction of negative ions with vortices in superfluid ^{3}He–B. *J. Phys. C: Solid State Phys.*, **17**, L181–6.

Mistura, L. and Sette, D. (1966). Shape of the coexistence curve in the critical region. *Phys. Rev. Lett.*, **16**, 268–70.

Mitchell, R. P. and Rayfield, G. W. (1972). Electron escape from negatively charged vortex rings incident on the free surface of He II. *Phys. Lett.*, **42 A**, 267–8.

Miyakawa, T. and Dexter, D. L. (1970). Interpretation of photoejection experiments and the well depth of electronic bubbles in liquid helium. *Phys. Rev. A*, **1**, 513–18.

Modena, I. (1963). Ionic mobility in liquid ^{3}He. In *Proceedings of the eighth international conference on low temperature physics* (ed. R. O. Davies), Volume 1, p. 51. Butterworth, London.

Modena, I. and Ricci, F. P. (1967). Mobility of electrons in liquid He^3 at the critical Point. *Phys. Rev. Lett.*, **19**, 347–50.

Moldover, M. R. (1969). Scaling of the specific heat singularity of He^4 near its critical point. *Phys. Rev.*, **182**, 342–52.

Moldover, M. R. and Little, W. A. (1965). Specific heat of He^3 and He^4 in the neighborhood of their critical points. *Phys. Rev. Lett.*, **15**, 54–6.

Mott, N. F. (1967). On electrons in disordered systems. *Adv. Phys.*, **16**, 49–144.

Mott, N. F. (1974). *Metal–insulator transitions.* Taylor & Francis, London.

Muirhead, C. M., Vinen, W. F., and Donnelly, R. J. (1984). The nucleation of vorticity by ions in superfluid ^{4}He. I. Basic theory. *Phil. Trans. Roy. Soc.*

(London), **311 A**, 433–67.

Muirhead, C. M., Vinen, W. F., and Donnelly, R. J. (1985). The nucleation of vorticity by ions in superfluid ^{4}He. II. Theory of the effect of dissolved ^{3}He. *Proc. Roy. Soc. (London)*, **402 A**, 225–43.

Muzikar, P. (1976). The internal structure of a de Gennes disgyration in ^{3}He–A. *J. Phys. Coll.*, **C6**, 53–4.

Nakanishi, K. and Suzuki, M. (1998). Absolute value of the surface tension of liquid ^{4}He. *J. Low Temp. Phys.*, **113**, 585–90.

Nakano, H. (1957). A method of calculation of electrical conductivity. *Prog. Theor. Phys.*, **17**, 145–61.

Namaizawa, H. (1978). A shell model of positive ions. *J. Physique Colloque (Paris)*, **39**(C-6), 72–3.

Nancolas, G. G., Bowley, R. M., and McClintock, P. V. E. (1985*a*). The breakdown of superfluidity in liquid ^{4}He. IV. Influence of impurities on the nucleation of quantized vortex rings. *Phil. Trans. Roy. Soc. (London)*, **313 A**, 537–606.

Nancolas, G. G., Bowley, R. M., and McClintock, P. V. E. (1986). Vortex nucleation in He II: An interesting distinction between the intrinsic and thermally activated nucleation mechanisms. *J. Phys. C: Solid State Phys.*, **19**, L37–41.

Nancolas, G. G., Ellis, T., McClintock, P. V. E., and Bowley, R. M. (1985*b*). A new form of energy dissipation by a moving object in He II. *Nature*, **316**, 797–9.

Nancolas, G. G. and McClintock, P. V. E. (1982). Quenching of the ion/vortex-ring transition in He II by intense electric fields. *Phys. Rev. Lett.*, **48**, 1190–2.

Nancolas, G. G., McClintock, P. V. E., Moss, F. E., and Bowley, R. M. (1981). Temperature-dependent vortex nucleation in isotopically pure superfluid ^{4}He. *J. Phys. C: Solid State Phys.*, **14**, L681–6.

Neeper, D. A. (1968). Vortex-ring formation by negative ions in He II under pressure. *Phys. Rev. Lett.*, **21**, 274–5.

Neeper, D. A. and Meyer, L. (1969). Ion motion and vortex-ring formation in pure liquid He^4 and He^3–He^4 solutions between 0.05 and 0.5 K. *Phys. Rev.*, **182**, 223–34.

Neppiras, E. A. (1980). Acoustic cavitation. *Phys. Rep.*, **61**, 159–251.

Neri, D. (1996). *Stati di extraelettroni e trasporto di carica in gas nobili densi.* Ph.D. thesis, University of Padua.

Neri, D., Borghesani, A. F., and Santini, M. (1997). Electron attachment to O_2^- molecules in dense helium and argon gases. *Phys. Rev. E*, **56**, 2137–42.

Nieminen, R. M., Vǎlimaa, I., Manninen, M., and Hautojǎrvi, P. (1980). Density-functional theory of positronium and electron bubbles in helium fluids. *Phys. Rev. A*, **21**, 1677–86.

Noble, J. D. and Bloom, M. (1965). Self-diffusion in ethane near the critical point. *Phys. Rev. Lett.*, **14**, 250–1.

Northby, J. A. and Donnelly, R. J. (1970). Detection of a vortex-free region in rotating liquid helium II. *Phys. Rev. Lett.*, **25**, 214–17.

Northby, J. A., Kim, C., and Jiang, T. (1994). Negatively charged helium microdroplets. *Physica B*, **197**, 426–34.

Northby, J. A. and Sanders, T. M. (1967). Photoejection of electrons from bubble state in liquid helium. *Phys. Rev. Lett.*, **18**, 1184–6.

Nummila, K. K., Simola, J. T., and Korhonen, J. S. (1989). Experiments with negative ions in ^{3}he superfluids. *J. Low Temp. Phys.*, **75**, 111–57.

Obara, K., Masutomi, R., Yamaguchi, A., and Ishimoto, H. (2000). Positive ion mobility in normal liquid ^{3}He under high magnetic fields. *J. Low Temp. Phys.*, **121**, 597–602.

Obara, K., Ueno, D., Yamaguchi, A., Efimov, V., and Ishimoto, H. (2001). Anomalous magnetic-field dependence of positive ion mobility in normal liquid ^{3}He. *Phys. Rev. Lett.*, **87**, 235301.

Ohmi, T., Tsuneto, T., and Usui, T. (1969). Phase separation in rotating helium. *Prog. Theor. Phys.*, **41**, 1395–400.

Ohmi, T. and Usui, T. (1969). Superfluid vortex trapping neutral impurities. *Prog. Theor. Phys.*, **41**, 1401–15.

O'Malley, T. F. (1963). Extrapolation of electron–rare gas atom cross sections to zero energy. *Phys. Rev.*, **130**, 1020–9.

O'Malley, T. F. (1980). Multiple scattering effects on electron mobilities in dense gases. *J. Phys. B: At. Mol. Phys.*, **13**, 1491–504.

O'Malley, T. F. (1992). Inhomogeneous gas model for electron mobility in high density neon gas. *J. Phys. B: At. Mol. Opt. Phys.*, **25**, 168–80.

O'Malley, T. F., Burke, P. G., and Berrington, K. A. (1979). R-matrix calculation of low-energy e–He scattering. *J. Phys. B: At. Mol. Phys.*, **12**, 953–65.

O'Malley, T. F. and Crompton, R. W. (1980). Electron–neon scattering length and S-wave phaseshifts from drift velocities. *J. Phys. B: At. Mol. Phys.*, **13**, 3451–64.

Onn, D. G. and Silver, M. (1969). Attenuation and lifetime of hot electrons injected into liquid helium. *Phys. Rev.*, **183**, 295–307.

Onn, D. G., Smejtek, P., and Silver, M. (1974). Cryogenic thin-film electron emitters. *J. Appl. Phys.*, **45**, 119–25.

Onsager, L. (1949). Comment on a paper of C. J. Gorter. *Nuovo Cimento*, **Suppl. 6**, 249–50.

Osheroff, D. D., Gully, W. J., Richardson, R. C., and Lee, D. M. (1972*a*). New magnetic phenomena in liquid He^3 below mK. *Phys. Rev. Lett.*, **29**, 920–3.

Osheroff, D. D., Richardson, R. C., and Lee, D. M. (1972*b*). Evidence for a new phase of solid ^{3}He. *Phys. Rev. Lett.*, **28**, 885–8.

Ostermeier, R. M. (1973). Pressure dependence of charge carrier mobilities in superfluid helium. *Phys. Rev. A*, **8**, 514–29.

Ostermeier, R. M. and Glaberson, W. I. (1974). The capture of ions by vortex lines at low temperatures. *Phys. Lett.*, **49 A**, 223–4.

Ostermeier, R. M. and Glaberson, W. I. (1975*a*). Motion of ions trapped on vortices in He II. *Phys. Rev. Lett.*, **35**, 241–4.

Ostermeier, R. M. and Glaberson, W. I. (1975*b*). Possible evidence for the condensation of ^{3}He atoms onto quantized vortex lines. *Phys. Lett.*, **51 A**, 403–4.

Ostermeier, R. M. and Glaberson, W. I. (1975*c*). The capture of ions by vortex lines in pure ^{4}He and ^{3}He–^{4}He solutions. *J. Low Temp. Phys.*, **20**, 159–79.

Ostermeier, R. M. and Glaberson, W. I. (1975*d*). The escape of positive ions from quantized vortex lines in superfluid helium. *Phys. Lett.*, **51 A**, 348–50.

Ostermeier, R. M. and Glaberson, W. I. (1976). The mobility of ions trapped on vortex lines in pure ^{4}He and ^{3}He–^{4}He solutions. *J. Low Temp. Phys.*, **25**, 317–51.

Ostermeier, R. M. and Schwarz, K. W. (1972). Motion of charge carriers in normal He^4. *Phys. Rev. A*, **5**, 2510–19.

Ostermeier, R. M., Yarmchuk, E. J., and Glaberson, W. I. (1975). Condensation of He^3 atoms onto quantized vortex lines in superfluid helium. *Phys. Rev. Lett.*, **35**, 957–60.

Ott-Rowland, M. L., Kotsubo, V., Theobald, J., and Williams, G. A. (1982). Two-dimensional plasma resonances in positive ions under the surface of liquid helium. *Phys. Rev. Lett.*, **49**, 1708–12.

Packard, R. E. and Sanders, T. M. (1969). Detection of single quantized vortex lines in rotating He II. *Phys. Rev. Lett.*, **22**, 823–6.

Packard, R. E. and Sanders, T. M. (1972). Observations on single vortex lines in rotating superfluid helium. *Phys. Rev. A*, **6**, 799–807.

Padmore, T. C. (1971). High-field behavior of ions in He II. *Phys. Rev. Lett.*, **26**, 63–6.

Padmore, T. C. (1972*a*). Motion of ions in helium II. *Phys. Rev. A*, **5**, 356–72.

Padmore, T. C. (1972*b*). Thermally assisted escape of ions from vortices in He II. *Phys. Rev. Lett.*, **28**, 469–72.

Padmore, T. C. and Cole, M. W. (1974). Free surface of He II and the electron bubble. *Phys. Rev. A*, **9**, 802–7.

Parks, P. E. and Donnelly, R. J. (1966). Ion trapping in rotating helium II. *Phys. Rev. Lett.*, **16**, 45–8.

Parshin, A. Ya. and Pereverzev, S. V. (1990). Direct observation of optical absorption by excess electrons in liquid helium. *Sov. Phys.–JETP Lett.*, **52**, 282–4.

Parshin, A. Ya. and Pereverzev, S. V. (1992). Spectroscopic study of excess electrons in liquid helium. *JETP*, **197**, 68–76.

Pauling, L. (1967). *The chemical bond.* Cornell University Press, Ithaca, NY.

Paulson, D. N., Kleinberg, R. L., and Wheatley, J. C. (1976). Zero-sound studies in superfluid ^{3}He. *J. Low Temp. Phys.*, **23**, 725–41.

Pereverzev, S. V. and Parshin, A. Ya. (1994). Spectroscopic study of excess electrons in liquid helium. *Physica B*, **197**, 347–59.

Phelps, A. V. and Brown, S. C. (1952). Positive ions in the afterglow of a low pressure helium discharge. *Phys. Rev.*, **86**, 102–5.

Phelps, A. V., Pack, J. L., and Frost, L. S. (1960). Drift velocity of electrons in helium. *Phys. Rev.*, **117**, 470–4.

Phillips, A. and McClintock, P. V. E. (1973). Evidence for roton creation in a superfluid field emission diode near 0.5 K. *Phys. Lett.*, **46 A**, 109–10.

Phillips, A. and McClintock, P. V. E. (1974*a*). Breaking the roton barrier: An experimental study of motion faster than the Landau critical velocity for roton creation in He II. *Phys. Rev. Lett.*, **33**, 1468–71.

Phillips, A. and McClintock, P. V. E. (1974*b*). The vortex ring nucleation rate due to negative ions travelling near the Landau velocity in He II. *J. Phys. C: Solid State Phys.*, **7**, L118–22.

Phillips, A. and McClintock, P. V. E. (1975). Field emission and field ionization in liquid ^{4}He. *Phil. Trans. R. Soc. London*, **A 278**, 271–310.

Pines, D. and Nozieres, P. (1966). *The theory of quantum liquids.* Volume *I: Normal Fermi Liquids.* Benjamin, New York.

Pitaevskii, L. (1959). Calculation of the phonon part of the mutual frictional force in superfluid helium. *Sov. Phys.-JETP*, **8**, 888–90.

Pitaevskii, L. P. (1960). BCS theory generalized to pairing with nonzero orbital momentum and spin. *Sov. Phys.-JETP*, **10**, 1267–71.

Plenkiewicz, B., Frongillo, Y., Plenkiewicz, P., and Jay-Gerin, J.-P. (1991). Density dependence of the conduction-band energy and of the effective mass of quasifree excess electrons in fluid neon and helium. *Phys. Rev. A*, **43**, 7061–63.

Plenkiewicz, B., Plenkiewicz, P., and Jay-Gerin, J.-P. (1989). Calculation of the conduction-band energy of excess electrons in fluid helium. *Chem. Phys. Lett.*, **163**, 542–4.

Plesset, M. S. (1949). The dynamics of cavitation bubbles. *J. Appl. Mech.*, **16**, 277–82.

Poitrenaud, J. (1970). Low temperature study by electron paramagnetic resonance of the susceptibility of the compound C_8K. *Rev. Phys. Appl.*, **5**, 275–81.

Poitrenaud, J. and Williams, F. I. B. (1972). Precise measurements of effective mass of positive and negative charge carriers in liquid helium II. *Phys. Rev. Lett.*, **29**, 1230–2.

Poitrenaud, J. and Williams, F. I. B. (1974). Erratum: Precise measurements of effective mass of positive and negative charge carriers in liquid helium II. *Phys. Rev. Lett.*, **32**, 1213.

Polishuk, A. Ya. (1983*a*). Quantum corrections to electron conductivity in a disordered medium of anisotropic scatterers. *J. Phys. B: At. Mol. Phys.*, **6**, 3853–58.

Polishuk, A. Ya. (1983*b*). Quantum corrections to the Boltzmann equation for electrons in a disordered medium. *J. Phys. B: At. Mol. Phys.*, **16**, 3845–51.

Polishuk, A. Ya. (1984). Theory of electron mobility in dense gases with small polarizability. *Physica C*, **124 C**, 91–5.

Pollock, E. L. and Alder, B. J. (1978). Charged particles in polarizable fluids. *Phys. Rev. Lett.*, **41**, 903–6.

Pollock, F. (1971). Analysis of the temperature dependence of the vortex core parameter in He II. *Phys. Rev. Lett.*, **27**, 303–4.

Pratt, W. P. and Zimmermann, W. (1969). Trapping lifetime of negative ions in rotating superfluid helium under pressure. *Phys Rev.*, **177**, 412–26.

Prokof'ev, N. V. (1993). Diffusion of a heavy particle in a Fermi-liquid theory. *Int. J. Mod. Phys. B*, **33**, 3327–51.

Prokof'ev, N. V. (1994). Theory of ion motion in normal liquid ^{3}He. *Physica B*, **194–196**, 747–8.

Prokof'ev, N. V. (1995). Effective masses of ions in superfluid ^{3}He–B. *Phys. Rev. Lett.*, **74**, 2748–51.

Rabinovich, V. A., Vasserman, A. A., Nedostup, V. I., and Veksler, L. S. (1988). *Thermodynamical properties of Ne, Ar, Kr, and Xe*. Hemisphere, Washington, DC.

Rae, A. I. M. and Vinen, W. F. (2001). Electrino collapse. *J. Low Temp. Phys.*, **123**, 1–5.

Rainer, D. and Vuorio, M. (1977). Small objects in superfluid ^{3}He. *J. Phys. C: Solid State Phys.*, **10**, 3093–106.

Raja Gopal, E. S. (1964). Oscillations of quantized vortices in rotating liquid helium II. *Ann. Phys. (NY)*, **29**, 350–65.

Rayfield, G. W. (1966). Roton emission from negative ions in helium II. *Phys. Rev. Lett.*, **16**, 934–6.

Rayfield, G. W. (1967). Evidence for a peeling model of vortex ring formation by ions in liquid helium. *Phys. Rev. Lett.*, **19**, 1371–3.

Rayfield, G. W. (1968*a*). Study of the ion–vortex-ring transition. *Phys. Rev.*, **168**, 222–33.

Rayfield, G. W. (1968*b*). Vortex-ring creation by negative ions in dilute mixtures of He^3–He^4. *Phys. Rev. Lett.*, **20**, 1467–8.

Rayfield, G. W. and Reif, F. (1963). Evidence for the creation and motion of quantized vortex rings in superfluid helium. *Phys. Rev. Lett.*, **11**, 305–8.

Rayfield, G. W. and Reif, F. (1964). Quantized vortex rings in superfluid helium. *Phys. Rev.*, **136**, A1194–208.

Rayfield, G. W. and Schoepe, W. (1971*a*). Thermal emission of electrons from liquid helium. *Z. Naturforsch.*, **A 26**, 1392–97.

Rayfield, G. W. and Schoepe, W. (1971*b*). Trapping and escape of negative ions from the surface of liquid ^{4}He. *Phys. Lett.*, **34 A**, 133–4.

Rayfield, G. W. and Schoepe, W. (1971*c*). Tunneling from electronic bubble states through the liquid–vapor interface of He II. *Phys. Lett.*, **37 A**, 417–18.

Reichert, J. F. and Dahm, A. J. (1974). Observation of electron spin resonance of negative ions in liquid helium. *Phys. Rev. Lett.*, **32**, 271–4.

Reichert, J. F., Jarosik, F., Herrick, R., and Andersen, J. (1979). Observation of electron spin resonance of negative ions in liquid ^{3}He. *Phys. Rev. Lett.*, **42**, 1359–61.

Reichert, J. F. and Jarosik, N. C. (1983). Magnetic-resonance studies of negative ions in liquid ^{3}He–^{4}He mixtures. *Phys. Rev. B*, **27**, 2710–21.

Reif, F. and Meyer, L. (1960). Study of superfluidity in liquid He by ion motion. *Phys. Rev.*, **119**, 1164–73.

Reut, L. S. and Fizher, I. Z. (1969). Adsorption of atomic admixtures on quantized vortexes in liquid helium II. *Sov. Phys.–JETP*, **28**, 375–84.

Rips, I. (1997). Cavitation model of electron solvation dynamics: Effect of the energy dissipation. *J. Chem. Phys.*, **106**, 2702–11.

Roach, P. D., Ketterson, J. B., and Roach, P. R. (1977*a*). Mobility of positive and negative ions in superfluid ^{3}He. *Phys. Rev. Lett.*, **39**, 626–9.

Roach, P. D., Ketterson, J. B., and Roach, P. R. (1977*b*). Positive ion mobility in normal ^{3}He. *Phys. Lett.*, **63**, 273–4.

Roach, P. R., Abraham, B. M., Ketterson, J. B., and Kuchnir, M. (1972). Absence of a quadratic term in the ^{4}He excitation spectrum. *Phys. Rev. Lett.*, **29**, 32–4.

Roach, P. R. and Douglass, D. H. (1966). Coexistence curve of He^4 near the critical point. *Phys. Rev. Lett.*, **17**, 1083–6.

Roach, P. R., Ketterson, J. B., and Roach, P. D. (1979). Positive-ion mobility in dilute solutions of ^{4}He in ^{3}He. *J. Low Temp. Phys.*, **34**, 169–95.

Roberts, P. H. and Donnelly, R. J. (1970). Dynamics of vortex rings. *Phys. Lett.*, **31 A**, 137–8.

Robertson, A. G. (1977). Drift velocities of low energy electrons in argon at 293 and 90 K. *Aust. J. Phys.*, **30**, 61–72.

Roobol, L. P., Remeijer, P., Steel, S. C., Jochemsen, R., Shumeiko, V. S., and Frossati, G. (1997). Shear viscosity of superfluid ^{3}He up to 15 Tesla. *Phys. Rev. Lett.*, **79**, 685–8.

Root, L. J., Bauer, J. D., and Skinner, J. L. (1988). New approach to localization: Quantum connectivity. *Phys. Rev B*, **37**, 5518–21.

Root, L. J. and Skinner, J. L. (1988). Localization phase diagram for the energetically and substitutionally disordered Anderson/quantum percolation model. *J. Chem. Phys.*, **89**, 3279–84.

Rosch, A. and Kopp, T. (1998). Giant mass and anomalous mobility of particles in a fermionic system. *Phys. Rev. Lett.*, **80**, 4705–8.

Rosenblit, M. and Jortner, J. (1995). Dynamics of the formation of an electron bubble in liquid helium. *Phys. Rev. Lett.*, **175**, 4079–82.

Rosenblit, M. and Jortner, J. (1997). Dynamics of excess electron localization in liquid helium and neon. *J. Phys. Chem. A*, **101**, 751–7.

Rudnick, I. and Shapiro, K. A. (1965). Velocity of sound in liquid helium at its lambda point. *Phys. Rev. Lett.*, **15**, 386–9.

Saam, W. F. and Laheurte, J. P. (1971). Theory of dilute solutions of ^{4}He in liquid ^{3}He at low temperatures. *Phys. Rev. A*, **4**, 1170–5.

Sadd, M., Chester, G. V., and Pederiva, F. (1999). Binding of impurities to a ^{4}He superfluid vortex. *Phys. Rev. Lett.*, **83**, 5310–13.

Sakai, Y., Bőttcher, E. H., and Schmidt, W. F. (1982). Excess electrons in liquid hydrogen, liquid neon, and liquid helium. *J. Electrostatics*, **12**, 89–96.

Sakai, Y., Bőttcher, E. N., and Schmidt, W. F. (1981). Excess electrons in liquid hydrogen and liquid neon: Photo injection and mobility. *J. Jpn. Inst. Electr. Eng.*, **61**, 499–503.

Sakai, Y., Schmidt, W. F., and Khrapak, A. G. (1992). High- and low-mobility electrons in liquid neon. *Chem. Phys.*, **164**, 139–52.

Salmelin, R. H. and Salomaa, M. M. (1987*a*). Anisotropy of ion mobility and the superfluid energy gap in ^{3}He–A. *J. Phys. C: Solid State Phys.*, **20**, L681–8.

Salmelin, R. H. and Salomaa, M. M. (1987*b*). Ion mobility along superfluid vortices with polar cores in ^{3}He–A. *J. Phys. C: Solid State Phys.*, **20**, L689–95.

Salmelin, R. H. and Salomaa, M. M. (1990). Resonant quasiparticle–ion scattering in anisotropic superfluid ^{3}He. *Phys. Rev. B*, **41**, 4142–63.

Salomaa, M., Pethick, C. J., and Baym, G. (1980*a*). Mobility tensor of negative ions in superfluid ^{3}He–A. *J. Low Temp. Phys.*, **40**, 297–356.

Salomaa, M., Pethick, C. J., and Baym, G. (1980*b*). Mobility tensor of the electron bubble in superfluid ^{3}He–A. *Phys. Rev. Lett.*, **44**, 998–1001.

Salomaa, M. M. and Volovik, G. E. (1983). Vortices with ferromagnetic superfluid core in ^{3}He–B. *Phys. Rev. Lett.*, **51**, 2040–3.

Salomaa, M. M. and Volovik, G. E. (1985*a*). Classification of axisymmetric vortices in ^{3}He–A. *Phys. Rev. Lett.*, **54**, 2127–30.

Salomaa, M. M. and Volovik, G. E. (1985*b*). Symmetry and structure of quantized vortices in superfluid ^{3}He–B. *Phys. Rev. B*, **31**, 203–27.

Salomaa, M. M. and Volovik, G. E. (1986). Vortices with spontaneously broken axisymmetry in ^{3}He–B. *Phys. Rev. Lett.*, **56**, 363–6.

Salomaa, M. M. and Volovik, G. E. (1987). Quantized vortices in superfluid ^{3}He. *Rev. Mod. Phys.*, **59**, 533–613.

Samuels, D. C. and Donnelly, R. J. (1991). Motion of charged vortex rings in helium II. *Phys. Rev. Lett.*, **67**, 2505–8.

Sanders, T. M. and Ihas, G. G. (1987). Nature of exotic negative carriers in superfluid ^{4}He. *Phys. Rev. Lett.*, **59**, 1722–25.

Scaramuzzi, F. (1963). Technical Report LNF 63/79, CNEN, Rome. Unpublished.

Scaramuzzi, F., Modena, I., Goodstein, D. L., and Cole, M. W. (1977*a*). Viscosity and positive-ion mobility near the melting transition in liquid ^{4}He. *Phys. Rev. B*, **16**, 3108–21.

Scaramuzzi, F., Savoia, A., Goodstein, D. L., and Cole, M. W. (1977*b*). Anomalous ion radius near the lambda transition. *J. Phys. C: Solid State Phys.*, **10**, L413–16.

Schappert, G. T. (1968). Mobility of a charged impurity in a Fermi liquid. *Phys. Rev.*, **168**, 162–6.

Schmidt, W. F., Illenberger, E., Khrapak, A. G., and Sakai, Y. (2001). Conjecture of the lifetime of delocalized electrons in liquid helium. *J. Chem. Phys.*, **115**, 10048–9.

Schmidt, W. F., Illenberger, E., Khrapak, A. G., Sakai, Y., and Yoshino, K. (2003). Electronic conduction and breakdown in liquid helium and liquid neon. *IEEE–TDEI*, **10**, 1012–21.

Schmidt, W. F., Volykhin, K. F., Khrapak, A. G., and Illenberger, E. (1999).

Structure and mobility of positive and negative ions in non-polar fluids. *J. Electrostatics*, **47**, 83–95.

Schoepe, W. and Probst, C. (1970). Evidence for a superfluid-induced surface barrier for electrons in liquid helium. *Phys. Lett.*, **31 A**, 490–1.

Schoepe, W. and Rayfield, G. W. (1973). Tunneling from electronic bubble states in liquid helium through the liquid–vapor interface. *Phys. Rev. A*, **7**, 2111–21.

Schwarz, K. W. (1968). Interaction of quantized vortex rings with quantized vortex lines in rotating He II. *Phys. Rev.*, **165**, 323–34.

Schwarz, K. W. (1970). Existence of ion mobility discontinuities in superfluid helium. *Phys. Rev. Lett.*, **24**, 648–51.

Schwarz, K. W. (1972*a*). Charge-carrier mobilities in liquid helium at the vapor pressure. *Phys. Rev. A*, **6**, 837–44.

Schwarz, K. W. (1972*b*). He^3-limited drift of charge carriers in superfluid He^4 $(0.3 < T < 1.0)\,°K$. *Phys. Rev. A*, **6**, 1947–49.

Schwarz, K. W. (1972*c*). Scattering of phonons by charge carriers in superfluid helium: The zero-velocity limit. *Phys. Rev. A*, **6**, 1958–66.

Schwarz, K. W. (1974). Spherical probes and quantized vortices: Hydrodynamic formalism and simple applications. *Phys. Rev. A*, **10**, 2306–17.

Schwarz, K. W. (1975). Mobilities of charge carriers in superfluid helium. *Adv. Chem. Phys.*, **33**, 1–49.

Schwarz, K. W. (1978). Anomalous electron mobilities in dense helium gas. *Phys. Rev. Lett.*, **41**, 239–42.

Schwarz, K. W. (1980). Electron localization in dense helium gas: New experimental results. *Phys. Rev. B*, **21**, 5125–36.

Schwarz, K. W. and Donnelly, R. J. (1966). Quantized vortex rings in rotating helium II. *Phys. Rev. Lett.*, **17**, 1088–90.

Schwarz, K. W. and Jang, P. S. (1973). Creation of quantized vortex rings by charge carriers in superfluid helium. *Phys. Rev. A*, **8**, 3199–210.

Schwarz, K. W. and Prasad, B. (1976). Dynamics of electron localization in dense helium gas. *Phys. Rev. Lett.*, **36**, 878–81.

Schwarz, K. W. and Stark, R. W. (1968). Phonon-limited drift of the electron bubble in superfluid helium. *Phys. Rev. Lett.*, **21**, 967–9.

Schwarz, K. W. and Stark, R. W. (1969). Scattering of positive ions by elementary excitations in superfluid helium. *Phys. Rev. Lett.*, **22**, 1278–80.

Senbetu, L. and Woo, C.-W (1979). Structure of negative ion in normal liquid ^3He. *J. Low Temp. Phys.*, **36**, 67–78.

Sengers, J. V. (1966). Behavior of viscosity and thermal conductivity of fluids near the critical point. In *Critical phenomena. Proceedings of a conference* (ed. M. S. Green and J. V. Sengers), pp. 242–406. National Bureau of Standards, US GPO, Washington, DC. Misc. Note No. 273.

Sengers, J. V. (1971). Transport properties of fluids near critical points. In *Critical phenomena* (ed. M. S. Green), Volume LI, Varenna Lectures, pp. 445–507. Academic Press, New York.

Sengers, J. V. and Levelt-Sengers, J. M. H. (1978). Critical phenomena in classical fluids. In *Progress in liquids physics* (ed. C. A. Croxton), pp. 103–74. Wiley, Chichester.

Sengers, J. V. and Levelt-Sengers, J. M. H. (1986). Thermodynamic behavior of fluids near the critical point. *Annu. Rev. Phys. Chem.*, **37**, 189–222.

Sheard, F. W. and Bowley, R. M. (1978). Influence of one-roton emission processes on the velocity-field characteristics of negative ions in liquid ^{4}He at very low temperatures. *Phys. Rev. B*, **17**, 201–3.

Sherman, R. H. (1965). Behavior of He^3 in the critical region. *Phys. Rev. Lett.*, **15**, 141–2.

Shih, Y. M. and Woo, C.-W. (1973). Structure of electron bubbles in liquid helium under pressure. *Phys. Rev. A*, **8**, 1437–9.

Shikin, V. B. (1970). Motion of helium ions near a vapor–liquid surface. *Sov. Phys.-JETP*, **31**, 936–40.

Shikin, V. B. (1973). Interaction between impurity excitations and negative ions in liquid helium. *Sov. Phys.-JETP*, **37**, 718–22.

Silver, M., Kumbhare, P., Smejtek, P., and Onn, D. G. (1970). Hot electron injection into liquid argon from a tunnel cathode. *J. Chem. Phys.*, **52**, 5195–99.

Silver, M., Onn, D. G., Smejtek, P., and Masuda, K. (1967). Hot electron injection into dense argon, nitrogen and hydrogen. *Phys. Rev. Lett.*, **19**, 626–30.

Simola, J. T., Nummila, K. K., Hirai, A., Korhonen, J. S., Schoepe, W., and Skrbek, L. (1986). Focusing of negative ions by vortices in rotating ^{3}He–A. *Phys. Rev. Lett.*, **57**, 1923–6.

Simola, J. T., Nummila, K. K., Skrbek, L., and Korhonen, J. S. (1987*a*). Fast negative ion thermometer for helium-3 superfluids. *Cryogenics*, **27**, 391–6.

Simola, J. T., Skrbek, L., Nummila, K. K., and Korhonen, J. S. (1987*b*). Two different vortex states in rotating ^{3}He–A observed by use of negative ions. *Phys. Rev. Lett.*, **58**, 904–7.

Simon, S. H., Dobrosavljevic', V., and Stratt, R. M. (1990). Semiclassical percolation approach to electronic states in simple fluids. *Phys. Rev. A*, **42**, 6278–81.

Sitton, D. M. and Moss, F. (1969). Trapping of negative ions in turbulent superfluid helium. *Phys. Rev. Lett.*, **23**, 1090–3.

Sitton, D. M. and Moss, F. (1971). Free ion mobility in He II using field emission points. *Phys. Lett.*, **34 A**, 159–60.

Sitton, D. M. and Moss, F. (1972). Observations on the escape of negative ions from pressurized, turbulent He II. *Phys. Rev. Lett.*, **29**, 542–5.

Sluckin, T. J. (1977). Positive ion structure in ^{3}He-rich liquid helium mixtures. *Phys. Lett.*, **64 A**, 211–14.

Smejtek, P., Silver, M., Dy, K. S., and Onn, D. G. (1973). Hot-electron injection into helium and other insulating liquids from a tunnel junction. *J. Chem. Phys.*, **59**, 1374–84.

Sobyanin, A. A. (1973). Superfluid component density distribution near the He

I–He II phase interface in external fields. *Sov. Phys.–JETP*, **63**, 1780–92.

Soda, T. (1975). Ion mobility in superfluid ^{3}He. *Prog. Theor. Phys.*, **53**, 903–4.

Soda, T. (1977). The motion and mobility of ions in superfluid ^{3}He. *Prog. Theor. Phys.*, **58**, 1096–101.

Soda, T. (1984). Theory of ion mobility below characteristic temperature for liquid (3) He and (3) He–B. *Prog. Theor. Phys.*, **72**, 387–401.

Sommer, W. T. (1964). Liquid helium as a barrier to electrons. *Phys. Rev. Lett.*, **12**, 271–3.

Soukoulis, C. M., Economou, E. N., and Grest, G. S. (1987). Localization in quantum percolation: Transfer-matrix calculations in three dimensions. *Phys. Rev B*, **36**, 8649—55.

Space, B. and Coker, D. F. (1991). Nonadiabatic dynamics of excited excess electrons in simple fluids. *J. Chem. Phys.*, **94**, 1976–84.

Space, B. and Coker, D. F. (1992). Dynamics of trapping and localization of excess electrons in simple fluids. *J. Chem. Phys.*, **96**, 652–63.

Spangler, G. E. and Hereford, F. L. (1968). Injection of electrons into He II from an immersed tungsten filament. *Phys. Rev. Lett.*, **20**, 1229–30.

Springett, B. E. (1967). Effect of pressure on the transport of ions through rotating helium II. *Phys. Rev.*, **155**, 139–50.

Springett, B. E. (1969). Ion sizes near the liquid–gas transition in helium. *Phys. Rev.*, **184**, 229–32.

Springett, B. E., Cohen, M. H., and Jortner, J. (1967). Properties of an excess electron in liquid helium: The effect of pressure on the properties of the negative ion. *Phys. Rev.*, **159**, 183–90.

Springett, B. E. and Donnelly, R. J. (1966). Pressure dependence of the radius of the negative Ion in helium II. *Phys. Rev. Lett.*, **17**, 364–7.

Springett, B. E., Jortner, J., and Cohen, M. H. (1968). Stability criterion for the localization of an excess electron in a non-polar fluid. *J. Chem. Phys.*, **48**, 2720–31.

Springett, B. E., Tanner, D. J., and Donnelly, R. J. (1965). Capture cross section for negative ions in rotating helium II. *Phys. Rev. Lett.*, **14**, 585–7.

Staas, F. A., Takonis, K. W., and Fokkens, K. (1960). Viscosity of liquid 3He–4He mixtures in the helium II region above 1 °K. *Physica*, **26**, 669–86.

Stamp, P. C. E., McClintock, P. V. E., and Fairbairn, W. M. (1979). Possible influence of thermal rotons on vortex nucleation by negative ions in pressurised He II below 1 K. *J. Phys. C: Solid State Phys.*, **12**, L589–93.

Stanley, H. E. (1971). *Phase transitions and critical phenomena.* The International series of Monographs on Physics. Oxford University Press.

Stauffer, D. (1985). *Introduction to percolation theory.* Taylor & Francis, London.

Stauffer, D. and Fetter, A. L. (1968). Distribution of vortices in rotating helium II. *Phys. Rev.*, **168**, 156–9.

Steingart, M. and Glaberson, W. I. (1970). Ion-mobility discontinuities in He II. *Phys. Rev. A*, **2**, 1480–2.

Steingart, M. and Glaberson, W. I. (1972). Quantized vortex ring dynamics and the vortex core radius in He II. *J. Low Temp. Phys.*, **8**, 61–77.

Stienkmeier, F., Higgins, J., Callegari, C., Kanorsky, S. I., Ernst, W. E., and Scoles, G. (1996). Spectroscopy of alkali atoms (Li, Na, K) attached to large helium clusters. *Z. Phys. D*, **38**, 253–63.

Storchak, V., Brewer, J. H., and Morris, G. D. (1996). Fast and slow electrons in liquid neon. *Phys. Rev. Lett.*, **76**, 2969–72.

Strayer, D. M. and Donnelly, R. J. (1971). Measurements of the probability of nucleation of a vortex ring in Hhelium II. *Phys. Rev. Lett.*, **26**, 1420–3.

Strayer, D. M., Donnelly, R. J., and Roberts, P. H. (1971). Motion of ions at finite velocities in helium II. *Phys. Rev. Lett.*, **26**, 165–9.

Strutt, J. W. (1917). On the pressure developed in a liquid during the collapse of a spherical cavity. *Philos. Mag.*, **34**, 94–8.

Surko, C. M. and Reif, F. (1968). Investigation of a new kind of energetic neutral excitation in superfluid helium. *Phys. Rev.*, **175**, 229–41.

Tabbert, B., Beau, M., Gűnther, H., Hǻussler, W., Hőnninger, C., Meyer, K., Plagemann, B., and zu Putlitz, G. (1995). Atoms and ions in superfluid helium. I. Optical spectra of atomic and ionic impurities. *Z. Phys. B*, **97**, 425–32.

Tabbert, B., Gűnther, H., and zu Putlitz, G. (1997). Optical investigation of impurities in superfluid ^{4}He. *J. Low. Temp. Phys.*, **109**, 653–707.

Takahashi, N., Shimoda, T., Fujita, Y., Itahashi, T., and Miyatake, H. (1995). Snowballs of radioactive ions—nuclear spin polarization of core ions. *Z. Phys.*, **B 98**, 347–51.

Takahashi, N., Shimoda, T., Miyatake, H., Mitsuoka, S., Mizoi, Y., Kobayashi, H., Sasaki, M., Shirakura, T., Ueno, H., Asahi, K., and Morinobu, S. (1996). Freezing-out of nuclear polarization in radioactive core ions of microclusters. "Snowballs" in superfluid helium. *Hyperfine Interactions*, **97/98**, 469–77.

Takken, E. H. (1970). Wave-radiation model for the onset of dissipation at the roton critical velocity in superfluid helium. *Phys. Rev. A*, **1**, 1220–39.

Tankersley, L. L. (1973). Energy barrier for electron penetration into helium. *J. Low Temp. Phys.*, **11**, 451–71.

Tanner, D. J. (1966). Negative-ion capture by vortex lines in helium II. *Phys. Rev.*, **152**, 121–8.

Taylor, R. D. and Dash, J. G. (1957). Hydrodynamics of oscillating disks in viscous fluids: Viscosities of liquids He^3 and He^4. *Phys. Rev.*, **106**, 398–403.

Tegeler, C., Span, R., and Wagner, W. (1997). Eine neue Fundamentalgleichung fűr das Fluide Zustandgebiet von Argon fűr Temperaturen von der Schmelzlinie bis 700 K und Drűcke 1000 MPa. VDI Forschungsberichte, VDI Verlag, Dűsseldorf.

Tempere, J., Silvera, I. F., and Devreese, J. T. (2001). Effect of pressure on statics, dynamics, and stability of multielectron bubbles. *Phys. Rev. Lett.*, **87**, 275301.

Tempere, J., Silvera, I. F., and Devreese, J. T. (2003). Fission of multielectron bubbles in liquid helium. *Phys. Rev. B*, **67**, 035402.

Theobald, J., Ott–Rowland, M. L., and Williams, G. A. (1981). Mobility of ions under the surface of liquid helium. *Physica B+C*, **108**, 957–8.

Thomas, J. E. and Schmidt, P. W. (1963). X-ray study of critical opalescence in argon. *J. Chem. Phys.*, **39**, 2506–16.

Thomson, W. (1873). On the ultramundane corpuscles of LeSage. *Philos. Mag.*, **45**, 321–32.

Thomson, W. (1880). Vibrations of a columnar vortex. *Phil. Mag.*, **10**, 155–68.

Thuneberg, E. V. (1986). Identification of vortices in superfluid ^{3}He–B. *Phys. Rev. Lett.*, **56**, 359–62.

Thuneberg, E. V., Kurkijärvi, J., and Rainer, D. (1981). Quasiclassical theory of ions in ^{3}He. *J. Phys. C: Solid State Phys.*, **14**, 5615–24.

Titus, J. A. and Rosenshein, J. S. (1973). Time dependence of vortex-ring creation in He II. *Phys. Rev. Lett.*, **31**, 146–9.

Tjerkstra, H. H. (1952). The influence of pressure on the viscosity of liquid helium I. *Physica*, **18**, 853–61.

Toennies, J. P. and Vilesov, A. F. (1998). Spectroscopy of atoms and molecules in liquid helium. *Annu. Rev. Chem.*, **49**, 1–41.

Tough, J. T. (1982). Superfluid turbulence. In *Progress in low temperature physics* (ed. D. F. Brewer), Volume 8, Chapter 3, pp. 133–219. North-Holland, Amsterdam.

Tough, J. T., McCormick, W. D., and Dash, J. G. (1963). Viscosity of liquid He II. *Phys. Rev.*, **132**, 2373–78.

Tough, J. T., McCormick, W. D., and Dash, J. G. (1964). Vibrating wire viscometer. *Rev. Sci. Instrum.*, **35**, 1345–48.

Trappeniers, N. J. and Oosting, P. H. (1966). Self-diffusion in gaseous and liquid methane. *Phys. Lett.*, **23**, 445–7.

Trappeniers, N. J., van der Gulik, P. S., and van den Hooff, H. (1980). The viscosity of argon at very high pressure, up to the melting line. *Chem. Phys. Lett.*, **70**, 438–43.

Trilling, L. (1952). The collapse and rebound of a gas bubble. *J. Appl. Phys.*, **23**, 14–17.

Tyndall, A. M. (1938). *The mobility of positive ions in gases.* Cambridge University Press.

Uehling, E. A. and Uhlenbeck, G. E. (1933). Transport phenomena in Einstein–Bose and Fermi–Dirac gases. I. *Phys. Rev.*, **48**, 552–61.

van Dijk, J. D., Postma, G. M., Wiebes, J., and Kramers, H. C. (1977). Ion escape from vortex rings in liquid He-II at various pressures. *Physica*, **85 B**, 85–103.

van Hove, L. (1954). Correlations in space and time and Born approximation scattering in systems of interacting particles. *Phys. Rev.*, **95**, 249–62.

van Vijfeijken, A. G., Walraven, A., and Staas, F. A. (1969). Energy and stability of vortex rings in liquid helium II; Critical velocities. *Physica*, **44**, 415–36.

Vicente, C., Yao, W., Maris, H. J., and Seidel, G. M. (2002). Surface tension of liquid ^{4}He as measured using the vibration modes of a levitated drop. *Phys. Rev. B.*, **66**, 214504.

Vinen, W. F. (1957). Mutual friction in a heat current in liquid helium II. III. Theory of the mutual friction. *Proc. Roy. Soc. (London)*, **A 242**, 493–515.

Vinen, W. F. (1958). Detection of single quanta of circulation in rotating helium II. *Nature*, **181**, 1524–25.

Vinen, W. F. (1961). The detection of single quanta of circulation in liquid helium II. *Proc. Roy. Soc. (London)*, **A 260**, 218–36.

Vinen, W. F. (1963). Critical velocities in liquid helium II. In *Liquid helium* (ed. G. Careri), Proceedings of the International School of Physics Enrico Fermi, Course XXI, pp. 336–55. Academic Press, New York.

Vinen, W. F. (1968). Topics in superfluid. In *Statistical physics, phase transitions and superfluidity* (ed. M. Chrètien, E. P. Gross, and S. Deser), Volume 2, Brandeis University Summer School in Theoretical Physics 1966, pp. 99–177. Gordon and Breach, New York.

Volykhin, K. F., Khrapak, A. G., and Schmidt, W. F. (1995). Structure and mobility of negative ions in dense gases and nonpolar liquids. *J. E. T. P.*, **81**, 901–8.

von Helmholtz, H. (1867). On the integrals of hydrodynamic equations to which vortex motions conform. *Philos. Mag.*, **33**, 485–512.

Vulovic, V. Z., Stein, D. L., and Fetter, A. L. (1984). NMR of textures in rotating ^{3}He–A. *Phys. Rev. B*, **29**, 6090–5.

Wallace, B. and Meyer, H. (1970). Equation of state of He^3 close to the critical point. *Phys. Rev. A*, **2**, 1563–75.

Walraven, A. (1970). Energy and motion of vortex rings in liquid helium II in the presence of various plane obstacles. *Phys. Rev. A*, **1**, 145–57.

Wang, S., Howald, C., and Meyer, H. (1990). Shear viscosity of liquid ^{4}He and ^{3}He–^{4}He mixtures, especially near the superfluid transition. *J. Low Temp. Phys.*, **79**, 151–87.

Wannier, G. (1966). *Statistical physics.* Dover, New York.

Watanabe, S. (1979). Effect of long-wavelength fluctuations on ion mobility in fluid. *J. Phys. Soc. Japan*, **46**, 1819–24.

Webeler, R. W. H. and Hammer, D. C. (1966). Viscosity coefficients for liquid helium-3 in the interval 0.36 to 2.6 K. *Phys. Lett.*, **21**, 403–4.

Welber, B. (1961). Damping of a torsionally oscillating cylinder in liquid helium at various temperatures and densities. *Phys. Rev.*, **119**, 1816–22.

Wőlfle, P. (1979). Low-temperature properties of liquid ^{3}He. *Rep. Prog. Phys.*, **42**, 269–346.

Wőlfle, P., Gotze, W., Kurkijärvii, J., and Smith, H. (1980). The mobility of a heavy impurity in a degenerate Fermi system. *J. Phys. C: Solid State Phys.*, **13**, 2461–84.

Wheatley, J. C. (1968). Experimental properties of liquid He^3 near the absolute zero. *Phys. Rev.*, **175**, 304–9.

Wheatley, J. C. (1975). Experimental properties of superfluid ^{3}He. *Rev. Mod.*

Phys., **47**, 415–70.

Whitney, W. M. and Chase, C. E. (1962). Velocity of sound in liquid helium at low temperatures. *Phys. Rev. Lett.*, **9**, 243–5.

Widom, B. (1972). Surface tension of fluids. In *Phase transitions and critical phenomena* (ed. C. Domb and M. S. Green), Volume 2, Chapter III, pp. 79–100. Academic Press, New York.

Wigner, E. and Seitz, F. (1933). On the constitution of metallic sodium. *Phys. Rev.*, **43**, 804–10.

Wilks, J. (1967). *The properties of liquid and solid helium.* The International Series of Monographs on Physics. Clarendon Press, Oxford.

Williams, C. D. H., Hendry, P. C., and McClintock, P. V. E. (1987). Production of "fast" and "exotic" negative ions in He II. *Jpn. J. Appl. Phys.*, **26**, 105–6.

Williams, G. A., DeConde, K., and Packard, R. E. (1975). Positive-ion trapping on vortex lines in rotating He II. *Phys. Rev. Lett.*, **34**, 924–6.

Williams, G. A. and Packard, R. E. (1974). Photograph of quantized vortex lines in rotating He II. *Phys. Rev. Lett.*, **33**, 280–3.

Williams, G. A. and Packard, R. E. (1975). Effect of ^{3}He impurities on the lifetime of ions trapped on quantized vortex lines. *Phys. Rev. Lett.*, **35**, 237–40.

Williams, G. A. and Packard, R. E. (1978). Ion trapping on vortex lines in rotating ^{3}He–^{4}He mixtures. *J. Low Temp. Phys.*, **33**, 459–80.

Williams, M. R. and Fetter, A. L. (1979). Textures in slowly rotating ^{3}He–A. *Phys. Rev. B*, **20**, 169–76.

Williams, R. L. (1957). Ionic mobilities in argon and helium liquids. *Can. J. Phys.*, **35**, 134–46.

Wisdom, J., Hartquist, T. W., and Lane, N. F. (1976). He (2^3S) bubble in liquid helium. *Phys. Rev. B*, **14**, 4205–8.

Wood, W. W. (ed.) (1967). *Shock and detonation waves.* Gordon Breach, New York.

Woods, A. D. B. and Hollis-Hallett, A. C. (1963). The viscosity of liquid helium II between 0.79 °K and the lambda point. *Can. J. Phys.*, **41**, 596.

Woolf, M. A. and Rayfield, G. W. (1965). Energy of negative ions in liquid helium by photoelectric injection. *Phys. Rev. Lett.*, **15**, 235–7.

Yakubov, I. T. (1970). Theory of the electrical conductivity of a weakly ionized imperfect plasma. *Sov. Phys.–JETP*, **30**, 567–70.

Yakubov, I. T. (1973). Mobility of electrons in moderately dense gases. *High Temp.*, **10**, 158–9.

Yakubov, I. T. and Polishuk, A. Ya. (1982). Quantum density corrections to mobility and dispersion law for electrons in a medium of disordered scatterers. *J. Phys. B: At. Mol. Phys.*, **15**, 4029–42.

Yamaguchi, A., Tajima, T., Obara, K., and Ishimoto, H. (2004). Negative ion mobility in superfluid ^{3}He under high magnetic field. *J. Low Temp. Phys.*, **134**, 643–8.

Yamaguchi, A., Ueno, D., Obara, K., Efimov, V., and Ishimoto, H. (2002).

Positive ion mobility in superfluid ^{3}He under high magnetic fields. *Phys. Rev. B*, **66**, 132502.

Yamaguchi, A., Ueno, D., Obara, K., Efimov, V., and Ishimoto, H. (2003*a*). Anomalous magnetic-field dependence of positive ion mobility in superfluid He-3. *Physica B*, **329**, 348–9.

Yamaguchi, A., Ueno, D., Obara, K., Efimov, V., and Ishimoto, H. (2003*b*). Positive ion mobility in superfluid He-3–A(1) and A(2). *Physica B*, **329**, 350–1.

Yarmchuk, E. J. and Packard, R. E. (1982). Photographic studies of quantized vortex lines. *J. Low Temp. Phys.*, **46**, 479–515.

Yarnell, J. L., Arnold, G. P., Bendt, P. J., and Kerr, E. C. (1959). Excitations in liquid helium: Neutron scattering measurements. *Phys. Rev.*, **113**, 1379–86.

Young, R. A. (1970). Phenomenological theory of the mobility of an electron in dense helium gas. *Phys. Rev. A*, **2**, 1983–6.

Yuan, D. Y. and Ebner, C. (1981). Theory of excess-electron states in classical rare-gas fluids. *Phys. Rev. A*, **23**, 285–93.

Zallen, R. and Scher, H. (1971). Percolation on a continuum and the localization–delocalization transition in amorphous semiconductors. *Phys. Rev. B*, **4**, 4471–9.

Zecca, A., Karwasz, G. P., and Brusa, R. S. (1996). One century of experiments on electron–atom and molecule scattering: A critical review of integral cross-sections. I. Atoms and diatomic molecules. *Nuovo Cimento*, **19**, 1–146.

Zieve, R. J., Mukharsky, Y. M., Close, J. D., Davis, J. C., and Packard, R. E. (1993). Investigations of quantized circulation in superfluid ^{3}He–B. *J. Low Temp. Phys.*, **91**, 315–39.

Ziman, J. M. (1960). *Electrons and Phonons.* Oxford University Press, Oxford.

Zimmerman, G. O. and Chase, C. E. (1967). Orthobaric density of ^{3}He in the critical region. *Phys. Rev. Lett.*, **19**, 151–4.

Zimmermann, P. H. and Reichert, J. F. (1977). Study of the electron spin resonance of negative ions field emitted into liquid helium. *Phys. Rev. B*, **15**, 2630–50.

Zinov'eva, K. N. (1956). The surface tension of liquid He^3 in the region of very low temperatures (1.0–0.35 °K). *Sov. Phys.-JETP*, **2**, 774–5.

Zinov'eva, K. N. and Peshkov, V. P. (1960). Phase diagram for Lliquid He^3–He^4 solutions. *Sov. Phys.-JETP*, **37**, 22–5.

Zipfel, C. L. (1969). *Structure of electron bubbles in liquid helium from photo-ejection under pressure.* Ph.D. thesis, University of Michigan, Chicago, IL.

Zoll, R. (1976). Study of the vortex-ring transition in superfluid ^{4}He. *Phys. Rev. B*, **14**, 2913–26.

Zoll, R. and Schwarz, K. W. (1973). New features of the vortex-ring transition. *Phys. Rev. Lett.*, **31**, 1440–3.

Index